D1518963

WATER WORKS ENGINEERING

Planning, Design, and Operation

Syed R. Qasim
The University of Texas at Arlington
Edward M. Motley
Guang Zhu
Chiang, Patel & Yerby, Inc.

In Cooperation with
CHIANG, PATEL & YERBY, INC.
Consulting Engineers • Planners • Project Managers
Dallas, Texas

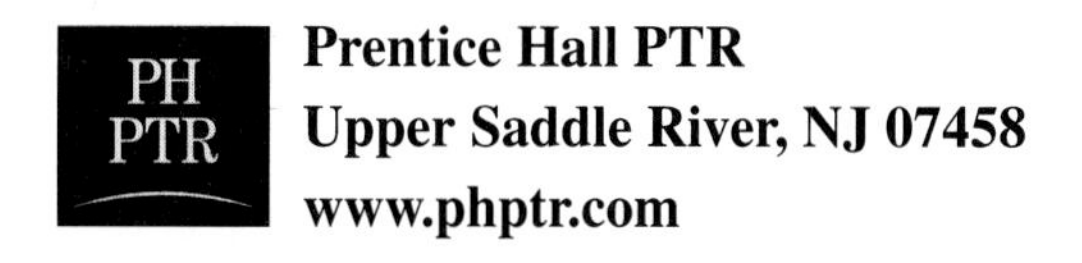

Prentice Hall PTR
Upper Saddle River, NJ 07458
www.phptr.com

Library of Congress Cataloging-in-Publication Data

Qasim, Syed R.
Water works engineering : planning, design, and operation / Syed R.Qasim, Edward M. Motley, Guang Zhu.
p. cm.
Incudes bibliographical references and index.
ISBN 0-13-150211-5
1. Water treatment plants--Design and construction. I. Motley, Edward M. II. Zhu, Guang. III. Title.

TD434.Q23 2000
628.1--dc21 00-024966

Editorial/production supervision: *Vincent Janoski*
Acquisitions editor: *Bernard Goodwin*
Marketing manager: *Lisa Konzellmann*
Manufacturing manager: *Alexis Heydt*
Cover design director: *Jerry Votta*

© 2000 by Prentice-Hall

Published by Prentice Hall PTR
Prentice-Hall, Inc.
Upper Saddle River, NJ 07458

Prentice Hall books are widely used by corporations and government agencies for training, marketing, and resale.

The publisher offers discounts on this book when ordered in bulk quantities. For more information, contact: Corporate Sales Department, Phone: 800-382-3419; Fax: 201-236-7141; E-mail: corpsales@prenhall.com; or write: Prentice Hall PTR, Corp. Sales Dept., One Lake Street, Upper Saddle River, NJ 07458.

All products or services mentioned in this book are the trademarks or service marks of their respective companies or organizations. Screen shots reprinted by permission from Microsoft Corporation.

All rights reserved. No part of this book may be reproduced, in any form or by any means, without permission in writing from the publisher.

Printed in the United States of America
10 9 8 7 6 5 4 3 2 1

ISBN 0-13-150211-5

Prentice-Hall International (UK) Limited, *London*
Prentice-Hall of Australia Pty. Limited, *Sydney*
Prentice-Hall Canada Inc., *Toronto*
Prentice-Hall Hispanoamericana, S.A., *Mexico*
Prentice-Hall of India Private Limited, *New Delhi*
Prentice-Hall of Japan, Inc., *Tokyo*
Pearson Education Asia Pte. Ltd.
Editora Prentice-Hall do Brasil, Ltda., *Rio de Janeiro*

Contents

Preface

The Safe Drinking Water Act Amendments of 1986 are the most sweeping legislative changes in the history of the public water supply field and will have more of an impact than the original act passed over two decades ago. These legislative changes reflect a perception by Congress that there is a great amount of public anxiety over the quality of our drinking water, as well as frustration with the manner in which the U.S. Environmental Protection Agency (USEPA) enforced the requirements of the original act. Practicing engineers and utility managers who are involved with the planning, design, and operation of water treatment plants must begin to evaluate current options and anticipate what lies ahead. At the present time, many programs in civil and environmental engineering at numerous universities are offering courses in the area of water treatment plant design.

Several excellent books have been written in recent years that present theory and principles of water treatment processes. The authors have observed during their years of experience in the water treatment field that no publication has been devoted entirely to water works engineering planning, design, and operation. The intent of the authors in developing this book is twofold: first, to consolidate the developments in design of water works engineering that have evolved as a result of technological advancement in the field and of the concepts and policies promulgated by the environmental laws and the subsequent guidelines; second, to develop step-by-step procedures for planning, design, and operation of a medium-size conventional water treatment plant.

This book has a total of nineteen chapters and four appendices. Chapters 1 through 4 are devoted to the basic facts of water engineering. Current and future trends in water treatment technology, water quality constituents, design factors, drinking water regulations, unit operations and processes, process combinations, and process trains are discussed in detail.

Chapter 5 is devoted to preparation of a predesign report. The general purpose of a predesign engineering report and an example of a model predesign report for a medium-size water supply project are presented. In this predesign report, procedures are presented for (1) estimation of population, water demand, plant capacity, and raw water quality; (2) evaluation and selection of a treatment process train; (3) coordination with distribution system; and (4) estimation of costs for proposed capital improvement project.

The remainder of the book is devoted to the design of the medium-size water treatment facility for which the predesign report is developed in Chapter 5. Step-by-step design calculations; equipment details; engineering drawings, plans, and specifications; and operation and maintenance of head works, raw water transport, treatment, and distribution systems: all are presented. These systems include raw water intake, screening, pump station, transport pipeline, flow measurement, rapid mix, flocculation, clarification, filtration, color, taste and odor control, disinfection, distribution, and residuals handling and disposal. Also, separate chapters have been devoted to plant layout, yard piping and hydraulics, instrumentation and automatic controls, design summary, and the avoidance of design errors. A separate chapter (Chapter 18) is provided to cover nonconventional treatment processes. In this chapter, the treatment processes for nitrate and fluoride removal, ion exchange, reverse osmosis, and heavy metals and organics removal are briefly presented.

The design procedures given in this book are for illustration and general information only and are not intended to be used as standard for water treatment plant designs. References made in this publication to specific methods, processes, and equipment do not constitute or imply an endorsement or recommendation. Equivalent or improved equipment may be obtained from many other manufacturers not mentioned in this publication.

This book will serve the needs of students, teachers, consulting engineers, equipment manufacturers, and technical personnel in city, state, and federal organizations who must review designs and specifications. In order to maximize the usefulness of this book, the material has been presented in a simplified and concise format. Many tables have been developed using a variety of sources. Those tables provide information used extensively in water treatment plant design. Basic properties of water, hydraulic design information, chemical properties and equipment selection, equipment manufacturers, cost equations, and unit conversions are arranged in four appendices.

It should be mentioned that a great deal of emphasis has been given to the predesign report and the design of conventional water treatment units. The authors are well aware of the fact that equal emphasis or in-depth design procedures for many other nonconventional water treatment processes such as denitrification, defluoridation, demineralization, and other specialized processes (briefly presented in Chapters 3 and 18) are not given elsewhere in this book. The reason is very simple. Covering design procedures for these processes would take another book of equal size. The authors strongly believe that the planning and design principles developed in this book can easily be extended to the design of many other treatment processes for a new plant or the upgrading of an existing facility. Therefore, the in-depth coverage and step-by-step design procedure for an entire conventional water treatment plant is the strongest feature of this publication.

This book is intended for a design course in water works engineering. Most of the programs in civil and environmental engineering are offering such courses at graduate and undergraduate levels. The material is developed in such a way that the normal prerequisites (fluid mechanics and an introductory course in water supply and wastewater treatment) would be sufficient. Furthermore, the basic material contained in this book can be covered in a three-semester credit course.

In this country, because the old plants will be upgraded in the future, the U.S. customary units will continue to be in use for some time to come. Therefore, where possible, both systems of units are used side-by-side, and proper conversion factors are provided. Complete conversion tables are given separately in an appendix.

Acknowledgments

A project of this size requires the cooperation and collaboration of a great many people. We are indebted to many persons who have helped us prepare this book and made constructive suggestions in process and equipment design. It represents a joint effort of the academic community and consulting engineers.

Chiang, Patel & Yerby, Inc. (CP&Y), a multi-disciplined consulting engineering firm in Dallas, Texas, has contributed a great deal to development of manuscript and preparation of artwork including complex AutoCAD drawings. In particular, we acknowledge our gratitude to W. Walter Chiang and Pete K. Patel for their stimulating discussions during the development of this book. We gratefully appreciate the assistance provided by the following CP&Y staff: Kimberly W. Brashear, Michael F. Graves, C. Kevin Chin, Gregory W. Johnson, Kathryn I. Neill, Susan M. Newsom, Dawn R. Anderson, and Marisa T. Vergara.

Many professionals, colleagues, and students reviewed various portions of this book, conducted literature searches, checked calculations, and worked out solutions to the problems. In particular, we thank Charles F. Anderson, Steve McCrary, Ron Tamada, Ernest C. Crosby, Max Spindler, Shih Pan Sutthipong Chanthikul, and Yuanchun Xu.

Many equipment manufacturers and their local representatives provided valuable information on equipment details and specifications. Fred Willms arranged for many photographs from equipment manufacturers. The Department of Civil and Environmental Engineering and the College of Engineering at The University of Texas at Arlington provided needed support.

Finally, we must acknowledge with deep appreciation the support, encouragement, and patience of our families.

Although portions of this book have been reviewed by professionals and students, the real test will not come until it has been used in classes and by professionals as a design guide. We shall appreciate it very much if all who use this book will let us know of any errors and of changes they believe would improve its usefulness.

Arlington, Texas

Syed R. Qasim
Edward M. Motley
Guang Zhu

CHAPTER 1

Introduction

The primary goals of a water treatment plant for over a century have remained the same: produce water that is biologically and chemically safe, is appealing to consumers, and is noncorrosive and nonscaling. Today, plant design has become very complex from discovery of unpronounceable and seemingly innumerable chemical substances, the multiplying of regulations, and trying to satisfy more discriminating palates. In addition to the basics, designers must now keep in mind all manner of legal mandates, as well as public concerns and environmental considerations, to provide an initial prospective of water works engineering planning, design, and operation. A brief review of the historical background, current status, and new directions in the area of water works engineering and water treatment plant design is presented in this chapter.

1.1 Historical Background

The desire to drink pure and wholesome water dates from ancient times. Early methods of treating foul water were boiling, exposing to sun, dipping a hot copper rod in repeatedly, and filtration.[1] The earliest water treatment practices were primarily in batch operations in individual homes. From the sixteenth century onward, centralized treatment facilities for large settlements were realized. By the eighteenth century, filtration of particles from water was established as an effective means of clarifying water.

The growth of community water supply systems in the United States started in the early 1800's. By 1860, over 400, and by the turn of the century, over 3000 major water systems had been built to serve the nation's major cities and towns. Many plants had slow sand filters. In the mid-1890's, the Louisville Water Company introduced coagulation with rapid sand filtration.

Although the first application of chlorine in potable water was introduced in the 1830's for taste and odor control, at that time diseases were thought to be spread by odors.[2] It was not until the 1890's and the advent of the germ theory of disease that the importance of disinfection in potable water was understood.[3] Chlorination was first introduced in 1908 and then became a common practice.

1.2 Federal Drinking Water Standards

Federal authority to establish standards for drinking water systems originated with the enactment by Congress in 1883 of the Interstate Quarantine Act.[4] The Act authorized the Director of the United States Public Health Services (USPHS) to establish and enforce regulations to prevent the introduction, transmission, or spread of communicable diseases. The brief history of advancements in water quality standards in the United States is summarized in Table 1-1.

Resource limitations have caused the United States Environmental Protection Agency (USEPA) to reassess schedules for new rules. Major changes to USEPA's current regulatory agenda are anticipated when the Safe Drinking Water Act (SDWA) is re-authorized.

1.3 Current Status and New Technologies

1.3.1 Current Status

A USEPA survey indicated that in 1987 there were approximately 202,000 public water systems in the United States. About 29 percent of these are community water systems, which serve approximately 90 percent of the population. Figure 1-1 provides a distribution of systems using surface or groundwater sources.[8] Of the 58,908 community systems that serve about 226 million people, 51,552 are classified as "small" or "very small." Each of these systems at an average serves a population of fewer than 3300 people. The total population served by these systems is approximately 25 million people

Small systems are the most frequent violators of federal regulations and accounted for almost 89 percent of the 43,000 violations posted in 1988. Microbiological violations accounted for the vast majority of cases, with failure to monitor and report. Among others, violations were due to exceeding SDWA maximum contaminant levels (MCLs). Bringing small water systems into compliance will require applicable technologies, operator ability, financial resources, and institutional arrangements.[9] The 1986 SDWA amendments authorized USEPA to set the best available technology (BAT) that can be incorporated in the design for the purposes of complying with the National Primary Drinking Water Regulations.[9,10] Current BAT to maintain standards are as follows:

- For turbidity, color and microbiological control in surface water treatment: filtration. Common variations of filtration are conventional, direct, slow sand, diatomaceous earth, and membranes.

Table 1-1 History and Advancements in Water Quality Standards in the United States

Year	Development
1912	First water-related regulation prohibiting the use of common drinking water cups on interstate carriers.
1913	Maximum level of bacterial contamination, 2 coliforms per 100 mL, was recommend
1914	Promulgation of standards by the Department of the Treasury; a basis for federal, state, and local cooperation was established.
1915	Federal commitment was made to review the drinking water regulations on a regular basis.
1925	Limit 1 coliform per 100 mL; also standards for lead, copper, zinc, and excessive soluble mineral substances were proposed.
1942	USPHS appointed an advisory committee for revision of the 1925 drinking water regulations. Significant new initiatives included bacteriological monitoring of water quality in the distribution system and maximum permissible concentration for heavy metals.
1946	Maximum permissible concentrations were published for heavy metals.
1962	The standards set mandatory limits for health-related chemicals and biological impurities. The standards covered 28 contaminants.
1970	A 1970 USEPA survey indicated that 41 percent of the systems surveyed did not meet the guidelines established in 1962.
1974	National Academy of Sciences (NAS) published trihalomethanes (THMs) in public water supplies and related health affects. The Safe Drinking Water Act (SDWA, PL 93-523) was signed, which required that national primary drinking water regulations be established.
1975-1980	Interim regulations were adopted, and SDWA amendments followed in 1977, 1979, and 1980. These amendments provided for reauthorization of the act and made a number of minor changes. Enforceable regulations were set for only 23 contaminants, most of which were interim standards. THMs list and best available technology (BAT) were published in 1979 and 1983, respectively.
1986-1999	SDWA amendments significantly altered the regulatory time table. USEPA was directed to set standards for 83 contaminants according to specific deadlines. Although most deadlines have not been met, the number of regulated contaminants has steadily increased to well above 83. Major regulations and standards revised and promulgated under SDWA amendments between 1986 through 1999 are fluoride standards, priority lists, Lead and Copper Rule, Phase I VOCs (volatile organic compounds), Phase II SOCs and IOCs (inorganic contaminants), Phase V SOCs and IOCs, Total Coliform Rule, Surface Water Treatment Rule, Enhanced Surface Water Treatment Rule, Information Collection Rule, Consumer Confidence Reports Rule, Radionuclides Rule, Disinfectants-Disinfection By-products (D/DBPs) Rule, Sulfate Rule, and Groundwater Rule.

Source: References 4–7.

- For inactivation of microorganisms: disinfection. Typical disinfectants are chlorine, chlorine dioxide, chloramines, and ozone.
- For organic contaminant removal from surface water: packed-tower aeration, granular activated carbon (GAC), powdered activated carbon (PAC), diffused aeration, advanced oxidation processes, and reverse osmosis.
- For inorganic contaminants removal: membranes, ion exchange, activated alumina, and GAC.
- For corrosion control: typically, pH adjustment or corrosion inhibitors.

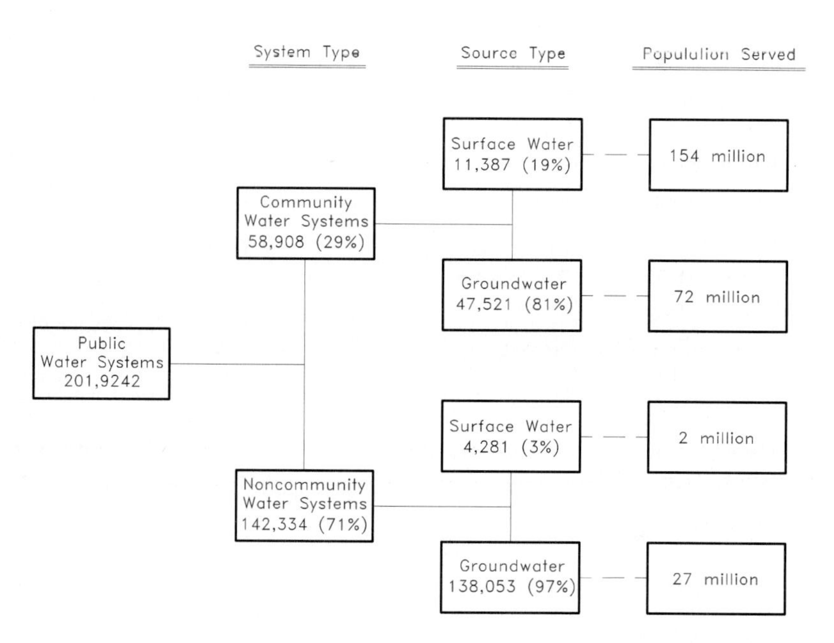

Figure 1-1 Distribution of public water systems by system type and source water in 1987.

1.3.2 New Technology

The implications of 1986 amendments to SDWA and new regulations have resulted in rapid introduction of new technologies for water treatment and monitoring. Until recently, the U.S. water industry showed little interest in *biological processes,* perhaps because of their most obvious drawback, the possible introduction of harmful microorganisms or of their by-products in the finished water. Its apparent effectiveness in removing biodegradable organic carbon that may sustain the regrowth of potentially harmful microorganisms in the distribution system, effective taste and odor control, and reduction in chlorine demand and DBP formation potential, has nonetheless made U.S. water suppliers and researchers slowly overcome their reluctance. Now research data in U.S. has shown that *biologically-active sand or carbon filters* provide more cost effective treatment of microcontaminants than do physicochemical processes. Other benefits reported are iron and manganese removal and conversion of ammonia by nitrification.[11,12] The process has the potential to upgrade existing conventional plants to a custom-designed new plant with this technology.

Over the past few years, *membrane technology* has been applied in drinking water treatment, partly because of affordable membranes and demand to removal many contaminants. *Microfiltration, ultrafiltration, nanofiltration*, and others have become common names in the water industry. Membrane technology is experimented with for the removal of microbes such as *Giardia* and *Cryptosporidium* and for selective removal of nitrate.[13,14] In other instances, membrane technology is applied for removal of DBP precursors, VOCs, and others.[15,16]

Many other treatment technologies that have potential for full-scale adoption are photochemical oxidation using ozone and UV radiation or hydrogen peroxide for destruction of refractory organic compounds.[17,18] One example of a technology that was developed outside North America and later emerged in the U.S. is the Haberer process. This process combines contact flocculation, filtration, and powdered activated carbon adsorption to meet a wide range of requirements for surface water and groundwater purification.[19,20]

In their quest to comply with multiple drinking water standards, utilities are seeking not only to improve treatment but also to monitor their supplies for microbiological contaminants more effectively. *Electro-optical sensors* are used to allow early detection of algal blooms in a reservoir and allow for diagnosis of problems and guidance in operational changes.[21] *Gene probe* technology was first developed in response to the need for improved identification of microbes in the field of clinical microbiology. Attempts are now being made by radiolabeled and nonradioactive gene-probe assays with traditional detection methods for enteric viruses and protozoan parasites, such as *Giardia* and *Cryptosporidium*. This technique has the potential for monitoring water supplies for increasingly complex groups of microbes.[22]

1.4 Water Works Engineering and Design

In spite of the multitudinous regulations and standards that an existing public water system must comply with, the principles of conventional water treatment process have not changed significantly over half a century. Whether a filter contains sand, anthracite, or both, slow or rapid rate, constant or declining rate, filtration is still filtration, sedimentation is still sedimentation, and disinfection is still disinfection. What has changed, however, are many tools that designers have at their disposal. *Computers* have bestowed the gifts of alacrity and accuracy in design and operation. Now engineers can compare the alternative processes and process trains with a speed that was not possible with a pencil and graph paper. Likewise, a supervisory control and data acquisition (SCADA) system can provide operators and managers with accurate process-control variables and operation and maintenance records. In addition to being able to look at the various options on the computer screen, engineers can conduct pilot plant studies of the multiple variables inherent in water treatment plant design. Likewise, operators and managers can utilize an ongoing pilot plant facility to optimize chemical feed and develop important information needed for future expansion and upgrading.

Water treatment plants should be designed so that water quality objectives can be met with reasonable ease and cost. The design should incorporate flexibility for dealing with seasonal changes, as well as with long-term changes in water quality and in future drinking water regulations. Good planning and design, therefore, must be based on five major steps:

1. characterization of the water source and finished water quality goals;
2. predesign studies, to develop alternative processes and selection of final process train;
3. detailed design of the selected alternative;
4. construction;
5. operation and maintenance of the completed facility.

Engineers, scientists, and financial analysts must utilize principles from a wide range of disciplines: engineering, chemistry, microbiology, geology, architecture, and economics, to carry out the responsibility of designing a water treatment plant.

1.5 Scope of Book

The objective of this book is to provide information for use by students and practicing engineers. Theory, design, operation and maintenance, troubleshooting, equipment selection, and specifications are integrated for each treatment process. The topics discussed include the following:[23]

- water quality criteria for raw and finished water;
- facility plan and headworks design;
- raw-water lifting and transport;
- theory, design, and layout of treatment processes;
- hydraulic profile;
- high-service pumping and distribution;
- instrumentation and controls.

References

1. Baker, M. N. *The Quest for Pure Water*, 2d ed., American Water Works Association, Inc., New York, 1981.
2. White, C. G. *The Handbook of Chlorination and Alternative Disinfectants*, 3d ed., Van Nostrand Reinhold, New York, 1992.
3. Pedden, T. M. *Drinking Water and Ice Supplies and Their Relations to Health and Disease*, G. P. Putnam's Sons, The Knickerbocker Press, New York, 1891
4. AWWA. *Water Quality and Treatment*, 4th ed., McGraw-Hill Book Co., New York, 1990
5. Pontius, F. W. "SDWA - A Look Back," *Jour. AWWA*, vol. 85, no. 2, pp. 22–24 & 94, February 1993.
6. Pontius, F. W. and Robinson, J. A. "The Current Regulatory Agenda: An Update," *Jour. AWWA*, vol. 86, no. 2, pp. 54–63, February 1994.
7. Pontius, F. W. "An Update of the Federal Drinking Water Regs," *Jour. AWWA*, vol. 90, no. 3, pp. 48–58, March 1998.

8. USEPA. *The National Public Water System Program*, FY 1988 Compliance Report, Office of Drinking Water, Cincinnati, OH, March 1990.
9. Goodrich, J. A., Adams, J. Q ., Lykins, B. W., and Clark, R. M. "Safe Drinking Water from Small Systems: Treatment Options," *Jour. AWWA*, vol. 84, no. 5, pp. 49–55, May 1992.
10. USEPA. *Technologies for Upgrading Existing or Designing New Drinking Water Treatment Facilities*, EPA/625/4-89/023, Office of Drinking Water, Cincinnati, OH, March 1990.
11. Le Chevallier, M. K., Becker, W. C., Schorr, P., and Lee, R. G. "Evaluating the Performance of Biologically Active Rapid Filters," *Jour. AWWA*, vol. 84, no. 4, pp. 136–146, April 1992.
12. Manem, J. A. and Rittmann, B. E. "Removing Trace-Level Organic Pollutants in a Biological Filter," *Jour. AWWA*, vol. 84, no. 4, pp. 152–157, April 1992.
13. Adam, S. S., Jacangelo, J. G., and Laine, J. M. "Low Pressure Membranes: Assessing Integrity," *Jour. AWWA*, vol. 87, no. 3, pp. 62–75, March 1995.
14. McCleaf, P. R. and Schroeder, E. D. "Denitrification Using a Membrane Immobilized Biofilm," *Jour. AWWA*, vol. 87, no. 3, pp. 77–86, March 1995.
15. Allgerier, S. C. and Summers, R. C. "Evaluating NF for DBP Control with RBSMT," *Jour. AWWA*, vol. 87, no. 3, pp. 87–99, March 1995.
16. Castro, K. and Zander, A. K. "Membrane Air-Stripping Effects of Pretreatments," *Jour. AWWA*, vol. 87, no. 3, pp. 50–61, March 1995.
17. Glaze, W. H. and Kang, J. W. "Advanced Oxidation Process for Treating Ground Water Contaminated with TCE and PCE: Laboratory Studies," *Jour. AWWA*, vol. 80, no. 5, pp. 57–63, May 1988.
18. Glaze, W. H., Kang, J. W., and Aieta, M. "Ozone-Hydrogen Peroxide Systems for Control of Organics in Municipal Water Supplies," *Proceedings of the Second International Conference in the Role of Ozone on Water and Wastewater Treatment*, TekTran International Ltd., Kitchener, Ontario, Canada, pp. 233–244, 1987.
19. Haberar, K. and Schmidth, S. N. "The Haberar Process: Combining Contact Flocculation, Filtration, and PAC Adsorption," *Jour. AWWA*, vol. 83, no. 9, pp. 82–89, September 1991.
20. Stukenberg, J. R. and Hesby, J. C. "Pilot Testing the Haberar Process in the United States," *Jour. AWWA*, vol. 83, no. 9, pp. 90–96, September 1991.
21. White, B. N., Kiefer, D. A., Morrow, J. H., and Stolarik, G. F. "Remote Biological Monitoring in an Open Finished-Water Reservoir," *Jour. AWWA*, vol. 83, no. 9, pp. 107–112, September 1991.
22. Richardson, K J., Stewart, M. H., and Wolfe, R. L. "Application of Gene Probe Technology to the Water Industry," *Jour. AWWA*, vol. 83, no. 90, pp. 71–81, September 1992.
23. James M. Montgomery, Inc. *Water Treatment Principles and Design*, John Wiley & Sons, New York, 1985.

CHAPTER 2

Water Quality

2.1 Introduction

Pure water is colorless, tasteless, and odorless. It is an excellent solvent that can dissolve most minerals that come in contact with it. This is the reason that there is really no such thing as "pure water" in nature; it always contains chemicals and biological impurities.

During precipitation or passage over and through land, water picks up a wide variety of dissolved and suspended impurities. As rain, snow, sleet, or hail falls through the atmosphere, it dissolves oxygen, nitrogen, carbon dioxide, and other gases. It also picks up dust, dirt, soot, bacteria, acid fumes, and other impurities. Nevertheless, *it is the purest form of natural water.* At the ground surface, the water picks up silt, microorganisms, and a wide variety of organic and inorganic compounds. The water in streams normally contains higher concentrations of suspended matter than the water in lakes. As water percolates into the ground, most of the suspended solids and bacteria are removed; however, the percolating water will dissolve various minerals that come in contact with it during its passage through the soil strata. The amount and character of dissolved minerals depend on the length of the underground strata and the chemical make-up of the geological formations traversed. In this chapter, the common impurities of surface and groundwater and their significance are presented. Appropriate treatment processes for removal of these impurities are discussed in Chapter 3.

2.2 Water Quality Constituents

Natural waters may contain suspended and dissolved inorganic and organic compounds and microorganisms. These compounds may come from natural sources and leaching of waste deposits. Municipal and industrial wastes also contribute to a wide spectrum of both organic and inorganic impurities. The inorganic compounds originate from weathering and leaching of rocks, soils, and sediments. Principally, the inorganic constituents are calcium, magnesium, sodium, and potassium salts of bicarbonate, chloride, sulfate, nitrate, and phosphate. Lead, copper, arsenic, iron, and manganese may also be present in trace amounts. The organic compounds originate from decaying plants and animal matters and from agricultural runoffs. The organic compounds range from natural humic materials to synthetic organics used as detergents, pesticides, herbicides, and solvents. These constituents and their concentrations influence the quality and uses of the natural water resource. Typical concentrations of common water quality constituents in water from several sources are given in Table 2-1. It is the purpose of the following sections to examine in some detail various impurities of water and their impacts upon the water quality and treatment

2.2.1 Inorganic Constituents

Principal Anions and Cations The principal constituents of ionic species and their distribution in natural waters vary greatly depending on the geographical formations and soil type. Important ionic species in all natural waters that influence water quality and represent the principal chemical constituents are:

Cation	Anions
Calcium (Ca^{2+})	Bicarbonate (HCO_3^-) and Carbonate (CO_3^{2-})
Magnesium (Mg^{2+})	Chloride (Cl^-)
Sodium (Na^+)	Sulfate (SO_4^{2-})
Potassium (K^+)	Nitrate (NO_3^-)
Iron (Fe^{2+})	Phosphate (PO_4^{3-})
Manganese (Mn^{2+})	Fluoride (F^-)

The principle of electroneutrality requires that the sum of cations and anions expressed as meq/L should be equal. The acceptable limit of difference in charge balance is as follows:

Total Ions Sum, meq/L	Acceptable %Difference
0–3.0	± 0.2
3.0–10.0	± 2
10–800	± 2–5

Calcium. Calcium is derived mostly from rocks. The greatest concentrations come from limestone, dolomite, gypsum, and gypsiferrous shale. Calcium is the second major constituent, after bicarbonate, present in most natural waters. It is required as a nutrient for plants and is a required mineral for humans and other animals. Suggested daily intake is 800 mg for humans. The deficiency of calcium may cause osteoporosis, and toxicity may include kidney stones.[1] The levels found in water are not great enough to contribute significantly to daily demand (approximately 24 percent). Concentration of calcium in natural waters may range between 10 and 100 mg/L. Waters with a calcium level between 40 and 100 mg/L are generally considered hard to very hard. Calcium is a primary constituent of water hardness.

Table 2-1 Typical Concentrations of Common Water Quality Constituents in Different Water Sources

Constituent	Typical Water Source				
	Rain	Tap	Surface	Ground	Sea
Biological:					
Coliform, MPN/100 mL	0	1	2,000	100	—
Virus, pfu[a]/100 mL	0	0	10	1	0
Chemical:					
BOD_5, mg/L	—	—	5	—	2
TOC, mg/L	0.2	0.5	3	0.5	1
NH_3, mg/L as N	0.05	—	0.2	0.1	—
TOTAL N, mg/L as N	—	—	3	<10	0.5
TOTAL P, mg/L as P	—	—	0.05	0.01	0.01
Hardness, mg/L as $CaCO_3$	25	90	90	120	—
Alkalinity, mg/L as $CaCO_3$	20	80	100	150	—
pH	7.0	8.0	7.5	7.5	7.9
Cation:					
Ca^{2+},mg/L	6	20	20	50	400
Mg^{2+},mg/L	2	3	3	5	1350
Na^{+},mg/L	5	20	20	5	10,500
K^{+},mg/L	—	—	2	2	350
Fe^{2+},mg/L	0.05	0.1	0.1	0.1	0.1
Anion:					
HCO_3^{-},mg/L	18	80	90	120	150
Cl^{-},mg/L	5	25	25	25	20,000
SO_4^{2-},mg/L	4	20	20	10	2800
SiO_2,mg/L	1	5	5	10	20
NO_3^{-},mg/L	0.1	1	0.5	<10	—
F^{-},mg/L	—	2	0.2	0.1	1
Physical:					
TDS, mg/L	25	200	150	250	35,000
Turbidity, NTU[b]	0	<0.5	10	<0.5	5
Color, units	—	<5	—	—	5

a Plaque forming units.

b Nephelometric Turbidity Unit.

Magnesium. Sources of magnesium include ferromagnesium minerals in igneous and metamorphic rocks and magnesium carbonate in limestone and dolomite. Magnesium salts are more soluble than calcium, but they are less abundant in geological formations. Magnesium concentration may range from 4 to 40 mg/L in natural waters[1]. Magnesium is an essential mineral for humans, with an acceptable adult daily intake level of 350 mg. Drinking water can provide only a small fraction of daily needs. Lappenbusch reported that magnesium deficiency may result in electrolytic imbalance of calcium and potassium[2]. At high concentrations in drinking water, magnesium salts may have laxative effects. They may also cause unpleasant taste at concentrations above 500 mg/L. For irrigation purposes, magnesium is a necessary plant nutrient as well as a necessary soil conditioner. Magnesium is associated with hardness of water, and is undesired in several industrial processes.[3]

Sodium. The major source of sodium in natural waters is from weathering of feldspars, evaporates, and clay. Sodium salts are very soluble and remain in solution. Typical sodium concentrations in natural waters range between 5 and 50 mg/L[4]. Sodium is required in limited amounts for most plant growth. High concentrations of sodium in soil may be toxic to plants, while wide fluctuations of sodium concentration may be toxic to fish and other aquatic animals. The actual adult intake of sodium is 5700 mg/d, so sodium intake through the water supply is a small dietary component (less than one percent). Excessive sodium intake is linked to hypertension in humans. A deficiency may result in hyponatremia and muscle fatigue.[2] The recommended EPA limit of sodium in drinking water supply is 20 mg/L.[5]

Potassium. Potassium is less abundant than sodium in natural waters. Its concentration rarely exceeds 10 mg/L in natural waters.[4] In highly cultivated areas, runoff may contribute to temporarily high concentrations as plants take up potassium and release it on decay.[1] It is an essential nutrient to plants, animals, and humans, although excessive quantities can be deleterious to any of them. A dose of 1–2 g is reported to be cathartic to humans.[1] From the point of view of domestic water supply, potassium is of little importance and creates no adverse effects.[2,6] There is presently no recommended limit in drinking water supplies.

Iron. Iron is present in soils and rocks as ferric oxides (Fe_2O_3) and ferric hydroxides ($Fe(OH)_3$). In natural waters, iron may be present as ferrous bicarbonate ($Fe(HCO_3)_2$), ferrous hydroxide, ferrous sulfate ($FeSO_4$), and organic (chelated) iron. Groundwaters containing iron in soluble form (ferrous) are usually clear and colorless when first drawn. Upon contact with air, they slowly cloud and finally deposit a yellowish to reddish brown precipitate of ferric hydroxide. Iron-containing waters stain porcelain fixtures and laundry. Iron bacteria (*Crenothrix* and *Gellionella*) utilizes ferrous iron as an energy source and precipitate ferric hydroxide. In iron-bearing waters, the growth of iron bacteria may cause pipe clogging.[6]

Iron is an essential element, with a suggested daily intake of 14 mg. Most people ingest around 20 mg per day. The deficiency of iron may result in decreased hemesynthesis and anemia. The drinking water contributes only a small fraction of daily iron needs.[1,2] The EPA secondary drinking water regulations limit iron to 0.3 mg/L, for reasons of aesthetics and taste .[7]

Manganese. Manganese is present in rocks and soils. In natural waters, it appears with iron. Common manganese compounds in natural waters are manganous bicarbonate ($Mn(HCO_3)_2$), manganous chloride ($MnCl_2$), and manganous sulfate ($MnSO_4$). Presence of manganese in higher concentrations causes staining, bad taste, and growth of microorganisms. It is an essential element for both plants and animals, with a typical daily intake for humans being 4 mg per day.[2] Less than two percent of the daily requirement comes from drinking water. A deficiency of Mn may result in decreased enzymatic reactions in carbohydrate metabolism. The toxicity of Mn may include neurobehavioral changes.[2] The U.S. EPA secondary standard for aesthetic reasons for Mn is 0.05 mg/L.[7]

Bicarbonate-Carbonate. Bicarbonate is the major constituent of natural water. It comes from the action of water containing carbon dioxide on limestone, marble, chalk, calcite, dolomite, and other minerals containing calcium and magnesium carbonate. The carbonate-bicarbonate system in natural waters controls the pH and the natural buffer system. The chemical species that compose a carbonate-bicarbonate system include gaseous and aqueous carbon dioxide (CO_2), carbonic acid (H_2CO_3), and solids containing bicarbonate (HCO_3) and carbonate (CO_3^{2-}). The carbonate and bicarbonate constituents are the major source of *alkalinity* responsible for the acid-neutralizing capacity of the water. The carbonate equilibrium system in water treatment and supply is important and may be found in References 1 and 8–11. Additional information on this subject is provided in Section 8.3.3.

The typical concentration of bicarbonate in surface waters is less than 200 mg/L as HCO_3^-. In groundwaters, the bicarbonate concentration is significantly higher.

Chloride. Chloride in natural waters is derived from chloride-rich sedimentary rock. In typical surface waters, the chloride concentration is less than 10 mg/L.[1,6] Chloride concentration may be high in arid or semiarid regions and in areas subject to seawater intrusion, flow from hot springs, brine disposal ponds, salt crystals originating from sea sprays, seawater trapped in sediments, and places where concentration by evaporation occurs. Concentration of chloride rarely diminishes in aquatic system because chloride salts are highly soluble.

Chloride is an essential element. Drinking water contributes to only a small fraction of daily intake.[2] In drinking water, chloride concentration exceeding 250 mg/L causes a salty taste. The EPA secondary standard provides a maximum limit of 250 mg/L.

Sulfate. Most sulfate compounds originate from the oxidation of sulfite ores and solution of gypsum and anhydride. Sulfate is also present in rainwater. The sulfate concentration in natural waters is generally less than 50 mg/L.[1] The bacteria reduction of sulfate produces H_2S as a by-product.

Sulfate is a non-essential contaminant.[2] High concentrations of sulfate in drinking water for individuals unaccustomed to consuming sulfate-rich water (above 300 mg/L) may cause laxative action. Tolerance to the laxative reaction can be developed over time. The taste threshold for sulfate in drinking water is around 300–400 mg/L, and medicinal or bitter taste is produced if the concentration exceeds 500 mg/L.[2] The EPA secondary maximum limit for sulfate is 250 mg/L.

Nitrate. Nitrogen is a complex element. In natural waters, it exists in many forms: organic nitrogen (proteins, amino acids, urea), ammonia (NH_3), nitrogen gas (N_2), nitrite (NO_2^-), and nitrate (NO_3^-).[8] In nature, nitrogen is cycled between organic and inorganic forms. Bacteria decompose organic matter (such as proteins) to ammonia, then to nitrite, and finally to nitrate. On the other hand, plants are responsible for the production of proteins from several inorganic forms of nitrogen, including NO_3^-, N_2, and NH_3. In addition to decaying organic matter, fertilizers are a major source of nitrate in water supplies. Surface waters generally contain nitrate concentrations below 1 mg/L unless polluted by municipal and industrial wastewaters, decaying organic matter, or agricultural chemicals.[1]

The major source of nitrogen in human diet is proteins. Nitrate is a non-essential contaminant with no minimum daily requirement.[2] In excessive amounts, it may cause infant *methemoglobinemia*. For this reason, the EPA primary standard of nitrate in drinking water is 10 mg/L as nitrogen.[5]

Phosphate. Phosphorus, like nitrogen, is of great significance in natural waters. They are the major nutrients and are the principal cause of *eutrophication* in natural waters. Serious water quality problems have resulted from uncontrolled algae and aquatic plant blooms in lakes and reservoirs that contained phosphorus in excess of 0.05 mg/L.[12] Phosphorus is an essential element. Suggested adult daily intake for phosphorus is 800 mg; the actual intake through diet is 1500 mg/day.[1,2] A deficiency in phosphorus may result in weakness, bone pain, and rickets.[2] Phosphorus contribution from water is insignificant, and EPA does not regulate it.

Fluoride. Fluoride occurs in few types of rocks and is only slightly soluble in water. Normally, fluoride concentration is less than 0.5 mg/L in natural waters; however, higher fluoride often occurs in aquifers.[6] Moreover, both surface waters and groundwaters may experience fluoride contamination from certain insecticides, from chemical wastes, and from airborne particles and gases from aluminum smelting plants.

Fluoride is an essential constituent and is utilized in the structure of bones and teeth. Suggested daily intakes for infants, children, and adults are 0.6, 1.0, and 2.7 mg, respectively.[1,2] A deficiency of fluoride may result in increased dental cavities. Large intakes of fluoride may cause *dental fluorosis (mottling)* and toxicity. A significant portion of fluoride comes from drinking water; therefore, its control in water supply is important. The addition of fluoride in water supply has been beneficial. The National Primary Drinking Water Standards limit the amount of fluoride on the basis of the annual average of the maximum daily temperature. This is because water intake in children and adults is related to air temperature. The concentration ranges from a low of 1.4 mg/L in hot areas to a maximum of 2.4 mg/L in cold regions.[5]

Minor or Trace Constituents There are many minor inorganic ionic species and nonionic minerals that may also be present in water. These constituents either are derived from the contact of water with various mineral deposits or are discharged by the industries. Many of these constituents are essential for life, while others are nonessential and toxic. The essential inorganics can also be toxic, depending upon the intake dosage. Basic information on many selected minor ionic species and nonionic constituents in terms of health effects and daily intake is listed in Table 2-2.[2–4]

Inorganic Water Quality Indicators Several water quality parameters are used to describe properties of water that are indicative of treatability and finished quality of water. These are alkalinity, hardness, total dissolved solids, conductivity, dissolved oxygen, turbidity, sodium adsorption ratio, and corrosion and scale formation.

Alkalinity. Alkalinity is a normal characteristic of natural waters that provides buffering capacity and maintains the pH of water within the range from 6.0 to 8.5. Alkalinity in natural waters is due primarily to the presence of bicarbonate and carbonate ions, although, in many cases, a small portion of the alkalinity may also be due to salts of silicate, borate, phosphate, organic acids, and hydroxides. Alkalinity of water varies in concentration with geographical location, depending particularly upon the character of the rocks and soils in the area. Sudden changes in alkalinity of streams are generally due to discharge of treated or untreated municipal and industrial wastes. For a healthy fresh-water aquatic habitat, the desirable bicarbonate alkalinity range is 30 to 130 mg/L as $CaCO_3$; the pH is well maintained within the range of 6.5 to 8.2. A minimum alkalinity of 30 mg/L as $CaCO_3$ is necessary for effective coagulation. Alkalinity results are needed to calculate lime and soda-ash dosage for water softening and to determine the corrosive or scaling action of water.[3,12]

Hardness. Hard water is generally associated with formation of scales in boilers, heaters, and hot-water pipes and on utensils and surfaces with which it comes into contact. Hard water also prevents formation of a lather until excessive soap is consumed and causes yellowing of fabrics and toughening of vegetables. Hardness in natural waters is due primarily to calcium and magnesium ions. Other ions such as strontium, iron, manganese, aluminum, copper, barium, zinc, lead, and other bivalent trace metals also cause hardness; however, these substances are found in very low concentrations in natural waters, so their effects generally are ignored. Areas that have limestone formations generally have hard water. Also, groundwater in a given area may have more hardness than surface water. Based on the hardness, the water is classified as soft, moderately hard, hard, very hard, and extremely hard. The classification criteria are summarized in the following table.[8,12]

Total hardness, mg/L as CaCO3	Classification
0–40	soft
40–100	moderately hard
100–300	hard
300–500	very hard
greater than 500	extremely hard

Table 2-2 Health Effects and Daily Intake of Many Minor Ionic Species and Nonionic Constituents in Water Sources Present Naturally or Added by Industrial Wastes

Constituent	Significance	Total Daily Intake, mg/d	Deficiency	Excess or Toxicity
Aluminum, Al^{3+}	Nonessential	30, less than 1% from water	None	Neuron disorder, (Alzheimer's disease)
Antimony, Sb^{2+}	Nonessential	None	None	Unknown
Arsenic, As^{+}	Not known	0.07, less than 5% from water	None	Skin disorder, cancer
Asbestos	Nonessential	None	None	Gastrointestinal irritation and cancer
Barium, Ba^{2+}	Nonessential	0.8, 11% from water	None	Cardiovascular disease and hypertension
Cadmium, Cd^{2+}	Nonessential	None	None	Kidney damage
Chromium, Cr^{6+}	Essential	0.12, 13% from water	Atherosclerosis	Tubular necrosis of kidney
Copper, Cu^{2+}	Essential	1.7, 12% from water	Anemia, loss of pigment, reduced growth and loss of arterial elasticity	Disorder of Cu Metabolism, and hepatic cirrhosis
Cyanide, CN^{-}	Nonessential	Unknown	None	Inhibition of cellular respiration and central nervous system, neuropathies
Iodine, I	Essential	0.51, 2% from water	Thyroid disorder, goiter (enlarged thyroid)	Goiter with possible hypothyroidism
Lead, Pb^{2+}	Nonessential	0.12, 50% from water	None	Anemia, nervous system impairment, mental retardation
Mercury Hg^{2+}	Nonessential	0.007, 27% from water	None	Kidney disease, and proteinuria methyl mercury affects nervous system
Molybdenum, Mo^{2+}	Essential	0.35, 3% from water	Reduced quantity of metalloenzymes	Elevated uric acid, gout, bone and joint deformities
Nickel, Ni^{2+}	Not known	0.34, 3% from water	Not known	Gastrointestinal irritation and dermatitis
Selenium, Se^{4+}/Se^{6+}	Essential	0.19, 8% from water	Liver injury	Growth inhibition, damage to liver and central nervous system, gastrointestinal disturbances and dermatitis
Silver, Ag^{+}	Nonessential	7% from water	None	Argyria
Vanadium, $^{2+}$	May or may not be essential	30% from water	Unknown	Gastrointestinal irritation, cramps and diarrhea
Zinc, Zn^{2+}	Essential	13, 3% from water	Reduced appetite and growth	Irritability, muscle stiffness and pain, loss of appetite and nausea

Source: Adapted in part from References 1, 2, 4, and 13.

Hard waters have had no harmful effects upon the health of people. Hardness over 100 mg/L in municipal water supplies has been found inconvenient by reason of excessive soap consumption and incrustation of utensils.[4] Recent studies, however, relate soft water to heart disease, simply because the soft water is corrosive and dissolves potentially harmful substances from the piping and plumbing.[13]

Total Dissolved Solids (TDS). Total dissolved solids in natural water are due to inorganic salts. In rivers and lakes, the concentration of TDS in water will vary depending on the time of year, local geological conditions, climate, and waste discharges. Classification of water into fresh, brackish, saline, and sea water is usually based on the content of TDS[a].[1] The TDS concentration in rivers and lakes is generally less than 200 mg/L.[1,8] TDS is not a measure of the safety or harmfulness of a water but is tied to a secondary or aesthetic standard. The EPA recommends a maximum of 500 mg/L TDS in drinking water supplies.[7] Above this level the water may have an increasingly salty taste. The recommended upper limit of TDS in irrigation water is 1500 mg/L.[3]

Conductivity. The electrical conductivity (EC) or specific conductance of water is related to TDS. It is a measure of the ability of the water to conduct an electrical current. It is reported into *microsiemens* per centimeter (also called *micromhos* per centimeter). Because the electrical current is transported by the ions in solution, the conductivity increases as TDS increases. The relationship is nearly linear, depending on TDS concentration, and must be determined for individual cases. The relationship is expressed by Eq. (2.1):

$$TDS = k \cdot EC \tag{2.1}$$

where

TDS = total dissolved solids, mg/L
EC = electrical conductivity, μmho/cm
k = constant

The value of k varies from 0.5 to 0.9 depending on TDS and water temperature. Typical value of k for most waters is 0.64.[8]

Dissolved Oxygen (DO). In nature, clean waters are saturated with DO, or nearly so. If organic wastes are discharged into the natural waters, microorganisms decompose these wastes and utilize DO. The level of DO in natural water is an indication of extent of pollution by oxygen-demanding wastes. Low DO concentrations are likely to be associated with low-quality water, and water treatment plants may face taste and odor problems. In treated water, the DO is one of the most important factors influencing the rate of corrosion of all the metals.[1]

a. Classification of brackish, saline, and sea water is based on TDS of 1000–5000, 5000–30,000, and 30,000–35,000 mg/L, respectively.[13]

Turbidity. Turbidity in water is caused by the presence of such suspended matter as clay, finely-divided organic and inorganic matter, and microorganisms.[14] Turbidity represents the optical property of a water which causes light to be scattered and absorbed rather than transmitted. Attempts to correlate turbidity with suspended solids concentration are impractical because the size, shape, and refractive index of the particulate materials are important optically but bear little direct relationship to the concentration and specific gravity of the suspended matter.[1]

The standard method for the determination of turbidity is based on the Jackson Candle Turbidimeter. However, the lowest turbidity value that can be measured directly on this instrument is 25 *Jackson turbidity units* (JTU). With turbidities of treated waters expected to fall within the range of 0 to 1 units, indirect secondary methods are needed. Nephelometers are now the standard instrument for measurement of low turbidities and give results in terms of the *nephelometric turbidity unit* (NTU)[b].[1,2,15]

The recent literature on the subject of water-borne diseases has stressed very strongly the need to reduce the turbidity as much as possible. Poor *cysts* removal of *Giardia, Cryptosporidium,* and *Entamoeba* has been noted when coagulation doses are low or less than optimal, when filter-water turbidities increase during filter runs, and during the filter ripening period. Specific health-related characteristics of turbidity include the association of microorganisms with particulate material; hosting of viruses, cysts, and other microorganisms by minute particulate matter in suspension; interference with disinfection; and distinct turbidity-related organic precursors and chlorine demand.[8] It is, therefore, important that safe and wholesome drinking water have a very low turbidity to qualify as safe water for municipal water supply.[1,8]

Particle Count. Although turbidity, the traditional parameter, is used extensively in water treatment, particle counts are currently gaining popularity to evaluate treatment efficiencies and finished water quality. Several types of particle counters are marketed that have sensors in different size ranges (such as 1.0–60 μm, or 2.5–140 μm). A particle count of 50–100 particles/mL (in the size range of 2.5–150 μm) represents finished water quality having turbidity units of 0.2–0.3 NTU[1]. One advantage of particle counters is that removal efficiency of particles in the specific size range of *Giardia* cysts can be measured.

Sodium Adsorption Ratio. In agricultural irrigation waters, the bulk of cations are Ca^{2+}, Mg^{2+} and Na^+; other ions are generally neglected in determining the risk of soil dispersion. The *sodium adsorption ratio (*SAR*)* is commonly used to predict water-permeability problems. SAR is calculated from Eq. (2.2), in which all cations are expressed in meq/L units.[3,8]

$$SAR = \frac{[Na^+]}{\sqrt{\dfrac{[Ca^{2+}] + [Mg^{2+}]}{2}}} \qquad (2.2)$$

b. NTU is a measure of light scattered by a formazine polymer. Approximately, 1 NTU is equal to 1 JTU.

Water having SAR values between 8 and 18 may have an adverse effect on the permeability of soils containing an appreciable proportion of clay because its use causes undesirable amounts of sodium to be adsorbed. Where used on sensitive crops, SAR values above 4 may be detrimental because of sodium phytotoxicity.[3]

Stability of Water. Water that is unstable will either deposit scale or corrode the pipes in the distribution system. *Stabilization* of water is the term applied to correction in the chemical nature of the water so that it will neither corrode the pipes through which it passes nor deposit an incrusting film of calcium carbonate. Stabilization is, therefore, adjustment of the pH and alkalinity of a water to its calcium carbonate saturation equilibrium value.

The application of some of the principles of physical chemistry, especially the law of mass action and carbonate equilibria, have provided means to interpret the stability of water.[17] Several procedures and indexes have been proposed over the years to help to predict and measure the tendency of finished water to deposit or dissolve calcium carbonate and, thus, to protect the distribution system against corrosion and scaling. Two of these procedures, the *marble test* and *Enslow's Stability Indicator*, are practical laboratory methods. The other indexes are largely related to and derived from mass law and stoichiometric equations for electrical neutrality in water. These indexes are: (1) *Langelier's Calcium Carbonate Saturation Index*, (2) *Ryznar's Stability Index*, (3) *The Momentary Excess (M.E.)*, and (4) *The Driving-Force Index (D.F.I.)*.[8,16] Use of some these indexes in achieving stable water that is fit for distribution to the consumer is presented in Chapter 13.

2.2.2 Organic Constituents

The organic compounds in natural waters are derived from natural decomposition of plants and animal material. The most commonly occurring natural organic matter (NOM) found in surface waters is humic, fulvic, and hymatomelanic acids, algae, and other microorganisms.[14,16] Organic compounds may also reach surface waters from municipal and industrial discharges and from runoffs from urban and agricultural lands. Concentrations of organics may range from none (in protected groundwaters) to 10–30 mg/L in naturally productive or contaminated surface waters.[1]

Organic compounds in natural waters present various types of problems. Among these problems are (1) turbidity and color formation, (2) dissolved oxygen depletion, (3) taste and odor problems, (4) interference with water treatment processes, and (5) formation of halogenated compounds when water is chlorinated.[1,8]

Natural Organic Matter Naturally occurring organic matter in an aquatic system originates from plant and animal residues. These organic compounds contain mainly proteins, carbohydrates, and lipids. Humification of these substances produces a broad category of organic compounds that are collectively called *humic substances*; the term simply denotes a group of compounds having similar properties. These substances (in water) are polymers with widely varying molecular weights ranging from less than 800 to as high as 50,000. The humic substances contain a variety of chemical groups such as hydroxyl, carboxyl, methoxyl, and quinoid groups, which give hydrophilic properties. The magnitude of charge is dependent on the degree of ionization and consequently on the pH of the water.[1,8,17]

Synthetic Organic Compounds More than 100,000 organic compounds have been synthesized since 1940. Many of these compounds are used widely. Their concentrations in both groundwaters and surface waters are increasing as a result of excessive use, discharges, spills, and intentional dumping. Many of these compounds, even at extremely low concentrations, are carcinogenic (oncogenic or tumor-causing), *mutagenic* (causing mutation in humans and other living forms), or teratogenic (causing disfiguring).[1,2,3,17] Principal classes of manufactured organic compounds of concern from the health, treatment, and ecological standpoint are surfactants, pesticides and herbicides, cleaning solvents, polychlorinated biphenyls, and disinfection by-products (trihalomethanes). Basic properties and representative members of many classes are given in Table 2-3.[1,2,8,18]

Measurement of Organic Content The concentration of organic matter in natural waters is generally small. Two types of measurements are commonly used: nonspecific, and specific.[1,8] Nonspecific measurements include BOD_5, COD, TOC, TOD, ThOD, color, ultraviolet absorbance, and fluorescence. Specific measurement involves quantification of certain compounds from the total organics in the water. The procedure involves gas chromatography, mass spectroscopy, and high-pressure liquid chromatography. Brief descriptions of these methods are provided in Table 2-4.[1,8,19]

2.2.3 Microorganisms

Water has long served as a mode of transmission of diseases. The most important of the water-borne diseases are those of the intestinal tract, including typhoid fever, paratyphoids, dysentery, infectious hepatitis, cholera, and some parasitic worm diseases. Many organisms cause bad taste and odor, corrosion, and slime production and, in general, are associated with "*nuisance*" in raw water transport, treatment, storage, and distribution. Major groups of organisms of interest in water supply and treatment are bacteria, viruses, fungi and mold, algae, protozoa, and helminths or parasitic worms. The purpose of this section is to introduce some basic concepts from microbiology and to present the importance of many microorganisms in water treatment and supply.

Basic Concepts Microorganisms are grouped into three kingdoms: *protista, plants,* and *animals*. The members of the kingdom protista are called "protists" and include *algae, protozoa, fungi, slime molds,* and *bacteria*. These organisms are distinguished from the plant and animal kingdoms on the basis of a lack of differentiation of cells and tissues. Each cell contains nucleic acid (deoxyribonucleic acid (DNA)), which contains genetic information that is considered vital for cell reproduction.

In algae, protozoa, fungi, and slime molds, the *nucleus* is clearly defined (eucaryotic cells). In bacteria and blue-green algae, the nucleus is poorly defined (prokayotic cells). Viruses

Table 2-3 Basic Properties and Representative Members of Different Classes of Synthetic Organic Compounds

Class	Description	Trade or Common Name
Synthetic detergents	Synthetic detergents, commonly called *syndets*, are a substitute for soap. They contain 20 to 30 percent *surfactant* and 70 to 80 percent builders. There are anionic sulfactants (sulfates and sulfonates), non-ionic syndets, and cationic syndets.	Alkyl benzene sulfonate (ABS), Linear alkyl sulfonate (LBS)
Pesticides	Includes insecticides, algicides, fungicide, and herbicides. Used extensively for control of objectionable insects, rodents, plants, weeds, and other forms of life. The major source of these chemicals is surface runoff from agricultural, residential, and park lands. These chemicals are toxic to aquatic organisms and are known or suspected carcinogens. These chemicals are grouped into four major categories based on molecular structure: chlorinated hydrocarbons, organophosphates, carbamates, and urea derivatives.	Chlorinated hydrocarbons: aldrin, chlordane, DDT, dieldrin, endrin, heptachlor, lindane, methoxychlor, toxaphane. Organophosphate: diazinon, malathion, parathion. Chlorophenoxys (herbicides): 2,4-D, 2,4,5-TP, silvex. Carbamate: captan, ferbam IPC, sevin. Ureas: fenuron
Cleaning solvents	Organic solvents are extensively used in the industry. Solvents have often leaked into underlying aquifers. Many solvents are suspected carcinogens.	Acetone, benzine, carbon tetrachloride, chlorobenzene, dichloromethane, trichloroethylene
Polychlorinated biphenyls (PCBs)	PCBs are mixtures of chlorinated biphenyls. These are found in transformers and capacitors. They persist in the ecosystem once released.	
Disinfection by-products (DBPs)	DBPs are produced during disinfection8 of treated water by the action of the disinfectant on organic compounds (known as *precursors*) in the water. Trihalomethanes (THMs) are the products with organics in water and chlorine (or other halogens such as bromine and iodine). THMs and other chlorinated hydrocarbons are suspected carcinogens.	Chloroform ($CHCl_3$), Bromoform ($CHBr_3$), Bromodichlromethane ($CHCl_2Br$), Chlorodibromomethane ($CHClBr_2$)

Source: Adapted in parts from References 1, 2, 8, and 18.

Table 2-4 Methods Commonly Used For Measurement of Organic Matter in Water Samples

Measurement	Description
Nonspecific	
BOD_5	Biochemical oxygen demand (5-d, 20°C), represents the amount of oxygen required to stabilize the organic matter by microorganisms under aerobic condition. It represents the biodegradable portion and is conducted over five days and at 20°C.
COD	Chemical oxygen demand represents the amount of oxygen required to oxidize the organic matter by a strong oxidizing chemical (potassium permanganate) under acidic condition.
TOC	Total organic carbon represents organic carbon present in the organic matter. It is measured by converting organic carbon to carbon dioxide in a high-temperature furnace in the presence of a catalyst. DOC (dissolved organic carbon) represents the soluble fraction of TOC.
TOD	Total oxygen demand measurement uses the conversion of organic compounds into stable oxides in a platinum- catalyzed combustion chamber.
ThOD	Theoretical oxygen demand measurement is based on a stoichiometric relation when the chemical formula of the organic matter is known
Color	Most organic matters in water cause color. Therefore, color measurement is used to quantify organics. Calorimetric analysis is performed to measure the color.
UV Absorbance (UV254mm)	Ultraviolet absorbance is used to quantify groups of organic compounds, such as aliphatic, aromatic, complex multi-aromatic, and multiconjugated humic substances.
Fluorescence	Some organic compounds absorb UV energy and then release energy at some longer wavelength. This phenomenon provides a basis for measurement of organics.
Specific Organic Compounds	
Gas chromatography	The sample is vaporized and swept by a carrier gas through a chromatographic column. The emergence of the compound is detected and measured.
Mass Spectroscopy	The sample is vaporized and the compounds are separated by gas chromatography. The bombardment of the organic molecule by the rapidly moving electrons breaks the organic molecule into charged fragments. The mass-to-charge ratio of each fragment provides quantitative analysis.
High-pressure liquid chromatography	The carrier stream is composed of a solvent (or a mixture of solvents) maintained under high pressure. The compounds are separated in a solid or liquid stationary phase and measured.

Source: Adapted in parts from References 1, 8, 12, 15, 18, and 19.

fall between living and nonliving. They are not complete organisms, being made up of a protein-protective coating surrounding a strand of nucleic acid.[8,20]

Microorganisms have different nutritional and environmental requirements, such as carbon source, energy source, oxygen, temperature, and the like. Basic classifications based on these requirements are given in subsequent subsections.

Autotrophic organisms derive their cell carbon from carbon dioxide. *Heterotrophic* organisms use organic carbon for cell synthesis. *Phototrophs* derive energy for cell synthesis from light. *Chemotrophs* derive their energy from chemical reaction. Based on oxygen requirements, the microorganisms may be grouped into *aerobic*, *anaerobic*, and *facultative*. Aerobic microorganisms require the presence of molecular oxygen (O_2) for their metabolism. Anaerobic microorganisms grow in absence of molecular oxygen. In fact, molecular oxygen is toxic to them. Facultative organisms can grow in the presence or absence of oxygen. The temperature plays an important role in the growth and survival of microorganisms. In accordance with the temperature range, the microorganisms may be classified as *psychrophilic* (or *cryophilic*), *mesophilic*, and *thermophilic*. The optimum temperature ranges for these groups of organisms are 12–18°C, 25–40°C, and 55–65°C, respectively. The pH of water also influences the growth of organisms. Most organisms have an optimum pH in the range from 6.5 to 7.5.[8,20]

Because of their importance in water quality, treatment, and distribution, (1) bacteria, (2) viruses, (3) fungi and molds, (4) algae, (5) protozoa, (6) worms, and (7) snails, crustaceans, and other nuisance-causing organisms are briefly discussed below.

Bacteria Bacteria are single-cell protists. Their importance in water supply is great, because they may cause disease, taste and odors, pipe corrosion, and pipe blockages. Most bacteria are grouped by form as spherical (cocci), rod-shaped (bacilli), curved rod-shaped (vibrios), spiral (spirilla), or filamentous. A schematic diagram of bacterial cells and of typical shapes of bacteria is shown in Figure 2-1. Most bacteria are harmless, and many are actually beneficial: some consume organic detritus, and some produce by-products that inhibit growth of many pathogens. In general, pathogens are best suited to the environment inside the human body and tend to die off rapidly in nature. There are some pathogens that are hardy and persistent. Some organisms form spores and are resistant to chlorination. Common bacterial diseases associated with contaminated waters are given in Table 2-5.

There are several groups of bacteria that are nonpathogenic but are undesirable because they are a nuisance. Actinomycetes are filamentous bacteria and cause musty and earthy tastes and odors in water supply. Hydrogen sulfide-producing bacteria cause a familiar rotten egg smell. Iron and manganese bacteria cause a severe clogging or corrosion of pipes (producing red water) associated with taste, coloration, and staining.

Viruses Viruses are the smallest of infectious agents, some being as small as a single protein molecule. They are obligate parasites and, as such, require a host. Common virus-caused diseases are given in Table 2-5. Viruses are more resistant to disinfection than bacteria, but in natural waters they are present in far fewer numbers than bacterial pathogens.

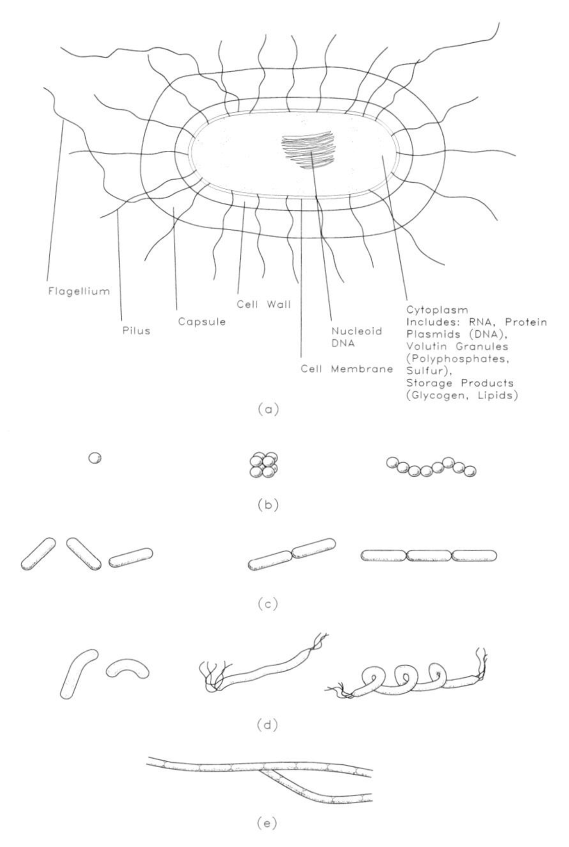

Figure 2-1 Bacterial cellular structure and typical forms of bacteria. (a) Schematic bacterial cellular diagram. (b) Spherical *(Cocci)*. (c) Rod *(Bacilli)*. (d) Curved rod *(Vibrios)* and spiral *(Spirilla)*. (e) Filamentous.

Fungi and Molds These are aerobic, single or multicellular, heterotrophic microorganisms. Most fungi and molds are saprophytes (obtaining food from dead organic matter). The fungi and molds frequently form dense, slimy mats that clog filters and other water treatment equipment. They also grow on the walls and weirs of water treatment units. They can also produce musty taste and odors, as well as color and turbidity.

Algae Algae are simple organisms that are autotrophic and photosynthetic and contain chlorophyll. Many algae also contain different pigments and, therefore, may have various colors. Algae produce their own food from sunlight and nutrients. Some algae produce slime that interferes with treatment processes. During algae blooms in reservoirs, they cause turbidity and color and interfere with coagulation and sedimentation, causing serious filter clogging. The chemical substances produced by algae are precursors and are also associated with different taste and odor

Table 2-5 Diseases Associated with Contaminated Water

Diseases	Causative Agent	Symptoms	Incubation Period
Viruses			
Hepatitis	Virus unknown	Fever, nausea, loss of appetite, fatigue, jaundice	30–35 days
Poliomyelitis[a]	Poliomyelitis virus	Headaches, nausea, vomiting, stiff neck, nasal voice, runny nose, difficulty in swallowing	3–21 days
Bacteria			
Cholera	*Vibrio comma*	Extremely heavy diarrhea, rice-water stool, vomiting, thirst, pain, dehydration, high death rate	A few hours to 3 days
Gastroenteritis	Bacteria, virus, chemicals, and toxins	Vague term for abdominal discomfort, characterized by mild to severe diarrhea, nausea, indigestion, vomiting, cramps, and possibly fever	8–12 hours
Leptospirosis (Weil's disease)	*Leptospira*	Fever, rigors, headaches, nausea, muscular pains, vomiting, prostration, jaundice	9–10 days
Paratyphoid Fever	*Salmonella paratyphi A*	Continued fever, diarrhea, occasional rose spots on body	1–10 days for gastroenteritis, 1–3 weeks for enteric fever
Salmonellosis	*Salmonella typhimurium*	Abdominal pain, diarrhea, vomiting and nausea, chills, fever	12–24 hours
Shigellosis (Bacillary dysentery)	*Shigella*	Diarrhea, fever, mucus and blood in stool	2–4 days
Typhoid fever	*Salmonella typhosa*	Continued fever, usually rose spots on body, abdominal discomfort	7–21 days

Table 2-5 Diseases Associated with Contaminated Water (cont'd)

Diseases	Causative Agent	Symptoms	Incubation Period
Protozoa			
Amebiasis (amoebic dysentery)	*Entamoeba histolytica*	Prolonged diarrhea, abdominal discomfort, blood and mucus in stool, abscesses of liver and small intestine	5 days or longer, averages 3–4 weeks
Giardiasis	*Giardia lamblia*	Mild to severe diarrhea, nausea, indigestion, flatulence	1–4 weeks
Cryptosporidiosis	*Cryptosporidium parvum*	Diarrhea, abdominal cramps, fever, nausea, and vomiting	Few days to weeks
Worms			
Ascariasis (Intestinal roundworm)	*Ascaris lumbricoides*	Worm in stool, abdominal pain, skin rash, protuberant abdomen, nausea, large appetite	About 2 months
Dracunculiaosis (Guinea worm)	*Dracunculus medinensis*	Abnormal redness of skin, itching, giddiness, difficulty in breathing (similar to asthma), vomiting, and diarrhea. Symptoms gradually manifest themselves over the entire incubation period and become pronounced a few hours before the appearance of the worm beneath the skin.	8–12 months
Paragonimiasis (Lung fluke)	*Paragonimus ringeri* and others	Chronic cough, clubbed fingers, dull pains, diarrhea	Variable
Schistosomiasis	*Schistosomes*	Worm enters blood stream. Itching of skin, dermatitis rash, body pain, dysentery, anemia, vigors. Snail is intermediate host. The disease is prevalent in tropics and Nile Deltas.	4–6 months, or longer

a Although there is no positive evidence of spread from contaminated water, there are certain facts from which health departments have made recommendations about water supply, waste disposal, swimming pools, and swimming in natural bodies of water during outbreaks.

problems.[1,20,21] Impact of algae on water quality and their control methods are covered in Chapter 11.

Protozoa Protozoa are a group of unicellular, nonphotosynthetic, aerobic organisms. Several protozoa cause disease. *Giardia lamblia, Entamoeba histolytica,* and *Cryptosporidium* are transmitted by drinking water (Table 2-5); *Nagleria fowleri* may enter by nasal inhalation exposure from swimming in polluted water and causes amoebic meningoencephalitis.

The outbreak of giardiasis, amoebic dysentery, and crytosporidiosis in the United States has been attributed to consumption of untreated or inadequately filtered drinking water. The cyst of these organisms have a long survival time, even in the presence of chlorine residual. It has been observed that poor cyst removal generally occurs when coagulant doses are less than optimum and when filtered water turbidities increase during filter runs and during filter ripening periods.[1,22] Also, turbidity interferes with the disinfection process; therefore, the best protection against cysts, viruses, and other pathogens in water supply is achieved with low turbidity through optimum coagulation and stable filter operation.

Helminths (Parasitic Worms) Parasitic worms cause many diseases. Common diseases are listed in Table 2-5. The transmission of worm and eggs is due to contaminated drinking water or vegetables and to body contact with polluted waters. Effective coagulation and filtration for turbidity control is a viable method of removing helminths from water supply.[1,8]

Nuisance-Causing Organisms Many organisms, snails (mollusca), crustaceans, and slime growth cause serious problems in raw-water conveyance. Snails and slime growth on the walls of the raw water lines result in reduced pipe capacity. Prechlorination of raw water has been used to reduce the growth on pipe walls. Another method of reducing slime growth is hydraulic pressure cleaning using an internal pipe scraper, commonly called a *pig*. Other devices include a bullet-shaped plastic ball with crisscrossed strips to scrape the interior pipe surface and a series of scraping and polishing blades of high-carbon spring steel mounted in a series on a rod.[23] Excessive growth of crustaceans and weeds in the reservoirs has caused clogging of pumping equipment.

Indicator Organisms In order to determine the presence of pathogenic organisms in natural and treated water, the microbiologist must have a reliable measurement technique. Unfortunately, the analytical procedures for detection of pathogenic organisms are not clear-cut. Hence, rather than to look for the specific pathogens, there is a need to find a group of indicator organisms to measure the potential of a water to transmit diseases. The ideal indicator organism should have the following characteristics:[1,8,20]

1. Detection should be quick, simple, and reproducible.
2. Results should be applicable to all waters; that is, the numbers should correlate with the degree of pollution (higher numbers in sewage, less in polluted waters, absent in unpolluted waters).

3. It should have greater or equal survival time and be present in larger numbers than the pathogen.
4. It should not grow in nature.
5. It should be harmless to man.

Naturally, no organism (or group of organisms) possesses all these characteristics. However, the *coliform* organisms currently used as indicator organisms have many of these qualifiers. These are nonpathogenic bacteria whose origin is in fecal matter. The presence of these bacteria in water is an indication of fecal contamination and probably unsafe water. The coliform bacteria are *gram-negative*, nonspore-forming bacilli capable of fermenting lactose with the production of acid and gas. The coliform bacteria are members of the *Entrobacteriacae* family, and include genera *Escherichia, Klebsiella, Citrobacter,* and *Entrobacter.* The *Escherichia coli* (*E. coli*) species appears to be most representative of fecal contamination. Fecal *Streptococci* and *Enterococci* are also used as indicator organisms under specific conditions.

The coliform organisms were originally believed to be entirely of fecal origin, but it has been shown that certain genera can grow in soil. Therefore, presence of coliform organisms in surface water may be due to fecal wastes from human and animal sources or from soil erosion. To separate fecal coliforms (*E.coli*) from possible soil types, special tests are generally run. Frequently, for this purpose, fecal *Streptococci* are also used as indicator organisms, as they also originate from the intestinal tract of warm-blooded animals.[15]

The standard techniques used to enumerate the coliform organisms are (1) multiple-tube fermentation and (2) the membrane filter. The multiple tube fermentation technique uses a lactose broth medium that is fermented by the coliform group. Gas bubbles collected inside an inverted inner vial are an indication of gas formation. The test may be carried out to presumptive, confirmed, or completed test levels. A series of dilutions in multiple tubes is used for statistical enumeration. The result is reported as *most probable number* per 100 mL (MPN/100 mL).[15] Standard MPN index tables of 95% confidence limits for a various number of tubes at dilutions of 10, 1.0, and 0.1 mL are provided in the Standard Methods.[15]

The membrane-filter technique is used to quantify the coliform organisms present in the water. This technique is faster and gives the actual number rather than the most probable number. In a standard filtration apparatus, a dilute sample is filtered through a membrane filter having a rated pore diameter of 0.45 μm. The filter is placed over an absorbent pad containing Endo-type selective media and incubated. The colonies with a golden-green metallic sheen are counted. The procedure is given in the Standard Methods.[15]

2.2.4 Radionuclides

Natural waters contain low levels of radioactivity. Increased radioactivity in water may be naturally-occurring or man-made. The naturally-occurring radioactivity may come from the earth's crust or from the cosmic-ray bombardment in the atmosphere. The man-made radioactivity may come from the mining and processing of ores, applications in industry and medicine, or disposal and testing of nuclear weapons. The radioactivity is due to release of *alpha*, *beta*, or *gamma* radiation. Radiation in all forms has different health effects.[2,8]

The units of radioactivity are the *curie* (ci) and the *rem*. One curie equals 3.7×10^{10} atom disintegrations per second (disintegrations per gram of radium). A common unit is the *picocurie* (pCi), which is equal to 10^{-12} curie. The background gross radioactivity in natural waters is 0.1 pCi/L. The objective of the water supply industry is to have gross radioactivity of less than 10 pCi/L, with a maximum permissible limit of less than 100 pCi/L.[1,2,8]

A *rem* (*Roentgen-equivalent-man*) is a measure of dosage of radiation exposure per unit of time.[18] The recommended *Maximum Permissible Dose* (MPD) for radiation workers is 5 *rem* per year and for nonradiation workers is ∫ rem per year.[1,2]

Problems and Discussion Topics

2.1 A water sample has Ca^{2+} and Mg^{2+} concentrations of 25 and 10 mg/L as Ca and Mg ions, respectively. Calculate total hardness in mg/L as $CaCO_3$.

2.2 Chemical analysis of a water sample gave the following results:

Ca^{2+} = 20 mg/L	CO_3^{2-} = 20 mg/L
Mg^{2+} = 5 mg/L	HCO_3^- = 13 mg/L
Na^+ = 5 mg/L	Cl^- = 15 mg/L
K^+ = 3 mg/L	SO_4^{2-} = 15 mg/L
Fe^{2+} = 1 mg/L	NO_3^- = 4 mg/L

Determine the completeness of the chemical-analysis data.

2.3 A water sample has a total hardness of 80 mg/L as $CaCO_3$ and a total alkalinity of 60 mg/L as $CaCO_3$. Calculate the carbonate and noncarbonate portions of the hardness.

2.4 The deficiency and excess of many ions in a water supply is associated with physiological disorders. Two lists are given below. Match the disorders in list B with the ions in list A.

List A	List B
___ Calcium	a. Mental retardation
___ Magnesium	b. Kidney damage
___ Sodium	c. Methomoglobinemia
___ Potassium	d. Osteoporosis
___ Iron	e. Gout
___ Manganese	f. Goiter
___ Nitrate	g. Hepatic cirrhosis
___ Fluoride	h. Laxative effect
___ Arsenic	I. Hypertension
___ Cadmium	j. Cathartic
___ Copper	k. Hemesynthesis
___ Iodine	l. Carbohydrate metabolism
___ Molybdenum	m. Mottling of teeth
___ Lead	n. Skin disorder

2.5 What is meant by the term "eutrophication"? What causes eutrophication?

2.6 Algae blooms in reservoirs cause many problems in the water supply. Explain.

2.7 What is the significance of turbidity in a municipal water supply? Why is water of low turbidity desired?

2.8 What are the major causes of alkalinity in natural waters? What is the significance of alkalinity in water source and treatment?

2.9 A groundwater sample has the following chemical analysis at 20°C.

Cations	Concentration, mg/L	Anions	Concentration, mg/L
Ca^{2+}	180	HCO_3^-	300
Mg^{2+}	65	CO_3^{2-}	40
Na^+	60	SO_4^{2-}	60
K^+	20	Cl^-	348
Fe^{2+}	0.5	NO_3^-	35
Cd^{2+}	0.2		

(a) Does the analysis appear complete?

(b) What is total hardness as $CaCO_3$?

(c) What is total alkalinity as $CaCO_3$?

(d) Would the water be acceptable for a domestic water supply? Why, or why not? List the treatment for removal of undesired constituents.

(e) If specific conductivity (EC) of the water sample is 2087 μmho/cm, calculate *k*.

2.10 Turbidity results in the water treatment field are expressed as NTU. Why are JTU units not used?

2.11 Discuss the significance of particle count in water treatment.

2.12 Calculate the SAR for two water supplies, for which Ca, Mg, and Na concentrations are given. Comment on the use of these water supplies for domestic and irrigation purposes.

Cations	Concentration, mg/L	
	Supply A	Supply B
Ca^{2+}	3.0	85
Mg^{2+}	1.2	29
Na^+	150	150

2.13 What is stability of water? Why is it important to supply stable water?

2.14 Review Chapter 16 and Reference 1. Discuss the sources and classification of naturally-occurring organic compounds and their significance in water treatment and supply.

2.15 What are the major health concerns with the synthetic organic compounds? Discuss the classification of synthetic organic compounds and give examples.

2.16 What are the major precursors? List some chlorine disinfection by-products in water supply. How can the formation of such compounds be reduced?

2.17 Calculate total carbonaceous and nitrogenous oxygen demand of a water sample that contains 5 mg/L organic compound having chemical formula $C_6H_6N_2O_2$. Assume that nitrogen is converted to ammonia and then to nitrate.

2.18 Contact the water treatment plant in your community and obtain the following information.

(a) What physical, chemical, and biological tests are run on finished water on a routine basis?

(b) What are the average values of these results?

(c) What method is used to determine organic matter in raw water (see Table 2-3)?

2.19 Two lists are given below. Match the terms.

List A	List B
____Cholera	a. Virus
____Typhoid fever	b. Ascariasis
____Dysentery	c. G. lamblia
____Hepatitis	d. Vibrio comma
____Worm	e. Entamoeba histolytica
____Giardiasis	f. Continued fever

2.20 Match the terms in lists A and B.

List A	List B
____Eucaryotic cell	a. Need oxygen
____Phototrophes	b. 12–18°C
____Procaryotic cell	c. Do not need oxygen
____Aerobic	d. Need organic carbon
____Heterotrophic	e. Genetic information
____Anaerobic	f. Nucleus clearly defined
____Psychrophilic	g. Sunlight
____Mesophilic	h. Nucleus poorly defined
____Actinomycetes	i. 25–40°C
____Protozoa	j. filamentous
____Giardiasis	k. Giardia lamblia

2.21 Why do raw-water lines need cleaning? Discuss various cleaning methods.

2.22 How does water play a role in transmission of diseases? Name common water-borne bacterial and viral diseases.

2.23 Define coliform organisms and the significance of the test. Why can coliform organisms be used as indicators of drinking water quality?

2.24 What is the difference between coliform group and fecal coliform?

2.25 Besides coliform organisms, are there other organisms that can be used as indicator organisms?

2.26 List the characteristics of an ideal indicator organism.

References

1. James M. Montgomery, Inc. *Water Treatment Principles and Design,* John Wiley & Sons, New York, 1985.
2. Lappenbusch, W. L. *Contaminated Drinking Water and Your Health,* Lappenbusch Environmental Health, Inc., VA, 1986.
3. National Technical Advisory Committee. *Water Quality Criteria*, Federal Water Pollution Control Administration, April 1968.
4. McKee, J. and Wolf, H. W. *Water Quality Criteria,* 2d ed., Publication No. 3-A, California State Water Quality Control Board, Sacramento, CA, 1971.

5. USEPA. "National Interim Primary Drinking Water Regulations," *Federal Register*, 59(126), 34320–34321, July 1, 1994.

6. Lehr, J. H., Gass, T. E., Pettyjohn, W. A., and DeMarre, J. *Domestic Water Treatment,* McGraw-Hill Book Co., New York, 1980.

7. USEPA. "National Secondary Drinking Water Regulations," *Federal Register,* 44(153), 42195–42202, July 19, 1979.

8. Tchobanoglous, G. and Schroeder, E. D. *Water Quality: Characteristics, Modeling and Modification,* Addison-Wesley Publishing Co., Reading, MA, 1985.

9. Butler, J. N. *Carbon Dioxide Equilibria and Their Applications,* Addison-Wesley Publishing Co., Reading, MA, 1982.

10. Fair, G. M., Geyer, J. C., and Okun, D. A. *Water and Wastewater Engineering, vol. 2, Water Purification and Wastewater Treatment and Disposal*, John Wiley & Sons, Inc., New York, 1968.

11. Benefield, L. D., Judkins, J. F., and Weand, B. L. *Process Chemistry for Water and Wastewater Treatment*, Prentice-Hall, Inc., Englewood Cliffs, NJ, 1982.

12. Viessman, W. and Hammer, M. J. *Water Supply and Pollution Control*, 6th ed., Addison-Wesley Longman, Inc., Menlo Park, CA, 1998.

13. National Academy of Sciences. "Safe Drinking Water Committee," *Drinking Water and Health,* National Academy of Sciences, Washington, D.C., 1977.

14. ASCE and AWWA. *Water Treatment Plant Design*, 2d ed., McGraw-Hill Publishing Co., New York, 1990.

15. APHA, AWWA, and WEF. *Standard Methods for the Examination of Water and Wastewater,* 19th ed., American Public Health Association, Washington, D.C., 1991.

16. AWWA. *Water Quality and Treatment,* 4th ed., McGraw-Hill Book Co., New York, 1990.

17. Schwartz, H. G. and Borzzel, J. C. *The Impact of Toxic Pollutants in Municipal Wastewater Systems,* USEPA Technology Transfer, Joint Municipal/Industrial Seminar on Pretreatment of Industrial Wastes, Dallas, Texas, July 12 and 13, 1978.

18. Sawyer, C. N. and McCarty, P. L. *Chemistry for Environmental Engineering,* McGraw-Hill Book Co., New York, 1978.

19. Qasim, S. R. *Wastewater Treatment Plants: Planning, Design, and Operation,* 2d ed., Technomic Publishing Co., Lancaster, PA, 1999.

20. Gaudy, A. F. and Gaudy, E. T. *Elements of Bioenvironmental Engineering,* Engineering Press Inc., San Jose, CA, 1988.

21. Kemp, L. E., Ingram, W. M., and Mackenthum, K. M. *Biology of Water Pollution, a Collection of Selected Papers on Stream Pollution, Wastewater, and Water Treatment*, 3d printing, U.S. Department of the Interior, Federal Water Pollution Control Administration, Washington, D.C., 1970.

22. Logsdon, G. S., Symons, J. M., and Arozarena, M. M. "Alternative Filtration Methods for Removal of Giardia Cysts and Cyst Models," *Jour. AWWA*, vol. 73, no. 2, pp. 111–118, February 1981.

23. Manuals Committee. *Manual of Water Utility Operation,* 7th ed., Texas Water Utilities Association, Austin, TX, 1979.

CHAPTER 3

Water Treatment Processes

3.1 Introduction

The principal objective of a water treatment plant is to produce water that satisfies a set of drinking water quality standards at a reasonable price to the consumers. A water treatment plant utilizes many treatment processes to produce water of a desired quality. These processes generally fall into two broad divisions: *unit operations* and *unit processes*. In the *unit operations*, the removal of contaminants is brought about by the physical forces. In the *unit processes*, however, the treatment is achieved by chemical and biological reactions.[1,2] Often the terms *unit operation* and *unit process* are used interchangeably, because many processes are integrated combinations of operations serving a single primary purpose. As an example, turbidity removal by coagulation combines chemical addition, mixing and dispersion, flocculation, and settling.

The collective arrangement of various unit operations and processes is called a *process train*, *flow sheet*, *process* or *flow diagram*, *flow schematic*, or *flow scheme*.[3] Basic considerations for developing a treatment process train depend upon the characteristics and seasonal variation in the raw water quality, regulatory constraints, site conditions, plant economics, and many other factors. The objective is to remove undesirable constituents from water supply and to process the residuals in a form that can be safely and easily disposed of or reused. In this chapter, an overview of various *unit operations* and *processes* used for water treatment and the processing of residuals is presented.

3.2 Raw Water Intake, Pumping, Conveyance, And Flow Measurement

Water treatment plants may utilize a raw water intake, pumping, conveyance, and flow measurement. Although these systems do not provide any treatment, they are necessary as a part of the overall treatment process train. A brief description of these systems is provided below.

3.2.1 Raw Water Intake

Raw water intakes are built to withdraw water from a river, lake, or reservoir over a predetermined range of pool levels. The structure can be from something as simple as a submerged intake pipe to a fairly elaborate tower-like structure that can house intake gates, screens, control valves, pumps, and chemical feeders. The intakes may be submerged, floating, or tower-like. Submerged and floating intakes are used for small water supply projects. Large projects utilize tower-like intakes that can be an integral part of the dam or can be a separate structure located elsewhere. Detailed discussion on various types of intake structures, site selection, design considerations, and design procedures are presented in Chapter 6.

3.2.2 Raw Water Pumping

Raw water pumping stations are generally located at the intake structure. The purpose is to lift or elevate the water from the source to an adequate height, from which the water can flow by gravity. The raw water pumping head is the sum of the static head, friction loss in the force main, and minor losses. Centrifugal pumps are most commonly used for water pumping. They may be suspended, submersible, or dry-well centrifugal pumps. The detailed discussion on types of pumps and pumping stations, pump selection, design procedure for raw water pumps, and operation and maintenance of pumping equipment is covered in Chapter 7.[3,4]

3.2.3 Raw Water Conveyance

The raw water conveyance or transport system is designed for the controlled movement of bulk quantities of water from the intake to the treatment plant. Generally, the treatment plant is located near the city; therefore, long raw water pipelines are required to transport the flow from the source. Topography, available head, construction materials and practices, economics, and water quality are the primary considerations in selecting suitable conduits and routes for water transport. The connecting conduit may be a canal, a flume, a grade aqueduct, grade tunnels, or a pressure pipeline, or a combination of these.[5,6] Hydraulics of open channel flow and pressure pipes and basic design considerations are provided in Chapter 7.

3.2.4 Flow Measurement

Flow measurement of raw water and finished water is essential for plant operation, process control, billing, and recordkeeping. The flow measurement device may be located in the raw-water line, in distribution mains after high-service pumps, or at any other location within a plant. Often more than one flow measurement device may be necessary at various treatment units. In general, the flow can be measured in pressure pipe and open channel. Flow of water through pipes under pressure is measured by mechanical or differential head meters, such as Venturi meters, flow nozzles, or orifice meters. Flow through an open channel is measured by a weir or a Venturi-type flume such as the Parshall flume. A discussion on flow measurement devices can be found in Chapter 7.[3,7]

3.3 Water Treatment Systems

3.3.1 Unit Operations and Processes

All water sources contain a broad spectrum of inorganic and organic constituents. Water treatment involves removal of undesirable constituents from the water and then disposal of them in the easiest and safest manner. To achieve this goal, a variety of treatment processes are utilized, which exploit various physical and chemical phenomena to remove or reduce the undesirable constituents from the water. Selection of the treatment processes is one of the keys to successful performance of a water treatment plant. Engineering decisions made during this stage can have a major impact on the total project cost. Errors in process selection may require extensive changes during operation to satisfy treated water quality standards. A summary of different unit operations and processes generally considered in the design of a water treatment plant is provided in Table 3-1.[3,5,8–13] A detailed description of each of the treatment processes is provided in different chapters of this book. Appropriate chapters are also referenced in Table 3-1.

It might be noted that the 25 treatment processes presented in Table 3-1 and discussed in various chapters of this book require techniques, understanding, and input from a wide range of disciplines, including engineering, chemistry, water quality, and microbiology. The order of presentation in Table 3-1 is a matter of personal opinion, but the authors feel that the arrangement follows a natural sequence, where information presented in this order is readily usable in the subsequent treatment process. Each process plays an important role at various stages of the treatment train. The predominant role, however, rests with the design engineers, who carry the responsibility for selecting and designing the treatment process in the overall treatment train.

3.3.2 Combination of Unit Operations and Processes in a Treatment Process Train

Many unit operations and processes summarized in Table 3-1 can be combined into a process train to achieve a desired level of treatment to meet the drinking water quality standards. The level of treatment may range from turbidity, taste, and odor removal to complete demineralization. Table 3-3 provides a general guideline for selection of treatment processes for desired applications. A number of water treatment process trains are shown in Figure 3-1. Choice of proper treatment processes and development of the process train is not a simple task. It requires understanding of the processes, performance, and operational capabilities. Experience acquired through treatment of the same or similar water source provide an excellent guide in preparing the process train. The following considerations, however, influence the selection of a treatment process train:[3,5,8,12,13]

1. ability to meet finished water quality objectives, considering both seasonal and long-term changes in raw water quality;
2. topography and site conditions, existing treatment facilities, land area available, and hydraulic requirements;

Table 3-1 Summary of Various Unit Operations and Processes Considered in the Design of a Water Treatment Plant

Unit No.	Unit Operation and Process[a]	Description and Principal Applications	Reference
A	Trash rack (UO)	Provided at the intake gate for removal of floating debris and ice.	Chap. 6
B	Coarse screen or fish screen (UO)	Mechanically cleaned screens provided at the intake gate or in the sump well ahead of pumps. Protect fish and remove small solids and frazil ice.	Chap. 6
C	Microstrainer (UO)	Removes algae and plankton from the raw water.	Chap. 6
D	Aeration (UP)	Strips and oxidizes taste and odor causing volatile organics and gases and oxidizes iron and manganese. Aeration systems include gravity aerator, spray aerator, diffuser, and mechanical aerator. Aeration in the reservoir helps destratification and T & O control.	Chap. 6
E	Mixing (UO)	Provides uniform and rapid distribution of chemicals and gases into the water.	Chap. 8
F	Pre-oxidation (UP)	Application of oxidizing agents such as ozone, potassium permanganate, and chlorine compounds in raw water and in other treatment units; retards microbiological growth and oxidizes taste, odor and color causing compounds.	Chap. 12
G	Coagulation (UP)	Coagulation is the addition and rapid mixing of coagulant resulting in destabilization of the colloidal particle and formation of pin-head floc.	Chap. 8
H	Flocculation (UO)	Flocculation is aggregation of destabilized turbidity and color causing particles to form a rapid-settling floc.	Chap. 8
I	Sedimentation (UO)	Gravity separation of suspended solids or floc produced in treatment processes. It is used after coagulation and flocculation and chemical precipitation.	Chap. 9
J	Filtration (UO)	Removal of particulate matter by percolation through granular media. Filtration media may be single (sand, anthracite, etc.), mixed, or multilayered.	Chap. 10

Table 3-1 Summary of Various Unit Operations and Processes Considered in the Design of a Water Treatment Plant (continued)

Unit No.	Unit Operation and Process[a]	Description and Principal Applications	Reference
K	Chemical precipitation (UP)	Addition of chemicals in water transforms specific dissolved solids into insoluble form. Removal of hardness, iron, and manganese, and many heavy metals is achieved by chemical precipitation.	Chap. 8
L	Lime-soda ash (UP)	Lime-soda ash is a chemical precipitation process used for water softening. Excess amounts of calcium and magnesium ions are precipitated from water.	Chap. 8
M	Recarbonation (UP)	Restores the chemical balance of water after softening by lime-soda process. Bubbling of carbon dioxide converts supersaturated forms of Ca and Mg into more soluble forms. The pH is lowered.	Chap. 8
N	Activated carbon adsorption (UP)	Removes taste and odor causing compounds, chlorinated compounds, and many metals. It is used as powdered activated carbon (PAC) at the intake or as a granular activated carbon (GAC) bed after filtration.	Chap. 11
O	Activated alumina (UP)	Removes certain species from water by hydrolytic adsorption. Fluoride, phosphate, arsenic and selenium have been effectively removed by activated alumina.	Chap. 18
P	Disinfection (UP)	Destroys disease-causing organisms in water supply. Disinfection is achieved by ultraviolet radiation and by oxidative chemicals such as chlorine, bromine, iodine, potassium permanganate, and ozone, chlorine being the most commonly used chemical.	Chap. 12
Q	Ammoniation (UP)	Ammonia converts free chlorine residual to chloramines. In this form, chlorine is less reactive, lasts longer, and has less tendency to combine with organic compounds, thus reducing taste and odors and THM formation.	Chap. 12
R	Fluoridation (UP)	Sodium fluoride, sodium siiicofluoride, and hydrofluosilicic acid may be added in finished water to produce water that has optimum fluoride level for control of tooth decay.	Chap. 12

Table 3-1 Summary of Various Unit Operations and Processes Considered in the Design of a Water Treatment Plant (continued)

Unit No.	Unit Operation and Process[a]	Description and Principal Applications	Reference
S	Biological denitrification (UP)	Nitrate in excessive concentration in drinking water may cause *methemoglobinemia* in infants. Nitrate is reduced to gaseous nitrogen by microorganisms in anaerobic environment. An organic source such as ethanol or sugar is needed to act as a hydrogen donor (oxygen acceptor) and to supply carbon for synthesis.	Chap. 18
T	Demineralization (UP)	Involves removal of dissolved salts from the water supply. Demineralization may be achieved by ion exchange, membrane processes, distillation, and freezing.	Chap. 18
T1	Ion exchange (UP)	The cations and anions in water are selectively removed when water is percolated through beds containing cation and anion exchange resins. The beds are regenerated when the exchange capacity of the beds is exhausted. Selective resins are available for hardness, nitrate, and ammonia removal.	Chap. 18
T2	Reverse osmosis (RO) and ultrafiltration (UO)	Semi-permeable membranes are used to permeate high-quality water while rejecting the passage of dissolved solids. RO is also used for nitrate and arsenic removal.	Chap. 18
T3	Electrodialysis (UO)	An electrical potential is used to remove the cations and anions through the ion-selective membranes to produce demineralized water and brine.	Chap. 18
T4	Distillation (UO)	Multiple-effect evaporation and condensation and distillation with vapor compression are common methods used for large-scale demineralization systems.	Chap. 18
T5	Freeze (UO)	Saline water is cooled to reach the freezing point. The ice contains pure water, while the dissolved salts remain in solution. The ice is removed and melted to recover fresh water. This method is used in colder climates.	Chap. 18

a UO = unit operation UP = unit process

Source: Adapted in part from References 3, 5, 8–13.

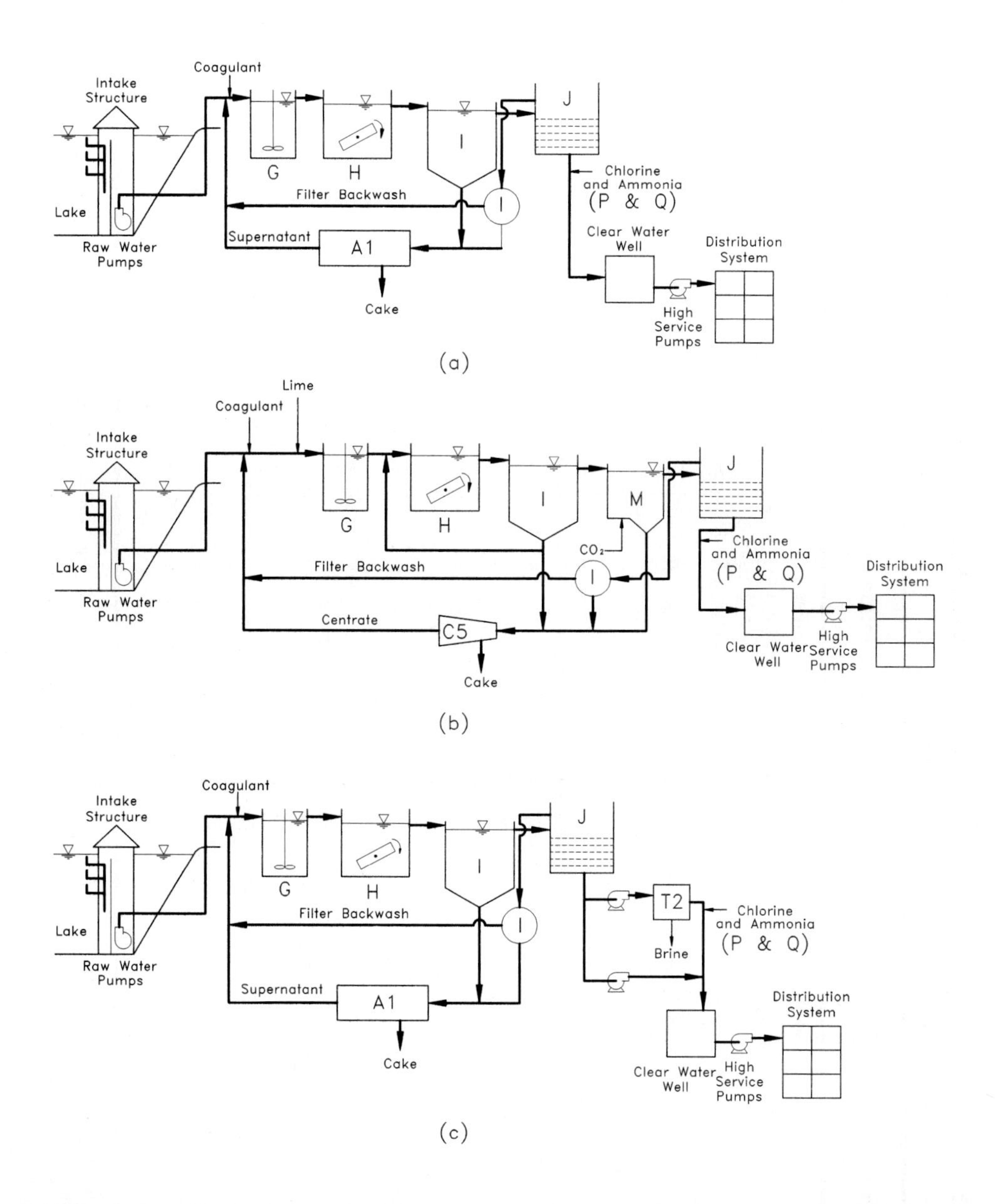

Figure 3-1 Water treatment process train. (See Table 3-1 and Table 3-3 for notation of unit operations and processes.) (a) Conventional water treatment plant treating surface water. Aeration may be applied in the lake or before coagulation. Preoxidation may be used at any desired location for control of T & O and undesired aquatic growth. (b) Conventional water treatment with partial softening. (c) TDS removal by split demineralization process.

Table 3-2 Treatment Desired and Process Selection (For explanation of process symbol, see Table 3-1.)

Contaminant Removal	Process	Remarks
Turbidity	a. In-line filtration (G + J)	Applicable to water having low turbidity and color.
	b. Direct filtration (G + H + J)	Applicable to water having low-to-medium turbidity and low-to-medium color.
	c. Conventional (G + H + I + J)	Applicable to low and high turbidity and color.
Algae and plankton	a. Microstrainer (C)	Microstrainers will not handle silt, sand, and other abrasive material.
	b. Conventional (G + H + I + J)	High population of algae and plankton are difficult to coagulate. They usually float.
Color	a.Oxidation (ozone, chlorine, chlorine dioxide, potassium permanganate) (F)	Applicable to water having low color (Chap. 11).
	b. Low pH coagulation (G + H)	Applicable to water having from low color to high. (Alum performs better than iron salts.) Optimum pH 5–6.
	c. Adsorption (N)	Applicable to water having low-to-moderate color episodes.
	d. Adsorption (N/GAC)	Applicable to water having moderate-to-low-level soluble color for routine color control.
	e. Ion-exchange (J + T1)	Synthetic resin bed after filtration removes soluble color of industrial origin.
Iron and manganese	a. Oxidation (D + F + I)	Iron and manganese are removed by oxidation and precipitation.

Table 3-2 Treatment Desired and Process Selection (For explanation of process symbol, see Table 3-1.) (continued)

Contaminant Removal	Process	Remarks
	b. Precipitation (D + K)	Iron and manganese are precipitated by aeration at high plumbing pH. Lime is generally used to raise pH.
	c. Conventional (G + H + I)	Iron and manganese are removed by conventional coagulation-flocculation.
	d. Ion exchange (T1)	Selective ion exchange resins are generally used for removal of Fe and Mn from groundwaters.
Taste and odor (T & O)		Source control prevents formation and reaching of T & O compounds into the reservoirs. Aquatic weeds and algae control reduce T & O causing compounds.
	a. Oxidation (D or F)	Aeration in the reservoir, at the head of plant, and in treatment units may reduce T & O. Oxidation with free chlorine may cause THM formation.
	b. Adsorption (N/PAC or GAC)	PAC is used for moderate and intermittent T & O episodes. GAC is used for prolonged T & O due to industrial sources.
Hardness	a. Precipitation (L)	Used for medium-to-hard water.
	b. Ion exchange (T1)	Zeolite beds remove divalent cations but add Na^+. Cation exchangers remove all cations.
Pathogens	a. Disinfection (P)	Free chlorine enhances THM formation potential.
THM	a. Enhanced coagulation (G + K+ H + I)	Efficient removal of precursors (organics) by low pH coagulation or in conjunction with softening.

Table 3-2 Treatment Desired and Process Selection (For explanation of process symbol, see Table 3-1.) (continued)

Contaminant Removal	Process	Remarks
	b. Adsorption (N)	PAC and GAC will remove TOC, precursors, and THM.
	c. Aeration (D)	Aeration removes organics.
	d. Preoxidation (F)	Preoxidation of precursors by O_3, H_2O_2, $KMNO_4$, chloramines, or chlorine dioxide. Avoids excessive free chlorine residual.
Nitrate	a. Denitrification (S)	Biological denitrification provides an effective method of NO_3 removal.
	b. Ion exchange (T1)	Selective resins are available.
	c. Demineralization (T2)	Demineralization with membranes removes NO_3 with other ions. Deionization with RO fluoride. High concentrations of fluoride in drinking water cause fluorosis and toxicity.
Fluoride	a. Activated alumina (O)	Activated alumina selectively removes F.
Arsenic	a. Enhanced (G + H + I)	Removes by precipitation of arsenate As(V), at low pH with coagulation and at high pH with partial softening with lime.
	b. Activated alumina (O)	Bed is effective in removing arsenic from groundwater.
	c. Demineralization (T2)	Arsenic is removed with other ions by membrane processes.

Source: Adapted in part from References 3, 5, 8, 12, and13.

3. overall system reliability;
4. flexibility and simplicity of operation;
5. capability of process to upgrade performance in case raw water quality or drinking water regulations change;
6. capability of process to meet the hydraulic peaks;
7. availability of skilled operation and maintenance personnel, major equipment, and chemical-delivery system;
8. state and federal requirements;
9. preference and experience of the design engineer;
10. ease of construction of facilities;
11. economics of construction and operation.

Laboratory and pilot plant studies are often necessary to develop design parameters for a given water source. These studies may include bench-scale study in the laboratory, pilot plant testing, and plant-scale simulation testing.[8]

3.4 Residuals Management

Water treatment residuals are materials that are removed in various treatment processes, along with large quantities of transport water. These residuals include turbidity and color-causing materials, organic and inorganic solids, algae, bacteria, viruses, and precipitated chemicals. These are by-products of chemical coagulation, iron and manganese precipitation, filter backwash, softening, regeneration brines, and microstrainer washwater. The liquid may constitute 3-10 percent of the volume of raw water processed. Solids concentration may vary from 0.1 to 4 percent, depending on the process and on raw water quality. These residuals cannot be discharged into the stream, river, or lake from which they originated. Currently, these wastes are considered industrial wastes, and disposal into surface waters falls under the jurisdiction of Federal Water Pollution Control Act Amendments of 1972 and the Clean Water Act of 1977.[14,15] For discharge into surface waters, a permit under National Pollutant Discharge Elimination System (NPDES) would be required.

3.4.1 Unit Operations and Processes

In general, the residuals-management processes include settling, thickening, conditioning, dewatering, drying, recovery, and disposal. Often, water treatment plant residuals are allowed to discharge into the sanitary sewers. Selection of any of these processes depends greatly on such factors as space available at the site, type and quantity of sludge, local weather conditions, cost of raw chemicals, and type of ultimate disposal available.[8,16,17] A summary of different residuals-management systems commonly used in an overall process train is provided in Table 3-3. These processes are discussed in Chapter 14.

3.4.2 Process Train for Residuals Management

Selection of a process train for sludge thickening, conditioning, dewatering, disposal, or recovery options is based on the size of the plant, construction cost, equipment, available land and plant location, and operational factors such as chemicals, power, labor, and reliability. The final selection of the most suitable treatment process train should be based on the merits and liabilities of each process and on the cost-effectiveness of the process train as a whole. There are many factors that must be considered in the final selection of the appropriate process train. These factors are (1) land requirements, (2) operation under adverse climatic conditions, (3) ability to handle flow variations, (4) process reliability, (5) ease of operation and maintenance, (6) prior requirements, and (7) quality of supernatant and sludge.

3.5 Treatability Studies

Because every water source is unique in terms of chemical makeup, it is often essential to conduct treatability studies to develop design and operational parameters of the processes and the overall process train. Such studies are especially important where new plants are proposed. Therefore, the design engineer must fully understand the general approach and methodology of (1) laboratory tests for assessing the treatability of the water to the desired level, (2) the procedure of laboratory and pilot-plant studies, and (3) the translation of experimental data into design and operational parameters.

The laboratory studies include batch and/or continuous-flow reactor studies. The classic *jar test* is almost essential to determine proper chemical dosages and physical conditions for coagulation.

Pilot plants are becoming a valuable tool for treatability studies to develop design parameters, to test new chemicals, to optimize dosages, and to evaluate processes to improve plant performance and reduce costs. There are a number of other cases where pilot studies may be needed. For example, a state-approved agency might, as a matter of policy, require a pilot study for innovative treatment processes. Another example would be a situation wherein a process cannot be modeled sufficiently to lead to design criteria, such as taste and odor episodes. In general, pilot plant studies in water works engineering are used for the following reasons:

- Test new processes
- Stimulate a process
- Predict process performance
- Document process performance
- Optimize system design
- Optimize system operation (chemical dosages, reaction periods, etc.)
- Satisfy regulatory agency requirements
- Satisfy legal requirements

Table 3-3 Unit Operations and Processes Used for Residuals Management (Detailed discussion on these unit operations and processes may be found in Chapter 14.)

Unit No.	Unit Operation and Process[a]	Description and Principal Applications
A	Sedimentation and thickening (UO)	The objective is to remove excess water and to concentrate solids. The liquid is usually recovered unless it contains taste and odors, algae, and other microorganisms.
A1	Sludge lagoon (UO)	Large open earthen or concrete reservoirs 3 to 4 m deep. Thickening of solids 5 to 10 percent can be achieved in one to three months with continuous decanting.
A2	Gravity (UO)	Circular tanks designed and operated similar to a solids-contact clarifier or sedimentation basin. Chemical conditioning may be needed.3
B	Conditioning	Sludge conditioning is done to aid in thickening and mechanical dewatering. The objectives are to improve the physical properties of the sludge, so that water will be released easily from solids. Conditioning is generally used for alum-coagulated sludge.
B1	Chemical (UP)	Polymers are the most commonly used chemicals for sludge conditioning. Lime and inert granular materials like fly ash have also been used.
B2	Freezing (UO)	Freezing destroys the gelatinous structure and thus improves thickening and dewatering. This process is usually applied only where natural freezing is possible.
B3	Heat treatment (UO)	Heating improves settling and dewatering. Because of energy cost, it is an undesirable method of sludge conditioning.
C	Dewatering (UO)	The process produces relatively dry sludge for further treatment or disposal.
C1	Drying bed (UO)	This is a gravity dewatering system where conditioned sludge may be applied without thickening. Sludge is applied on a filling and drying cycle over filter beds or lagoons that have an underdrain system. Water is removed by filtration and by decanting.
C2	Centrifuge (UO)	Conditioned sludge is dewatered in a solid-bowl or basket centrifuge. Both capital and operating costs are high.
C3	Vacuum filter (UO)	Rotary drum vacuum filter with traveling media or precoat media filters are used for dewatering. Sludge conditioning is generally needed.
C4	Filter press (UO)	Also known as plate-and-frame or leaf filter. It is an effective method of sludge dewatering. It is a batch process and uses chemically conditioned sludge.

Table 3-3 Unit Operations and Processes Used for Residuals Management (Detailed discussion on these unit operations and processes may be found in Chapter 14.) (continued)

Unit No.	Unit Operation and Process[a]	Description and Principal Applications
C5	Belt filter press (UO)	Provides continuous operation. The sludge is squeezed between two belts as it passes between various rollers. Sludge conditioning is necessary.
D	Recovery	Recovery of water, coagulants, and magnesium bicarbonate is possible from the sludge.
D1	Coagulants (UP)	Recovery of aluminum and iron can be accomplished by adding acid (sulfuric acid) to solubilize the metal ions from the sludge. Accumulation of heavy metals, manganese, and other organic compounds is possible.
D2	Lime (recalcination) (UP)	Recovery of lime from calcium carbonate sludge is achieved by recalcination. The dewatered sludge is dried and heated to about 1000 degrees C. CO_2 is driven off, which may be used for recarbonation.
D3	Magnesium	Magnesium recovery is possible from a sludge containing calcium carbonate and magnesium hydroxide. Bubbling CO_2 produces soluble magnesium bicarbonate, while calcium carbonate remains insoluble.
E	Disposal	Disposal of residuals may be achieved on land, in sanitary sewer, in surface waters, and by deep-well injection.
E1	Land disposal	Dewatered, semi-liquid, and liquid residuals are disposed of by land-filling or land-spreading governed by the Resource Conservation and Recovery Act of 1976.[18] Cost of transportation may be significant.
E2	Sanitary sewer	Direct discharge into sewer system is an attractive option if residuals do not adversely affect the operation of the wastewater treatment plant.
E3	Surface water	NPDES permit is required for disposal of residuals into surface water. The permit requirements vary with the types of residuals and with the type of surface water.
E4	Deep-well injection	Brines from ion exchangers and membrane processes may be injected into deep wells. Such injection disposals are controlled by local environmental regulations subject to geology and groundwater hydrology.

a UO = unit operation UP = unit process.

It is believed that pilot plant studies at municipal water treatment plants are valuable in developing designs and process trains that can improve performance of new and existing plants and reduce operating costs. Details of pilot plant design and the conducting of treatability studies in water treatment engineering are given in References 8, 13, 19–24. The process train and some photographs of a pilot plant designed, constructed, and operated by the authors are provided in Figures 3-2 and 3-3, respectively.

Problems and Discussion Topics

3.1 Various water treatment unit operations and unit processes are given in list A. Applications of these processes are given in list B. Match the processes by writing against these applications in list B. Some processes may have more than one application.

List A (Processes)	List B (Applications)
(a) Coagulation/sedimentation	_____ Hardness removal
(b) Aeration	_____ THM removal
(c) Reverse osmosis	_____ Sludge concentration
(d) Ion exchange	_____ Color removal
(e) Recarbonation	_____ Taste and odor removal
(f) Carbon adsorption	_____ Turbidity removal
(g) Thickening	_____ Iron and manganese removal
(h) Filtration	_____ Pathogen removal
(i) Disinfection	_____ TDS removal
(j) Ammoniation	_____ Undesired-growth control
(k) Precipitation	_____ Combined chlorine residual
(l) Centrifugation	_____ Algae removal
(m) Prechlorination	_____ Reduction in THM formation potential
	_____ Restore chemical balance after softening

3.2 Describe the purpose of a microstrainer in water treatment and the most desirable location for it in a process train.

3.3 List four types of aerators. What is the purpose of aeration in water treatment, and what physical and chemical changes does it induce for removal of various compounds?

3.4 Describe mixing, and give five examples of mixing in a water treatment plant.

3.5 Why is water softening necessary in water treatment, and how can it be achieved?

3.6 Describe the difference between coagulation and flocculation.

3.7 What is the purpose of recarbonation in water treatment?

3.8 Write basic chemical reactions of iron and manganese precipitation when water is aerated.

3.9 Water is withdrawn from a reservoir that has a heavy algae bloom. Draw a process train to treat the water. Give special consideration to taste and odors and to THM problems.

3.10 Visit the water treatment plant in your community. Draw the process diagram and indicate the common problems that often develop in water treatment. What steps must be taken by the operators to control these problems?

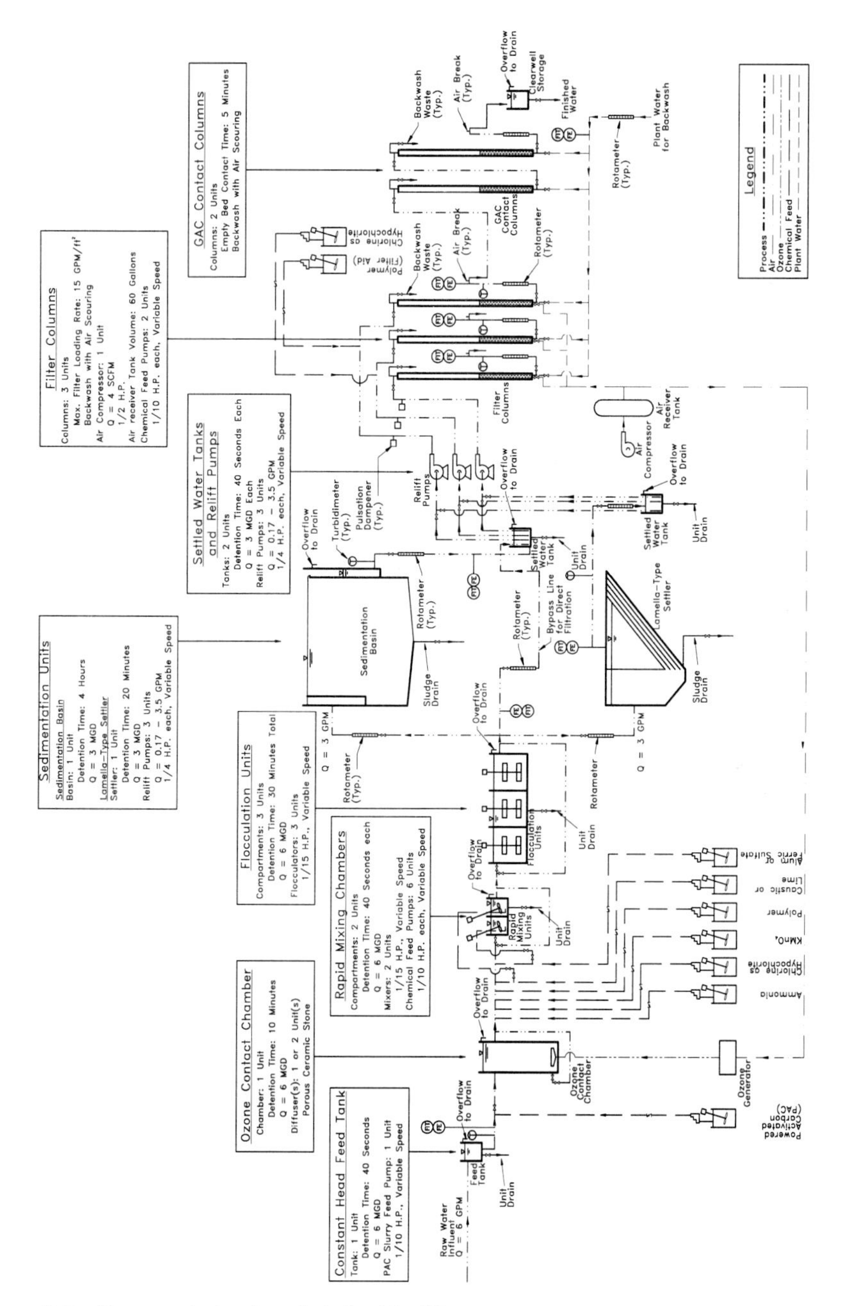

Figure 3-2 Process train of a pilot plant facility.

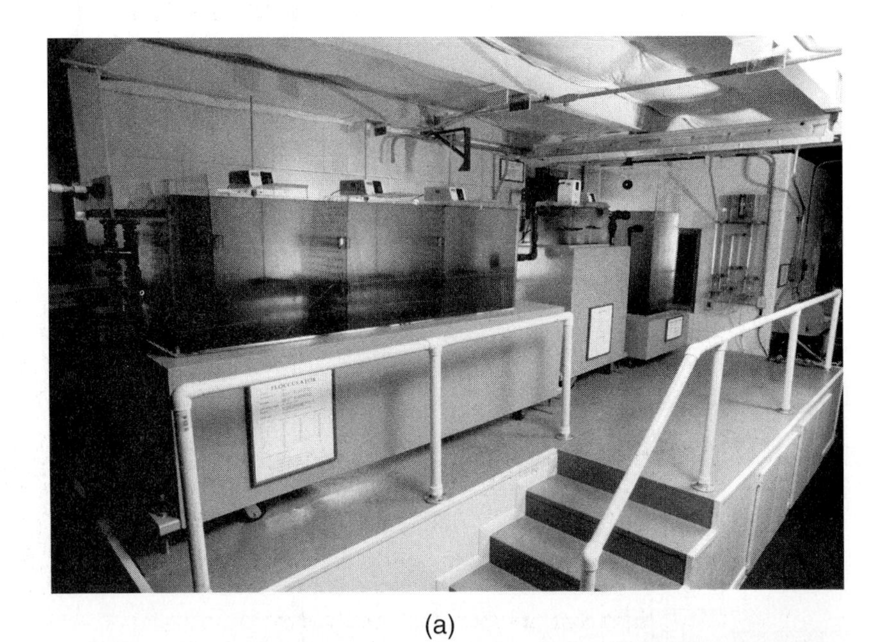

(a)

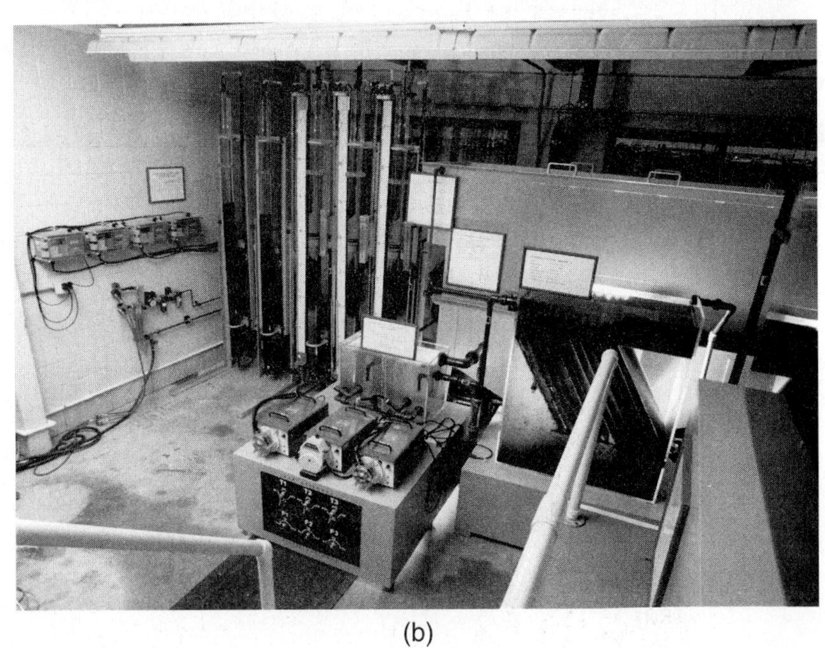

(b)

Figure 3-3 Photographs showing process train of pilot plant in Figure 3-2. (a) Process train from constant-head tank to flocculator. (b) Process train from sedimentation tanks to GAC columns.

References

1. Rich, Linvil G. *Unit Operations of Sanitary Engineering,* John Wiley and Sons, Inc., New York, 1961.
2. Rich, Linvil G. *Unit Processes of Sanitary Engineering,* John Wiley and Sons, Inc., New York, 1963.
3. Qasim, S. R. *Wastewater Treatment Plants: Planning Design and Operation,* 2d ed., Technomic Publishing Co., Lancaster, PA, 1999.
4. McGhee, T. J. *Water Supply and Sewerage*, 6th ed., McGraw-Hill Book Co., New York, 1991.
5. Fair, G. M., Geyer, J. C., and Okun, D. A. *Water and Wastewater Engineering, Vol. 1 and 2*, John Wiley and Sons, New York, 1966.
6. Wiessman, W. and Hammer, M. J. *Water Supply and Pollution Control,* Addison-Wesley, Menlo Park, CA, 1998.
7. Metcalf and Eddy, Inc. *Wastewater Engineering: Treatment Disposal and Reuse,* 3d ed., McGraw-Hill Book Co., New York, 1991.
8. James M. Montgomery, Inc. *Water Treatment Principles and Design,* John Wiley & Sons, New York, 1985.
9. American Water Works Association. *Water Quality and Treatment,* 4th ed., McGraw-Hill, Inc., New York, 1990.
10. Sanks, R. L. *Water Treatment Plant Design for the Practicing Engineer,* Ann Arbor Science, Ann Arbor, MI, 1980.
11. ASCE and AWWA. *Water Treatment Plant Design,* 2d ed., McGraw-Hill Publishing Co., New York, 1990.
12. Qasim, S. R., Hashsham, S., and Ansari, N. I. "TOC Removal by Coagulation and Softening," *Jour. of Envir. Engrg. Division, Am. Soc. of Civ. Engrs.*, vol. 118, no. 3, pp. 432–437, June 1992.
13. Tchobanoglous, G. and Schroeder, E. D. *Water Quality: Characteristics, Modeling, and Modification,* Addison-Wesley Publishing Co., Reading, MA, 1985.
14. U.S. Congress. *Federal Water Pollution Control Act Amendments of 1972* (PL 92-500), 92nd U.S. Congress, October 18, 1972.
15. U.S. Congress. *Clean Water Act* (PL 92-217), 95th U.S. Congress, December 27, 1977.
16. Cornwell, David A., et. al. *Handbook of Practice: Water Treatment Plant Waste Management,* American Water Works Association Research Foundation, Denver, CO, 1987.
17. Koorse, S. J. "The Role of Residuals Disposal Laws in Treatment Plant Design," *Jour. AWWA*, vol. 85, no. 10, pp. 57–62, October 1993.
18. U.S. Congress. *The Resource Conservation and Recovery Act of 1976* (PL 94-580), 94th U.S. Congress, October 21, 1976.
19. Kawamura, S. "Hydraulic Scale-Model Simulation of the Sedimentation Process," *Jour. AWWA*, vol. 73, no. 7, pp. 372–378, July 1981.
20. Letterman, R., Quon, J., and Gemmell, R. "Influence of Rapid Mix Parameters on Flocculation," *Jour. AWWA*, vol. 65, no. 11, pp. 716–722, November 1973.
21. Lang, J. "Selection of Pilot Plant Equipment," *Paper presented at Annual AWWA Conference*, Miami, FL, 1982.
22. Qasim, S. R., Chiang, W. W., and Sawey, R. "Innovative Pilot Plant Design for Water Treatment Studies," *Proceedings of the International Conference and Exhibition on Water and Wastewater*, pp. 21-30, Beijing, China, July 12-16, 1994.
23. Qasim, S. R., Clines, W. A., and Hossain, M. I. *Construction and Startup of Pilot Plant at Rolling Hills Water Treatment Plant*, Final Report submitted to the City of Fort Worth, Texas, p. 108, Department of Civil Engineering, The University of Texas at Arlington, Arlington, TX, August 1990.
24. Qasim, S. R., Stone, F. R., and Miller, R. *Pilot Plant Operation at Rolling Hills Water Treatment Plant*, Final Report submitted by City of Fort Worth, Texas, Department of Civil Engineering, The University of Texas at Arlington, Arlington, TX, October 1992.

CHAPTER 4

Basic Design Considerations

4.1 Introduction

Water treatment projects vary widely in scope: source of water supply, raw water transmission, treatment, storage, and distribution. Extensive engineering and economic studies are needed to evaluate alternatives dealing with sources of raw water and treatability, intake location, plant siting, plant capacity, process selection and design, and storage and distribution. Many decisions are complex and can raise political and social issues. Predesign studies are necessary to address and to resolve the specific elements of the project. Predesign studies are covered in the next chapter. The objective of this chapter is to provide a general discussion on many design factors that must be considered in the planning and predesign studies and are finally implemented into the design of the water treatment plants. Basic design factors covered in this chapter include the following:

1. water quality objectives;
2. regulatory requirements;
3. initial and design years;
4. service area;
5. plant capacity;
6. raw water quality and source selection;
7. process selection and treatment process train;
8. plant siting, layout, and hydraulics;
9. equipment selection;
10. energy and resource requirement;
11. plant economics;

12. environmental-impact assessment;
13. state requirements.

Although most of these factors are considered in the predesign studies and have been presented in greater detail in other chapters, the information given below provides an overview of the design factors necessary for preparation of predesign reports and design plans and specifications.

4.2 Water Quality Objectives

The objective of every water supply project is to produce safe water that meets federal, state, and local drinking water quality standards. It should also be free of taste and odor and aesthetically pleasing. The finished water quality goals are generally governed by the existing water quality regulations. It is expected that, in the future, the number of regulated constituents in the finished water may be increased or made tighter. It is also expected that the raw water quality of supply sources will deteriorate with time. Therefore, the water quality objectives should also include flexibility and innovation in the design, to upgrade the plant economically and to handle new situations that may develop in the future.

4.3 Regulatory Requirements

Water quality standards are generally promulgated on federal, state, and local levels. Currently, the water quality standards emphasize low-level contaminants that can have chronic health effects (such as higher incidence of cancer). A great deal of controversy occurs during the setting of such standards. As improved detection techniques for many new contaminants in water supply (and greater information on their chronic health effects) become available, it is believed that the drinking water quality standards will continue to become more and more complex and controversial.[1]

With the passage of the *Safe Drinking Water Act (SDWA) of 1974 and its Amendments of 1986 and 1996*, the Federal Government, through USEPA, was given the authority and timetable to set the standards for drinking water quality.[2] This Act provides for direct federal involvement in the quality of the drinking water supplied to the communities. The *National Academy of Sciences* recommended a systematic approach to establishing the quantitative criteria for carcinogenic compounds based on risk assessment.[3–5] The USEPA Office of Drinking Water, using the National Academy of Sciences criteria, has developed advisory documents collectively known as *health advisories*.[1] For individual compounds, *suggested no-adverse-effect response levels* (SNARLs) for short-term exposure and incremental lifetime-cancer-risk concentrations for long-term exposure to cancer-causing compounds have been established[a].[1] Individual state agencies have been given discretion in the use of these criteria.[1] The regulatory provisions derived from the 1986 amendments use the following terms to describe controls for water contaminants:

Maximum Contaminant Levels (MCLs) are maximum permissible levels of contaminants delivered to a user of public water supply.

Maximum Contaminant Level Goals (MCLG) are non-enforceable limits that indicate the level of the contaminant that does not cause any known or anticipated adverse effect on human health. They were formerly termed Recommended Maximum Contaminant Levels (RMCLs).

Secondary Maximum Contaminant Levels (SMCLs) are non-enforceable goals for preserving the aesthetic qualities of drinking water.

The Safe Drinking Water Act requires two types of standards: primary and secondary.[5–7] *Primary drinking water standards* protect the public, to the extent feasible, using the treatment technology. The standards include *maximum contaminant levels*, or MCLs, and/or treatment requirements. *Secondary drinking water regulations* pertain to those contaminants that affect the aesthetic quality of drinking water (color, taste, odor, etc.).

The amendments authorized USEPA to set treatment-technique requirements in lieu of MCLs when it is not economically or technically feasible for water suppliers to determine the level of contaminants. In addition, the amendments require USEPA to set *best available technology* (BAT) for purposes of complying with National Primary Drinking Water Regulations. In addition to the quantitive provisions, there are several far-reaching provisions on monitoring, filtration, disinfection, and use of lead materials.

There are three primary entities involved with the regulatory effort:

- USEPA, with the primary roles of national primary and secondary drinking water regulations, designating BATs, and overseeing State programs and enforcement;
- States, with the primary responsibility for implementation, program administration, and enforcement;
- utilities, which will have to increase monitoring, install new treatment processes, and increase public awareness of contamination problems.

The regulatory effort is divided into the following five phases:

- Phase I: Volatile organic compounds
- Phase II: Synthetic organic compounds, inorganic compounds, and unregulated contaminant monitoring
- Phase III: Radionuclide contaminants
- Phase IV: Disinfectant and oxidant by-products
- Phase V: Inorganic compounds and synthetic organic compounds

a. An incremental life-time cancer risk of 10^{-6} for 0.29 μg/L chloroform concentration in drinking water means that if a person consumed 2 L/d water containing 0.29 μg/L chloroform for life time (70 years), then his cancer risk would increase by 1 in a million; alternatively, in a population of one million, one person would get cancer in his life-time due to water containing 0.29 μg/L chloroform concentration who otherwise would not have.

Some specific details of regulatory efforts are presented in Table 1-1.

4.3.1 Primary Drinking Water Regulations

The Safe Drinking Water Act required a series of steps and timetables for developing and revising national primary and secondary drinking water regulations and periodic review and update. The National Interim Primary Drinking Water Regulations (NIPDWRs), first established in 1975, have been amended several times.

The existing National Primary Drinking Water Regulations (NPDWRs) set limits for public water systems (PWSs)[b] on those organics, inorganics, radionuclides, and microbial contaminants known to be important to public health in drinking water, and they establish sampling schedules to measure the concentration of these contaminants in public water systems.[5,7] Additional regulations established limits on trihalomethanes(THMs) and their monitoring requirements and limits on sodium levels and on corrosivity of finished water.[8–12] In 1982 and 1983, USEPA published a list of 83 contaminants to be controlled by establishing maximum concentration levels (MCL)[c].[13] By 1985, recommended maximum contaminant level (RMCL) for volatile synthetic organic chemicals and for microorganisms, detection limits for man-made beta particle and proton emitters, and final MCL for fluoride were established.[13–18] By the end of 1995, the list of contaminants in the standard had grown to over 102. The USEPA Primary Drinking Water Standards for regulated contaminants are given in Table 4-1.

4.3.2 Secondary Drinking Water Regulations

The secondary drinking water regulations were promulgated in 1979 to regulate those contaminants that can adversely affect the aesthetic quality. These contaminants generally relate to taste, odors, and color. The secondary drinking water regulations present MCLs that are considered a reasonable goal for water supply. They are not federally enforced. Each state may establish appropriate levels in accordance with its needs, as long as the health and welfare of the consumers is fully protected. By 1986 the secondary MCL for fluoride and *National Secondary Drinking* water regulations were revised.[19] The National Secondary Drinking Water Regulations are summarized in Table 4-2.

b. Public water systems (PWSs) are defined as having at least 15 service connections or serving at least 25 people. Public water system are divided into two classifications, depending on the type of population they serve. A *community water system* is a PWS that serves residential year-round population. A noncommunity water system is a PWS that serves a transient population, such as restaurants and schools. The sampling and contaminant level requirements are different for the two types of public water supplies.

c. Maximum concentration levels (MCLs) are the enforceable standards that USEPA must set as close to the maximum contaminant level goals (MCLGs) as feasible. MCLGs are numerical limits set for each contaminant at a level at which no adverse health effects on humans can be expected, with an adequate margin of safety. Decisions on the levels of an MCL that is feasible are based on a consideration of the best technology, treatment techniques, and laboratory analyses that are available, taking cost into consideration.

Table 4-1 USEPA Primary Drinking Water Standards for Regulated Contaminants

Contaminant	MCLG, mg/L	MCL, mg/L
Fluoride Rule		
Fluoride	4.0	4.0
Phase I Volatile Organic		
Benzene	Zero	0.005
Carbon tetrachloride		
p-Dichlorobenzene	Zero	0.005
1,2-Dichloroethane	0.075	0.075
1,1-Dichloroethylene	Zero	0.005
Trichloroethylene	0.007	0.007
1,1,1-Trichloroethane	Zero	0.005
Vinyl chloride	0.2	0.2
Total Coliform Rule and Surface Water Treatment Rule		
Giardia Iamblia	Zero	TT[a]
Legionella	Zero	TT
Heterotrophic plate count	N/A	TT
Total coliform	Zero	<5%[b]
Turbidity	N/A	TT
Viruses	Zero	TT
Phase II Rule Inorganics		
Asbestos (>10 μm)	7 MFL[c]	7 MFL
Barium	2	2
Cadmium (total)	0.005	0.005
Chromium (total)	0.1	0.1
Mercury (inorganicl)	0.002	0.002
Nitrate (as N)	10	10
Nitrite (as N)	1	1
Nitrate + Nitrite (as N)	10	10
Selenium	0.05	0.05
Phase II Rule Organics		
Acrylamide	Zero	TT
Alachlor	Zero	0.002
Atrazine	0.003	0.003
Carbofuran	0.04	0.04

Table 4-1 USEPA Primary Drinking Water Standards for Regulated Contaminants (continued)

Contaminant	MCLG, mg/L	MCL, mg/L
Chlordane	Zero	0.002
Chlorobenzene	0.1	0.1
2,4-D	0.07	0.07
o-Dichlorobenzene	0.6	0.6
cis-1,2-Dichloroethylene	0.07	0.07
trans-1,2-Dichloroethylene	0.1	0.1
Dibromochloropropane	Zero	0.0002
1,2-Dichloropropane	Zero	0.005
Epichlorohydrin	Zero	TT
Ethylbenzene	0.7	0.7
Ethylene dibromide	Zero	0.00005
Heptachlor	Zero	0.0004
Heptachlor epoxide	Zero	0.0002
Lindane	0.0002	0.0002
Methoxychlor	0.04	0.04
Pentachlorophenol	Zero	0.001
PCBs	Zero	0.0005
Styrene	0.1	0.1
Tetrachloroethylene	Zero	0.005
Toluene	1	1
Toxaphene	Zero	0.003
2,4,5-TP	0.05	0.05
Xyenes (total)	10	10
Lead and Copper Rule		
Lead	Zero	TT[d]
Copper	1.3	TT[e]
Phase V Rule Inorganics		
Antimony	Zero	0.006
Beryllium	0.004	0.004
Cyanide	0.2	0.2
Thallium	0.0005	0.002
Phase V Rule Organics		
Adipate [di(2-ethyhexyl)]	0.4	0.4

Table 4-1 USEPA Primary Drinking Water Standards for Regulated Contaminants (continued)

Contaminant	MCLG, mg/L	MCL, mg/L
Dalapon	0.2	0.2
Dichloromethane	Zero	0.005
Dinoseb	0.007	0.007
Dioxin	Zero	$3 \infty 10^{-8}$
Diquat	0.02	0.02
Endothall	0.1	0.1
Endrin	0.002	0.002
Glyphosate	0.7	0.7
Hexachlorobenzene	Zero	0.001
Hexachlorocyclopentadiene	0.05	0.05
Oxamyl (Vydate)	0.2	0.2
Phathalate, [di(2-ethylhexyl)]	Zero	0.006
Picloran	0.5	0.5
Polyaromatic hydrocarbons [benzeo(*a*) pyrene]	Zero	0.0002
Simazine	0.004	0.004
1,2,4-Trichlorobenzene	0.07	0.07
1,1,2-Trichloroethane	0.003	0.005
Other Interim (I) and Proposed (P) Standards		
Sulfate (P)	500	500[f]
Arsenic (I)		0.05
Interim (I) and Proposed (P) Standards for Radionuclides		
Beta/photon emitters	Zero (I), zero (P)	4mrem/yr (I), 4 mrem/yr (P)
Alpha emitters	Zero (I), zero (P)	15 pCi/L (I), 15 pCi.L (P)
Radium 226 + 228	Zero (I)	5 pCi/L
Radium 226 (P)	Zero (P)	20 pCi/L
Radium 228 (P)	Zero (P)	20 pCi/L
Uranium (P)	Zero	0.02
Interim (I) and Proposed (P) Standards for Disinfection By-Products		
Bromate (P)	Zero	0.010
Bromodichloromethane (P)	Zero	see TTHMs
Bromoform (P)	Zero	see TTHMs
Chloral hydrate (P)	0.04	TT

Table 4-1 USEPA Primary Drinking Water Standards for Regulated Contaminants (continued)

Contaminant	MCLG, mg/L	MCL, mg/L
Chlorite (P)	0.08	1.0
Chloroform (P)	Zero	see THHMs
Dibromochloromethane (P)	0.06	See TTHMs
Dichloroacetic acid (P)	Zero	see HAA5
Haloacetic acids (HAA5)[g]	Zero	0.060 (P stage 1), 0.030 (P stage 2)
Trichloroacetic acid (P)	0.3	See HAA5
Total trihalomethanes (TTHMs)	Zero	0.10 (I), 0.080 (P stage 1), 0.040 (P stage 2)

a TT–treatment technique requirement.

b For water systems analyzing at least 40 samples per month, no more than 5.0 percent of the monthly samples may be positive for total coliforms. For systems analyzing fewer than 40 samples per month, no more than one sample per month may be positive for total coliforms.

c MFL–million fibers per liter.

d Action level = 0.015 mg/L.

e Action level = 1.3 mg/L.

f Alternatives allowing public water systems the flexibility to select compliance options appropriate to the protection of the population served were proposed.

g Sum of the concentrations of mono-, di-, and trichloroacetic acids and mono- and dibromoacetic acids

Source: Adapted in part from Reference 6.

4.3.3 State Drinking Water Regulations

The goal of the Safe Drinking Water Act is to assure the provision of safe drinking water to Americans by combining the efforts of Federal and State authorities. Under the Act, the USEPA has the responsibility for establishing regulations defining safe drinking water quality for public water systems and of assuring that all public water systems provide water meeting this definition. States that adopt regulations at least as stringent as those established by USEPA and adopt appropriate administrative and enforcement procedures are delegated primary enforcement responsibility ("primacy") for administration of the program.[20]

USEPA supports the efforts of primacy States with technical assistance and with financial assistance in the form of program grants. Where States do not assume primacy, the USEPA must administer the regulations directly. States with primacy can delete contaminants from the list if USEPA approves and can add to the list without USEPA approval.

4.3.4 Long-Term Changes

The National Primary Drinking Water Regulations will be reviewed and revised intermittently. The 1996 SDWA amendments require USEPA to review and revise as appropriate each

Table 4-2 National Secondary Drinking Water Regulations

Contaminant	Level, mg/L, unless specified
Chloride	250
Color	15 color units
Copper	1
Corrosivity	Noncorrosive
Foaming Agents	2.5
Fluoride	2.0
Iron	0.3
Manganese	0.05
Odor	3 threshold odor
pH	6.5 to 8.5
Sulfate	250
Total dissolved solids	500
Zinc	5

Source: Adapted in part from Reference 19.

NPDWR at least each six years. Any revision of an NPDWR is to be promulgated in accordance with the SDWA as amended in 1996. It is expected that USEPA's drinking water regulatory activity will increase in the next few years as the agency works to meet the various mandates and deadlines imposed by the 1996 SDWA amendments.

4.4 Initial and Design Years and Staging Periods

A water treatment plant is generally designed and constructed to serve the needs of a community for a number of years in the future. The initial year is the year when the construction is completed and the initial operation begins. The design or planning year is the year when the facility is expected to reach its full designed capacity and further expansion may become necessary.[21] Selecting the design year is not a simple task. It requires sound judgement and skills in developing future population-growth estimates from the past social and economic trends of a community. Several factors are considered in making these decisions.[1,21,22]

- Useful life of the treatment units, taking into account wear and tear, development in water treatment technology, and process obsolescence
- Convenience of future expansion
- Anticipated changes in the drinking water quality requirement
- Anticipated changes in raw water quality
- Growth pattern of the community, the service area, and the region (including shifts in community)

- Performance of the treatment facility during early years when it is over-sized
- Trends in interest rates, cost of present and future construction, and availability of funds

It is generally desirable to divide the design period into two or three *staging periods*. Initially, those facilities that are not conveniently or economically added are constructed for the ultimate capacity. Examples of these facilities are intake structure, pump stations, raw water lines, main conduits, channels, appurtenances, chemicals and control buildings, clearwells, and storage facilities. The facilities that are constructed in stages are: pumping equipment, chemical feeders, and treatment units such as rapid mix, flocculation, sedimentation basins, and filters. Some designers use a modular concept for the staging units by providing a complete process train in each phase.[1] The number of staging periods depends on the rate of growth of the service area and the financial ability of the utility. It is, however, desirable to allow sufficient time between the staging periods (at least 5 years) for engineering investigations, design, financing, and construction.[1,21]

Another approach to the concept of design period and staging periods is the consideration of strategically located water treatment plants at more than one location. This concept is beneficial if additional water supply sources are available and large geographical areas with shifting population centers and communities of different pressure zones are encountered.[1]

4.5 Service Area

Service area (also called water district) is defined as the total land area that will be eventually served by the proposed water treatment plant. The area may be based on geographical and topographical characteristics, on political boundaries, or on both. Areas outside the city limit that night become incorporated into the city at future dates should also be considered in the service area.

It is important that the design engineer and the project team become familiar with the service area of the proposed project. Site visits and review of engineering data on topography, geology, hydrology, climate, ecological elements, and social and economic conditions may be necessary. Existing zoning regulations and land use or zoning plans and probable future changes that can affect both developed and undeveloped lands should be studied. Such efforts should be carefully coordinated with the state, regional, and local planning agencies and should be in conformance with the *water master plan* of the area.[21] Under certain geographical, topographic, and regional development conditions, providing more than one water treatment plant night economically and efficiently meet the needs of the water utility.

4.6 Plant Capacity

The design capacity of a water treatment plant is generally based on the maximum day demand that is expected to occur at the design year. This involves forecasting of the demand in the future. Forecasting of water demand is a complex problem: the total demand includes residential, commercial, industrial, fire-fighting, public uses, and water lost or unaccounted for. All these components vary greatly, depending on such factors as the following: (1) climate; (2) geographic location; (3) size, population, and economic conditions of the community; (4) degree of industrialization; (5) metered water supply; (6) cost of water; (7) supply pressure; and (8) water conservation efforts of the general public and of regulatory agencies.[21,22]

The most common approach to forecasting the plant capacity is based on population projections of the service area to the design year and projection of the rate of water demand.

4.6.1 Population Projections and Population Density

Population Projections There are many mathematical and graphical methods that are used to project past population data to the design year. Some widely employed methods are the following:

1. arithmetic growth;
2. geometric growth;
3. decreasing rate of increase;
4. mathematical or logistic curve fitting;
5. graphical comparison with similar cities;
6. ratio method;
7. employment forecast;
8. birth cohort.

All these methods utilize different assumptions and, therefore, give different results. Selection of any one method depends on the amount and type of data available and whether the projections are made for the short or long term. In general, the population and rate of growth greatly depend on the following: (1) birth and death rates, (2) migration and immigration rates, (3) annexation, (4) urbanization and commercialization, and (5) industrialization. The arithmetic, geometric, decreasing rate of increase, and logistic curve-fitting methods are summarized in Table 4-3. The remaining methods are presented in Table 4-4 and Figure 4-1, Figure 4-2, and Figure 4-3. Several excellent references may be consulted on population forecasting.[23–27]

Population Density The average population density for the entire city rarely exceeds 7500–10,000 persons per km^2 (30–40 per acre). Often it is important to know the population density in different parts of the city in order to estimate the flows and to design the collection network. Density varies widely within a city, depending on the land use. The average population densities based on land use characteristics are summarized in Table 4-5.

4.6.2 Municipal Water Demand

Municipal water demand includes the following: (1) domestic or residential, (2) commercial, (3) industrial, (4) public service, and (5) unaccounted system losses and leakage. For design of waterworks projects, it is necessary to estimate accurately the amount of water that is needed.

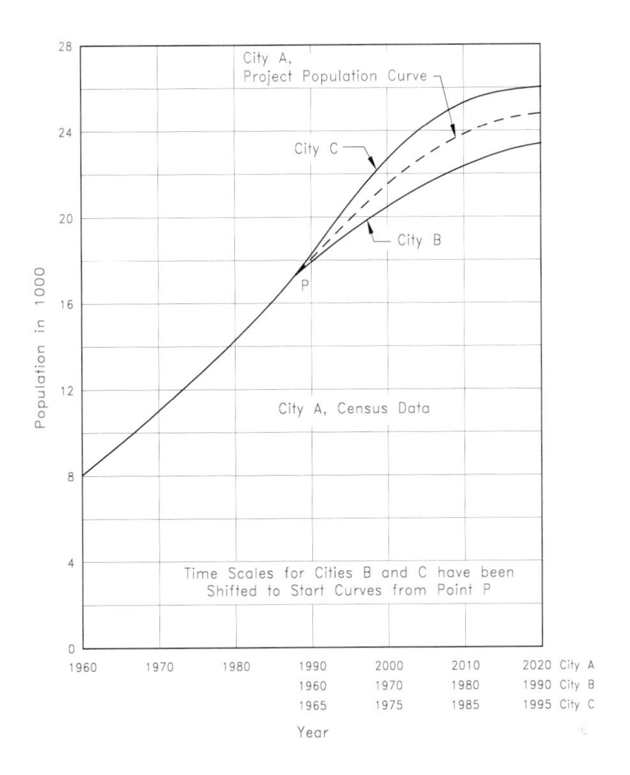

Figure 4-1 Population estimate by graphical comparison. *(Reprinted from Reference 21, with permission from Technomic Publishing Co, Inc., copyright 1999.)*

This involves estimating the population of the service area, per capita water consumption, and those factors that affect the water demand.

It is customary to express water consumption in liters per capita per day (Lpcd) or gallons per capita per day (gpcd). This figure represents the average daily amount of water required per person during a period of one year, and it generally is estimated by dividing the annual-average daily water consumption by the total population served. The per capita demand typically includes all treated water that is used by residential, commercial, industrial, and public service and system losses. Per capita demand may be further classified as usage under normal conditions or drought conditions. There is a wide variation in the per capita water withdrawal rate in public water supplies in the United States. The average water withdrawal rate, however, is 628

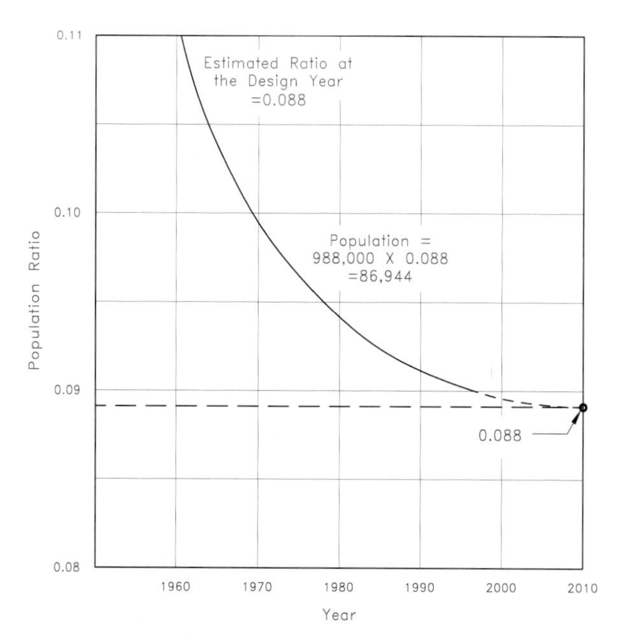

Figure 4-2 Population estimate by ratio method. *(Reprinted from Reference 21, with permission from Technomic Publishing Co., Inc., copyright 1999.)*

Lpcd (166 gpcd).[27–29] The water withdrawal rates for 50 states in the United States and other countries and cities are summarized in Table 4-6.[30, 31]

Components of Municipal Water Demand Various components of municipal water demand are residential, commercial, industrial, and public water uses and unaccounted system losses. These components are illustrated in Figure 4-4.

Residential Water Use. The residential or domestic water demand is the portion of the municipal water supply that is used in homes. Residential water use may vary greatly.[31–34] Typical use is given in Table 4-7.[21] It includes toilet flush, cooking, drinking, washing, bathing, watering lawn, and other uses. The average residential water demand varies from 100 to 450 Lpcd, while most commonly used numbers are 200–300 Lpcd.[23,25,29] Typical water use for various household devices is summarized in Table 4-8.[21]

Commercial Water Use. Commercial establishments include motels, hotels, office buildings, shopping centers, service stations, movie houses, airports, and the like. The commercial water demand depends on the type and number of commercial establishments. The commercial water demand is about 10–20 percent of the total water demand.[23,25] Based on floor area, the typical value ranges from 10 to 15 $L/m^2 \cdot d$.[22] The water demands in various types of commercial establishments may vary greatly.[21,33,35,36] Typical values are given in Table 4-9.

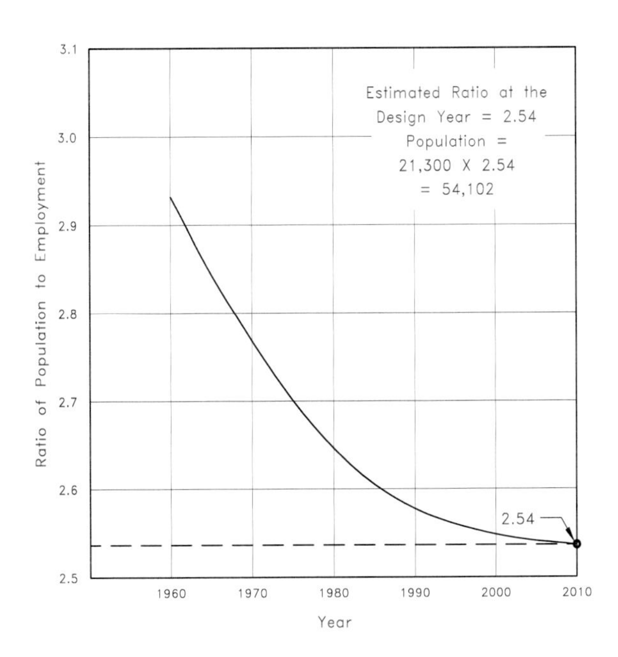

Figure 4-3 Population estimate by employment forecast. *(Reprinted from Reference 21, with permission from Technomic Publishing Co., Inc., copyright 1999.)*

Industrial Water Use. Industrial water demands are very large in the United States. Generally, large industries develop their own water supply systems. Only small industries purchase water and, therefore, impose demand on local municipal systems. Industrial water demand may be estimated on the basis of proposed industrial zoning and of the types of industries most likely to develop within the city. Typical water demand of small industries is 25–35 percent of total municipal water demand. Some industrial water-demand data are summarized in Table 4-10.[21,33,35,36]

Public Water Use. Water used in public buildings (city halls, jails, schools, etc.) as well as water used for public services (including fire protection, street washing, and park irrigation) is considered public water use. Public water use accounts for 5–10 percent of total municipal water demand.[23,25]

Unaccounted System Losses and Leakage. In a water supply system there is a certain amount of water that is lost or unaccounted for because of meter and pump slippage, leaks in mains, faulty meters, and unauthorized water connections. In municipal supply systems this may be 8-18 percent of the total water demand.[23,25]

Fluctuations The municipal water demand discussed above is based on annual average daily demand. There are wide variations in seasonal, daily, and hourly water demands. Some of the general observations of municipal water demands can be summarized as follows:

Table 4-3 Population Projections Using Arithmetic, Geometric, Decreasing Rate of Increase, and Mathematical Curve Fittings

Method	Description	Basic Equations or Procedure[a]	Calculated Values[b]
Arithmetic method	Population is assumed to increase at a constant rate. Average value of proportionality constant over several decades may also be used. The method is commonly used for short-term estimates (1-5 yrs.).	$\frac{dY}{dt} = K_a; Y_t = Y_2 + K_a(T - T_2)$ $K_a = \frac{Y_2 - Y_1}{T_2 - T_1}$	$K_a = 300/\text{yr}$ $Y_t = 21{,}000$
Geometric method	Population is assumed to increase in proportion to the number present. Average value of proportionality constant over several decades may also be used. The method is commonly used for short-term estimates (1-5 yrs.).	$\frac{dY}{dt} = K_p Y; \ln Y_t = \ln Y_2 + K_p(T - T_2)$ $K_p = \frac{\ln Y_2 - \ln Y_1}{T_2 - T_1}$	$K_p = 0.0182$ $Y_t = 21{,}600$

Table 4-3 Population Projections Using Arithmetic, Geometric, Decreasing Rate of Increase, and Mathematical Curve Fittings (continued)

Method	Description	Basic Equations or Procedure[a]	Calculated Values[b]
Decreasing rate of increase	Population is assumed to reach some limiting value or saturation point.	$\frac{dY}{dt} = K_d(Z - Y); Y_t = Y_2 + (Z - Y_2)(1 - e^{(-k_d)(T - T_2)})$ $Z = \frac{2Y_0Y_1Y_2 - (Y_1^2)(Y_0 + Y_2)}{Y_0Y_2 - Y_1^2}$ $K_d = -\frac{1}{T_2 - T_1}\ln\left(\frac{Z - Y_2}{Z - Y_1}\right)$	$Z = 20{,}000$ $K_d = 0.09163$ $Y_t = 19{,}200$
Mathematical or logistic curve fitting	It is assumed that the population growth follows a logistical mathematical relationship. Most common relationship is an S-shaped curve.	$Y_t = \frac{Z}{1 + ae^{(b)(T - T_0)}}$ $a = \frac{Z - Y_0}{Y_0}$ $b = \left(\frac{1}{n}\right)\ln\left[\frac{Y_0(Z - Y_1)}{Y_1(Z - Y_0)}\right]$	$Z = 20{,}000$ $a = 1.00$ $b = 0.1099$ $n = 10$ $Y_t = 19{,}287$

a dY/dt = rate of change of population with time. Y_0, Y_1, Y_2 = populations at time T_0, T_1, and T_2. Y_t = estimated population at the year of interest. Z = saturation population. K_a, K_p, and K_d are proportionality constants. a and b = constant. n = constant interval between T_0, T_1, and T_2 (generally 10 yrs.).

b Population for 2000 is estimated using the following census results: $T_0 = 1970$, $Y_0 = 10{,}000$, $T_1 = 1980$, $Y_1 = 15{,}000$, $T_2 = 1990$, $Y_2 = 18{,}000$.

Source: Reprinted from Reference 21, with permission from Technomic Publishing Co., Inc., copyright 1999.

Table 4-4 Population Projections by Using Graphical Comparison, Ratio Method, Employment Forecast, and Birth Cohort

<table>
<tr><th>Method</th><th>Description</th><th>Problem Definition</th><th>Estimated Population</th></tr>
<tr><td>Graphical comparison</td><td>The procedure involves the graphical projection of the past population data for the city being studied. The population data of other similar but larger cities are also plotted in such a manner that all the curves are coincident at the present population value of the city being studied. These curves are used as guides in future projections.</td><td>Estimate the population of City A by using graphical comparison with cities B and C. The design year is 2020.
<table>
<tr><th>Year</th><th colspan="3">Population in 1000s</th></tr>
<tr><th></th><th>City A</th><th>City B</th><th>City C</th></tr>
<tr><td>1960</td><td>8.0</td><td>18.0</td><td>16.0</td></tr>
<tr><td>1970</td><td>11.0</td><td>20.3</td><td>20.0</td></tr>
<tr><td>1980</td><td>14.0</td><td>22.0</td><td>25.0</td></tr>
<tr><td>1990</td><td>18.0</td><td>23.2</td><td>25.6</td></tr>
</table>
The procedure is illustrated in Figure 4-1.</td><td>$Y_{2020} = 24{,}800$</td></tr>
<tr><td>Ratio and correlation</td><td>In this method, the population of the city in question is assumed to follow the same trends as that of the region, county, or state. From the population records of a series of census years, the ratio is plotted and then projected to the year of interest. From the estimated population of the region, county, or state and the projected ratio, the population of the city concerned is obtained.</td><td>Estimate the population of a city using the ratio method. The design year is 2010. The estimated population of the region in the year 2010 is 988,000.
<table>
<tr><th>Year</th><th colspan="3">Population in 1000s</th></tr>
<tr><th></th><th>City</th><th>Region</th><th>Ratio</th></tr>
<tr><td>1960</td><td>50</td><td>455</td><td>0.110</td></tr>
<tr><td>1970</td><td>61</td><td>623</td><td>0.098</td></tr>
<tr><td>1980</td><td>72</td><td>766</td><td>0.094</td></tr>
<tr><td>1990</td><td>77</td><td>850</td><td>0.091</td></tr>
</table>
The procedure is illustrated in Figure 4-2.</td><td>$Y_{2010} = 86{,}944$</td></tr>
</table>

Table 4-4 Population Projections by Using Graphical Comparison, Ratio Method, Employment Forecast, and Birth Cohort (continued)

<table>
<tr><th>Method</th><th>Description</th><th>Problem Definition</th><th>Estimated Population</th></tr>
<tr><td>Employment forecast or other utility connections forecast</td><td>The population is estimated using the employment forecast. From the past data of population and employment, the ratio is plotted and population is obtained from the projected employment forecast. This procedure is similar to that of the ratio method. Similar procedure can be utilized from the forecast of various utility service connections such as telephone, electric, gas, and sewers. Utility companies conduct studies and develop reliable forecasts on the future connections. Forecasts of postal and newspaper service points have also been used in population estimates.</td><td>Estimate the population of a city using employment forecast. The design year is 2010. Use the following data. Employment forecast for the year 2010 is 21,300.
<table>
<tr><th>Year</th><th>Population in 1000s</th><th>Employment in 1000s</th><th>Ratio Population employment</th></tr>
<tr><td>1960</td><td>20</td><td>6.80</td><td>2.94</td></tr>
<tr><td>1970</td><td>30</td><td>10.79</td><td>2.78</td></tr>
<tr><td>1980</td><td>39</td><td>14.77</td><td>2.64</td></tr>
<tr><td>1990</td><td>46</td><td>17.83</td><td>2.58</td></tr>
</table>
The procedure is illustrated in Figure 4-3.</td><td>$Y_{2010} = 54{,}102$</td></tr>
<tr><td>Birth cohort</td><td>A birth cohort is defined by demographers as a group of people born in a given year or period[a]. The existing populations of males and females in different age groups are determined from the past records. From birth and death rates of each group and population migration data, the net increase in each group is calculated. The population data are then shifted from one group to the other until the design period is reached. This procedure is discussed in Reference 11.</td><td></td><td></td></tr>
</table>

a Demography is that branch of anthropology which deals with the statistical study of the characteristics of human population with reference to total size, density, number of deaths, births, migration, and so forth.

Source: Reprinted from Reference 21, with permission from Technomic Publishing Co., Inc., copyright 1999.

Table 4-5 Range of Population Densities in Various Sections of a City

Land Use	Population Range, Persons per km^2	Persons per acre
Residential Areas		
Single-family dwellings		
Large lots	1200–4000	5–16
Small lots	3700–8700	15–35
Multiple-family dwellings		
Small lots	8700–25,000	35–100
Apartment or tenement houses	25,000–250,000	100–1000 or more
Commercial areas	3700–7500	15–30
Industrial areas	1250–3700	5–15
Total, exclusive of parks, playgrounds, and cemeteries	2500–13,000	10–53

Source: Adapted in part from References 21 and 27.

- Working days have higher demand than holidays.
- Hot and dry days have more demand than wet or cold days.
- Maximum months are typically July or August (summer).
- Within a day, there are two demand peaks . One peak is in the morning, as the day's activities start; the other peak is in the evening.
- The minimum water demand normally occurs around 4:00 a.m.

The average demand represents the average daily demand over a period of one year. It is often referred to as *annual average day demand.* The *maximum day demand* represents the amount of water required during the day of maximum consumption in a year. This information is required to analyze the peak capacity of the production and treatment facilities. The *maximum demand* is particularly sensitive to the weather conditions. For an extended dry, hot summer (drought condition), the peak day demand tends to be much higher, on account of increased lawn watering and bathing. Conversely, for a wet, cool summer, the maximum day demand is generally lower.

Peak hour demand represents the amount of water required during the maximum consumption hour in a given day. This information is used to analyze the peak capacity required of the distribution system, elevated reservoirs, and high-service pumps, to be able to deliver the peak water demand during the peak hour of the day.

Table 4-6 Average Municipal Water Demand by State, Country, and Major City

State	Lpcd[a]	gpcd[b]	State	Lpcd	gpcd	Country	Lpcd	gpcd
Alabama	806	213	Nebraska	636	168	Austria	291	76
Alaska	1790	473	Nevada	1154	305	Baghdad, Iraq	172	45
Arizona	787	208	New Hampshire	485	128	Belgium	119	31
Arkansas	503	133	New Jersey	526	139	Bombay, India	165	43
California	685	181	New Mexico	772	204			
Colorado	746	197	New York	609	161	Columbo, Sri Lanka	229	60
Connecticut	541	143	North Carolina	644	170			
Delaware	700	185	North Dakota	477	126	Italy	281	74
Florida	617	163	Ohio	594	157	Netherlands	160	42
Georgia	946	250	Oklahoma	492	130	Spain	167	44
Hawaii	746	197	Oregon	712	188	Sweden	340	90
Idaho	897	237	Pennsylvania	685	181	Switzerland	408	107
Illinois	772	204	Rhode Island	462	122	Tel Aviv-Yufo, Israel	306	81
Indiana	534	141	South Carolina	916	242			
Iowa	466	123	South Dakota	549	145	Tokyo, Japan	294	77
Kansas	587	155	Tennessee	488	129			
Kentucky	314	83	Texas	587	155	United Kingdom	271	71
Louisiana	545	144	Utah	1113	294			
Maine	553	146	Vermont	553	146	United States	628	166[b]
Maryland	515	136	Virginia	420	111			
Massachusetts	530	140	Washington	1200	317	West Germany	173	46
Michigan	636	168	West Virginia	568	150			
Minnesota	473	125	Wisconsin	587	155			
Mississippi	507	134	Wyoming	746	197			
Missouri	485	128	Washington, D.C.	799	211			
Montana	829	219	Puerto Rico	326	86			

a Average is based on total municipal water demand divided by the population served.

b 1 gallon = 3.785 liters.

Source: Adapted in part from References 30 and 31.

Annual average, maximum day, and peak hour demands are important for planning and design of municipal water supply systems. These variations are conveniently expressed as a ratio to the mean average daily flow. These ratios vary greatly for different cities; therefore, a careful study for each city must be made from the past data to develop these fluctuations. In the absence of water-demand data, Eq. (4.1) is used to estimate the ratios of maximum month, week, day, and hour demand.[22,25]

Table 4-7 Typical Breakdown of Residential Water Uses

Types of Water Uses		Percentage
Toilet flush, including toilet leakage		33
Shower and bathing		28
Wash basin		11
Kitchen		9[a]
Drinking, cooking	(2–6 percent)	
Dishwashing	(3–5 percent)	
Garbage disposal	(0–6 percent)	
Laundry and washing machine		16[b]
Lawn sprinkling and miscellaneous		3

a Dishwasher is 3% and remaining is faucet use.

b Higher percentages because of increased use of washing machines and reduced total demand.

Source: Adapted in part from References 21, 37, and 38.

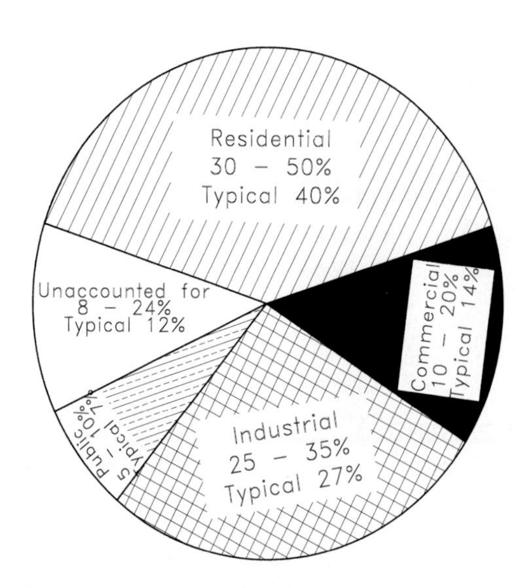

Figure 4-4 Components of municipal water demand.

Table 4-8 Typical Rates of Water Use from Various Devices

Device	Range of Flow
Household Faucet	10-20 L/min
Wash basin	4-8 L/use
Shower head	90-110 L/use; 19-40 L/min
Tub bath	60-190 L/use
Toilet flush, tank-type	19-27 L/min
Toilet flush, valve-type	90-110 L/min
Dishwasher	15-30 L/load
Washing machine	110-200 L/load
Lawn sprinkler	6-8 L/min
Continues flowing drinking fountain	4-5 L/min
Garbage disposal	600-7500 L/wk 4-8 L/ person per day
Dripping or leaky faucet	10-1000 L/d

Source: Adapted in part from References 21 and 36–38.

$$p = 180t^{-0.10} \tag{4.1}$$

where

p = percent of the annual average demand for time, t

t = time, d. The time t in days varies from 2/24 to 365

Typical peak demand ratios and those calculated from Eq. (4.1) are summarized in Table 4-11. It may be mentioned that peak hourly demands in residential areas may be very high because of lawn-watering and other water uses. In general, the ratio is higher for subdivisions that have a lower population density. Also, the commercial and industrial demands reduce the peaking ratio. The accepted value for peak hourly demand is 150 percent of maximum day demand.[22] Typical daily variation in municipal water demand is shown in Figure 4-5.[21]

Fire Demand Water demand for firefighting must be considered in the design of water distribution systems. Although the annual volumes of water required for fire fighting are small, the demand may be exceedingly high during the periods of need. Therefore, fire flows must be considered while designing the water distribution systems.

In the past, the fire flow requirements for a community were established by the *National Board for Fire Underwriters* (NBFU). Eqs. (4.2) and (4.3) were used to estimate the fire demand required in downtown business districts.[22,25,26]

Table 4-9 Average Water Demand in Residential, Institutional, and Commercial Establishments

Source	Unit	Flow (L/unit·d)
Residential		
Single-family		
Low-income	Person	270
Medium-income	Person	310
High-income	Person	380
Summer cottage	Person	190
Trailer park	Person	150
Apartment	Person	230
Hotel, motel	Unit	380
Camp	Person	133
Resort (day or night)	Person	190
Institutional		
Hospital	Bed	950
Rest home	Bed	380
Prison	Inmate	450
School		
Boarding	Student	300
Day	Student	76
Commercial		
Country club		
Resident	Member	380
Nonresident	Member	95
Restaurant	Customer	30
Cafeteria	Customer	6
	Employee	40
Bar	Customer	8
	Employee	50

Table 4-9 Average Water Demand in Residential, Institutional, and Commercial Establishments (continued)

Source	Unit	Flow (L/unit·d)
Coffee shop	Customer	20
	Employee	40
Dance hall	Person	8
Store	Toilet room	1520
	Employee	40
Department store	m^2 Floor area	8
	Employee	40
Shopping center	m^2 Floor area	6
	Employee	40
Office building	Employee	65
Barber shop	Chair	210
Beauty salons	Station	1026
Laundry		
Laundromat	Machine	2200
Commercial	Machine	3000
Service station	First bay	3800
	Additional bays	1900
	Employee	190
Theater [a]		
Drive-in	Car space	19
Movie	Seat	8
Airport [a]	passenger	10
Car wash[a]	car washed	209
Industrial building	Employee	55
Factories		
With showers	Employee-shift	133
Without showers	Employee-shift	95

a Does not contain per day unit. Multiply unit per day to obtain total per day flow.
Source: Adapted in part from References 21, 33, 35, and 36.

Table 4-10 Typical Industrial Water Demand

Industrial Use	Quantity
Canning	30–60 m^3/metric ton
Milk, dairy	2–3 m^3/metric ton
Meat packaging	15–25 m^3/metric ton
Cattle	40–50 L/head·d
Dairy	70–80 L/head·d
Chicken	30–40 L/head·d
Pulp and paper	200–800 m^3/metric ton
Steel	260–300 m^3/metric ton
Tanning	60–70 m^3/metric ton raw hides processed

Source: Adapted in part from References 21, 33, 35, and 36.

$$Q = 3.86\sqrt{P}(1 - 0.01\sqrt{P}) \tag{4.2}$$

where

Q = fire flow rate, m^3/min

P = population in thousands

$$Q' = (1020\sqrt{P})(1 - 0.01\sqrt{P}) \tag{4.3}$$

Q' = fire flow rate, gpm

Table 4-11 Typical Fluctuations in Water Demand in Municipal Water Supply Systems

Condition	Percent of Annual Average Day		From Eq. (4.1)	
	Range	Typical	t,d	Percent of Annual Avg Day
Daily average in maximum month	110–140	120	30	128
Daily average in maximum week	120–170	140	7	148
Maximum day in a year	160–220	180	1	180
Peak hour within a day	225–320	270[a]	2/24[b]	231

a 1.50 ∞ maximum day demand from Eq. (4.1).

b Peak hour demand is distributed over two hours.

Source: Adapted in part from References 22 and 28.

More recently, the fire flow requirements for a community have been established by the *Insurance Service Office* (ISO). The required fire flow for a neighborhood depends upon the size of the area, type of construction, occupancy, and exposure of the buildings within and surrounding the block or group complex.[22, 25]

The Insurance Services Office uses Eq. (4.4).[22,25,39]

$$F = 320C(A)^{0.5} \tag{4.4}$$

where

F = fire flow required, m^3/d

A = total floor area in m^2, excluding the basement of the building. For fire-resistive buildings, the six largest successive floor areas are used if the vertical openings are unprotected, but, where the vertical openings are properly protected, only the three largest successive floor areas are included.

C = coefficient related to the type of construction. The C values for different types of construction follow:

1.5 for wood-frame construction
1.0 for ordinary construction
0.8 for noncombustible construction
0.6 for fire-resistive construction

Interpolation between these values is used for construction that does not fall into the above four categories. The computed value is then adjusted up or down for (1) occupancy, (2) sprinkler protection, and (3) exposure. Regardless of the value calculated from the above formula, the fire flow shall not exceed 30.3 m^3/min (8000 gpm) for wood frame and ordinary construction, and 22.7 m^3/min (6000 gpm) for noncombustible or fire-resistive construction. Additional flow may be required to protect nearby buildings. The fire flow shall not be less than 1.9 m^3/min (500 gpm). For groupings of single-family and small two-family dwellings of two stories at most, the fire flow is obtained from Table 4-12.

The fire-flow values obtained from Eq. (4.3) must be adjusted for special conditions: (a) reduced up to 25 percent for occupancies having a light fire loading;[d] (b) increased up to 25 percent for occupancies having a high fire loading; (c) reduced up to 50 percent with automatic sprinkler protection; and (d) increased if within 46 m (150 ft) of the fire area (25 percent for one side exposed, to a maximum of 75 percent for all sides exposed). After all corrections, the fire flow shall not be less than 1.9 m^3/min (500 gpm) and shall not be greater than 45.4 m^3/min (12,000 gpm). The fire flow should be available from 2 to 10 hr, as summarized in Table 4-13.[39,40]

d. Light fire loadings are occupancies of low hazard, such as churches, hospitals, schools, offices, museums, and other public buildings. High fire-hazard loadings are commercial and industrial establishments that involve processing or storage of oils, flammables, paints, solvents, explosives, and so forth.

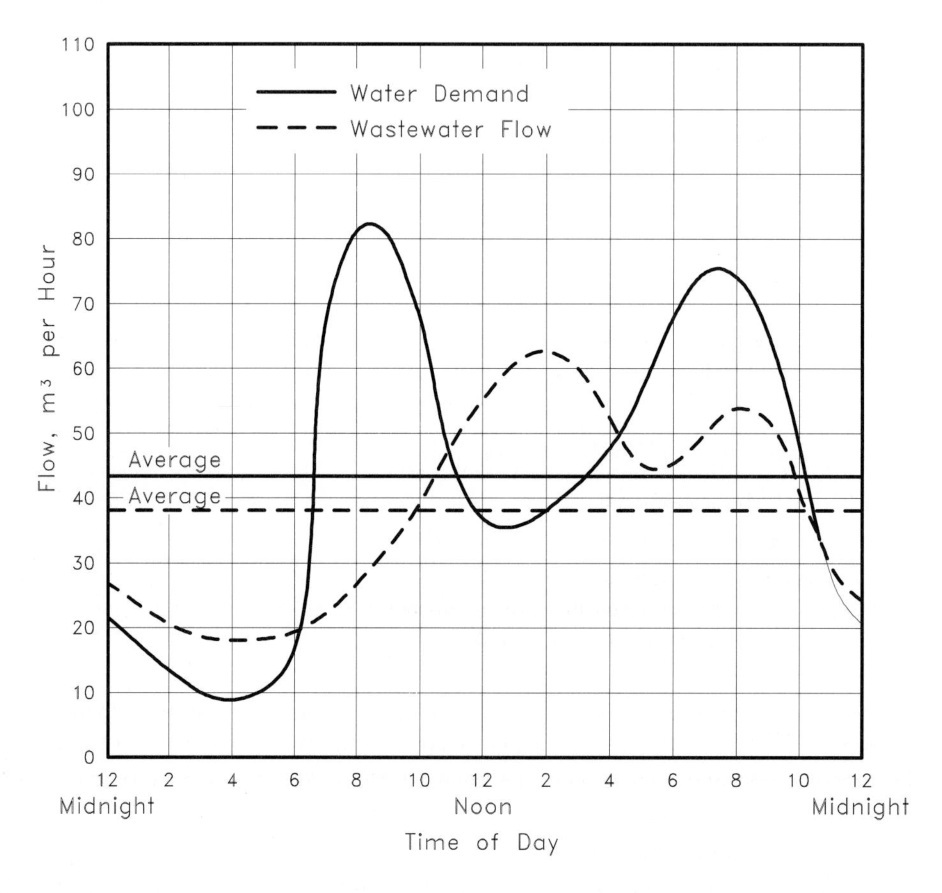

Figure 4-5 Typical variation in municipal water demand and wastewater flow. *(Reprinted from Reference 21, with permission from Technomic Publishing Co., Inc., copyright 1999.)*

The required fire flow must be available in addition to the coincident maximum daily flow rate. For the distribution system, the design flow rate is the larger of the peak hourly flow or maximum day flow, plus fire. The fire hydrant pressure during a fire should generally exceed 137.9 kN/m^2 (20 psi) where motor pumpers are used. Because a city will be penalized in its fire insurance rates if the needed flows cannot be met for specified durations, most cities provide emergency storage to meet fire demands.

The guide for determination of required fire flow are developed by the *Insurance Service Office* and other fire-rating organizations. Considerable knowledge and experience in fire protection engineering are necessary for detailed application of the guidelines. Many states and cities specify minimum fire flows for residential, commercial, and industrial areas and minimum storage requirements for checking the design of distribution systems. This topic and additional

Table 4-12 Required Fire Flows for Single- and Two-Family Residential Areas not Exceeding Two Stories in Height

Distance Between Dwelling Units		Required Fire Flow,	
m	ft	m³/min	gpm
Over 30.5	over 100	1.9	500
9.5–30.5	31–100	2.8–3.8	750–1,000
3.4–9.1	11–30	3.8–5.7	1,000–1,500
3.1 or less	10 or less	5.7–7.6	1,500–2,000
continuous	continuous	9.5	2,500

Source: Adapted in part from Reference 39.

details on fire-flow estimation, fire-hydrant location, and pressure distribution are covered in Chapter 13. The following examples provide some basic information.[22,25,39]

Example 4-1

A 4-story wooden-frame building has each floor of area 509 m^2. This building is adjacent to a 5-story building of ordinary construction with 900 m^2 per floor. Determine the fire flow and duration for each building and that for both buildings (assuming they are connected).

Solution

C for wooden construction is 1.5, and for ordinary construction is 1.0.

Area occupied by 4-story building = $4 \infty 509\ m^2 = 2036\ m^2$

$F_1 = 320 \infty 1.5 \infty (2036)^{0.5} = 21{,}659\ m^3/d = 15.0\ m^3/min$

The duration from Table 4-12 is 4 hrs.

Area occupied by 5-story building = $5 \infty 900\ m^2 = 4500\ m^2$

$F_2 = 320 \infty 1.0 \infty (4500)^{0.5} = 21{,}466\ m^3/d = 14.9\ m^3/min$

The duration from Table 4-12 is 4 hrs.

The fire flow for the both buildings is obtained from the fractional area occupied by each building.

Total area = $2036\ m^2 + 4500\ m^2 = 6536\ m^2$

Fractional area of 4 story building = $2036\ m^2/6536\ m^2 = 0.31$

Fractional area of 5 story building = $4{,}500\ m^2/6536\ m^2 = 0.69$

$F = 320\ [0.31 \infty 1.5(2036)^{0.5} + 0.69 \infty 1.0(4500)^{0.5}] = 21{,}526\ m^3/d = 14.95\ m^3/min$

The duration from Table 4-13 is 4 hrs.

Example 4-2

Estimate the water requirements for a community that will reach a population of 120,000 at the design year. The estimated municipal water demand for the community is 610 Lpcd. Calculate fire flow, design capacity of the water treatment plant, and design capacity of the water distribution system. Use NBFU formula for fire flow.

Solution

$$\text{Average day water demand} = \frac{610 \text{ L/person/d} \times 120{,}000 \text{ persons}}{1000 \text{ L/m}^3 \times 1440 \text{ min/d}} = 50.83 \text{ m}^3\text{/min}$$

Fire flow from Eq. (4-2) $= 3.86\sqrt{120}(1 - 0.01\sqrt{120}) = 37.65 \text{ m}^3\text{/min}$

The maximum day demand from Eq. (4.1) or Table 4-10 = 1.8 ∞ 50.83 m^3/min = 91.5 m^3/min

Peak hour demand from Table 4-10 = 1.5 ∞ 91.5 m^3/min = 137.25 m^3/min

Design capacity of water treatment plant = 91.50 m^3/min = 131,760 m^3/d

Design capacity of water distribution system is larger of:

(a) Peak hour flow of 137.25 m^3/min

(b) Fire + coincident maximum day demand
= 37.65 m^3/min + 91.50 m^3/min = 129.15 m^3/min

The peak hour demand of 137.25 m^3/min governs the design of the distribution system.

Water Conservation Water conservation in homes, commercial establishments, and industries is gaining popularity. Water conservation has the beneficial effect of reducing water and sewer bills. A community awareness program is necessary for the general public to understand the importance of water conservation and the cost and benefits of water-saving devices. A number of inexpensive devices are now available as part of most faucet combinations. Water savings as high as 20–40 percent can be achieved by use of these flow reduction devices and by practicing simple water-conservation measures in homes, businesses, and industries.[32,41–43] Principal flow-reduction devices and systems and the percent reduction that can be achieved over conventional devices are summarized in Table 4-14.[21] Many water-saving ideas for homes are listed in Table 4-15.[21,34,43] Water-conservation efforts require public cooperation and the efforts of the local government in the form of plumbing codes.

Projection of Water Demand The design criteria for most states require that the treatment capacity of the water treatment plant must always be in excess of the maximum expected draft of any day of the year. Therefore, for planning of the water treatment facilities, the maximum day drought condition should be used for projection of the treatment plant capacity and that of the water distribution system. The following procedure is generally used;

1. Estimate the future population of the service area for the initial and design years and for the staging periods.
2. Using the historical water usage and billing information, develop annual water-consumption data and population served. Determine average per capita daily water demand (Lpcd). In many cases, it may be desirable to delete from this analysis the demand exerted by large

Table 4-13 Required Duration for Fire Flow

Required Fire Flow,		Duration,
m^3/m	gpm	h
7.6	2000 or less	2
11.3	3000	3
15.1	4000	4
18.9	5000	5
22.7	6000	6
26.5	7000	7
30.2	8000	8
34.0	9000	9
37.8	10,000 and greater	10

Source: Adapted in part from Reference 39.

industries that can account for disproportionately large percentages of the industrial water demand. If industrial demand is not disproportionately high for a few industries, it may not be necessary to separate industrial demand.

3. Plot annual average per capita demand versus the year. Prepare curves representing drought and normal conditions. Generally, a drought condition is represented by the curve enveloping the high points. Such analysis is given in Chapter 5 (Figure 5-3).
4. Considering water-conservation efforts and other general trends in water demand, establish for the desired future years the annual average per capita water demands under drought and normal conditions.
5. Develop the ratio of maximum day to average day from the maximum day (of the year) usage record and the annual average day data developed in step (2). Be sure to include the adjustments made for industrial water usage.
6. Establish a desirable maximum day/average day ratio for future projections.
7. Establish the ratio of peak hour to maximum day demand. For this purpose, a study must be undertaken during the high-demand hours, to develop the maximum system withdrawal rate. Information such as pump discharge rates and pressures, flow-meter readings, and elevated storage tank levels and their rate of decrease are necessary.
8. The average day demand and the peaking ratios of maximum day and peak hour developed above should be used to determine the projected average day, maximum day, and peak hour demands for drought and normal conditions. Proper adjustments to the projected industrial demands should also be made if deleted in the above analysis. Examples of such analysis and step-by-step calculations are presented in Section 5.4.3 and Section 5.4.4.

4.7 Raw Water Quality and Source Selection

Type of treatment depends on the quality of the source and on the desired quality of the finished water. In general, raw water should have quality high enough to provide finished water quality with a minimum number of treatment processes and at the least treatment cost. It is therefore essential that careful consideration and extensive engineering evaluations, including a water resource development investigation, be conducted as part of source selection. Factors such as quantity, quality, reliability, safety of source, water rights, and environmental impacts, along with capital and operation costs of the project, enter into the decision making.

Types of sources fall into three general categories: (1) rain water, (2) groundwater sources, principally wells, and (3) surface water sources such as rivers, lakes, and impoundments on rivers and streams. Individual homes have utilized cisterns to catch rain water for their domestic needs. Groundwater sources are used extensively by individual homes and by small-to-medium-sized communities. Groundwater sources tend to be uniform in quality and free of turbidity and color but to contain greater amounts of dissolved solids. Water quality of a typical rain-water and groundwater source is given in Chapter 2 (Table 2-1). Consideration of groundwater for a supply source should require studies to include aquifer safe yield, permissible drawdown, MCLs of regulated contaminants, saltwater intrusion, other sources of groundwater contamination, and water rights.

Surface waters tend to be variable in quality in terms of turbidity, color, and taste- and odor-causing substances. Surface waters are also exposed to waste inputs, including accidental spills, natural causes, and man-made activity that may change the quality in a short time or over the lifetime of the project. Therefore, surface water sources should be carefully selected. The following should be considered:

1. safe watershed yield during the drought years to meet the projected demands; this may include statistical analysis of precipitation, evapotranspiration, runoff, and stream flow.
2. urbanization and land development in the watershed;
3. proposed impoundments on tributaries;
4. water quality, based on short-term, seasonal, and long-term changes expressed as physical, chemical, microbiological, and radiological indicators;
5. assessment of reliability in terms of possible disruptions due to flood damage, earthquake, sabotage, or accidental spills of toxic substances;
6. requirements for construction of reservoir, collection headworks, conveyance system, treatment plant, and distribution system;
7. economics of the project, including cost of construction and operation of the facility;
8. environmental impacts of the project.

Information sources on surface water quality may be obtained from local, state, and federal departments and officials. Concentrations of common constituents in the selected surface water source should be such that USEPA drinking water standards could be met by a conven-

Table 4-14 Principal Flow-Reduction Systems and Devices, and Percentage Reduction Over Conventional Devices

Flow-Reduction System/device	Description	Percentage Reduction in Water Use over Conventional System/Device
Pressure-reducing valves (PRV)	Installed to reduce the city water supply pressure; thus household pressure and flow are reduced. This eliminates unnecessary waste due to spurts from faucet, leaks, and drips.	20–30
Sink faucet		
Faucet aerator	Designed to provide smooth and even flow of water. Aerated jet of water increases rinsing power of water, thus reducing amount of washwater used in kitchen and bathroom sinks.	1–2
Limiting flow valve	Restricts flow to a constant rate independent of supply pressure.	1–2
Shower		
Flow-limiting shower head	Reduces water consumption by restricting and concentrating water jet.	10–12
Limiting flow valves	Restricts flow to a fixed rate independent of supply pressure.	8–10
Washing machine		
Level controller or water-efficient washer	Level setting uses water in accordance to wash load.	2–4
Dishwasher		
Water-efficient dishwasher	Machine designed to use less water.	4–10
Toilet		
Shallow trap water closet	Shallow-trap toilets are similar to conventional toilets. Shallow trap requires less water per flush. These are cost-effective for new homes.	30–40
Dual-cycle toilet inserts	Toilet insert devices convert conventional toilets to dual-cycle operation (separate for urine or fecal waste transport).	15–18
Dual-cycle toilet	New types of toilets that have two flush cycles, one for urine and the other for fecal transport. Recommended for new construction.	25–30

Table 4-14 Principal Flow-Reduction Systems and Devices, and Percentage Reduction Over Conventional Devices (continued)

Flow-Reduction System/device	Description	Percentage Reduction in Water Use over Conventional System/Device
Reduced-flush device	Toilet tank inserts that reduce the volume of flush by occupying a portion of the tank or prevent the tank from completely emptying.	12–18
Flush valve		
single-flush valve and dual-flush valve	The flush valves operate directly from the water supply lines. Controlled amount of water is used for flushing. The mechanism can also be used with shallow-trap toilets. Recommended for new homes.	10–14; 20–30
Immersed bottle or brick in toilet tank	Plastic bottle filled with water and weighted with pebbles occupy tank space and reduce water used per flush (obstruction to float must be avoided). Bricks are also used, but they flake and may clog tubes and valves.	2–5
Toilet-leak detector	Tablet that dissolves in tank and releases dye to indicate the leak	0
Vacuum-flush toilet system	The toilets use air and small amount of water or foam to transport waste.	30–40
Recirculating toilet system	The flush water is recirculated for a certain period. The system operates on a closed-loop principle in which waste is accumulated in a holding tank. Chemicals are used to suppress the microbiological activity in recirculating water. Other innovations utilize mineral oil as flushing medium. The holding tank is emptied periodically by vacuum trucks.	25–45
Washwater reuse	Wastewater stream from sinks, bathtubs, showers, and laundry may be treated on-site and reused for toilet flushing and lawn sprinkling.	30–50
Urinals	Use of urinals in homes reduces water consumption. Wall-type urinals with flush valves for home use 5–6 L per flush.	10–15

Source: Adapted in part from References 21, 32, 34, and 41–43.

tional water treatment plant, and by BAT for regulated contaminants (Table 4-1). Surface water quality criteria for public water supplies utilizing conventional water treatment processes were developed over 30 years ago.[44] At that time, the concepts of permissible criteria and desirable criteria were introduced. A water source meeting *permissible criteria* will allow production of

Table 4-15 Water-Saving Ideas for Homes

Location	Water-Saving Ideas
Bathtub	A full bathtub holds 190 L. One can bathe adequately in a one-quarter-full bathtub.
Shower	Water consumption in a quick shower is less than bathing (40–70 L). Turn off water while soaping. Use light spray.
Toilet	Do not flush more than necessary. Use water-saving devices discussed in Table 4-14. Water leaks or waste can be detected by adding a few drops of food coloring in the tank. Coloring appears in the toilet if there is a leak.
Washing machine	Use load selector for large or small loads, if there is one. Otherwise, wash only full loads. Using cold water saves energy. Buy a machine that uses less water and energy per unit weight of wash. Use suds-saver attachment and less detergent.
Dishwasher	Preclean dishes with used paper napkins. Wash full load. Use less detergent.
Utility sink	Preclean dishes with used paper napkins. Soak overnight with small quantities of low-sudsing detergent. Save rinse water for next soak. Plan ahead to thaw frozen foods in air, not running water or microwave.
Drinking water	Do not run tap waiting for cold water. Use ice cubes or keep water pitcher in refrigerator. Use paper cups at water fountain to avoid waste.
Garbage disposal	Avoid using garbage grinders. Use leftover food for feeding pets. Start a compost pile.
Bathroom sink	Turn off water while shaving or brushing teeth.
Faucet	Check all faucets, including outside hose connections, for leaks. Replace worn 0 rings, and packing and faulty fixtures. A pinhole leak can waste up to 1000 L of water per day. In many areas, leaks cause about 95 percent of complaints about excessive water bills.
Lawn, garden	Water slowly, thoroughly, and as infrequently as possible. Water at night to minimize evaporation. Keep close watch on wind shifts. Select hardy species and native plants that do not need as much watering. Mulch heavily. Let grass grow higher in dry weather—it saves burning, and less water is needed. Recycle water from bathtub, kitchen sinks, etc. A 1.25-cm garden hose under normal water pressure pours out more than 2300 L/h. A 2-cm hose uses almost 7000 L/h.
Backyard pool	Cover when not in use to prevent evaporation. Keep clean, and reduce algae growth. Recycle wading pool water for plants, shrubs, lawns.
Carwash	Try to wash car near hedges, shrubs, or lawn. Wash car in short spurts from hose. Use a commercial carwash that recycles water.

Source: Adapted in part from References: 21, 29, 41– 43.

safe water by a *conventional treatment plant. Desirable criteria* represent higher-quality water in all respects, and produce water with a greater factor of safety or at less cost than is possible with waters meeting permissible criteria. AWWA statement of policy on public water supply matters suggests that the raw water supply should come from the highest-quality source of supply available and be appropriately treated to meet regulatory water supply criteria.[45] A comprehensive water quality management program must be instituted for watershed protection and source monitoring.

4.8 Process Selection and the Treatment Process Train

Sound understanding of raw water quality, finished water quality, and performance levels of various unit operations and processes are essential to develop process alternatives. Water treatment plants utilize a number of treatment processes to achieve the desired degree of treatment. In addition to this, the design engineer must evaluate numerous other important factors in the selection of the treatment processes. These factors include raw water and finished water quality standards, State design criteria, constituents treated, topography and geology, hydraulic requirements, sludge disposal, energy requirements, and plant economics. The collective arrangement of various treatment processes is called a *flow scheme*, *flow sheet*, *process diagram*, or *process train*.

Choice of proper treatment processes and development of the process train is a complex task. It requires understanding of the unit operations and processes, operational capabilities, and compatibility of various treatment components that are arranged to develop the process train. Laboratory and pilot plant studies are often necessary to conduct treatability tests. Chapter 3 is devoted entirely to process selection and the process train. Readers are advised to study this chapter for valuable information on process selection.

4.9 Plant Siting, Layout, and Hydraulics

During early planning and design stages, careful consideration must be given to the site selection of the proposed water treatment plant. Conditions such as natural and man-made disasters, topography, available land area and land acquisition, access roads, flood conditions, available head, economy of conveying water to and from the plant, availability of electric power, and disposal of residuals should all be considered in site selection and plant layout. Plant layout requires centralization of operations and control, short chemical feed lines, and coordination of plant elements to provide efficient operation and flow by gravity. Chapter 15 is devoted exclusively to the topics of plant siting, layout, and hydraulics. Design engineers should study this chapter in order to include the basic considerations relative to plant siting, layout and hydraulics during the preliminary design and process selection.

4.10 Equipment Selection

Every water treatment plant requires manufactured equipment or materials. In fact, most designs are governed by the dimensions and installation requirements of the selected equipment. The design engineer has the responsibility for selecting the treatment processes and corresponding equipment for achieving the desired treatment. To do this, a review of the design standards, design procedure, and design assumptions, preliminary design calculations, and careful study of the manufacturers' catalogs may be necessary in advance. The manufacturers of the appropriate equipment and their local representatives should be requested to furnish the needed details. It may often be necessary for the designer to work closely with the equipment supplier and provide as much information as possible to ensure that the equipment selection is best for the intended applications. Equipment selection must depend, to an extent, on the type of process, reliability, redundancy, and parts inventory required.

The request for information on equipment should not impose any obligation on the part of the engineer to use the equipment or even to include it in the specifications; however, the equipment data should not be requested unless there is some specific application in the design. It may often be necessary to obtain equipment data from more than one supplier. In no case, however, should the request be sent as a routine to all equipment manufacturers. Preparing the needed information on equipment costs money, and supplying such information without obligation can be justified only if there is a prospect for sale.[21,46]

4.11 Energy and Resource Requirement

Because of the recently increased concern about the limited resources available to meet our energy needs, the project planning and design must also utilize energy-conservation concepts. Primary energy is the energy used in the operation of the facility; secondary energy is needed to manufacture chemicals, other consumable materials, and construction material such as concrete and steel. Designers should encourage techniques that reduce total energy requirements. Such efforts should be utilized throughout the planning, the project formulation, and the preliminary engineering design.[21]

4.12 Plant Economics

As an integral part of the water treatment plant planning and design, a cost-effective approach should be utilized to ensure that the construction and operation and maintenance (O&M) costs are reasonable and appropriate for the planned level of treatment and process train. A cost-effective solution is one which will minimize total costs of the resources over the life of the treatment facility. Resources costs include capital (land plus construction), operation, maintenance, and replacements. Many publications are available that can be used in making preliminary cost estimates.[47,48] When preparing cost estimates, proper assumptions, cost curves, and cost indexes should be used. Discussion on cost equations and cost indexes is provided in Appendix E. Procedures for cost estimation and cost-effective evaluation are given in Chapter 5.

4.13 Environmental-Impact Assessment

The National Environmental Policy Act of 1969 (NEPA) was enacted to ensure that federal agencies consider environmental factors in the decision-making process and utilize an interdisciplinary approach in evaluating these issues.[49] The environmental-impact assessment must evaluate all impacts, "beneficial" and "adverse," "primary" and "secondary," that may result from the construction of a water treatment facility. The primary impacts are those directly associated with construction and operation of the waterworks project. The secondary impacts are indirect, resulting from the growth or change in land use induced or facilitated by the construction of associated water lines. A detailed environmental-impact assessment report should address the following areas:

1. environmental setting or existing conditions;
2. the environmental impact of the proposed action;
3. any adverse environmental effects that cannot be avoided should the proposal be implemented;
4. alternatives to the proposed action;
5. the relationship between local short-term uses of the human environment and the maintenance and enhancement of long-term productivity;
6. any irreversible and irretrievable commitments of resources that would be involved in the proposed action should it be implemented.

The environmental-impact assessment for a water treatment project is covered in Chapter 5. In-depth coverage and preparation of environmental-impact reports may be found in References 49–51.

4.14 State Requirements

Most states require the approval of plans and specifications for public water supply projects before construction or expansion begins. Most states have developed design criteria or minimum design requirements or standards.

The design engineer should meet with the representatives of the state's review agency in the early stages of planning, to obtain necessary design details and requirements of the state and local conditions. The design engineer should also consult with local planning boards, commissions, and government agencies and become familiar with local laws, regulations, and building and construction codes.[52]

Problems and Discussion Topics

4.1 It is believed that the National Drinking Water Regulations will continue to become more complex and controversial in the future. Do you support such notion? Discuss.

4.2 Using References 5 through 24, prepare a table to summarize the changes in the National Primary Drinking Water Regulation.

4.3 Visit the water treatment plant in your community. Mark on a map the service area and the plant location. Briefly summarize the history of expansion of the plant and the distribution system, including the major events that helped to bring about improvements. Draw the process train.

4.4 Review the engineering report on the most recent expansion of the water treatment facility in your community. List the major issues addressed in the report.

4.5 What factors must be considered in selection of a design year and staging periods?

4.6 Discuss various factors that may influence the population growth of a community.

4.7 Estimate the population of a service area in the design year 2020. Also estimate the populations in staging years 2010 and 2015. Use arithmetic, geometric, decreasing rate of increase, and logistic curve-fitting methods. Use the following census data

Year	Population in thousands
1970	30.0
1980	40.0
1990	62.0
2000	67.0

4.8 Estimate the population in the year 2010 for a community using employment forecast. The employment projection for 2010 is 9,850.

Year	Population	Employment
1970	20,000	6400
1980	21,100	7385
1990	22,800	8664
2000	23,500	9400

4.9 Estimate the average, maximum day, and peak hour water demand for a community of 55,000 population. Also calculate the design capacity of the water-distribution system and the water treatment plant. Use Eq. (4.2) to estimate the fire demand and Eq. (4.1) and Table 4-11 to estimate the fluctuations in water demand. Assume average water demand of 170 Lpcd.

4.10 A 3-story wooden-frame building has each floor of area 900 m^2. Determine the fire flow and the duration for fire flow.

4.11 A municipal water system has a 20-mgd average water demand. It is estimated that 10 percent of the flow is lost through leaks in the water mains at 60 psi pressure. Assume that these leaks behave as orifices; how much water can be saved from leaking out if the main pressure is reduced to 35 psi.

4.12 A community has a population of 80,000 in the year 2000, and is expected to reach 110,000 in the year 2010. The average water demand in 2000 was 150 gpcd. It is expected that the water consumption rate will be reduced to 135 gpcd by the year 2010. Assuming that the population increase and decrease in consumption rate have a linear relationship, calculate the year when the treatment plant will reach an average treatment capacity of 15 mgd.

4.13 A 200-home subdivision with single-family dwellings is proposed. Calculate the annual water saving that can be achieved over conventional fixtures if all homes are provided with suitable water-saving devices. The suitable water-saving devices are faucet aerator, water-efficient washing machine and dishwashers, and shallow-trap water closet. Use average saving values from Table 4-14. Also use the typical breakdown of residential water demand given in Table 4-7. Assume that the average residential water demand is 380 Lpcd.

4.14 A home has installed a pressure-reducing valve (PRV) at the line supplying water into the home. If water usage is 350 Lpcd and there are four residents, calculate percent water saving. Typical breakdown of residential water uses is given in Table 4-7. Average savings from PRV is found in Table 4-14.

4.15 In a home, total average water loss caused by toilet leaks, faucet drips, and leakage from the lawn sprinkler is approximately 4 percent of total water consumed. The water supply pressure is 345 kPa. A PRV is installed, which drops the supply pressure in the home to 110 kPa. Assuming that water consumption is 304 Lpcd and all leaks behave like orifices, calculate the water saving that will be achieved from these sources after the PRV is installed.

4.16 A city of 60,000 residents has an average water demand of 350 Lpcd. The institutional and commercial and the industrial average areas in the city are 200 and 300 ha, respectively, and water demand expected is 20 and 23 m^3 per ha per day. The public water use and water unaccounted for are 10 and 6 percent of total municipal water demand, respectively. Calculate total municipal demand and each component as a percent of total municipal demand.

4.17 A small subdivision of a city is being developed. The ultimate zoning plan shows the following residential, institutional, and commercial establishments.

Single-family population	
In low-income units	500
In medium-income units	800
In high-income units	1500
Apartment residents	500
One hotel/ motel	500 units
One hospital	300 beds
One rest home	150 beds
One boarding school	1500 students
Five restaurants serving	1900 customers
One bar having	150 customers
Shopping center	250 employees
Office building	1500 employees
Three barber shops	30 chairs
Two beauty salons	20stations
One commercial laundry	20 machines
One service station	10 bays
1 movie theater	500 seats

Calculate the residential water demand and the institutional and commercial water demand. What is the average water demand for the entire community?

4.18 Obtain the following information for the water treatment facility in your community:

(a) design year, estimated population, and average demand;

(b) initial year, population, and average flow;

(c) service area in km^2.

Determine: (1) design period, (2) population density at the design year (person/km^2), (3) staging periods, and (4) whether the estimated population growth pattern is similar to the actual population trends of the community.

References

1. James M. Montgomery, Inc. *Water Treatment Principles and Design*, John Wiley & Sons, New York, 1985.
2. U. S. Congress. *Safe Drinking Water Act* (PL 93-523), 93rd U. S. Congress, December 16, 1974.
3. USEPA. *Technologies for Upgrading Existing or Designing New Drinking Water Treatment Facilities*, EPA/625/4-89/023, Cincinnati, OH, March 1990.
4. National Academy of Sciences, Safe Drinking Water Committee. *Drinking Water and Health*, National Academy of Sciences, Washington, D.C., 1977.
5. National Academy of Sciences, Safe Drinking Water Committee. *Drinking Water and Health*, Vol. 2 and 3, National Academy Press, Washington D.C., 1980
6. Pontius, F. W. "New Horizons in Federal Regulations," *Jour. AWWA*, vol. 90, no. 3, pp. 38–50, March 1998.
7. U. S. Congress. *Safe Drinking Water Act Amendments of 1986* (L.99-339), 99th U. S. Congress, June 19, 1986.
8. Pontius, F. W. "An Update of the Federal Drinking Water Regulations," *Jour. AWWA*, vol. 87, no. 2, pp. 48–58, March 1995.
9. USEPA. "National Primary Drinking Water Regulations; Amendments, Corrections," *Federal Register*, 47 (49), p. 10998, March 12, 1982.
10. USEPA. "National Interim Primary Drinking Water Regulations; Trihalomethanes," *Federal Register*, 48 (40), p. 8406, February 28, 1983.
11. USEPA. "National Primary Drinking Water Regulations; Synthetic Organic Chemicals and Inorganic Chemicals; Monitoring for Unregulated Contaminants; National Primary Drinking Water Regulations Implementation. Final Rule," *Federal Register*, 56(20), p. 3526, January 30, 1991.
12. USEPA. "National Primary Drinking Water Regulations; Synthetic Organic Chemicals and Inorganic Chemicals, Final Rule," *Federal Register*, 57 (138), p. 31776, July 13, 1992.
13. USEPA. "National Primary Drinking Water Regulations; Interim Enhanced Surface Water Treatment Rule," *Federal Register*, 62 (212), p. 59485, November 3, 1997.
14. USEPA. "National Primary Drinking Water Regulations; Interim Enhanced Surface Water Treatment Rule, Notice for Data Availability, Proposed Rule," *Federal Register*, 62 (212), p. 59485, November 3, 1997.
15. USEPA. "National Primary Drinking Water Regulation Synthetic Organic Chemicals; Final Rule," *Federal Register*, 52 (130), p. 25689, July 8, 1987.
16. USEPA. "Drinking Water Regulations; Public Notification; Final Rule," *Federal Register*, 52 (208), p. 41535, October 28, 1987.
17. USEPA. "Drinking Water, National Primary Drinking Water Regulation, Total Coliforms, Proposed Rule," *Federal Register*, 52(212), p. 42223, November 1987.
18. USEPA. "National Primary Drinking Water Regulations, Filtration and Disinfection; Turbidity, Giardia Lamblia, Virus, Legionnella, and Heterotrophic Bacteria, Proposed Rule," *Federal Register*, p. 42177, November 1987.
19. USEPA. "National Secondary Drinking Water Regulations; Final Rule," *Federal Register*, 56(20), p. 3526, January 30, 1991.
20. USEPA. *Commentary on the Safe Drinking Water Act*. EPA Region VI, Revised February 4, 1988.
21. Qasim, S. R. *Wastewater Treatment Plants: Planning, Design, and Operation*, 2d ed., Technomic Publishing Co., Inc., Lancaster, PA, 1999.
22. Tchobanoglous, G. and Schroeder, E. D. *Water Quality: Characteristics, Modeling, and Modification*, Addison-Wesley Publishing Co., Reading, MA, 1985.
23. Viesseman, W. and Hammer, M. J. *Water Supply and Pollution Control*, 6th ed., Addison-Wesley Longman, Inc., Menlo Park, CA 1998.
24. Metcalf and Eddy, Inc. *Wastewater Engineering: Treatment Disposal and Reuse*, 3d ed., McGraw-Hill Book Co., New York, 1991.
25. McGhee, T. J. *Water Supply and Sewerage*, 6th ed., McGraw-Hill Inc., New York, 1991.

26. Turk, A., Turk J., and Wittes, J. T. *Ecology, Pollution, Environment*, W. B. Saunders Co., Philadelphia, PA, 1972.
27. Fair, G. M., Geyer, J. C., and Okun, D. A. *Water and Wastewater Engineering*, Vol. 1, *Water Supply and Wastewater Removal*, John Wiley & Sons, New York, 1966.
28. Murray, C. R. and Reeves, E. B. *Estimated Use of Water in the United States in 1970*, U.S. Geological Survey, Department of the Interior, Geological Survey Circular 676, Washington, D.C., 1972.
29. Metcalf & Eddy, Inc. *Wastewater Engineering: Treatment, Disposal, and Reuse*, 3d ed., McGraw-Hill Book Co., New York, 1991.
30. *The International Drinking Water Supply and Sanitation Decade Directory*, 2d ed., World Water Magazine, March 1984.
31. Dufor, C. N. and Becker, E. *Public Water Supplies of the 100 Largest Cities in the United States*, U.S. Geological Survey, Water Supply Paper 1812, p. 35, 1964.
32. Bailey, J. R. et al. *A Study of Flow Reduction and Treatment of Wastewater from Households*, Water Pollution Control Research Series Report, No. 11050 FKE 12/69, USEPA, December 1969.
33. Salvato, J. A. "The Design of Small Water Systems," *Public Works*, vol. 91, no. 5, pp. 109–133, May 1960.
34. ECOS, Inc. *Water Wheel: Your Guide to Home Water Conservation*, USEPA, Public Information Center (PM-215), Washington, D.C., 1977.
35. Culligan International Co. *Engineering Data Handbook*, 8181-70, 1970.
36. Hubbell, J. W. "Commercial and Institutional Wastewater Loadings," *Jour. WPCF*, vol. 34, no. 9, pp. 962–968, September 1962.
37. U.S. Department of Housing and Urban Development. *Residential Water Conservation,* Summary Report, June 1984.
38. U.S. Department of Housing and Urban Development. *Water Saved by Low-Flush Toilets and Low-Flow Shower Heads,* March 1984.
39. *Guide for Determination of Required Fire Flow*, 2d ed., Insurance Service Office, New York, 1974.
40. Hammer, M. J. *Water and Wastewater Technology*, John Wiley & Sons, Inc., New York, 1975.
41. Qasim, S. R., Drobny, N. L., and Cornish, Alan. "Advanced Waste Treatment Systems for Naval Vessels," *Journal of Environmental Engineering Division, ASCE 99 (EES)*, pp. 717, October 1973.
42. Qasim, S. R. "On-Site Wastewater Treatment Technology for Domestic and Special Applications," *The Encyclopedia of Environmental Science and Engineering*, Gordon & Breach, Science Publishers, New York, pp. 1150–1162, 1998.
43. Sharp, W. E. "Selection of Water Conservation Devices for Installation in New and Existing Dwellings," *Proceedings, National Conference on Water Conservation and Municipal Wastewater Reduction*, November 28 and 29, 1978, Chicago, EPA/430/9-79-015, USEPA, Cincinnati, OH, August 1979.
44. *Water Quality Criteria*, Report of the Committee, Federal Water Pollution Control Administration, U.S. Department of the Interior, Washington, D.C. April 1, 1968.
45. AWWA. *Statement of Policy on Public Water Supply Matters*, Officers and Committee Directory, pp. 154, 1993–1994.
46. Hardenburgh, W. A. and Rodie, E. B. *Water Supply and Waste Disposal*, International Textbook Co., Scranton, PA, 1966.
47. Qasim, S. R., Lim, S., Motley, E. M., and Heung, K. G. "Estimating Costs for Treatment Plant Construction," *Jour. AWWA*, vol. 84, no. 8, pp. 56–62, August 1992.
48. Gumerman, R. C., Culp, R. L., and Hansen, S. P. *Estimating Water Treatment Costs*, EPA/600/2-79-162, Vol. 1–4, USEPA, Cincinnati, OH, August 1979.
49. Canter, L. W. *Environmental Impact Assessment*, McGraw-Hill Book Co., New York, 1977.
50. Garing, Taylor & Associates, Inc. *A Handbook Approach to the Environmental Impact Report*, Garing, Taylor & Associates, Inc, Arroyo Grand, CA, 2d ed., 1974.
51. Ortolano, Lo. *Environmental Planning and Decision Making*, John Wiley & Sons, New York, 1984.
52. ASCE and AWWA. *Water Treatment Plant Design*, 2d ed., McGraw-Hill Publishing Co., New York, 1990.

CHAPTER 5

Predesign Report and Problem Definition for the Design Example

5.1 Introduction

A predesign study is a systematic planning process to define the specific elements of the project , so that detailed design can proceed smoothly and efficiently.[1] The study involves input from clients, design engineer, regulatory agencies (local, state, and federal), civic groups, and the public. Often, an interdisciplinary effort utilizing expertise from many disciplines, including hydrology, water resources, hydraulics, process engineering, and urban planning and architecture and from legal and financial consultants is involved.[2,3] The purpose of the predesign study is to determine the most economical, dependable, and safe water supply for the community. The study usually results in one or more reports outlining the needs of the community and the recommended steps that the community should undertake to meet its future needs. These reports are called the planning-phase report, the predesign-phase report, the preliminary-phase report, and the engineering-phase report.[2]

The purpose of this chapter is to provide the basic guidelines and considerations involved in the preparation of a predesign report for a water supply project. Also, a sample predesign report is prepared in this chapter, to present a generalized format and to develop necessary information needed for the preparation of the final design plans and specifications.

5.2 Significance of a Predesign Report

Predesign studies are essential, because they document data and major engineering decisions for the project. Most of the data developed in a predesign report are used in the preparation of the design plans, specifications, and cost estimates of the project. Therefore, the predesign report is essential for the following reasons:

1. The predesign report represents only a small portion of the project cost (∫ to 1 percent)[1]. Decisions made in the predesign phase may realize substantial savings in the design, construction, and operational phases of the project.[1,4]
2. The predesign report provides major project decisions, design criteria, alternatives, and a basis for treatment-process selection.
3. The predesign report documents major engineering and architectural features, equipment, and operation and maintenance requirements within the overall scope of the project.
4. Based on the design information developed during the predesign phase, reliable capital and operation and maintenance-cost estimates of the project are established.
5. The predesign report provides the client and concerned agencies with a great deal of information for a limited amount of spending. At this stage, the available resources and financial strategy can be thoroughly and informatively evaluated and fully agreed upon or committed to by all concerned parties. Decisions made at this phase will eliminate expensive changes and delays during the design phase. Overall, the project will realize substantial savings in the design, construction, and operational phases of the project through the use of best available treatment technology.
6. After review, revisions, and final approval of the predesign report, the design team can proceed with preparation of design plans and specifications.

5.3 Contents of a Predesign Report

Unlike similar studies for wastewater treatment projects, predesign reports for water supply projects usually do not have a government-dictated format or guidelines. Because predesign reports must be approved by the state, a close working relationship with the state representatives is needed during the preparation stages. The format for such reports is usually established by the engineer in charge of conducting the study or by company policy. Most predesign reports for water supply projects, however, contain discussion on the following areas:[2] (1) existing water production facilities and need for expansion; (2) design and staging period; (3) population projections, water-use projections, and plant capacity; (4) water quality objectives and source selection; (5) site selection for intake, treatment plant, and conveyance system; (6) alternatives to expansion or design of water-production facilities; (7) treatability studies and selection of process train; (8) coordination with distribution study; (9) preliminary design and cost estimates; (10) recommended capital-improvement plan; and, (11) environmental-impact assessment. Often, a predesign report supported by many appendices is prepared. Many times, a predesign study results in more than one report. Whatever the outcome, in all cases, the purpose is to develop the most economical, safe, and dependable water supply system for the community.

5.4 Model Predesign Report

In the following section, a model predesign report for a water-production project is developed. A medium-sized community has been selected, and efforts have been made to simplify the problem and to keep the contents of the predesign report realistic and brief. The generalized format presented below is one that the authors have used with success in the past. The report is used for problem definition and to develop the needed design information for preparing the design plans and specifications in Chapters 6 through 17.

The names of the town and state (and all other information used to prepare this predesign report) are not real. The information is meant only to be illustrative. It is neither complete nor definitive.

TITLE PAGE OF THE
MODEL PREDESIGN REPORT

WATER PRODUCTION FACILITY DESIGN
for
THE CITY OF MODELTOWN
ANY STATE
Prepared by
ENGINEERING DESIGN CONSULTANTS
November 1999

5.4.1 Introduction

The City of Modeltown has enjoyed an adequate supply of quality groundwater throughout its history. The City has experienced steady growth over the past thirty years, and continual growth will place an increasing demand on the city's water system. The present groundwater supply does not have sufficient capacity to meet the increasing demand. If the present growth rate continues, the water supply for Modeltown is expected to become inadequate within five years. Therefore, the surface water supply sources for Modeltown must be considered. The Big River Reservoir is the only reliable water supply source for the region and is being considered for the project. To utilize the water from the Big River Reservoir, the city must construct a new raw water intake and pump station, a raw water pipeline, a new water treatment plant, a clearwater well, a high service pump station, transmission pipelines, and distribution storage reservoirs.

With the objective of undertaking a capital-improvement plan for development of the most economical alternative and of insuring a continued supply of quality water, the City of Modeltown retained Engineering Design Consultants (EDC), a consulting engineering firm in 1998, to perform the predesign study. Following is the predesign report that was approved by all concerned parties.

5.4.2 Current and Future Economic Demographic and Land Use

The Modeltown water system presently serves a 30.65 km^2 (11.8 sq. mile) area located on the west bank of the Big River. The economy of the town is based on agriculture, allied enterprise, and retail business. The major industries include one slaughter house, one meat-processing and packaging plant, one dairy, one canning plant, one brewery, and one farm-implement manufacturing company. The town also has large business and commercial establishments. There are 12 modern public grade schools and high schools and one college that also serves the surrounding farming area.

The topography of the service area is best described as gently rolling plains. The region is subject to frequent precipitation, with an average annual rainfall of approximately 94 cm (37 in.) and with high-intensity showers occurring during the late summer and early fall. The climate is mild, with maximum, minimum, and average temperatures of 35°, –2° and 27°C, respectively. There are no surface water impoundments in the planning area. Groundwater is the only current source of water supply for the town and is fully utilized.

The existing land use within the city has not changed substantially over the past twenty years. There have been some changes via transition of vacant land to single-family dwellings, and the service area has slightly increased. The continued growth in the population served has been due to more industrial development in the city's industrial park. At the present time, the city has plans to expand its service area in the north, which is anticipated to grow in the future. The land-use plan for the year 2000 and that for the design year 2020 are illustrated in Figure 5-1.

5.4.3 Historical and Projected Population

The planning department of the City of Modeltown, on the basis of census data, has estimated the current and future populations of the town. The population forecasts were coordinated and compared with the information from other sources. These sources included *State Regional Planning Commission, State Department of Water Commission, area-wide wastewater management* population forecasts established under the Clean Water Act and population projections established for the county and region. Historical and projected populations for the City of Modeltown are summarized in Table 5-1. Figure 5-2 graphically depicts the historical and projected population of the city. The following important observations may be made.

The City of Modeltown has experienced a continual growth in population since its founding in 1890. Some of this increase is attributed to general population growth in the entire region. The U.S. Census Bureau reported that, in 1990, a total of 47,000 persons resided in the City of Modeltown. This number represents a 47 percent increase in population from the 1980 census figures. The population projections show a decreasing rate of increase following a logistic S-shaped curve. The projected increases in population from 1990 to 2000 and from 2000 to 2010 are, respectively, 32 and 15 percent, indicating that the city is fast approaching a saturation point.

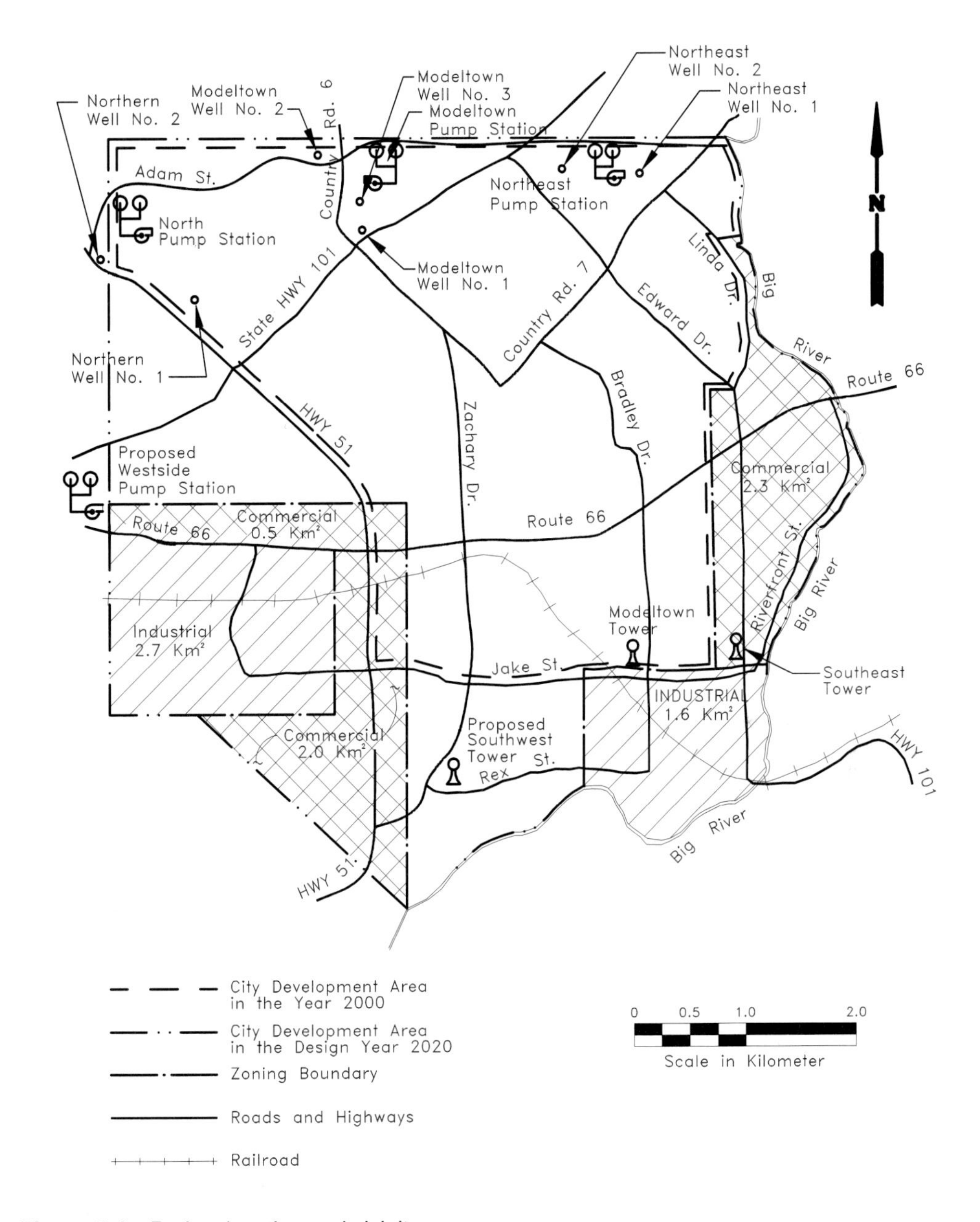

Figure 5-1 Project location and vicinity map

Table 5-1 Historical and Projected Populations of Modeltown

Year	Population	Basis	Significance
1960	15,000	Census	
1970	22,000	Census	
1980	32,000	Census	
1990	47,000	Census	
1999	59,600	Estimated	Completion of Predesign Report
2000	62,000	Estimated	Completion of design plans and contract documents
2005	67,000	Estimated	Initial year, completion of construction and operation begins
2010	71,200	Estimated	
2015	74,200	Estimated	
2020	76,200	Estimated	Design year, facility will reach full capacity and further expansion would be needed.

5.4.4 Historical Water Usage

The water demand for a community varies by the hour, day, month, and year. A detailed analysis of the historical pattern and extent of variations from average or normal conditions to maximum or drought conditions is required as a basis for predicting future water requirements. Future demands are generally linked to population, per capita use, industrial demands, and the supplying of customers outside the city limits. The water requirements for a community are generally expressed in terms of per capita *average day, maximum day, and peak hour demands.* These terms, in the overall context of the predesign planning, have been discussed in Section 4.6.2. Readers should review this section.

The City of Modeltown maintains an excellent set of water-production records, from which an analysis of water usage for average and peak conditions can be made. The historical water-usage record for the period from 1970 through 1998 is used to develop residential, commercial, and industrial water demands, average per capita municipal water demand, maximum day and peak hour demands, and projected future water demand.

A: Residential, Commercial, and Industrial Water Demand The residential, industrial, and commercial water usage is included in the total water-production quantities of finished water given in Table 5-2. Approximately 49 percent of the total water demand consists of the residential consumption. The industrial and commercial demands are 22 and 17.7 percent, respectively. The remaining 11.3 percent includes water used for unmetered public uses such as pipes and street flushing and that lost through leaks or unaccounted system losses. These uses are typical of most municipal water systems.

The industrial and commercial water usage throughout the entire design period is anticipated to remain at the same proportionate value of the total water produced. The six major indus-

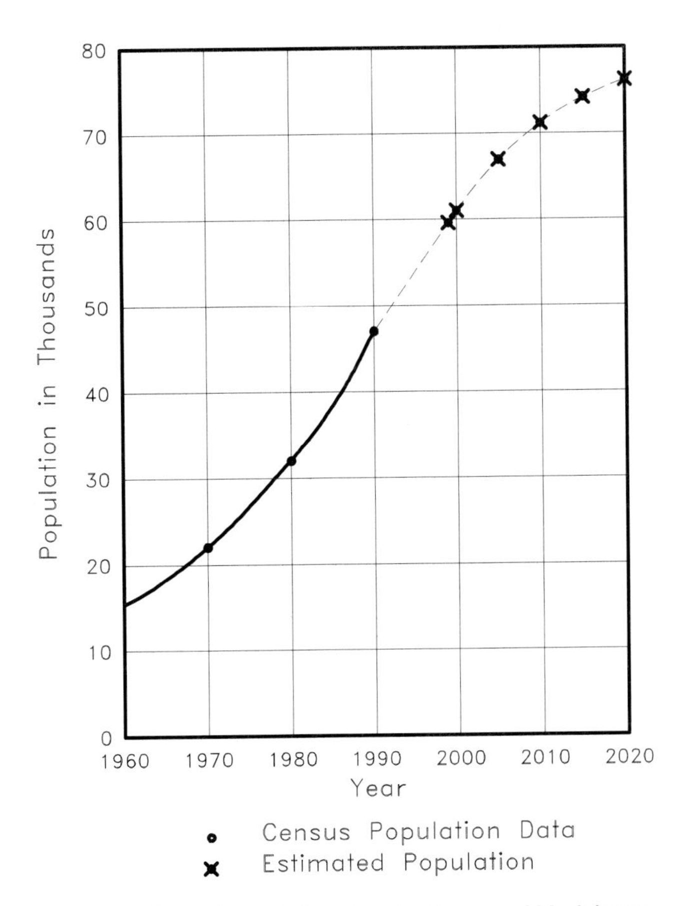

Figure 5-2 Census and estimated population data for the city of Modeltown.

tries mentioned above will continue to withdraw the major share of the industrial water demand, and their combined demand has historically followed the same trend as the total water demand for the City.

B: Per Capita Demand The annual average day water usage and annual average per capita water usage for Modeltown from 1970 through 1997 are summarized in Table 5-2. The average daily per capita water usage has steadily increased from a low of 417 Lpcd (110 gpcd) in 1970 to a high of 680 Lpcd (179 gpcd) in 1997. However, it may be noted that the rate of increase in per capita water usage has declined during this period. The stabilization of water-use rate in recent years has reflected both a slowdown of changes in the prevailing standard of living and the water-conservation efforts gaining popularity among the general public and the regulatory agencies.

Table 5-2 Historical Water-Usage Data for the Modeltown Service Area

Year	Population Served, Thousands	Annual Average Day Water Use, m^3/d	Annual Average Day Per Capita Water Use, Lpcd	Ratio of Max. Day to Avg. Day Demand
1970	22.0	9170	417	1.60
1971	23.0	12,030	523[a]	1.90
1972	24.0	11,280	470	1.65
1973	25.0	14,030	560[a]	1.89
1974	26.0	13,000	500	1.6
1975	27.0	13,200	489	1.58
1976	28.0	17,300	618[a]	1.93
1977	29.0	16,050	553	1.68
1978	30.0	17,280	576	2.05
1979	31.0	17,050	550	1.60
1980	32.0	21,570	674[a]	1.96
1981	33.5	20,300	606	1.79
1982	35.0	21,320	584	1.75
1983	36.5	25,550	700[a]	1.96
1984	38.0	21,620	569	1.68
1985	39.5	25,440	644	1.75
1986	41.0	29,070	709[a]	1.96
1987	42.5	27,290	642	1.70
1988	44.0	29,830	678	2.00
1989	45.5	30,030	660	1.74
1990	47.0	35,630	758[a]	1.95
1991	49.0	32,590	665	1.75
1992	51.0	38,250	750[a]	1.90
1993	52.5	35,700	680	1.78
1994	54.0	36,180	670	1.76
1995	55.8	37,950	680	1.75
1996	57.0	43,600	765[a]	1.95
1997	58.5	39,780	680	1.70

a Represents drought condition. (See Figure 5-3.)

Water conservation has the beneficial effect of reducing the per capita water demand. There are many popular *water-saving devices* that are effective in flow reduction. Many devices are commonly manufactured as part of most faucet combinations and are readily available from most appliance suppliers. These devices and the savings achieved by their use are discussed in Chapter 4. The social acceptability and the ease with which many of the water-saving devices can be installed in the existing and new systems and public cooperation toward use of water-saving devices will all affect the flow reduction; however, it is expected that, in spite of all available water-saving techniques and all public-awareness programs, there will not be any significant reduction in water demand. It is safe to assume, however, that implementation of all water-conservation efforts will prevent a net (Lpcd) increase in residential and commercial flow rates.

The per capita water usage rate is also sensitive to weather conditions. The data indicate that the annual water use rate for the Modeltown service area under the drought condition[a] is significantly higher than that under normal weather conditions. The annual average demand for normal weather and for drought conditions can be separated by plotting the per capita consumption values, as shown in Figure 5-3. The curve passing through the bulk of the data points represents normal weather conditions, while the curve enveloping the normal usage points represents the drought condition. It may be noted that the annual average water usage rates for the Modeltown service area under normal weather conditions is 675 Lpcd (178 gpcd) and that for the drought condition is 760 Lpcd (200 gpcd). These demand conditions occur in 1990 and onward.

C: Maximum Day and Peak Hour Demands The historical daily record on maximum day water consumption for Modeltown was used to calculate the ratio of maximum day[b] to annual average day demand for the period between 1970 and 1997. These ratios are obtained from maximum day water usage data obtained from the water demand data divided by the annual average day usage. For example, in 1997 the population was 58,500, and the maximum day was August 4 with a total demand of 67,626 m^3. (This information is not given elsewhere.) From these data, the maximum day per capita demand is 1156, and the ratio of maximum day to average day is 1156/680, or 1.70. These ratios are provided in Table 5-2, which ranged from 1.58 to 2.05. The maximum ratio for the drought condition is 1.96 (Table 5-2). In the study area, the likelihood that the high-day demand will occur simultaneously with the drought condition is very real and may happen for several consecutive days. Therefore, a maximum day to average day ratio of 1.96 is adopted for projection of the future maximum day demand condition.

The peak hour rate is the measure against which the adequacy of the distribution system, high-service pumps, and elevated storage tanks are established. A study of pump-discharge rates, discharge pressures, elevated storage-tank levels, and demand was conducted in July of 1996, during a record water usage month. The ratio of peak hour to maximum day demand dur-

a. Drought condition represents high water demand (during the growing season or during the summer months) due to exceptionally dry weather conditions.

b. The maximum day water usage data is not provided in Table 5-2.

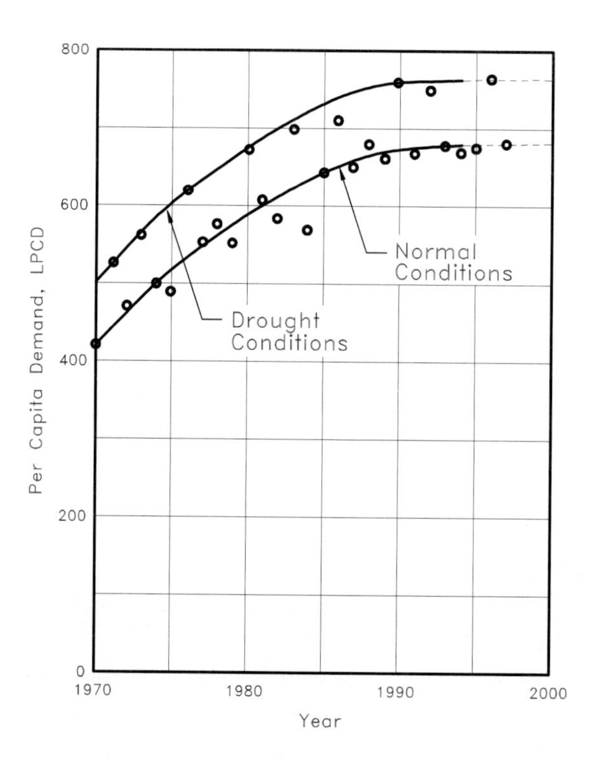

Figure 5-3 Historical annual water demand per capita under normal weather and under drought conditions.

ing the three high-use days ranged from 1.48 to 1.70. Therefore, a factor of 1.6 times the potential maximum day demand was established to project the potential peak hour water usages in the service area.

D: Projected Water Demand The *State Department of Health Design Criteria* require that the treatment capacity of the water treatment plant must always be in excess of the *maximum expected draft* of any day of the year. Therefore, for the planning of the water treatment facilities the maximum day drought condition will be considered as the *maximum expected draft*. The projected water demands for the Modeltown service area under normal conditions and under drought conditions are summarized in Table 5-3. The per capita use of water under normal and drought conditions, as developed earlier, is 675 and 760 Lpcd, respectively. The average daily water demand is derived by multiplying the projected population by the per capita consumption. The maximum day water demand is derived by multiplying the average day demand by the factor 1.96. The peak-hourly demand is derived by multiplying the maximum day demand by the factor 1.60. The annual average day, maximum day, and peak hour demands for normal and drought conditions for several key years are summarized in Table 5-3.

It may be noted that the flows corresponding to the drought condition are used to design the facility and equipment. The flows under normal conditions are often used to develop the operating and maintenance costs, particularly chemicals and power.

5.4.5 Justification of the Selected Design Period

The design period proposed for the capital-improvement project for the Modeltown service area is 15 years. As discussed in Section 4.4, the design or staging period is generally selected with many factors in mind. The most important factors are population growth patterns, plant capacity, availability of initial capital investments, the useful life of the treatment units, and process obsolescence. The 15-year design period proposed for this project is considered appropriate for the following reasons:

1. The Modeltown service area has undergone a rapid population growth in the past; however, the growth rate is following a decreasing rate of increase after the year 2000, and the city will reach a saturation population soon after the year 2020. Further plant expansion beyond the year 2020 will be possible by process modification to increase the capacity and to improve the operation.
2. The City of Modeltown has plans to finance the capital-improvement project through the sale of municipal revenue bonds. These bonds will be retired by revenues generated by the sales of water over a period of 18 to 20 years.[5] Assuming that 3 to 5 years are needed for completion of the construction phase of the project, EDC concluded that a 15-year design period will provide the necessary revenues over the life of the bonds.

5.4.6 Evaluation of Existing Facilities

A: Groundwater Supply The City of Modeltown currently operates a total of seven ground wells located in the northern part of the city and surrounding countryside. The approximate location, date of completion, and estimated capacity of each well is listed in Table 5-4. The location of these wells are also shown in Figure 5-1. The existing Modeltown Well 1 and Modeltown Well 2 are founded in the *Grey Sand Aquifer*. The quality of water from this aquifer (the shallowest local aquifer) has deteriorated over the past 10 years; as a result, these wells are used only periodically. The remaining wells are found in the *Red Sand Aquifer*. This water, although somewhat high in iron, has remained adequate in quality over the past 25 years.

In 1997, the combined average pumping rate for all wells provided the needed average service-area water demand of 39,800 m^3/d. In the same year, the total available capacity of all wells combined was 80,000 m^3/d. The maximum day demand of 67,600 m^3/d was therefore met by the combined capacity of the wells. The Red Sand Aquifer has been used heavily by agricultural users over the past 20 years. As a result of this, the level of the aquifer has dropped significantly over several decades. Indications are that this trend will continue and will result in a decreased

Table 5-3 Projected Water Demand for the Modeltown Service Area

			Normal Condition				Drought Condition			
Year	Significance	Population	Per Capita Consumption, Lpcd	Ave. Day, $\times 10^3$ m^3/d	Max. Day, $\times 10^3$ m^3/d	Peak Hourly, $\times 10^3$ m^3/d	Per Capita Consumption, Lpcd	Ave. Day, $\times 10^3$ m^3/d	Max Day, $\times 10^3$ m^3/d	Peak Hourly, $\times 10^3$ m^3/d
1999	Completion of pre-design report	59,600	675	40.2	78.8	126.1	760	45.3	88.8	142.1
2005	Initial year	67,000	675	45.2	88.6	141.8	760	50.9	99.8	159.7
2020	Design year	76,200	675	51.4	100.7[a]	161.1[b]	760	57.9	113.5[c]	181.6[d]

a Maximum day demand = average day demand × 1.96.

b Peak hour demand = maximum day demand × 1.60.

c Maximum day demand = average day demand × 1.96.

d Peak hour demand = maximum day demand × 1.60.

Table 5-4 Existing Groundwater Wells Located in the Modeltown Service Area

Name of Well	Location	Aquifer	Year Completed	Capacity, $\times 10^3$ m^3/d
Modeltown Well No. 1	North quadrant of Country Road 6 and Highway 101	Grey sand	1935	1.2
Modeltown Well No.2	Northeast of Adam Street and Country Road 6	Grey sand	1940	2.3
Modeltown Well No. 3	Southwest of Adam Street and Country Road 7	Red sand	1952	5.3
Northeast Well No. 1	Southwest of Adam Street and Country Road 7	Red sand	1960	5.7
Northeast Well No. 2	Southeast of Adam Street and Highway 101	Red sand	1965	6.1
Northern Well No. 1	Northwest of Highway 51 and Highway 101	Red sand	1968	7.5
Northern Well No. 2	Southeast of Adam Street and Highway 51	Red sand	1970	11.7
Total average daily pumping in 1997:				39.8
Total available pumping capacity of all wells in 1997:				80.0

total capacity for the wells founded in the Red Sand Aquifer. Therefore, it is clear that, within a period of from eight to ten years, all the wells, collectively, will not be able to meet the maximum day demand for the Modeltown service area.

B: Storage The City of Modeltown has three ground-storage and pump station sites, with a total capacity of 12.0 million liters. The city also has two elevated tanks, with a total capacity of 6.0 million liters. The location and capacity of each reservoir is listed in Table 5-5, and the location is shown in Figure 5-1.

At present, all of the city's storage reservoirs are in excellent condition; however, as the peak demand increases, additional storage will be required. The estimated future storage requirements are given in Table 5-6. The locations of future storage requirements are also shown in Figure 5-1.

5.4.7 Proposed Capital-Improvement Project

The proposed capital-improvement project for Modeltown involves abandonment of all of the groundwater supply wells that have served the needs of the community for so long. A new state-of-the-art conventional water treatment plant will be constructed to treat the surface water available from the Big River Reservoir. The treated surface water will be supplied through the existing groundwater storage and distribution system, which will need a major overhaul and expansion. Thus, the proposed capital-improvement project will involve: (a) acquisition of water rights, (b) construction of a new raw water intake structure to supply surface water from the Big River Reservoir, (c) construction of a new raw water pump station, (d) construction of a new raw water pipeline, (e) construction of a new conventional water treatment plant and high-service pump station, and (f) coordination with the existing water distribution system. Each of these basic elements of the proposed capital-improvement project is briefly presented below. Details are provided in subsequent chapters.

Table 5-5 Storage Reservoir Sites, Type, and Capacity

Name	Location	Type	Capacity, m^3
Modeltown pump station	Southeast of Adam Street and Country Road 6	Ground	3000
Northeast pump station	Southwest of Adam Street and Country Road 7	Ground	4000
North pump station	Southeast of Adam Street	Ground	5000
Modeltown tower	North of Intersection of Bradley and Jake Streets	Tower	2000
Southeast tower	East of Riverfront Street and Jack Street	Tower	4000
Total existing storage			18,000

A: Acquisition of Water Rights from the Big River Reservoir The U.S. Army Corps of Engineers completed the construction and filling of the Big River Reservoir approximately 15 years ago. The reservoir is formed by the construction of an earth-filled dam that is located approximately 15 miles southwest of the center of Modeltown. The dam creates a conservation pool covering 9421 ha (23,280 acres) and containing 536 million m^3 (436,000 acre-ft) of conservation storage space for municipal and industrial purposes. The reservoir is a multi-purpose project for flood control, water conservation, and recreation purposes. The dam controls a drainage area of 4300 km^2 (1660 sq. miles) of the Big River watershed.

Modeltown and several surrounding cities participated as local sponsors in the Big River Reservoir project, with Modeltown paying 3.7 percent of the local share of costs while other cities paid 75 percent. An unallocated 21.3 percent share remained. Modeltown holds water rights in the Big River Reservoir under Permit 168, issued in 1985, which allowed Modeltown to utilize 19.8 million m^3 of conservation storage per annum for municipal water supply purposes (an average daily rate of 54,300 m_3/d). The expected average annual withdrawal rate at the design year is 18.8 million m^3 (51,400 m^3/d), leaving one million m^3 reserve capacity (see Table 5-3). Under the terms of the contract with the Federal Government, the City of Modeltown is entitled to use its allocated conservation storage in return for payment of the proportional construction and maintenance and operation costs.

Table 5-6 Future Storage Reservoir Sites, Type and Capacity, and Existing Reservoir Capacity

Name	Location	Type	Capacity, m^3
New westside pump complex	Near intersection of Highway 101 and Route 66	Ground	4000
New tower	Crossing of Rex Street and Zachary Drive	Tower	2000
Total new storage			6000
Total existing storage	Locations are given in Table 5-5 and Figure 5-1	Three ground storages and two towers	18,000
Total future storage			24,000

The water rights for Modeltown from the Big River Reservoir are slightly in excess of the future municipal water supply; however, in order to utilize water from the Big River Reservoir, the project will require construction of: (1) an intake structure, (2) a raw water pump station, (3) a raw water pipeline, (4) a water treatment plant and a high-service pump station, and (5) a treated water transmission line. An overview of all these components of the capital-improvement project are shown in Figure 5-4.

B: Raw Water Intake Structure The intake structure and raw water pump station will be constructed at the Big River Reservoir. An ideal site for an intake structure and raw water pump station must possess several distinct features. Among them are (1) allowing access to deep water close to shore, (2) allowing withdrawal of best quality water from multiple elevations, (3) providing flexibility for future expansion, (4) minimizing the length of water transmission lines, access roads, and electrical power, and (5) minimizing adverse environmental impacts, including those on recreation, navigation, aquatic life, and aesthetics. Detailed discussion of site selection and about design of a raw water intake structure can be found in Chapter 6.

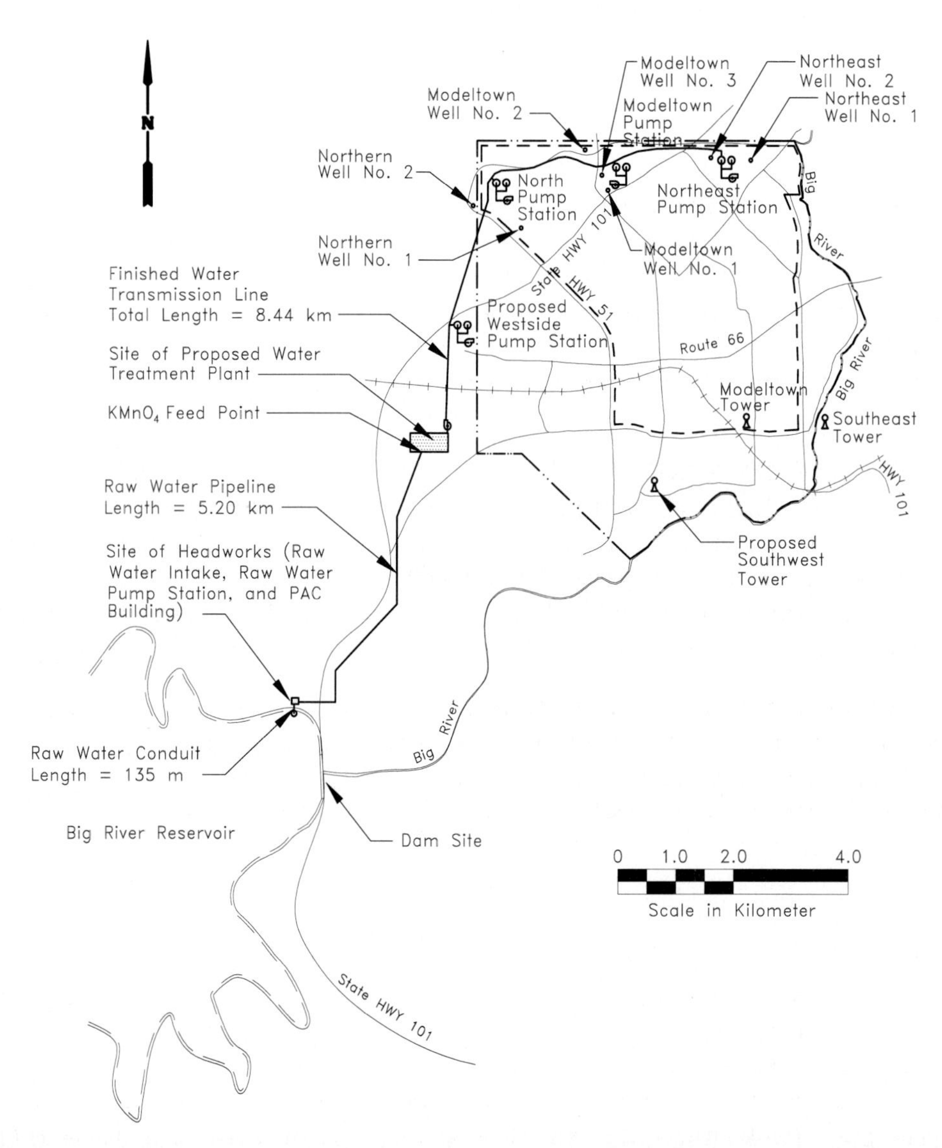

Figure 5-4 An overview of all components of the capital improvement project: Big River Reservoir, raw water intake, raw water pump station, raw water transmission line, water treatment plant, and treated water transmission line.

A two-ha (5-acre) tract of land is available on the northeast shore of the Big River Reservoir, approximately 0.8 km (0.5 miles) from the dam site. This site has many desirable features, and it has been selected for construction of the raw water and intake pump station. The intake structure will be located approximately 100 meters from the shore. It will be a dry tower structure, of 8-m × 8-m inside dimensions, with two 3-m × 3-m gates fitted with bar racks and located at each of four elevations, to withdraw water of the best quality available at the desired level. A conduit 152 cm diameter (60 inches) in diameter and 135 m long connects the intake structure with the raw water pump station. Design details of the raw water intake structure are presented in Chapter 6.

C: Raw Water Pump Station The raw water pump station houses a mechanically-cleaned traveling fine screen, six identical vertical-turbine submersible pumps with variable-speed drives, and a powdered activated carbon (PAC) storage and feed system. Six pumps will be installed with one pump as a stand-by unit. Thus, the maximum day flow will be provided by five pumps in service. All pumps discharge into a common header 107 cm (42 inches) in diameter. The design details of the raw water pump station are provided in Chapter 7.

D: Raw Water Line The 122-cm (48 inches) cast-iron raw water line will transport water from the pump station to the treatment plant. The raw water transport line is about 5.2 km (3.2 miles) long and is designed to allow insertion of a mechanical cleaning *pig* at the pump station and at two intermediate locations along its length. There are also two intermediate pig-retrieval locations and one at the treatment plant. The pipeline is approximately 1 to 2 m below the ground surface and follows the natural terrain. The raw water line finally discharges in Junction Box A, located at the south property boundary of the water treatment plant (Figure 5-4).

E: Water Treatment Plant and High Service Pump Station Careful consideration was given to site selection for the proposed water treatment plant. Conditions such as topography, available land area and land acquisition, access roads, flood conditions, economy of conveying water to and from the plant, availability of electric power, disposal of residuals, and natural and man-made disasters were the major considerations. One very important factor was compatibility with the distribution system. Because the existing distribution system utilizes several groundwater wells and ground and elevated storage, it is necessary that the treatment plant conveniently and economically supply the reservoirs and the other points of input to the distribution system.

Several candidate sites were evaluated for plant location. After a tedious site-ranking analysis, a 20-ha (50-acre) tract on the western end of the town was selected for the plant site. State Highway 101 passes by the site, and the site is also accessible from Jake Street. This site provides most of the features desirable, in particular, topography and geology, access roads, electrical power, and land-acquisition requirements. It is high ground and is currently in use as pasture land. Other desirable features include close proximity to the major existing and proposed storage and a location approximately 4 km (2.5 miles) from the downtown business area. The city has approved this site for plant location and has initiated proceedings to acquire the land.

The proposed water treatment plant is a state-of-the-art conventional water treatment plant with special features for control of taste and odor episodes, iron and manganese, and THM formation. Chemical storage and feed facilities, clear water wells, and a high-service pump station will also be integrated with this plant.

F: Coordination With the Distribution System The existing water distribution system has three ground-storage and pumping systems and two elevated storage facilities. The currently available combined capacity of the first three is 128,000 m^3/d. Under the peak hour demand, the existing water distribution system utilizes three groundwater storage basins and distribution pumping; the two elevated storage facilities are to meet the peak hourly demand. The expansion will involve construction of a new ground-storage and pumping facility and one of elevated storage. The existing distribution network will be expanded by connecting several nodes with new pipes. Several old pipes will be replaced with new pipes that have larger diameters. A new treated-water transmission pipeline 90-cm in diameter will begin at the high-service pump station at the treatment plant. The transmission line will feed all four ground-storage distribution reservoirs. The diameter of the pipeline will decrease as it feeds different ground-storage reservoirs. Eventually, it will terminate at the Northeast pump station and ground-storage facility. The modified system will expand distribution capacity from 128,000 m^3/d to 181,700 m^3/d.

5.4.8 Evaluation of Raw Water Quality and Treatment Processes

Water quality characteristics of the raw water source could affect the selection of the treatment processes and operation of the plant. Investigations were made by reviewing the historical data available from the U.S. Geological Survey, the State Department of Water Commission, and other sources on the Big River Reservoir. Water quality samples were also collected from the vicinity of the proposed intake structure during a winter month (February) and a summer month (August) and analyzed by a water quality testing laboratory to supplement the existing data.

The raw water conditions and characteristics that were significant with respect to the water treatment process development and plant design are the following:

1. reservoir stratification;
2. chemical and physical water quality process-related parameters;
3. algae;
4. THM formation potential and raw water chlorine demand.

A: Reservoir Stratification The historical data indicate that temperature stratification normally exists in the Big River Reservoir from April through October. The intensity of the temperature stratification is greatest during the summer months, when an obvious thermocline[c] develops from around 8 meters downward. This period is accompanied by a marked dissolved oxygen (DO) stratification, with low DO in the hypolimnion[d]. The water in the hypolimnion also has a lower temperature and a lower pH.[6] The upper-level intake gates should be used

c. The transition zone comprising a layer of water in which water temperature changes rapidly.

d. Stagnant zone lying below the thermocline and characterized by lack of mixing.

exclusively during this period, to assure that the raw water is drawn from the epilimnion[e] during stratification.

B: Chemical and Physical Water Quality

Process-Related Parameters Water quality samples were collected during summer and winter months from several depths near the vicinity of the proposed intake structure. These samples were preserved upon collection and taken to a water quality laboratory for analysis. The analyses performed included all parameters listed in the primary and secondary drinking water standards (Table 4-1 and Table 4-2) and other process-related parameters, such as temperature, pH, alkalinity, and hardness.

The analytical results indicated that none of the primary standards was exceeded except that for turbidity. Two of the secondary parameters, iron and manganese, exceeded the secondary maximum contaminant levels. These results were consistent with information obtained from other sources.

The historical average monthly turbidity values ranged from 3 to 17 NTU. The highest turbidity was in the spring, the lowest in the summer, and for nine months of the year the turbidity averaged less than 10 NTU. The total iron and total manganese concentrations in unfiltered samples varied from 0.2 to 0.7 mg/L and from 0.05 to 0.4 mg/L, respectively. It is believed that iron and manganese are converted to their reduced soluble species under the anaerobic conditions present in the hypolimnion. Solubilization of these species induces their release from the bottom sediments into the upper layers.[7] Iron and manganese concentrations will remain below MCL as long as water is withdrawn from the top layers and the reservoir remains stratified. If, however, the lower intakes are used to draw water from the hypolimnion, the iron and manganese concentrations will likely exceed acceptable levels. Other reduced chemicals species, such as hydrogen sulfide, will also be of concern.

The ranges of such process-related parameters as temperature, pH, total alkalinity, total hardness, and TDS were developed from the treatability studies. The temperature ranged from an average low of 5°C in January to an average high of 30°C in August. The pH value ranged from 7.9 to 8.2, with the higher values occurring in the warmer months (most likely because of the influence of algae. Total alkalinity, hardness, and TDS remained fairly constant, ranging from 80 to 110 mg/L as $CaCO_3$, from 100 to 120 mg/L as $CaCO_3$, and from 200 to 250 mg/L, respectively.

C: Algae Available data on algae show that the number and types of algae in the Big River Reservoir are highly variable. The general cycle is: (1) the highest number of algae occurred in late summer (August), with blue-green algae predominating, (2) the lowest algae counts were in the late spring (March), when the green algae and diatoms predominate, and (3) blue-green algae dominate in the winter months, with their numbers significantly lower than those in the summer.

e. Circulation zone lying above the thermocline and characterized by good mixing and uniform temperature and density.

Not enough data exist to indicate whether a substantial algae bloom occurs in the fall during reservoir destratification. Currently, other municipal water treatment plants that treat raw water from the Big River Reservoir face occasional taste and odor episodes and filter clogging during late summer months.

D: THM Formation Potential and Chlorine Demand THM are not present in any of the raw water supplies but are the byproducts of chlorination. Free chlorine reacts with organic precursors present in water to form THMs. The laboratory studies with raw water samples from the Big River Reservoir indicated the following:

1. The THM formation potential is dependent on chlorine dose applied, pH, temperature, and reaction time.
2. The THM formation potential is developed upon TOC concentration. Generally, the THM formation potential, lower in the winter months and higher in the summer, is due to prevailing TOC concentrations.
3. In summer months, at a chlorine dosage of 10 mg/L, the concentration of total THMs exceeded 100 μg/L within eight hours of contact time. Also, there was no significant difference in the THM formation potential between the water from the epilimnion and that from the hypolimnion.
4. Chlorine dosage in the raw water transport line will cause THM formation unless ammonia feed is maintained to insure that chlorine occurs in the combined state.

The chlorine dose, state, and demand influence the THM formation potential of the raw water. Chlorine demand for raw water from the Big River Reservoir was highest in the summer, averaging about 4 mg/L in a 12-h contact time at a 10-mg/L chlorine dosage.

5.4.9 Process Design Criteria Development

The review of the important raw water quality characteristics of the Big River Reservoir makes it necessary that the water treatment processes be designed to accomplish the following:

1. turbidity removal;
2. filtration;
3. disinfection and THM control;
4. taste and odor control and iron and manganese removal;
5. corrosion and scaling control;
6. residuals management.

At times, it may be necessary to control taste and odors as well as iron and manganese. Seasonal algae blooms may impose a constraint on the turbidity removal process.

A: Turbidity Removal Jar Tests The heart of the treatment process for the Modeltown Water Treatment Plant will be turbidity removal. As a result, extensive bench-scale studies (jar tests) were performed to optimize the design of the turbidity removal process. Process design criteria were established that included (1) the choice of coagulants, optimum dose and pH, (2) the best coagulant aid (polymer) and its point of addition, (3) rapid-mix time and energy, (4) total flocculation time, optimum energy, and compartmentation, and (5) the design overflow of the settling basin.

Most of the turbidity removal processes were developed during the winter jar test program, because cold-water coagulation conditions are generally more sensitive than warm-water conditions, because of the relative difference in floc and water density. The optimum conditions developed during the winter test program were then confirmed during the summer program. The emphasis of the jar testing was placed on the reduction of THM formation and on the potential for direct filtration during the summer months, when the THM formation potential was the highest and turbidity the lowest. The TOC concentration in the raw water was always below 3 mg/L. The D/DBP Rule requires TOC removal of 30 percent based on source water TOC and alkalinity. (See Table 12-14.) Removal of TOC is accomplished by using *enhanced coagulation.* This topic is presented in Sections 8.2.3, 8.2.4, and 12.2.4.

Settling data were developed by conducting flocculent settling tests in a column 15 cm (6 in) in diameter and 3.1 m (10 ft) tall. The optimum turbidity removal conditions developed from jar tests are summarized here:

Choice of primary coagulants:

Liquid aluminum sulfate

Optimum dose = 20 mg/L as $Al_2(SO_4)_3$

Optimum pH range = 6.4–7.1

Turbidity remaining < 0.5 NTU

Liquid Ferric sulfate

Optimum dose 25 = mg/L as $Fe_2(SO_4)_3$

Optimum pH range = 8.2–9.0

Hydrated-lime dosage = 15 mg/L as $Ca(OH)_2$

Turbidity remaining < 0.5 NTU

Polymers

Coagulant aid is high molecular weight cationic polymer. The optimum dose is 0.05 mg/L. The point of adding coagulant aid is after coagulant has dispersed and pin-head floc has been produced.

Rapid Mix

Detention time = 20–30 s

Velocity gradient = 950/s

Flocculation

Compartmentation—three-stage tapered flocculation

Detention time—total 30 min, each stage 10 min

Velocity gradient—overall 30/s, first, second and third stages, 60, 30, and 15/s, respectively

Settling Basin

Overflow rate = 35 $m^3/m^2 \cdot h$ (0.60 gpm/ft^2 or 860 gpd/ft^2)

Detention time = 4 h

Side-water depth—not less than 5 m (16.4 ft)

Detailed discussion on coagulation, flocculation, and sedimentation can be found in Chapters 8 and 9.

B: Filtration Development of an efficient filter design depends on four interrelated design factors: (1) filter rate control, (2) filtration rate and surface area, (3) filter-media selection, and (4) filter backwash, media support, and underdrain system. Generally, pilot testing is necessary if the filter design of a plant is to be optimized. Without the benefit of pilot-test results, the design factors are developed from experience and are made conservative to accommodate unknown effects. The selection criteria for key design factors is briefly summarized here:

Filter Rate Control. Three practical filter rate control systems could be used at the Modeltown Water Treatment Plant. These are (1) effluent-controlled constant-rate or constant-level, (2) influent-flow-splitting constant-rate, and (3) variable declining-rate.

Comparison of the three filter rate control systems led EDC to propose an effluent-controlled constant-rate control for the Modeltown Water Treatment Plant. The reasons for this selection are (1) superior filtrate quality throughout a filter run is expected as the filtration rate is kept constant by gradually reducing the head loss at the controller as the filter becomes clogged, (2) a longer filter run is expected, because head loss through the underdrain system is kept constant, (3) a nonionic organic polymer will be added for enhanced productivity, and finally, (4) filter-design procedure is much simpler with constant rate filtration.

Filtration Rate And Surface Area. Recent trends in filter-aid media selection, the use of filter polyelectrolytes or nonionic polymers, and backwashing methods have resulted in the successful design and operation of full scale plants with a rate of filtration as high as 24.6 $m^3/m^2 \cdot h$ (10 gpm/ft^2). The AWWA Direct Filtration Subcommittee recommended that the nominal filtration rates for direct filtration should range from 9.6 to 15 $m^3/m^2 \cdot h$ (4 to 6 gpm/ft^2).[8] Direct filtration is not expected for this design, so a filtration rate of 10 $m^3/m^2 \cdot h$ (4.1 gpm/ft^2) will be used to design the filters. The estimated time between the backwashes is 24 hours for each filter, with a maximum head loss of 2.5 m (8.2 ft) through the dirty media.

Filter Media Selection. Selection of the optimum filter media for any plant should be based on pilot test results collected over a wide range of raw water conditions. Without the pilot test results, selection of the filter media is also made from experience.[9–12]

For both high-rate and conventional filter designs, a medium-grained sand layer overlaid by a coarser, but lighter, anthracite coal layer has been used with success. The effective size, uniformity coefficient, and specific gravity of the sand and coal should be specified in such a way that the two media are compatible. An anthracite layer with sufficient depth over a sand layer should be provided for an adequate degree of protection and storage volume to accommodate solids for conventional filtration. Provision for feeding filter aid should be made to enhance filtration during periods of poor settled-water quality.

Filter aid is a high-molecular-weight nonionic polymer. The optimum dose is 0.02 mg/L, and the application point is at the exit of the central collection channel, before free fall into the effluent box of each sedimentation basin.

Filter Backwash, Media Support, and Underdrain. The selected backwash method should be capable of removing a large quantity of solids in a reasonable length of time. Also, ferric salt or alum and synthetic polymers will be used in the conventional filtration mode. It is possible that the solids removed during filtration will be "sticky" and will have a tendency to adhere to the media, thus increasing the potential for *mudball* formation. Therefore, it is necessary that auxiliary scour systems, such as surface-wash scour, be provided in the design. The auxiliary surface wash should be used during conventional mode. This arrangement will provide the necessary flexibility for fluidization of a dual-media bed.

The media support and underdrain system will consist of *double-reverse* graded gravel over a precast underdrain designed for backwash. Such system may be a manufactured underdrain or a custom-designed filter bottom. It is estimated that a backwash rate of 36 to 42 $m^3/m^2{\cdot}h$ (15 to 17 gpm/ft^2) will be required for a minimum of 10 minutes.

C: Disinfection and THM Control The disinfection process will be closely interrelated with THM control at the water treatment plant. Use of free chlorine as the primary disinfectant in the plant (to meet CT requirement), followed by chloramines for residual disinfection in the distribution system, will produce water that will consistently meet the MCL for the THMs. The general disinfection scheme was also shown to be most cost-effective for disinfection, THM control, and maintenance of desired residual in the distribution system. Multiple chlorine and ammonia feed points should be provided to control undesired slime growth and allow flexibility for meeting CT requirements, lower THM formation potentials, and chlorine demand. The chlorine and ammonia feed points should, therefore, include (1) raw water at the flash mix, (2) settled water prior to filtration, and (3) filtered water prior to storage. Additional details on chlorine and ammonia chemistry, equipment, and unit design may be found in Chapter 12.

D: Taste and Odor Control, and Iron and Manganese Removal It is expected that seasonal taste and odor episodes and high iron and manganese problems will arise. PAC may be used for control of seasonal taste and odors. The anticipated dosage of PAC is 4 to 8 mg/L (typical 5 mg/L), and it must be applied at the *raw water pump station*. PAC will reduce precursors, free-chlorine demand, and possibly THM formation.

Potassium permanganate ($KMnO_4$) will be used to oxidize iron and manganese when their levels exceed the secondary MCLs. Best point of $KMnO_4$ application is in the raw water line at the plant property. It is expected that at 1 to 4 mg/L (typical 3 mg/L), $KMnO_4$ dosage will oxidize and precipitate both iron and manganese. In addition to this, there may be slight reduction in taste and odor causing compounds, chlorine demand, and THM formation potential.

E: Corrosion and Scaling Control Production and supply of non-corrosive and non-scaling water is essential for water quality control and to prolong the life of the water distribution system. Stabilization of water may be necessary to adjust pH and alkalinity to its calcium carbonate-saturation equilibrium value. *Langelier Saturation Index*, *Ryznar's Stability Index*, or other tests must be conducted on treated water and samples obtained from the distribution system to determine the chemical stability of the water. In-depth discussion on this subject can be found in Chapter 13.[13–15]

F: Residuals Management There are two major residual sources that must be controlled at the water treatment plant. These include the filter-backwash water from the filters and sludge from the sedimentation basins. The filter backwash will be settled for recovery of water that will be returned to the head of the plant. The solids withdrawn from the sedimentation basins and filter-backwash recovery system will be thickened in a gravity thickener followed by dewatering and in-situ disposal in sludge lagoons. The dewatering sludge lagoons will be used in such a way that, while one is being filled, others will be drying. The thickener supernatant and decant/percolate from the lagoon will be returned to the coagulation flocculation unit prior to the filter-backwash recovery system.

5.4.10 Process Selection and Process Train

In the previous sections, it has been established that the Modeltown water treatment plant must be capable of removing turbidity and provide complete disinfection, THM control, and elimination of seasonal episodes of taste and odors and of iron and manganese.

Optimum turbidity removal can be achieved by adding a primary coagulant in a rapid mix and a polymer aid in a Parshall flume, then by three-stage flocculation, sedimentation, and filtration. The most cost-effective disinfection and THM control method will involve maintaining free chlorine residual in a chlorine contact channel upstream of the clearwells, followed by chloramines in the clearwells and distribution system for desired residuals. Taste and odor episodes will be controlled by a powdered activated carbon feed at the raw water pump station. If the iron and manganese MCLs are exceeded periodically, potassium permanganate in the raw water line at the plant property will be used for oxidation and conversion into insoluble forms for removal with turbidity. Filter backwash will have a backwash recovery system. Sludge from clarifiers and solids from filter backwash recovery system will be thickened in gravity thickeners and dewatered in drying lagoons. The thickener overflow and other decants will be recycled through the plant, while dewatered solids will be disposed of in-situ in the sludge drying lagoons. A generalized process train selected for the conventional water treatment plant for Modeltown is illustrated in Figure 5-5.

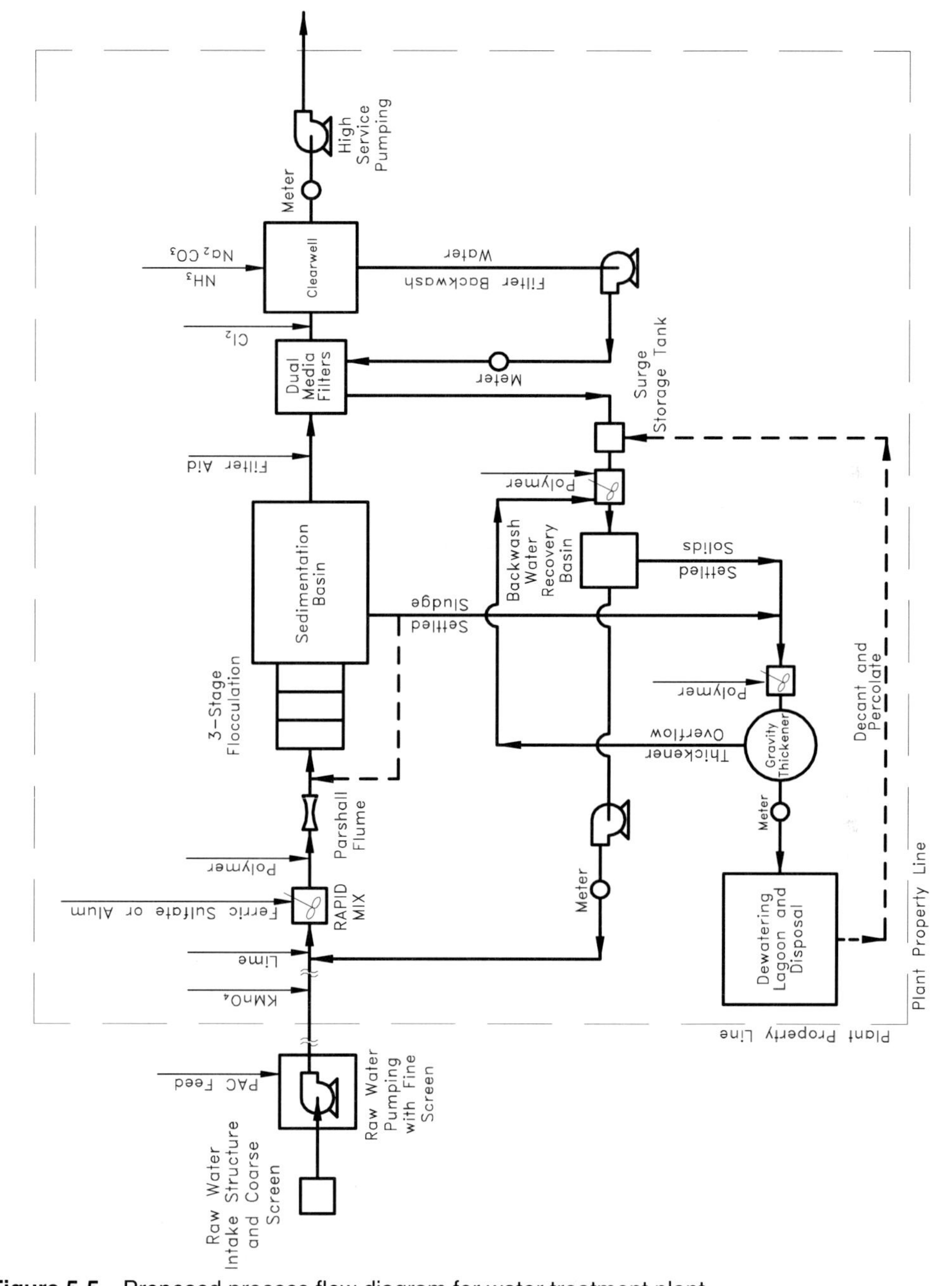

Figure 5-5 Proposed process flow diagram for water treatment plant

5.4.11 Description of Water Treatment Plant

The previous sections have established that the Modeltown water treatment project will involve construction of an intake structure in the Big River Reservoir, a raw water pump station, and a raw water line. The treatment plant must be capable of removing turbidity and provide complete disinfection, control of THMs, and control of possible seasonal episodes of taste and odors and of iron and manganese. The project also involves construction of high-service pumps, a major treated water transmission pipeline, and coordination with the distribution system. A summary of the design criteria and preliminary unit sizing is an essential part of a predesign report. Such information is given in Chapter 17. Design analysis and design computations are covered in Chapters 6 through 16 of this book.

A: Plant Capacity The water treatment will be designed for the year 2020. The design capacity of the plant will be 113,500 m^3/d (30 mgd). This capacity is based on the maximum day drought condition and is essentially the maximum draft of any day of the year over the design period.

B: Water Treatment Plant The water treatment process train selected is essentially a conventional treatment, including coagulation, flocculation, sedimentation, filtration, and clear-water storage, with capability for direct filtration. The residuals management involves filter-backwash recovery, gravity thickening, dewatering lagoons, and in-situ disposal of solids in dewatering lagoons.

All processes used at the treatment plant are interrelated. As a result, a change in process operation in any one process unit may cause a change in other processes. In addition, the performance and optimum process design parameters of all processes are related to the raw water quality. The operational flexibility under variable raw water conditions will be achieved by providing multiple alternative chemical feed points and equipment with the versatility to accommodate variable raw water conditions. The process layout utilizes two identical process trains arranged in parallel. This essentially gives two plants, each providing half of the total capacity. Each process train is further subdivided into two process trains with piping interconnected. This arrangement will provide a total of four units each of rapid mixes, flocculators, and sedimentation basins and 8 units of filters. The plant can thus be operated at 3/4 capacity with any single unit out of service. Furthermore, the plant can also be operated at 1/4 and 1/2 capacity, if necessary. Two storage reservoirs with a total storage capacity of 28,800 m^3 are provided at the plant site. The treated water will be delivered to the distribution system by means of a high service pump station containing five pumps with a combined station capacity of 120 percent of the maximum water treatment plant capacity of 113,500 m^3/d. The process layout of the water treatment plant and the capacity of each process are given in Figure 5-6. The process flow diagram and the chemical-feed points are illustrated in Figure 5-5.

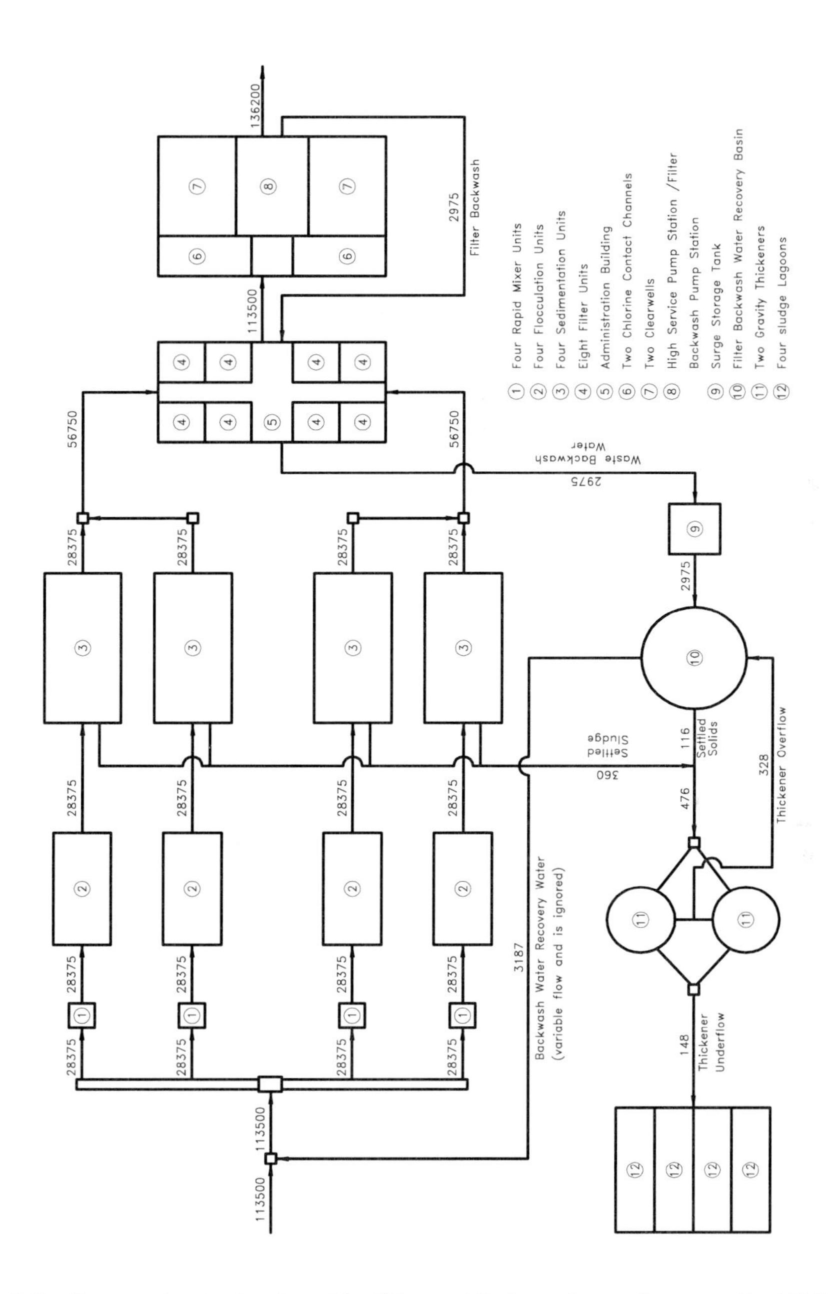

Figure 5-6 Proposed water treatment facilities and their maximum day capacity. (All flows are in m^3/d.)

5.4.12 Operation and Maintenance

A: Personnel The treatment facility will be operated by trained licensed operators on a 7-days-per-week basis. Supporting staff, including mechanics and electricians, will be provided on a 24-hour basis. The laboratory facility for water quality and process control and the administration building and equipment maintenance shops shall be fully staffed.

B: Electrical Power The power supply in the area is very reliable. Interruption records for 10 years at the Big River Dam and the water treatment plant site have revealed only five power outages, ranging in durations from 10 minutes to one hour; prolonged power failure is unlikely. Auxiliary power generators and multiple electrical primary-power lines from two different sources are considered unnecessary.

C: Flood Hazard The entire areas of the proposed raw water pumping station and water treatment plant are well above the 100-year flood level. Both sites are fully accessible under all weather conditions.

5.4.13 Construction and Maintenance Costs

The construction, operation, and maintenance costs of the water supply project in this example are developed using detailed engineering cost estimates. Various components of the capital costs are (1) site preparation and earthwork, (2) structure, (3) pipeline, (4) equipment and installation, (5) electrical and instrumentation, and (6) site improvement and project enclosure. These cost components are based on actual quantities of material and on manufacturer's data for the equipment.[16–19] The estimated annual O&M costs include energy, parts and material, and labor. Table 5-7 summarizes these estimated construction, operation, and maintenance costs for the entire project. It may be noted that these costs represent actual estimates based on quantities, current prices, and parts and labor.

In many cases, an alternate but less reliable cost estimate is made by using several computer programs and published cost curves and equations.[20–25] This procedure gives only preliminary cost data but is good enough for comparison of alternatives. Actual costs (similar to those in Table 5-7) must be developed for the selected alternative after the engineering plans and specifications are developed.

5.4.14 Present Worth and Equipment Annual Costs

The procedure for the annual cost estimate includes determination of the *present worth* and *equivalent annual costs*.[2,25,26] The present worth may be thought of as the sum which, if invested now at a given interest rate, would provide exactly the funds required to make all necessary expenditures during the planning period. Equivalent annual cost is the expression of a nonuniform series of expenditures used as a uniform annual amount to simplify calculations of the present worth. Detailed procedures for making these calculations are well known and are explained in many textbooks.[27,28] The interest rate for formulation and evaluation of construction-grant projects is published by the Water Resources Council; these are *the Discount Rate* and

Table 5-7 Estimated Construction, Operation, and Maintenance Costs of the Water Treatment Plant

Item No.	Item	Construction Costs—1999, thousands of $	Annual O & M Costs—1999, thousands of $
1	Intake structure and raw water pump station	$7441	$357
2	Raw water transmission pipe	$5255	$207
3	Rapid mixer units (including junction structures, rapid mixers and housing, and Parshall flume)	$752	$73
4	Flocculation units	$1258	$80
5	Sedimentation basins	$4263	$103
6	Filter units	$5262	$403
7	High-service/filter backwash pump station (including chlorine contact channel and clearwells)	$7945	$282
8	Chlorine/ammonia facilities	$481	$49
9	Chemical building	$1012	$167
10	Backwash water recovery system (including elevated backwash water storage tank)	$2527	$88
11	Sludge disposal (including the meters)	$1304	$74
12	Administration building (including maintenance and warehouse, plant vehicles, electrical equipment, control system, plant process piping, plant sewer system, plant pavement, and plant site preparation and improvements)	$4919	$670
13	Finished water transmission pipe	$6414	$312

Table 5-7 Estimated Construction, Operation, and Maintenance Costs of the Water Treatment Plant(continued)

Item No.	Item	Construction Costs—1999, thousands of $	Annual O & M Costs—1999, thousands of $
14	New ground storage and distribution pump storage (including new distribution pipes, new elevated storage, existing ground storage and distribution pump station improvements, existing distribution network improvements, existing elevated storage improvements)	$5923	$866
15	Subtotal (actual project cost)	$54,805	$3733
16	Preparation of predesign engineering reports (1% of project cost)	$548	—
17	Preparation of design plans and specifications (6% of project cost	$3288	
18	Engineering and construction supervision (15% of project cost)	$8221	
19	Contingencies (15% of project cost)	$8221	
20	Total project cost	$75,083	$3733

are periodically revised.[29] The discount interest rate used in this example is 7.625 percent.

The present worth and equivalent annual costs of the water treatment project are summarized in Table 5-8. These costs must be updated after completion of the design plans and specifications.

5.4.15 Financial Status and Users Charge

The city has the legal, institutional, managerial, and financial capability to ensure adequate funds for construction, operation, and maintenance of the project proposed in this predesign report. The city is currently charging a fee to users of water and sewer service. The water-service rates are based on water used by the customers. The users' charge schedule shall be made in accordance with the state guidelines. The financial and statistical information about the city is provided in an attachment to the predesign report.

Table 5-8 Present Worth and Equivalent Annual Cost of the Proposed Water Treatment Plant

Item	Cost, Millions of $
Total capital—1999	$75.083
Total annual O & M—1999	$3.733
Present worth of annual O & M	$32.697[a]
Present worth	$107.780[b]
Equivalent annual cost	$8.575[c]

a Interest Rate 7.625 percent, planning 15 years, present value of annuity = $[1 + i)^{-n}]/i$ = $(1 + 0.07625)^{-15}]/0.07625$ = 8.759. Present Worth of O & M Cost = 3.733 × 8.759 = 32.697 millions of dollars.

b Present Worth = 32.697 + 75.083 = 107.780 millions of dollars.

c Capital Recovery Factor = $i(1 + i)^n/[(1 + i)^n - 1]$ = $0.07625\ (1+ 0.07625)^{15}/[(1+ 0.07625)^{15} - 1]$ = 0.1142. Equivalent Annual Cost = 75.083 × 0.1142 = 8.575 millions of dollars.

5.4.16 Environmental-Impact Assessment

An environmental-impact assessment report normally is prepared as part of the predesign report. Only basic information pertinent to the environmental-impact assessment is presented in the main text of the predesign report. Other supporting material is provided in several attachments.

The proposed water treatment plant is expected to produce high-quality water and serve the needs of Modeltown for many years. The project also reduces impact on groundwater resources. It is expected that the proposed project will not have any significant primary adverse impacts on the following areas of environmental concern:

- navigation and recreational use (Permit 404 has been issued);[f]
- plant and animal communities;
- ecosystem;
- endangered or locally threatened species;
- unique or vulnerable environmental features;
- unique archeological, historic, scientific, or cultural areas, parks, wetlands, or stream corridors;
- community growth pattern and land-use pattern.

f. Further information may be found in Section 6.2.2.

The city owns a small tract of land that is proposed for construction of the raw water pump station. Approximately 12 hectare (30 acres) will be allocated to the plant property. Some relocation of people, but no relocation of existing structures, will be required for the project. Also, no change in the value of the adjacent property is anticipated as a result of this project.

The minor adverse effects that cannot be avoided are those normally associated with the existence and operation of water treatment facilities. Some degree of disruption of the environment and of inconvenience to citizens during construction of water mains is unavoidable, but it will be reduced in severity by proper construction scheduling and techniques.

Electrical power, chemicals, and fuel will be utilized during construction and operation of the facilities. The construction materials used for the project will remain permanently. Chemicals such as chlorine, lime, alum, ferric chloride, and organic polymers will be needed for normal plant operation. The land for treatment facilities will be permanently lost to agricultural production; however, the productivity of the area in current use as pasture land is minimal.

5.4.17 Public Participation Program

During the entire project-planning process and predesign report-preparation period, an active public participation program was maintained. The program utilized public meetings, mail-outs, and press releases to keep the public aware of the progress and to receive their comments before major decisions were made. Various appointed environmental advisory committees held public meetings to review the work.

The final meeting will be a public hearing, and all public comments will be included in the engineering and design phase of the project.

5.4.18 Cost Breakdown and Time Schedule for Project Milestone

The estimated cost for preparation of the predesign report, the design plans and specifications, and the estimated construction cost of the project are summarized in Table 5-7. The project milestones for preparing design plans and specifications and construction programs are generally developed for specific project using computer software.[30] Thus, the scheduling for starting and completion dates for all elements of the project can be established.

5.4.19 Attachments

The following information should normally be submitted as separate attachments or appendices to the predesign report:

1. design analysis and summary of design computations;
2. Permit 404 from U.S. Army Corps of Engineers;
3. Permit 201 from U.S. Army Corps of Engineers;
4. statement demonstrating applicant's necessary legal, technical, financial, institutional, and managerial resources;
5. correspondence from agencies;

6. review comments or approval of state and local agencies;
7. public notifications.

Problems and Discussion Topics

5.1 Figure 5-2 and Table 5-1 represent the historical and projected populations of Modeltown. Using population values for the years 1990, 2000, and 2010, calculate saturation population. Also, check the populations of the years 1992, 1995, and 2005. Use the logistic-curve-fitting method.

5.2 The procedure used to establish the projected water demand for the Modeltown service area (Table 5-3) is somewhat different from the general procedure presented in Table 4-10 and Example 4-2. Compare the procedures and discuss the results. Use drought conditions in both cases.

5.3 The removal of iron and manganese is achieved by treatment with potassium permanganate. Write the chemical reactions that are involved. List other chemicals that can also be used for the same purpose. Write the chemical equations.

5.4 In the process design for Modeltown WTP, PAC will be used for taste and odor control. Discuss alternate methods of taste and odor control.

5.5 In the process flow diagram (Figure 5-5), several chemicals are used before rapid mix, filters, and clearwell. Discuss the purpose of each chemical feed.

5.6 The suggested method of THM control in the process diagram is by conversion of free chlorine to combined chlorine residual before clearwell. List the advantages and disadvantages of this method. Suggest other methods of THM control.

5.7 Direct filtration has been proposed in the treatment process train. Discuss why such capability is desirable.

5.8 What is the purpose of the auxiliary scour system? Discuss surface wash and air scour systems.

5.9 Describe (a) effluent-controlled constant-rate or constant-level, (b) influent-flow-splitting constant-rate, and (c) variable-declining-rate filter-rate control systems.

5.10 Visit your local water treatment plant. Determine which year the predesign report was prepared. Review a copy of the predesign report and identify the following:
   (a) initial and design years;
   (b) population served and design capacity;
   (c) process diagram.

References

1. James M. Montgomery, Inc. *Water Treatment Principles and Design*, John Wiley and Sons, New York, 1985.
2. Qasim, S. R. *Wastewater Treatment Plants: Planning Design and Operations*, 2d ed., Technomic Publishing Co., Lancaster, PA, 1999.
3. ASCE and AWWA. *Water Treatment Plant Design*, 2d ed., McGraw-Hill Publishing Co., New York, 1990.
4. Sanks, R. L. *Water Treatment Plant Design for the Practicing Engineers*, Ann Arbor Science, Ann Arbor, MI, 1980.
5. USEPA. Grants for Construction of Treatment Works, Federal Register, 43(188), p. 44022–44097, September 27, 1978.
6. Fair, G. M., Geyer, J. C., and Okum, D. A. *Elements of Water Supply and Wastewater Disposal*, 2d ed., John Wiley and Sons, New York, 1971

7. Ficek, K. G. "Potassium Permanganate for Iron and Manganese Removal and Taste and Odor Control," *Water Treatment Plant Design*, R. L. Sanks (Ed.), Ann Arbor Science, Ann Arbor, MI, 1980.

8. AWWA Committee Report. "The Status of Direct Filtration," *Jour. AWWA*, 72(7), vol. 72, no. 7, pp.405–411, July 1980.

9. Hudson, H. E. "Declining Rate Filtration," *Jour. AWWA*, vol. 51, no. 11, pp. 1455–1463, November 1959.

10. Cleasby, J. L. et. al. "Backwashing of Granular Filter," *Jour. AWWA*, vol. 69, no. 2, pp. 115–126, February 1977.

11. Arboleda, J. "Hydraulic Control Systems of Constant and Declining Flow Rate in Filtration," *Jour. AWWA*, vol. 66, no. 2, pp. 87–94, February 1974.

12. Culp, R. L. "Direct Filtration," *Jour. AWWA*, vol. 64, no. 7, pp. 375–378, July 1977.

13. Tchobanoglous, G. and Schroeder, E. D. *Water Quality: Characteristics, Modeling and Modification*, Addison-Wesley Publishing Co., Reading, Massachusetts, 1985.

14. *Drew Principles of Industrial Water Treatment*. Drew Chemical Corporation, Boonton, N.J., 1977.

15. AWWA. *Water Quality and Treatment*, 4th ed., McGraw-Hill Book Co., New York, 1990.

16. RSMeansCo., Inc. *Building Construction Cost Data*, 57th Annual Edition, RSMeansCo., Inc., Kinston, MA, 1999.

17. RSMeansCo., Inc. *Heavy Construction Cost Data*, 12th Annual Edition, RSMeansCo., Inc., Kinston, MA, 1999.

18. RSMeansCo., Inc. *Mechanical Cost Data*, 22nd Annual Edition, RSMeansCo., Inc., Kinston, MA, 1999.

19. RSMeansCo., Inc. *Site Work and Landscape Cost Data*, 18th Annual Edition, RSMeansCo., Inc., Kinston, MA, 1999.

20. Gumerman, R. C., Culp, R. L., and Hansen, S. P. *Estimating Water Treatment Costs*, Vol. 1, Summary, Municipal Environmental Research Laboratory, EPA-600/2-79-162 a, NTIS PB80-139817, USEPA, Cincinnati, OH, August 1979.

21. Gumerman, R. C., Culp, R. L., and Hansen, S. P. *Estimating Water Treatment Costs,* Vol. 2, *Cost Curves Applicable to 1 to 200 mgd Treatment Plants*, EPA-600/2-79-162b, NTIS PB80-139827, USEPA Municipal Environmental Research Lab., Cincinnati, OH, August 1979.

22. Hansen, S. P., Gumerman, R. C., and Culp, R. L. *Estimating Water Treatment Costs,* Vol. 3, *Cost Curves Applicable to 2,500 gpd to 1 mgd Treatment Plants*, EPA-600/2-79-162c, NTIS PB80-148455, USEPA Environmental Research Lab, Cincinnati, OH, August 1979.

23. Lineck, T. S., Gumerman, R. C., and Culp, R. L. *Estimating Water Treatment Costs,* Vol. 4, *Computer User's Manual for Retrieving and Updating Cost Data*, EPA-600/2-79-162d, NTIS PB80-225979, USEPA Municipal Environmental Research Lab, Cincinnati, OH, August 1979.

24. Gumerman, R. C., Burris, B. E., and Hansen, S. P. *Small Water System Treatment Costs*, Noyes Data Corporation, Park Ridge, NJ, 1986.

25. Qasim, S. R., Lim, S. W., Motley, E. M., and Heung, K. G. "Estimating Costs for Treatment Plant Construction," *Jour. AWWA*, vol. 84, no. 8, pp. 1156–1162, August 1992.

26. USEPA. *Guidance for Preparing a Facility Plan*, EPA-430/9-76-015, Construction Grants Program Requirements, USEPA, Washington, D.C., May 1975.

27. Grant, E. L. and Jreson, R. S. L. *Principles of Engineering Economy*, 8th ed., Prentice Hall Press, Englewood Cliffs, NJ, 1990.

28. Thuesew, G. J. and Fabrycky, W. J. *Engineering Economy*, 8th ed., Prentice Hall Press, Englewood Cliffs, NJ, 1993.

29. American Consulting Engineering Council. *The Last Word*, vol. XVIII, no. 10, p.1, April 5, 1996.

30. Pyron, T., Guthrie, B., Nelson, S., Orvis, W. J., and Valentine, K. *Using Microsoft Project 4 for Windows*, Que Corp., Indianapolis, IN, 1994.

CHAPTER 6

Raw Water Intake, Screening, and Aeration

6.1 Introduction

Raw water intakes withdraw water from a river, lake, or reservoir over a predetermined range of pool levels. Screens remove large floating objects from the water. In order to protect pumping equipment, screens are generally located at the intake structures, to remove large objects before they enter the pump station. Aeration is a unit process that removes gases and volatile organics that may cause taste and odor problems in the water supply. Aeration is also used to oxidize certain dissolved metals by forming insoluble oxides. This unit process may be located either at the intake structure or just ahead of the treatment plant.

In this chapter, an overview of design and selection considerations for intake structure, for screening, and for aeration equipment is presented. Step-by-step design calculations for intake structure and screens are also provided in the design example.

6.2 Raw Water Intake Structures

Raw water intake structures are used to control withdrawal of raw water from a surface water source. Their primary purpose is to selectively withdraw the best-quality water while excluding fish, floating debris, coarse sediment, and other objectionable suspended matter. These structures may be simple submerged intake pipes or elaborate tower-like structures that house intake gates, screens, control valves, pumps, chemical feeders, flow meters, offices, and machine shops. They may be an integral part of a dam or may be a separate structure located elsewhere.

6.2.1 Types of Intake Structures

Several different types of intake structures are in common use. These are floating, submerged, tower, shore, and pier structures. Selection of an intake structure depends on the nature of the water source and the quantity and quality of the water being withdrawn. The operational characteristics, advantages, and disadvantages of these types of intake structures are compared in Table 6-1. Additional discussion is given below.

Floating Intakes Floating intake structures offer a relatively inexpensive method of constructing an intake in an existing lake or reservoir. Floating intake structures are particularly suitable in water sources that have unsuitable geological conditions and where relatively little difference in water-surface elevation over time occurs. They consist of a barge-type structure that floats on the water surface and supports pumps, screens, valves, electrical switches, gears, and other equipment. The structure is anchored by at least two mooring piers. These piers each consist of a steel or concrete column that is embedded in the lake or reservoir bottom and extends through a specially designed sleeve attached to the structure. This type of mooring arrangement anchors the structure in place but allows it to freely float on variable water levels. The structure is connected to the shore by a bridge that supports the raw water discharge pipe and electrical cables and provides access to the structure. Figure 6-1 illustrates a typical floating intake structure.

Submerged Intakes Submerged intakes are used to withdraw water from streams or lakes that have relatively little change in water-surface elevation throughout the year. These intakes are constructed as *cribs* or *screened bellmouth.*[1] They may consist of a simple concrete box or of a rock-filled timber crib to support the influent end of the withdrawal pipe. The withdrawal pipe discharges either into a sump at the shore where pumping equipment is installed or into a gravity conveyance system. The top opening is covered with cast iron or mesh grating. A fine screen may be installed in the sump at the pumping station or at the treatment plant. In lakes with heavy siltation, the intake opening is raised about 1 to 2 meters above the bottom. Figure 6-2 illustrates a typical submerged intake.

Exposed or Tower Intakes Exposed intake gatehouse or tower structures are generally used for large projects on rivers or reservoirs with large water-level fluctuations. They may be built either sufficiently near the shore to be connected by a bridge walkway or at such distance that they can be reached only by boat. Gate-controlled openings (also called ports) are generally provided at several levels in order to permit selection of the best-quality water. There are two major types of exposed or tower intakes: *wet-intake towers*, and *dry-intake towers*.

Wet-Intake Towers. Wet-intake towers consist of a concrete circular shell filled with water up to the reservoir level, with an inner well that is connected to the withdrawal conduit. Ports are provided into the outer concrete shell and into the inner well. The ports leading to the inner well have gates that regulate the flow of water into the well. The purpose of the water-filled outer shell is to provide weight needed to balance the buoyant forces when the inner well is dry. General details of a wet-intake tower are given in Figure 6-3(a).

Table 6-1 Comparison of Types of Intake Structures

Intake Type	Relative Cost	Operational Flexibility	Advantages	Disadvantages
Floating	Low to moderate	Can withdraw water from a fixed depth below the water surface	Relatively low cost, can be fabricated off-site and assembled on-site. Can operate over a small range of water surface elevations.	Must be securely anchored to prevent damage from wind and wave action. Can only draw water from a fixed depth below the water surface.
Submerged	Low	Can withdraw water from a fixed elevation.	Simple, easy, and relatively inexpensive to construct.	Can draw water only from a fixed elevation near the bottom where poor quality water is usually located. Grating is inaccessible, so it is difficult to remove lodged debris and perform normal maintenance.
Tower	Moderate to high	With multiple gates can withdraw water from different elevations to obtain optimum water quality.	Can withdraw optimum water quality through multiple gates. Can be located in deep water when constructed prior to reservoir completion. Gates and water conduit can be dewatered and maintained.	More expensive to construct. Located offshore and may be less accessible than shore intakes.
Shore	Moderate to high	With multiple gates can withdraw water from different elevations to obtain optimum water quality.	Accessible for maintenance. If constructed with multiple gates, optimum water quality can be selected. Relatively easy and inexpensive to construct in existing water bodies.	More expensive than floating or submerged structures. May require dredging or excavating a channel to deep water.
Pier	Moderate	Can withdraw from a fixed elevation only, unless pumps are set at different depths.	Relatively easy to construct in existing water bodies. Accessible for maintenance.	Can only withdraw water from an elevation fixed by setting depth of pump. May not be able to obtain optimum water quality.

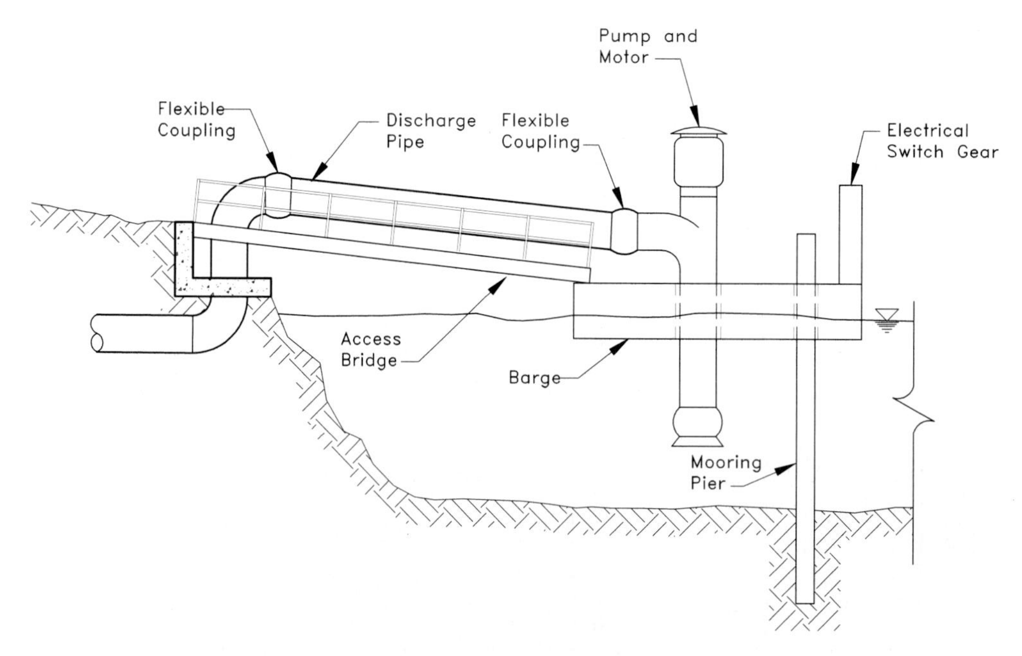

Figure 6-1 Typical floating intake structure and pump station.

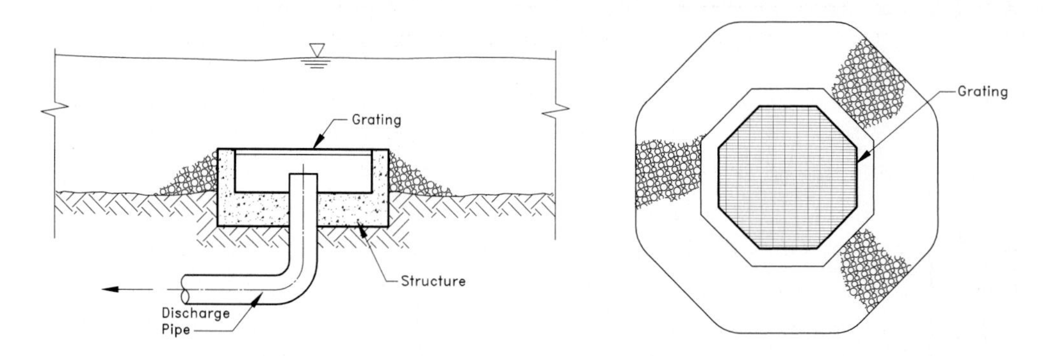

Figure 6-2 Typical submerged intake structure.

Dry-Intake Tower. In a dry-intake tower, water enters the well through gate-controlled ports. In the absence of a water-filled outer shell, the dry-intake tower must be heavier in construction than a wet-intake tower, to resist the buoyant forces when empty; however, these intakes are beneficial, because water can be withdrawn from any selected level by using the proper port. Details of the dry-intake tower are given in Figure 6-3(b).

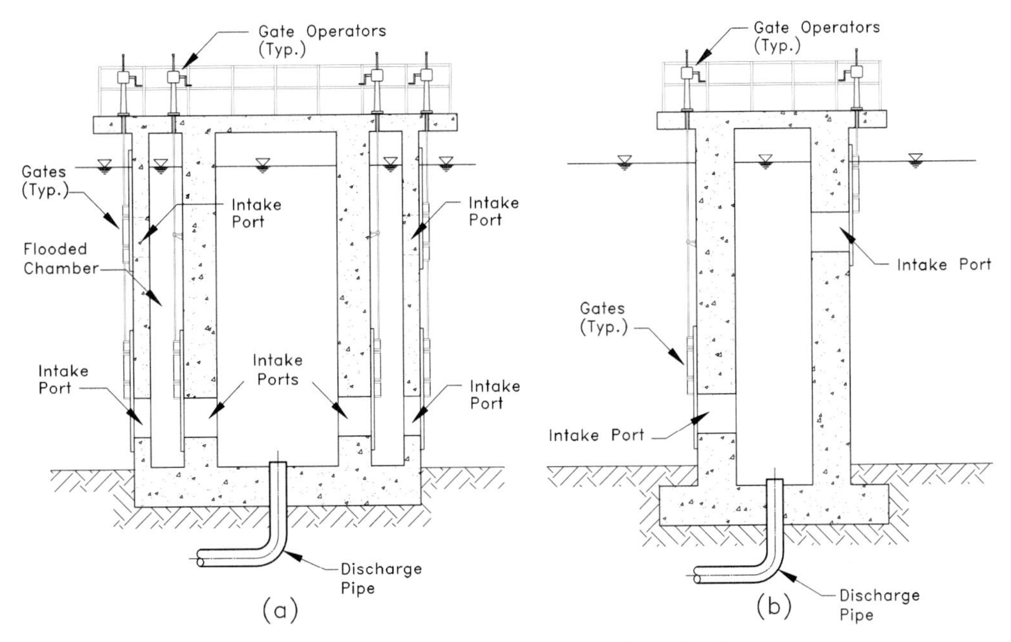

Figure 6-3 Typical intake towers. (a) Wet-intake tower. (b) Dry-intake tower.

Shore-Intake Structures Shore-intake structures are used on rivers and lakes with a near-constant level or on lakes and reservoirs with deep shorelines. Typically, they are concrete structures open on the water side. The structures may be constructed to be integral with a pump station. A shore-intake structure is illustrated in Figure 6-4.

Pier Intakes Pier-intake structures are used on lakes and rivers where the water depth at the shoreline is too shallow for a shore-type structure. They are also used when an intake must be constructed in an existing water body. This type of structure consists of a structural steel or concrete platform resting on steel piles or concrete piers. The platform supports the pumps, piping, valves, electrical switchgear, and other equipment. A bridge connects the platform to the shore, providing access and a support for the raw water pipeline. A pier-intake structure is illustrated in Figure 6-5.

6.2.2 Intake-Site Selection

Engineers must carefully consider numerous factors in site selection for a raw water intake. These factors include water quality, water depth, stream or current velocities, foundation stability, access, power availability, proximity to water treatment plant, hazards to navigation, and environmental impact. Each factor is briefly discussed below.

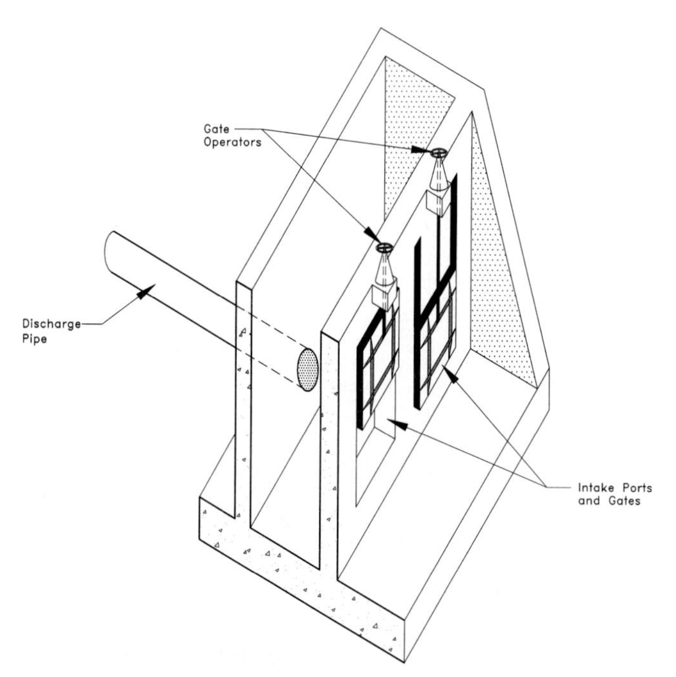

Figure 6-4 Typical shore intake structure.

Water Quality The intake-structure site should be selected to yield the best-quality water possible. Intakes should not be located in dead areas (areas with little or no water circulation), near wastewater outfalls, near large inflows of poor-quality water, or in areas susceptible to hazardous chemical spills.

Water Depth The intake structure should be located in such a way that water can be withdrawn from the full range of water levels. The range of water levels should include the lowest expected drought level and the extreme flood level in the water source.

Stream or Current Velocities Both the direction and magnitude of the stream or current velocities can have an impact on the operation of intakes. Water currents flowing past intake structure and ports at velocities greater than 0.6 m/s (2 fps) can cause eddy currents that will affect the hydraulic function of the intake.[2] Water velocity also affects the lateral stability and foundation stability of the structure.

Foundation Stability Typically, intake structures are tall, narrow structures that often have extreme lateral loads caused by water velocity. Therefore, the foundation must be carefully designed to resist the resulting overturning moments.

Access Intakes must be accessible under all weather conditions, to allow operators to inspect and maintain the intake equipment, which includes gates and screens and often pumps.

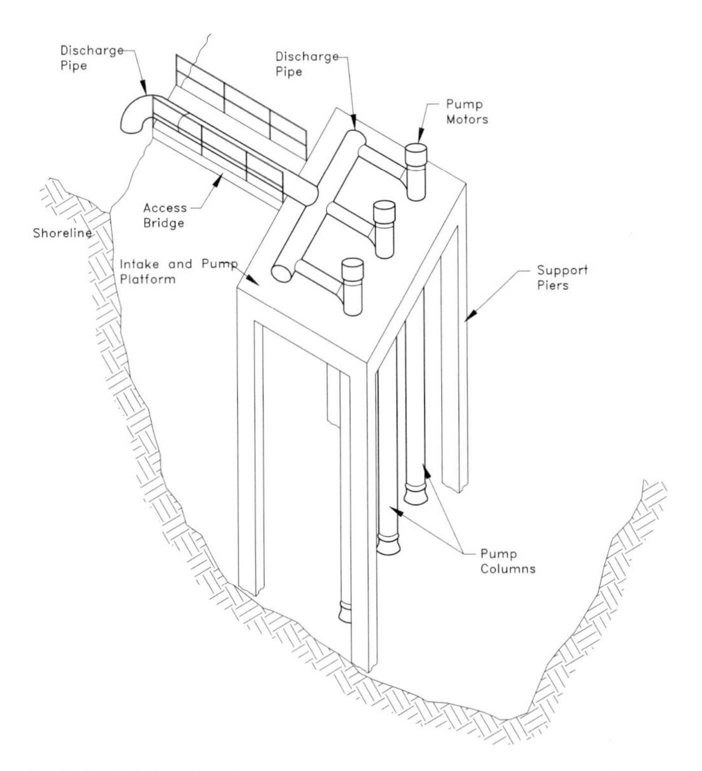

Figure 6-5 Typical pier intake structure.

For this reason, an all-weather road that is not subject to flooding should be provided to the intake structure.

Power Availability Electrical power is always required for intake structures, to operate equipment such as gates and screens and for hazard lighting. If raw water pumps are not located at the intake, the power load may be small; therefore, in some cases, electrical power can be provided by an on-site generator.

Proximity To Water Treatment Plant The cost of facilities to convey water from the source to the treatment plant is related directly to the distance between the source and the treatment plant. Therefore, during selection of an intake site, the length of the conveyance system must be given serious consideration.

Environmental Impact Environmental issues must be carefully considered when one is selecting intake sites. Usually, the greatest impact of raw water intake is on recreational uses in a body of water. Typically, a clear zone of 61 m (200 ft) or more is required around intake sites, to protect boaters from the hazards of the intake and to protect the water quality near the intake. In many cases, this zone severely limits recreational use in a small body of water. Other environmental issues that must be considered are impacts on wildlife habitats, endangered species, and historical or archaeological sites.

Construction in wetland areas of the United States that results in the discharge of dredged materials requires a permit from the U.S. Army Corps of Engineers. These permits, known as *Wetlands Permits* and *Section 404 Permits,* respectively, are required for many intake construction projects. Information on these permits can be obtained from the nearest district office of the U.S. Army Corps of Engineers.

Hazard to Navigation Intakes of the floating, submerged, tower, or pier type require structures to be located in the waterway. In many cases, particularly in major rivers, the water body may also serve as a navigation channel. In these cases, intake structures must be located and designed so that they do not pose a hazard to navigation. Particular attention in this regard must be given to submerged intakes, because they may pose a hidden hazard and are more susceptible to accidents.

Construction of structures within navigable waters of the United States requires a permit from the U.S. Army Corps of Engineers. This permit, called a *Section 10 Permit,* is applied for in conjunction with the Section 404 Permit. Design engineers should consult the nearest U.S. Army Corps of Engineers district office for information regarding these permits.

6.2.3 Intake-Design Considerations

The design of intake structures is generally site-specific. Rarely can a standard design be adapted for a given site without major modifications. Some design considerations that must be addressed in the design of each structure are listed here.

Intake Velocities The velocity of water entering the intake port is the single most important design value to be selected by an engineer. High intake velocities increase head loss, entrain suspended matter, trap fish and other aquatic animals, and compound ice problems. Low velocities require the intake port to be larger and so add to the cost of the structure. Experience has shown that a velocity below 8.0 cm/s. (0.3 fps) allows fish and other aquatic animals to escape, reduces frazil ice obstruction, and minimizes the entrainment of suspended matter.[1]

Intake-Port Location Another important design consideration for intake structures is the location of intake ports. Lakes and reservoirs tend to be stratified. As a result, water quality in each stratum may vary. Properly designed intake structures should provide water treatment plant operators the flexibility to draw water from the stratum with the best water quality. In order to achieve this, multiple intake ports set at various levels are generally provided.[1]

Ideally, the vertical location of ports should be determined by testing of water quality from various depths at the proposed intake site. These tests should be performed during all seasons, which different lake or reservoir levels and climatic conditions. In the absence of such testing, intake ports should be provided along the entire height of the structure.

The top intake port should be located not less than 2 m (6 ft) below the normal water surface so that floating debris will not be drawn into the port. The bottom port should be located at least 1 m (3 ft) above the bottom to prevent the suspension of bottom sediments.[1]

Gates Gates are used in intake structures to control inflow of water from the raw water source into the water-conveyance system. Gates are used to select water from a stratum that has the best-quality water. Gates are also used to prevent water from entering the system altogether; however, gates should not be used to regulate the flow into the system.

Gates typically used in intake structures are sluice gates. These are large cast-iron gates that slide vertically on a guide track. The gate is moved by a threaded rod (stem) that extends to the top of the structure. The gate and its frame are attached to a cast-iron casting (thimble) that is cast in the structure wall. Details of a typical sluice gate are given in Figure 6-6.

Figure 6-6 Typical sluice gates. *(Courtesy of Rodney Hunt Company.)*

Control of Ice In colder regions, ice accumulation around and in the intake structure can become a severe problem during winter. Icing, known as anchor ice, may form on metal surfaces such as screens, gates, valves, and pumps. Frazil ice in the shape of needles, flakes, or slush may also obstruct the flow by attaching to metallic surfaces. In cold climates, ice problems may be reduced if intake ports are kept sufficiently below the water surface (below 8 m).[1] Compressed air can be used to remove ice accumulations.[1] Other preventive measures include steam piping and space heaters within the intake structure, arranged so that exposed metal surfaces are kept at a temperature slightly above the freezing point of water.[1]

6.3 Screening

Screening is a unit operation that removes suspended matter from water. Screens may be classified as coarse, fine, or microstrainer, depending on the size of material removed. Screens may be located at the intake structure, raw water pump station, or water treatment plant.

6.3.1 Coarse Screen or Trash Rack

Intake ports should be equipped with a coarse screen or bar rack to prevent large objects from entering the conveyance system. These screens consist of vertical flat bars, or, in some cases, round pipes spaced with 5 to 8 cm (2 to 3 in) of clear opening.

Screens should be installed outside (on the water side) of any sluice gate or stop log slot, to prevent debris from interfering with their operation. If the intake is located such that cleaning currents sweep debris from the gates, and low intake velocities are encountered though the gate, there should not be large accumulations of debris on the bar racks; therefore, mechanical cleaning may not be required. Nonetheless, the design should provide access for equipment to remove any debris that does accumulate. Often, boats with cleaning apparatus are used for this purpose. The velocity through the coarse screen is generally less than 8.0 cm/s (0.3 fps).

6.3.2 Fine Screen

Fine screens are used to remove smaller objects that may damage pumps or other equipment. They may be located either at the intake structure or at the raw water pump station. In the case of gravity conveyance systems, the fine screen may be provided at the water treatment plant. These screens consist of heavy wire mesh with 0.5 cm (1/4 in) square openings or of circular passive screens with similar opening widths. The screen area efficiency factor is 0.5 to 0.6, and the typical design velocity through the effective area is in the range of 0.4 to 0.8 m/s. Design velocities are much lower if fish protection is required.

Screens with small openings will retain a large quantity of material; therefore, they may require automatic cleaning. Traveling water screens, as illustrated in Figure 6-7(a), have been used extensively in the past at water intakes.[3,4] The screen assembly consists of a number of screens that move slowly upward on a chain system catching debris. At the top, a system of water jets removes the debris from each screen before the screen travels back down for a repeat of the cycle.[5]

Over the past few years, a passive screening system such as that shown in Figure 6-7(b), has come into widespread use. This system consists of a circular screen, with small bars spaced at the desired spacing. Debris is removed from the screen either by air purge or by allowing water to flow backward through the screen.[6]

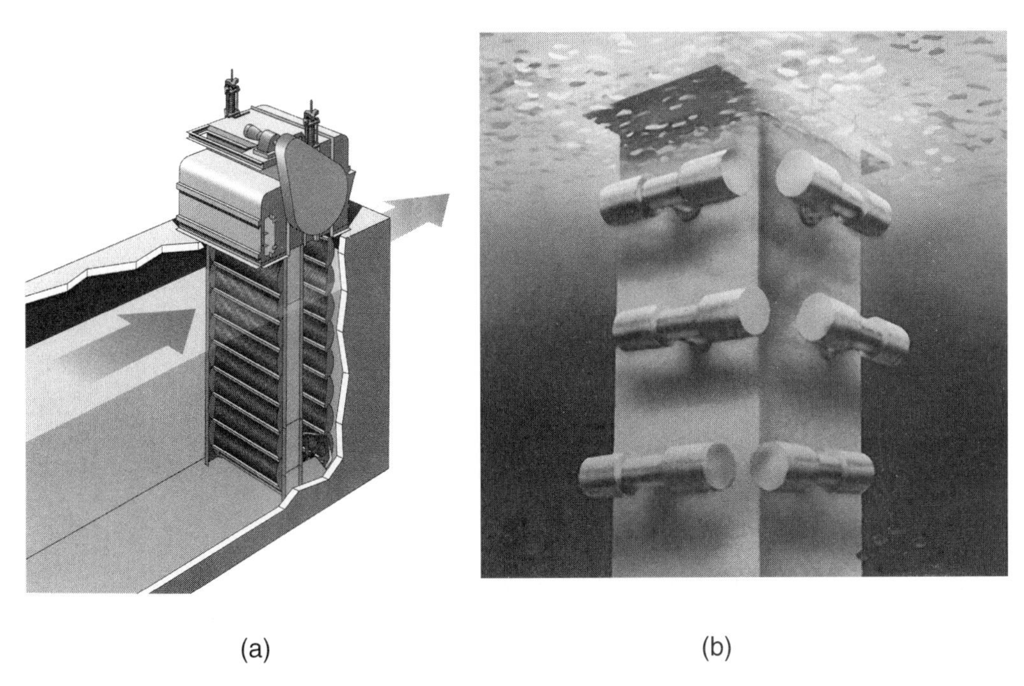

(a) (b)

Figure 6-7 Typical fine screens. (a) Traveling screen. *(Courtesy of U.S. Filter/Envirex.)* (b) Passive screen installation. *(Courtesy of U.S. Filter/Johnson Screens.)*

6.3.3 Microstrainer

Microstrainers are used for the removal of plankton and algae from impounded waters. Raw water containing a heavy population of algae and plankton is difficult to coagulate. Algae and plankton usually float, because their specific gravity is less than 1.0. Therefore, microstrainers installed before chemical coagulation will often improve the performance of clarifiers.

If raw water is pre-chlorinated, certain species of algae produce intense taste and odor problems by forming substitution products. Removal of algae prior to prechlorination and coagulation has beneficial effects on taste and odor control and on reduction in THM formation and other chlorine by-products. Also, chlorine dosage for prechlorination is significantly reduced after microstraining.

A microstrainer consists of a rotating cylindrical frame covered with fine wire-mesh fabric. Water enters the cylinder and moves radially out. Deposited solids are removed by a jet of water from the top and discharged into a trough. Typical head loss across a microstrainer is less than 15 cm (6 in). The volume of wash water varies according to the microfabric used and the algae loading, but it is usually between 1 and 3 percent of the microstrainer capacity.[7] Microstrainers are available up to 1.32 m^3/s capacity (30 mgd).[3] Operation and equipment details of a microstrainer are shown in Figure 6-8.[3]

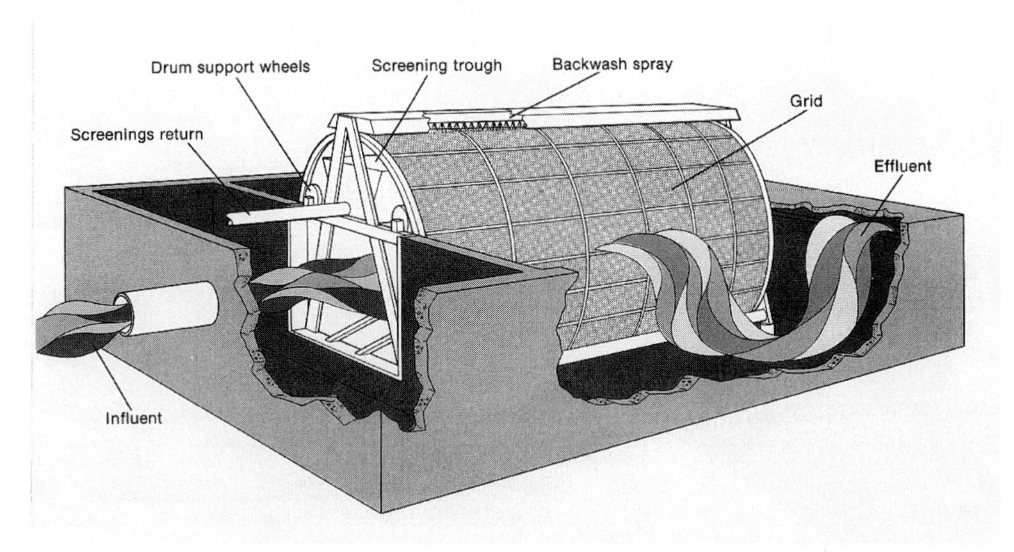

Figure 6-8 Typical microstrainer. *(Courtesy of U.S. Filter/Envirex.)*

Microstrainers are designed to remove plankton and algae; but they will not handle sand, silt, or other abrasive material. Microstrainers occasionally experience operational problems due to buildup of slime on the fabric. Occasional slime growth can be corrected by washing the fabric with chlorine solution. Frequent slime build-up can be corrected by installing ultraviolet *irradiation* equipment over the drum, to inhibit the microbiological slime growth.

6.4 Aeration

Aeration involves bringing air or other gases in contact with water to transfer volatile substances from the liquid to the gaseous phase and to dissolve beneficial gases into the water. The volatile substances that may be removed include dissolved gases, volatile organic compounds, and various aromatic compounds responsible for tastes and odors. Gases that may be dissolved into water include oxygen and carbon dioxide. The purposes of aeration in water treatment are (1) to reduce the concentration of taste- and odor-causing substances, such as hydrogen sulfide and various organic compounds, by volatilization or oxidation, (2) to oxidize iron and manganese, rendering them insoluble, (3) to dissolve a gas in the water (examples: addition of oxygen to groundwater and addition of carbon dioxide after softening), and (4) to remove those compounds that may in some way interfere with or add to the cost of subsequent water treatment (examples: removal of hydrogen sulfide before chlorination and removal of carbon dioxide prior to softening).

6.4.1 Types of Aerators

Four types of aerators are in common use: (1) gravity aerators, (2) spray aerators, (3) diffusers, and (4) mechanical aerators.[1,8] A major design consideration for all types of aerators is to provide maximum interface between air and water at a minimum expenditure of energy.[1] A comparison of the performances of the various types of aerators is given in Table 6-2. A brief description of each type of aerator is provided here. Design information on aeration devices may be found in References 1, 9, and 10.

Table 6-2 Design and Operational Characteristics of Aerators

Aerator	Removal Rates	Design Specifications	Equipment
Gravity aerators			
Cascade	20–45% CO_2	Head: 1.0–3.0 m Area: 85–105 $m^2/m^2 \cdot s$ Flow velocity of approx. 0.3 m/s	Pump and concrete, or other corrosion-resistant flow surface baffles. Energy required to raise water
Packing Tower	>95% VOC[a] >90% CO_2	Max. column diameter 3 m Hydraulic loading: 2000 $m^3/m^2 \cdot d$	Tower structure, packing material, pump, and air blowers. Energy required to raise water and provide counter-current air flow.
Tray	>90% CO_2	Water rate: 0.8–1.5 m^3/m^2/min. Air requirement: 7.5 m^3/m^3 water Tray spacing: 30–75 cm Area: 50–160 $m^2/m^3 \cdot s$	Corrosion-resistant trays, tray media, pump, air blowers. Energy required to raise water and provide counter current air flow.
Spray	70–90% CO_2, 25–40 H_2S	Head: 1.2–9 m Nozzle diameter: 2.5–4.0 cm Nozzle spacing: 0.6–3.6 m Nozzle discharge: 5–10 L/s Basin area: 105–320 $m^2/m^3 \cdot s$ Spray pressure: around 70 kPa	Bronze or cast iron nozzles, piping, meters and valves. Energy required to operate pressurized spray pump.
Diffused aerators	80% VOC[a]	Detention time: 10–30 min. Air: 0.7–1.1 m^3/m^3 water Tank: 2.7–4.5 m deep, 3–9 m wide Width/depth < 2 Maximum volume = 150 m^3 Diffuser orifice: 2–5 mm diameter	Diffusers, air piping, air compressor, air filter, valves, meters, basin with drains. Energy for air compressor around 11–45 $kW/m^3 \cdot s$.
Mechanical aerators	50–80% CO_2	Detention time: 10–3 0 min. Tank: 2–4 m deep	Motor-driven impellers turbine aerators or draft tube aerators.

a Volatile Organic Compounds

Gravity Aerators Gravity aerators utilize weirs, waterfalls, cascades, inclined planes with riffle plates, vertical towers with updraft air, perforated tray towers, or packed towers filled with contact media such as coke or stone. Various types of gravity aerators are shown in Figure 6-9(a).

Spray Aerators Spray aerators spray droplets of water into the air from moving or stationary orifice or nozzles.[1] The water rises either vertically or at an angle and falls onto a collecting apron, a contact bed, or a collecting basin. Spray aerators are also designed as decorative fountains. To produce an atomizing jet, a large amount of power is required, and the water must be free of large solids. Losses from wind carryover and freezing in cold climates may cause serious problems. A typical spray aerator is shown in Figure 6-9(b).

Diffused-Air Aerators Water is aerated in large tanks. Compressed air is injected into the tank through porous diffuser plates, or tubes, or spargers. Ascending air bubbles cause turbulence and provide opportunity for exchange of volatile materials between air bubbles and water. Aeration periods vary from 10 to 30 min.[8] Air supply is generally 0.1 to 1 m^3 per min per m^3 of the tank volume.[8] Various types of diffused-aeration systems are shown in Figure 6-10(a).

Mechanical Aerators Mechanical aerators employ either motor-driven impellers or a combination of impeller with air-injection devices. Common types of devices are submerged paddles, surface paddles, propeller blades, turbine aerators, and draft-tube aerators. Various types of mechanical aerators are shown in Figure 6-10(b).

Aeration may be achieved in a reservoir by forced-draft tube or by turbine aerators. This is the first stage in treatment of stagnant-water or ice-covered reservoirs during winter months. These systems are also used to raise the warmer water from the bottom during winter and to prevent the surface from freezing.

6.4.2 Mechanics of Gas Transfer

Equilibrium Gases dissolved in a liquid will seek an equilibrium condition. The concentration of a gas dissolved in a liquid at equilibrium is called the saturation value (C_s). The saturation value of a gas depends upon the temperature of the liquid, the partial pressure of the gas, and the dissolved-solids concentration in the liquid. The saturation value is directly proportional to the partial pressure and inversely proportional to the temperature and dissolved-solids concentration.

The difference between the saturation value and the actual concentration provides the driving force for the exchange of gases from the gaseous state to the dissolved state and vice versa. The rate of exchange is directly proportional to the difference between the actual concentration and the saturation value.

Gas Transfer The rate of gas transfer across a liquid-gas interface is commonly expressed by Eq. (6.1).[8] Integration of Eq. (6.1) yields Eq. (6.2).

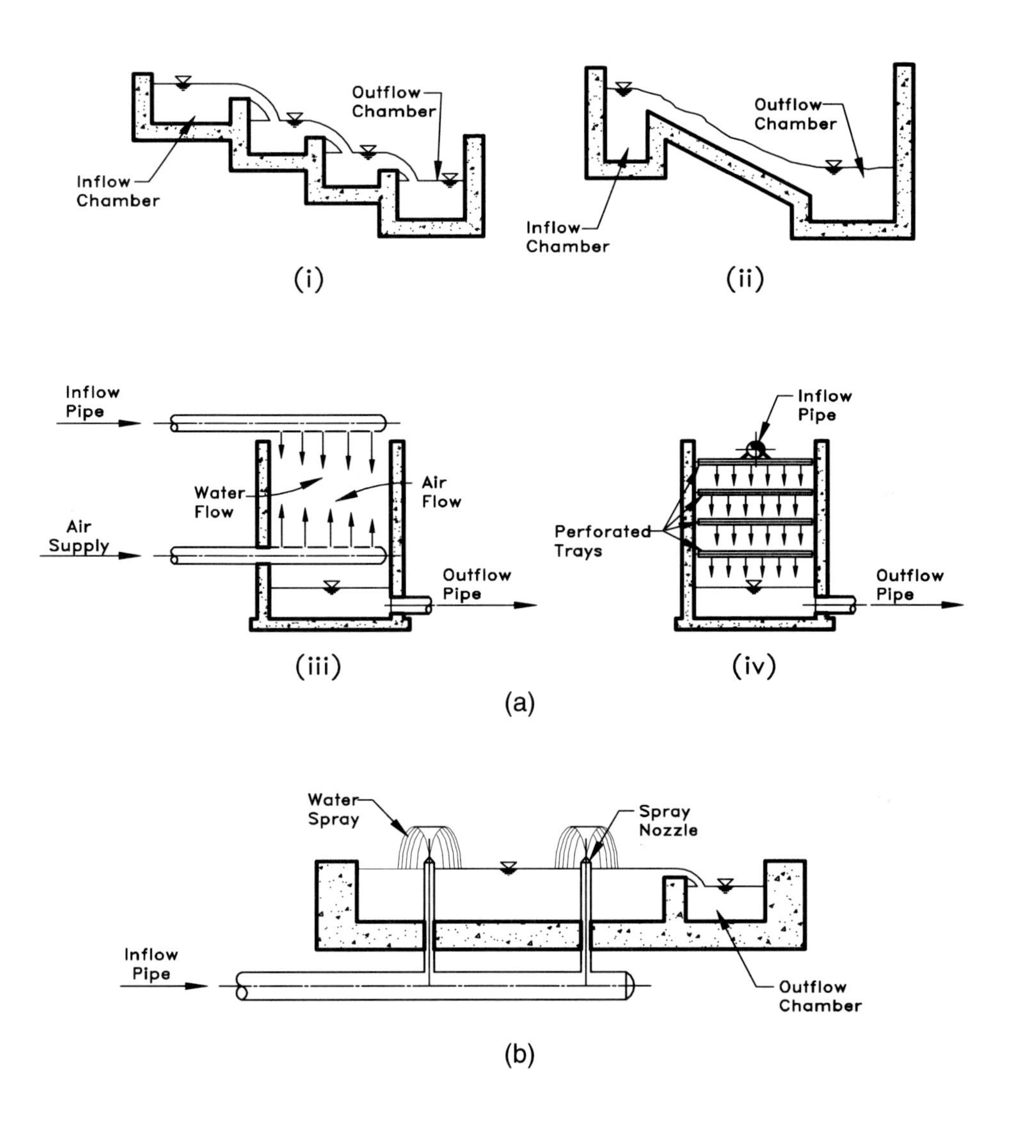

Figure 6-9 Various types of gravity and spray aerators. (a) Gravity aerators: (i) cascade, (ii) inclined apron possibly studded with riffle plate, (iii) tower with counter-current flow of air and water, (iv) stack of perforated pans possibly containing contact media. (b) Spray aerator.

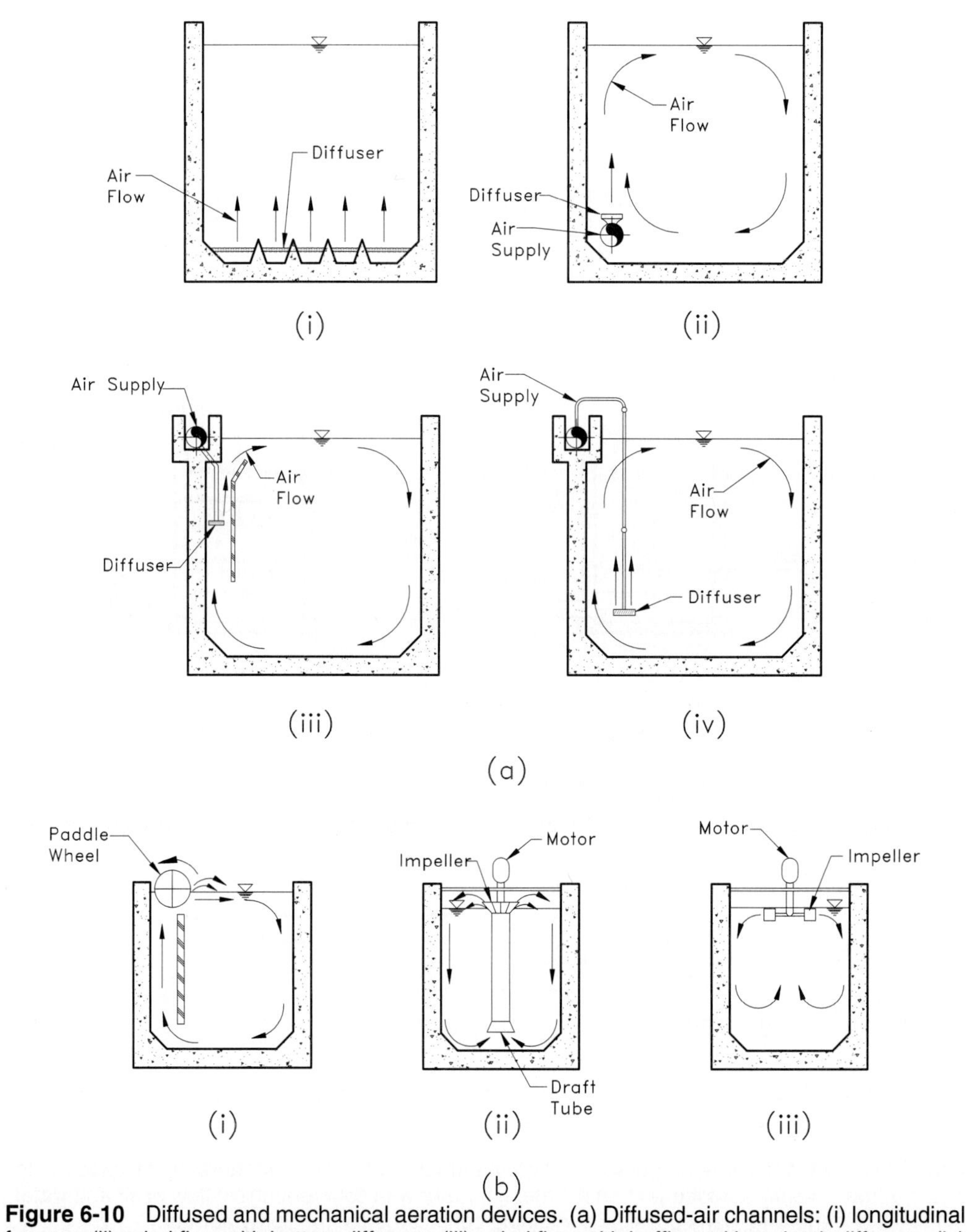

Figure 6-10 Diffused and mechanical aeration devices. (a) Diffused-air channels: (i) longitudinal furrows, (ii) spiral flow with bottom diffusers, (iii) spiral flow with baffle and low-depth diffusers, (iv) swing diffusers. (b) Mechanical aeration units: (i) surface paddles, (ii) draft-tube turbine type, (iii) turbine aerator.

$$\frac{dC}{dt} = K_L a(C_s - C_0) \tag{6.1}$$

$$(C_s - C) = (C_s - C_0)e^{-K_L a \cdot t} \tag{6.2}$$

where

dC/dt = rate of change in concentration (mg/L·s)
$K_L a$ = overall mass-transfer coefficient, 1/s
C_s = saturation concentration, mg/L
C = concentration at any time t, mg/L
C_0 = initial concentration at t = 0, mg/L

The gas-transfer coefficient $K_L a$ is a variable value and depends upon many complex interrelationships, including temperature, area through which gas is diffused, volume of liquid in contact, and coefficient of gas diffusion. It is best determined experimentally for a given gas and aeration condition. The temperature correction for $K_L a$ is expressed by Eq. (6.3).

$$(K_L a)_T = (K_L a)_{20} \times \theta^{T-20} \tag{6.3}$$

where

$(K_L a)_T$ = gas-transfer coefficient at temperature T, 1/s.
$(K_L a)_{20}$ = gas-transfer coefficient at temperature 20°C, 1/s.
θ = temperature-correction coefficient that can vary with test condition (See Problem 6.6.)

From Eqs. (6.1) and (6.2), the following key design considerations for aerators are developed.

1. The gas-transfer rate is directly proportional to the exposure area per unit volume. An ideal aeration device will maximize the area of exposure. In diffused aerators, for a given air supply, fine bubbles offer a greater area of exposure than coarse bubbles. In cascade aerators, a larger length of overflow weir increases the area. In spray aerators, the nozzle that produces a finer spray gives the larger area of exposure.
2. The rate of gas transfer is also directly proportional to the time of exposure; therefore, an aeration device should maximize exposure time. For example, the rising velocity of air bubbles in a diffused-aeration system is generally twenty percent (one-fifth) of the falling velocity in spray aerators. If all other factors were equal, the rate of oxygen transfer would be five times greater for the diffused aerators than for the spray aerators.
3. The rate of gas transfer is directly proportional to the difference between the saturation value and the initial concentration $(C_s - C_0)$. The saturation value depends upon the partial pressure of gas, temperature of the liquid, and dissolved-solids concentration in the liquid, so a change in any of these parameters will affect the rate of gas transfer. For example, by

increasing the percentage of oxygen in the air mixture, the partial pressure is increased, resulting in a higher saturation value. This situation is encountered with pure-oxygen or high-oxygen diffuser systems.

4. If C_0 is greater than C_s (typical case of groundwater supersaturated with CO_2 or H_2S), the right side of Eq. (6.2) will yield a negative value. This represents desorption, or stripping action. Thus, Eq. (6.2) can also be applied to desorption.

6.4.3 Application of Aeration

Taste and Odor Removal Aeration has a limited value in taste and odor removal. Most taste- and odor-causing compounds have a high solubility in water. Therefore, with some exceptions, aeration is generally less efficient in removal of tastes and odors than other methods, such as chemical oxidation or adsorption.[8,9,11]

Iron and Manganese Removal Iron and manganese removal is typically accomplished by oxidizing the iron and manganese ions into insoluble oxides. The application of aeration in this process is to provide a sufficient amount of oxygen for this reaction to occur. This process is used most commonly in groundwater applications, where the dissolved oxygen of the water is low. Therefore, aeration in this application will result in oxidation of iron and manganese, causing their precipitation and an increase in the dissolved-oxygen concentration.[12] Manganese often cannot be readily oxidized at normal pH levels. Increasing the pH to 8.5 will enhance manganese oxidation, particularly if a packed-tower aerator with coke-bed packing is used.[13] The iron and manganese oxidation reactions are given in Section 8.3.3.

Volatile Organic Compounds Removal The 1986 amendments to the Safe Drinking Water Act imposed new limitations on a class of organic compounds known as *volatile organic compounds (VOCs)*. Most of these chemicals can readily be removed by aeration. Consideration must be given in the design of aeration units to removal of VOCs in the exhaust air; some of these compounds are also subject to maximum ambient air-concentration limitations.

Carbon Dioxide Removal Carbon dioxide can readily be removed by aeration. Carbon dioxide has a low solubility in water; therefore, aeration is very efficient in its removal. This process is usually applied prior to softening of groundwater, because high concentrations of carbon dioxide are generally encountered in groundwater. Since high concentration of carbon dioxide increases the chemical requirements for softening, so there are benefits to carbon dioxide removal prior to softening.[8,9,11]

Hydrogen Sulfide Removal Hydrogen sulfide is an important taste- and odor-causing compound that can be treated effectively by aeration. The mechanism of treatment in this case is primarily oxidation of hydrogen sulfide, resulting in water and free sulfur. In some cases, free sulfur can interfere with other treatment processes.

Hydrogen sulfide can be more effectively removed by initially aerating the water with a high concentration of carbon dioxide. In this variation of the process, conditions are more favorable for the removal of hydrogen sulfide.[8,9,11]

6.5 Design of Intakes and Screens

The hydraulic considerations in intake structure design are energy losses due to the acceleration and deceleration of water at bar racks, intake ports, and fine screens. The low velocities of a properly designed intake typically make these losses very small; however, a designer should perform calculations to assure that the system will function properly. The losses through the intake port can be calculated by using the orifice equation (Eq. (6.4)).[7,14–16] The orifice area is the submerged effective gate area.

$$h_L = \frac{1}{2g}\left(\frac{v}{C_d}\right)^2 \tag{6.4}$$

where

h_L = head loss, m (ft)
g = acceleration due to gravity, m/s^2(ft/s^2)
v = velocity, m/s (fps)
C_d = coefficient of discharge for orifice (usually 0.6–0.9)[16]
A = effective submerged orifice area, m^2(ft^2)

Losses through screens can be calculated using Eq. (6.5) for either clean or dirty screens.[7,14,15]

$$h_L = \frac{(v^2 - v_v^2)}{2g} \times \frac{1}{0.7} \tag{6.5}$$

where

h_L = head loss through the screen, m (ft)
v = velocity through the screen opening, m/s (fps)
v_v = velocity upstream of the screen (0 in most cases), m/s (fps)
g = acceleration due to gravity, m/s^2 (ft/s^2)

Losses through a fine screen can also be calculated from Eq. (6.5) or from the orifice equation (Eq. (6.4)).

6.6 Equipment Manufacturers of Intake, Screening, and Aeration Devices

A list of manufacturers of gates, screens, microstrainers, aeration, and other equipment for intake structures is provided in Appendix C. The equipment selected for water supply projects should meet the project requirements for flexibility of operation, maintainability, and durability, while meeting the primary design requirements for the unit process.

6.7 Information Checklist for Design of Raw Water Intake and Screen

The following information should be obtained and necessary decisions made before the design engineer begins designing a raw water intake and screening facilities.

1. Information to be obtained from pre-design report:

   a. design flow: maximum day demand;
   b. location of the proposed intake;
   c. minimum and maximum water surface elevations of the raw water source.

2. Information to be obtained from the preliminary design; the following information may be available as a part of the pre-design report often submitted to the client for review and approval:

   a. preliminary site plan of the raw water intake structure;
   b. preliminary layout of the intake structure;
   c. type of screening equipment;
   d. wetlands, Section 404, and/or Section 10 Permit requirements, as applicable.

3. Information to be obtained from the plant owner and operator:

   a. any particular preferences with regard to equipment types and manufacturers;
   b. any particular preferences with regard to structural, mechanical, and electrical design of the raw water intake.

4. Information to be obtained from field investigations:

   a. topography of the proposed site;
   b. water current patterns and velocities in the reservoir or river near the intake structure;
   c. geological conditions at the intake structure (this information should include the level of groundwater and location of rock formations).

5. Information to be developed by the design engineering team:

   a. design criteria established by the regulatory authorities;
   b. types, sizes, and limitations of available equipment.

6.8 Design Example

6.8.1 Design Criteria Used

The following design criteria are used to develop the design of the raw water intake and the screening facility.

1. Hydraulic data

   a. Design flow (maximum day demand) = 113,500 m^3/day (30 mgd). Although the maximum raw water flow in Figure 5-6 is 110,313 m^3/d, this value is obtained by subtracting the return flow from the backwash-recovery basin from the design maximum day demand. The return flow is a variable flow and cannot be relied upon. Therefore, the intake structure should be designed for maximum day demand.
   b. Minimum reservoir elevation = 70.00 m (mean sea level [msl]).
   c. Maximum reservoir elevation = 90.00 m (msl).
   d. Normal water surface elevation = 85.00 m (msl).
   e. Bottom elevation = 60.00 m (msl).

2. Location and configuration of headworks

   The proposed raw water intake is located in the lake and connected by a bridge to the on-shore pump station. The headworks arrangement is shown in Figure 6-11.
3. General design guidelines

   a. The raw water intake shall be located in the lake such that it will withdraw water from maximum and minimum elevations of 90.00 m and 70.00 m.
   b. The raw water intake shall be a dry-tower design, with several gates to allow the water to be selectively withdrawn from the level or levels that yields the most desirable quality of water.
   c. A coarse screen shall be provided at each gate to prevent large objects from entering the intake. The velocity through the rack should be less than 8 cm/s (0.3 fps).
   d. A mechanically-cleaned fine screen shall be provided at the pump station in order to prevent fish and trash from entering the wet well of the pump station. The velocity through the fine screen shall be less than 0.2 m/s (0.6 fps).
   e. Aeration of raw water at the head works is not required.

6.8.2 Design Calculations

Step A: Unit Arrangement and Layout The preliminary layout of the raw water intake, raw water conduit, fine screens, and raw water pump station are shown on Figure 6-11. Calculations for the raw water conduit and pump station are provided in Chapter 7. The following designs and calculations are for the raw water intake.

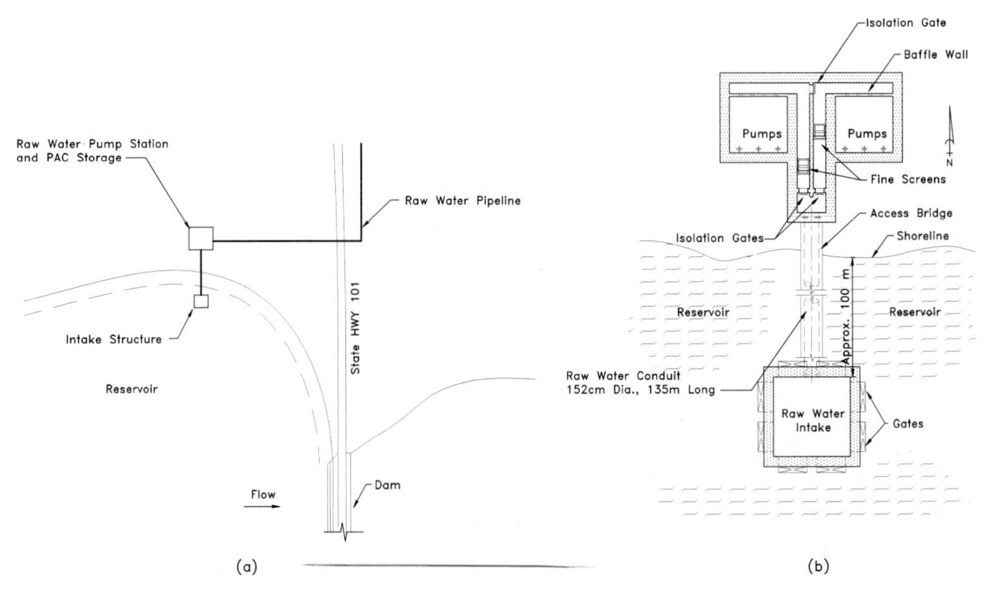

Figure 6-11 Location and layout of intake structure. (a) Location of intake structure. (b) Connecting conduit, screens, and raw water pump station. (See Figure 7-12 for additional details.)

Step B: Design of Intake Structure

1. Select the size of the intake gate.

The dimensions of the gates will set the minimum requirements for the size of the intake tower. The gates are sized such that the entire maximum flow of 113,500 m^3/day can be withdrawn from a single level at a maximum velocity of 0.08 m/s (0.3 fps). Water withdrawn from the selected level should provide uniform and best raw water quality to be treated.

$$Q = 113,500 \text{ m}^3/\text{day} = 1.31 \text{ m}^3/\text{s}$$

$$A = \frac{1.31 \text{ m}^3/\text{s}}{0.08 \text{ m/s}} = 16.38 \text{ m}^2$$

This is too large for a single gate, so select two equal-size square gates.

$$\text{Width} = (16.38 \text{ m}^2/2)^{1/2} = 2.86 \text{ m}$$

Select the next-larger standard-size gate from the manufacturer's catalog. The selected gate has:

$$\text{Width and Height} = 3.00 \text{ m (120 in)}$$

$$\text{Velocity through the gate} = \frac{1.31\ \text{m}^3/\text{s}}{3.0\ \text{m} \times 3.0\ \text{m}} = 0.07\,\text{m/s}$$

2. Determine the layout of the intake gates.

   Set the highest gate with its top two meters below the normal water-surface elevation of 85.00 m, resulting in a centerline elevation of 81.50 m. Likewise, set the lowest gate at a centerline elevation of 65.00 m, providing 3.5 m of head at the 100-year drought elevation of 70.00 m. At this elevation, the bottom of the lowest gate is 3.5 m above the reservoir bottom. In order to provide the flexibility to withdraw water from intermediate elevations, provide additional gates at two levels equally spaced over the 16.5 meter range (81.50 m–65.00 m).

$$\text{Spacing} = \frac{16.5\ \text{m}}{3\ \text{spaces}} = 5.5\ \text{m/space}$$

   Therefore, gates will be provided at centerline elevations of 81.50, 76.00, 70.50, and 65.00 m.

   Locate two gates on each side of the intake tower. Each gate is slightly more than 3 m wide, so set the width of the tower wall at 8 m. This will provide approximately 0.5 m between the gate and the wall, and approximately 1.0 m between gates (Figure 6-12). The overall dimensions of the tower are 10 m × 10 m. The gate elevations are staggered, so that the two gates on each side of the structure will be set at different elevations. This will result in less weakening of the structure by a large opening in a single plane. Figure 6-12 illustrates the proposed layout of the intake tower. The centerline elevations of each gate are summarized in Table 6-3.

Step C: Design of Coarse Screen:

1. Select layout of the coarse screen.

   The coarse screen will be located at the intake ports, slightly projected away from the intake gate. This will prevent debris from interfering with the operation of the gate. The coarse screen will be constructed of flat bars attached to a concrete seat projected from the intake tower. The seat projects from the wall a sufficient distance to allow for operation of the gate. Figure 6-13 illustrates the layout of the gate and coarse-screen arrangement.

2. Select bar arrangement.

   Use 13 mm (0.5 in) square edge bars, 4.8 m long, spaced at approximately 8 cm (3.1 in) on centers. This provides a clear opening of 6.7 cm (2.6 in). This size opening will preclude large debris from entering the intake. Small suspended debris will enter the gate and will be removed at the fine screen. The number of bars required is computed as follows. The bars cover 3.6 m width over the gate.

$$\text{Number of spaces} = \frac{360\ \text{cm}}{8.0\ \text{cm/space}} = 45\ \text{spaces}$$

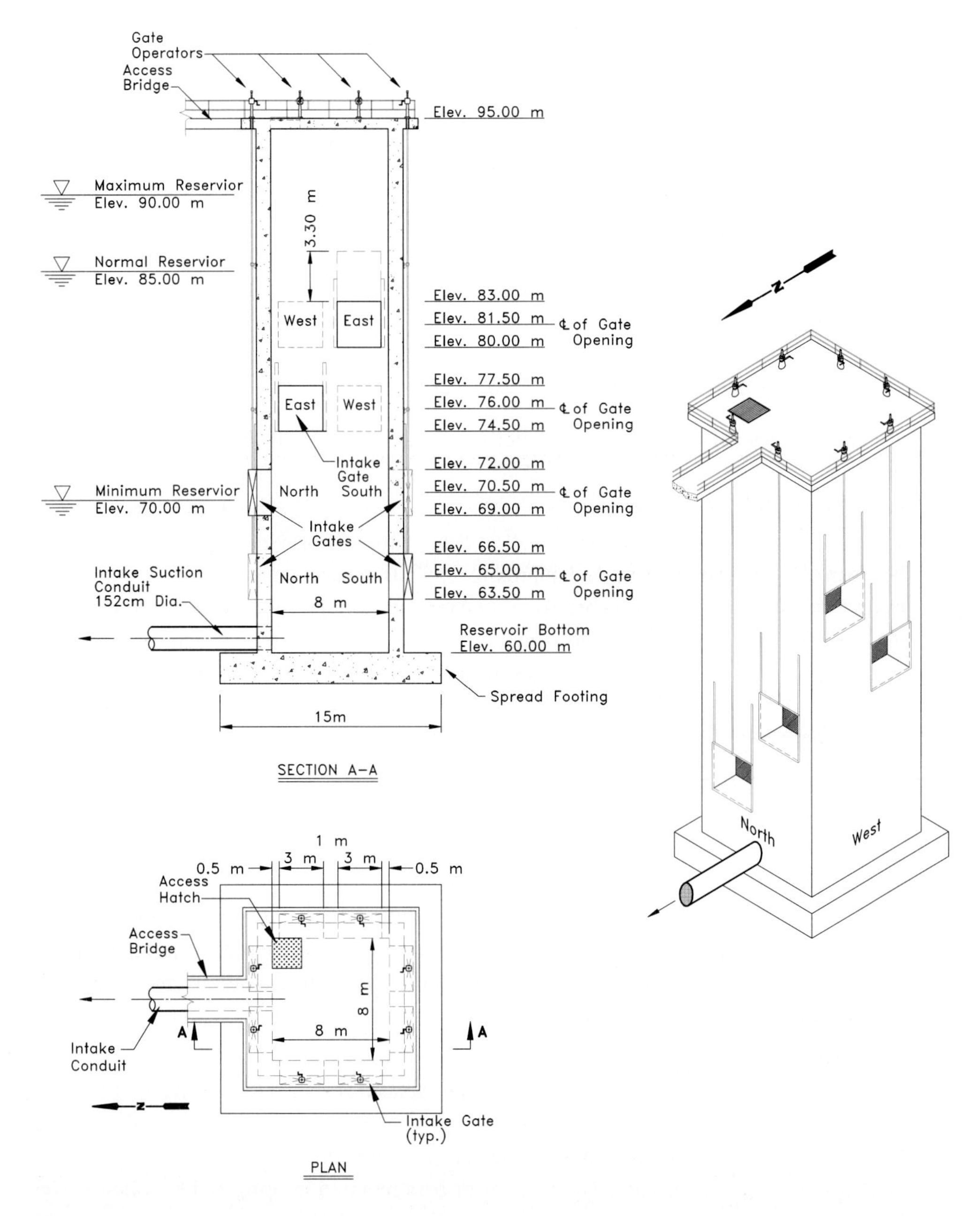

Figure 6-12 Example intake-structure layout.

Table 6-3 Intake-gate Layout

No.	Location	Centerline Elevation, m
1	East	81.50
2	West	81.50
3	East	76.00
4	West	76.00
5	North	70.50
6	South	70.50
7	North	65.00
8	South	65.00

The number of bars is equal to the number of spaces minus 1, or 44. The bar arrangement is shown in Figure 6-13.

3. Calculate the velocity through the bar rack.

The open area through the rack is the total area, minus the area of the bars. Refer to Figure 6-13.

$$\text{Area of the rack} = 3.6 \text{ m} \times 4.8 \text{ m} = 17.28 \text{ m}^2$$

$$\text{Area of the bars} = 44 \text{ bars} \times 0.013 \text{ m} \times 4.8 \text{ m} = 2.75 \text{ m}^2$$

$$\text{Open area} = 17.28 \text{ m}^2 - 2.75 \text{ m}^2 = 14.53 \text{ m}^2 \quad (4.8 \text{ m} \times 3.028 \text{ m})$$

Maximum flow through the rack is half of the design flow because there are two gates at each level.

$$\text{Velocity} = \frac{1.31 \text{ m}^3/\text{s}}{2} \times \frac{1}{14.53 \text{ m}^2} = 0.0451 \text{ m/s (0.15 fps)}$$

This is within the acceptable velocity.

Step D: Design of Fine Screen

1. Select the location of the fine screen.

The fine screens are located at the pump station. This location provides better access for maintenance and removal of screenings than does the location at the intake tower. The design procedure of fine screens is covered in this section to keep the necessary details of coarse and fine screens together. The information developed here on fine screens is used in Chapter 7.

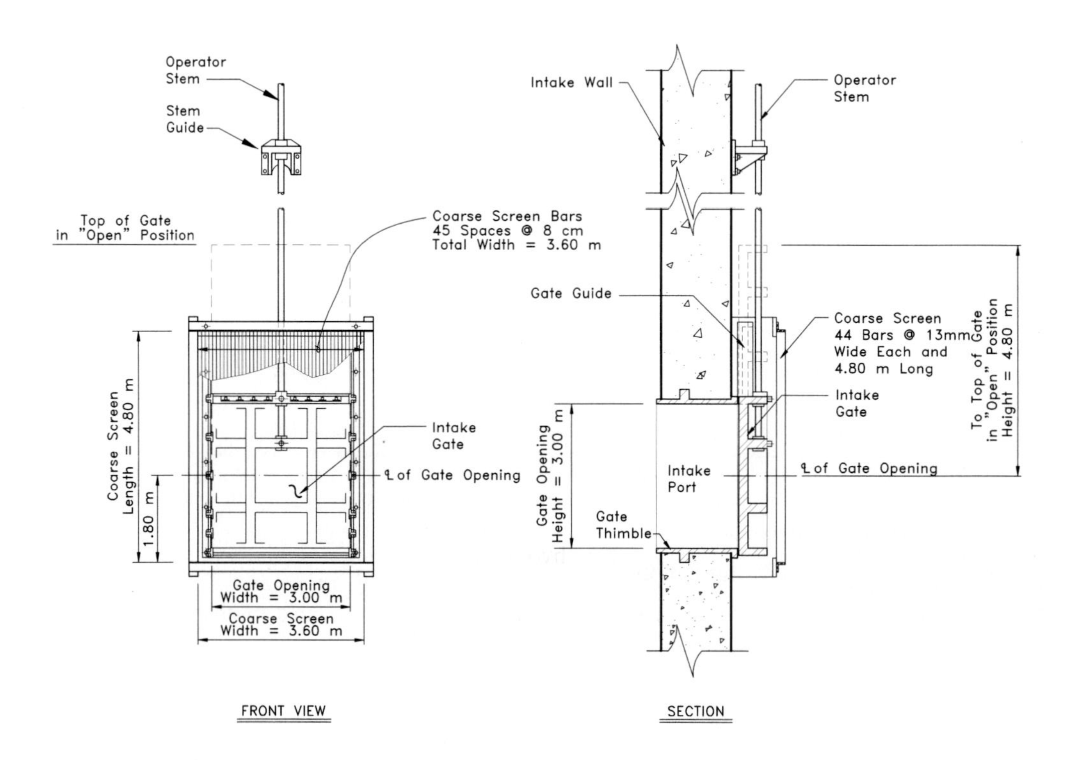

Figure 6-13 Coarse-screen dimensions.

2. Select screening equipment.

   Provide two fine screens, each with a 9.5-mm (3/8-in) opening. This size of opening will remove objects that may harm the pumps but will not result in a heavy accumulation of solids. The depth of flow at the screen chamber is 9.53 m at the minimum reservoir level of 70.00 m. (See Figure 7-11 for details.) A maximum velocity of 0.2 m/s (0.6 fps) through the screen is assumed. (See Design Criteria.) The recommended screen efficiency factor is 0.56 for stainless-steel screens. The design flow through one screen is 0.66 m^3/s under normal conditions.

3. Calculate width of screen.

$$\text{Width} = \frac{0.66 \text{ m}^3/\text{s}}{9.53 \text{ m} \times 0.2 \text{ m/s} \times 0.56} = 0.62 \text{ m}$$

   a. From the manufacturer's catalogs, the smallest screen available is 0.9 m (3 ft). Use two 0.9 m screens.[4] The width of the screen channel will be 1.3 m, to allow for guide rack. The fine screen is provided at the pump station, so the layout of the screens is included with the layout of the wet well of the pump station (Chapter 7). General arrangement is shown in Figure 6-11(b).

b. Calculate the actual velocity through the screen.

$$\text{Velocity through the screen} = \frac{0.66\ \text{m}^3/\text{s}}{9.53\ \text{m} \times 0.9\ \text{m} \times 0.56} = 0.14\ \text{m/s}$$

This velocity is acceptable through the fine screen under normal conditions when fish protection is desired. Under normal conditions with two screens, the velocity through a screen will be 0.14 m/s.

4. Description of mechanical cleaning equipment.

The traveling mechanical cleaning screen removes floating debris on an endless series of baskets.[4] The operation could be intermittent or continuous on water entering the baskets that are mounted on two strands of chain. As the screen revolves, the debris is lifted by the upward travel of the baskets. At the head of the screen, the debris is removed from the basket by powerful jet sprays of water. The debris is washed into a trough and collected in a container.

Step E: Hydraulic Calculations The hydraulic calculations presented in this chapter include the head losses through the coarse screens, intake ports, and fine screens. Calculations of other head losses, including the raw water conduit and pump station losses, are presented in Chapter 7.

1. Calculate the head loss through the coarse screen.

The approach velocity in the reservoir is assumed to be zero. The head loss through the rack is calculated from Eq. (6.5).

$$h_L = \frac{(0.0451\ \text{m/s})^2 - (0^2)(1/0.7)}{0.7 \times 2 \times (9.81\ \text{m/s}^2)} = 0.0001\ \text{m}$$

2. Calculate the head loss through the intake port.

The head loss through the intake port can be calculated by the orifice equation, Eq. (6.4). The velocities are extremely small, so the effective area is assumed equal to the total port-opening area.

The flow will be one-half of the total flow of 1.31 m^3/s or 0.66 m^3/s. The orifice coefficient is assumed to be 0.6.

$$h_L = \frac{1}{2 \times (9.81\ \text{m/s}^2)} \times \left(\frac{0.66\ \text{m}^3/\text{s}}{0.6 \times 9\ \text{m}^2}\right)^2 = 0.0008\ \text{m}$$

3. Calculate the head loss through the fine screens.

The head loss through the fine screens will also be calculated from the orifice equation, Eq. (6.4).

The total flow of 1.31 m^3/s will be used for maximum head loss calculations. This condition will develop when one screen is out of service, five pumps are running, and the other screen must pass the entire flow.

The coefficient of discharge (C_d) is assumed to be 0.6. The manufacturer's area reduction factor is 0.56 for screens with 6.4 mm openings.[4]

$$h_L = \frac{1}{2 \times (9.81 \text{ m/s}^2)} \times \left(\frac{1.31 \text{ m}^3/\text{s}}{(0.6 \times 9.53 \text{ m} \times 0.9 \text{ m} \times 0.56)} \right)^2 = 0.011 \text{ m}$$

4. Determine the total head loss through port and screens.

The head loss through ports and coarse and fine screens, as calculated above, totals

$$0.0001 \text{ m} + 0.0008 \text{ m} + 0.011 \text{ m} = 0.0119 \text{ m}$$

It may be noted that the major head loss occurs through the fine screen and that the total head loss is small.

Step F: Stability Calculations The hydraulic stability of the intake tower must be checked. The tower displaces a large amount of water. To ensure structural stability, the weight of the structure must be greater than the weight of water displaced. The worst condition for the hydraulic stability will occur when the tower is empty and the reservoir is at its maximum level of 90 m elevation. The outside and inside dimensions of the tower are 10 m and 8 m square, respectively, and the height of the tower is 35 m (95.00 m – 60.00 m). The water depth is 30 m (maximum reservoir elevation, 90.00 m, minus minimum reservoir elevation, 60.00 m). The base slab is 15 m × 15 m × 2.5 m (thick).

1. Calculate the uplift forces.

$$\text{Volume of water displaced by tower and base slab} = 10 \text{ m} \times 10 \text{ m} \times 30 \text{ m} + 15 \text{ m} \times 15 \text{ m} \times 2.5 \text{ m} = 3563 \text{ m}^3$$

$$\text{The weight of water displaced} = 1000 \text{ kg/m}^3 \times 3563 \text{ m}^3 = 3.56 \times 10^6 \text{ kg}$$

The weight of the structure must be more than 3.56×10^6 kg in order to provide stability.

2. Calculate the weight of structure (unit weight of concrete is 2308 kg/m^3 (144 lb/ft^3)).

$$\text{Weight of the side walls} = [(10 \text{ m} \times 10 \text{ m} \times 35 \text{ m}) - (8 \text{ m} \times 8 \text{ m} \times 35 \text{ m})] \times (2308 \text{ kg/m}^3) = 2.91 \times 10^6 \text{ kg}$$

$$\text{Weight of the foundation slab} = (15 \text{ m} \times 15 \text{ m} \times 2.5 \text{ m}) \times 2308 \text{ kg/m}^3 = 1.30 \times 10^6 \text{ kg}$$

$$\text{Total downward weight} = 2.91 \times 10^6 \text{ kg} + 1.30 \times 10^6 \text{ kg} = 4.21 \times 10^6 \text{ kg}$$

3. Calculate the factor of safety.

$$\text{Safety factor} = \frac{4.21 \times 10^6 \text{ kg}}{3.56 \times 10^6 \text{ kg}} = 1.2$$

In order to ensure stability, the factor of safety should be in the range of from 1.5 to 2.0. The factor of safety of this structure can be increased by adding additional weight to the structure or by anchoring the footing by the use of piles or drilled piers embedded into a strong foundation material such as bed rock.

Step G: Instrumentation and Control Ultrasonic transducers will be provided to measure the water-surface elevations in the lake, inside the intake tower, and upstream and downstream of the fine screen. Each intake position will be monitored by limit switches. The fine-screen rake will have time-clock operation with high-water level override. The rake motor will have a torque monitor. Additional information on instrumentation and control may be found in Chapter 16.

6.9 Operation, Maintenance, and Troubleshooting for Raw Water Intake and Screens

Raw water intake, gates, and screens are mechanically simple, for easy maintenance. If properly maintained, these systems will operate efficiently with few mechanical problems.

6.9.1 Common Operational Problems and Troubleshooting Guide

There are many common operational problems that can occur at the raw water intake, the coarse screen, and the fine screen. Important problems and corrective procedures are listed here.

1. Raw Water Intake and Coarse Screen

   The purpose for the raw water intakes is to selectively withdraw optimum-quality water from a surface water source. Typically, intakes consist of a structure and gates to control the withdrawal of water.

   a. Odorous or foul-tasting water may be due to poor-quality water. Test other strata for better-quality water and operate intake gates accordingly. Reservoir stratification is most severe during the summer months. Typically, the thermocline develops at an approximate depth of 8 m. During this condition, the dissolved oxygen (DO) concentration is low in the hypolimnion (lower stratum). During summer months, the lower intake gates should not to be opened to withdraw water from the hypolimnion.

   b. Logs and debris lodged near the intake indicate excessive intake velocities. Some accumulation may be unavoidable in most cases, but it can be minimized by reducing intake velocities. This can be achieved by opening additional gates.

c. Extreme force required to operate gates indicates corrosion of gate guides. To prevent jamming, operate gates through one cycle monthly.
d. If icing clogs the intake port, heat the metal bars with steam or other heat sources. An optional solution is to utilize a deeper intake port that may receive warmer water at, typically, 4°C.

2. Fine Screen

Screens remove suspended material from water. Screens, if not properly maintained, can cause serious operational problems and can result in damage to the pumping equipment. Fine screens are provided ahead of the pumps at the raw water pump station.

a. Excessive head loss is indicative either of improper cleaning or of slime growth on the screen. Check automatic cleaning system for proper operation and slime build-up.
b. Large debris can jam traveling-screen mechanisms. These mechanisms are typically equipped with overload protection, which will inactivate the screen during overload conditions. To correct this problem, remove debris and reset motor.
c. Broken links in the traveling screen will cause the screen to be inoperative even though the motor continues to run. Inspect chain and linkages.
d. Excessive slime build-up will result in excessive head loss and can impart a taste and/or odor to the water. If slime build-up is persistent, an ultraviolet-light illumination device will inhibit further build-up.

6.9.2 Routine Operation and Maintenance

1. Intake Structure and Coarse Screen

The following preventive-maintenance procedures are necessary to ensure satisfactory operation.

a. Operate gates through one cycle at least once in a month.
b. Clean debris from the intake area at least four times a year. Clean the area more often if accumulation is excessive.
c. Inspect foundation for scour at least annually.

2. Fine Screens and Microstrainers

a. Monitor head loss through screens.
b. Lubricate bearing as recommended by equipment manufacturers.
c. Check cleaning system periodically to ensure proper operation.
d. Remove screenings daily and dispose of in an approved landfill.

6.10 Specifications

Brief specifications for the raw water intake and screens designed above are presented below and are provided to describe the design more fully. Many details have been omitted; therefore, these specifications should be used only as a guide to preparing more detailed specifications. Manufacturers' representatives should be consulted during the preparation of every equipment specification.

6.10.1 Intake

The contractor shall construct a dry-tower raw water intake with eight 3.0 × 3.0-meter gates. The structure shall be constructed of concrete. The foundation shall have spread footings, to provide lateral stability to the structure. The structure shall be designed to resist an uplift force of 3.0×10^6 kg with a safety factor of at least 1.5.

The gates shall be cast-iron sluice gates with bronze guides, seats, and wedges. The gate stem shall be stainless steel and designed with sufficient diameter and stem guides to prevent buckling. The gate operator shall be a floor-stand type with a hand crank and be located on the operator platform on top of the intake structure.

6.10.2 Coarse Screen

Coarse screens shall be installed outside of each intake port. Each screen shall consist of 39 13mm-square American Society of Testing Materials (ASTM) A36 steel bars. The bars shall be equally spaced and installed vertically. They shall be welded to a 13mm × 40mm A36 steel frame bolted to the structure with stainless steel bolts.

6.10.3 Fine Screen

Install two 0.9-meter-wide traveling water screens. The screens shall have 6.4-mm-square openings made from 0.27-mm galvanized steel wire. The chain shall be cast-iron, with hardened steel bushings. The chain sprockets shall be made of cast iron, with bronze bearings. The screen shall be powered by a 440-volt three-phase motor. The screen shall be equipped with a spray-wash system and a fiberglass head house. The screening shall be discharged into a removable hopper.

Problems and Discussion Topics

6.1 A water treatment plant has a design flow rate of 37,800 m^3/day (10 mgd). Design an aeration tank with dual chambers, 13.0 m depth, a 2-to-1 length-to-width ratio, and a 15-minute detention time. Estimate the total air requirements to supply 0.5 m^3/min. of air per cubic meter of the tank.

6.2 Estimate the head loss through a 2-m-wide (6.56 ft) vertical-bar screen with seventy-eight 9.5 mm (0.375 in) square-edge bars, if the flow rate is 113,500 m^3/day (30 mgd) and the depth of flow is 10.0 m. Assume that the approach velocity is zero and the coefficient of discharge is 0.68. Use Eq. (6.4).

6.3 In the design example, an intake tower was designed for a maximum flow rate of 113,500 m^3/day (30 mgd). Determine the impact of increasing the peak flow rate to 170,250 m^3/day (45 mgd). Discuss the alternatives available to implement this increase.

6.4 In the design example, the fine screen was designed for a maximum flow rate of 113,500 m^3/day (30 mgd). Determine the impact of increasing the peak flow rate to 170,250 m^3/day (45 mgd). Discuss the alternatives available to implement this increase.

6.5 A raw water source has 10 μg/L of a chemical compound. The allowable concentration of the compound in the finished water is 5 μg/L. The table below provides the results of a pilot test conducted on the removal of the compound. Determine the capacity of the aeration basin required to remove the compound to meet the allowable limit. What size basin would you recommend? Explain your answer.

Time min	Concentration μg/L	Percent Removal
0	10	0.0
1	9	10.0
2	8.2	18.0
3	7.7	23.0
4	7.2	27.0
5	7.0	30
7.5	6.3	37
10	5.5	45
15	4.5	55
20	4.0	60
30	3.0	70

6.6 An aeration study was conducted to establish the removal of VOC from groundwater. The water temperature at the time of study was 23°C, and initial concentration of VOC in the groundwater was 12 μg/L. The results of the study are summarized in the table below. Calculate K_La at 23°C and at 20°C, and the aeration period to achieve a residual concentration of 6 μg/L of VOC in the aerated water. Assume C_s at 23°C is 3 μg/L. The temperature correction coefficient, θ, is 1.026.

t, min	*C*, μg/L
0	12.0
2.5	11.0
5.0	10.0
7.5	9.5
10.0	8.4
15.0	7.2
20.0	6.2
30.0	5.0
40.0	4.3

6.7 Visit a local water treatment plant. Determine the type and size of the raw water intake, screening, and aeration facilities. Prepare sketches of the facility layout. Ask the operator to explain any operational problems experienced with these facilities.

6.8 The Design Example utilized a dry-type tower. Ballast for stability was provided by structure weight. Discuss alternatives to relying on weight of the structure for stability.

6.9 Calculate the flotation forces on a dry-tower intake with outside dimensions of 15 m × 8 m and a maximum water depth of 17 m.

6.10 Design an intake tower to meet the following requirements:

- Maximum water surface elevation = 60 m
- Minimum water surface elevation = 45 m
- Reservoir bottom elevation = 42 m
- Number of gates at each elevation = 2
- Number of gate levels = 2
- Flow rate = 20,000 m^3/day

Calculate intake port size, location of ports, and head loss through the intake port. Make a sketch illustrating your design. Normal water surface elevation is 57.50 m.

6.11 Design a fine screen to meet the following requirements:

- Flow rate = 20,000 m^3/day
- Screen opening = 9.5 mm
- Area reduction factor = 0.35
- Maximum velocity = 0.15 m/s
- Depth of flow = 6 m
- Discharge coefficient = 0.6

Calculate screen width and head loss through the screen.

References

1. Fair, G. M., Geyer, J. C., and Okum, D.A. *Water Purification and Wastewater Treatment and Disposal, Vol. II*, John Wiley & Sons, New York, 1968.
2. Hydraulic Institute, Inc. *American National Standard for Pump Intake Design*, ANSI/HI 9.8-1998, Hydraulic Institute, Inc., Parsippany, NJ, 1998.
3. American Public Works Association. *The 1971 Environmental Waste Control Manual and Catalog File*, Public Works Journal Corp., Ridgewood, NJ, 1971.
4. U.S. Filter/Envirex. *Rex Water Intake Screens*, Bulletin 315-331, 1/90-3M, Envirex, Waukesha, WI, Undated.
5. Kawamura, S. *Integrated Design of Water Treatment Facilities*, John Wiley and Sons, Inc., New York, 1991.
6. U.S. Filter/Johnson Screens. *Johnson Surface Water Screens*, Product Literature, 1987.
7. Qasim, S. R. *Wastewater Treatment Plants Planning, Design and Operation,* 2d ed., Technomic Publishing Co., Lancaster, PA, 1999.
8. ASCE and AWWA. *Water Treatment Plant Design*, 2d ed., McGraw-Hill Book Co., New York, 1990.
9. Scott, A. R. *Aeration in Water Quality and Treatment*, 3d ed., McGraw-Hill Book Co., New York, 1971.
10. Tchobanoglous, G. and Schroeder, E. D. *Water Quality: Characteristics, Modeling and Modification,* Addison-Wesley Publishing Co., Reading, MA, 1985.
11. AWWA. *Water Quality and Treatment*, 4th ed., McGraw-Hill Book Co., New York, 1990.
12. Peavy, H. A., Rowe, D. R., and Tchobanoglous, G. *Environmental Engineering,* McGraw-Hill Book Co. New York, 1985.
13. James M. Montgomery, Inc. *Water Treatment Principles and Design*, John Wiley & Sons, New York, 1985.
14. Metcalf & Eddy Inc. *Wastewater Engineering: Treatment, Disposal and Reuse*, 3d ed., McGraw Hill Book Co., New York, 1991.
15. Joint Task Force of the Water Environment Federation and American Society of Civil Engineers. *Design of Municipal Wastewater Treatment Plants,* Vol I, MOP/8, Water Environment Federation, Alexandria, VA, 1992.
16. Brater, E. F. and King, H. W. *Handbook of Hydraulics*, 6th ed., McGraw-Hill Book Co., New York, 1976.

CHAPTER 7

Water Conveyance, Flow Measurement, and Pumping

7.1 Introduction

Water-conveyance systems include the conduits needed to transport water from one point to another. In most cases, water conveyance is necessary from its source to the water treatment facility and from the treatment facility to the consumers. Pumping is used to impart the energy required to move water through the conveyance system. Pump stations are often located at the raw water intake, at the water treatment plant, and in the water distribution system. Flow measurement provides information regarding the quantity of water passing through a point in the water-conveyance system. Flow meters are generally provided at the following locations: at the raw water intake, ahead of the water treatment plant, at individual filter units, at the high service pump station, and at various points in the water treatment and distribution systems.

This chapter presents an overview of design considerations for water-conveyance, pumping, and flow measurement systems. Design calculations and operation and maintenance procedures for the raw water pump station are provided in the Design Example.

7.2 Water Conveyance Systems

The purpose of the water conveyance system is to provide a conduit to move water from the raw water intake to the treatment plant and from the treatment plant to the point of use.

7.2.1 Types of Conveyance Systems

Various types of conduits are used for transporting water. Topography, available head, construction materials and practices, economics, and water quality are the primary considerations in selecting suitable conduits for a water conveyance system. Water conduits are classified as *open channels* or *pressure conduits*. Open channels have a free water surface in contact with the atmosphere; pressure conduits have a confined water surface. Under each of these general classifications there are several types of conveyance systems in common use.

Open Channels

Canals. Canals are earthen channels excavated through the ground. They may be lined or unlined, depending on soil conditions, bottom slope, and the cost of water. They are usually trapezoidal in cross section, but may also be rectangular, parabolic, or of some other geometric shape.

Flumes and Aqueducts. Flumes and aqueducts are channels supported above the ground to transport water across valleys and depressions. They are typically constructed from concrete, masonry, metal, or wood.

Gravity Conduits. Gravity conduits or grade aqueducts are usually buried pipes. However, they have a free water surface, as canals do. Gravity conduits are constructed of concrete, brick, or stone, and are typically of rectangular, circular, or horseshoe cross-sections.

Grade Tunnels. Grade tunnels are built through mountains or other extreme elevations, either to shorten the route or to conserve head. They are typically circular sections, but other geometric shapes are also used. They are usually excavated by special boring machines. The tunnel may be lined or unlined. In unstable rock, lining may be essential. Lining materials are typically cast-in-place concrete, precast concrete, or steel pipe sections. Circular pipe may also be an open channel if the flow is by gravity.

Pressure Conduit. Pressure conduits are generally circular in cross section and are made of precast reinforced concrete, cast iron, steel, or plastic. The water flows under pressure due to hydrostatic or pumping heads. Pressure conduits (also called pressure aqueducts or pressure pipelines) generally follow the ground profile and are buried 0.6 to 1.5 m below the ground surface. The depth of cover usually allows for typical frost depth and protects the conduit from surface disturbances. When pipeline drops beneath a valley, stream, and other depression, it is called a *sag*, *depressed pipe*, or *inverted syphon*. Pressure pipelines can also penetrate through hills and other high elevations as a *pressure tunnel* and can be supported above ground as a *pressure flume* or *aerial crossing*.

7.2.2 Design Considerations for Conveyance System

Open Channel Systems The important design considerations for open channel systems are turbulence control at hydraulic transitions, protection of water quality, erosion control, and minimization of seepage and evaporation losses.

Hydraulic Transitions. Flow conditions in an open channel are classified as either *supercritical* or *subcritical*, depending on the relative velocity and amount of turbulence. Velocity is controlled by the channel slope and cross-section. Mild slopes result in lower velocities and less turbulence and are classified as subcritical flow. Steeper slopes result in higher velocities, and more turbulence and are classified as supercritical flow. Changes in slope or cross sectional area can cause the flow to transition from one type to the other. When this occurs, extreme turbulence can damage the channel lining. The procedure for predicting type of flow and design details for subcritical, transition-section, and supercritical flows may be found in Reference 1.

Water Quality. Open channel systems are often open to the atmosphere. This creates the possibility of accidental, intentional, or natural contamination of the water. Accidental contamination could occur where vehicular traffic crosses over a canal or other open conduit. In such cases, a traffic accident can result in spillage of hazardous chemicals into the water supply. Accidental contamination can also occur if contaminated surface water runoff is discharged into the conduit. To prevent contamination, gravity conduits should be isolated from all sources of contamination. Intentional contamination is a rare occurrence, but its possibility should be considered. Limiting access to open conduits reduces the potential for intentional contamination.

Natural contamination occurs when the water quality degrades through natural processes. The most common example is algae blooms and rooted plants in canals. The algae impart a taste and odor to the water that is difficult to remove. Natural contamination can be reduced by nutrient control (through watershed management practices), by the elimination of surface water inflow into the canals, and by controlling the growth of rooted plants.

Erosion Control. Earthen canals must be protected from erosion by limiting the velocity and turbulence of the water. References 1 to 4 have excellent discussions on various methods of determining the appropriate velocity in canals. One method discussed is the *permissible velocity method.* In this method the velocity is limited to a value below which erosion will not occur. For straight channels, this value ranges from 0.45 m/s (1.5 fps) for canals lined with fine sand with colloidal binders, to 1.14 m/s (3.75 fps) for stiff clays, to 1.22 m/s (4 fps) for coarse gravels with no colloids.[1,2]

Seepage and Evaporation Losses. Canals may often be routed through arid areas and/or through areas with soils of high permeability. If either of these conditions exists, high losses can occur that are due to evaporation or seepage. To control evaporation, designers should limit the exposed water surface area in arid regions. If seepage is anticipated, the canal may require lining with clay, concrete, or some other impermeable material to reduce seepage.

Pressure Conduit. The important design considerations for pressure systems are the control of air and the control of transient pressure waves.

Air Control. Typically, pressure pipelines follow the ground terrain. As a result, there are numerous hills and valleys in the pipeline. Any air trapped in the pipeline will tend to accumulate at the high points, and, if not removed, will increase the friction losses in the pipeline at that point. If air bubbles become entrained in the water, they may restrict the flow or release explosively at the exit point. Air is best controlled by installing automatic air relief valves designed to allow air to escape at high points in a pipeline. These valves, called *air release valves*, are discussed in Section 7.2.4.

Transient Pressure Wave. When the velocity changes in an enclosed conduit, the kinetic energy of the water is transformed into pressure waves that move rapidly through the piping system. These waves are often called *water hammer, surge waves*, or *transient pressures*. They can exert pressures of high magnitude, oscillating between positive (higher pressure) and negative (lower pressure) pressures. These waves are generally caused by a sudden change in velocity or direction of water flow caused by fast-opening or -closing valves or by rapid starting or stopping of pumps. If uncontrolled, these waves can result in extensive damage, such as the collapse, rupture, or displacement of the piping system.

These transient waves are controlled by reducing the rate of velocity change, so that the magnitude of the waves are reduced. This is done by installing standpipes called surge towers along the pipeline to provide a pressurized source of water to slow the rate of velocity change. Specially designed surge-control valves can also be installed at critical points to reduce high pressures or to allow air to enter the pipeline, preventing vacuum conditions during low pressure waves.[3,4]

The *transient analysis* is a technology used by the engineers to conduct engineering evaluations of pipeline-transient conditions. This analysis is usually performed using suitable computer software.[5,6] The software can simulate the pipeline-transient conditions for different operational scenarios. The surge-analysis module demonstrates the pipeline profiles graphically, showing important engineering data such as maximum and minimum values of head and location and time of transient occurrence. These data can then be used by engineers to optimize the types and locations of surge-protection devices along the pipeline. The *real time control* (RTC) function of some software can also provide operator guidance to optimize the operation and to prevent severe damage to the piping system. Further discussion of transient pressure waves is beyond the scope of this text. Designers should recognize that this phenomenon is a potential problem in every pipeline. It is recommended that designers consult appropriate literature and experienced professionals to develop proper procedures for predicting and controlling transients. References 2–4 and 6 provide excellent discussion of transients. Valves for controlling transient pressure waves are discussed in Section 7.2.3.

7.2.3 Hydraulic Considerations for Conveyance Systems

The hydraulic analysis and design of a water transport system is carried out by using the basic principles of open-channel and pressure-conduit hydraulics. It is assumed that the readers have the understanding of basic concepts in fluid mechanics. References 1–4 and 7–11 may be used in conjunction with the basic design information presented below.

Open Channel Systems Open channel flow occurs under gravity. The hydraulic grade line coincides with the water surface and is generally parallel to the bottom of the conduit. Equations proposed by Manning, Chezy, Gangrullet, Kutter, and Scobey are used to predict the hydraulic gradient of gravity systems. These equations will, under most circumstances, predict the flow characteristics (velocity, depth of flow, hydraulic slope, and dimensions) of a gravity system. Of these equations, the Manning equation has received the most widespread application. Various forms of Manning's equations are expressed as follows: [8–10]

$$v = \frac{1}{n} r^{2/3} S^{1/2} \quad \text{(SI unit)} \tag{7.1}$$

$$Q = \frac{0.312}{n} D^{8/3} S^{1/2} \tag{7.2}$$

$$v = \frac{1.486}{n} r^{2/3} S^{1/2} \quad \text{(U.S. customary unit)} \tag{7.3}$$

$$Q = \frac{0.464}{n} D^{8/3} S^{1/2} \quad \text{(circular pipe flowing full, U.S. customary unit)} \tag{7.4}$$

where

v = velocity in the channel or pipe, m/s (fps)
r = hydraulic mean radius (area/wetted perimeter), m (ft)
D = diameter of circular pipe, m (ft)
Q = flow in pipe flowing full, m^3/s (ft^3/s)
S = slope of energy grade line or channel bottom, m/m (ft/ft)
n = Manning's roughness coefficient

The hydraulic characteristics for different cross-sections of the channel and pipe are given by Eqs. (7.5) - (7.11).[9]

$$R = \frac{D}{4} \quad \text{(for circular pipe flowing full)} \tag{7.5}$$

$$r = \frac{D}{4}\left(1 - \frac{360 \sin\theta}{2\pi\theta}\right) \quad \text{(for circular pipe flowing partially full)} \tag{7.6}$$

$$\cos\left(\frac{\theta}{2}\right) = 1 - \frac{2d}{D} \tag{7.7}$$

$$a = \frac{D^2}{4}\left(\frac{\pi\theta}{360} - \frac{\sin\theta}{2}\right) \tag{7.8}$$

$$p = \frac{\pi D\theta}{360} \tag{7.9}$$

$$r = \frac{b \times d}{2d + b} \quad \text{(for rectangular channel)} \tag{7.10}$$

$$r = \frac{bd + zd^2}{b + 2d\sqrt{z^2 + 1}} \tag{7.11}$$

where

R = hydraulic mean radius for circular pipe flowing full, m (ft)

θ = central angle, degree (θ<180° if d< D/2, θ=180° if D=d/2, and 180°<θ<360° if d>D/2)

d = depth of flow, m (ft)

a = area of cross-section of circular pipe under partial-flow condition, m^2 (ft^2)

p = wetted perimeter of circular pipe under partial-flow condition, m (ft)

b = width of open channel at the bottom, m (ft)

z = horizontal fraction of slope per unit vertical rise, dimensionless

The various terms in Eqs. (7.5)–(7.11) are illustrated in Figure 7-1. It may be noted that it is a common practice to use the *d*, *v*, *q*, *a*, and *p* notations for depth of flow, velocity, discharge, area, and wetted perimeter under partial-flow condition and open channel, while *D*, *V*, *Q*, *A*, and *P* are used for pipelines flowing full or under pressure. Roughness coefficient of *n* depends on the material and age of the conduit. Commonly used values of *n* for different materials are given in Table 7-1.[1,2,9,12]

Table 7-1 Common Values of Roughness Coefficient Used in the Manning Equation

Material	Commonly Used Values of *n*
Concrete	0.011–0.015
Verified clay	0.011–0.015
Cast iron	0.011–0.015
Brick	0.013–0.017
Corrugated metal pipe	0.022–0.025
Asbestos cement	0.013–0.015
Plastic	0.011–0.015
Earthen channels	0.030–0.035

Various types of nomographs have been developed for solution of problems involving flow through open channels.[1,9–12] A nomograph based on Manning's equation for circular pipe flowing full and variable *n* values is provided in Appendix B. Hydraulic elements of circular pipes under partially full flow conditions and of rectangular channels are also given in Appendix B.

Manning's and the other equations mentioned in this section will predict flow characteristics only under *normal flow* conditions (long channel without flow-control points). Obstructions

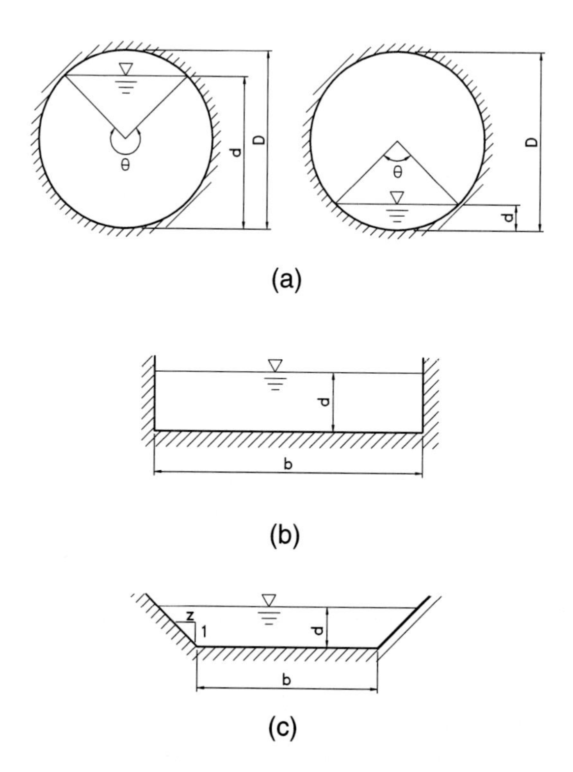

Figure 7-1 definition of variables for different cross sections. (a) Circular. (b) Rectangular. (c) Trapezoidal.

in the channel will cause the water to flow slower and deeper than predicted by these normal flow equations. Conversely, a free discharge will cause the velocity to be higher and the depth lower than those predicted by normal flow equations. Both of these conditions affect the velocity and water depth up to several hundred meters above the control point, leaving the normal flow equations invalid for those reaches. Reference 1 provides methods of calculating flow characteristics under these circumstances.

The water surface profile for open-channel flow is developed from water depth, bed slope of the channel, and head losses encountered in the appurtenances. The minor losses in an appurtenance are due to (1) entrance, (2) change in direction, (3) contractions and expansions, (4) transitions, and (5) exit conditions. The procedure may involve applying an energy equation at two sections and solving for velocity and depth of flow by a trial-and-error procedure. A simplified method is to combine all losses in terms of velocity head, as given by Eq. (7.12).

$$h_m = \frac{kv^2}{2g} \tag{7.12}$$

where

h_m = minor losses due to entrance, exit, change in direction, m (ft)
v = velocity of flow, m/s (fps)
g = acceleration due to gravity, 9.81 m/s^2 (32.2 ft/s^2)
K = head loss coefficient

The value of K for different conditions of open channel flow and appurtenances are summarized in Appendix B.

Pressure Conduits Pressure conduits are enclosed pipes. Water flows at a pressure higher than the atmospheric pressure. In other words, the hydraulic grade line is above the crown of the conduit; therefore, pressure pipes can follow the natural ground surface, provided the ground surface is below the hydraulic grade line. The velocity in the conduit is flow divided by the area.

The friction losses in the pipe are typically calculated from Darcy–Weisbach or Hazen–Williams equations (Eqs. (7.13) and (7.14)). Since the Hazen-Williams equation is most commonly used in water works engineering, this equation in various forms is given below (Eqs. (7.14)–(7.20)).[8]

Darcy–Weisbach:

$$h_f = f\frac{L}{D}\frac{V^2}{2g} \tag{7.13}$$

Hazen–Williams:

$$h_f = 6.81\left(\frac{V}{C}\right)^{1.85}\left(\frac{L}{D^{1.167}}\right) \quad \text{(SI units)} \tag{7.14}$$

$$V = 0.850CR^{0.63}S^{0.54} \quad \text{(SI units)} \tag{7.15}$$

$$V = 1.318CR^{0.63}S^{0.54} \quad \text{(U.S. customary units)} \tag{7.16}$$

$$V = 0.355CD^{0.63}S^{0.54} \quad \text{(SI units)} \tag{7.17}$$

$$V = 0.550CD^{0.63}S^{0.54} \quad \text{(U.S. customary units)} \tag{7.18}$$

$$Q = 0.278CD^{2.63}S^{0.54} \quad \text{(SI units)} \tag{7.19}$$

$$Q = 0.432CD^{2.63}S^{0.54} \quad \text{(U.S. customary units)} \tag{7.20}$$

where

h_f = total friction head loss in suction or discharge pipes, m (ft)
V = velocity in the pipe, m/s (fps)
Q = flow rate, m^3/s (ft^3/s)

f = coefficient of friction (Darcy-Weisbach)
C = coefficient of roughness (Hazen-Williams)
D = diameter of the pipe, m (ft)
L = length of pipe, m (ft)
S = slope of energy grade line (h_f/L), dimensionless

The value of f in the Darcy–Weisbach equation depends on the Reynolds number and the relative roughness and diameter of the pipe; it may range from 0.01 to 0.10. The procedure for calculating f is discussed in References 2, 10 and 12. The roughness coefficient C in Hazen–Williams equations depends on the material and age of the pipe. Common values are summarized in Table 7-2.[7]

The minor losses in pressure pipes are due to entrance, bends, valves, fittings, exits, contractions, or expansions. These losses are expressed in terms of velocity head by Eq. (7.12). The common values of the head loss coefficient K for pressure pipes are summarized in Appendix B.

Table 7-2 Hazen–Williams Friction Coefficient

Description of Pipe	C
Smooth and straight cast iron	
New	140
5 years old	130
10 years old	120
15 years old	110
20 years old	90–100
30 years old	75–90
Concrete or cement lined	120
Welded steel, new pipe	120
Asbestos cement	120–140
Plastic	150

7.2.4 Valves in Conveyance System

Water-conveyance systems utilize valves for numerous purposes. The type of valve used depends on the size of the pipeline and its purpose.[6]

Isolation Valves Isolating segments of a pipeline is the most common purpose for installing valves in water conveyance systems. Requirements for these valves include providing a tight seal and having low head loss when fully open. The valve should be relatively simple and inexpensive. Gate valves are the most common type of valve used for isolation. Butterfly, ball, and globe valves are other types of valves generally found in this use. Sluice gates are commonly used to isolate gravity flow systems such as intake and canal. Check valves are used to prevent flow from occurring in the reverse direction.

Valves for Regulating Flow Valves are often used to regulate or throttle the rate of flow. Such valves are used to control the rate of flow through filters from a raw water source into a gravity conveyance system and of flow between the pressure planes of a distribution system. The requirements for this type of valve are a tight seal when closed, low cavitation potential, and a near-linear relationship between head loss and percent opening. Butterfly and globe valves are commonly used to regulate flow. Specially designed ball and plug valves are also used in some applications.

Air-Relief Valves Air entrained in the water often accumulates at high points in pipelines. When this occurs, the flow may be restricted and excessive head loss is experienced. If the air escapes through the outlet of the pipeline, it may do so explosively, causing damage to the outlet works. Therefore, air-relief valves are normally installed at high points of pipelines to automatically release the accumulated air.

Surge-Control Devices Change in pumping rate or closure of a valve on a pressure pipeline may result in rapid velocity changes. These velocity changes result in transient pressure waves that travel at the speed of sound through the piping system and create transient pressure waves. One method of controlling transient pressure waves is the installation of special valves that mitigate rapid velocity changes. These are quick-opening and slow-closing air valves; they prevent the propagation of secondary pressure waves. Surge-control valves are also installed to minimize this situation. These are typically globe valves or special plug valves with elaborate controls to ensure rapid opening and slow closure. There are other methods of controlling transient pressure waves, such as surge towers and hydropneumatic tanks. Surge towers are typically standpipes that extend above the maximum hydraulic gradient of the pipeline. These towers provide a reservoir of water that acts as a buffer, slowing velocity changes, and thus reducing the magnitude of the pressure waves. Figure 7-2(a) illustrates a typical surge-tower installation. The hydropneumatic tanks used for surge control are usually located near the pump station and typically connected to the pump discharge header. Figure 7-2(b) illustrates the schematic of a typical hydropneumatic surge-control system.

7.3 Flow Measurement

Flow measurement is a critical element of a water supply system. These measurements are essential to control process operations such as filtration and backwash rates, chemical feed, and operation of the distribution system and to maintain records for billing and future expansions. Flow meters are, therefore, provided, typically at raw water intakes, treatment plant influent pipes, flow-division pipes, filter controls, pump stations, and various critical points in the distribution system.

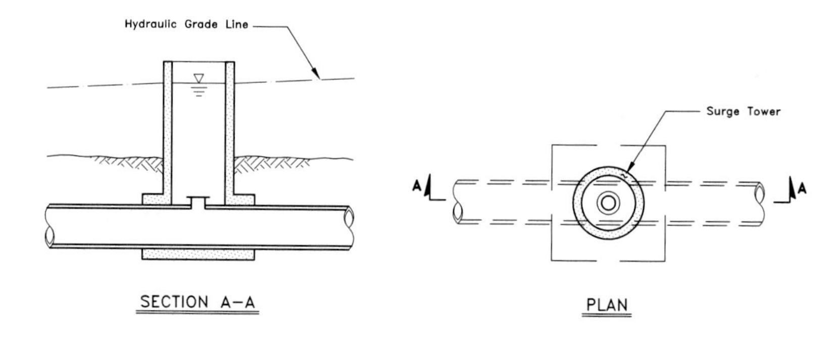

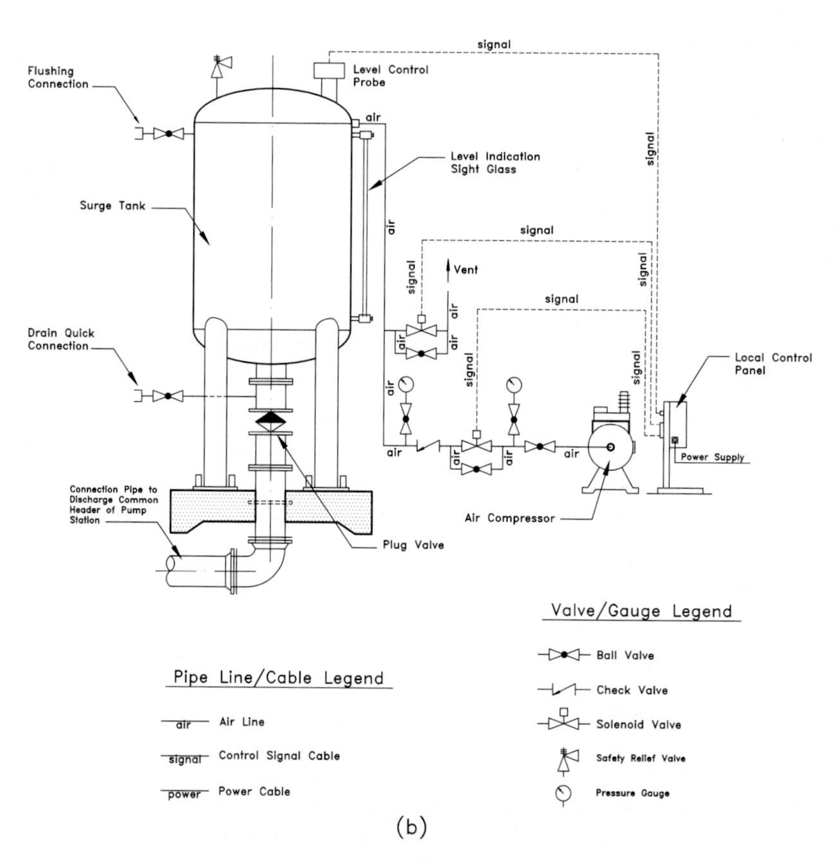

Figure 7-2 Typical surge control devices. (a) Typical surge-tower installation. (b) Schematic of typical hydropneumatic surge-control system.

7.3.1 Elements of Flow Measurement System

Flow measurement systems are composed of primary and secondary elements. Primary elements produce a head, pressure, electrical current, or other measurable parameter that is proportional to the flow rate. The secondary element measures the parameter produced by the primary element and provides an indication of the flow rate. These indications may be visual, such as an analog or digital readout, or a telemetry signal. Secondary elements are an integral part of the control system. Applications of primary and secondary elements are presented in Chapter 16. Numerous types of primary flow elements are commonly used in the water industry. A list of many commonly used primary flow elements and their basic characteristics are summarized in Table 7-3.

Selection of a primary flow element requires consideration of several factors. These factors are range of flow, accuracy required, head loss constraints, and cost. The accuracy of a primary flow element is generally excellent over a narrow range of flow rate. The selected flow element should meet the accuracy requirements over the entire range of the expected flow rates. If the range of flow is too wide for a single flow element to measure accurately, multiple flow elements should be considered. The head or pressure loss should also be considered in the flow element selection. Higher head loss results in additional energy cost. Many simple, less expensive flow elements, such as weirs and orifices, have high head loss across them. On the other hand, acoustic, magnetic, or ultrasonic meters exhibit very low head loss. Cost is also a major factor in selection of primary flow elements. The cost should include the capital cost and the present worth of annual operation and maintenance costs.

7.3.2 Flow Rate Controllers

A *flow rate controller* is a device used to maintain a preset constant rate of flow from a treatment unit. This is accomplished by attaching a valve or gate in the line with the flow meter. The secondary flow element adjusts the valve or gate setting to achieve a constant flow. The common applications for constant flow rate controllers are to maintain (1) a constant influent flow into a process or a treatment plant, (2) a constant flow from a filter, and (3) flow into a ground storage tank.

7.4 Pumping

Pumping is needed to transport water, chemicals, and residuals streams. Pumping involves imparting energy to water to raise its head, causing it to flow from a lower elevation to a higher elevation.

7.4.1 Terminology Used in Pumping

Many terms are commonly used in pumping and pump applications. These terms are briefly define and discussed below.

Table 7-3 Description of Primary Flow Elements

Element Type	Open Channel	Pressure Conduit	Operating Range	Accuracy Range, %	Description
Flumes Parshall Palmer-Boulers	Yes	No	20 to 1	±5	Critical depth through the flume is measured. Free fall is needed to prevent backwater effects. Upstream approach section equal to 2 times the throat width is needed. Turbulence downstream of flume can be used for rapid mixing of chemical. The device can be installed in pipe, channel, and appurtenance.
Weirs Rectangular Triangular Trapezoidal	Yes	No	20 to 1	±5	Critical depth over the weir is measured. Free fall is needed to prevent backwater effects. This device is generally less accurate than flumes. It is generally installed in a channel.
Acoustic	Yes	Yes	20 to 1	±1	Both velocity and depth of flow in the channel and velocity in a pressure pipe are measured by reflected sound waves.
Venturi tube	No	Yes	10 to 1	±1	Change in pressure between the "approach" section and the smaller diameter throat section is measured. Requires 10 pipe diameters of straight pipe upstream of the meter. Downstream piping has minimal effects. Typical head loss is approximately 10 percent of differential pressure.
Magnetic meter	No	Yes	10 to 1	±1	Voltage created by a conductor (water) that passes through a magnetic field is measured. Requires 5 pipe diameters of straight pipe upstream. No head loss is encountered. Electrode fouling from solids may often be a problem in sludge or raw water applications.
Ultrasonic meter	No	Yes	10 to 1	±1	Change in velocity of sound waves transmitted diagonally across the pipe is measured. Water should be free of bubbles and solids.
Orifice meter	No	Yes	4 to 1	±0.5	The differential pressure across a small diameter orifice in pipe is measured. Requires 10 pipe diameters of straight pipe upstream of the meter for accurate measurement.
Propeller	Yes	Yes	10 to 1	±1	Flowing water causes rotation of a meter propeller, and the speed of propeller rotation is measured. Requires clean water and 5 diameter of straight piping upstream of the meter.
Chemical	Yes	Yes	Varies	Varies	Chemical tracer is injected into water stream at a constant rate. Concentration of substance is measured downstream after proper mixing. Concentration is inversely proportional to flow rate.

Head Head describes hydraulic energy, expressed as the height of a column of liquid above a datum. The minor losses, *kinetic energy* or *velocity head*, are expressed by Eq. (7.21).[10] *Potential energy* or *static head* is expressed by Eq. (7.22).[10]

$$h_m = \frac{KV^2}{2g} + h_v \tag{7.21}$$

$$H_{stat} = H_D + H_S \text{ (for lift suction head)} \tag{7.22}$$

$$H_{stat} = H_D - H_S \text{ (for flooded suction head)} \tag{7.23}$$

where

h_m = minor losses due to entrance, exit, change in direction, etc., m(ft). The value of K for different valves, fittings and specials for pressure pipes are given in Appendix B.

V = velocity in the pipe, m/s (fps).

h_v = velocity head at a given point, m (ft). It is equal to $V^2/2g$. If water is discharged from a pipe into a reservoir, the velocity head h_v (or $V^2/2g$) is lost.

H_{stat} = total static head (difference in elevations of free water surface at discharge and suction reservoirs of the pump), m (ft)

H_D, H_S = discharge and suction static heads measured at water surface levels referenced to the centerline of the pump impeller, respectively; m (ft)

Figure 7-3 illustrates the relationships of various static head terms used in a pump system.

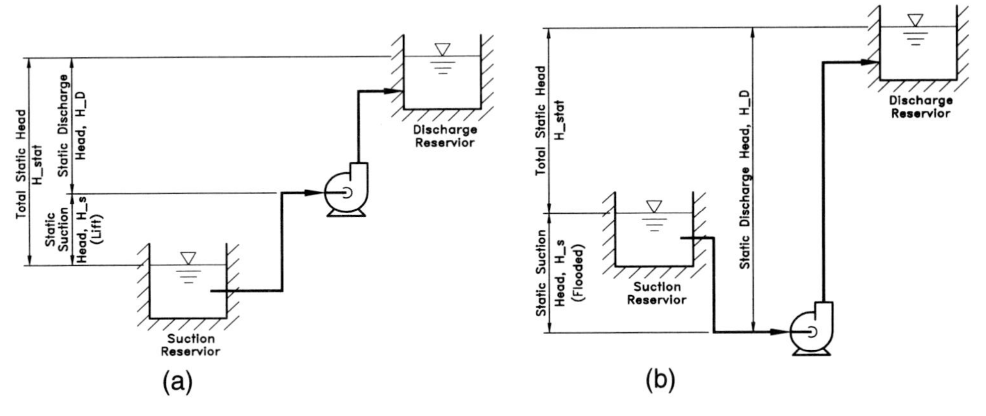

Figure 7-3 Definition sketches for determination of pump total static heads. (a) Lift suction head. (b) Flooded suction head.

The Total Dynamic Head (TDH) is the total energy barrier that must be overcome before the water can be lifted by a pump. The TDH for a pump is the net energy imposed into the water by the pump. It can be calculated from Eq. (7.24).

$$TDH = (\text{discharge energy} - \text{suction energy})$$

$$= \left(\frac{p_D}{\gamma} + \frac{V_D^2}{2g} + Z_D\right) - \left(\frac{p_S}{\gamma} + \frac{V_S^2}{2g} + Z_S\right) \quad (7.24)$$

where

p_D and p_S = gauge pressures on discharge and suction nozzles of the pump, respectively; N/m^2, (lb/ft^2)

γ= density of fluid, N/m^3 (lb/ft^3)

V_D, V_S = velocity in discharge and suction nozzles of the pump, respectively; m/s (ft/s)

Z_D, Z_S = elevation of gauges on discharge and suction nozzles of the pump above datum, respectively; m (ft)

The values of TDH at different flow rates for a pump is usually provided by the pump manufacturer based on their pump-testing results. These values are then used in pump selection. The TDH provided by the pump is simply utilized to counter the total head loss encountered in the pumping system due to the static head and other head losses through the system. Therefore, the TDH requirement by the pumping system is determined from Eq. (7.25).

$$TDH = H_{stat} + h_f + h_m \quad (7.25)$$

All variables in Eq. (7.25) have been defined earlier.

Capacity Capacity is the volumetric rate of flow of a liquid through a pump. The capacity of a pump is dependent on the total dynamic head and the pump characteristics. The relationship between the pump capacity and *TDH* is expressed in a *pump characteristic curve.* These curves are discussed in detail in Section 7.4.3.

Power and Efficiency The power produced by a pump (output power) is given by Eq. (7.26).[13] The output power of a pump is often referred to as its *water power.* The input power is the power applied to the pump by a driver and always exceeds the output power. This is also called *break power.* Input power and output power are related by the *pump efficiency* which is given by Eq. (7.27).[9,14]

$$P_W = K'Q(TDH)\gamma \quad (7.26)$$

$$E_p = \frac{P_w}{P_p} \times 100 \quad (7.27)$$

where

P_w = power output of the pump (water power), kW (hp).

K' = constant depending on the units of expression

(TDH = m, $Q = m^3/s$, $\gamma = 9.81\ kN/m^3$, P_w = kW, K' = 1 kW/kN·m/s)

(TDH = ft, $Q = ft^3/s$, $\gamma = 62.4\ lb/ft^3$, P_w = hp (horse power), K' = (1/550) hp/ft·lb/s)

E_p = pump efficiency, usually 70–90 percent

P_p = power input to the pump (brake power), kW (hp)

Net Positive Suction Head The net positive suction head (*NPSH*) is the absolute pressure of the fluid at the pump centerline or impeller eye as it enters the pump suction. Two values of *NPSH* are important in pump selection. These are *NPSH* required ($NPSH_{req}$) and *NPSH* available ($NPSH_{av}$).

NPSH required is the absolute fluid pressure required by the pump for smooth, efficient operation and is determined experimentally by the pump manufacturer. *NPSH* available is the absolute pressure of the liquid as it enters the pump suction. The value of *NPSH* available is mainly depends on the head and layout of pump suction piping system, and is given by Eq. (7.28).[12]

$$NPSH_{av} = H_{abs} + H_s - h_L - H_{vp} \tag{7.28}$$

where

$NPSH_{av}$ = available net positive suction head, m (ft)

H_{abs} = absolute pressure on the surface of the liquid in the suction well or reservoir (usually atmospheric pressure), m (ft)

H_s = suction head at the pump suction, m (ft)

h_L = total head loss due to friction, entrance, valves, and fittings in the suction piping, m (ft)

H_{vp} = vapor pressure of fluid at the operating temperature, m (ft).

In Eq. (7.28), H_s is the elevation difference between the surface in the suction well or reservoir and the pump centerline. The value of H_s is positive under flooded suction condition and negative under suction lift conditions. (See Figure 7-3.)

For any installation, it is recommended that the *NPSH* available should exceed the *NPSH* required by one meter or more.[8,15] If the *NPSH* available is less than the *NPSH* required or these two values are too close to each other, no water can be lifted by the pump. Under this condition, the pump station design must be corrected by implementing the following adjustments, as applicable:

- raise water surface level in the suction well or reservoir;
- lower pump centerline elevation;
- reduce head loss by increasing suction-piping size;
- select different model or type of pump.

Furthermore, the *NPSH* required of a pump should be evaluated over the entire head and capacity range for which the pump is expected to operate.

Cavitation Cavitation is a phenomenon that occurs when the absolute pressure of a fluid reaches the fluid vapor pressure. When this occurs, a cavity is formed in the fluid and the liquid literally boils. Cavitation occurs in pumps, control valves, and other high-velocity locations. It results in a distinctive rattling or pinging noise. If this condition is allowed to continue, severe pitting of the metallic surfaces may occur. Cavitation can be corrected by making the following adjustments:

- increase the diameter of the pump-suction piping;
- decrease the pump speed;
- increase the static head on suction side;
- decrease the flow rate.

Specific Speed The specific speed is a term that applies to centrifugal pumps and is defined as the rotational speed at which the pump discharges a unit flow at a unit head and at maximum efficiency.[8] It is expressed by Eq. (7.29).

$$N_s = \frac{NQ^{1/2}}{(TDH)^{3/4}} \tag{7.29}$$

where

N_s = pump specific speed[a]
N = pump rotational speed, rev/min
Q = flow at optimum efficiency, m^3/s (gpm)

7.4.2 Types of Pumps

Pumps are generally classified as either kinetic energy or positive displacement pumps.[13–18] Under these two general classifications, there are numerous types of pumps. Each pump type is specifically suited for a certain application. Table 7-4 lists various types of pumps commonly used in water treatment and supply systems. Figure 7-4 illustrates typical applications of pumps and selection guide in water works engineering.

7.4.3 Centrifugal Pump Characteristics

Centrifugal pumps are the most commonly used pumps in water supply systems. Because of their frequent application in water supply systems, the remainder of this discussion is devoted to the design and application of centrifugal pumps.

a. N_s in SI units × 51.6 = N_s in U.S. customary units.

Table 7-4 Pumps Commonly Used in Water Work Applications[13]

Classification	Type	Description
Kinetic	Centrifugal	An impeller rotates inside an enclosed casing. Head is developed principally by the centrifugal force created by the impeller rotation.
	Peripheral or recessed impeller	A type of centrifugal pump used in high-solids application. The impeller is recessed on the side of the casing entirely out of the flow stream. The fluid rotates due to viscous drag. These pumps are less efficient than centrifugal pumps.
Positive displacement	Plunger or piston	A piston oscillates in a cylinder and displaces a fixed amount of fluid with each stroke. Typically, these pumps are used to pump sludge and chemical solutions.
	Diaphragm	Similar to a piston pump but a flexible diaphragm is placed over the cylinder. Typically these pumps are used to pump and meter corrosive liquid and chemical solutions.
	Rotary	Consist of a fixed casing with rotating gears, cams or vanes. The fluid is pushed around the casing into the discharge pipe.
	Screw	A spiral screw rotates in an inclined casing. Fluid is pushed up the casing as the screw rotates.
	Airlift	Air is bubbled in a partially submerged tube. Air bubbles reduce the unit weight of the fluid causing it to be forced up by the heavier fluid below.

Categories of Centrifugal Pumps Centrifugal pumps are categorized into radial-flow, mixed-flow, and axial-flow. The categories are made according to specific speed and pumping action. Table 7-5 lists these major differences in the pump categories.[9,13] Centrifugal pumps are also classified according to the details of the pump construction. Table 7-6 lists some of these classifications.[9,13]

Pump Characteristic Curves Pump manufacturers conduct tests and publish a series of graphs or curves that represent the performance of a pump under various conditions. These are known as pump characteristic curves. Typical characteristic curves are illustrated in Figure 7-5. A pump casing can accommodate impellers of several sizes; therefore, a particular pump casing may be used for different head and flow applications. This feature is helpful because an existing pump can be expanded simply by replacement of the impeller and the motor. Following is a description of various curves.

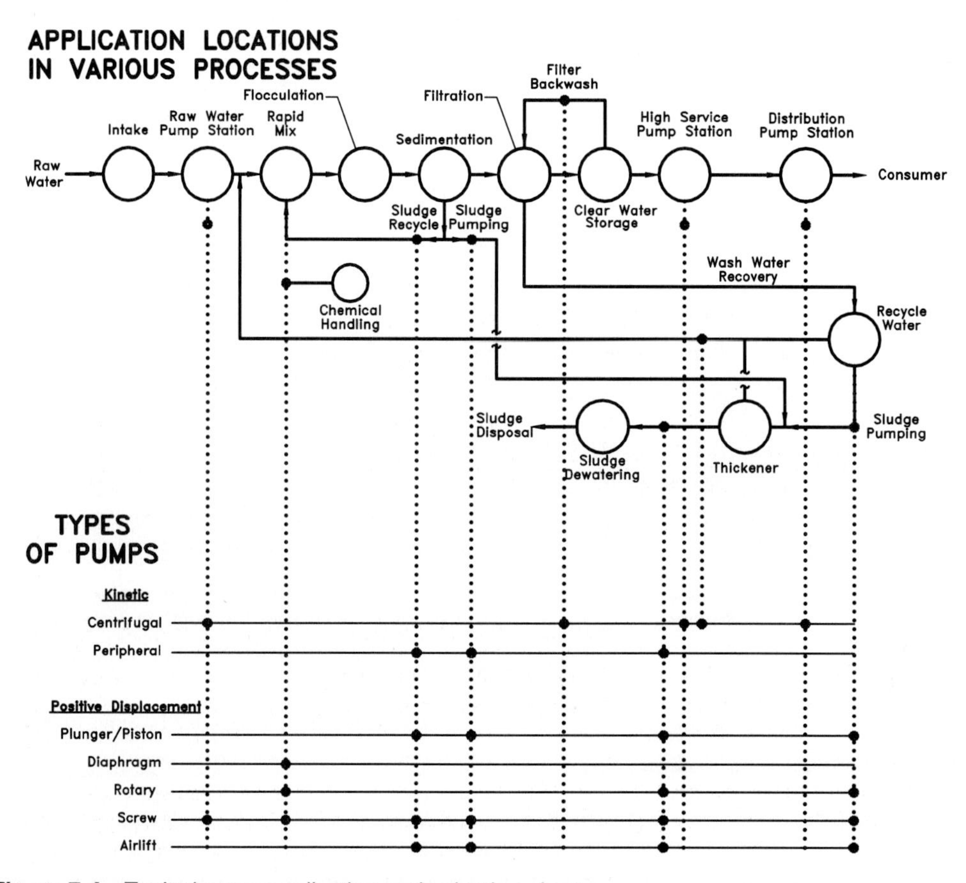

Figure 7-4 Typical pump application and selection chart.

Head-Capacity Curve. This is the most important of the pump characteristic curves. It demonstrates the pump discharge or capacity at a given head condition. The shape of the head-capacity curve may be rising-head, flat, steep, or dropping-head. Such curves are a function of specific speed and pump design. The point where the curve intersects the y-axis (zero discharge) is known as the *shut-off head.* The point where the curve ends on the right side is known as the *run-out point.* If a pump is operated near the shut-off point, excessive heat will build up; at the run-out point, the pump will cavitate. Pumps should not be allowed to operate near either of these points for an extended period of time. Such operational conditions will result in damage to the pump. An example of typical head-capacity relationship from the manufacturer is shown in Figure 7-5.

Table 7-5 Categories of Centrifugal Pumps

Category	Radial Flow	Mixed Flow	Axial Flow
Specific speed	Less than 80 for single suction (4200 U.S. customary units)	80–120 (4200–6250 U.S. customary units)	More than 200 (10,000 U.S. customary units)
Discharge direction from impeller	Radial to the axis of rotation	Partially radial and partially parallel to the axis of rotation	Parallel to the axis of rotation
Pumping action	Centrifugal force and lifting action	Centrifugal	Lifting action
Head ranges	High to medium (more than 16 m)	Medium to low (7–16 m)	Low (less than 10 m)
Capacity	Low to medium	Medium to high	High

NPSH Curve. This curve demonstrates the $NPSH_{req}$ for a given discharge. It is given by the manufacturer. The $NPSH_{av}$ is calculated from Eq. (7.28) and should exceed the $NPSH_{req}$ by one meter or more.

Power. The power curve demonstrates the power required to operate the pump at a given discharge. Typically, radial-flow pumps have a power curve that climbs from left to right. Therefore, these pumps are often started against a closed valve to reduce the starting torque. The valve is slowly opened after the pump starts reducing the effects of a surge or water hammer phenomenon. Mixed-flow and axial-flow pumps have power curves that drop from the left to the right, with a sharp increase near the y-axis. These pumps should not be started against a closed valve.

Efficiency. The efficiency curve demonstrates the efficiency of a pump at various discharge rates. Typically, this curve will have an inflection point or peak. Ideally, the pump should be selected to operate near its peak efficiency point.

Effects of Speed on Centrifugal Pumps The rotational speed of an impeller affects the operating characteristics of the pump. Eqs. (7.30)–(7.32) give the relationship of pump discharge, head, and power output with respect to impeller rotational speed for two conditions.

$$\frac{Q_1}{Q_2} = \frac{N_1}{N_2} \tag{7.30}$$

$$\frac{TDH_1}{TDH_2} = \frac{N_1^2}{N_2^2} \tag{7.31}$$

Table 7-6 Centrifugal Pump Classifications

Type	Variation	Description
Suction	Single	A pump with a single suction opening in the impeller.
	Double	A pump with two suction openings in the impeller.
Stages	Single stage	A pump where total head is developed by one impeller.
	Multi-stage	A pump with two or more impellers in series.
Shaft position	Horizontal	Pump shaft is in the horizontal plane.
	Vertical	Pump shaft is in the vertical plane.
Submergence	Dry pit	Pump is located in a dry pit or other location out of the water.
	Submersible	Pump is immersed in the water. Motor is suspended above water, or is an integral part of the pump and is immersed in water.
Casing	Volute	A pump made of a casing in the form of a spiral or volute; kinetic energy is transformed into pressure by gradually increasing the area of water passage.
	Circular	A pump made up of a circular casing of constant cross-section. Curved vanes are used to convert kinetic energy into pressure.
	Diffusion	A pump that transforms kinetic energy into pressure (turbine) by diffusion vanes.
Impeller	Enclosed	Pump with impeller vanes enclosed in a shroud.
	Open or (semi-enclosed)	Pump with impeller vanes attached to one side of the shroud. One side of the shroud is omitted; therefore, the vanes are exposed.

$$\frac{P_1}{P_2} = \frac{N_1^3}{N_2^3} \tag{7.32}$$

where

N_1, N_2 = rotational speed of the pump for conditions 1 and 2, respectively.
Q_1, Q_2 = discharge corresponding to speeds N_1 and N_2, respectively.
TDH_1, TDH_2 = total dynamic heads corresponding to speeds N_1 and N_2, respectively
P_1, P_2 = power output corresponding to speeds N_1 and N_2, respectively

Effects of Impeller Geometry on Pump Performance The shape and size of the impeller has an effect on the operating characteristics of a pump.[14,18] Eqs. (7.33)–(7.35) give the

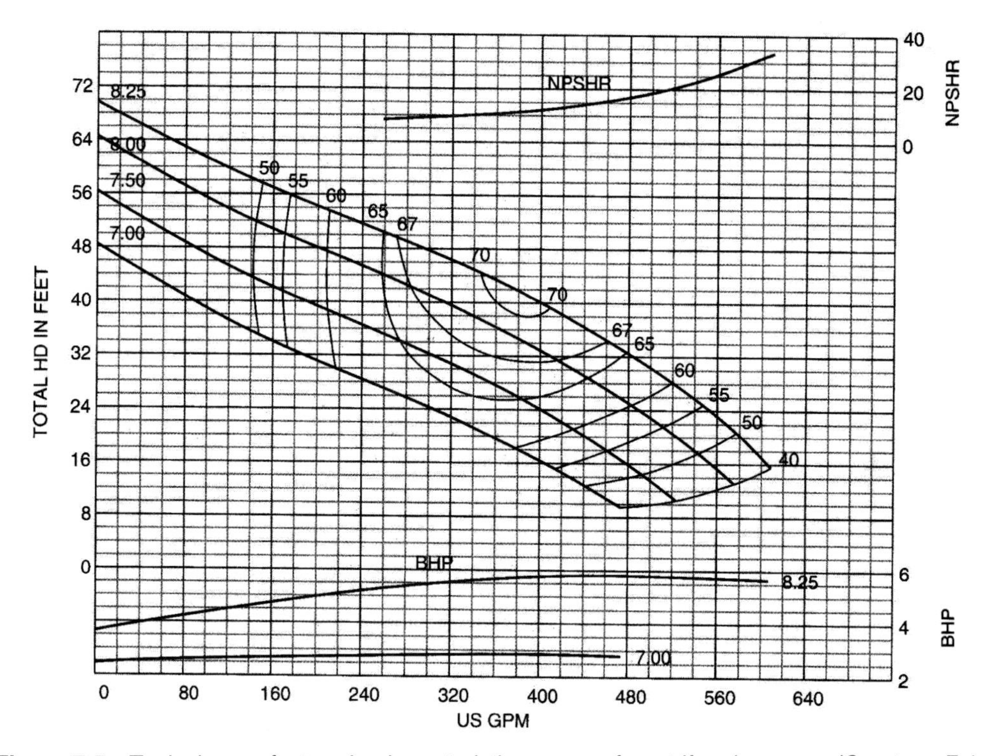

Figure 7-5 Typical manufacturer's characteristic curves of centrifugal pumps. *(Courtesy Fairbanks Morse Pump Corp.)*

relationship for two conditions of head, capacity, and power of a pump with respect to impeller diameter.

$$\frac{TDH_1}{TDH_2} = \frac{D_1^2}{D_2^2} \tag{7.33}$$

$$\frac{Q_1}{Q_2} = \frac{D_1^3}{D_2^3} \tag{7.34}$$

$$\frac{P_1}{P_2} = \frac{D_1^5}{D_2^5} \tag{7.35}$$

D_1, D_2 = impeller diameter for conditions 1 and 2, respectively. Other terms have been defined earlier.

Pumps in Combination It is a common practice to use multiple pumps in water supply applications. The two possible pump combinations are *series* and *parallel.* In general, to develop the characteristics of pumps in series, the pump heads are added for respective discharge rates. This procedure is shown on Figure 7-6(a). For pumps in parallel, the pump discharges are added for respective heads. This procedure is shown on Figure 7-6(b).

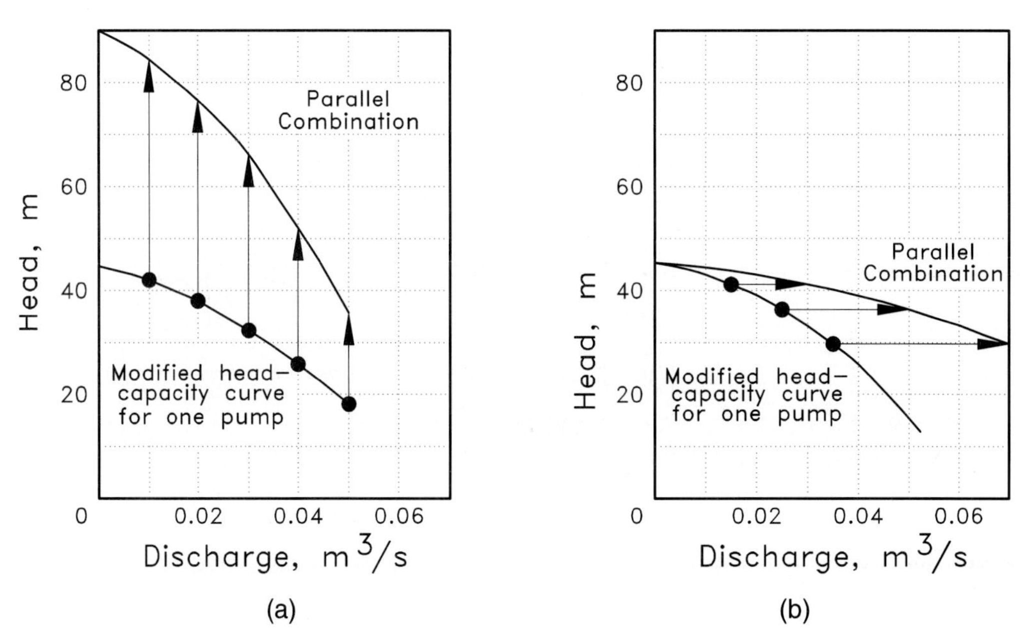

Figure 7-6 Pump combination. (a) Two identical pumps operating in series combination. (b) Two identical pumps operating in parallel combination.

7.4.4 Pump Selection

One of the most important tasks in the design of water supply systems is selecting pumps properly. Pumps must be carefully selected to operate near peak efficiency at the design flow rate and to operate smoothly throughout the entire range of expected operating conditions. For this purpose, several design considerations are important. Many of these considerations are discussed below.[14–18]

System Head-Capacity Curve The system head-capacity curves are used to define the head-capacity characteristics of the piping system. An example of a system head curve is given in Figure 7-7(a). This curve is developed by calculating the system head losses, including friction and minor losses for various discharge rates. Only losses common to all pumps from the start of the discharge header to the delivery point are included in the system head curve. The system head losses are then added to the static head to determine the system dynamic head. These points are plotted with respect to the flow rates to obtain the system head curve.

Some pumping systems have variable static-head conditions. A variable-level suction and discharge reservoir may be an example of such a condition. In this case, the system head curve is calculated for both extreme head conditions, to yield a band of possible operating conditions between the maximum and minimum suction and discharge reservoir conditions. An example of such a curve is shown on Figures 7-3 and 7-7(b).

Modified Pump Head-Capacity Curve When multiple pumps operate in parallel, each pump typically has a separate discharge pipe prior to discharging into a common pipe known as a *discharge header.* The effects of each individual discharge pipe on the overall operating characteristics of a pump station with multiple pumps must be considered. A typical procedure dealing with this situation is the use of a modified head-capacity curve. In this procedure, the head losses associated with the suction and discharge piping of each pump are subtracted from the manufacturer's head-capacity curve. These losses are also called *station losses.* The intersection of the modified head-capacity curve with the system head curve indicates the operating point of the pump. An example of a modified head-capacity curve is shown in Figure 7-7(c). Figure 7-7(d) illustrates how the pump operating point is determined.

Other Pump Selection Considerations Many other considerations are also important for design and selection of pumping equipment. Some of these considerations are provided here.[16–18]

Suction Head. The *NPSH* required for the selected pump should be at least one meter less than the *NPSH* available. Furthermore, if the installation requires a suction lift, the pump must be primed each time it is started. This is typically achieved by use of a vacuum pump connected to the pump controls. The vacuum pump is used to evacuate air from the pump casing and allow water to enter.

Efficiency. The pump should be selected to operate near its peak efficiency point. In typical water supply applications, pumps operate over a band of head conditions. Therefore, they cannot operate at their peak efficiency all the time. Nonetheless, efforts should be made to achieve peak efficiency at the normal operating conditions.

Shut-off and Run-out. Pumps should be selected to operate in the middle of the head-capacity curve. Operations too near the left-hand side of the curve results in low flows. Continued operation in this condition will result in damage to the pump from overheating. Axial-flow and mixed-flow pumps usually have a sharp rise in discharge head and require more power to operate near the left-hand side of the curve. The higher discharge head can damage pipes and valves from excessive pressure, and the higher power requirements can overload the pump driver. Radial-flow pumps, however, can be started against a closed valve to take advantage of lower starting torques. If this is done, the valve should be opened soon after the pump accelerates to full speed, to prevent heat damage. Operation of pumps too far to the right-hand side of the curve

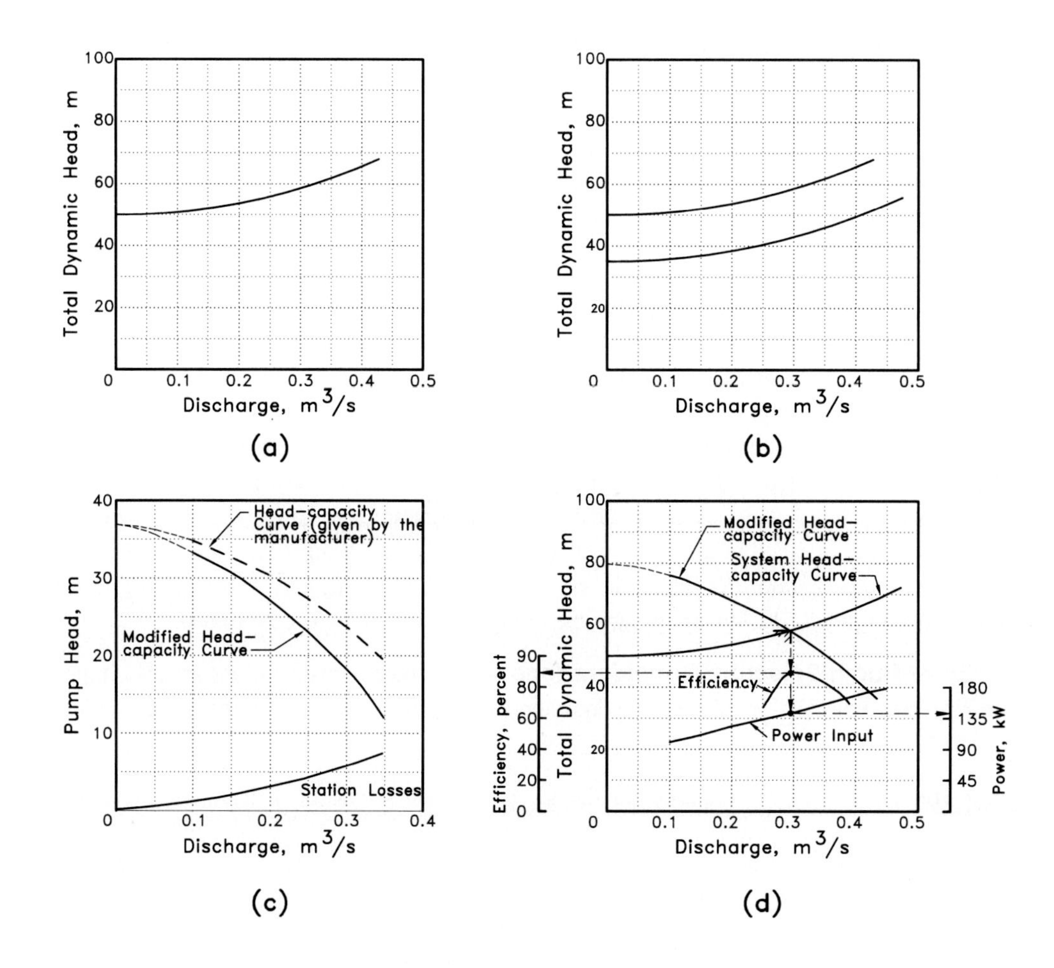

Figure 7-7 Typical pump and system curves. (a) System head-capacity curve for constant static head. (b) System head-capacity curve for variable static head. (c) Modified head-capacity curve. (The station losses are subtracted from the manufacturer's head-capacity curve to obtain the modified head-capacity curve.) (d) Pump operating point is obtained from intersection of system head-capacity curve and modified head-capacity curve. For best pump selection, the pump operating point should be near maximum efficiency point.

results in discharge rates in excess of the suction capacity. Continued operation in this condition will result in damage to the pump from cavitation.

Unstable Operation. Some axial-flow and mixed flow pumps have an inflection point or knee in the head-capacity curve. An example of such a curve is illustrated in Figure 7-8. Such pumps should always be operated at a head that is lower than the minimum head of the knee. Operation near the knee will result in oscillations or unstable flow and will ultimately damage the pump.

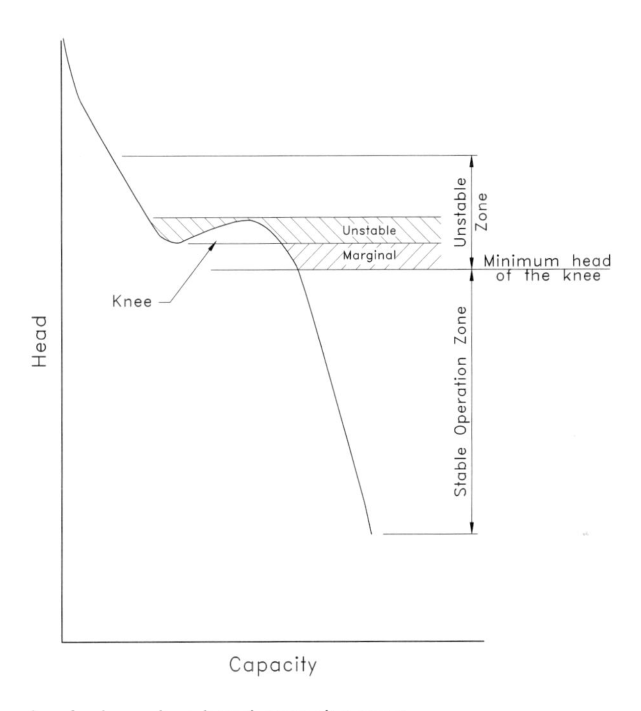

Figure 7-8 Example of a knee in a head-capacity curve.

Solids. Pumps should be selected to handle the largest solids they may be required to pass. Water supply pumps typically are designed to pass only very small solids. Therefore, screens must be used to remove solids prior to pump intakes.

Pumping Sludges One of the most difficult materials to pump is *solids suspension* or "sludge." Pump selection is generally dependent upon viscosity of the sludge. Sludges can be classified as low-viscosity or high-viscosity suspensions. Table 7-7 lists typical viscosity ranges for different types of sludges from water treatment plants.[19]

Low viscosity sludges behave much like water and can easily be pumped with centrifugal sludge pumps. High-viscosity sludges require a high-suction lift pump such as a positive-displacement sludge pump.[14] Sludge is also often gritty from sand and silt particles in it; therefore, sludge pumps must be selected to perform reliably under such service.

Table 7-7 Typical Viscosity Ranges for Water Treatment Sludges

Sludge	Solids Content, percent	
	Low-Viscosity Sludge	High-Viscosity Sludge
Lime sludge	Less than 10	Greater than 10
Alum or ferric sludge	Less than 2	Greater than 2

The head loss in the pipelines transporting sludge is generally calculated from the Hazen-Williams equation (Eq. (7.14)). The Hazen–Williams coefficient C is decreased, because sludge is more difficult to pump than water. The suggested values of Hazen–Williams coefficient for different types of sludges are summarized here.

- Chemically coagulated sludge (total solids up to 1 percent), $C = 90$
- Compacted sludge (total solids up to 4.5 percent), $C = 55$
- Thickened sludge (total solids up to 8 percent), $C = 35$

If the sludge is pumped long distances, more accurate head loss calculations are needed to design the pumping equipment. Friction losses should be calculated using thixotropic behavior, rheological concept, and other flow properties of sludge. The procedure involves non-Newtonian fluid and turbulent conditions. Both the Reynolds number and the Hedstrom number are needed to calculate the Darcy–Weisbach friction factor. The procedure may be found in References 11 and 20– 22.

It is a good design practice to provide sludge piping larger than 15 cm (6 in.) in diameter. For small sludge flows, the pumps are operated on a time cycle, to achieve larger flows and velocities during the pumping cycle. A liberal number of cleanouts should be provided in the piping for cleaning the stoppages.

7.4.5 Pump Drivers

Pump drivers convert electrical or thermal energy into the rotating kinetic energy required to drive a pump. They include electric motors, internal combustion engines, gas turbines, and steam turbines. The selection of a pump driver must be based on evaluation of available types, reliability and cost of the energy sources, reliability and cost of the driver, characteristics of the load demand, and power output of the driver. Centrifugal pumps generally have a relatively low starting-torque requirement and a smooth running torque. Reciprocating pumps often have a higher starting torque and a more cyclic or pulsating running torque.[14]

Electric Motors Electric motors are the most common pump drivers in use today.[15] Their common use as pump drivers is due to their high efficiency in energy conversion, their simplicity and reliability, and the widespread availability of reliable electric power. Typical electric-motor efficiencies are given in Table 7-8. Electric motors used as pump drivers are primarily alternating current (a-c) motors; however, some direct-current (d-c) motors are also used.[13] Two general types of a-c motors in common use are *induction* and *synchronous*.[15–18]

Table 7-8 Typical Electrical Motor Efficiencies

Motor Power Rating, kW	Typical Efficiency, percent
1–5	70–80
5–7.5	80–85
7.5–20	85–88
20 and above	88–92

Source: Reference 26.

Induction Motors. The most common electric a-c motor is the induction motor. This type of motor is commonly used for lower-power applications, less than 750 kW (1000 hp), or for higher-speed applications, more than 600 rpm. The primary disadvantage of induction motors is their *power factor.* In simple terms, the voltage applied to the motor leads the current, resulting in a power factor of less than 1.0. Some electric utilities assess additional charges for low power factors, so it may be desirable to increase the power factor for large motors.

Synchronous Motors. One way to increase the power factor is to use a synchronous motor. These motors can be operated over a range of power factors from less than 1.0 to greater than 1.0. In simpler terms, synchronous motors can be operated with either leading or lagging current. Therefore, these motors tend to correct the deficiency of induction motors by increasing the power factor.[23,24] Synchronous motors are more complex than induction motors and typically are cost-effective only for high-power applications (above 750 kW) or low-speed applications (less than 500 rpm).

Variable Speed Drives. The a-c electric motors are best suited for constant-speed applications. If variable-speed operation is desired, an auxiliary variable-speed drive unit is required. The two most common variable-speed units for pumps driven by a-c electric motors are *eddy-current* drives and *variable-frequency* controllers.

Eddy-current drives are magnetic devices. An electromagnetic input rotor is rotated around an electromagnetic output rotor. The speed of the output is adjusted by increasing and decreasing the current applied to the magnets. Increasing the current increases the magnetic field, reduces slip between the magnets, and increases the output speed. Conversely, decreasing the current decreases the magnetic field, increases slip, and decreases output speed.

Variable frequency controllers adjust the speed of an a-c motor by varying the frequency of the a-c current applied to the motor. The speed of an a-c motor is directly proportional to the frequency of the current. In variable-speed control panels, special control panels are used to change the current frequency to meet the speed requirements of the motor.

Internal Combustion Engines Internal combustion engines operate on natural gas, gasoline, or diesel fuel. The engines are similar to common automobile or truck engines but are designed for continuous high-power output service. Internal combustion engines are most often used as standby power for electric motor-driven pumps. They are also used to drive pumps in remote areas where electricity is not available.

Gas Turbines Gas-turbine engines are similar to jet engines used on aircraft. They operate at very high speeds and require gear reducers to drive pumps. They are very seldom used as pump drivers.

Steam Turbines At the turn of the 20th century, steam was a common power source for pumps, but steam-driven pumps require boilers and other auxiliary equipment to generate steam. Unless steam is available from other sources (such as waste heat generated from solid-waste and sludge incineration), steam turbines are usually not feasible alternatives for pump drivers.

7.5 Pump Stations

Pump stations are the structures that house pumps, piping, equipment, and chemicals. They may be large, architecturally designed facilities or simple and small structures. In either case, the design must ensure that the pump station meets the functional requirements of the pumping facility. In addition, the structure of the pump station houses and protects pumps and auxiliary facilities from weather, vandals, and other undesirable elements. Many design considerations regarding pump stations are provided here.[15–18]

7.5.1 Types of Pump Stations

In general, pump stations can be classified as *wet-pit* or *dry-pit*. These classifications are based on the location of the pumps relative to the wet well or dry pit.

In wet-pit pump stations, the pumps are located in the wet well, as illustrated in Figure 7-9(a). In dry-pit pump stations, pumps are located in a dry enclosure separated from the wet well, as illustrated in Figure 7-9(b). Selection of a pump station is usually based on the pumping equipment selected. For example, submersible and vertical pumps require wet-pit structure; horizontal pumps usually require dry-pit structure.

7.5.2 Site Selection of Pump Station

Many of the criteria for site selection of a pump station are same as those for an intake-site selection. Key considerations in site selection include power, access, and flood protection. These are briefly discussed here.

Power Pump stations require a significant amount of power to operate pumps. In order to continue operation during emergencies, a second source of power should be provided. The second source of power may be either a totally independent electrical circuit, a motor generators

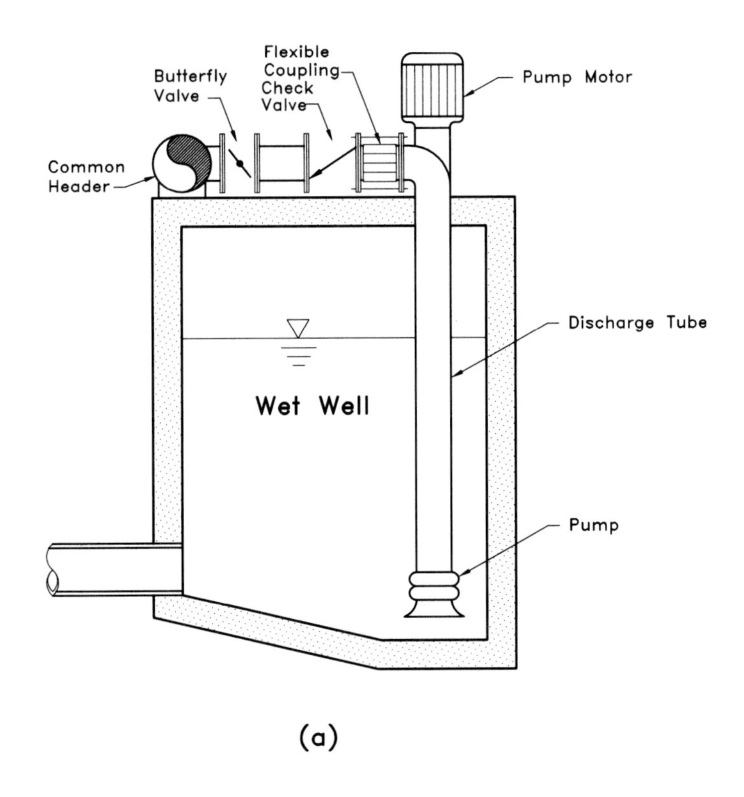

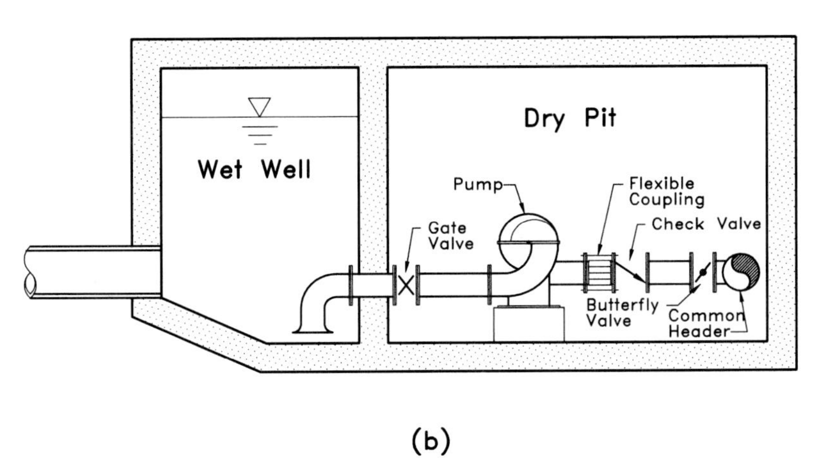

Figure 7-9 Common pump station design. (a) Wet-pit pump station. (b) Dry-pit pump station. (*Reprinted from Reference 9, with permission from Technomic Publishing Co., Inc., copyright 1999.*)

using natural gas or diesel, or secondary engine drivers for the pumps. Pumps with secondary engine drivers have both an electric motor and an internal combustion engine connected to the pump drive shaft. A clutch mechanism is installed on each driver, so that either can be used to drive the pump.

Access Pump stations require a great deal of maintenance and attention. Therefore, all-weather access to the structure is vital.

Flood Protection The equipment contained in a pump station can be severely damaged by flood waters. Therefore, pump stations must be located away from flood-prone areas or protected by flood-protection levees.

7.5.3 Design Considerations For Pump Stations

Design of pump stations involves the layout of the structure and equipment for efficient operation and maintenance. Necessary facilities are required to house pumps and pump controls, spare parts and accessories, and basic facilities for operation and maintenance personnel.

Suction Well and Suction Piping One of the most critical aspect of pump station design is the pump-suction well and suction piping. The important consideration in design of the suction well is to provide a smooth transition; turbulence in the pump suction can adversely affect pump performance.

Suction Wells. Proper design of suction wells includes consideration of the pump-suction placement relative to the water surface, suction-well bottom and walls, and other pumps in the suction well. Velocities and direction of flow in the suction well are other important considerations. References 13–18 and 25 provide an excellent discussion on wet well design. *Hydraulic Institute* has provided design standards for suction wells.[13] Figure 7-10 provides a summary of these recommendations. Common problems associated with poorly designed existing suction wells and recommendations for correcting them are discussed in *Hydraulic Institute Standard* ANSI/HI 9.8-1998.[25]

If site-specific limitations dictate variances from these recommendations, suction-model tests should be conducted. These tests will verify the operational features of the proposed design of the suction well. Design modifications based on such tests will provide considerable cost savings and trouble-free operation.

Suction Piping. Important design considerations for design of pump-suction piping include prevention of air pockets and reduction of turbulence in the suction piping. To prevent air pockets, suction piping should be leveled or sloped slightly upward toward the pump. If reducers are required, they should be eccentric-type for horizontal piping and concentric for vertical piping. To prevent turbulence, suction piping should be straight and no smaller than the pump suction nozzle. If bends are required, they should have long radius. Many recommended designs of suction piping and undesired conditions are given in References 13 and 25.

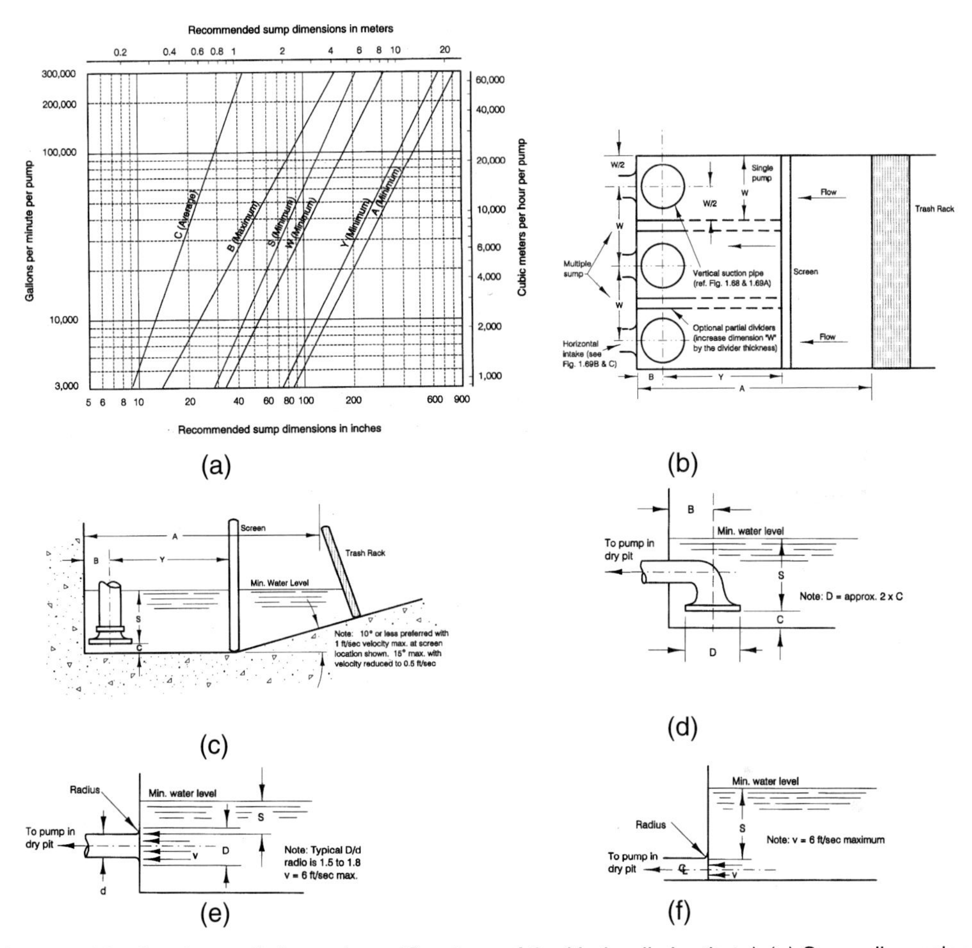

Figure 7-10 Suction-well dimensions *(Courtesy of the Hydraulic Institute).* (a) Sump dimensions for different flows. (b) Sump dimensions—plan view. (c) Sump dimensions—section view for wet-pit pump. (d) Sump dimensions—section view for dry-pit pump with turned-down bellmouth elbow. (e) Sump dimensions—section view for dry-pit pump with circular horizontal intake. (f) Sump dimensions—section view for dry-pit pump with noncircular horizontal intake.

Flexible couplings should be installed on the pump-suction as well as discharge piping to reduce vibration and prevent temperature effects on piping from stressing the pump casing. When flexible couplings are provided, the pump foundation and base plate must be designed to resist the thrust forces created by the flexible coupling.

Pump Station Ventilation and Cooling Pumping equipment generates a great deal of heat during operation. This heat is typically generated from lost efficiency of the pump driver. If this heat is not removed from the pump station, temperatures will become excessive and may damage the equipment. The heat load generated by electric motors is calculated from Eq. (7.36).[23,24]

$$H = \frac{P_m UL(1-E)}{E} \tag{7.36}$$

where

H = heat load, kW
P_m = motor-power rating, kW. The motor rating is the power output of the motor.
U = motor-use factor, 1.0 or decimal fraction. The motor-use factor may be applied when motor use is intermittent, with significant nonuse during all hours of operation. For conventional applications, its value could be 1.0.
L = motor load factor, 1.0 or decimal fraction. The motor load factor is the fraction of the load being delivered under the estimated cooling condition, [L = (pump power requirement)/(motor rated power)]
E = motor efficiency, decimal fraction.

Pump stations typically are cooled by forced-air ventilation. This method requires forcing a sufficient quantity of outside air through the pump station to remove the heat generated from the pump operation. Eq. (7.37) is typically used to estimate heat removed by air flow.[23]

$$Q_a = \frac{H_r}{(C_p\rho)(T_i - T_0)} \tag{7.37}$$

where

Q_a = air-flow rate required to remove heat, m^3/s (ft^3/hr)
H_r = quantity of heat removed, kW (Btu/hr)
C_p = specific heat of air; 1.02 kW-s/kg-°C or 1.02 kJ/kg-K (0.245 Btu/lb-°F)
ρ= density of air, 1.2 kg/m^3 (0.075 lb/ft^3)
T_i = desired temperature of indoor air, °C (°F)
T_0 = maximum temperature of outside air, °C (°F)

Ventilation for cooling of switch-gear rooms should also be designed using the above equations. The heat load for such designs can be calculated from an empirical equation (Eq. (7.38)).

$$H = 0.07P_s \tag{7.38}$$

P_s = total power load on the switchgear, kW

Condensate Removal During some periods of the year, the water temperature will be less than the dew point of the ambient air. When this occurs, condensation will form over pumps, piping, and other metallic surfaces in contact with water. This can be prevented by (1) installing air driers to reduce the dew point of the ambient air, or (2) insulating all metallic surfaces in contact with the water. Usually, neither of these options can be justified economically. Therefore, engineers must make provisions in pump-room floors to collect and convey condensation or any other water drips that may occur.

Equipment Access The layout of pump stations should provide access to all pumps, valves, piping, and miscellaneous equipment for maintenance. Lifting equipment, such as overhead bridge cranes, is often installed in larger pump stations, to facilitate repair and removal of equipment.

Miscellaneous Equipment and Facilities

Electrical Switchgear. Sufficient room for electrical switchgear cabinets should be provided in the pump station. The space required will depend on the power rating of the motors, voltages, and complexity of the control system. The reader should seek assistance from experienced electrical engineers for information on switchgear space requirements for specific installations.

Control Facilities. Space should be provided for instrumentation and control equipment. These facilities may range from a simple relay cabinet to computer-control facilities. In larger installations, a dedicated room for control equipment is often provided. An instrumentation and control engineer should be consulted for information on space requirements for control facilities.

Potable Water and Sanitary Facilities. Pump stations should provide potable water and sanitary services for operations and maintenance personnel. A clean water supply is often required for pump-seal and bearing lubrication. For lubrication needs, an alternative potable water supply, perhaps from a groundwater source, should be considered.

Pump stations should be provided with toilet, shower, and locker facilities for operator comfort. Additionally, all floor drains should be discharged into sanitary sewers. If sanitary sewers are not available, the sanitary waste treatment and disposal systems should be properly located and designed to ensure protection of raw and treated water supplies.

7.6 Manufacturers of Water Conveyance, Pumping, and Flow-Measurement Equipment

A list of the manufacturers of water-conveyance, pumping, and flow-metering equipment is provided in Appendix C. General guidelines for review and selection of the equipment are also provided in Appendix C. The equipment selected should provide flexibility of operation, maintainability, and durability, while meeting the primary design requirements for selected applications.

7.7 Information Checklist For Design of Raw Water Conveyance, Pump Stations, and Flow Meters

The following information should be obtained, and necessary decisions should be made, before the design engineer begins designing raw water conveyance, pumping, and flow-metering systems.

1. Information to be obtained from the predesign report:

   a. design flow: peak, average, and minimum;
   b. location of the proposed pump station and water treatment plant;
   c. minimum and maximum water surface elevations of the raw water source;
   d. water surface elevation of the first treatment unit at the water treatment plant;
   e. route of the raw water pipeline.

2. Information to be obtained from the preliminary design:

   As a part of the predesign report, the design engineer generally develops preliminary design information that may include the following:

   a. preliminary site plan of the raw water pump station;
   b. preliminary layout of the pump station;
   c. type of pumps and valves;
   d. size of the raw water pipeline.

3. Information to be obtained from the plant owners and operator:

   a. preferences with regards to the type of pumping and flow measuring equipment, and the manufacturer;
   b. preferences with regards to any existing pump stations and flow measuring devices in operation at the facility.

4. Information to be developed by the design engineer.

   a. minimum design parameters and design criteria established by the concerned regulatory authorities;
   b. manufacturers' catalogs containing information on types, sizes, and limitations of available equipment;
   c. site plan, piping schematic, and hydraulic profile from the reservoir to the treatment plant. A great deal of engineering judgment and experience is required in establishing these elevations.
   d. type of pumping station and suction conditions. This may include limits of submergence, suction head, suction lift, and available *NPSH*.

e. system head-capacity curve;

f. initial pumping-unit selection, to include types and number of pumps, constant- or variable-speed drive, specific speed, and so forth;

g. drive unit (electrical motor or engine) and expected power requirements.

7.8 Design Example

7.8.1 Design Criteria Used

1. Hydraulic data

   a. Design flows (from Chapter 5)

   Maximum flow rate = 113,500 m^3/d (30 mgd)

   Average flow rate, normal condition (see Table 5-3) = 51,400 m^3/d (13.6 mgd)

   Minimum average flow rate, initial year (see Table 5-3) = 45,200 m^3/d (11.9 mgd)

   b. Reservoir elevations (from Chapter 6)

   Maximum reservoir elevation = 90.00 m above mean sea level (msl)

   Minimum reservoir elevation = 70.00 m (msl)

   c. Raw water line (from Chapter 5)

   Hydraulic gradient at Junction Box A[b] = 110.00 m (msl)

   Raw water pipeline size = 122 cm (48 in)

   Raw water pipeline length = 5200 m

   Suction conduit connecting intake tower and wet well—diameter and length = 152 cm (60 in.) and 135 m.

2. Location and configuration of raw water pump station

   The proposed raw water pump station has two levels and is located at the shore of the Big River Reservoir. A suction conduit 135 m long connects the pump station with the intake structure. The raw water pipeline is routed across the countryside to the water treatment plant. Figure 6-11 illustrates a preliminary layout of the pump station and intake. Figure 5-4 illustrates a preliminary layout of the raw water transmission pipeline.

3. General design guidelines

   a. Pumps shall be multiple-stage, vertical, mixed-flow type.

   b. Pump station shall be wet-pit type, with two separate wet wells to facilitate maintenance.

b. Junction Box A is the first box at the treatment property in which the raw water transmission line is ended.

c. Each wet well shall have three pumps, with a total of six pumps in the pump station. Five pumps will meet the full design capacity of the treatment plant. One pump will be a stand-by unit.

d. Maximum flow from each wet well shall be attained when all three pumps in the wet well are in operation.

e. Fine screens shall be located at the pump station. A brief discussion of and design criteria for the fine screen is provided in Section 6.8.2, Step D.

f. PAC storage and feed facility shall be provided in a building adjacent to the raw water pump station. The design of PAC feed system is covered in Chapter 11.

7.8.2 Design Calculations

Step A: Unit Arrangement and Layout The pump station has two levels. A general schematic flow sequence through the lower level of the pump station is illustrated in Figures 6-11 (plan view) and 7-11 (section view). This layout provides two influent channels with fine screens, as discussed in Section 6.8.2, Step D. Each screen is dedicated to three pumps. A gate is provided in the front of each screen to allow isolation for necessary maintenance of the screen and suction well. Ultrasonic transducer monitors liquid level in the influent channel, and limit switches indicate the position of the gates. The raw water, after passing through the fine screen, flows through a distribution channel that is separated from the suction well by a baffle wall. This wall directs the flow towards the pumps and reduces eddy currents in the suction well.

Figure 7-12(a) illustrates the lower level (suction well) of the pump station. The arrangement of gates, screen, distribution channel, baffle wall, and pump suction is shown. A schematic layout of the pump-discharge piping is shown in Figure 7-12(b). This is the upper level layout and illustrates pump motors and a separate discharge pipe with a check valve and isolation plug valve for each pump. Each pump-discharge pipe connects to a common discharge header that connects to the raw water transmission pipeline. A flow meter is installed in a meter vault at the pump station site. The raw water transmission pipeline has a diameter of 122 cm (48 in). The total length of the pipeline is about 5.2 km (3.2 miles) from the raw water pump station to Junction Box A, located near the south property boundary of the water treatment plant. Motor Control Center (MCC) is located in a separate building that also houses electrical switchgear and instrumentation panels. Two transformer pads are provided near the MCC building. A PAC storage and feed facility is provided adjacent to the raw water pump station. Detailed dimensions of the pipeline are shown in Figure 7-12(c). The general layout of the head works, with MCC building, PAC facility, the sanitary sewer system, and service roads, is shown in Figure 7-13.

Step B: PAC Feed System The PAC storage and feed facility houses two PAC-slurry bulk-storage tanks, carbon delivery and dust collection, recirculation pumps, volumetric feeder, feed pumps, and other equipment. The structure has a direct access road, to enable easy unloading of PAC by the delivery trucks. Design of various components of the PAC facility and layout are covered in Section 11.6.2, Step B.

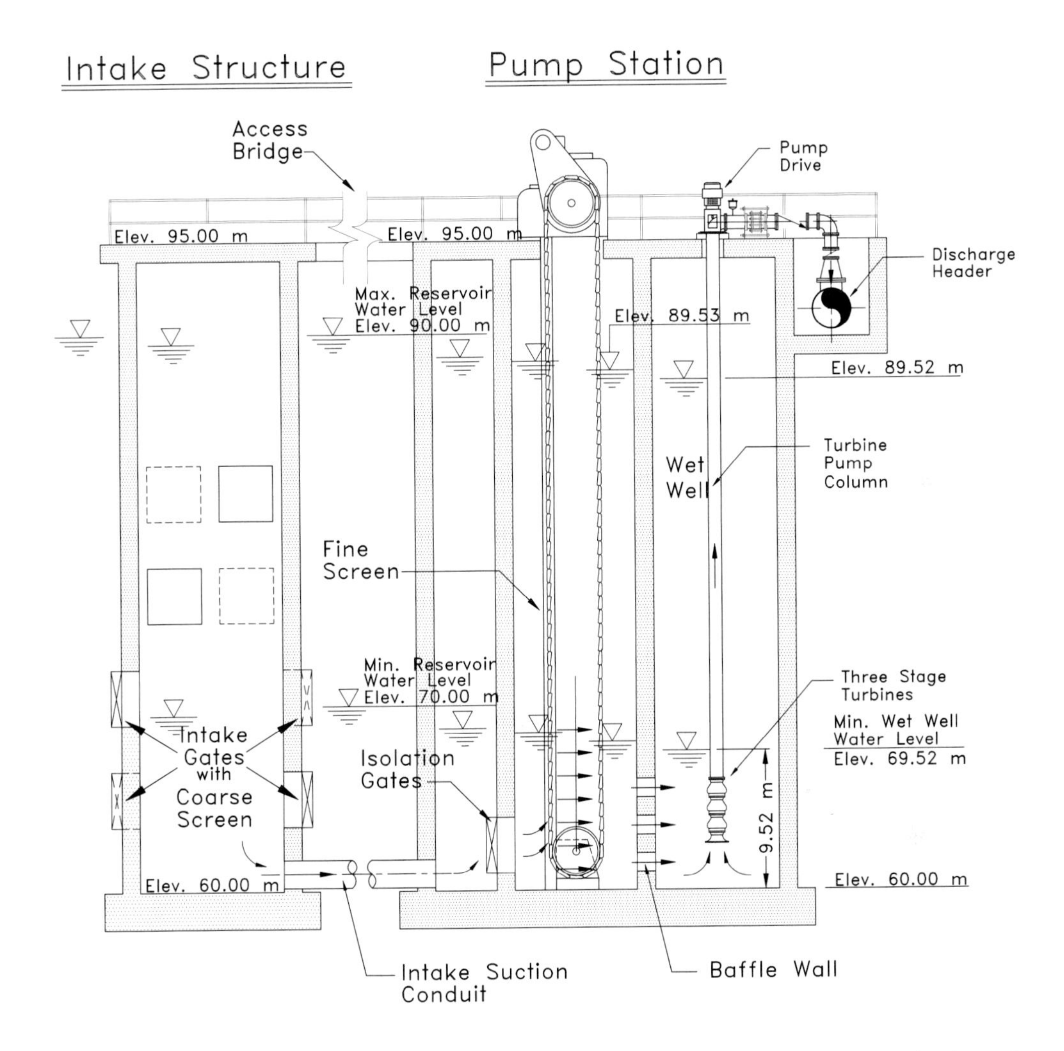

Figure 7-11 Schematic flow sequence in the raw water pump station. (The flow train shows intake structure, suction conduit, isolation gate, fine screen, baffle wall, and pump arrangement.)

Step C: Wet Well Design The Hydraulic Institute Standard (Reference 13) is used to design the wet well. The Hydraulic Institute Standards are revised regularly. The last Standard for Pump Intake Design was published in 1998.[25] Designers should contact the Hydraulic Institute for the updated copy of the Standard before designing a wet well or a pump station.

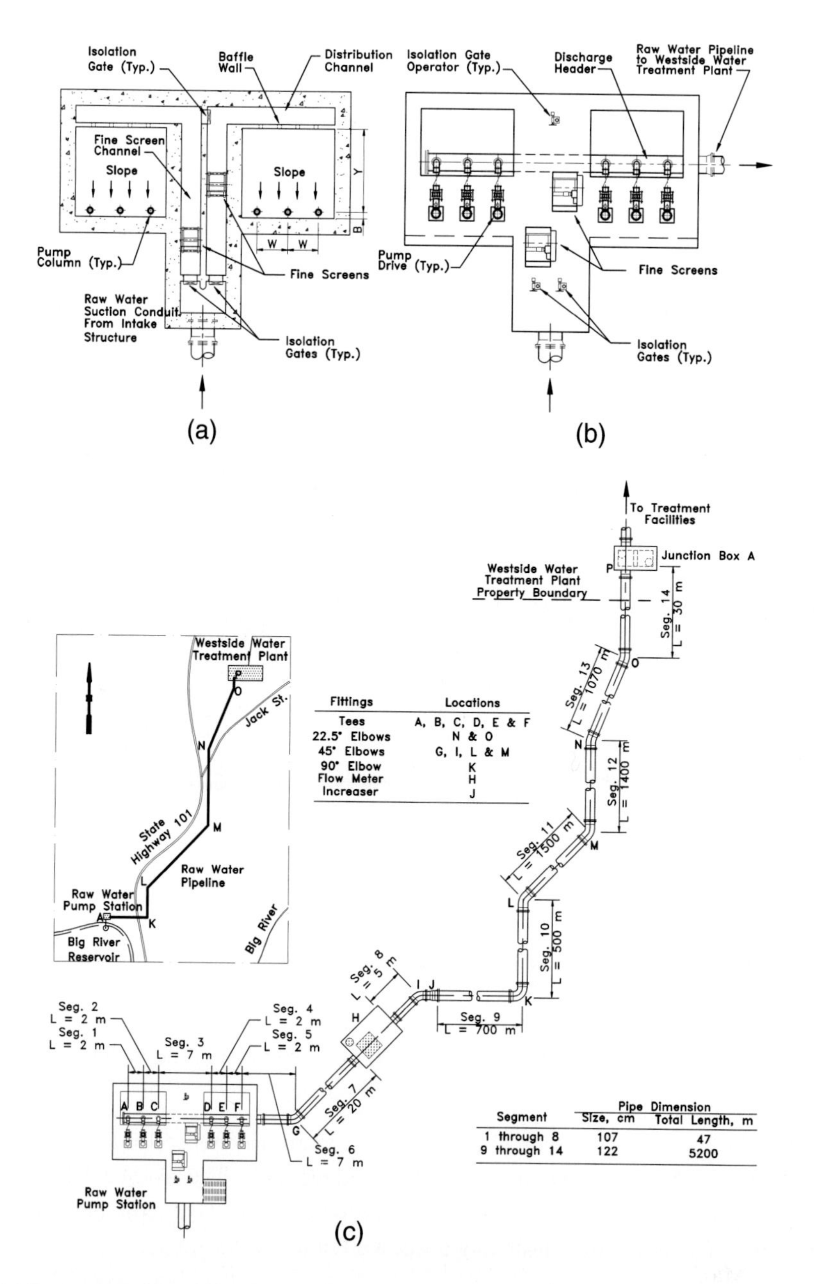

Fittings	Locations
Tees	A, B, C, D, E & F
22.5° Elbows	N & O
45° Elbows	G, I, L & M
90° Elbow	K
Flow Meter	H
Increaser	J

Segment	Pipe Dimension Size, cm	Total Length, m
1 through 8	107	47
9 through 14	122	5200

Figure 7-12 Schematic layout of pump station and raw water transmission pipeline. (a) Distribution channel and wet well (lower level). (b) Pumps, fine screens and discharge piping (upper level). (c) Raw water transmission pipeline.

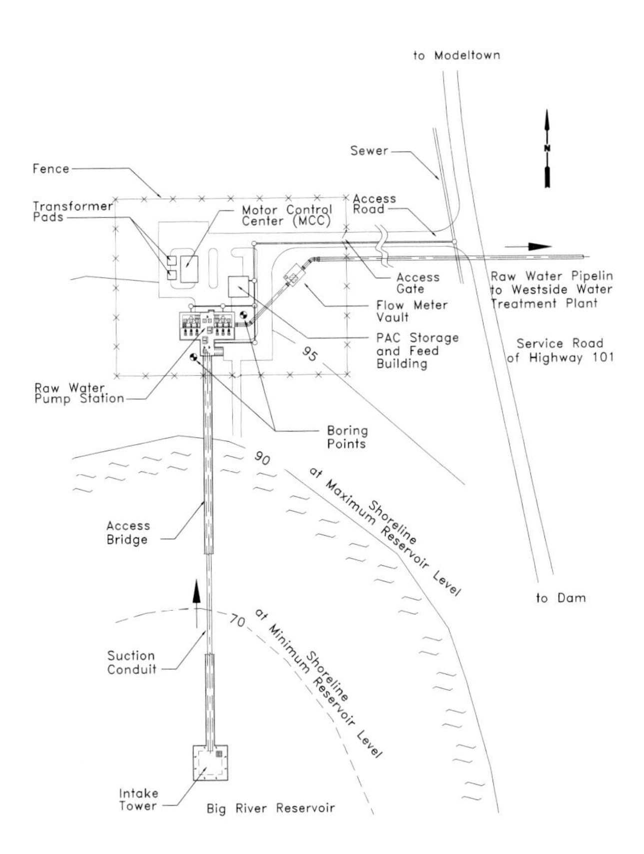

Figure 7-13 Site plan for head works.

1. Calculate individual pump discharge based on five pumps.

   Six pumps are provided, but one pump is a stand-by unit. Therefore, the maximum flow through each pump is as follows:

$$\frac{113{,}500\ \text{m}^3/\text{d}}{5\ \text{pumps}} = 22{,}700\ \text{m}^3/\text{d} = 0.263\ \text{m}^3/\text{s} = 15.78\ \text{m}^3/\text{min}$$

2. Calculate maximum flow through each wet well.

   Maximum flow through each wet well will occur when three pumps on one side are in operation.

$$\text{Maxiumum flow} = 3 \times 0.263\ \text{m}^3/\text{s} = 0.789\ \text{m}^3/\text{s}$$

3. Determine required dimensions of suction well from Figure 7-10(a) at a flow of 15.78 m^3/min = 947 m^3/h) per pump.

   Dimension Y (minimum distance from baffle wall to pump center line) = 2.15 m

Dimension S (minimum submergence) = 0.8 m

Dimension B (maximum distance between pump center line and the opposite wall) = 0.4 m

Dimension A (minimum) = 2.7 m (Not applicable in the design)

Dimension W (minimum spacing between pumps) = 1.0 m

Dimension C (average clearance between the suction bed and the floor) = 0.25 m

The recommended minimum dimensions in Figure 7-10(a) are based on the hydraulic requirements to optimize the pump performance. The actual dimensions are much larger. The main reasons to provide much large space are: (1) provide extra room to accommodate the actual pumping equipment sizes, (2) provide extra space per recommendations by the manufacturers, and (3) maintain the necessary clearances for routine operation and maintenance needs. Therefore, all dimensions must be rechecked against all three conditions stated above, after pumps are selected. In this design, provide the following dimensions. (See Figures 7-12(a), 7-14, and 7-15.)

Dimension (baffle wall to pump center line) Y = 5.25 m

Dimension (suction bell to water surface) S = 9.27 m

Dimension (center line of pumps to opposite wall) B = 0.4 m

Dimension (spacing between pump center lines) W = 2.0 m

Dimension (suction bell to floor) C = 0.25 m

Bottom slope of wet well = 7.5°

The wet well dimensions, $(Y+B) \times (3W)$, are 5.65 m × 6.0 m

4. Select isolation gate dimensions.

Assume that the maximum velocity through the gate is 1.0 m/s. The kinetic energy due to higher entrance velocity will be dissipated by the fine screen before the water will enter the distribution channel.

$$\text{Area of the isolation gate} = \frac{0.789\ \text{m}^3/\text{s}}{1.0\ \text{m/s}} = 0.789\ \text{m}^2$$

Keep the width of the isolation gate the same as the width of the fine-screen opening (0.9 m).

$$\text{Height of the isolation gate} = \frac{0.789\ \text{m}^2}{0.9\ \text{m}} = 0.88\ \text{m}$$

Select from the manufacturer's catalog the available standard isolation gate closest to 0.9 m × 0.9 m. Preliminary layout of the pump station suction well is illustrated in Figure 7-12(a).

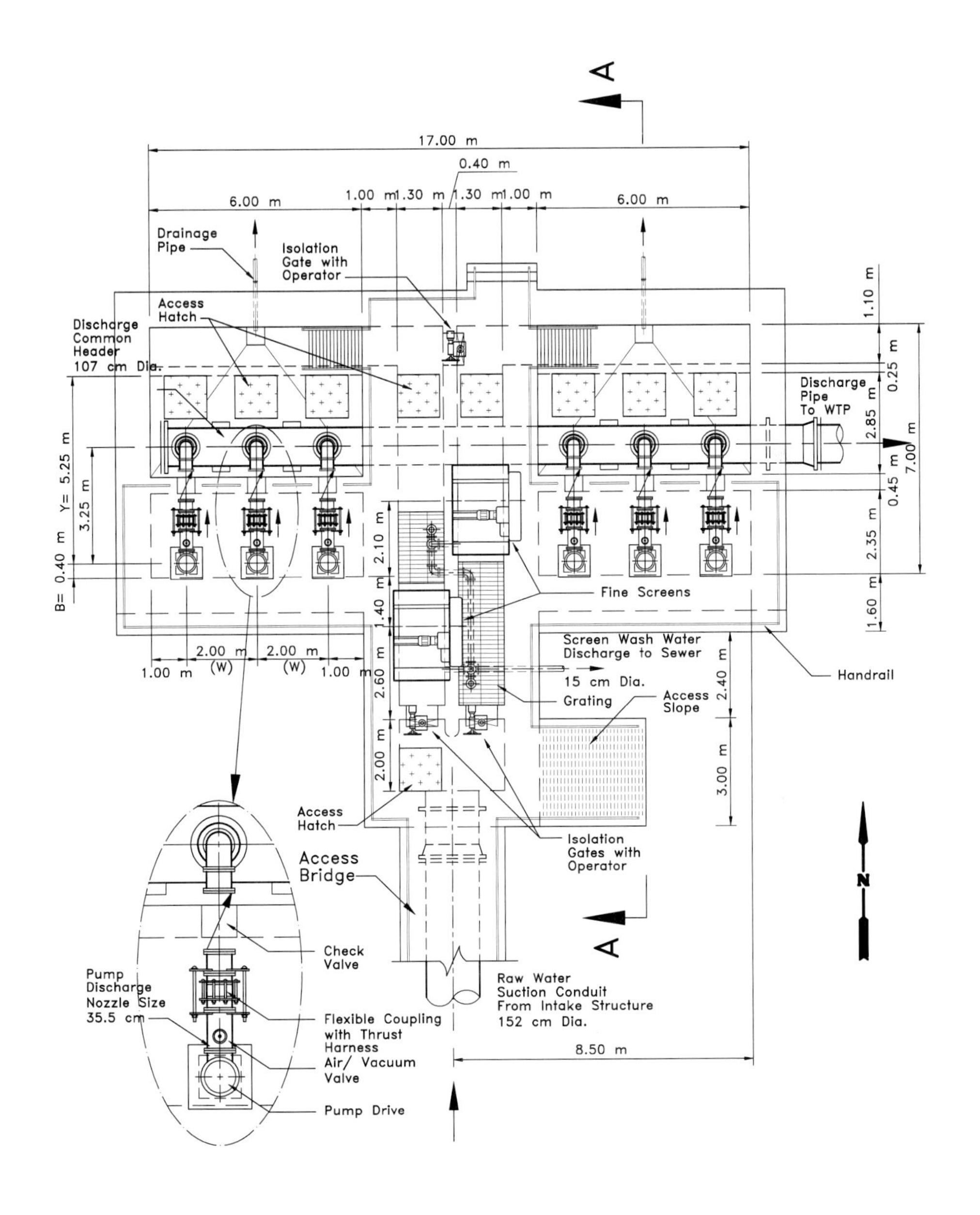

Figure 7-14 Design dimensions of raw water pump station—plan view. (Section A-A is shown in Figure 7-15.)

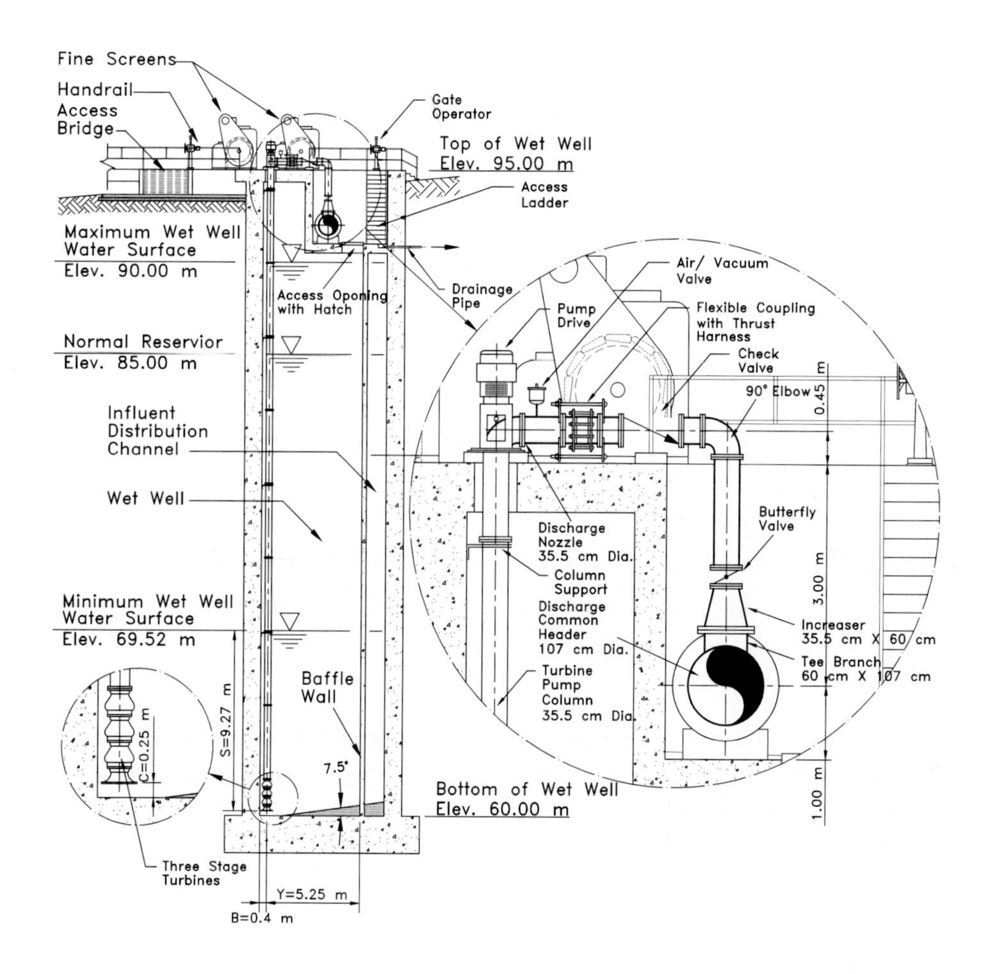

Figure 7-15 Design dimensions of raw water pump station—section A-A. (See section A-A location in Figure 7-14.)

5. Design fine screen and channel.

   Two fine screens are provided in the wet well of the raw water pump station. These screens are 0.9 m wide and have 9.5-mm (3/8-inch) clear openings. The width of the screen channel is 1.3 m. The isolation of a screen is possible by closing the gate ahead of the screen, and the gate on the center common wall after the screen. Design calculations for the fine screens are covered in Section 6.8.2, Step D.

6. Check water velocity in the distribution channel.

   The width of the distribution channel is 1.1 m and the water depth is 9.53 m at the minimum reservoir level of 70.0 m.

$$\text{The velocity in the distribution channel} = \frac{0.789 \text{ m}^3/\text{s}}{1.1 \text{ m} \times 9.53 \text{ m}} = 0.075 \text{ m/s}$$

   This is acceptable; it is much lower than the maximum allowable distribution-channel velocity of 0.6 m/s typically used in wet well design.

7. Design baffle wall.

   The most important design consideration for a baffle wall is the size of the orifice openings that will create a head loss at-least equal to the upstream kinetic energy. The thickness of the baffle wall should be at least equal to the diameter of the orifice opening to fully develop a horizontal jet. Also, a large number of orifice openings should be used.

   a. Calculate the kinetic energy (h_v).

      The velocity in the distribution channel is 0.075 m/s. The upstream kinetic energy, h_v, is calculated as follows:

$$h_v = \frac{v^2}{2g} = \frac{(0.075 \text{ m/s})^2}{2 \times (9.81 \text{ m/s}^2)} = 0.0004 \text{ m}$$

      It is desirable to use head loss through the orifice that is at least four times the upstream kinetic energy, to ensure proper flow distribution. For this example, use a higher value, because floc break-up is not a concern on raw water (as it is in a sedimentation basin). The higher value will improve distribution even more. Use head loss across the baffle port equal to 0.01 m.

   b. Calculate the velocity through the orifice. The velocity is calculated from orifice equation (Eq. (6.4)). For a head loss of 0.01 m,

$$\text{Velocity}, v = C_D(h_L 2g)^{1/2}$$

      Assume $C_d = 0.6$.

$$v = (0.6)[(0.01 \text{ m})(2 \times 9.81 \text{ m/s}^2)]^{1/2} = 0.27 \text{ m/s}$$

   c. Calculate the total area of the openings needed to achieve this velocity.

$$A = \frac{Q}{v} = \frac{0.789 \text{m}^3/\text{s}}{0.27 \text{m/s}} = 2.92 \text{m}^2$$

   d. Calculate the total number of ports if 25.4-cm (10-inch) ports are provided.

$$\text{Area for each port} = \frac{\pi}{4}(0.254 \text{m})^2 = 0.051 \text{m}^2$$

$$\text{Total number of ports} = \frac{2.92 \text{ m}^2}{0.051 \text{ m}^2/\text{opening}} = 58$$

Provide a total of 56 openings, to maintain symmetry.

e. Calculate opening spacing.

Provide ports at a center to center vertical spacing of 0.65 m. Use the configuration shown in Figure 7-16. The ports are arranged in 14 rows and 4 columns; this arrangement gives the desired number, 56 ports.

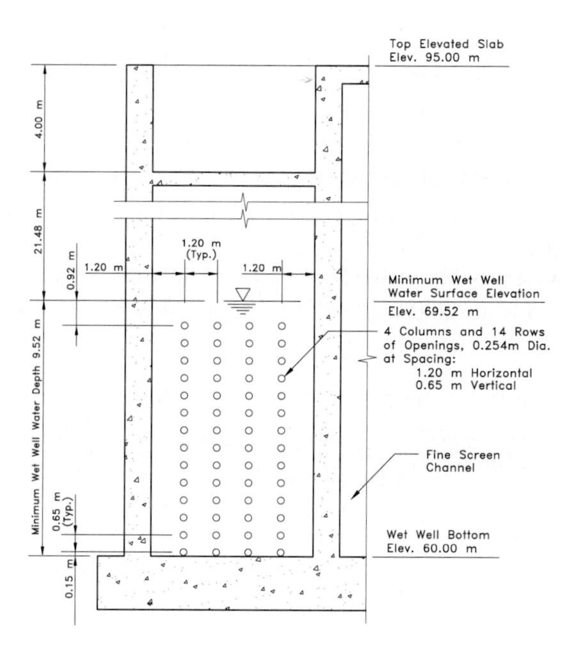

Figure 7-16 Design detail of baffle wall in the wet well.

Step D: Develop System Head Curve

1. Components of System Head Loss.

The total system head loss is basically the sum of all head losses from the wet well to the point of discharge in Junction Box A, which is located at the south boundary of the water treatment plant property (Figure 7-12(c)). In this example, vertical mixed-flow pumps are utilized and are located in the wet well. The wet well is connected to the intake structure through a suction conduit and several appurtenances. A drawdown of water-surface level in the wet well will be created during pumping. The total static head will increase because of this drop of water level. The system head curve calculations, therefore, will utilize the static head and head losses through the transmission pipeline.

2. Calculate the head losses from the reservoir to the wet well.

The head loss calculations from the reservoir to the wet well are needed to develop the drawdown curve during the pumping operation. Head losses are encountered in the following locations: coarse screen at the intake port, intake gate, suction conduit, isolation gate, fine screen, distribution channel, and baffle wall. The flow schematic (and locations where many of these head losses are encountered) are shown in Figure 7-11. This drawdown is relatively small in comparison with the static head and the head loss through the transmission pipeline, so the calculation procedure can conservatively be simplified by using a constant drawdown independent of flow rate and reservoir level. This simplified value is determined by the head losses from the reservoir to the wet well at the maximum flow and minimum reservoir-level condition. The calculation procedures for the head losses (or the drawdown) are listed in Table 7-9. The minor loss equation (Eq. (7.12)) is used for all losses except in screens, port, pipe, and baffle wall. The orifice equations (Eq. (6.4) or Eq. (6.5)) are used to calculate the head losses through the coarse-screen rack, fine screen, and baffle wall. Hazen–Williams (Eq. (7.14)) and Manning (Eq. (7.1)) equations are used respectively to calculate the losses through pipe and channel. A constant drawdown of 0.48 m (Table 7-9) is therefore used in the calculation of total system head loss.

The water surface level in the wet well varies in response to seasonal change in the reservoir levels. By subtracting constant drawdown from the reservoir levels, the design water surface level in the wet well is 89.52 m at the maximum reservoir level of 90.0 m and 69.52 m at the minimum reservoir level of 70.0 m, respectively.

3. Calculate the maximum and minimum static head.

Because of the variation of water surface in the wet well, there is a maximum and a minimum static head. They are determined by the difference in water surface between the minimum and maximum water surface level in the wet well, respectively, and the design water level in Junction Box A at the water treatment plant. By using a relatively constant design water level of 110 m in Junction Box A, the maximum and minimum static heads will be as follows:

Maximum Static Head

= Water Level in Junction Box A – Minimum Wet-Well Level

= 110 m – 69.52 m

= 40.48 m

Table 7-9 Calculations of Head Losses from Reservoir to Wet Well

Appurtenance	Equation Used	Flow Rate through Appurtenance, m/s	Dimension of Appurtenance	Velocity, m/s	Head Loss, m
Coarse screen at intake[a]	Eq. (6.5)	0.657	$H = 4.8$ m, $W_{net} = 3.028$ m	0.045	0.0001
Intake gates[b]	Eq. (6.4), $C_d = 0.6$	0.657	$H = 3.0$ m, $W = 3.0$ m	0.073	0.0008
Suction-conduit entrance[c]	Eq. (7.12), $K = 0.5$	1.314	$D = 1.52$ m	0.724	0.0134
Suction conduit[c]	Eq. (7.14), $C = 120$	1.314	$D = 1.52$ m, $L = 135$ m	0.724	0.0442
Suction-conduit exit[c]	Eq. (7.12), $K = 1.0$	1.314	$D = 1.52$ m	0.724	0.0267
Isolation gate[d]	Eq. (6.4), $C_d = 0.6$	1.314	$H = 0.9$ m, $W = 0.9$ m	1.622	0.3724
Fine screen[d]	Eq. (6.5), $C_d = 0.6$	1.314	$H = 9.53$ m, $W = 0.9$ m	0.274	0.0106
Screen and distribution channels[e]	Eq. (7.1), $n = 0.013$	0.789	$H = 9.53$ m, $W_{ave} = 1.2$ m $L_{approx} = 15$ m	0.070	0.0000
Baffle wall[f]	Eq. (6.4), $C_d = 0.6$	0.789	$D = 0.254$ m	0.278	0.0109
Total head loss (or drawdown)					0.4791

a The flow is divided between two ports; therefore, head loss calculations are based on half flow. The net open area of coarse screen is 14.53 m^2. (See Section 6.8.2, Step C.3 for detailed calculation.)

b The flow is divided between two gates. (See Section 6.8.2, Step D2 for detailed calculation.)

c Head loss calculations are based on 152-cm-diameter pipe. It is desirable to limit the maximum velocity in the suction conduit to less than 1.0 m/s to limit head loss and turbulence.

d The maximum flow through one side of the channel is 0.789 m^3/s if one fine screen is out of service. Under this condition, the isolation gate upstream of the screen will be closed, and the inner isolation gate on the common wall between the two distribution channels will be open. Therefore, the maximum design flow of 1.314 m^3/s will pass through one side of the distribution channel. An area-reducing factor of 0.56 is used in the calculation.

e The fine screen and distribution channels are shown in Figure 7-12(a). The average width of these channels is 1.2 m. The approximate length of 15 m is used for the channels downstream of the fine screen. Eq. (7.1) is rearranged $h_L = L_{approx}\,[(Q/A_{ave})/R^{2/3}\,)]^2$. $A_{ave} = 9.53\text{m} \times 1.2\text{m} = 11.4\ \text{m}^2$. Manning's $n = 0.013$. $R = 9.53\ \text{m} \times 1.2\ \text{m}/((2 \times 9.53\ \text{m}) + 1.2\ \text{m}) = 0.56$ m. The frictional head loss is 0.00002, which is assumed to be zero.

f The head loss is obtained from Eq. (6.4). The maximum flow through one side of the wet well is 0.789 m^3/s. A total of 56 openings are in the baffle wall of each wet well.

Minimum Static Head

= Water Level in Junction Box A – Maximum Wet-Well Level

= 110 m – 89.52 m

= 20.48 m

4. Calculate the head loss through the transmission pipeline.

The total head losses in the transmission pipeline are the sum of (1) minor head losses due to the turbulence created at various valves and fittings and (2) the friction head losses through the straight pipe lengths from the start of the common header to the discharge point in Junction Box A. The pipe arrangement and various valves, turns, fittings, and specials are shown in Figures 7-12(c), 7-14, and 7-15. The head loss calculations are made by using proper diameters of pipeline sections and K coefficients for various gates, valves, turns, fittings and specials. This important design information are summarized in Table 7-10. The head loss computations for each appurtenance are then made for different flow conditions, to cover the entire anticipated flow ranges. A suitable spreadsheet computer program is usually required to perform these calculations. The calculation results at flow rates up to 200,000 m^3/d (2.315 m^3/s) are tabulated in Table 7-11. Example calculations at design flow of 1.314 m^3/s (maximum daily demand) are given below to illustrate the procedure.

a. Velocity for each pipe size

$$\text{Velocity in pipe segments 1–8 } (D = 107 \text{ cm}) = \frac{4 \times 1.314 \text{ m}^3/\text{s}}{\pi \times (1.07 \text{ m})^2} = 1.461 \text{ m/s}$$

$$\text{Velocity in pipe segments 9–14 } (D = 122 \text{ cm}) = \frac{4 \times 1.314 \text{ m}^3/\text{s}}{\pi \times (1.22 \text{ m})^2} = 1.124 \text{ m/s}$$

b. Minor head losses due to turns, fittings and specials

The minor head losses are calculated from Eq. (7.12):

$$\text{Minor head losses in pipe segments 1–8 } (D = 107 \text{ cm}) = \frac{3.45 \times (1.46 \text{ m/s})^2}{2 \times 9.81 \text{ m/s}^2} = 0.375 \text{ m}$$

$$\text{Minor head losses in pipe segements 9–14 } (D = 12 \text{ cm}) = \frac{2.05 \times (1.124 \text{ m/s})^2}{2 \times (9.81 \text{ m/s}^2)} = 0.132 \text{ m}$$

c. Venturi meter selection and head loss

The Venturi meter selected should provide accurate metering over the entire flow range from 0.3 m^3/s to 1.3 m^3/s (an operating range of 1 to 4.3). At the lower end of the range, the differential should be at least 10 cm, because this is the lower range of the measurable differential pressures. At the high end, the differential should be as

Table 7-10 Dimensional Information and Head-Loss Coefficients of Transmission Pipelines

Pipe Segment[a]	Size, cm	Length, m	*C* Coefficient	Minor Head Loss Coefficient *K*				
				Type	Number	Location	Value of *K*	Total
Pipe Segments 1–8	107	47	120	Tee, runs	5	Between Segments 1&6	0.6	3.0
				45° Elbows	2	Between Segments 6&7, Segments 8&9	0.225	0.45
				Total *K* for Segments 1–8				3.45
Venturi meter	48	N/A	N/A	N/A	1	Between Segments 7&8	0.098	0.098
Increaser	107 x 122	N/A	N/A	Divergent	1	Between Segments 8&9	0.078	0.078
Pipe Segments 9–14	122	5200	120	22.5° Elbows	2	Between Segments 12&13, Segments 13&14	0.15	0.3
				45° Elbows	2	Between Segments 10&11, Segments 11&12	0.225	0.45
				90° Elbows	1	Between Segments 9&10	0.3	0.3
				Total *K* between Segments 9–14				1.05

a Pipe segments, bends, and fittings are shown in Figure 7-12 (c).

low as possible while still meeting the minimum differential of lower flows. The approach sections should be 107 cm, to match the discharge piping of the Venturi meter.

The design information about a Venturi meter is obtained from the manufacturer's catalog.[26] A 107-cm Venturi meter with a 48-cm throat will meet these requirements. The differentials are calculated from the Eq. (7.39).

$$\frac{h_1}{h_2} = \left(\frac{Q_1}{Q_2}\right)^2 = \left(\frac{v_1}{v_2}\right)^2 \tag{7.39}$$

where

Q_1 and Q_2 = flow rates for two conditions, say condition 1 and 2, respectively; m^3/s (ft^3/s)
h_1 and h_2 = differentials corresponding to Q_1 and Q_2, m (ft)

The head loss through a Venturi meter is typically 10 percent of the differential. The manufacturer generally provide the differential value at unit flow. For the selected meter the differential at a flow of 1.0 m^3/s is 1.52 m. From these values, K value in Eq. (7.12) is calculated:

$$\text{Velocity through Venturi throat } (D = 48 \text{ cm}), \text{ flow } 1.0 \text{ m}^3/\text{s} = \frac{4 \times 1.0 \text{ m}^3/\text{s}}{\pi \times (0.48 \text{ m})^2} = 5.526 \text{ m/s}$$

$$\text{Head loss through Venturi meter, flow } 1.0 \text{ m}^3/\text{s} = 1.52 \text{ m} \times 0.1 = 0.152 \text{ m}$$

Thus,

$$0.152 \text{ m} = \frac{K(5.526 \text{ m/s})^2}{2 \times (9.81 \text{ m/s}^2)}$$

$$\text{or } K = 0.098$$

$$\text{Velocity through Venturi meter throat, flow } 1.314 \text{ m}^3/\text{s} = \frac{4 \times 1.314 \text{ m}^3/\text{s}}{\pi \times (0.48 \text{ m})^2} = 7.261 \text{ m/s}$$

Therefore,

$$\text{Minor head loss through Venturi meter, flow } 1.314 \text{ m}^3/\text{s} = \frac{0.098 \times (7.261 \text{ m/s})^2}{2 \times (9.81 \text{ m/s}^2)} = 0.263 \text{ m}$$

The meter is placed downstream of the long straight section of the pipe. The manufacture recommends that the length of the straight pipe section be 10 times greater than the diameter of the pipe.[26] In this case, 20 meters of straight piping (more than 18 times the pipe diameter) is provided. The Venturi meter and vault layout are shown in Figure 7-17.

Table 7-11 Calculations of Head Losses Through Transmission Pipelines

Flow,[a]		Velocity, m/s			Head Loss, m						
					Minor				Friction		
m^3/d	m^3/s	Pipe Segs. 1–8	Pipe Segs. 9–14	Venturi Meter	Pipe Segs. 1–8	Pipe Segs. 9–14	Venturi Meter	Increaser	Pipe Segs. 1–8	Pipe Segs. 9–14	Total
0	0.000	0.000	0.000	0.000	0.000	0.000	0.000	0.000	0.000	0.000	0.000
20,000	0.231	0.257	0.198	1.279	0.012	0.004	0.008	0.000	0.003	0.200	0.228
40,000	0.463	0.515	0.396	2.558	0.047	0.016	0.033	0.000	0.012	0.722	0.830
60,000	0.694	0.772	0.594	3.838	0.105	0.037	0.074	0.001	0.026	1.528	1.770
80,000	0.926	1.030	0.792	5.117	0.186	0.066	0.131	0.002	0.045	2.602	3.031
100,000	1.157	1.287	0.990	6.396	0.291	0.102	0.204	0.003	0.067	3.931	4.599
113,500	1.314	1.461	1.124	7.261	0.375	0.132	0.263	0.003	0.085	4.969	5.828
120,000	1.389	1.545	1.188	7.675	0.420	0.147	0.294	0.004	0.094	5.508	6.447
140,000	1.620	1.802	1.386	8.955	0.571	0.201	0.401	0.005	0.125	7.326	8.629
160,000	1.852	2.059	1.584	10.234	0.746	0.262	0.523	0.007	0.161	9.379	11.077
180,000	2.083	2.317	1.782	11.513	0.944	0.332	0.662	0.009	0.200	11.662	13.808
200,000	2.315	2.574	1.980	12.792	1.165	0.410	0.817	0.011	0.243	14.172	16.818

a Only selected flow calculations are tabulated. To draw smooth curves small, flow intervals are required.

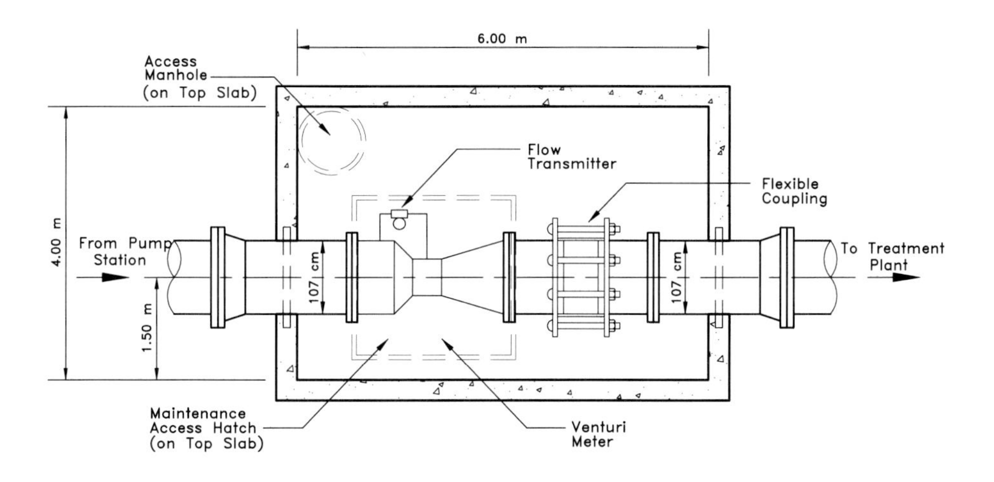

Figure 7-17 Venturi-meter vault layout.

d. Increaser selection and head loss

The head loss in an increaser is calculated from Eq. (7.40).

$$h_L = \frac{(K)(v_1^2 - v_2^2)}{2g} \tag{7.40}$$

where

h_L = minor head loss in increaser, m (ft)

v_1 and v_2 = velocity in smaller and larger sections of the enlarger, m/s (ft^3/s)

K = headloss constant

Select an increaser 107 cm × 122 cm in the force main. Using a 10-degree cone angle and $K = 0.078$, the head loss at flow rate of 1.314 m^3/s is as follows:

$$\text{Minor head loss in increaser } (h_L) = \frac{0.078 \times [(1.461 \text{ m/s})^2 - (1.124 \text{ m/s})^2]}{2 \times 9.81 \text{ m/s}^2} = 0.003 \text{ m}$$

e. Friction head losses through the pipes

The friction head loss in a pipe is calculated from Eq. (7.14). The pipe segments 1 through 8 have a pipe diameter D = 107 cm and a total length L = 47 m.

$$h_f = 6.82 \times \left(\frac{1.461 \text{ m/s}}{120}\right)^{1.85} \times \left(\frac{47 \text{ m}}{(1.07 \text{ m})^{1.167}}\right) = 0.085 \text{ m}$$

The pipe segments 9 through 14 have a pipe diameter $D = 122$ cm and a total length $L = 5200$ m.

$$h_f = 6.81 \times \left(\frac{1.124 \text{ m/s}}{120}\right)^{1.85} \times \left(\frac{5200 \text{ m}}{(1.22 \text{ m})^{1.167}}\right) = 4.971 \text{ m}$$

f. Total head losses in the pipes

The total head loss in the transmission pipeline is the sum of all losses.

$$\text{Total head loss} = (0.375 \text{ m}) + (0.132 \text{ m}) + (0.263 \text{ m}) + (0.003 \text{ m}) + (0.085 \text{ m}) + (4.971 \text{ m}) = 5.83 \text{ m}$$

This value of head loss may also be found in Table 7-11.

5. Prepare system head curves.

The system head curves are developed by adding all losses in the transmission pipeline (including Venturi meter and increaser) to the static heads. The summary computations of total dynamic head under maximum and minimum static-head conditions used for preparation of the system head curve are provided in Table 7-12. Figure 7-18 shows the system head curves under maximum and minimum static-head conditions.

Table 7-12 Total Dynamic Head Data Used to Develop System Head Curves

Flow,		Static Head[a], m		Total Head Losses through Transmission Pipeline[b], m	Total dynamic head, m	
m^3/d	m^3/s	Maximum	Minimum		at Maximum Static Head	at Minimum Static Head
0	0.000	20.48	40.48	0.000	20.48	40.48
20,000	0.231	20.48	40.48	0.228	20.71	40.71
40,000	0.463	20.48	40.48	0.830	21.31	41.31
60,000	0.694	20.48	40.48	1.770	22.25	42.25
80,000	0.926	20.48	40.48	3.031	23.51	43.51
100,000	1.157	20.48	40.48	4.599	25.08	45.08
113,500	1.314	20.48	40.48	5.828	26.31	46.31
120,000	1.389	20.48	40.48	6.447	26.95	46.95
140,000	1.620	20.48	40.48	8.629	29.11	49.11
160,000	1.852	20.48	40.48	11.077	31.56	51.56
180,000	2.083	20.48	40.48	13.808	34.29	54.29
200,000	2.315	20.48	40.48	16.818	37.30	57.30

a See Section 7.8.2 Steps D.2 and D.3 for detailed calculations.

b From Table 7-11.

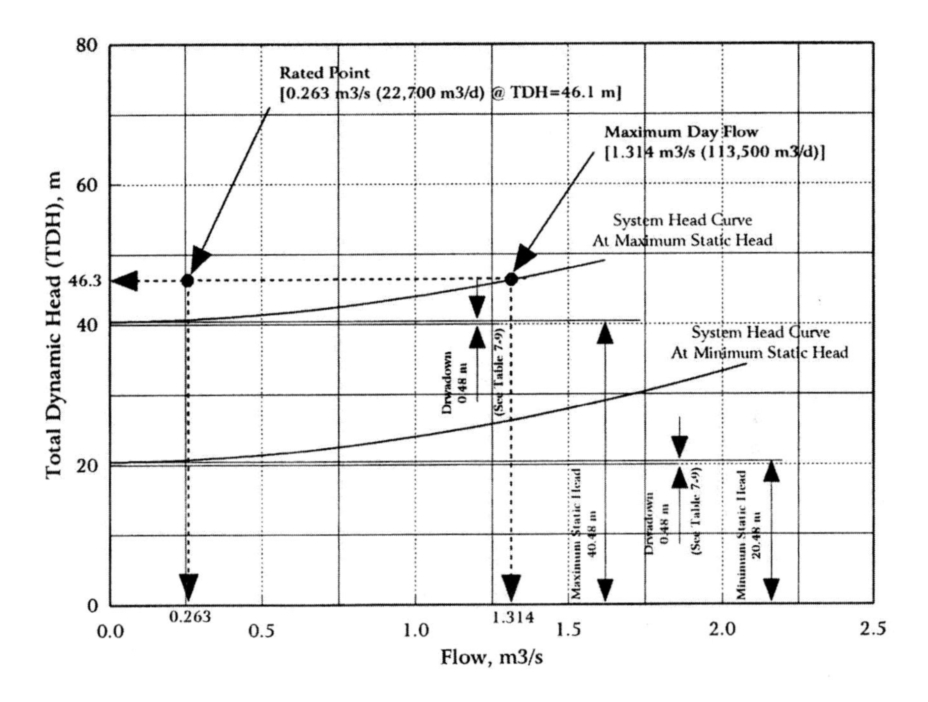

Figure 7-18 System head capacity curves under maximum and minimum static head conditions.

6. Determine pump rated point

The pump rated point is one of the most important values used for pump selection. The rated point corresponds to a required head and discharge condition that must be provided by a single pump in parallel operation delivering the station capacity. The required head equals the *TDH* value on the system head curve at maximum static head and the pump station design capacity. The required discharge is the proportional share of station capacity attributed to each single pump only. In this design example, the rated point is established from system head curve at maximum static head and the maximum day flow (Figure 7-18).

$$\text{Rated total dynamic head} = 46.3 \text{ m}$$

$$\text{Rated flow rate} = \frac{1.314 \text{ m}^3/\text{s}}{5 \text{ pump units}} = 0.263 \text{ m}^3/\text{s}$$

The procedure for pump selection using this pump rated point is shown next.

Step E: Develop Modified Pump and Select Pump Curves

1. Calculate the station head losses.

   The manufacturer head-capacity curve is generally developed from the gauge readings at the suction and discharge lines. The gauges are typically located prior to the suction bell and after the discharge nozzle. In a pump station, each pump has additional suction and discharge piping and appurtenances from the suction bell to the discharge common header. It is therefore necessary to account for the losses encountered. These head losses are commonly called *station head losses* and are calculated from the suction bell to the common header. It may be noted that the manufacturers, while developing the head-capacity curve of a vertical turbine pump, commonly include the suction bell, vertical riser pipe, and horizontal nozzle in the test piping. Therefore, the head losses due to these items should not be included. In a vertical turbine pump, the station head losses thus involve only the losses from the pump discharge nozzle to the common header. The dimensions, friction, and minor head loss coefficients for the discharge piping, valves, fittings, and specials are given in Table 7-13. The pump station piping layout and appurtenances are shown in Figure 7-15. The station head loss calculations are then conducted at selected flows to cover the anticipated flow ranges. The calculations for selected flows are provided in Table 7-14. The relationship between the station head losses and flows is presented in the next section.

2. Select the pump.

   The pump selection for a desired application is essentially a trial-and-error procedure. The design engineer must review the manufacturers' catalogs to identify a pump that will meet the rated point as well as other specific design criteria required for the raw water pumps. After a careful review of several manufacturers' catalogs, a vertical turbine pump is selected. The pump is manufactured by Ingersoll-Dresser Pumps (pump Series APW, 880 rpm).[27] The pump has three stages, with a 376-mm (14.8 in) impeller in each stage[c]. The characteristic curve of this pump is shown in Figure 7-19. From this curve, the total dynamic head values for single stage is obtained for various flow rates. These readings are included in Table 7-14 and are plotted as Curve (b) in Figure 7-20. Additionally, the station losses for these selected flows (Table 7-14) are used to develop the station head loss curve. This curve is shown as Curve (a) in Figure 7-20. The original multistage pump head-capacity curve is obtained by multiplying the manufacturer's single-pump head value by the number of stages (three stages in this case), and then plotting against the corresponding pump discharges. This curve for a three-stage pump is shown as Curve (c) in Figure 7-20, and the expected performance data are also included in Table 7-14. This pump has a shut-off head of 21.9 m and approximate run-out point of 8.5 m on the sin-

c. It is a good practice to select initially a pump with a small impeller. In the future, this impeller can be replaced by a larger impeller to increase the capacity. (See Section 7.4.3.)

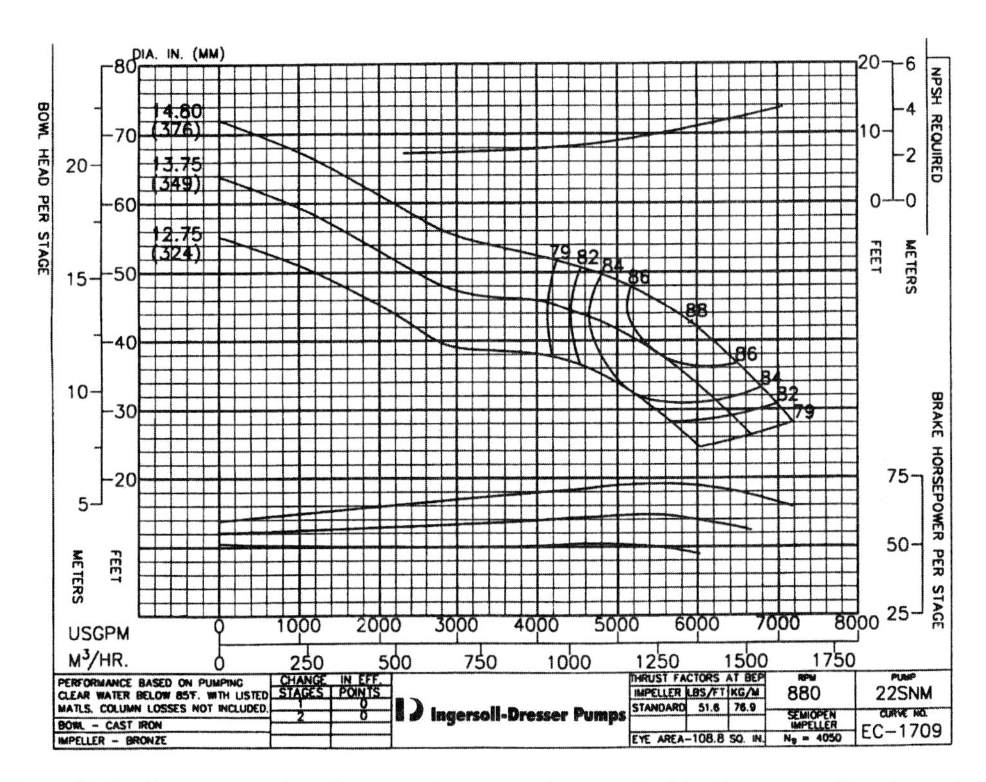

Figure 7-19 Pump characteristic curves for a one-stage vertical turbine pump. *(Courtesy of Ingersoll-Dresser Pumps.)*

gle-stage pump curve (Curve (b)). The maximum and minimum total dynamic heads of 65.7 m and 25.5 m will be developed by the selected pump with three-stage pumping (Curve (c)).

3. Prepare modified pump head-capacity curves.

   The total dynamic head for the modified pump curve is prepared by subtracting the calculated station head losses (Curve (a) in Figure 7-20) from the original three-stage pump head-capacity curve. This modified curve is also shown as Curve (d) in Figure 7-20, and the calculation results are also provided in Table 7-14. It is suggested that the design engineer should consult the manufacturer's representative to verify the operating range, pump rated point, and other features. Additional information on preparation of modified pump curves is presented in Section 7.4.4.

4. Prepare parallel combination curves.

   The parallel-pump combination curves are prepared using the modified head capacity curve for one pump with three impellers in a series (three stages). The procedure for pre-

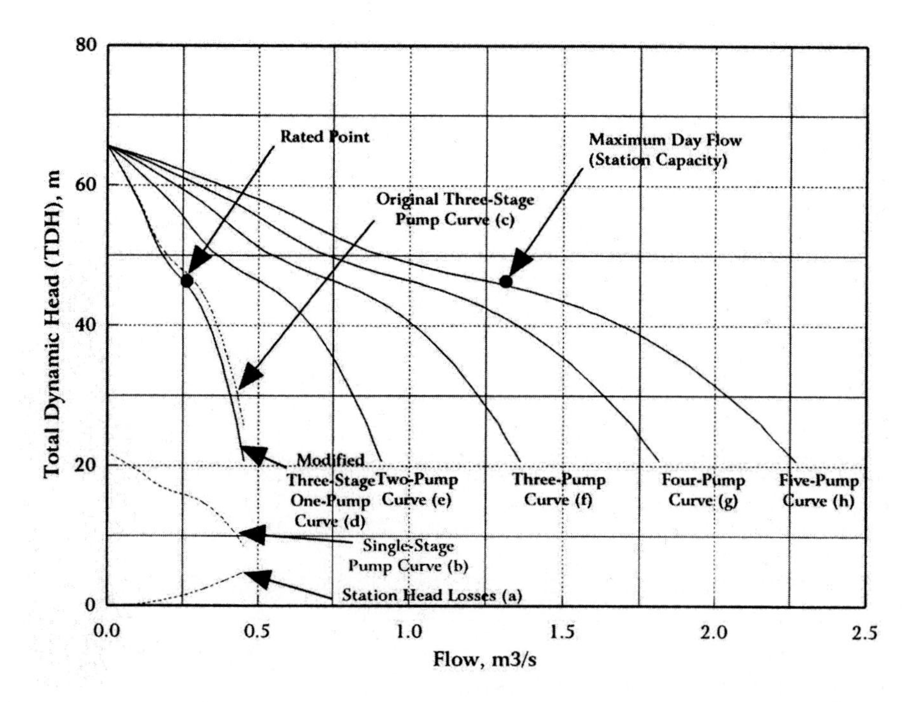

Figure 7-20 Pump head-capacity curves. The following curves are shown: (1) station head loss; (2) single-stage pump curve, (3) three-stage pump curve; (4) modified three-stage one-pump curve; and (5) curves for two, three, four, and five pumps in parallel combination.

paring parallel-pump combination curves is discussed in Section 7.4.4. The parallel combinations with two, three, four, and five pumps are shown as Curves (e) through (h) in Figure 7-20.

Step F: Evaluate Pump Station Design

1. Evaluate pumping capacities.

   The pump operating head and capacity under different pumping conditions are obtained by combining the system head curves and pump-combination head-capacity curves. The procedure is illustrated in Figure 7-21. The intersection points of system curves and pump combination curves are the *operating points* and can be used to determine the pumping capacity at different operating conditions. The raw water pumping capacities are evaluated under six different operating conditions. (See operating points A through F in Figure 7-21.) Information about these operating points is summarized in Table 7-15.

Table 7-13 Dimensional Information and Coefficients of Pump-Discharge Piping

				Minor Head-Loss Coefficient K				
Pipe Segment[a]	Size, cm	Length, m	C Coefficient	Type	Number	Location	Value of K	Total
Pipe segment from pump discharge disc to increaser	35.5	5.25	120	Check valve	1	Between flexible coupling and 90° elbow	2.5	2.5
				90° Elbow	1	Vertical bend	0.3	0.3
				Butterfly valve	1	Ahead of increaser	1.2	1.2
				Total K for this segment				4.0
Increaser	35.5 x 60	N/A	N/A	Divergent	1	Between butterfly valve and tee branch	0.08	0.08
Tee branch	60 x 107	N/A	N/A	Tee, branch to run	1	Between increaser and common header	1.8	1.8

a Pipe segments, valves, bends, and fittings are shown in Figure 7-15.

Table 7-14 Calculations of Station Head Losses and Modified Pump Head-Capacity Curves

Flow,		Velocity, m/s		Station Head Loss, m					Total Dynamic Head, m		
					Minor		Friction			Multistage	
m^3/d	m^3/s	Pipe Size 35.5 cm	Pipe Size 60 cm	Pipe Size 35.5 cm	Pipe Size 60 cm	Increaser	Pipe Size 35.5 cm	Total	Single Stage	Original	Modified
0	0.000	0.000	0.000	0.00	0.000	0.000	0.000	0.000	21.90	65.70	65.70
3,000	0.035	0.351	0.123	0.025	0.001	0.000	0.002	0.029	21.10	63.30	63.27
6,000	0.069	0.702	0.246	0.100	0.006	0.002	0.009	0.116	20.20	60.60	60.48
9,000	0.104	1.052	0.368	0.226	0.012	0.004	0.019	0.261	19.30	57.90	57.64
12,000	0.139	1.403	0.491	0.401	0.022	0.007	0.032	0.462	18.15	54.45	53.99
15,000	0.174	1.754	0.614	0.627	0.035	0.011	0.048	0.721	17.10	51.30	50.58
18,000	0.208	2.105	0.737	0.903	0.050	0.015	0.068	1.036	16.50	49.50	48.46
21,000	0.243	2.456	0.860	1.229	0.068	0.021	0.090	1.408	16.02	48.08	46.74
22,700	0.263	2.654	0.929	1.436	0.079	0.025	0.104	1.644	15.75	47.25	45.76
24,000	0.278	2.806	0.982	1.606	0.089	0.027	0.115	1.837	15.60	46.80	44.96
27,000	0.313	3.157	1.105	2.032	0.122	0.035	0.143	2.322	15.00	45.00	42.53
30,000	0.347	3.508	1.228	2.509	0.138	0.043	0.174	2.864	14.00	42.00	39.06
33,000	0.382	3.859	1.351	3.036	0.167	0.052	0.208	3.463	12.55	37.65	34.56
36,000	0.417	4.210	1.474	3.613	0.199	0.062	0.244	4.118	10.85	32.55	28.58

Table 7-15 Raw Water Pumping Capacity Evaluation

Capacity Condition	Operating Point	Static Head Condition	Number of Pumps in Operation	Pumping Capacity, m^3/s	Modified Three-Stage TDH, m
Maximum-day flow	A	Maximum	5	1.29	46.1
	B	Minimum	3	1.29	26.1
Average-day flow	C	Maximum	2	0.634	42.0
	D	Minimum	2	0.884	23.3
Maximum capacity available	E	Minimum	5	1.96	32.8
Maximum $NPSH_{req}$ point	F	Minimum	1	0.451	21.3

2. Evaluate pump efficiency and determine power input.

The pump performance at rated point and at any other operating point can be evaluated by the information provided in pump characteristic curves by the manufacturer (Figure 7-19). These curves typically include the pump efficiency, power output, power input, and motor power. When multiple pumps are used at a pump station, the following procedure is necessary to obtain the desired information from the original pump characteristic curves and to evaluate the pump performance for a given pumping condition (Figure 7-21):

a. Obtain the capacity of an individual pump on the modified curve corresponding to the same total dynamic head when five pumps are in service (point A to A´, Figure 7-21).

b. Obtain total dynamic head from the operating point on the modified curve to the multistage head-capacity curve at the same capacity value (point A´ to A´´, Figure 7-21).

c. Obtain total dynamic head from the operating point on the multistage curve to the manufacturer's original single-stage head-capacity curve at the same capacity value (point A´´ to A´´´, Figure 7-21).

d. Compare the pump efficiency and power requirements in the specification against the manufacturer's recommended values corresponding to this point (point A´´´).

The preceding procedure is used to establish the efficiency and power requirements at each operating point. These values are provided in Table 7-16.

The maximum power input typically occurs at one of the operating points under minimum static-head condition. The maximum required motor output power (also called brake horsepower) per stage is 162 kW (217 hp) at operating points C and E. The required motor

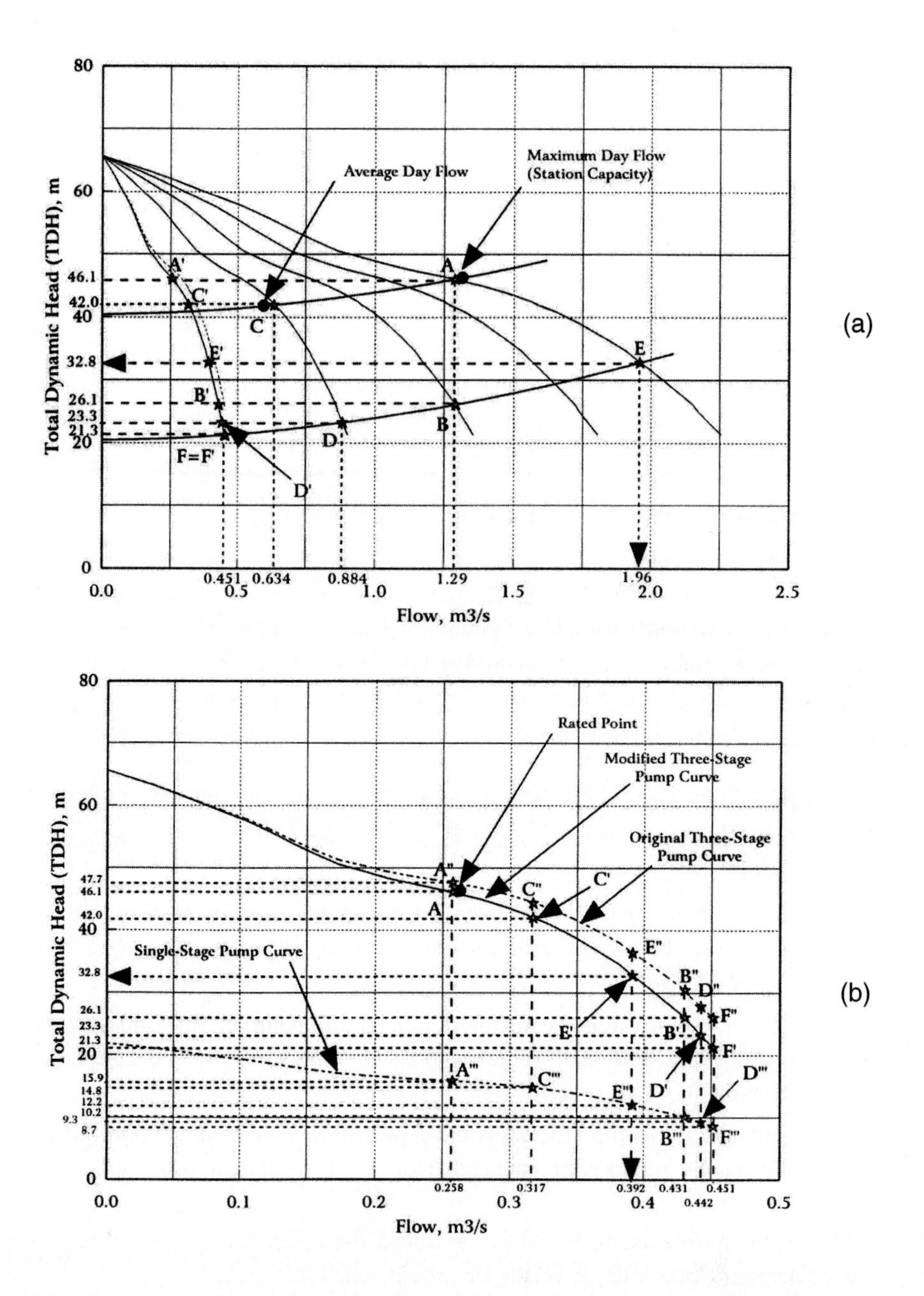

Figure 7-21 Raw water pump station operating points. (a) Operating points on parallel-combination head-capacity curves. (b) Operating points on single-stage, multistage, and modified head-capacity curves.

Table 7-16 Single-Stage Pump Operating Conditions and Characteristics

Operating Point	Each Pump Discharge Flow Rate, m3/s[a]	*TDH*, m			Characteristics				
							Power Requirements, kW		
		Modified Three-Stage Pump Curve (A′-F′)[a]	Original Three-Stage Pump Curve (A″-F″)[a]	Single-Stage Pump Curve (A‴-F‴)[a]	Pump Operating Efficiency, [b]%	Required NPSH,[b] m	Pump Output from Eq. (7.2 6)[c]	Pump Input or Motor Output from Eq. (7.2 7)[d]	Motor Input or Wire Power[e]
A	0.258	46.1	47.7	15.9	77	2.4	121	157	194
B	0.431	26.1	30.5	10.2	84	4.0	129	154	190
C	0.317	42.0	44.4	14.8	85	2.7	138	162	200
D	0.442	23.3	27.9	9.3	82	4.1	121	148	183
E	0.392	32.8	36.5	12.2	87	3.5	141	162	200
F	0.451	21.3	26.1	8.7	80	4.3	115	144	178

a See Figure 7-21 and Table 7-15 for more information.

b Pump efficiency and $NPSH_{req}$ are obtained from the manufacturer's curves (see Figure 7-19).

c Example calculations for pump output-power requirement corresponding to operating point A: 15.9 m × 0.258 m^3/s × 9.81 kN/m^3 × 3 (stages) = 121 kW.

d Example calculations for motor output-power requirement corresponding to operating point A: 121 kW/0.77 = 157 kW. In this design example, the maximum motor output-power requirement of 162 kW in this column occurs at operating points C and E, respectively. This value is used to select the motor.

e Motor wire power is obtained by assuming motor efficiency of 82 percent. Example calculations for wire-power requirement corresponding to operating point A: 157 kW/0.82 = 191 kW.

input power is therefore 200 kW (268 hp), for a motor efficiency of 81%, which is equivalent to a wire-to-water efficiency of 70%.

The maximum motor-output power of 162 kW (217 hp) is required when the pump is operated between points C‴ and E‴ on the single-stage pump curve (Figure 7-21). The motor selection is based on this value. The nearest available motor size to meet this power requirement is 168 kW (225 hp).

3. Check for available *NPSH* value.

   The available net positive suction head ($NPSH_{av}$) is calculated from Eq. (7.28) and is compared against the required *NPSH* value from the manufacturer. The maximum required *NPSH* value occurs at the operating point F, where a single pump is in service under minimum static-head condition. When vertical turbine pumps are used, the calculated *NPSH* value is much higher than the *NPSH* value required. Therefore, it is not calculated here. Procedure for calculating $NPSH_{av}$ may be found in References 8, 9, and 11 and the Example.

4. Details of pumping station and pump installation.

   The pump arrangement in the suction well should be carefully selected. Proper clearances between the pumps, baffles, and walls are essential. Figure 7-12 and Figure 7-14 show the dimensions of the wet well, distribution channels, screens, and isolation gates. Figures 7-12(b) and 7-14 show the pump base, pump discharge, valves and fittings, and the common header. Other details of the selected pump are provided in Figure 7-22.

   Each pump will have a pressure switch to monitor seal and water pressure, for pump protection. Pressure indicators will monitor discharge pressures of the pumps. Motor monitors (contacts provided in the motor control center) will monitor the on/off condition of each motor and any motor malfunction. The transformers in the motor control center will measure the current withdrawn by the motor, to monitor its efficiency. All pump motors will have a resistance temperature device to measure temperature of the motor bearings, for protection.

7.9 Operation, Maintenance, and Troubleshooting for Conveyance Systems, Pump Stations, and Flow Meters

7.9.1 Conveyance Systems

Common Operational Problems, Troubleshooting, and Suggested Solutions Conveyance systems are mechanically simple and do not require a great amount of maintenance. Following are some common problems and suggested corrective actions.

1. Excessive pumping heads at pump stations (or low flow rates) can be the result of high head losses in pressure conduits. The reasons may be (a) partially closed valve, (b) air accumulation at high points, or (c) excessive growth of slime in the pipe interior, scale

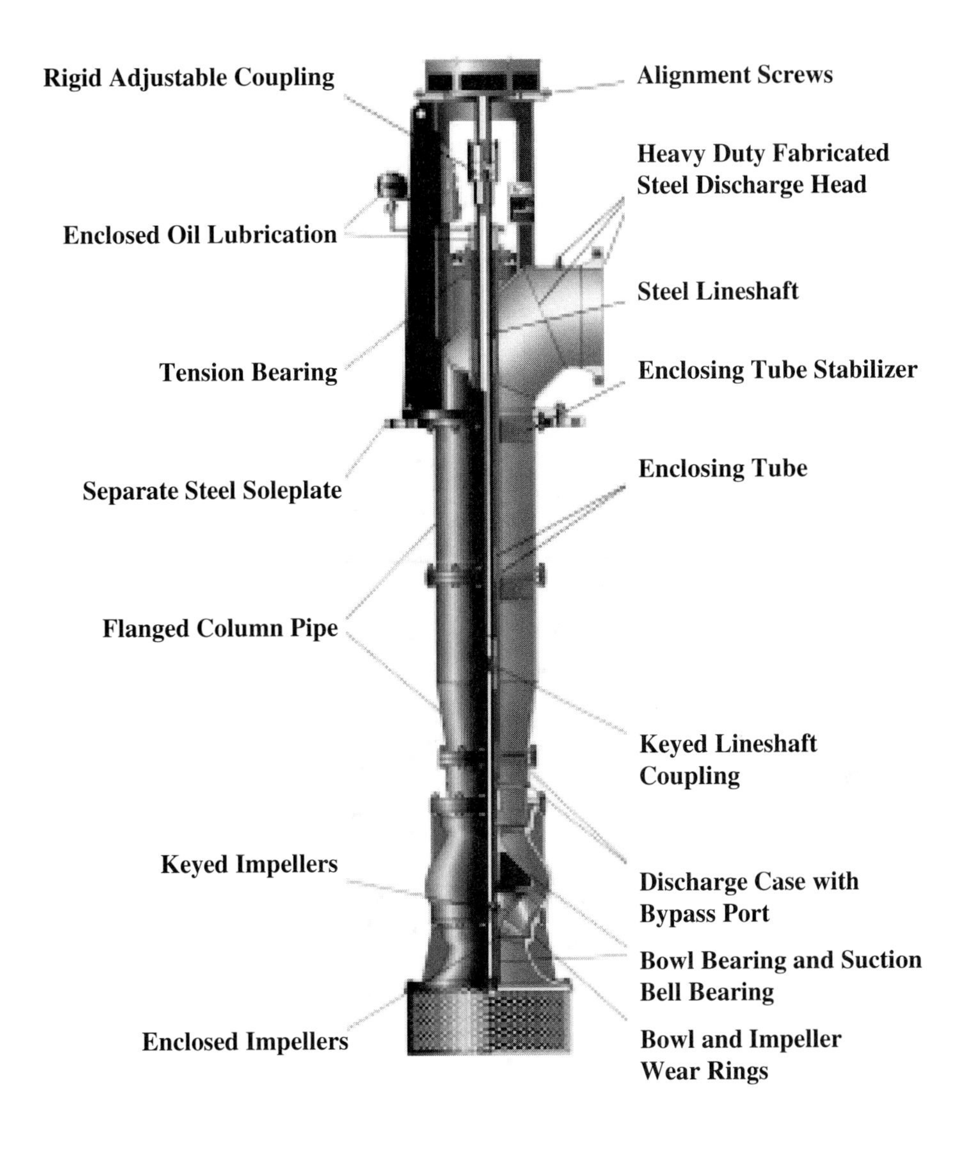

Figure 7-22 Construction details of selected raw water vertical turbine pumps. *(Courtesy of Ingersoll-Dresser Pumps.)*

formation, or other restrictions. The corrective actions include opening the valves fully, release of accumulated air, or installation of an air-relief valve if this problem persists. The pipe cleaning is achieved with pigs (large flexible plugs used to clean pipelines).

2. Excessive loss of water due to leakage from the water-conveyance system will reduce the flow. The pipelines may have leaks. Inspect alignment for existence of leaks and repair them. Also check for unauthorized connections.
3. Corrosion of concrete pipes is the result of dissolution of calcium components of the concrete. Check the stability of water, and add chemicals if necessary to reduce the corrosivity. This topic is covered in Chapter 13. Also, use lining materials that are inert to soft-water attack. These include concrete with high lime content, cement and silicious aggregate, plastic coatings, and plastic liners.

Operation and Maintenance Following are recommended operation and maintenance procedures for conveyance systems.

1. Maintain a record of pumping head versus flow rates. This will provide an indication of pipeline deterioration.
2. Regularly (monthly) inspect conveyance system alignments for (a) illegal or unauthorized connections, (b) right-of-way encroachments, (c) construction activity near pipelines, (d) slime growth in the pipes, (e) leaks, and (f) damaged structures.
3. Check the operation of air-release and vacuum valves bimonthly.
4. Inspect pipeline interiors annually for slime accumulation and for deterioration of lining material. Flush the lines to clean the deposits.

7.9.2 Pump Stations

Common Operational Problems and Troubleshooting Following are some common operational problems of pump stations:

1. Absence of pump discharge may be due to (a) insufficient water level of suction head to maintain prime, (b) motor running too slowly, (c) discharge head too high, (d) impeller or suction piping clogged, (e) rotation in wrong direction, (f) air leak in suction pipe, (g) impeller loose on shaft, and (h) closed suction or discharge valve. If any other problem is identified, proper corrective action shall be taken.
2. Low discharge of pumps may result from (a) air leak in suction pipe of pump-shaft seals, (b) discharge head too high, (c) suction head too low ($NPSH_{av}$ too low), (d) impeller worn, (e) wear rings worn out, (f) pipelines partially plugged, or (g) wrong rotation direction. Once the problem is identified, corrective actions can be taken.
3. Low discharge pressure of the pump is a result of (a) speed too low, (b) impeller worn out, (c) impeller too small, (d) system head lower than anticipated, and (e) wrong direction of rotation.

Operation and Maintenance Following are recommended operation and maintenance procedures for pump stations:

1. Inspect pump stations daily, to (a) determine pump condition, discharge head, motor and bearing temperature, and unusual noise, (b) clean pump station, (c) check valve positions, (d) check switchgear selector switch positions, and (e) note any pumps that are out of service or in need of maintenance.
2. Inspect pumps annually to determine the following: (a) impeller wear, (b) wear ring clearance, (c) bearing condition, and (d) seal condition.
3. Maintain logs on each pump, noting performance and any maintenance performed.

7.9.3 Venturi Meters

Operation and Maintenance Before plant start-up, inspect the Venturi meter for proper installation. The pressure-differential probe sensors shall be checked for proper connection. All tubings shall be checked frequently for clogging and slime growth. Erratic reading on Venturi meter may be due to clogged pressure taps. Flush the taps. Inspect flow meters regularly for visible problems, leaks, corrosion, and loose connections. Calibrate flow meters quarterly.

7.10 Specifications

Brief specifications of the raw water conveyance, raw water pump station, and flow meters just designed are presented next. These specifications are provided to describe the design more fully. Many details are omitted; therefore, these specifications should be used only as a guide to prepare more detailed specifications. Manufacturers' representatives should be consulted during the preparation or equipment specifications.

7.10.1 Conveyance System

Pipe The raw water pipeline shall be a 122 cm (48-inch) diameter reinforced concrete pressure pipe conforming to AWWA C30 with rubber gasket-type joints. It shall be designed for a maximum operating head of 50 m. (164 ft) and a maximum surge head of 70 m (230 ft). It shall be installed approximately 1.5 m deep. The backfill around the pipe shall be of sand material.

Valves Valves shall be butterfly type. They shall conform to AWWA C504 and be designed for a maximum head of 50 m. The valve seat shall be a resilient material and shall be mechanically adjustable. The seating surface shall be stainless steel. The valve body and disc shall be designed to close and open the valve under all head conditions.

7.10.2 Pump Stations

Structure The pump station structure shall be a well-type structure constructed of concrete. It shall have two wet wells, each of which can be isolated without affecting the operation of the other.

Pumps The pumps shall be three-stage vertical turbine pumps with appurtenances to meet the requirements. The pump is to deliver a maximum discharge of 0.451 m^3/s against a three-stage head of 21.3 m and a minimum capacity of 0.258 m^3/s against a three-stage head of 46.1 m.

The distance from the base of the discharge header to the bottom of the strainer shall not be less than 35 m. Pump speed shall not exceed 880 rpm. The pump will be installed in a sump having a minimum of 9.52 m. The efficiency at the design point shall be at least 80 percent. The pump shall have three bronze impellers, enclosed-type and statically balanced. They shall be securely fastened to the carbon-steel shafting. The suction case, intermediate bowls, and discharge bowl shall be of cast iron or steel. The pump impeller shaft shall be stainless steel, supported by bearings above and below each impeller. A galvanized basket strainer shall be furnished and have a net inlet area of not less than four times the suction inlet area to the impeller section.

Motor The pump motor shall be a minimum 162-kW, 3-phase, 480-volt electric motor. It shall have grease-lubricated antifriction bearings. The motor shall be equipped with a ratchet device to prevent reverse rotation of the pump and motor.

7.10.3 Venturi Meter

The Venturi meter shall have a 107-cm approach section, a 48-cm throat, and a full-diameter 107-cm recovery section. It shall be constructed of cast iron with a stainless steel throat section. Four pressure taps shall be provided at both the approach and throat sections. The Venturi meter shall be capable of measuring flows within the range indicated. The flow sensors shall measure the differential pressure, translate the differential pressure signals into an output signal, and transmit the signal by a transmitter to a central panel for recording.

Problems and Discussion Topics

7.1 Calculate the slope required for a 41-cm pipe to carry a flow of 12,000 m^3/d. Use (a) Manning's equation (Eq. (7.2)) with a 0.013 friction factor. (b) Hazen–Williams equation (Eq. (7.12)) with a 120 friction factor. Compare the results.

7.2 Calculate the diameter of a pipe line to carry 50L/s when flowing 60 percent full. The velocity in the line at partial-full condition must be 0.6 m/s (n=0.013).

7.3 A 60-cm diameter (24-in) pipeline is designed to carry a peak design flow of 0.01 m^3/s at a slope of 0.0008 m/m. Calculate the actual depth of flow and velocity. Assume n=0.013.

7.4 Calculate the size of circular pipe required for a flow of 9000 m^3/d, slope of 0.2 percent and a friction factor n=0.013 (use Manning's equation Eq. (7.2)). (a) Assume the pipe is flowing full. (b) Assume the pipe is flowing ∫ full.

7.5 Calculate the normal depth of flow in a trapezoidal canal with a slope of 0.1 percent, 2 meter bottom width and side slopes of 3 horizontal to 1 vertical. The average flow in the canal is 32 m^3/s. Assume n=0.015.

7.6 In the design example, the pump station was designed for the projected maximum design flow in the design year. In the future, the maximum design flow will increase and the pump station will need expansion. Discuss the alternatives available to expand the pumping capacity by about 20 percent.

7.7 Develop a system head-capacity curve for the pipeline shown in Figure 7-12(c). Use the following design information.

Maximum design flow—7800 m^3/d

Pipeline size —30 cm

7.8 A centrifugal pump is operating at a speed of 1200 rpm and discharge 2.5 m/min. The total head of 120 kPa is measured on a pressure gauge located in the discharge pipe. The power required is 7.0 kW. Calculate (a) the pump efficiency and (b) the discharge, head, and power, if the pump speed is changed to 1800 rpm.

7.9 A pump with a Francis–Vane impeller is selected for operation at a best efficiency of 87 percent. The operating capacity is 0.2 m^3/s against a total head of 16 m. Determine the operating speed and the specific speed.

7.10 A centrifugal pump is operating at an elevation of 1829 m above sea level. The pump requires 30 kPa net positive suction head (*NPSH*) when delivering water at 20 °C. What is the allowable suction lift of the pump if entrance and friction losses in the suction lines are 15 kPa? Assume the reduction in barometric pressure caused by weather change is 4.1 kPa.

7.11 Water is pumped from reservoir A to B. The water-surface elevations in the reservoirs A and B are 300 m and 318 m, respectively. The suction line is 9.5 cm, which is short enough that the head losses may be neglected. The pressure pipe is 400 m long and 79 cm in diameter. There are three 90° and two 45° bends. Calculate the annual power bill for pumping 0.30 m^3/s. Unit power cost is 5 cents per kW-h. Assume that C=100 and combined pump and motor efficiency is 78 percent.

7.12 The data for characteristic curves of a variable-speed pump supplied by a manufacturer are given below. Draw the modified curve and head-capacity curve for two pumps operating in parallel. Also draw the system head-capacity curves. Use the following data (piping, valves, fittings and specials from wet well to common header are given below):

Pipe size = 46 cm

Pipe straight length is short and negligible

Two 46-cm gate valves, $K = 0.19$

One 46-cm 90° elbow, $K = 0.30$

Two 46-cm ∞ 31-cm reducers, $K = 0.25$

One 46-cm check valve, $K = 2.5$

One 46-cm tee, $K = 1.80$

One 62-cm entrance suction bell, $K = 0.04$

(piping, valves, fittings and specials from common header to discharge are given below):

Pipe size = 76 cm

Pipe straight length = 65 m, $C = 100$

Two 76-cm 90° elbows, $K = 0.30$

Two 76-cm 45° elbows, $K = 0.20$

One Venturi meter with throat diameter of 38 cm, $K = 0.14$

One 76-cm plug valve, $K = 1.00$

One 76-cm Wye branch, $K = 0.38$

The pump volute elevation is 300.00 m. Maximum and minimum water-surface elevations in the wet well are 303.00 m and 301.00 m. The water-surface elevation in the discharge unit is 315.00 m. The maximum and minimum station discharges are 0.65 and 0.20 m^3/s. Determine (a) operating head and capacity of the pump station when both pumps are operating, (b) operating head and discharge of each pump, and (c) corresponding head and discharge given on the manufacturer's head-capacity curve.

Pump Characteristic Curve Data Supplied by the Manufacturer

Head, m	Discharge, m^3/s
20	0
15	0.30
10	0.45
5	0.50

7.13 A centrifugal pump delivers 0.15 m^3/s at 30 m *TDH* when operating at a speed of 2000 rpm. Determine the discharge, total dynamic head, and power input if the impeller diameter is reduced from 30 cm to 24 cm.

7.14 Determine the available *NPSH* of a pump that delivers water at 20 °C. The pump is operating at an elevation of 40 m above sea level. The water-surface elevation in the wet well is 38 m, and the water-surface elevation in the discharge unit is 60 m. The capacity of a single pump is 0.2 m^3/s. The details of piping, valves, fittings, and specials from wet well to common pump header are as follows:

Pipe size = 25 cm

Pipe straight length = 10 m, $C = 100$

One 25-cm gate valve, $K = 0.10$

Two 25-cm 45° elbows, $K = 0.20$

Two 25-cm 90° elbows, $K = 0.25$

One 46-cm entrance suction bell, $K = 0.04$

7.15 The modified pump head-capacity curve and system head-capacity curve at the minimum water-surface level in the wet well is as follows. Using these data, determine the operating heads and capacities at minimum and maximum wet-well levels for one pump, two pumps in parallel, and two pumps in series. The difference between maximum and minimum wet-well elevations is 2.0 m.

Pump Head-Capacity Data		System Head-Capacity Data	
Head, m	Capacity, m^3/s	Head, m	Capacity, m^3/s
40	0	30	0
39	0.02	32	0.02
36	0.04	34	0.04
29	0.06	37	0.06
20	0.08	40	0.08
8	0.10	44	0.10

7.16 Calculate the differential pressure for a Venturi meter at 10,500 m^3/d. The differential pressure at 7500 m^3/d is 120 cm.

7.17 A Venturi meter is 30 cm × 20 cm. The differential gauge is deflected 5.52 cm when flow is 0.04 m^3/s. The specific gravity of the gauge liquid is 1.25. Determine the meter coefficient K and discharge coefficient C_I.

7.18 A Venturi meter with a 10-cm throat is installed in a pipe that is inclined upward at an angle of 45° to the horizontal. The diameter of the pipe is 20 cm. The distance between the pressure taps along the pipe is 1.5 , and the differential pressure is 69 kPa. Estimate the flow of water in the pipe.

7.19 A Venturi meter has a 30-cm throat. Calculate the head loss if $K = 0.12$ and the discharge is 0.1 m^3/s.

7.20 The Francis equation (Eq. (8.24)) is generally used to calculate the discharge over a rectangular weir. This equation takes into account the contracted width of the weir

$$Q = \frac{2}{3}C_d L' \sqrt{(2gH^3)}$$

where

Q = discharge flow, m^3/s
L' = contracted width, and $L' = L - 0.1nH$
n = number of end contractions
H = head over weir, m

A sharp-crested rectangular weir is used to measure flow in a rectangular channel that is 3 m width. The weir is 1 m above the floor of the channel and extends across the channel width. Calculate the depth in the channel upstream of the weir if discharge in the channel is 1.2 m^3/s. Assume $C_d = 0.6$.

7.21 The discharge over a triangular weir is expressed by the following equation (Eq. (9-16))

$$Q = \frac{8}{15}C_d\sqrt{2g}\tan\frac{\theta}{2}H^{5/2}$$

where

Q = discharge flow, m^3/s
H = head over V-notch, m
θ = angle of the notch, degree.

An effluent weir plate contain 25 90° V-notches. Calculate the discharge if head over the weir notches is 20 cm. Assume $C_d = 0.6$.

7.22 Visit a local water treatment plant and pump station. Determine the type and size of raw water pump station raw water pipeline and flow meter facilities. Prepare sketches of facility layouts. Ask the operator to explain any operational problems he experiences with these facilities.

References

1. Chow, V. *Open Channel Hydraulics*, McGraw Hill Book Co. New York, 1959.
2. Brater, E. F. and King, H. W. *Handbook of Hydraulics*, 6th ed., McGraw-Hill Book Co., New York, 1976.
3. Fox, J. A. *Hydraulic Analysis of Unsteady Flow in Pipe Networks*, The MacMillan Press Ltd, London, U.K., 1979.
4. Wylie, E. B. and Streeter, V. L. *Fluid Transients*, FEB Press, Ann Arbor, MI., 1983.

5. Wood, D. J. and Funk, J. E. *SURGE5 Addendum*, University of Kentucky, Lexington, KY, 1993.
6. Jullis, J. P. *Hydraulics of Pipelines: Pumps, Valves, Cavitation, Transients*, John Wiley and Sons, New York, 1989.
7. Hammer, J. *Water and Wastewater Technology*, 2d ed., John Wiley and Sons, New York, 1986.
8. American Society of Civil Engineers, Committee on Pipeline Planning, Pipeline Division. *Pipeline Design for Water and Wastewater*, American Society of Civil Engineers, New York, 1975.
9. Qasim, S. R. *Wastewater Treatment Plants Planning, Design and Operation*, 2d ed., Technomic Publishing Co., Lancaster, PA, 1999.
10. Daughtery, R. and Franzini, J. *Fluid Mechanics with Engineering Applications*, McGraw-Hill Book Co., New York, 1965.
11. Metcalf and Eddy, Inc. *Wastewater Engineering: Collection and Pumping of Wastewater*, McGraw-Hill Book Co., New York, 1981.
12. McGee, T. J. *Water Supply and Sewerage*, 6th ed., McGraw-Hill Book Co., New York, 1991.
13. Hydraulic Institute, Inc. *American National Standards for Centrifugal Pumps*, 15th ed., Hydraulic Institute, Inc., Parsippany, NJ, 1994.
14. Hicks, T. G. and Edwards, T. W. *Pump Application Engineering*, McGraw-Hill Book Co., New York, 1971.
15. Walker, R. *Pump Selection: A Consulting Engineers Manual*, Ann Arbor Science Publishers, MI, 1972.
16. Karassik, I. J., Krutzsch, W. C., Fraser, W.H., and Messina, J. P. *Pump Handbook*, 2d ed., McGraw-Hill Book Co., New York, 1986.
17. Stewart, H. L. *Pumps*, 2d ed., Macmillan Publishing Book Co., New York, 1986.
18. Bartlett, R. E. *Pumping Stations for Water and Sewage*, Applied Science Publishers, Ltd., London, U.K., 1977.
19. ASCE and AWWA. *Water Treatment Plant Design*, 2d ed., McGraw-Hill Publishing Co., New York, 1990.
20. Metcalf and Eddy, Inc. *Wastewater Engineering: Treatment, Disposal, and Reuse*, 3d ed., McGraw-Hill Inc., New York, 1991.
21. Carthew, G. A., Goehring, C. A., and Van Teylingen, J. E. "Development of Dynamic Head Loss Criteria of Raw Sludge Pumping," *Journal of Water Pollution Control Federation*, vol. 55, no. 5, pp. 472–483, 1983.
22. Sanks, R. L.,Tchobanoglous, G., Newton, D., Bosserman, B. E., and Jones, G. M. (editors). *Pumping Station Design*, Butterworths, Stoneham, MA, 1989.
23. Avallone, E. A. and Baumeister, T. (editors). *Mark's Standard Handbook for Mechanical Engineers*, 10th ed., McGraw-Hill Book Co., New York, 1996.
24. American Society of Heating, Refrigeration and Air Conditioning Engineers, Inc. *ASHRAE Handbook Fundamentals, I-P Edition*, Atlanta, GA, 1993.
25. Hydraulic Institute, Inc. *American National Standard for Pump Intake Design*, ANSI/HI 9.8-1998, Hydraulic Institute, Inc., Parsippany, NJ, 1998.
26. BIF, A Unit of General Signal. *Flow Meters Differential Producers and Flow Controllers*, Manufacturer's Literature, Re. No. 180. 21–1, 1983.
27. Ingersoll-Dresser Pump Co. *Water Resources Pump Manual*, Section 6, Vertical Turbine Pumps, Liberty Corner, NJ, Sheet 2330, June 1992.

CHAPTER 8

Coagulation, Flocculation, and Precipitation

8.1 Introduction

Water from a natural source usually contains many dissolved and suspended solids. Large suspended particles, such as sand, behave as "discrete" particles and can readily be removed by sedimentation and/or filtration processes. Suspended particles at the lower end of the size spectrum do not readily settle. These are colloidal particles and can be removed by sedimentation and filtration only after physical and chemical conditioning. Chemical conditioning of colloids is known as *coagulation* and involves the addition of chemicals that modify the physical properties of colloids to enhance their removal. Physical conditioning is known as *flocculation.* This process involves gently mixing the suspension to accelerate interparticle contact, thus promoting agglomeration of colloidal particles into larger floc for enhanced settling.

Some dissolved minerals or solids in water can be precipitated as suspended solids by chemically or physically modifying the solution. Examples of such minerals are calcium and magnesium, which can be converted into an insoluble state by the addition of lime and soda ash, thus forming a precipitate of the mineral. The resulting precipitate can then be removed by conventional coagulation, sedimentation, and filtration processes. This process of removal is known as *precipitation.*

In this chapter, the theory and design considerations of coagulation, flocculation, and precipitation processes are presented. The design example provides typical design calculations and details of coagulation and flocculation units. Operation and maintenance recommendations and equipment specifications for coagulation/flocculation systems are also covered in the Design Example.

8.2 Suspended Solids

Solids suspended in water include sand, soil, organic material, bacteria, virus, and other particulate material. Some of these materials pose a nuisance in water treatment. Characteristics of particulate pollutants in raw water sources are discussed in greater detail in Chapter 2. Typical size variations of particulates found in surface water are listed on Table 8-1.[1] Particles at the higher end of the spectrum (greater than 1 micron) will usually settle in quiescent water. Smaller particles will not settle readily. A suspension of particles that will not settle is known as a stable suspension. The particles that make up these suspensions are known as colloids.

Table 8-1 Particle Sizes Found in Water Treatment

Material	Particle Diameter Micron $\mu^{a,b}$
Viruses	0.005– 0.01
Bacteria	0.3– 3.0
Small colloids	0.001– 0.1
Large colloids	0.1– 1
Soil	1– 100
Sand	500
Floc particle	100–2000

a μ = micron or micrometer; $\mu = 10^{-6}$ m or 10^{-3} mm.
b $m\mu$ = millimicron = 10^{-6} mm.

8.2.1 Characteristics of Colloids

Colloidal particles are defined by size. Their size range is generally considered as being from 0.001 micron (10^{-6} mm) to one micron (10^{-3} mm).[2] Particles found within this size range are (1) inorganic particles, such as asbestos fibers, clays, and silts, (2) coagulant precipitates, and (3) organic particles, such as humic substances, viruses, bacteria, and plankton.[1] Colloidal dispersions have light-scattering properties; true solutions scatter very little light. The property of scattering light is generally measured in *turbidity* units. Colloidal solutions may be classified, by their affinity for the dispensing medium (water), as either *hydrophobic* or *hydrophilic*. Colloidal particles are hydrophobic (water-hating) when a weak affinity for water exists and hydrophilic (water-loving) when there is a strong affinity for water.[2] The unique behavior of colloidal particles is a result of surface phenomena. The colloids have a very large ratio of surface area to mass. Their mass is so small that the gravitational force has little effect on their behavior. The principal phenomena controlling the behavior of colloids are *electrostatic forces*, *van der Waals forces*, and *Brownian motion.*

Electrostatic Forces Electrostatic force is the principal force contributing to the stability of the colloidal suspensions. Most colloids are electrically charged. The nature of this charge varies somewhat, depending on the nature of the colloid. Metallic oxides are generally positively charged, while nonmetallic oxides and metallic sulfides are generally negatively charged. The result of this electrical charge is that colloids of similar charge will repel each other. Typically, negatively charged colloids predominate in natural waters.[2]

The surface charge on the colloids attracts ions of opposite charge, known as *counter ions.* These ions, which include hydrogen and other cations, form a dense layer adjacent to the particle known as the *stern layer.* Water molecules are also attracted to the colloidal particle. The attraction of water is due to the asymmetric electrical charge of water molecules. A second layer of ions, known as the *diffused layer*, is also attracted to the colloid. In this layer, ions of both electrical charge are attracted, but counter ions predominate.[3] The two layers together are often referred to as the *double layer.* A model of a colloidal particle and its double layer is illustrated in Figure 8-1.[4] In the diffused layer, the molecules of water are sufficiently bound to create a shear surface. The water molecules inside this shear surface will behave as if attached to the colloid; water molecules outside the shear surface will behave as if independent of the colloid.

As shown in the model, the electrical potential at the shear surface is known as the *zeta potential.*[2,4] The zeta potential is often measured in water treatment, to give operators an indication of the stability of the colloidal system or of the effectiveness of the coagulation process.

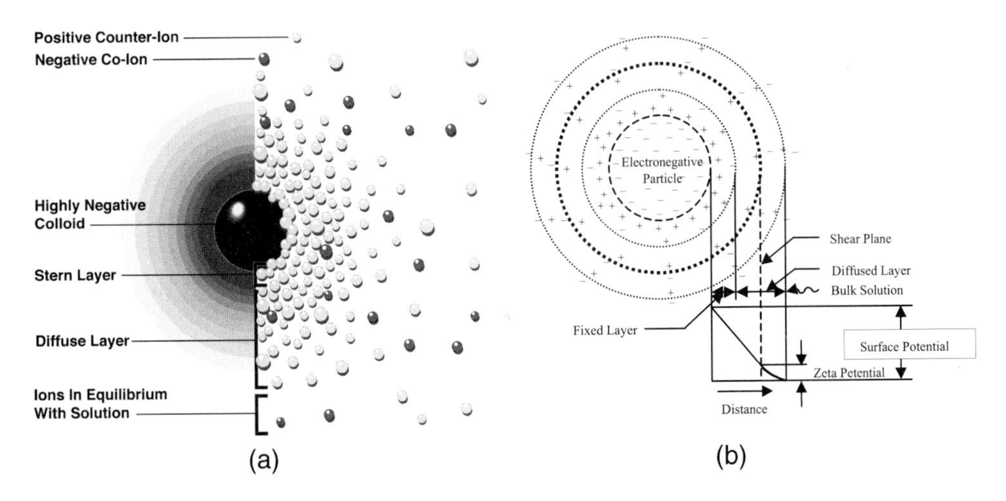

Figure 8-1 Model of a colloidal particle and double layer. (a) A negative colloidal particle with its electrostatic field. *(Adapted from Reference 4; with permission of L.A. Ravina and Zeta-Meter, Inc.)* (b) Distribution of electrostatic potential, shear plane, and zeta potential.

van der Waals Forces A force of attraction exists between any two masses. The magnitude of this attraction is a function of the mass of the two bodies and the distance between them. This attraction is known as the *van der Waals* forces.[2] In colloidal chemistry, van der Waals forces are the antithesis of electrostatic forces. As indicated in Figure 8-2, the force of repulsion due to the electrical charge will normally repel the colloids before they can move close enough for van der Waals forces to become significant.[2] If the magnitude of the electrostatic forces could be reduced, the particles would move close enough for van der Waals forces to predominate.

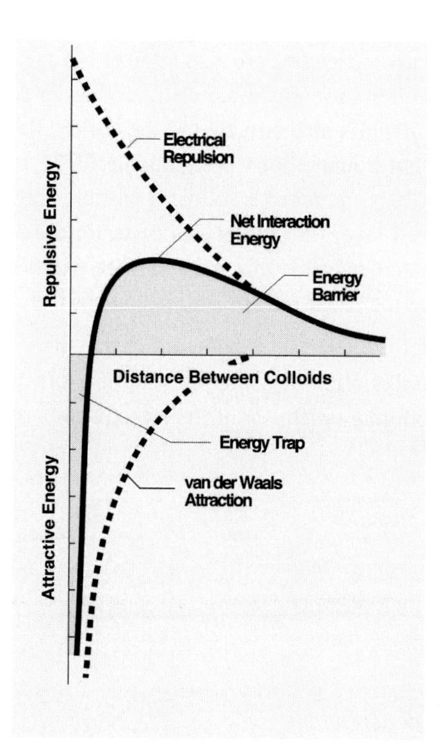

Figure 8-2 Forces acting on colloids. *(Adapted from Reference 4; with permission of L.A. Ravina and Zeta-Meter, Inc.)*

Brownian Motion Another destabilizing force acting on colloids is *Brownian motion.* Colloids have a sufficiently small mass that collisions with molecular-size particles will cause movement of the colloid. Molecules in water are in constant motion, with the intensity of this motion being dependent upon water temperature. This molecular motion causes random collisions with colloids, resulting in what appears to be the random movement of the colloid. This phenomena is known as *Brownian motion.*[5,6]

In some instances, Brownian motion can cause two colloids to move sufficiently close for van der Waals forces to predominate over electrostatic repulsion. If this occurs, the particles could agglomerate. Such instances rarely occur, and the destabilizing influence of Brownian motion is insignificant.

8.2.2 Coagulation of Colloidal Suspensions

Coagulation is a chemical treatment process used to destabilize colloidal particles. In the coagulation process, chemicals are added to the water that either break down the stabilizing forces, enhance the destabilizing forces, or both. Traditionally, such metal salts as aluminum sulfate (alum), ferric sulfate, ferric chloride, and ferrous sulfate have been utilized as coagulants. In recent years, however, polymers (long-molecular-chain organic compounds) have been used in conjunction with, or in lieu of, metal salts to enhance the coagulation process.

Metal salts, when added to water that has sufficient alkalinity, will hydrolyze into complex metal hydroxides of the form $Me_q(OH)_p$ (Me = metallic ion).[2] The actual hydroxide formed is dependent upon the chemical makeup of the water, particularly its pH, and the coagulant dosage. Figure 8-3 illustrates a typical equilibrium diagram for both iron and aluminum hydroxides in water.[7] Figure 8-3 also illustrates dosage rates and pH ranges that are typically used in coagulation. (See Section 8.3.3 for aqueous chemistry of iron and aluminum salts.)

Coagulants typically destabilize colloids by a combination of five mechanisms: *compression of the double layer, counterion adsorption and charge neutralization, interparticle bridging, enmeshment in a precipitate*, and *heterocoagulation*.[1,2,8–11] Each of these mechanisms is discussed below.

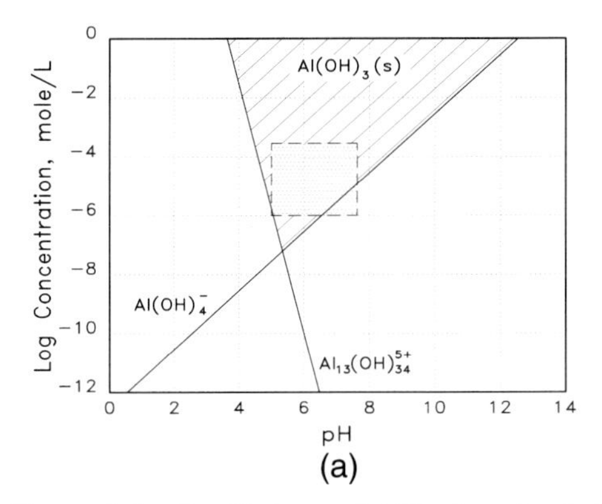

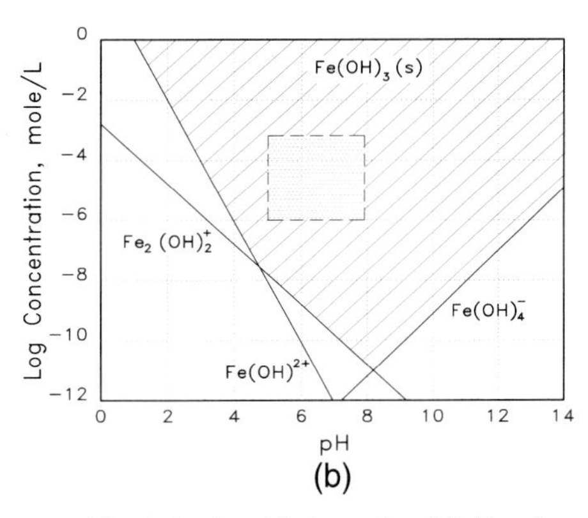

Figure 8-3 Equilibrium solubility domain of aluminum and ferric hydroxide in water. (a) Aluminum hydroxide. (b) Ferric hydroxide. Note: Shaded area is concentration and pH range of water treatment process.

Compression of the Double Layer (DLVO Theory) Compression of the double layer can be accomplished by the addition of a coagulant having a positive charge (to counteract negatively charged colloids). As the concentration of counter ions in the solution increases, the counterions cause the net charge in the diffused layer to neutralize and result in the compression of this layer. This compression affects the thickness of the entire double layer and so allows colloids to come closer together. If the colloids can be caused to come close enough together for van der Waals forces to predominate, the colloids will agglomerate into a floc.[2] This phenomenon is illustrated in Figure 8-4.

Singley cited that the Schulze–Hardy rule relates coagulating powers of counterions to their valence state. This rule states that divalent ions are approximately 50 to 70 times more effective, trivalent ions approximately 600 to 700 times more effective, than monovalent ions.[12]

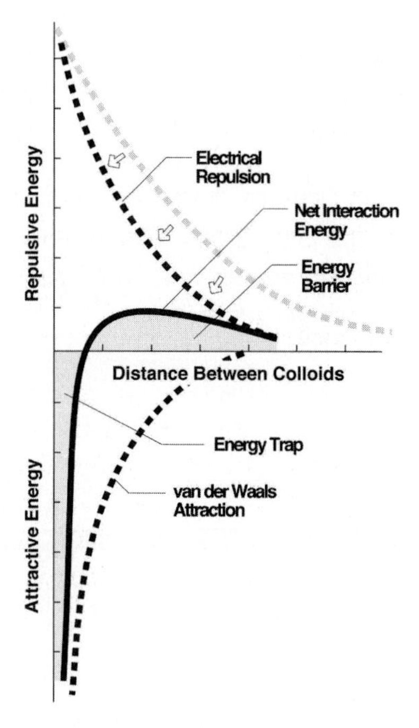

Figure 8-4 Forces acting on a colloid after compression of the double layer. *(Adapted from Reference 4; with permission of L.A. Ravina and Zeta-Meter, Inc.)*

Counterions Adsorption and Charge Neutralization The counterions from the coagulant can also be adsorbed onto the surface of the colloidal particles. In this way, the repulsive charges on the surface of the particles may be fully neutralized by the charges carried by the counterions. Therefore, the destabilized colloidal particles can adhere to each other to form colloidal-colloidal complexes, by van der Waals attractions or by further adsorption of counterions; however, an excess addition of counterionic electrolyte may result in restabilization by a charge reversal; the net charge on the particles may be reversed by the adsorption of an excess of counterions.

Interparticle Bridging If a synthetic organic polymer is utilized as the flocculation aid, the interparticle bridging begins with adsorption of the polymer onto specific sites on the surface of colloidal and/or coagulant particles. The resulting structure grows into a single particle several times larger than any of its individual constituents. In order for this bridging to occur, the segments of a polymer chain must be adsorbed onto sites on more than one particle. The particle-polymer-particle aggregates are typically formed through this mechanism, in which the polymer services as a bridge. Figure 8-5(a) and (b) illustrate normal coagulation and flocculation by interparticle bridging.[12] An excess dosage of polymer may cause restabilization (of destabilized particles) due to surface saturation or sterical stabilization. Figure 8-5(c) illustrates how particles can be restabilized if there is an insufficient number of particle collisions during flocculation. Figure 8-5(d) illustrates how particles can be restabilized if there is an insufficient number of colloidal particles available for bridging to occur. The destabilization and restabilization of a negatively charged dispersion by an anionic polyelectrolyte is commonly considered, not an interparticle bridging, but counterion adsorption and charge neutralization.

Enmeshment in a Precipitate The dosage of metal salts used in coagulation is usually slightly in excess of the amount required for reduction of the zeta potential. The excess metal salts hydrolyze into the form $Me_q(OH)_p$. These hydroxides are extremely insoluble in water. As the hydroxide precipitate forms and accumulates, the small colloidal particles are entrapped or enmeshed in the hydroxide floc structures.[1] This phenomenon is known as *enmeshment in a precipitate* and is also termed *sweep-floc coagulation.*[10]

Heterocoagulation In water treatment practice, the surface charges on the surface of some naturally occurring particles may not be uniform. Oppositely charged sites may exist on the surface of the same particle, such as on plate-like clay particles. The coagulation of these colloidal particles can therefore occur via simple electrostatic interaction between these oppositely charged sites. This mechanism is termed *heterocoagulation.* It may possibly play an important role in the process that is especially involved with sweep-floc coagulation.[11]

8.2.3 Enhanced Coagulation

In surface water treatment practice, the coagulation process is one of the processes most commonly used in all size applications. The essential purpose of using this process is to improve the efficiency of removal of small turbidity-causing particles in the sedimentation basin, which is followed by filters for polishing purposes. In this process, the destabilization of colloids is primarily targeted, and, therefore, the turbidity removal is the indicator of process efficiency. This process is called *conventional coagulation.*

In recent years, the coagulation process has also been broadly utilized to remove, not only turbidity, but also other undesirable organic and inorganic contaminants from the raw water. These objectives can be achieved by use of an *enhanced coagulation process,* in which an elevated coagulant dosage is usually required. The term *enhanced coagulation* was originally defined in conjunction with the Enhanced Surface Water Treatment Rule and the Disinfection and Disinfection By-Products Rule (D/DBPR), which were proposed in 1994. Depending on the initial total organic carbon (TOC) and alkalinity in the raw water, a certain level of total organic

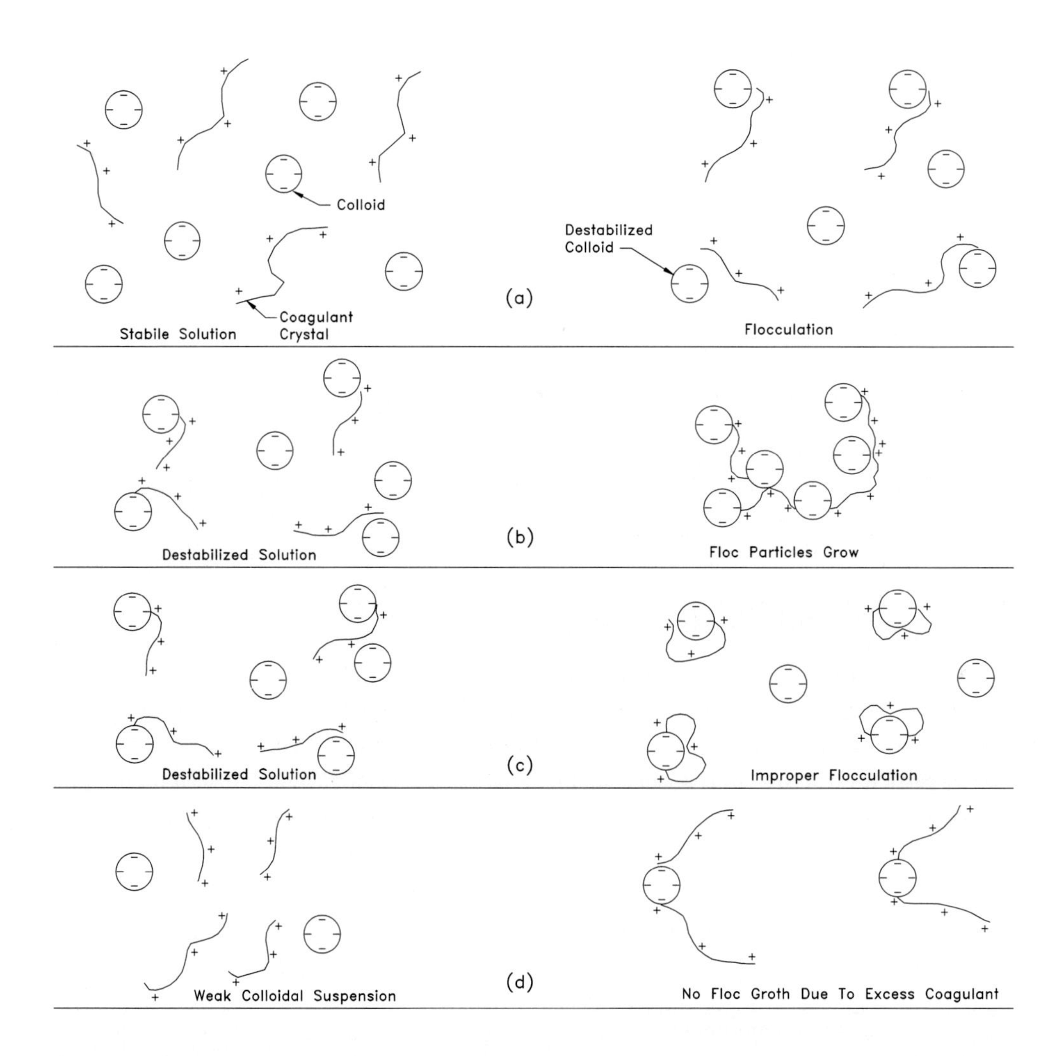

Figure 8-5 Schematic representation of the destabilization and restabilization of colloids by polymers. (a) Initial adsorption at optimum polymer dosage. (b) Floc formation. (c) Secondary adsorption. (d) Initial adsorption with excess polymer dosage.

carbon (TOC) removal is required by the regulations. (See also Section 12.2.4, for additional details.) A higher coagulant dosage than that in a conventional process is recommended if a higher TOC removal is required. Most recently, the applications of enhanced coagulation are no longer limited to TOC removal but also applied to many other impurities. Therefore, the term enhanced coagulation is broadly defined as the addition of an excess coagulant dosage for improved removal of such contaminants as normal organic matters (NOMs), color, arsenic, and other heavy metals from the water.[13–15] As a result, enhanced coagulation is not limited to five destabilization mechanisms described in Section 8.2.2.

Important mechanisms such as *precipitation* and *adsorption* also play a vital role in the enhanced coagulation process. Precipitation of NOM refers to the formation of an aluminum or iron humate or fulvate having a lower solubility product. Low solubility product in a substance will cause its precipitation from the solution into a solid form. The adsorption is the immobilization of *adsorbate* (i.e., the dissolved contaminants) on the *adsorbent* (i.e., the hydrous metal oxide). It may be the result of a simple van der Waals interaction or include the formation of chemical bonds between the adsorbate and the adsorbent. The bonds involved in adsorption of material can include hydrogen bonds, hydrophobic bonds, ionic bonds, and dipole interactions. The most important adsorption phenomenon involved in enhanced coagulation process is *coprecipitation*. The surface of coagulation precipitates is generally covered by some surface functional groups, such as ≡S-OH, ≡S-SH, and ≡S-COOH (≡S represents the metal ion on the surface of the hydrous metal oxide). Among these groups, only the so-called type A *surface hydroxyl groups* (i.e., ≡S-OH) are the actual "reactive" or "exchangeable" sites. Adsorption can occur effectively through chemical bonds (covalent or ionic) between the surface hydroxyl group and the adsorbate ions (cations or anions). During coprecipitation, the soluble organic material is adsorbed onto the reactive sites on the surfaces of growing hydrous metal oxide crystals. This type of adsorption mechanism can also be called *surface complex formation*. The removal of arsenic species, NOMs, and many other contaminants by enhanced coagulation can be explained by the coprecipitation process. The site-specificity, pH dependence, and surface of the hydrous metal oxide are the most important factors affecting the efficiency of coprecipitation.[16,17]

Extensive jar testing is necessary to establish optimum coagulant dosage and pH condition. A high coagulant dosage is typically not necessary for precipitation and adsorption. The only purpose of higher coagulant dosage in an enhanced coagulation process is simply to achieve sweep-floc coagulation. This is necessary to effectively remove the fine hydrous metal oxide crystals on which the contaminants have been successfully adsorbed. The coagulation pH usually affects both the hydrous metal oxide species and the level of dissociation of contaminant species, such as fulvic and humic acids and arsenate and arsenite acids. For instance, at lower pH values, the surface of an amorphous ferric hydroxide precipitate is positively charged, while the predominant arsenate species are (negatively charged) $H_2AsO_4^-$ and $HAsO_4^{2-}$. As a result, adsorption becomes more favorable and the required coagulant dosage decreases.[18] It is therefore likely that the mechanisms of charge neutralization and coprecipitation by adsorption are enhanced.[19]

Implementation of enhanced coagulation in a conventional water treatment plant may raise the following issues.[19–23]

- Increased coagulant dose causes higher sludge production. Existing sludge-handling facilities may not be sufficient.
- Chemical storage and feed facilities may not be adequate.

- Optimum turbidity removal may not coincide with the enhanced coagulation required for removal of specific constituents.
- Overall cost will increase, because of the cost of the chemicals used for coagulation and for pH adjustment before and after coagulation.

8.2.4 Coagulation Process Design Considerations

The design of a coagulation process involves (1) selection of proper coagulant chemicals and their dosages and (2) design of rapid-mix and flocculation basins. Designs for rapid-mix and flocculation basins are provided later in this chapter.

Selection of proper coagulants and their dosages are best determined experimentally for each raw water source. Operating experience from other treatment plants treating similar waters is a second common method of developing design parameters. It may be noted that the chemical make-up of any water may vary to such a degree that the design of a coagulation facility cannot be rigidly established. The most important consideration in coagulant process design is to provide a flexibility in chemical feed equipment and types of chemicals. Such flexibility provides the operator an ability to adjust proper dosages as the chemical make-up of raw water changes with time. In conventional coagulation, the Fe^{3+} dose is in the range of 2 to 4 mg/L; the Fe^{3+} dose may reach as high as 11 mg/L in enhanced coagulation.

8.3 Dissolved Solids And Chemical Precipitation

The dissolved solids in natural water include ions of such dissolved minerals as calcium, magnesium, sodium, iron, and manganese, in combination with such anions as bicarbonates, carbonates, chlorides, sulfates, nitrates, phosphates, and fluorides. Most of these compounds are beneficial; others may be harmful in higher concentrations. A discussion on the effects of various dissolved solids may be found in Chapter 2.

8.3.1 Characteristics of Dissolved Solids

Dissolved solids in water may be either ionic compounds, such as calcium bicarbonate ($Ca(HCO_3)_2$), or molecular compounds, such as sucrose ($C_{12}H_{22}O_{11}$). Dissolved ionic compounds in water form a strong electrolyte; molecular compounds typically form a non-electrolyte solution.

Many ionic compounds such as those of iron, manganese, calcium, and magnesium, are removed from raw water mostly by chemical precipitation; molecular compounds, on the other hand, are not removed by precipitation. Because of the importance of ionic compounds and the precipitation of many cations, the remainder of the discussion on chemical precipitation of dissolved solids is devoted exclusively to ionic compounds.

8.3.2 Solubility Product

When an ionic compound dissolves in water, the ions that compose the compound dissociate. For example, $Ca(HCO_3)_2$ dissociates into Ca^{2+} and two HCO_3^- ions. Both ions are free to react with other ions in the solution. The compound will continue to dissolve up to a limiting value where the solution becomes saturated. The solubility product (K_{sp}) is used to calculate the solubility of an ionic compound and is expressed by Eqs. (8.1) and (8.2).[2]

$$MX \Leftrightarrow M^+ + X^- \tag{8.1}$$

$$K_{SP} = [M^+] \times [X^-] \tag{8.2}$$

where

K_{sp} = solubility product constant of compound MX
$[M^+]$ = molar concentration of the cation
$[X^-]$ = molar concentration of the anion

The solubility products of various ionic compounds of interest in water treatment are listed in Table 8-2. If the product of the molar concentrations of ions of a compound is less than the solubility product, the solution is undersaturated. On the other hand, if the product of the molar concentrations of the ions is greater than the solubility product, the solution is supersaturated and a precipitate of the compound will form until the product of the molar concentrations equals the solubility product.[24,25]

8.3.3 Precipitation of Dissolved Solids

In the precipitation process, the physical or chemical properties of a solution are modified to the point that the solution becomes supersaturated. In water treatment, the precipitation process is used for softening (removal of the hardness caused by calcium and magnesium) and for removal of iron and manganese.

Aqueous Chemistry of Iron and Aluminum Salts Coagulant reactions are carried out by the addition of a *coagulant*, usually a metal salt, to water. Commonly used coagulants are ferric sulfate [$Fe_2(SO_4)_3$], ferric chloride ($FeCl_3$), and alum [aluminum sulfate, $Al_2(SO_4)_3 \bullet 14H_2O$]. The overall reactions of these coagulants in water are described by Eqs. (8.3)–(8.5).

Ferric Sulfate:

$$\underset{400}{Fe_2(SO_4)_3} + \underset{\substack{3 \times 100 \\ (\text{as } CaCO_3)}}{3Ca(HCO_3)_2} \Leftrightarrow \underset{2 \times 107}{2Fe(OH)_3(s)} + \underset{3 \times 136}{3CaSO_4} + \underset{6 \times 44}{6CO_2} \tag{8.3}$$

Table 8-2 Solubility-Products Constants of Ionic Compounds

Equilibrium Equation	K_{sp} at 25°C	Significance in Water Treatment
$CaCO_3 \Leftrightarrow Ca^{2+} + CO_3^{2-}$	$5 \infty 10^{-9}$	Hardness removal, scaling
$Ca(OH)_2 \Leftrightarrow Ca^{2+} + 2OH^-$	$8 \infty 10^{-6}$	Hardness removal
$MgCO_3 \Leftrightarrow Mg^{2+} + CO_3^{2-}$	$2 \infty 10^{-5}$	Flue gas desulfurization
$CaSO_4 \Leftrightarrow Ca^{2+} + SO_4^{2-}$	$4 \infty 10^{-5}$	Hardness removal, scaling
$Mg(OH)_2 \Leftrightarrow Mg^{2+} + 2OH^-$	$9 \infty 10^{-12}$	Hardness removal, scaling
$Al(OH)_3 \Leftrightarrow Al^{3+} + 3OH^-$	$1 \infty 10^{-32}$	Coagulation
$Fe(OH)_3 \Leftrightarrow Fe^{3+} + 3OH^-$	$6 \infty 10^{-38}$	Coagulation, iron removal, corrosion
$Fe(OH)_2 \Leftrightarrow Fe^{2+} + 2OH^-$	$5 \infty 10^{-15}$	Coagulation, iron removal, corrosion
$Mn(OH)_3 \Leftrightarrow Mn^{3+} + 3OH^-$	$1 \infty 10^{-36}$	Manganese removal
$Mn(OH)_2 \Leftrightarrow Mn^{2+} + 2OH^-$	$8 \infty 10^{-14}$	Manganese removal
$Cu(OH)_2 \Leftrightarrow Cu^{2+} + 2OH^-$	$2 \infty 10^{-19}$	Heavy metal removal
$Zn(OH)_2 \Leftrightarrow Zn^{2+} + 2OH^-$	$3 \infty 10^{-17}$	Heavy metal removal
$Ni(OH)_2 \Leftrightarrow Ni^{2+} + 2OH^-$	$2 \infty 10^{-16}$	Heavy metal removal
$Cr(OH)_3 \Leftrightarrow Cr^{3+} + 3OH^-$	$6 \infty 10^{-31}$	Heavy metal removal

Source: Adapted in part from References 1 and 25.

Ferric Chloride:

$$\underset{2 \times 162.5}{2FeCl_3} + \underset{\substack{3 \times 100 \\ (\text{as } CaCO_3)}}{3Ca(HCO_3)_2} \Leftrightarrow \underset{2 \times 107}{2Fe(OH)_3(s)} + \underset{3 \times 111}{3CaCl_2} + \underset{6 \times 44}{6CO_2} \tag{8.4}$$

Alum

$$\underset{594}{Al_2((SO_4)_3 \cdot 14H_2O)} + \underset{3 \times 100}{3Ca(HCO_3)_2} \Leftrightarrow \underset{2 \times 78}{(2Al(OH)_3)(s)} + \underset{3 \times 136}{3CaSO_4} + \underset{6 \times 44}{6CO_2} + \underset{14 \times 18}{14H_2O} \tag{8.5}$$

The following observations can be made on these reactions (Eqs. (8.3), (8.4), and (8.5)):

- 1 mg of ferric sulfate will produce approximately 0.54 mg of insoluble $Fe(OH)_3$ precipitates and will consume approximately 0.75 mg of alkalinity (expressed as $CaCO_3$).
- 1 mg of ferric chloride will produce approximately 0.66 mg of $Fe(OH)_3$ precipitates and will consume approximately 0.92 mg of alkalinity (expressed as $CaCO_3$).

- 1 mg alum [$Al_2(SO_4)_3 \cdot 14H_2O$] will produce approximately 0.26 mg of insoluble $Al(OH)_3$ precipitates and will consume approximately 0.51 mg of alkalinity (expressed as $CaCO_3$).

Because of the consumption of alkalinity, CO_2 is produced during coagulation. The pH value may also be lowered after the coagulation process, depending on the amount of coagulant applied and the initial alkalinity in the raw water. The stoichiometric equations can be used to estimate the coagulant quantity requirement; however, the actual amounts of coagulant required for destabilization of colloids may depend, not only on the reaction stoichiometry, but also on other operational conditions such as ionic species, pH value, temperature, type and properties of particles, mixing-energy input, and the effective content of metal ions in the coagulant. The metal-ion content in a liquid commercial coagulant is typically in the range of 10–15% by liquid weight. The physical and chemical properties of commonly used coagulants are summarized in Table 8-3.

Water Softening Hardness in water is caused mainly by calcium and magnesium ions. These cations typically cause carbonate hardness if associated with alkalinity-causing anions in water (hydroxide (OH^-), carbonate (CO_3^{2-}), and bicarbonate (HCO_3^-). Calcium and magnesium may also be combined with other anions, such as sulfate (SO_4^{2-}), chloride (Cl^-), and nitrate (NO_3^-). Hardness in this form is known as non-carbonate hardness. As indicated in Table 8-2, calcium carbonate and magnesium hydroxide are relatively insoluble ($K_{sp} = 5 \infty 10^{-9}$ and $9 \infty 10^{-12}$), while other products of hardness are quite soluble. In the softening process, the chemical reactions increase the concentrations of calcium carbonate and magnesium hydroxide, thus removing calcium and magnesium as precipitates.

Figure 8-6 illustrates the relationship between carbon dioxide (CO_2) and various forms of alkalinity. Eqs. (8.6) and (8.7) represent the chemical reactions between carbon dioxide and alkalinity ions.

$$CO_2 + H_2O \Leftrightarrow H_2CO_3 \Leftrightarrow H^+ + HCO_3^- \tag{8.6}$$

$$CO_2 + OH^- \Leftrightarrow HCO_3 \Leftrightarrow H^+ + CO_3^{2-} \tag{8.7}$$

At lower pH values (below 4.5), carbon dioxide is in equilibrium and no alkalinity exists. At moderate pH values (between 4.5 and 8.3), the reaction represented by Eqs. (8.3), (8.4), and (8.5) shifts to the right. Carbon dioxide concentrations begin to reduce and bicarbonate (HCO_3^-) ions are formed. At high pH values (above 8.3), bicarbonates are converted into carbonate (CO_3^{2-}) ions, with the maximum concentration occurring at a pH value of 10.5.[16]

These relationships between carbon dioxide and alkalinity are utilized in water softening to remove hardness and alkalinity. The pH of the solution is raised to increase the concentrations of carbonate and hydroxide ions. These ions combine with calcium and magnesium to form calcium carbonate and magnesium hydroxide precipitates.

The pH can be raised by adding any ionic compound containing hydroxide ions. The chemical most commonly used is either lime (CaO) or hydrated lime ($Ca(OH)_2$). The physical

Table 8-3 Commonly Used Coagulants and Limes in Water-Treatment Practice

Chemical Name	Synonyms	Chemical Formula	Molecular Weight	Appearance	Commercial-Grade Qualities: Bulk Density, kg/m^3	Specific Gravity	Solubility in Water, kg/m^3	Chemical Content % w/w	Water Content % w/w	pH of Solution
Aluminum sulfate	Alum	$Al_2(SO_4)_3 \cdot 14.3H_2O$	599.77	White to light-tan solid	1000–1096	1.25–1.36	Approx. 872	Al: 9.0-9.3	–	approx. 3.5
	liquid alum	$Al_2(SO_4)_3 \cdot 49.6H_2O$	1235.71	White or light- gray to yellow liquid	–	1.30–1.34	Very soluble	Al: 4.0–4.5	71.2–74.5	–
Ferric chloride	Iron (III) chloride, Iron trichloride	$FeCl_3$	162.21	Green-black power	721–962	–	Approx. 719	Fe: approx. 34	–	–
	Liquid ferric chlorine	$FeCl_3 \cdot 6H_2O$	270.30	Yellow-brown lump	962–1026	–	Approx. 814	Fe: 20.3–21.0	–	–
		$FeCl_3 \cdot 13.1H_2O$	398.21	Reddish-brown syrupy liquid	–	1.20–1.48	Very soluble	Fe: 12.7–14.5	56.5–62.0	0.1–1.5
Ferric sulfate	Iron (III) sulfate, Iron persulfate	$Fe_2(SO_4)_3 \cdot 9H_2O$	562.02	Red-brown power	1122–1154	–	–	Fe: 17.9–18.7	–	–
	Liquid ferric sulfate	$Fe_2(SO_4)_3 \cdot 36.9H_2O$	1064.64	Reddish-brown syrupy liquid	–	1.40–1.57	Very soluble	Fe: 10.1–12.0	56.5–64.0	0.1–1.5
Ferrous sulfate	Copperas	$FeSO_4 \cdot 7H_2O$	278.02	Green crystal lump	1010–1058	–	–	Fe: approx. 20	–	–
Calcium oxide	Lime, Quick lime	CaO	74.09	Off-white powder or lump	561–801	–	Approx. 1.3	CaO: approx. 95	–	12.6
Calcium hydroxide	Hydrated lime	$Ca(OH)_2$	56.08	Off-white (faintly) powder	–	–	Approx. 1.8	CaO: approx. 71	approx. 24	12.6

Note: kg/m^3 ∞ 0.00835 = lb/gal
Source: Adapted in part from References 24–26.

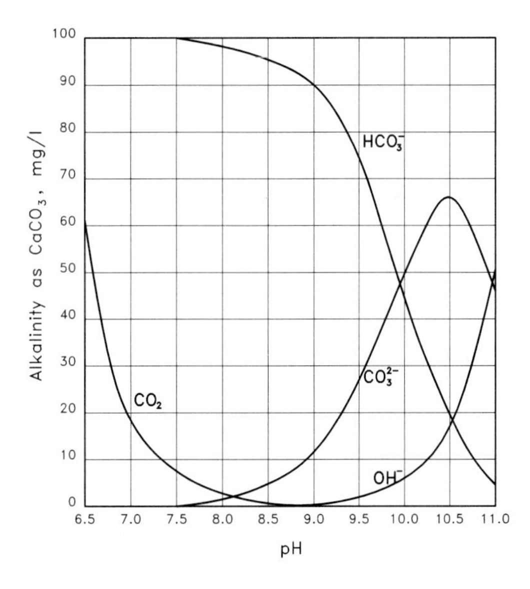

Figure 8-6 Equilibrium concentrations of CO_2 and alkalinity with respect to pH. (Note: The temperature is 25°C, and the total alkalinity is 100 mg/L as $CaCO_3$.)

and chemical properties of commercial lime products are included in Table 8-3. The chemical reactions of lime softening with hydrated lime are illustrated in Eqs. (8.8) through (8.10).

$$CO_2 + Ca(OH)_2 \Leftrightarrow CaCO_3(s) + H_2O \tag{8.8}$$

$$Ca(HCO_3)_2 + Ca(OH)_2 \Leftrightarrow 2CaCO_3(s) + 2H_2O \tag{8.9}$$

$$Mg(HCO_3)_2 + 2Ca(OH)_2 \Leftrightarrow 2CaCO_3(s) + Mg(OH)_2(s) + 2H_2O \tag{8.10}$$

If noncarbonate hardness is to be removed, the lime softening process alone will be unsuccessful in the removal of calcium and magnesium. In this case, soda ash (Na_2CO_3) is used. The chemical reactions involved in lime–soda processes are represented by Eqs. (8.11)–(8.13).[8]

$$CaSO_4 + Na_2CO_3 \Leftrightarrow CaCO_3(s) + Na_2SO_4 \tag{8.11}$$

$$MgSO_4 + Na_2CO_3 \Leftrightarrow MgCO_3 + Na_2SO_4 \tag{8.12}$$

$$MgCO_3 + Ca(OH)_2 \Leftrightarrow Mg(OH)_2(s) + CaCO_3(s) \tag{8.13}$$

To calculate the chemical dosage requirements for the softening reactions, the concentrations of calcium, magnesium, total alkalinity, and carbon dioxide are required. The amount of lime required is expressed by Eq. (8.14).[1]

$$[Ca(OH)_2] = [CO_2] + [HCO_3^-] + [Mg^{2+}] + [\text{excess lime}] \tag{8.14}$$

where

$[Ca(OH)_2]$ = amount of $Ca(OH)_2$ required for softening in me/L

$[CO_2]$ = concentration of CO_2 in me/L

$[HCO^{3-}]$ = concentration of HCO^{3-} in me/L

$[Mg^{2+}]$ = concentration of Mg^{2+} in me/L

[excess lime] = additional amount of lime needed to maintain an elevated pH value, usually 1 me/L.

The dosage of soda ash required is expressed by Eq. (8.15).[1,27]

$$[Na_2CO_3] = [Ca^{2+}] + [Mg^{2+}] - [Alkalinity] \tag{8.15}$$

where

$[Na_2CO_3]$ = amount of Na_2CO_3 required for noncarbonate softening in me/L

$[Ca^{2+}]$ = concentration of Ca^{2+} in me/L

[Alkalinity]= concentration of alkalinity in me/L

The concentrations in Eqs. (8.14) and (8.15) are in me/L. To calculate the amount of lime or soda ash required in mg/L, multiply by their respective equivalent weights as shown by Eqs. (8.16) and (8.17).[1,27]

$$\text{Lime required in mg/L} = [Ca(OH)_2] \times 37.0 \tag{8.16}$$

$$\text{Soda ash required in mg/L} = [Na_2CO_3] \times 53 \tag{8.17}$$

The use of Eqs. (8.14)–(8.17) is illustrated in the following example.

Example: Calculate the amount of lime and soda ash required for water-softening. The concentration of hardness and alkalinity-causing components are given below. The excess lime is 1.0 me/L.

Components	Concentration	Molecular Weight	Equivalent Weight	Equivalent Conc., me/L
CO_2	10 mg/L	44	22	0.45
Ca^{2+}	70 mg/L	40	20	3.50
Mg^{2+}	40 mg/L	24	12	3.33
HCO_3^-	250 mg/L	61	61	4.10
CO_3^{2-}	2 mg/L	60	30	0.07
OH^-	0.02 mg/L	17	17	0.0012

Solution: Calculate the amount of lime required.

$$[Ca(OH)_2] = [CO_2] + [HCO_3^-] + [Mg^{2+}] + [\text{excess lime}]$$
$$= 0.45 + 4.10 + 3.33 + 1.00$$
$$= 8.88 \text{ me/L}$$

Lime required = 8.88 me/L ∞ 37 = 328 mg/L as $Ca(OH)_2$

Calculate the amount of soda ash required.

$$[Na_2CO_3] = Ca^{2+} + Mg^{2+} - [Alk]$$
$$= [Ca^{2+}] + [Mg^{2+}] - ([HCO_3^-] + [CO_s^{2-}] + [OH^-])$$
$$= 3.50 + 3.33 - 4.10 + 0.07 + 0.0012$$
$$= 2.66 \text{ me/L}$$

Soda ash required = 2.66 me/L ∞ 53 = 141 mg/L as Na_2CO_3

The chemical dosages calculated from Eqs. (8-12) and (8-14) will result in softening of the water to approximately 35 mg/L of $CaCO_3$ and 25 mg/L of $Mg(OH)_2$ (expressed as $CaCO_3$).[1,27] These concentrations are more than the theoretical concentrations calculated by the solubility products but represent the lowest concentrations achievable by chemical precipitation. Removal of hardness to this level is, however, adequate for most domestic water uses.

The lime–soda softening process leaves the water saturated with $Mg(OH)_2$ and $CaCO_3$ and with an elevated pH. A process known as recarbonation is used to lower the pH into the 9.2–9.7 range, thus restabilizing $CaCO_3$.[1,27] In this process, carbon dioxide is added to the water. This lowers the pH and converts $CO_3{}^{2-}$ to more stable $HCO_3{}^-$ ions.

Iron and Manganese Removal

Iron and manganese are often present in groundwater supplies or in water drawn from the hypolimnion of a reservoir. Under anaerobic conditions, bacteria reduce iron and manganese to the Fe^{2+} and Mn^{2+} valence state, a more soluble state. Likewise, when waters containing Fe^{2+} and Mn^{2+} ions are exposed to air, these ions are oxidized to a less soluble Fe^{3+} and Mn^{3+} valence state that causes red or brown stains on plumbing fixtures and laundry. Therefore, removal of iron and manganese is desirable. Eqs. (8.18) and (8.19) represent the chemical reactions of iron and manganese oxidation.

$$4Fe^{2+} + O_2 + 10H_2O \rightarrow 4Fe(OH)_3 + 8H^+ \tag{8.18}$$

$$2Mn^{2+} + O_2 + 2H_2O \rightarrow 2MnO_2 + 4H^+ \tag{8.19}$$

Removal is accomplished by oxidizing the Fe^{2+} and Mn^{2+} ions and removing the precipitates by sedimentation and filtration. One method of accomplishing oxidation is by aeration. The aeration process is discussed in Chapter 6. The aeration process is most successful in removing iron and manganese when the pH of the solution is above 8.5 and stacked-tray aerators with coke beds are employed.[27] The aeration process is usually unsuccessful at removal of iron in the chelated form.[5] In this form, the iron ion is bound to a larger molecule that is also dissolved in the water. Iron in the chelated form must be removed by chemical oxidation, as described next.

In the chemical oxidation process, a strong oxidant such as ozone, permanganate, hydrogen peroxide, chlorine, or chlorine dioxide is used.[27,28] Iron is effectively removed at a neutral pH by any of these oxidants. Manganese can also be removed by permanganate oxidation at neutral pH levels. The remaining chemicals require a pH in excess of 8.3 for effective removal of manganese.[27,28]

Iron and manganese precipitates are also removed by coagulation and filtration. These precipitates coat the filter and enhance the removal of colloids, but, if a chlorine residual is not maintained across the filters, bacterial growth in the precipitate coating will reduce the iron and manganese back to soluble Fe^{2+} and Mn^{2+} state and causes resolubilization.[5]

8.4 Rapid Mix

Coagulation and precipitation processes both require the addition of chemicals to the water stream. The success of these processes depends on rapid and thorough dispersion of the chemicals. The process of dispersing chemicals is known as *rapid mix* or *flash mix.*

8.4.1 Types of Mixers

Rapid-mixing units can be classified according to the method of agitation (mechanical or static) and type of flow pattern (plug-flow or complete-mixed). A mechanically agitated rapid mixer utilizes a mechanical mixer with an impeller or propeller to create turbulence in the mixing chamber. Examples of impellers and propellers typically used in water treatment are shown in Figure 8-7. These impellers and propellers are generally classified in accordance with the type of flow produced. Impellers that force water outward at right angles to the axis of rotation, in a manner similar to that of a radial-flow pump, are called *radial-flow impellers* or turbine impellers (Figures 8-7(a) and (b)). The impellers shown in Figure 8-7(c) tend to force water parallel to the axis of rotation, much as an axial-flow pump does; hence, they are called *axial-flow pitched-blade impellers.* Propellers are also high-speed axial-flow devices that have pitched blades. These are illustrated in Figures 8-7(d) and (e). The type of impeller used in a rapid mixer is dependent upon the geometry of the basin and the flow pattern desired within the basin. Static mixers create turbulence by the use of hydraulic jumps, baffles, turbulent flow in a pipeline or channel, or contractions or enlargements in a pipeline. Table 8-4 compares the advantages and disadvantages of mechanical and static mixers.

Plug-flow mixing units are designed so that the fluid particles pass through the unit in the same sequence in which they enter. There is minimal longitudinal dispersion, and particles retain their identity and remain in the unit for a time equal to the theoretical hydraulic detention time. In trace-profile study, an initial slug or continuous feed of dye tracer concentration C_0 in a plug-flow unit would result in dye distributions at the outlet as shown in Figure 8-8(a).[18] The theoretical hydraulic detention time t_0 is obtained by dividing the unit volume by the flow rate. An

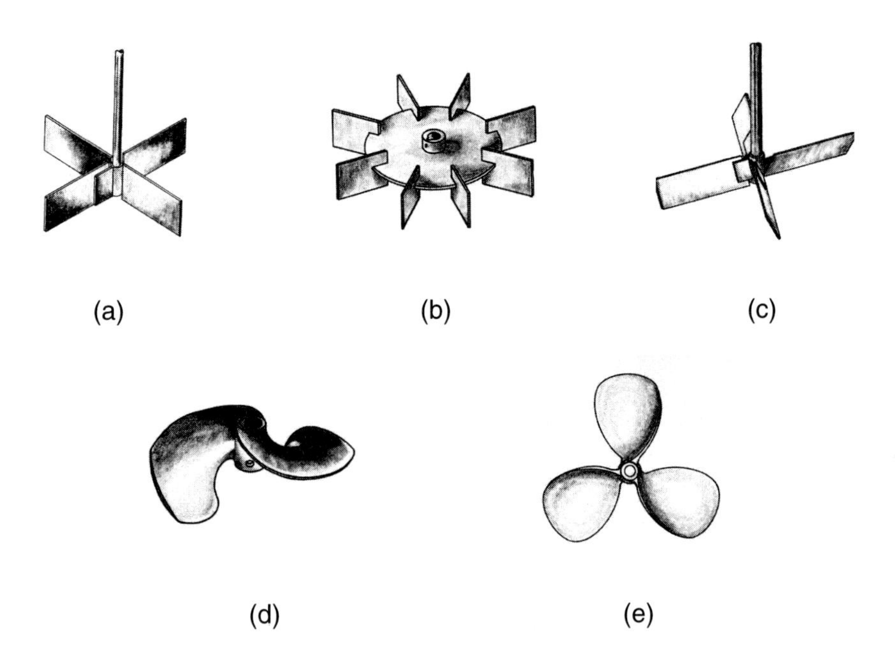

Figure 8-7 Radial- and axial-flow impellers. *(Courtesy of Aqua-Aerobic System, Inc.)* (a) Straight-blade or flat-blade radial-flow turbine impeller. (b) Flat-blade disk impeller. (c) Pitched-blade axial-flow impeller. (d) Nonclog pitched-blade propeller-type axial-flow impeller. (e) Marine-type pitched-blade axial-flow impeller.

Table 8-4 Comparison of Mechanical and Static Mixers

Mixer Type	Advantages	Disadvantages
Mechanical mixers	• Agitation independent of flow rate • Agitation is adjustable • High flexibility in operation	• Additional equipment required for maintenance • Reliability subject to equipment failure
Static mixers	• Little or no maintenance • Very reliable	• Agitation dependent on flow rate • High head loss • Less flexibility in operation

excellent example of plug flow would be water flowing through a pipeline or in a series of completely mixed reactors. Due to their nature, static mixers are typically plug-flow type.

Complete-mixed units are designed such that the contents of the tank are completely mixed. A slug or continuous dye tracer injected at the inlet of a complete-mixed unit would result in a dye distribution at the outlet as shown in Figure 8-8(b).[18] Due to their nature, mechanically mixed rapid mixers tend to have complete-mixing characteristics or those of an arbitrary-flow system (Figure 8-8(c)). In an arbitrary-flow unit, partial mixing between that of a plug-flow unit and that of a complete-mixed unit is encountered.

8.4.2 Agitation Requirements

Rapid mixers should provide sufficient agitation to disperse the chemicals thoroughly in the water stream. Reactions of coagulation chemicals are rapid; softening reactions occur more slowly.[27,31,32] In either case, it is desirable to disperse the chemicals quickly, before the reactions are complete. In order to quickly and completely homogenize the coagulant or softening chemicals, the rapid mixer should be designed to provide a short period of violent agitation, with the chemicals being added at the point of greatest turbulence. Traditionally, in water treatment, the degree of agitation in a mixing unit is measured by *velocity gradient*. For mixing equipment, the value of the velocity gradient is given by Eq. (8.20).[2]

$$G = \sqrt{(P/V\mu)} \tag{8.20}$$

where

G = velocity gradient, 1/s (G = 700 to 1000/s)
P = power imparted to the water, N-m/s or Watt, W (lb-ft/s)[a]
V = volume of the basin, m^3 (ft^3)
μ = absolute viscosity of the fluid, N-s/m^2 (lb-s/ft^2)

One important point to be remembered is that Eq. (8.20) represents the power imparted to the water. The motor power of the mixer is the power to drive the speed reduction gears. The power imparted to the water by a mixer is calculated from Eq. (8.21).

$$P = 2\pi nT \tag{8.21}$$

where

n = impeller speed, revolutions per second (rps)
T = impeller shaft torque, N-m (lb-ft)

Other expressions for the power imparted to the water are given by Eqs. (8.22) and (8.23).[29,33,34]

$$P = N_p \mu n^2 d^3 \tag{8.22}$$

$$P = N_p \rho n^3 d^5 \quad (\text{or } (\gamma)/g N_p n^3 d^5) \quad (\text{or } N_p (\gamma)/g n^3 d^5) \tag{8.23}$$

a. Watt (W) = N-m/s

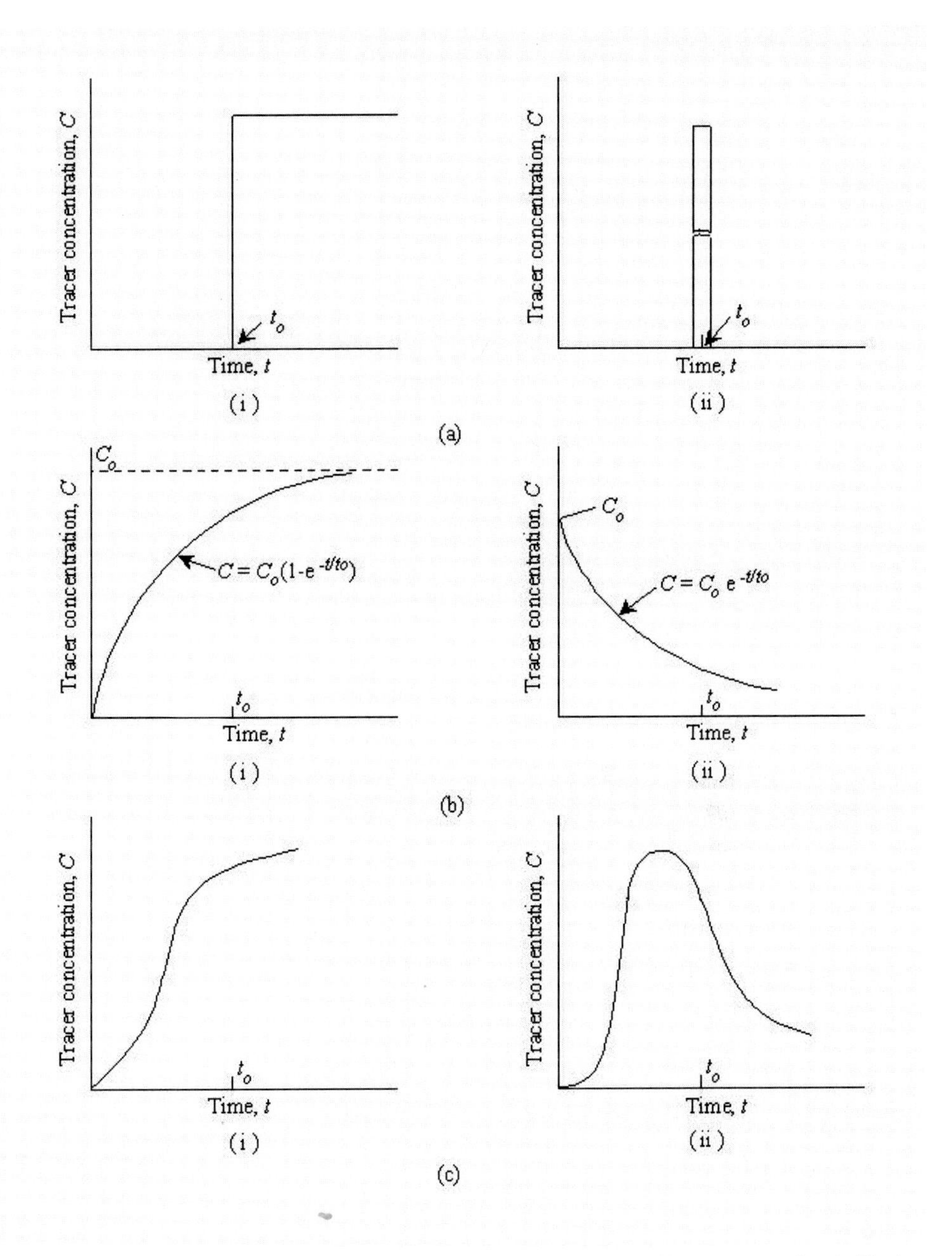

Figure 8-8 Dye-tracer slug test in rapid-mix units. (a) Plug-flow unit: (I) continuous feed, (ii) slug feed. (b) Complete-mixed unit: (i) continuous feed, (ii) slug feed. (c) Arbitrary-flow unit, (i) continuous feed, (ii) slug feed.

where

N_p = power number of the impeller (power numbers for different types of impellers are given in Table 8-5).

d = impeller diameter, m. (ft)

ρ = mass density of fluid, kg/m3 (slug/ft3)[b]

γ = water density, N/m^3 (lb/ft^3)

g = acceleration due to gravity, m/s^2 (ft/s^2)

μ = dynamic viscosity of water, N-s/m^2(lb-s/ft^2)

Eq. (8.22) is used for the laminar-flow range (Reynolds number N_R<10), and Eq. (8.23) is used for the turbulent-flow range (Reynolds number N_R>10,000). The Reynolds number for rapid mixers is given by Eq. (8.24).[14]

$$N_R = \frac{d^2 n\rho}{\mu} \text{ (or } d^2 n\gamma/\mu g) \tag{8.24}$$

Other terms are previously defined.

The velocity gradient for a mixing basin utilizing flow-induced turbulence can be calculated from Eq. (8.25).[2]

$$G = \left[\frac{\gamma\sqrt{h_1}}{t\mu}\right] = \left[\frac{g\rho\sqrt{h_L}}{t\mu}\right] \tag{8.25}$$

where

h_L = total head loss through the mixer, m

t = detention time, s

In U.S. Customary units, Eq. (8.25) can be simplified and is expressed by Eq. (8.26) at 4°C.

$$G = 178\sqrt{\left(\frac{h'_L}{t'}\right)} \tag{8.26}$$

b. $$\rho = \frac{\gamma}{g} = \frac{\text{N/m}^3}{g} = \frac{(\text{kg·m/s}^2)}{\text{m/s}^2} \times \frac{(1/\text{m}^3)}{\text{m/s}^2} = \text{kg/m}^3$$

$$\text{N} = \text{mass} \times \text{acceleration (kg·m/s}^2)$$

$$\text{Watt (W)} = \text{N·m/s} = (\text{kg·m/s}^2) \times (\text{m/s}) = \text{kg·m}^2/\text{s}^2$$

$$\mu = \frac{\text{N·s}}{\text{m}^2} = \frac{(\text{kg·m/s}^2) \times \text{s}}{\text{m}^2} = \text{kg/m·s}$$

where

h'_L = total head loss, ft.
t' = detention time, min.

The design value of the velocity gradient G is dependent on the detention time in the mixing unit, the coagulant dosage rate, and the geometry of the mixing unit. Velocity gradients normally range from 100 to 1000/s.

Table 8-5 Power Numbers of Various Rapid-Mix Impellers

	Power Number, N_p
Radial flow	
Straight blade turbine	
4 blade (w/d = 0.15)[a]	2.6
4 blade (w/d = 0.2)	3.3
Disc turbine	
4 blade (w/d = 0.25)	5.1
6 blade (w/d = 0.25)	6.2
Axial flow	
Propeller 1:1 pitch	0.3
Propeller 1.5:1 pitch	0.7
45° Pitched blade	
4 blade (w/d = 0.15)	1.36
4 blade (w/d = 0.2)	1.94

a w/d = blade width-to-diameter ratio.
Source: Adapted in part from References 2, 5, 27, and 28.

The velocity gradient selected for rapid-mixer design should be developed through the use of bench-scale or pilot plant testing. An alternative to laboratory testing is to utilize the experience of existing plants treating the same (or similar) water. In general, a short duration of high-intensity mixing often gives the best results.

8.4.3 Detention Time in Rapid-Mix Basin

The detention time in rapid mixers should provide sufficient time for complete homogenization of the chemicals with the water and also provide sufficient time for the floc to reach particle-size equilibrium. *Particle-size equilibrium* refers to the condition where no additional rapid mixing will result in any further turbidity removal by settling alone. Such rapid mixing time can be determined only through laboratory tests or through experience with an existing treatment plant treating water with a similar chemistry and utilizing the same coagulant chemicals. Typical detention time[c] for rapid mixers ranges from 10 s to 5 min.[31]

The average detention time in a rapid mixer is determined from Eq. (8.27).

$$t = V/Q \tag{8.27}$$

where

t = average detention time, min

Q = flow rate, m^3/min. (ft^3/min)

V = volume of the tank, m^3 (ft^3)

Shorter detention times require higher velocity gradients to achieve effective mixing. Conversely, longer detention times permit lower velocity gradients. The term Gt is often used to match the proper velocity gradient with detention time. Typical Gt values range between 30,000 and 60,000.[2,19] Optimum Gt values vary a great deal with the chemicals and dosage rates.[29] Optimum design values are best determined experimentally.

8.4.4 Geometry of Rapid-Mix Basin

The geometry of the rapid mixer is the most important aspect of its design. The primary concern in the geometric design is to provide uniform mixing for the water passing through the mixer and to minimize dead areas and short-circuiting.

Rapid mixers utilizing mechanical mixers are usually square and have a depth-to-width ratio of approximately 2. The size and shape of the mixer impeller should be matched to the flow desired through the mixer. Mixing units with vertical flow patterns utilizing radial-flow mixers tend to minimize short-circuiting effects. Figure 8-9 illustrates the flow pattern from such a mixer.[23] Round or cylindrical mixing chambers should be avoided for mechanical mixers. A round cross section tends to provide little resistance to rotational flow (induced in the tank by the mixer) resulting in reduced mixing efficiencies. Baffles can be employed to reduce rotational motion and increase efficiencies.[33,35]

A channel with fully turbulent flow of sufficient length to yield the desired detention time, followed by a hydraulic jump, has been used successfully. Figure 8-10 illustrates a typical rapid mixer utilizing a hydraulic jump.[1,36] Also, pipe reducers and increasers with a sufficient length of pipe develop fully turbulent flow condition and may give the desired detention time. Figure 8-11 illustrates a typical design of such a rapid-mix basin utilizing pipe contraction.[31] In all cases, the coagulant chemical should be added to the water stream immediately prior to the point of greatest turbulence. Such a design ensures rapid homogenization of the chemical in the water.

c. Other terms using average detention time are theoretical, nominal, or hydraulic detention time.

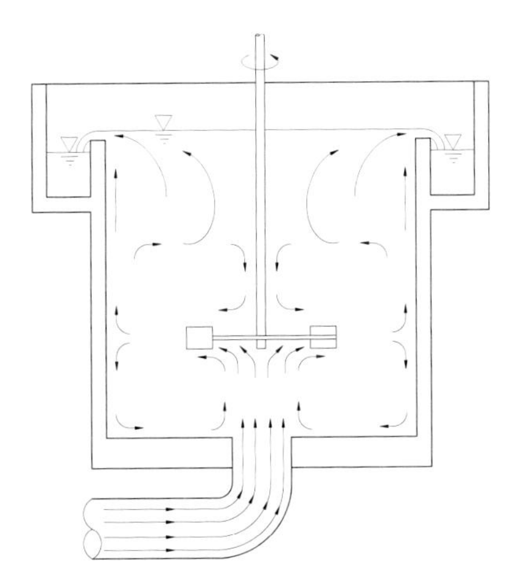

Figure 8-9 Typical flow pattern in radial flow mechanically mixed unit.

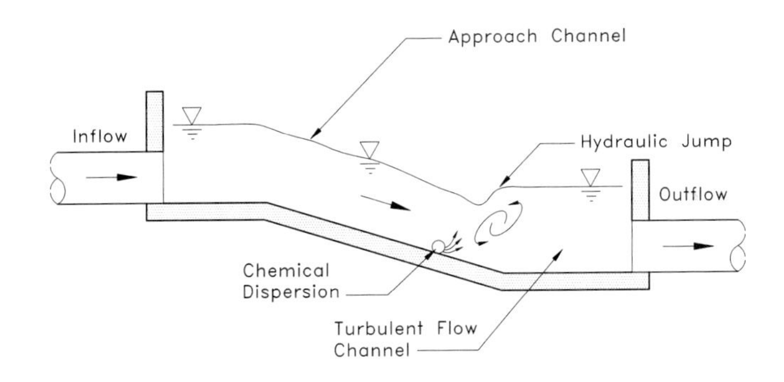

Figure 8-10 Typical rapid mixing utilizing a hydraulic jump.

8.5 Flocculation

The coagulation process chemically modifies the colloidal particles so that the stabilizing forces are reduced. To insure that a maximum amount of turbidity is removed, mixing condition and energy input must be properly provided after rapid mixing, to allow the aggregation of destabilized particles. The coagulated water must be gently stirred to promote the growth of the floc. This process is known as *flocculation*. Flocculation is also important in precipitation processes. The precipitate initially forms into small particles that cannot readily be settled or filtered. In the flocculation process, the mixture is gently stirred to promote the growth of the floc to a size that can be removed by sedimentation and filtration. The typical floc size is in the range from 0.1 to 2.0 mm.

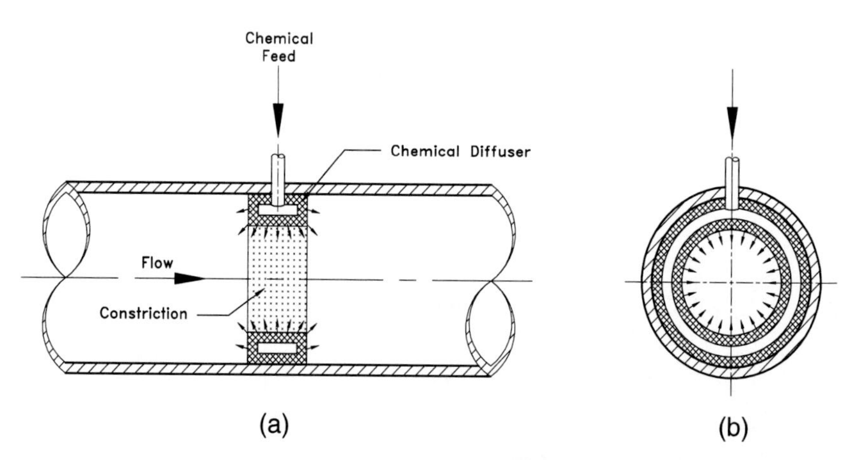

Figure 8-11 Typical rapid mixing utilizing a pipe contraction. (a) Longitudinal section. (b) Cross section.

The terms *parakinetic* and *orthokinetic* are often used in describing the coagulation flocculation process. *Parakinetic* refers to the growth of particles as a result of interparticle contacts due to *Brownian motion. Orthokinetic* refers to particle growth as a result of interparticle contacts due to fluid motion.[2] *Parakinetic* coagulation alone is typically inefficient for turbidity removal. Only in extremely high solids concentrations will a sufficient number of particle collisions occur due to *Brownian motion.* In most water treatment processes, *orthokinetic* flocculation predominates as a mechanism to promote particle growth.

8.5.1 Types of Flocculators

Flocculation units are often divided into two general groups: (1) hydraulic flocculators, and (2) mechanical flocculators. The hydraulic flocculators simply utilize cross-flow baffles or 180° turns to produce the required turbulence. The critical design objective in hydraulic flocculators is to achieve gentle, uniform mixing that will not shear the floc. These types of flocculators are effective only if the flow rate is relatively constant. They are rarely used in medium- and large-sized water treatment plants, because of their sensitivity to flow changes.[37] In mechanical flocculators, any of the mixers described in Section 8.4.1 can also be used (at a reduced speed) for flocculation. The mixers typically used in flocculation basins are horizontal-shaft *paddle-wheel* flocculators (Figure 8-12(a)). The turbine types are axial- and radial-flow vertical and horizontal flocculators. In addition to mixer types, other common flocculators are the *walking-beam* type and the *oscillating* type. These flocculator types are shown in Figure 8-12(b) and (c).[38,39]

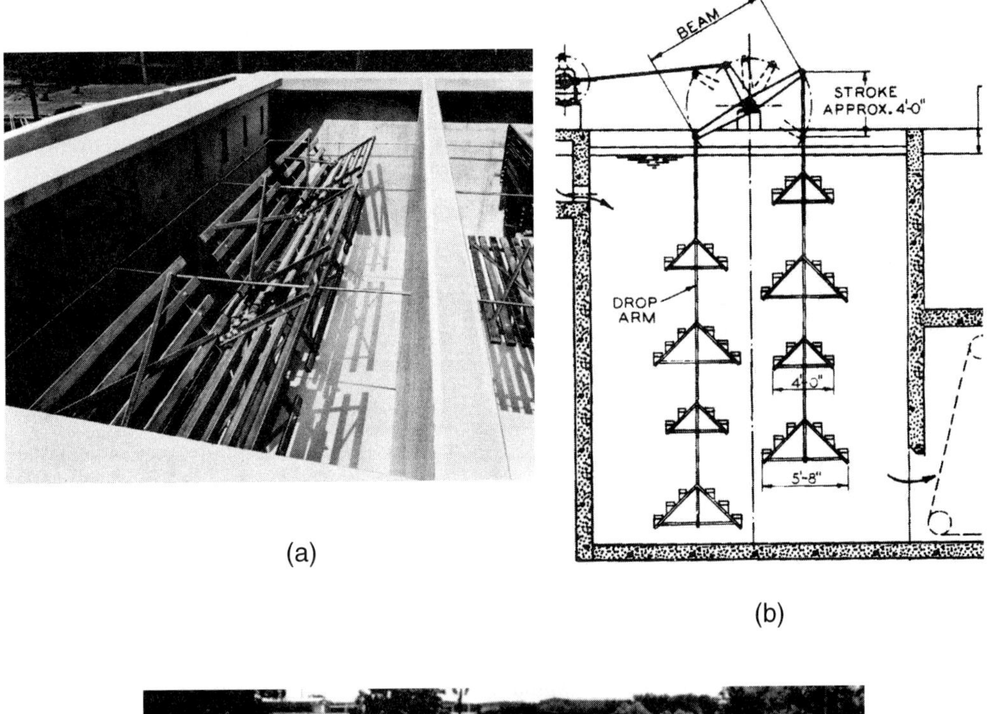

(a)

(b)

(c)

Figure 8-12 Typical flocculators. (a) Paddle wheel. *(Courtesy of Envirex/U.S. Filter)* (b) Walking beam. *(Courtesy of JDV Equipment Corp.)* (c) Oscillating paddle. *(Courtesy of Baker Process)*

8.5.2 Agitation Requirements

The degree of agitation employed in flocculators is much less than that used for rapid mixing. The purpose of flocculation is to cause particle contact, while not creating sufficient turbulence to break up the floc particles already formed. Typical velocity gradients (G) for flocculators range from 15 to 60/s.[1,2,27] Flocculation basins are normally designed with multiple mixing compartments in a series, with velocity gradients successively lower in each compartment.[2] This type of design is called *tapered flocculation* and has been found to produce a uniform and tough floc that will settle readily.

The higher G value in the first compartment causes a rapid transformation of the primary particles into higher-density floc; the lower G values in subsequent compartments cause the buildup of progressively larger-size floc, for better settling. Compartmentalization and tapering improve the process of flocculation significantly.[27] The velocity gradient for flocculators employing flow-induced turbulence is computed by Eqs. (8.25) and (8.26). The velocity gradient for mechanical mixers can be calculated from Eq. (8.20). In the case of paddle-wheel mixers, the water power is given by Eq. (8.28).[2]

$$P = \frac{C_D A \rho v^3}{2} \quad \left(\text{or } \frac{C_D A \gamma v^3}{2g}\right) \tag{8.28}$$

where

P = power imparted to water, W (ft-lb/s)
C_D = coefficient of drag, which varies with the length-to-width ratio (L/W) of the paddle blades. The values of C_D for different L/W ratios are summarized in Table 8-5.
A = area of paddles, m^2 (ft^2)
v = velocity of the paddle relative to the water, m/s (ft/s).

The paddle velocity relative to the water is approximately 75 percent of the absolute peripheral velocity of the paddle. The average absolute peripheral velocity of the paddle, v, is $2\pi n \infty$ (distance from the center of the paddle to center of the shaft), where n is the rotational speed of the shaft.

The peripheral speed of a paddle-wheel mixer should be limited to 0.40 m/s for alum coagulation and 0.50 m/s for lime softening flocculation.[36] Mechanical mixers should have variable-speed drives, to enable the operator to vary the velocity gradients to match variations in the raw water conditions. The ratio of maximum to minimum mixer rotational speeds normally ranges between 2 and 4.

8.5.3 Detention Time in Flocculation Basin

The detention time in flocculation basins is much higher than that in rapid-mix basins. Detention times from 20 to 60 minutes are common.[1] The key design factor in a flocculation basin is the value of Gt (velocity gradient × detention time), because the number of particle collisions within the basin is directly proportional to the value of Gt. Typical Gt values range from

10,000 to 150,000.[2,3,27] As with rapid mixing, the design parameters for a flocculation facility are best determined experimentally or from experience at water treatment plants treating similar water.

Table 8-6 Coefficient of Drag (C_D) for Paddle-Wheel Flocculator, Based on Length-to-Width Ratio of the Paddle

Length-to-Width Ratio (L/W)	C_D
5	1.20
20	1.50
∞	1.90

8.5.4 Geometry of Flocculation Basin

Short-circuiting in a flocculation basin should be minimized. In order to achieve this, multiple compartments are used in a series. When properly designed, three or four compartments have been found enough to minimize short circuiting effectively.[2] Proper design dictates that each flocculation compartment be separated by a baffle wall.[2,31,37] Square compartments with maximum dimensions of 6 m ∞ 6 m (20 ft ∞ 20 ft) and a depth of 3 to 5 m (10 to 16 ft) are usually used for vertical flocculators. When horizontal-shaft reel or paddle flocculators are used, the compartments are typically 6 to 30 m (20 to 100 ft) long and 3 to 5 m (10 to 16 ft) wide.[40] Openings on the baffle wall shall be designed to create a velocity gradient no greater than that in the preceding basin. The velocity gradient through a baffle-wall opening is given by Eqs. (8.25) and (8.26) or (8.27). The head loss through these openings is calculated from Eq. (6.4).

Typical velocity criteria used in flocculation unit design are as follows:

- Typical velocity in conduits or flume from the rapid-mixing unit to the flocculation basins is from 0.45 to 0.9 m/s (1.5 to 3.0 ft/s). Tapered shape, either in width or depth, is sometimes used to achieve constant velocity in the flocculation-basin influent-distribution channel.[40]
- The flocculation basins should be designed to have a velocity through the basin between 0.15 and 0.45 m/min (between 0.5 and 1.5 feet per minute).[41] Velocities greater than 0.5 m/min may result in shearing of the floc particles.[31] Flow through the basin should be uniform across the cross-section, to ensure that no dead spots exist in the basin.
- The baffles are typically designed to have a velocity of 0.3 to 0.45 m/s (1.0 to 1.5 ft/s). The baffle-wall opening ratio is usually 3 to 6 percent. Staggered slots (typically 0.1–0.15 m high ∞ 0.4–0.6 m long) are also provided in the bottom, for cleaning.[40]
- A velocity in the range from 0.15 to 0.45 m/s (from 0.5 to 1.5 ft/s) is typically used in design if a pipe or conduit is used to transfer coagulated water to the sedimentation basin.[41,42] When a diffusion wall (or *end baffle*) is provided between the flocculation and sedimentation basins, a velocity of 0.15 m/s (0.5 ft/s) is typically used to determine the orifice's opening area. A small port at the bottom and a small submerged section at the top

of the diffusion wall are often provided, to allow sludge and scum to pass to the sedimentation basin.[40]

8.6 Manufacturers of Coagulation and Flocculation Equipment

A list of manufacturers of mixing, flocculation, and chemical-handling equipment is provided in Appendix C. Information regarding the type and quantity of chemicals to be used, flow rates, basin geometry, velocity gradients, and equipment operating ranges are necessary to make proper equipment selection. Manufacturers can assist the design engineer in selecting and specifying the proper equipment for the proposed application.

8.7 Information Checklist for Design of Coagulation and Flocculation Facilities

The following information must be obtained and necessary decisions made before the design engineer can begin to design the coagulation and flocculation facilities.

1. Information to be obtained from the predesign report:

   a. design flows—peak, average, minimum;
   b. chemicals to be used and dosage rates;
   c. rapid-mix time and velocity gradients;
   d. flocculation time and velocity gradients.

2. Information to be obtained from the preliminary design

   a. preliminary plant layout, showing existing facilities and the general size, location, and geometry of the proposed facilities;
   b. number of proposed units.

      i. Chemical feeders
      ii. Rapid mixers
      iii. Flocculators

   c. type and location of other chemical-treatment units that are used in conjunction with coagulation—examples include powdered carbon, oxidant feed (potassium permanganate), chemicals for partial softening, and disinfectant;

      i. Application point
      ii. Dosage rate
      iii. Chemical handling and storage requirements

d. preliminary hydraulic profile of the proposed unit illustrating head loss through the various treatment units being designed;
e. type of rapid-mixing equipment;
f. type of flocculation equipment.

3. Information to be obtained from the plant owner and operator:

a. preferences with regard to the equipment type and manufacturer;
b. preferences with regard to the design of any existing coagulation and flocculation facility in operation at the plant.

4. Information to be developed by the design engineer:

a. minimum design parameters established by the regulatory authorities concerned;
b. types, sizes, and limitations of available chemical-feed, coagulation, and flocculation equipment.

8.8 Design Example

8.8.1 Design Criteria Used

The following criteria are used in the Design Example.

1. Flow Rates.

a. Maximum flow rate = 113,500 m^3/d (30 mgd)
b. Average flow rate = 57,900 m^3/d (15.3 mgd)

2. Raw Water Quality. As discussed in Section 5.4.8, the raw water quality is expressed by the following conditions. This information is presented in more detail in Section 5.4.8.

a. Turbidity: 3–17 NTU
b. Seasonal total iron and manganese concentration: 0.2 to 0.7 mg/L and 0.05 to 0.4 mg/L, respectively
c. pH: 7.9 to 8.2
d. Total alkalinity: 80 to 110 mg/L as $CaCO_3$
e. Total hardness: 100 to 120 mg/L as $CaCO_3$
f. Various metering and instrumentation shall be provided at strategic locations, to continuously monitor and record temperature, turbidity, pH, and specific conductance of the raw water reaching the plant.

3. Chemicals

a. Coagulant: ferric sulfate, optimum dosage 25 mg/L, maximum feed rate 60 mg/L[d]. Application point is rapid-mix basin.
b. Coagulant aid: cationic polymer, optimum dosage 0.05 mg/L, maximum feed rate 0.08 mg/L. Application point is the Parshall flume ahead of the flocculation basin.
c. Filter aid: nonionic polymer; optimum and maximum dosages 0.02 and 0.04 mg/L, respectively. Application point is the exit of central effluent collection channel, before free fall into the effluent box of each sedimentation basin.
d. pH adjustment: Hydrated lime $Ca(OH)_2$ is required to adjust the pH 8 to 9 units. Lime dose of 15 mg/L as $Ca(OH)_2$ is sufficient to provide optimum pH for coagulation. Application point is in the lime-mixing chamber ahead of primary coagulant in the rapid mixer.
e. Potassium permanganate and powdered activated carbon (PAC) are added for iron and manganese precipitation and for taste and odor control[e]: Potassium permanganate ($KMnO_4$) is added into the raw water at Junction Box A inside the south plant property boundary. (See Figure 5-4 for feed point.) Maximum dosage is 4 mg/L. (Calculations for $KMnO_4$ requirements are discussed in Chapter 11.) PAC will be added at the raw water pump station. (See Figure 7-1 for PAC storage and feed building.) The PAC dose is 4 to 8 mg/L.

4. Rapid-Mix Basin—Design Parameters

a. Number of units: four
b. Number of stages: one
c. Detention time: 20–30 s
d. Velocity gradient (G): 950/s

5. Flocculation Basin—Design Parameters

a. Number of basins: four
b. Number of stages: three

d. The values of ferric sulfate are generally recorded as ferric sulfate ($Fe_2(SO_4)_3$). This is a dry form, used by many plant operators because it makes equipment adjustments easier. If liquid chemicals are available, the plant operators prefer liquid feed rate. Generally, 50 percent (by weight) solutions of ($Fe_2(SO_4)_3$) and ($Al_2(SO_4)_3$) are supplied by vendors. A more accurate method is to express coagulant feed as Fe^{3+}. This removes the inaccuracies caused by variations in water of hydration or dilution water. This approach, however, is confusing to some operators.

e. PAC is added at the raw water pump station (Chapter 7). PAC feeders and storage facility are designed in Chapter 11.

c. Detention Time: 30 min total, 10 min in each stage

d. Velocity gradient (G): first stage 60/s, second stage 30/s, third stage 15/s. The overall process G is 35/s.

6. Unit Arrangement and General Design Guidelines

The general layout of the rapid-mix and flocculation facilities is illustrated in Figure 8-13. The rapid-mix basins are located adjacent to the chemical feed facility. Chemical handling and delivery are often among the major maintenance problems in a water treatment plant. This layout will minimize the potential for such problems by reducing the length of the piping between the chemical-handling equipment and the application points.

The 122-cm-diameter conduit will lead the raw water influent from Junction Box D into a forebay. The raw water then will flow over the weir into the lime-mixing chamber, where hydrated lime will be added for pH adjustment. The pH-adjusted water will then be split into two common influent channels before the rapid-mix basins. Short retention period (approximately 2 minutes at the maximum flow rate) is provided for lime slurry dissolution in the lime-mixing chamber and the influent channel before the rapid-mix basins. From each influent channel, two totally separate process trains are provided for coagulation, flocculation, and sedimentation. This layout offers the capability of totally shutting down one-fourth of the process train for maintenance while continuing to process water in the other three process trains. This capability is often desirable when one treatment plant is the only water supply source for a community. Such cost may not be warranted in areas where more than one water supply source is available.

a. Rapid Mixers: A total of four mixing basins shall be operated in parallel, each process train designed for 25 percent of the maximum flow. Each mixing basin shall utilize a radial-flow mechanical mixer. The coagulant will be added in the rapid-mix basins. Coagulant aid (a polymer) will be added in the Parshall flume at the head of flocculation basins. At this application point, pin-head floc will have already formed. Turbidimeters, pH meters, streaming-current monitors, and chlorine residual meters will be installed at several strategic locations through the process train to provide information for chemical feed-rate dosage and treatment-process optimization. Potassium permanganate will be added to the raw water at Junction Box A inside the plant property at the south boundary, so the oxidation of iron and manganese will be initiated several minutes prior to coagulation reaction in the rapid mixer. This design consideration will effectively improve iron and manganese removal.

b. Flocculation Basins: The flocculation basins shall be attached to the sedimentation basins. The basins shall be separated by a diffusion wall (or end-baffle wall). Four flocculation basins shall be provided, operated in parallel. Each basin shall be designed for 25 percent of the design flow and shall have three-stage paddle-wheel mechanical mixers operating in a series. The coagulated water is fed evenly through the influent distribution channel and weir. The cross-flow pattern is provided through the basin. Two baffle walls perpendicular to the flow direction are used to separate the flocculation stages. A diffusion wall with numerous orifices separates the flocculation and sedimentation basins.

c. Chemical Building: The chemical building shall house the chemical feed and storage facilities. Chemicals that will be handled in this building are ferric sulfate, hydrated lime, coagulant and filter aids, and potassium permanganate. The building is multi-storied and is climate-controlled to a temperature of 5°C–40°C. An air-conditioned electrical switchgear room is provided on the ground floor. Chemical delivery is from the unloading area located on the east side of the building. A maintenance shop is also provided on the ground floor in the northeast part of the building adjacent to the coagulant-handling area. A covered passage over the street is provided between the chemical building and the administration/control building.

8.8.2 Design Calculations

Step A: Rapid-Mix Basin Design

1. Unit dimensions

a. Calculate design flow for each process train.

$$\text{Maximum design flow} = 113.5 \text{ MLD} = 113,500 \text{ m}^3/\text{day}$$

Each process train receives one-fourth of maximum design flow.

$$\text{Design flow for each process train} = \frac{113,500 \text{ m}^3/\text{day}}{4} = 28,375 \text{ m}^3/\text{day}$$

$$= 0.328 \text{ m}^3/\text{s}$$

b. Calculate basin volume for 30-second detention time at the design flow rate.

$$\text{Volume required} = Q \times t = 0.328 \text{ m}^3/\text{s} \times 30 \text{ s} = 9.84 \text{ m}^3$$

Use a square basin with a depth-to-width ratio of 1.5. Experience has shown that a depth-to-width ratio of 1.5 yields excellent performance with turbine mixers.

$$\text{Volume} = w \times l \times d$$
$$= w \times w \times 1.5w$$
$$= 1.5w^3$$

Solve for w.

$$w = \left(\frac{\text{Volume}}{1.5}\right)^{1/3} = \left(\frac{9.84 \text{ m}^3}{1.5}\right)^{1/3} = 1.9 \text{ m}$$

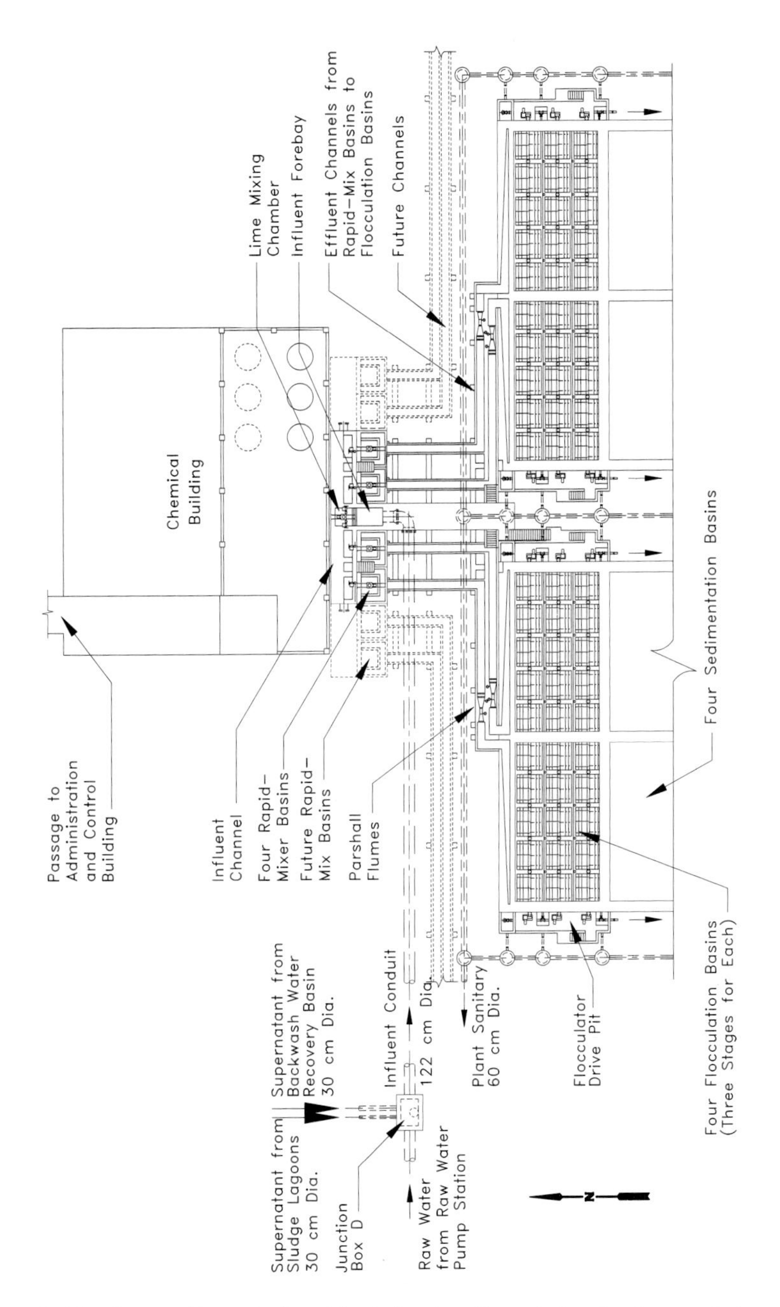

Figure 8-13 General layout of rapid-mix basins, flocculation basins, and chemical building.

$$\text{Water depth required} = 1.5 \times w = 1.5 \times 1.9 \text{ m} = 2.85 \text{ m}$$

Provide four basins, with dimensions of 1.9 m wide ∞ 1.9 m wide ∞ 2.9 m deep (from bottom to the top of effluent weir) and with volume $V = 10.5 \text{ m}^3$. Figure 8-14 illustrates the details of the rapid-mix basins.

2. Influent Structure

The influent structure consists of forebay, lime-mixing chamber, and common influent channel. The forebay receives raw water near the bottom from a 122-cm-diameter conduit. The water flows upward through the forebay and is discharged from a 2-m-long straight weir into the lime-mixing chamber. The square-shaped lime-mixing chamber has dimensions of 2 m ∞ 2 m. The pH-adjusted water is then split through two isolation gates near the bottom and flows into two common influent channels on either side. The selected sluice gates have dimensions of 0.9 m × 0.9 m. Thus, two rapid-mix basins are fed by a common channel on both sides of the lime-mixing chamber. The width of the influent channel is 2 m. Each rapid-mix basin is connected underneath by a 60-cm-diameter pipe. The pipe discharges at the bottom of the rapid mix basin, directly below the turbine impellers. The influent end of the influent pipe has an influent sluice gate that is used to regulate the flow into the rapid mix unit or to isolate it from others as necessary. Details of the influent structure are shown in Figure 8-14.

3. Effluent Structure

The effluent structure of the rapid-mix basins consists of a straight weir on three sides, two 0.6-m-wide effluent-collection flumes, and a 0.85-m-wide exit channel that conveys the coagulated water into the Parshall flume at the head of the flocculation basin. The width of the effluent channel is equal to the required entrance width for a standard Parshall flume. (See Figure 8-17 for Parshall flume dimension requirements.) The overall layout of the rapid-mix and flocculation basins and the details of the influent and effluent structures of the rapid mix are shown in Figures 8-13 and 8-14, respectively.

4. Equipment design

a. Rapid mixer

i. Calculate mixer power requirements. Design each mixer for a velocity gradient (G) of 950/s at a flow rate of 0.328 m^3/s, that is, one-fourth of the maximum design flow rate. The sustained temperature of raw water is expected to be in the range from 5°C to 28°C. The lowest temperature will present the critical condition in the mixer design.

ii. Calculate mixer power.

Rearranging Eq. (8.20), to have the following:

$$P = G^2 V \mu$$

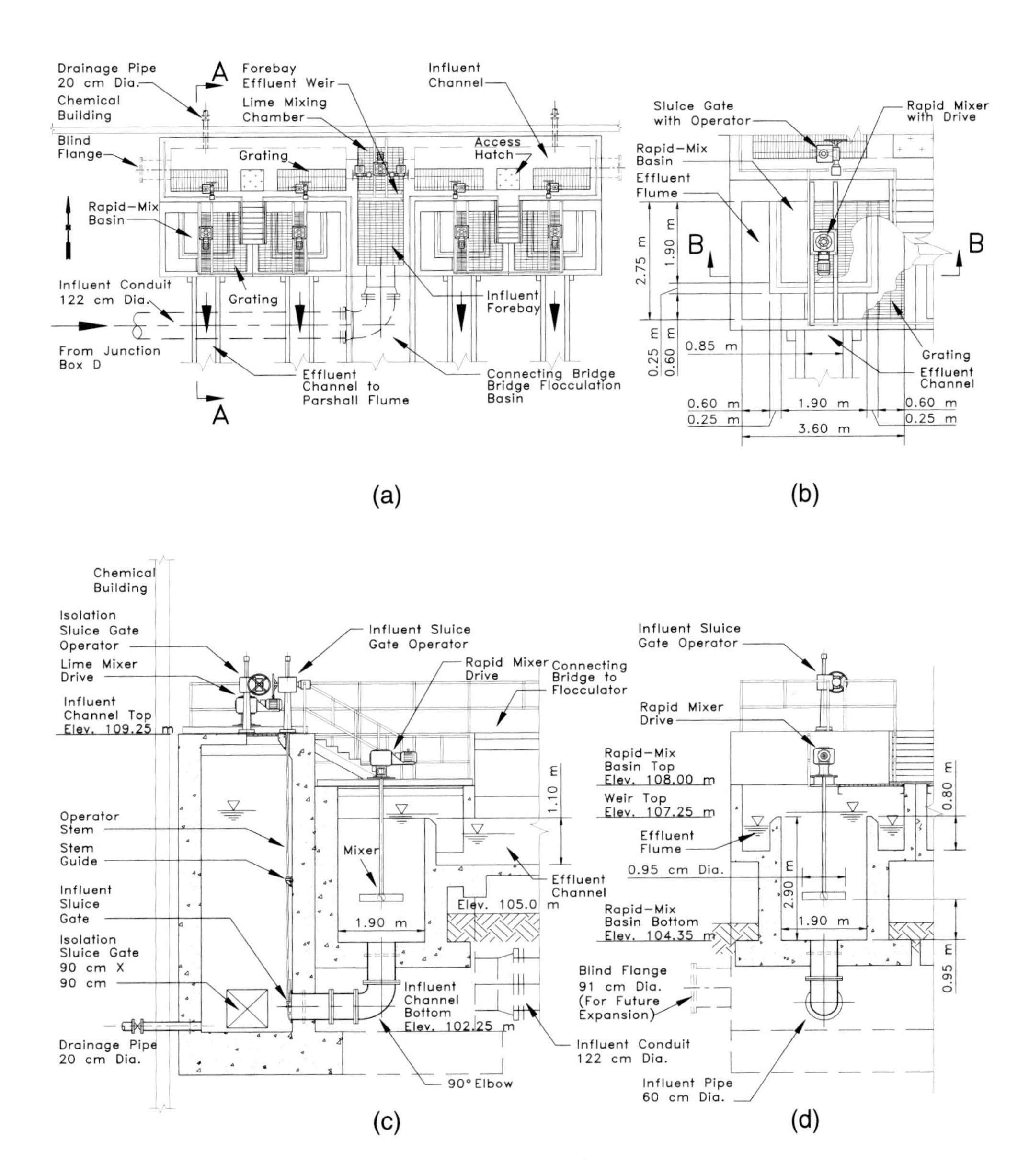

Figure 8-14 Details of rapid-mix basins. (a) Overall layout. (b) Plan view of rapid-mix basin. (c) Section A-A. (d) Section B-B.

$$\mu = 1.518 \times 10^{-3} \text{ N-s/m}^2 \text{ at } 5° \text{ C}$$

$$V = \text{basin volume} = 10.5 \text{ m}^3$$

$$P = (950/\text{s})^2 \times 10.5 \text{ m}^3 \times 1.518 \times 10^{-3} \text{N-s/m}^2$$
$$= 1.44 \times 10^4 \text{N-m/s}$$
$$= 14.4 \text{ kW}$$

P is the power imparted to the water. The power of the driver (P') is calculated by dividing P by the efficiency of the gearbox, which is typically around 90 percent.

$$P' = \frac{P}{0.9}$$
$$= \frac{14.4 \text{ kW}}{0.9}$$
$$= 16 \text{ kW (12hp)}$$

Each mixer motor will be provided with motor contacts, to monitor operating status and malfunction.

iii. Calculate impeller size and rotational speed. The rapid-mix basin will be an "up flow" type. Experience has shown that radial-flow mixers perform better than axial-flow mixers in a vertical-flow basin. For this example, use a four-blade turbine mixer with a w/d ratio of 0.16. Interpolating from Table 8-4, Np = 2.75.

Rearranging Eq. (8.23), to have the following:

$$n = \left(\frac{P}{\rho N_p d^5}\right)^{1/3}$$

Here, using ρ = 1000 kg/m^3 at 5°C and

P = 14.4 kW (14.4 kN-m/s or 14,400 kg m^2/s^3)

Use a turbine diameter equal to ∫ the basin width or 0.95 m, and locate the mixer blade one diameter above the bottom of the basin. Experience has shown that such a configuration provides good mixing currents.

$$n = \left(\frac{14.4 \text{ kN-m/s} \times 1000 \text{ N/kN} \times \text{kg·m/s}^2\text{·N}}{1000 \text{ kg/m}^2 \times 2.75 \times (0.95 \text{ m})^5}\right)^{1/3}$$

$$= 1.89 \text{ rps} = 113 \text{ rpm}$$

Check Reynolds number for turbulent flow (Eq. (8.24)).[f]

$$N_R = \frac{(0.95\ \text{m})^2 \times 1.89/\text{s} \times 1000\ \text{kg/m}^3}{1.518 \times 10^{-3}\ \text{N-s/m}^2 \times \text{kg·m/s}^2\text{·N}}$$

$$= 1.12 \times 10^6 > 10,000$$

Therefore, Eq. (8.23) is valid.

5. Head loss calculations and hydraulic profile through the rapid mix basin.

   In this section, the hydraulic calculations will be conducted for the head losses from the forebay to the Parshall flume. The head losses through the 122-cm diameter conduit between Junction Box A and the forebay are calculated in Chapter 15.

   a. Calculate head loss at the influent structure.

      The head losses at the influent structure are due to (1) free-fall from the forebay into the lime-mixing chamber, (2) loss through the isolation gate, (3) loss in the common influent channel, and (4) loss through the influent pipe to the rapid-mix basin. The water surface elevation in the forebay is controlled by the weir elevation. At least 0.6 m of free fall at the weir is considered desirable when preparing the hydraulic profile through the influent structure. The water surface in the rapid-mix basin is controlled by the effluent structure.

      i. Calculate the head over the forebay-discharge weir. The head over a weir can be calculated from Francis equation (Eq. (8.29)).

$$Q = \frac{2}{3}C_d L' \sqrt{(2gH^3)} \qquad (8.29)$$

where

Q = flow over weir, m^3/s
H = head over weir, m
C_d = coefficient of discharge, assume $C_d = 0.6$
$L' = L - 0.1\ nH$: where n = number of horizontal end contractions (0 , 1 or 2), and
L = length of weir, m.

f. If modified equation ($n = (Pg/\gamma N_p d^5)^{1/3}$) is used, γ=9.81 kN/m^3 at 5°C and g=9.81 m/s^2.

$$n = \left[(14.4\ \text{kN·m/s} \times 9.81\,\text{m/s}^2)/(2.75 \times 9.81\,\text{kN/m}^3 \times 0.95\,\text{m})^5\right]^{1/3} = 1.89\,\text{rps}$$

$$N_R = [(0.95\ \text{m})^2 \times 1.89/\text{s} \times 9810\ \text{N/m}^3]/[1.518 \times 10^{-3}\ \text{N·s·m}^2 \times 9.81\,\text{m/s}^2] = 1.12 \times 10^6$$

In this design, $n = 0$, and $L' = L = 2$ m, and the total maximum design flow rate $Q = 113{,}500$ m3/day/ 86,400 s/day = 1.314 m^3/s.

Substitute these values in Eq. (8.24), and solve for H.

$$H = \left(\frac{1.314\ \text{m}^3/\text{s} \times 3/2}{0.6 \times 2\ \text{m} \times \sqrt{2 \times 9.81\ \text{m/s}^2}}\right)^{2/3}$$

$$= 0.52\ \text{m}$$

A free-fall of 0.95 m is provided at the weir.

ii. Calculate the head loss through the isolation gate. The head loss through the isolation gate is calculated using the orifice equation (Eq. (6.5)). One half of the total maximum flow rate goes through each gate.

$$Q = \frac{1.314\ \text{m}^3/\text{s}}{2} = 0.657\ \text{m}^3/\text{s}$$

$$h_L = \frac{1}{2 \times 9.81\ \text{m/s}^2} \times \frac{0.657\ \text{m}^3/\text{s}^2}{0.6 \times (0.9\ \text{m})^2} = 0.10\ \text{m}$$

iii. Calculate the head loss through the common influent channel. The influent channel is 2 m wide, and an average water depth of at least 5 m is maintained. The channel has a variable flow. The maximum flow of 0.657 m^3/s enters the head of each channel, and each rapid-mix basin receives its distribution of one-half of this flow (0.328 m^3/s). The maximum velocity at the head of each channel will be 0.07 m/s, and the velocity will decrease gradually. The head loss in the influent channel is considered small and is ignored.

iv. Calculate the head loss through the influent pipe of the rapid-mix basin. The 60-cm-diameter influent pipe receives a flow of 0.328 m^3/s.

$$\text{Velocity in pipe} = \frac{4 \times 0.328\ \text{m}^3/\text{s}}{\pi \times (0.60\ \text{m})^2} = 1.16\ \text{m/s}$$

The friction loss is small because of the small length. Only minor head losses are considered. The minor head losses are encountered due to the entrance (K = 0.5), one 90° elbow ($K = 0.3$), and an exit loss ($K = 1.0$). The head loss is calculated from Eq. (7.12).

$$h_m = (0.5 + 0.3 + 1.0) \times \frac{(1.16\ \text{m})^2}{2 \times 9.81\ \text{m/s}^2} = 0.12\ \text{m}$$

v. Calculate the total head losses through the effluent structure.

- Free fall provided at the forebay weir 0.95 m
- Head loss through the isolation gate 0.10 m
- Head loss through the influent pipe 0.12 m

Total = 1.17 m

b. Calculate head loss at the effluent structure.

The head losses at the effluent structure are due to (1) free-fall from the weir into the effluent flume, (2) loss in the effluent flume, and (3) loss in the effluent channel. The water surface elevation in the effluent channel is controlled by the Parshall flume provided at the influent end of the flocculator. With normal flow conditions in the effluent channel, the head-loss calculations can be performed as follows:

i. Calculate the head over the effluent weir. As in the calculation for the influent forebay weir, the head over the effluent weir is calculated from the weir equation (Eq. (8.24)). Substitute $n = 0$, $L = 5.7$ (3 ∞ 1.9 m), and $Q = 0.328\ m^3/s$ in the equation, and solve for H. The head loss over the effluent weir $H = 0.10$ m. A free-fall of 0.33 m is provided at the effluent weir.

ii. Calculate the depth of flow in the effluent channel. The calculations are started from the Parshall flume. The required water depth at the entrance of the flume is 0.71 m. (See Section 8.8.2, Step B.) Assuming a normal flow condition in the channel, the required slope is calculated from Eq. (7.1) on the basis of the following design values:

n (Manning′s roughness coefficient) = 0.013

d (water depth) = 0.71 m

B (width of effluent channel) = 0.85 m

$$v = \frac{0.328\ m^3/s}{0.85\ m \times 0.71\ m} = 0.54\ m/s$$

$$r = \frac{0.85\ m \times 0.71\ m}{2 \times 0.71\ m + 0.85\ m} = 0.266\ m$$

Substitute these values in Eq. (7.1).

$$0.54\ m/s = \frac{1}{0.013} \times 0.266^{2/3} \times S^{1/2}$$

Solve for S (slope of effluent channel).

$$S = 0.0003$$

iii. Calculate the depth of water at the upstream end of the effluent flume. The depth of flow in the upstream end of the effluent flume is calculated from Eq. (8.30).

$$y_1 = \sqrt{y_2^2 + \frac{2Q^2}{(gb^2 y_2)}} \tag{8.30}$$

where

y_1 = water depth at the upstream end of the flume, m
y_2 = water depth at the downstream end of the flume, m
Q = total flow at the downstream end of the flume, m^3/s
b = width of the flume, m

Eq. (8.30) is widely used by designers to calculate the water surface profile in a flume receiving flow from a free-falling weir. This equation provides an approximate solution and was originally developed for flumes with level inverts and parallel sides; channel friction is neglected, and the drawdown curve is assumed parabolic. The calculation for a more accurate water surface profile throughout the entire length of the flume under variable flow conditions is presented in Chapter 9.

In this design example, the water surface elevation at the exit point in the effluent flume is the same as that in the effluent channel, and the invert of the flume is 0.3 m above the upstream invert of the effluent channel. Therefore, $y_2 = 0.71 - 0.30 \text{ m} = 0.41 \text{ m}$.

The length of each effluent flume

$$= \text{half the length of the effluent weir} - \text{half the width of the effluent channel}$$

$$= \frac{(5.7 \text{ m} - 0.85 \text{ m})}{2}$$

$$= 2.43 \text{ m}$$

Total flow at downstream end of each effluent flume is as follows:

$$= \text{total flow per rapid mixer} \times \frac{\text{length of each effluent flume}}{\text{total length of effluent weir}}$$

$$= 0.328 \text{ m}^3/\text{s} \times \frac{2.43 \text{ m}}{5.7 \text{ m}}$$

$$= 0.140 \text{ m}^3/\text{s}$$

The width of flume $b = 0.6$ m. Substituting these values in Eq. (8.30), the calculated value of $y_1 = 0.44$ m.

Add 30 percent for losses due to friction, turbulence, and 90° turn in the flume.

The depth of the water at the upstream end of the flume = (0.44 m ∞ 1.30) = 0.57 m. The head loss through the effluent flume is therefore 0.57 m – 0.41 m = 0.16 m.

iv. Calculate the head loss in the effluent channel. The total length of the effluent channel from the effluent flume to the entrance of Parshall flume is approximately 25 m. Total head loss due to the slope of the channel (friction losses) = 25 m ∞ 0.0003 = 0.01m.

At each 90° bend in the effluent channel leading to the flocculator, minor head losses will be encountered from turbulence and from change in flow direction. At the downstream section of the bend, the free water surface will rise and will eventually reach the normal depth of 0.71 m.

v. Calculate the total head losses through the effluent structure.

• Free fall provided at the effluent weir	0.33 m
• Head loss through the effluent flume	0.16 m
• Head loss through the effluent channel	0.01 m
Total =	0.50 m

c. Prepare hydraulic profile through the rapid-mix basin.

In the design of water treatment plants, the upstream water surface elevation in a unit normally is determined with respect to the water surface elevation in the downstream facility. Today, various types of computer software are commonly used by the design engineers to prepare the hydraulic profile. A computer hydraulic model is generally developed. Such a program allows flexibility, to perform hydraulic calculations for the upstream facilities simultaneously while any change in design conditions such as elevations, flows, and dimensions of the downstream facility is being made. Thus, the effects of such changes on the hydraulics of the upstream facilities can be effectively and accurately assessed. In this Design Example, a hydraulic model is developed using a computer spreadsheet to prepare the hydraulic profile through the treatment units. The hydraulic profile from the influent forebay through the rapid-mix basin to the Parshall flume located at the influent end of the flocculation basin is computed by this program. The results are shown in Figure 8-15. The calculations for key design elevations in the hydraulic profile are presented below.

Starting from the effluent channel before the Parshall flume, the water surface elevation (WSEL) at the entrance of the Parshall flume is 106.85 m. This value is obtained from the flocculation basin design. (See Section 8.8.2, Step B.5.e.ii, and Figure 8-17(b)).

i. Calculate WSEL and other major design elevations in the rapid-mix basin and effluent structure.

The WSEL in the rapid-mix basin = ((downstream WSEL in the effluent channel at the entrance of the Parshall flume) + (total head losses through the effluent structure)) = (106.85 m + 0.50 m) = 107.35 m.

The top elevation of the effluent weir = ((WSEL in the rapid-mix basin) – (head over the effluent weir)) = (107.35 m – 0.10 m) = 107.25 m.

The bottom elevation of rapid-mix basin = ((top elevation of the effluent weir) – (design water depth)) = (107.25 m – 2.9 m) = 104.35 m.

ii. Calculate WSEL and other major design elevations in the influent structure.

The WSEL in the forebay = ((WSEL in the rapid-mix basin) + (total head losses through the influent structure)) = (107.35 m + 1.17 m) = 108.52 m.

The top elevation of the forebay weir = ((WSE in the forebay) – (head over the weir)) = (108.52 m – 0.52 m) = 108.00 m.

With a forebay-weir height of 5.75 m, the floor elevation of the forebay = ((top elevation of the forebay weir) – (provided forebay-weir height)) = (108.00 m – 5.75 m) = 102.25 m.

Step B: Flocculation-Basin Design

1. Unit Dimensions

a. Calculate the required volume.

The required design flow for each basin is one-fourth of the maximum day flow, and a total detention time of 30 min is provided for three-stage flocculation basin.

$$\text{Volume} = 0.328\ \text{m}^3/\text{s} \times 60\ \text{s/min} \times 30\ \text{min} = 590\ \text{m}^3$$

$$\text{The volume of each stage of flocculation basin} = \frac{590\ \text{m}^3}{3} = 197\ \text{m}^3$$

b. Calculate basin dimensions.

In this design example, a flocculator axis perpendicular to the flow direction is provided in the flocculation basin. The width of the basin parallel to the flocculator axis is equal to the width of the sedimentation basin, which is 18.4 m. (See Chapter 9.) The length of each stage perpendicular to the axis of the flocculator is equal to the water depth. There are three flocculation stages. Therefore, the total length of three stages = 3d.

$$\text{Volume of each stage of the flocculator } (V) = 18.4 \times d \times d$$

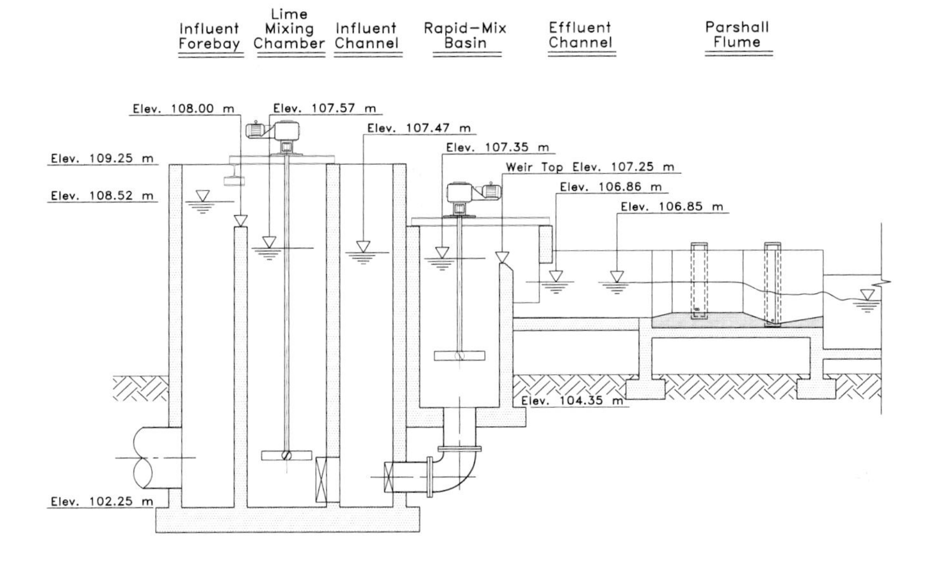

(a)

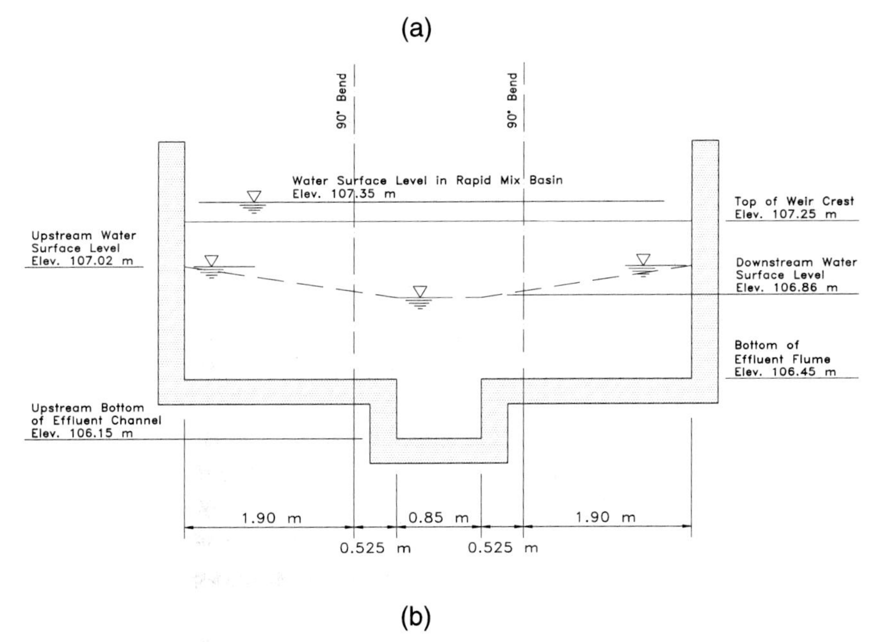

(b)

Figure 8-15 Hydraulic profile through the rapid-mix basin. (a) Hydraulic profile from the influent forebay to the entrance of the Parshall flume. (b) Conceptual longitudinal section in rapid-mix basin effluent flume.

$$d = \left(\frac{197 \text{ m}^3}{18.4 \text{ m}}\right)^{1/2}$$

The required water depth $d = 3.27$ m.

With a water depth of 3.3 m, the dimensions for each stage are 3.3 m long ∞ 18.4 m wide ∞ 3.27 m deep, for a volume of 199 m^3. The water flows from one stage into the other through a baffle wall. The total length of three stages plus 0.1 m for two baffle walls is 10 m. Therefore, the dimensions of each of the four flocculation basins are 10 m long ∞ 18.4 m wide ∞ 3.27 m deep. The total volume of each basin (including the volume of baffle walls) is 596 m^3. The general layout of the flocculation basins is shown in Figure 8-13. Details of the flocculation-basin design are illustrated in Figure 8-16.

2. Influent structure. The influent structure of each flocculation basin consists of a Parshall flume, an influent channel, an influent-distribution channel, and influent-distribution weirs. The coagulated water is conveyed by the 0.85-m-wide effluent channel from the rapid-mix basin to the Parshall flume. A free fall is provided after the Parshall flume. The metered water flows into the influent-distribution channel through a short influent channel with two 90° turns. The influent channel is designed with a tapered section, to provide a more nearly constant velocity in the channel. Five straight-side weirs parallel to the channel are provided, to distribute water evenly into the first stage of the flocculation basin.
3. Effluent structure. The effluent structure of each flocculation basin consists of simply a diffusion wall. The flocculated water enters the sedimentation basin through the orifices evenly arranged on the diffusion wall. The velocity through the orifices is low enough to prevent the floc structure from breaking up. These flocs are formed during three stages of flocculation and are quite fragile.
4. Parshall flume and equipment design.

a. Design of Parshall flume.

The Parshall flume is located at the head of the influent-distribution channel of the flocculation basin. The purpose of the Parshall flume is to provide the plant operator with flow data at the entry of each process train, to ensure that the total plant capacity is equally shared by the four process trains. The turbulence created by the hydraulic jump at the free fall after the flume will provide energy for mixing of a cationic polymer (coagulant aid) with the coagulated water.

A Parshall flume (Figure 8-17(a)) consists of three basic sections: (1) a converging upstream section, (2) a constant-width throat section, and (3) a diverging downstream section. A Parshall flume can be used as an accurate flow measurement and control device if the discharge from the flume is unrestricted or under free-fall conditions. The unrestricted flow condition is commonly determined by the ratio of the flow depth H_b near the end of the throat to the flow depth H_a in the upstream con-

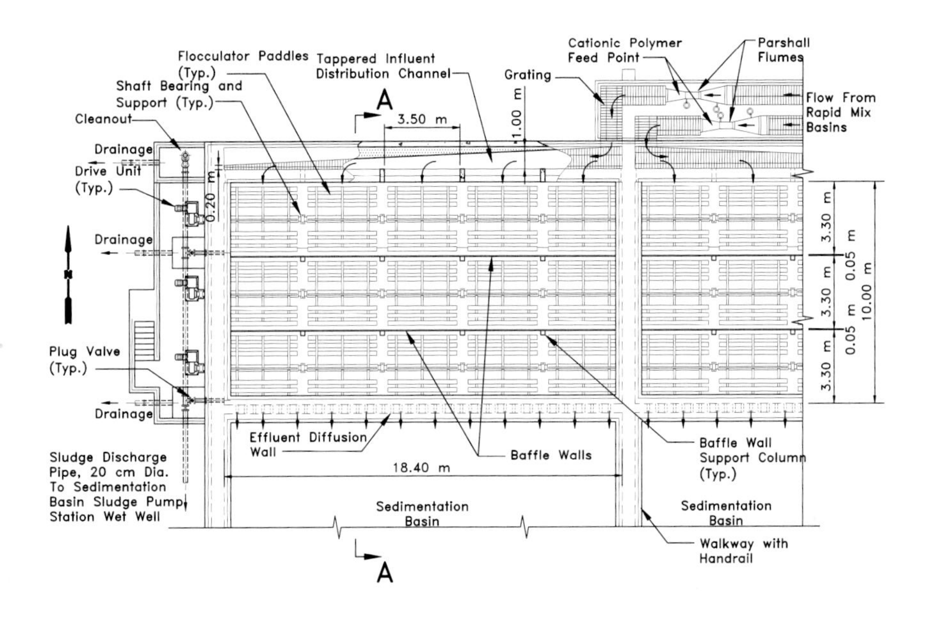

(a)

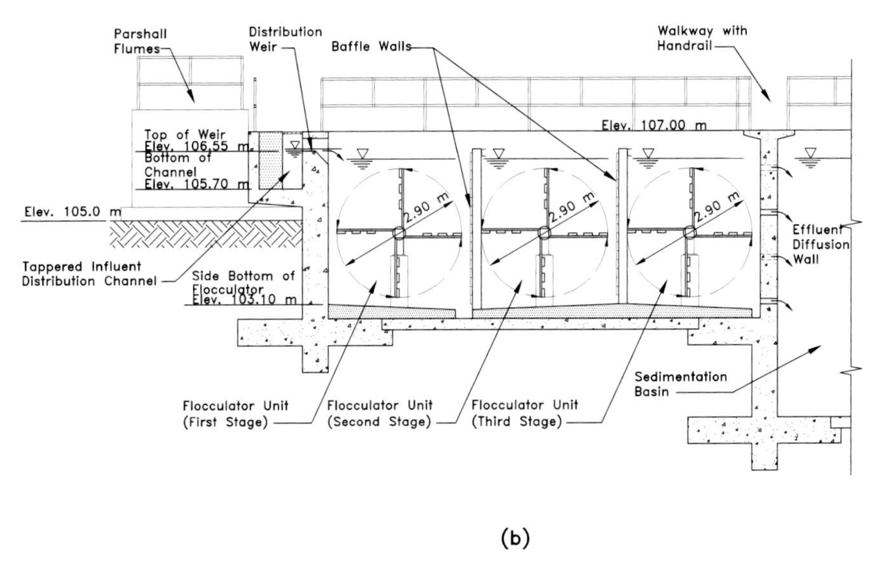

(b)

Figure 8-16 Details of flocculation basins. (a) Flocculation basin plan (typical of four). (b) Section A-A.

verging section. The free-fall condition in a Parshall flume is dependent upon the H_b/H_a ratio at different throat widths. The following criteria are commonly used to check the free-fall condition.[43]

W, throat width	H_b/H_a ratio
3-9 in. (76-229 mm)	< 0.6
1-8 ft. (0.30-2.44 m)	< 0.7
0-50 ft. (3.05-15.2 m)	< 0.8

Under unrestricted flow conditions, the discharge through a Parshall flume is can be determined from Eq. (8.31) by using the reading of flow depth H_a.[44]

$$Q = 4WH_a^{1.522W^{0.026}} \tag{8.31}$$

where

Q = free-flow discharge, cfs

W = throat width, ft.

H_a = depth of water at upstream gauging point, ft.

If the ratio of H_b/H_a exceeds the limit given above, the Parshall flume is operating under the submerged condition. The discharge will be retarded under this flow condition. The flow can be obtained only from both H_a and H_b readings, by using calibration curves that must be developed.

In this example, the design parameters for the Parshall flume are as follows:

Maximum flow $Q = 0.328\ m^3/s$ (11.5 cfs)

Throat width $W = 0.30$ m (1.0 ft.)

Submergence H_b/H_a ratio < 70 percent

A free-flow Parshall flume having throat width of 0.3 m (1.0 ft.) has minimum and maximum flows of 0.01 and 0.456 m^3/s (0.35 and 16.1 cfs).[43] The dimension requirements of the Parshall flume used in this example are given in Figure 8-17(a).[45,46] A calibration curve for the selected 0.3-m Parshall flume is also developed from Eq. (8.31). This curve is shown in Figure 8-18.

b. Flocculator-paddle design

Design of flocculator paddles involves selection of mixer speed, paddle size, power requirements, number of paddles, paddle-wheel size, and layout.

i. Calculate flocculator power requirements.

The velocity gradient in the first stage of the flocculator $G = 60/s$. The basin volume $V = 199\ m^3$ and $\mu = 1.518 \times 10^{-3}\ N\text{-}s/m^2$ at 5°C. The power imparted to the water is calculated from Eq. (8.20).

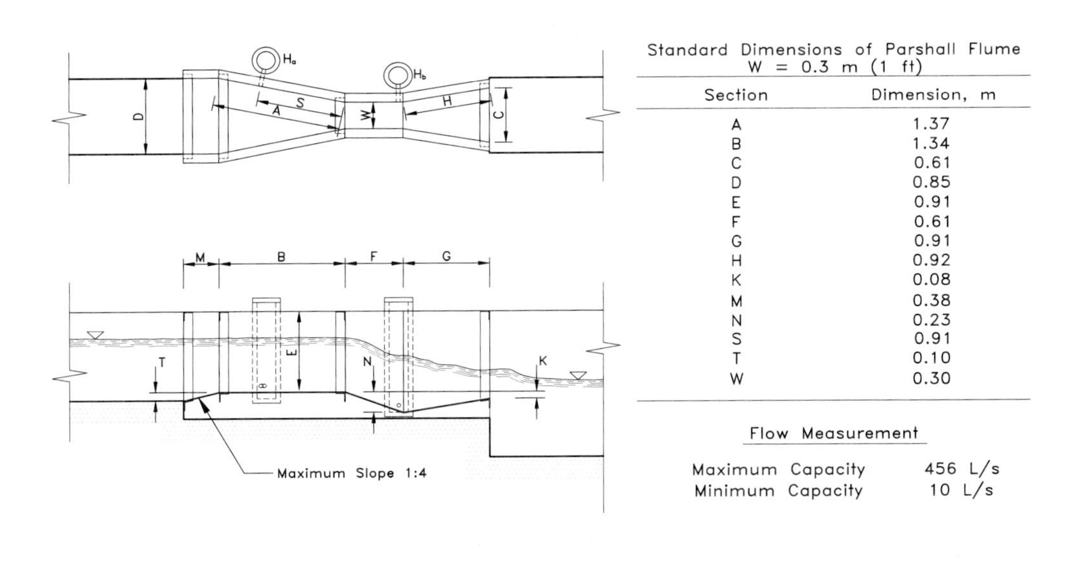

Standard Dimensions of Parshall Flume
W = 0.3 m (1 ft)

Section	Dimension, m
A	1.37
B	1.34
C	0.61
D	0.85
E	0.91
F	0.61
G	0.91
H	0.92
K	0.08
M	0.38
N	0.23
S	0.91
T	0.10
W	0.30

Flow Measurement

Maximum Capacity	456 L/s
Minimum Capacity	10 L/s

(a)

Top of Parshall Flume Wall Elev. 107.50 m

Elev. 106.85 m

Downstream Bottom of Rapid-Mix Basin Effluent Channel Elev. 106.14 m

0.71 m

0.10 m

H_a = 0.61 m

H_b = 0.43 m (Maximum Allowable Water Depth)

0.08 m

0.46 m

Elev. 106.60 m

Top of Flocculation Basin Wall Elev. 107.00 m

0.25 m (Head Loss)

0.90 m

Bottom of Flocculator Influent Box Elev. 105.70 m

(b)

Figure 8-17 Design details of Parshall flume. (a) Dimension requirements for a standard W=0.3-m (1 ft) Parshall flume. (b) Hydraulic profile through the Parshall flume.

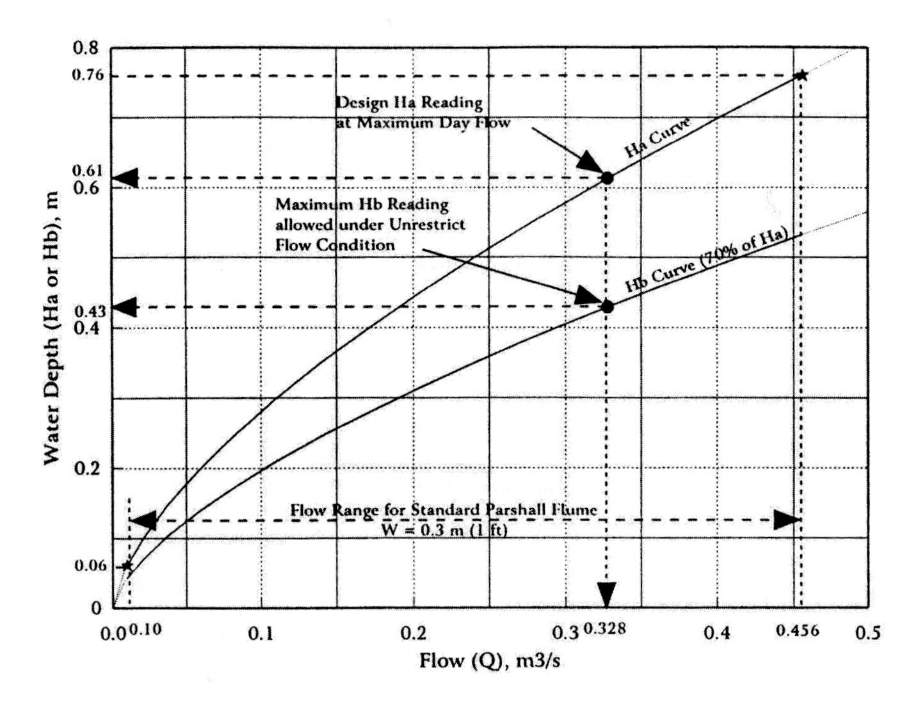

Figure 8-18 Calibration curve of the 0.3-m throat Parshall flume.

$$P = (60/\text{s})^2 \times 199 \text{ m}^3 \times 1.518 \times 10^{-3} \text{ N-s/m}^2 = 1087 \text{ N-m/s} = 1.10 \text{ kW}$$

The motor power requirement is calculated from the efficiencies of the gearbox and the flocculator bearings.

$$P' = \frac{P}{E_{gears} \times E_{bearings}}$$

where

E_{gears} = efficiency of gears, assume 90 percent
$E_{bearings}$ = efficiency of bearings, assume 70 percent

Assume $E_{gears} = 0.90$ and $E_{bearings} = 0.7$; then the motor power requirement is as follows:

$$P' = \frac{1.10 \text{ kW}}{0.9 \times 0.7} = 1.75 \text{ kW (2.3 hp)}$$

Similarly, the power imparted for the second and third flocculator stages at a G value of 30/s and 15/s have been determined. These values are 0.27 kW and

0.07 kW, respectively. The motor power requirements are 0.43 kW (0.58 hp) and 0.11 kW (0.15 hp), respectively. Each flocculator motor will have motor contacts to monitor motor stats and malfunctions. The rotational speed of the paddles will also be monitored.

ii. Determine the paddle size, number of paddles, and paddle-wheel layout.

The paddle wheels shall be constructed in segments to facilitate construction, installation, and shipping. (See Figure 8-16). Each segment will be 3.2 m long. The paddle-wheel diameter will be approximately 85 percent of the water depth in the flocculation basin. This will ensure complete mixing within the flocculation chamber.

Each stage of the flocculation basin is provided with 5 segments, and each segment has a total of twelve paddle blades. (See Figure 8-16(a).) Each paddle blade is 20 cm wide and 3.2 m long. The space between two adjacent blades is 15 cm. The arrangement is illustrated in Figure 8-19.

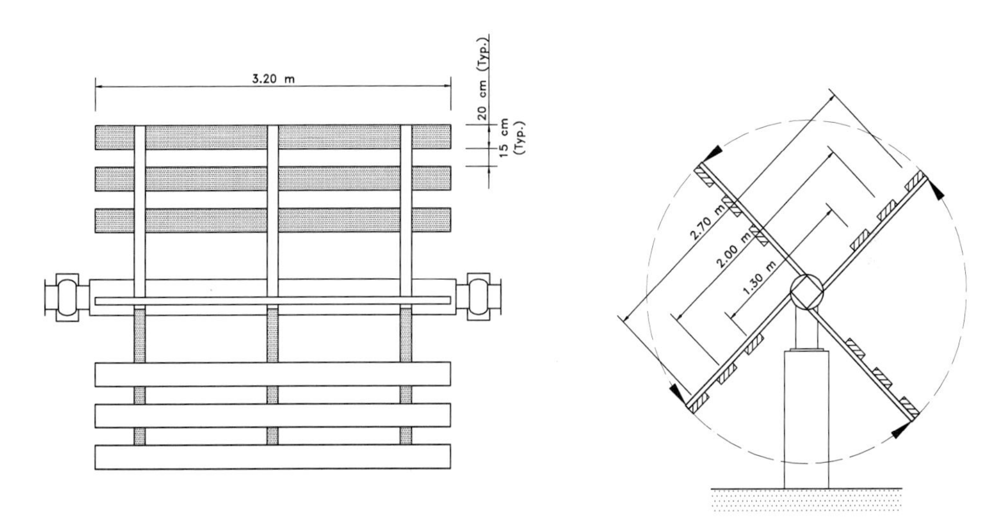

Figure 8-19 Flocculator paddle details.

iii. Calculate flocculator speed requirements.

In the first stage, a total of five segments or paddle wheels is provided. Each paddle wheel contains 12 paddles, 3.2 m long ∞ 0.20 m wide. As illustrated in Figure 8-19, four paddles are placed at each location, 2.7 m, 2.0 m, and 1.3 m center–center. The rotational speed is calculated from Eq. (8.28) after appropriate modifications.

$$P = \frac{C_D \rho}{2}(A_1 v_1^3 + A_2 v_2^3 + A_3 v_3^3)$$

where

P (power imparted to the water) = 1095 N-m/s for the first stage

C_D (coefficient of drag) = 1.35. (See Table 8-5 for blade length-to-width-ratio $L/W = 3.2/0.2 = 16$.)

A_1, A_2, and A_3 = total area of the paddles at each location per stage. There are five segments in the first stage and a total of four paddles at each radius per segment, so the total paddle area at each location is equal and is calculated as follows:

$$A = A_1 = A_2 = A_3 = 5 \text{ paddle-wheel segments per stage}$$
$$\times 4 \text{ paddle blades at each radius per segment}$$
$$\times 3.2 \text{ m (length of paddle)} \times 0.2 \text{ m (width of paddle)}$$
$$= 12.8 \text{ m}^2 \text{ total paddle blade area per location}$$

v_1, v_2, and v_3 = the average velocity of each paddle blade relative to the water. Assume that the paddle-blade velocity relative to water is 75 percent of the absolute peripheral velocity of the blade[g].

$$v_1 = \pi n_{rel} d_1 = 0.75 \pi n_{abs} d_1$$

$$v_2 = \pi n_{rel} d_2 = 0.75 \pi n_{abs} d_2$$

$$v_3 = \pi n_{rel} d_3 = 0.75 \pi n_{abs} d_3$$

where

d_1, d_2, and d_3 = diameter measured from center to center of each paddle blade, m

n_{abs}= absolute rotational speed of the blades, rev/s

Substitute these values in Eq. (8.28).

$$P = \frac{C_D \rho A}{2}(0.75 \pi n_{abs})^3 (d_1^3 + d_2^3 + d_3^3)$$

g. The paddles impart a velocity to the water, so the absolute velocity of the paddle (n_{abs}) must exceed the relative velocity (n_{rel}). Experience has shown that the relative velocity of the water is 75 percent of the rotational speed of the blades (i.e., $n_{rel} = 0.75\ n_{abs}$ or $n_{abs} = 1.33\ n_{rel}$).

$$1087\text{N-m/s} = \frac{1.35}{2} \times 1000\ \text{N/m}^4 \times 12.8\ \text{m}^2 \times (0.75 \times \pi \times n_{abs})$$
$$\times [(2.7\ \text{m})^3 + (2.0\ \text{m})^3 + (1.3\ \text{m})^3]$$

Solve for $n_{abs} = 0.069$ rev/s or 4.1 rpm.

The rotational speeds of the second- and third-stage flocculators are calculated in a similar manner. Table 8-7 lists the design parameters of the flocculators. The peripheral velocity ranges of paddle blades in the first, second, and third stages are 0.28–0.58 m/s, 0.18–0.37 m/s, and 0.11–0.23 m/s, respectively. Lower peripheral velocity in the third stage will not only meet the requirement for lower power input but also effectively prevent break-up of the floc structure due to high local peripheral velocities.

5. Head losses and hydraulic profile.

In this section, the hydraulic calculations are conducted for the head losses from the Parshall flume at the entrance of the flocculation basin to the end of the diffusion wall between the flocculation and sedimentation basins.

a. Head loss at the Parshall flume.

i. Calculate the flow depth H_a. Flow depth H_a at maximum day flow of 0.328 m^3/s (11.5 cfs) is calculated from Eq. (8.31).

$$11.6 = 4 \times 1 \times H_a^{1.522(1)^{0.026}}$$

Solving this equation, $H_a = 2.0$ ft (0.61 m). At submergence < 70 percent, H_b < 0.7 ∞ H_a or $H_b < 1.4$ ft (0.43 m) above the bottom of the throat section.

ii. Calculate head loss through the Parshall flume. The minimum head loss that must be provided to maintain unrestricted flow condition is obtained form Figure 8-20.[45] At maximum day flow of 0.328 m^3/s (11.6 cfs) and 70 percent submergence, the minimum head loss that must be provided is 0.24m (0.8 ft). In other words, the maximum allowed downstream water surface level shall be $(H_a - 0.24\ \text{m}) = (0.61 - 0.24\ \text{m}) = 0.37$ m above the bottom of the throat section. In this design, a free-fall head loss of 0.25 m is provided at the Parshall flume.

b. Head loss at the influent structure downstream of the Parshall flume.

The head loss at the influent structure below the Parshall flume is due to the losses in the influent channel, in the distribution channel, and at the influent weir. (See Figure 8-16.)

i. Calculate the head in the influent channel. Provide a water depth of approximately 0.9 m in the influent channel. The channel width in the first segment

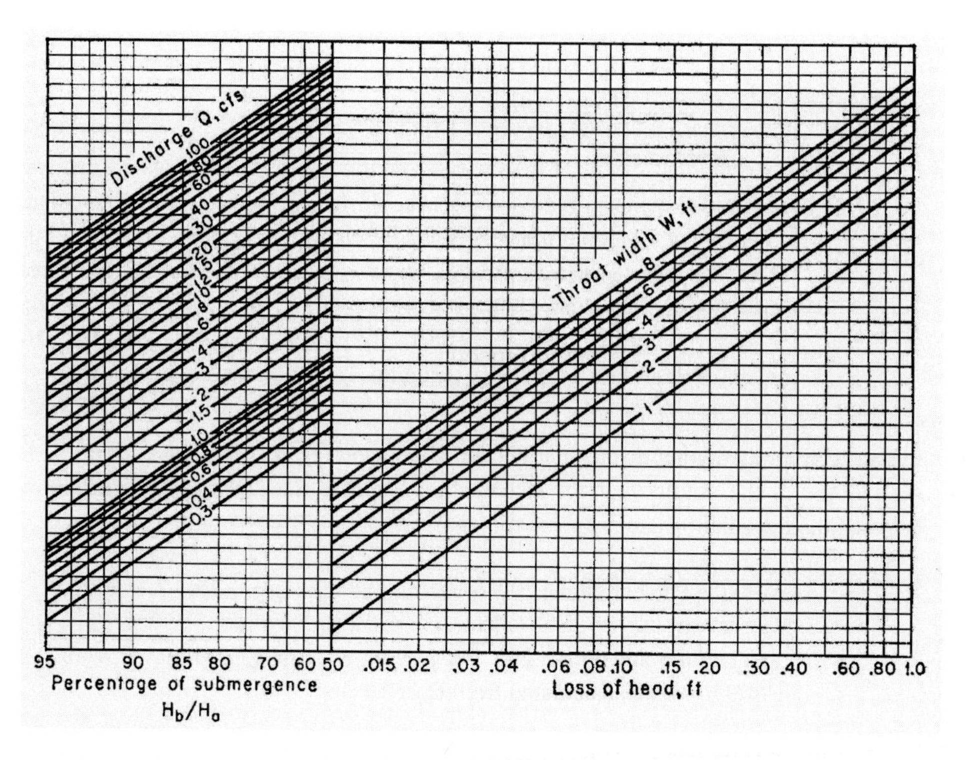

Figure 8-20 Minimum head loss requirement through a standard Parshall flume. *(Adapted from Reference 45.).*

before the first 90° turn is maintained at 0.85 m, like that before the Parshall flume. The width is then increased to 1 m in the segment between the first and second 90° turns. This width is also used as the initial width of the tapered influent-distribution channel. Using an average width of 0.9 m and a length of approximately 4 m, the head loss at the maximum day flow is calculated from Eq. (7.1): $h_L = 0.00002$ m, which is also assumed zero. The average velocity in the channel at the maximum day flow is as follows:

$$v_{avg} = \frac{0.328 \text{ m}^3/\text{s}}{0.9 \text{ m} \times 0.9 \text{ m}} = 0.40 \text{ m/s}$$

This velocity is slightly lower than the typical range from 0.45 to 0.9 m/s recommended for the water-conveyer channel between the rapid-mix and the flocculation channel. The minor head losses at the two 90° turns are also ignored because of the low-velocity head.

ii. Calculate the head loss through the influent-distribution channel. The width of the distribution channel is tapered to achieve better distribution of water along the length of the channel. The initial and final widths of the channel are 1 m and 0.2 m, respectively. The head loss can be calculated from Eq. (7.1), like that of the influent channel. This head loss h_L = 0.00004 m, which is also assumed zero.

iii. Calculate the head over the influent-distribution weirs. Five straight-side weirs are provided. The length of each weir is L = 3.5 m. A 0.225-m-wide support column is provided between two weirs. The head over a weir is calculated from the weir equation (Eq. (8.29)) with C_d = 0.6.

Table 8-7 Design Parameters of the Flocculator

	Power Requirement			Speed. rpm		
Stage	Velocity Gradient, *G*, 1/s	Motor[a]	Water, kW	n_{rel}	n_{abs}	Actual Speed *n*
1	60	1.75	1.10	3.1	4.1	2–10
2	30	0.43	0.27	1.9	2.6	1–6
3	15	0.11	0.07	1.2	1.6	0.5–4

a Although the motor-power requirement for each flocculator is different, provide identical motors for all three units having interchangeable parts. The motor power specified is 0.86 kW (2.5 hp).

It is assumed that each weir receives one fifth of the maximum design flow.

$$Q = \frac{0.328 \text{ m}^3\text{/s}}{5} = 0.0656 \text{ m}^3\text{/s}$$

Assume n = 2, and calculate the head over the weir by trial and error. The final results are H = 0.05 m and L' = 3.40 m. Provide a free fall of 0.23 m at the weir.

iv. Calculate the total head losses through the influent structure after the Parshall flume.

• Head loss through the influent channel	0.00 m
• Head loss through the distribution channel	0.00 m
•Free fall provided at the distribution weirs	0.23 m
Total =	0.23 m

c. Head losses in the flocculation basin.

The head losses through the flocculation basin and across the baffle walls between the stages are small and are ignored.

d. Head loss at the effluent structure.

The head loss at the effluent is encountered at the ports on the diffusion wall between the flocculation and sedimentation basins. (See Figure 8-16.)

i. Diffusion-wall arrangement. The diffusion wall separating the flocculation and sedimentation basins is of concrete, with circular ports. The flocculation tank is designed as part of the sedimentation basin, so this baffle wall located between the tanks is called the *diffusion wall.* The important design consideration for the diffusion wall is to distribute the flow evenly into the sedimentation basin. To distribute the flow evenly into the sedimentation basin, the kinetic energy of the water flowing through the flocculation basin must be dissipated. To achieve this, the diffusion wall will be provided with the following design features. These design features have been extensively discussed in References 31 and 45–47. The basic design requirements are given below:

- The ports should be uniformly distributed throughout the baffle wall.
- In order to minimize the length of the jets and the dead zones between ports, a maximum number of ports should be provided.
- The optimum head loss through the ports should be 2–3 mm.
- The maximum flow velocity through the ports should be approximately 150 mm/s, in order to prevent floc breakup.
- The port configuration should be such as to assure that the discharge jets will direct the flow toward the basin outlet. Timber walls tend to distort the direction of flow vertically.
- The most effective type of diffusion wall should have uniformly distributed 125-mm ports with an opening ratio of 6 to 8 percent. Some designers prefer keeping the diameter of the port less than the thickness of the baffle wall.[31]

ii. Calculate the total port area. Assume the velocity through the port is 0.15 m/s.

$$A = \frac{0.328\ \text{m}^3/\text{s}}{0.15\ \text{m/s}} = 2.187\ \text{m}^2$$

Provide 12.5 cm (5 in.) diameter ports.

$$\text{Area of each port} = \frac{\pi}{4} \times (0.125\ \text{m})^2 = 0.0123\ \text{m}^2$$

$$\text{Number of ports} = \frac{2.187\ \text{m}^2}{0.0123\ \text{m}^2/\text{opening}} = 178$$

Provide a total of 178 ports in four rows. The design details of the diffusion wall are provided in Figure 8-21(a).

iii. Calculate the head loss through the ports on the diffusion wall.The head loss across the port is calculated from orifice equation (Eq. (6.4)) with $C_d = 0.6$: $h_L = 0.0032$ m. This head loss is small and is also assumed zero.

e. Prepare the hydraulic profile through the flocculation basin.

A procedure similar to that for the rapid-mix basin is used. The hydraulic profile is developed with respect to the water surface elevation in the sedimentation basin. The WSEL of 106.37 m is calculated in the sedimentation basin design. (See Section 9.6.2, Step A.5.e and Figure 9-21.) The head loss calculations through the influent and effluent structures have been presented earlier. The head losses through the flocculation basin are ignored.

i. Calculate the WSEL and other major design elevations in the flocculation basin and influent and effluent structures.

The WSEL in the flocculation basin = ((WSEL in the sedimentation basin) + (head loss through the effluent structure)) = (106.37 m + 0.00 m) = 106.37 m.

The bottom elevation of the flocculation basin = ((WSEL in the flocculation basin) – (design water depth)) = (106.37 m – 3.27 m) = 103.10 m.

The WSEL in the influent distribution channel = ((WSEL in the flocculation basin) + (free fall provided at the distribution weirs)) = (106.37 m + 0.23 m) = 106.60 m. The top elevation of the distribution weir = ((WSEL in the influent distribution channel) – (head over the weir)) = (106.60 m – 0.05 m) = 106.55 m.

The bottom elevation of distribution channel = ((WSEL in the distribution channel) – (design water depth)) = (106.60 m – 0.90 m) = 105.70 m.

The head losses from the Parshall flume to the distribution channel are ignorable, so the WSEL and bottom elevation in the channel after the Parshall flume are 106.60 m and 105.70 m, respectively.

The hydraulic profile through the flocculation basin is shown in Figure 8-21(b).

ii. Calculate WSEL and other major design elevations in the Parshall flume.

The WSEL at the throat section = ((WSEL after the Parshall flume) + (head loss through the Parshall flume)) = (106.60 m + 0.25 m) = 106.85 m.

The bottom elevation of the throat section = ((WSEL at the throat section) – H_a) = (106.85 m – 0.61 m) = 106.24 m.

The crest elevation at the end of the flume = ((bottom elevation of the throat section) – K (See Figure 8-17)) = (106.24 m – 0.08 m) = 106.16 m. The height from the bottom of the downstream channel to the crest = ((crest elevation) –

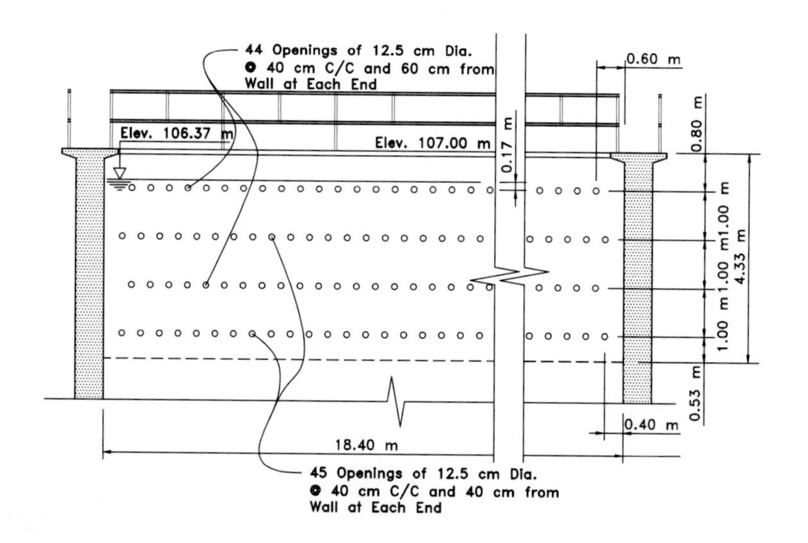

(a)

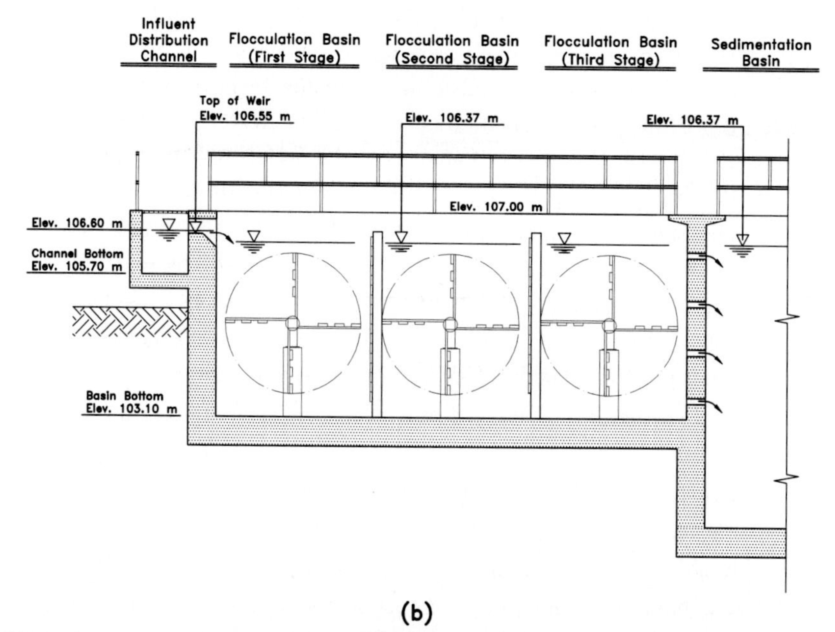

(b)

Figure 8-21 Flocculation-basin effluent structure and hydraulic profile. (a) Details of diffusion wall. (b) Hydraulic profile through the flocculation basin.

(bottom elevation in the downstream channel)) = (106.16 m – 105.70 m) = 0.46 m.

The bottom elevation of upstream channel = ((bottom elevation of the throat section) – (T (See Figure 8-17)) = (106.24 m – 0.10 m) = 106.14 m. The water depth in the upstream channel = ((WSEL at the throat section) – (bottom elevation of upstream channel)) = (106.85 m – 106.14 m) = 0.71 m.

The hydraulic profile through the Parshall flume is provided in Figure 8-17(b).

Step C: Chemical Storage and Feed Systems

1. Chemical system

   a. Chemical storage and feed equipment is housed in a chemical building. Multiple chemical-storage tanks and feed pumps with necessary backup capability are provided for each chEmical. A schematic of a proposed storage and feed system is shown for each chemical in Figure 8-22.

   b. Liquid ferric sulfate. Liquid ferric sulfate storage is provided for a 30-day supply. Three storage tanks and five feeders are included. Space for three additional tanks and five additional feeders for future expansion are also considered. Each feeder is designed to feed 25 percent of the maximum feed rate. The piping for the feeders is arranged in such a pattern that one feeder feed ferric sulfate into a separate rapid-mix basin while the remaining feeder is capable of feeding into any of the four rapid-mix basins. (See Figure 8-22(a).) Diaphragm-type metering pumps are used as chemical feeders. Ultrasonic level transducers will monitor the storage level. The meter contacts will monitor the pump feed status. The diaphragm positioner will be used to adjust and control the feed rate.

   c. Polymer. Two separate polymer-feed systems have been provided. These are coagulant aid and filter aid. These polymers are stored and delivered from 200-L (55-gallon) drums. Storage for a 30-day supply is provided. The concentrated solution of polymer is diluted in solution tanks, then pumped to the application points. Four separate pumps deliver coagulant aid (a cationic organic polymer) into the Parshall flumes, and four pumps deliver filter aid (a nonionic organic polymer) into the effluent channel of each sedimentation basin. The details of the polymer-feed system are shown in Figure 8-22(b).

   d. Lime. Provide 30-day storage in three storage silos for hydrated lime ($Ca(OH)_2$).[h] The hydrated lime is fed into three slurry tanks by dry-chemical feeders. Two slurry

h. Hydrated lime ($Ca(OH)_2$) is a common form of commercial lime product. The operational cost is normally high when hydrated lime is used. When large quantities are required, pebble lime (CaO) is often less expensive than the hydrated lime; however, the pebble lime needs slaking. Care must be taken when storing either hydrated or pebble lime. It is necessary to maintain a dry atmosphere. If water comes in contact with pebble lime, it will begin to hydrate in the storage bin and a great amount of heat is released. If hydrated lime becomes moist, it will cake up or form large chunks in the bin and clog the feeding mechanism.

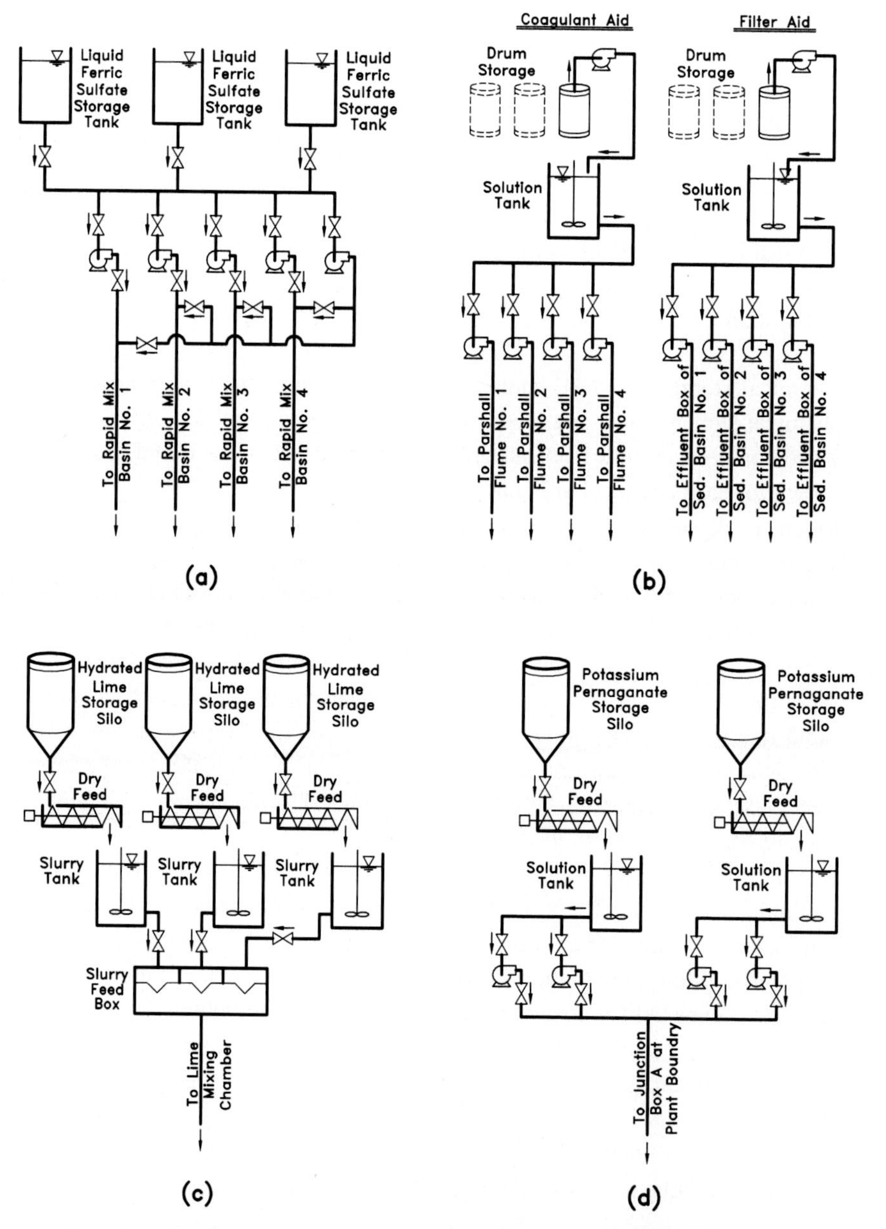

Figure 8-22 Chemical feed system schematic. (a) Ferric sulfate feed system. (b) Polymer feed system. (c) Lime feed system. (d) Potassium permanganate feed system.

tanks are kept in service, and each slurry tank is capable of feeding 50 percent of the maximum feed rate. A mixer is provided in each tank to keep the slurry in suspension. The lime slurry from each tank flows into a distribution box by gravity and is further diluted to the desired concentration. The distribution box then distributes the combined flow into the lime-mixing chamber by gravity. A schematic of the lime-feed system is shown in Figure 8-22(c). Ultrasonic level transducers will monitor storage levels, and galvanometric feeders will monitor and adjust chemical feed.

e. Potassium permanganate. Provide two potassium permanganate storage and feed systems in the chemical building. Each system shall consist of a dry-chemical storage tank, a dry-chemical volumetric feeder, a solution tank, and two chemical metering pumps. The chemical shall be delivered by hopper truck and blown into the storage silos. A dust filter shall be provided to prevent fine particles from escaping during delivery. The solution shall be pumped to the raw water influent at Junction Box A near the plant property line. (See site plan Chapter 15.) Free fall of water in this box will provided enough energy for the mixing of potassium permanganate solution and the raw water. The design details of the potassium permanganate ($KMnO_4$) storage feeder, solution, and feed pumps are provided in Section 11.6.2. The schematic flow diagram is shown in Figure 8-22(d). The motor controls will monitor the pumps. Weight cells will monitor storage drum weight, and a diaphragm positioner will be used to adjust and control the feed rate.

2. Chemical quantities.

a. Liquid ferric sulfate storage.

A volume for 30 days of storage is required in this design. Assume that bulk liquid ferric sulfate contains 50 percent by weight $Fe_2(SO_4)_3$. Use optimum feed rate and maximum flow rate to calculate the storage requirements. This will provide a minimum of 30 days storage during peak-summer water-demand conditions.

$$\text{Optimum feed rate} = 25 \text{ mg/L} = 0.025 \text{ kg/m}^3 \text{ as } Fe_2(SO_4)_3$$

$$\text{Maximum flow rate} = 113,500 \text{ m}^3\text{/d}$$

$$\begin{aligned}\text{Maximum daily feed rate}\\ &= 0.025 \text{ kg/m}^3 \times 113,500 \text{ m}^3\text{/d}\\ &= 2840 \text{ kg/d of } Fe_2(SO_4)_3\end{aligned}$$

For 30 days' storage, 2840 kg/day ∞ 30 day = 85,200 kg of $Fe_2(SO_4)_3$.

49 percent by weight of the liquid ferric sulfate is $Fe_2(SO_4)_3$, so the mass of liquid ferric sulfate for 30 days' storage required is 85,200 kg ÷ (0.49) = 173,900 kg or 174,000 kg of ferric sulfate liquid.

Assume that bulk ferric sulfate liquid has a mass density of 1122 kg/m^3; then the total volume of the storage facility = (174,000 kg/1122 kg/m) = 155 m^3 (say, 156 m^3).

There are three separate systems in the chemical building, so provide three 52-m^3 storage tanks for liquid ferric sulfate. The storage tanks shall be made of fiberglass to resist chemical corrosion by the ferric sulfate. Each tank is 4 m in diameter and has a maximum liquid depth of 4.15 m.

The tanks and metering pumps are located on the ground floor of the chemical building. The final discharge point of the chemical-feed piping shall be above the maximum level of liquid chemical in the storage tank and will provide a positive head on the chemical-feed pumps. This will prevent inaccurate metering of the chemical and the siphoning of chemical solution through the pump.

b. Other Chemical Storage.

Other chemical-storage facilities include polymers, lime, and potassium permanganate. The quantities of these chemicals and storage needs are computed similarly. The results of these calculations are summarized in Table 8-8.

3. Layout of chemical building

The layout of the chemical storage is shown in Figure 8-23. The ferric sulfate storage and feed system and the two polymer-feed systems are located on the ground floor of the building. The polymer drums are stored in the chemical-storage room. The potassium permanganate solution tanks and feed pumps are also located on the ground floor, while the storage silos are located on the upper balcony. The lime-slurry tanks are located on the balcony, so that lime slurry can be fed into the lime-mixing chamber by gravity. The lime-storage silos are located on the roof of the building, directly above the slurry tanks.

8.9 Operation, Maintenance, and Troubleshooting for Coagulation and Flocculation Facilities

Coagulation and flocculation facilities are the first in a series of treatment processes used to remove turbidity, hardness, iron, and manganese from water in a conventional water treatment plant. The results obtained in the coagulation and flocculation processes are critical to the success of subsequent treatment processes.

8.9.1 Common Operational Problems and Troubleshooting

Operational problems associated with coagulation and flocculation processes typically relate to either equipment failure or process inefficiencies. Problems associated with equipment operations are specific to the installed equipment and are not discussed here. Problems associated with the coagulation process are typically indicated by high-turbidity water in the sedimentation-basin effluent and/or the filtered water. Following are some of the common results of poor performance of coagulation and flocculation facilities.

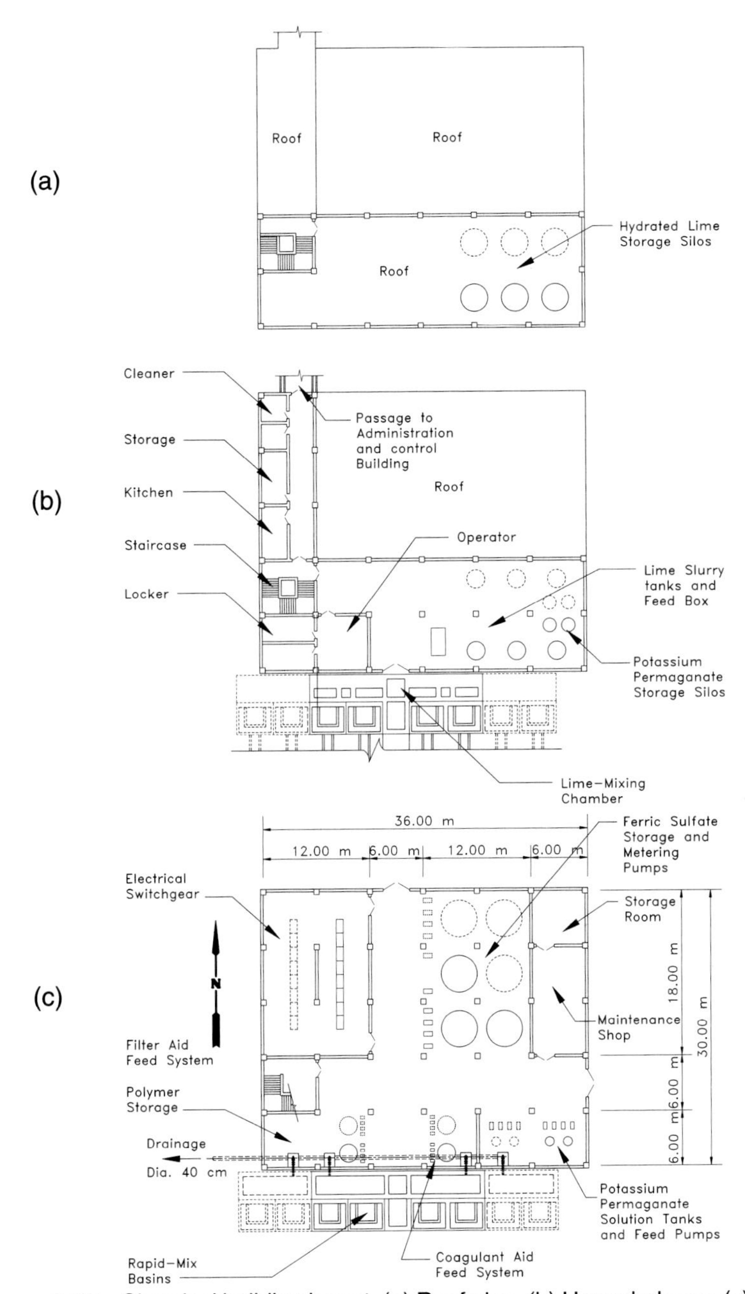

Figure 8-23 Chemical building layout. (a) Roof plan. (b) Upper balcony. (c) Ground floor with underground pit.

1. High effluent turbidity, with no floc carryover, can be the result of too little coagulant or of incomplete dispersion of the coagulant. Perform jar tests with varying coagulant dosages, coagulant dilutions, and rapid-mix intensities. Make adjustments as indicated by the tests.
2. Unsatisfactory effluent turbidity can also result from raw water that has low initial turbidity. An insufficient number of particle collisions during flocculation will inhibit floc growth. Increase flocculation intensity, recycle sludge, or add bentonite to provide a nucleus for floc formation.
3. High effluent turbidity with floc carryover is an indication of a poor-settling floc. High flocculation intensity will often shear floc and result in poor settling. Lower the flocculation intensity, or add a coagulant aid that will toughen the floc and make it more readily settleable.
4. Too much coagulant will often result in restabilization of the colloids. If unsatisfactory performance is obtained, run a series of jar tests with various coagulant dosages and adjust feed rates accordingly.
5. Calcium carbonate precipitate will often accumulate on lime-feed pipes. Flush lime pipes with an acid solution periodically, to dissolve the scale.
6. Improper feed rate of coagulant through positive-displacement metering pumps can be the result of siphoning through the pump. Locate pumps in such a way that a positive head is present at all times on the pump discharge. An alternative correction method is to install a back-pressure valve on the pump discharge.

8.9.2 Operation and Maintenance

The following preventive maintenance procedures are necessary for the satisfactory operation of rapid-mix and flocculation facilities.

1. Perform jar tests on raw water samples daily when significant raw water quality changes are experienced. Adjust coagulant dosages and mixer speeds accordingly.
2. Clean accumulated precipitate and sludge from rapid-mix and flocculation basins annually.
3. Calibrate chemical feeders every month.
4. Check the chemical analysis of each delivery of coagulant. Adjust feed rates as indicated by the analysis and jar tests.

Table 8-8 Chemical-Storage Tank-Size Calculation

Chemical	Dosage (mg/L)	Average Strength[a] (percent)	Chemical Density[a]	Pure Chemical daily Usage (kg/d)	Bulk Chemical Monthly Usage (kg/month)	Storage Required (m^3)	No. of Tanks	Required Size of Each Tank (m^3)
Ferric sulfate, $Fe_2(SO_4)_3$	25.0	49 (Liquid)	1125	2840	174,000	155	3	52
Cationic polymer (Coagulant Aid)	0.05	90 (liquid)	1050	5.7	190	0.18	4[b]	N/A
Nonionic polymer (filter aid)	0.02	90 (liquid)	1050	2.3	77	0.07	4[b]	N/A
Hydrated lime, $Ca(OH)_2$	15 as Ca(OH)[c]	71 (Dry)	800	1700	72,000	90	3	30
Potassium permanganate, $KMnO_4$	4.0	90 (Granular)	1500	454	15,100	10	2	5.0

a The purity and density of chemical products varies within manufacturers.

b Store cationic polymer in the drums in which the chemical is delivered.

c If pebble lime is used, the feed rate as CaO = 15 mg/L $(Ca(OH_2)$/ 37 eq wt of Ca $(OH)_2 \infty$ 28 eq wt of CaO = 11.4 mg/L as CaO.

5. Lubricate flocculator and mixer gear boxes and bearings as specified by the manufacturer.
6. Inspect rapid-mix impellers and flocculator paddles annually. Remove any accumulations of floc or calcium carbonate scale. Inspect more frequently if build-up is severe.

8.10 Specifications

Brief specifications for the rapid-mix and flocculation units designed above are presented in this section. These specifications are provided to describe the design better. Many details have intentionally been omitted from these specifications, to keep the contents brief and simple. Therefore, these specifications should be used only as a guide to prepare more detailed specifications. Manufacturers' representatives should be consulted during the preparation of specifications for all equipment.

8.10.1 Rapid Mixers

The contractor shall construct four rapid-mix basins, as indicated on the drawings. Each unit shall be 1.9 m square in plan and 2.9 m deep. The basins shall be constructed of reinforced concrete. Each basin shall be equipped with a four-blade turbine-type rapid mixer with a power number of 2.7 to 2.8. The mixer shall be equipped with a drive unit consisting of a 16 kW, 480-volt, 3-phase electric motor. The maximum speed of the motor shall be 1170 revolutions per minute (rpm). Provide a gear reducer to reduce the mixer shaft speed to the required 113 rpm.

The mixer units shall be constructed of the following materials:

Table 8-9 Rapid-Mixer Materials

Steel Plate and Shapes	ASTM[a] A36
Shafts	
Solid	Carbon steel, machined and polished
Torque Tube	Schedule 80 seamless carbon steel pipe with closed ends
Impeller Assemblies and Couplings	Carbon Steel

a ASTM—American Society of Testing Materials.

Each drive unit shall consist of a combination of speed and gear reducer with an attached alternating-current, squirrel-cage motor. The drive units shall be designed for mixer service and shall be suitable for 24-hours-per-day operation under moderate shock-loading conditions. Lifting lugs shall be provided on all drive units to permit lifting of the mixer and drive unit.

All gear-reducer and motor bearings shall be oil- or grease-lubricated, antifriction type. Thrust bearings shall be designed to carry all shafting and impeller loads plus an additional 25 percent. No bearings shall be located below the bottom of the supporting platform. Gear reducers shall be selected, designed, and rated in accordance with the appropriate AGMA standards.[i] The gearing shall be designed with a life factor equivalent to at least 100 million cycles, with service factor of 1.5 (based on the sum of the continuous loads plus any transient loads other than starting loads). Bearings shall have an AGMA L10 rating of at least 200,000 hours.[j]

Each mixer shall be designed and mounted so that it can be removed from its basin as a complete unit after the impeller is removed from the shaft. Each mixer shall be assembled on a rigid base. Vertical shafting for the impeller assembly shall be of ample size and design for the service intended. The impeller assembly shall be braced so that the unit will operate without excessive vibration. The impeller-shaft diameter shall be no less than that of the gear reduction unit of the output shaft. Each motor shall have contacts provided in the motor control center to monitor the operating status and malfunctions.

8.10.2 Flocculation Units

The contractor shall construct four flocculation basins, as indicated on the drawings. Each basin shall be 10 m long (overall) by 18.4 m wide, with a water depth of 3.27 m. The basins shall be an integral part of the sedimentation basins. A concrete diffusion wall shall be installed to separate the flocculation basin from the sedimentation basin. Each diffusion baffle wall shall have 180 circular ports, each with a 12.5 cm diameter.

Each basin shall be equipped with three paddle-wheel flocculators meeting the following requirements:

Motor power for all three units shall be identical, with interchangeable parts. The specified motor power is 0.86 kw. Each motor shall have a motor contact in the motor control center to monitor flocculator running status and malfunctions.

Paddle blades shall be made of clear hard-grade redwood lumber. Paddle blades shall be bolted to and supported by steel angle arms. Paddle arms shall be steel shapes welded or bolted to the paddle shaft. Each half of each complete paddle arm shall be stiffened with suitable diagonal cross-bracing members. Paddle shafts may be either solid cold-rolled steel or tubular steel. If tubular paddle shafting is furnished, solid shafting shall be installed in the dry well, through each stuffing box and bearing. All tubular shafting shall be fabricated from Schedule 80 steel pipe. Keyed cast-iron compression couplings shall be provided for connecting all sections of continuous solid shafting.

Shaft connections shall be made with forged or cast-steel flange couplings. Flanges for couplings shall be ANSI B16.5, 150 welding neck-type flanges.[k] Flange couplings shall be bored to slip over the shaft and babbitted for accurate alignment. Flanges shall be turned and faced square with the shaft after being keyed or welded in place. Each flanged coupling shall be bolted together with stainless steel machine bolts, washers, and locknuts.

i. AGMA—American Gear Manufacturers Association

j. AGMA L_{10} life—Number of hours of operation by which 10 percent of bearings will fail.

k. ANSI—American National Standards Institute

Each shafting assembly shall be machined, faced, and matched so that it will be straight and in rotational balance throughout its entire length when assembled on properly aligned bearings and bearing supports. Shafting shall be furnished in sections that can be conveniently handled and installed. Shafting in power-transmission equipment and through stuffing boxes and paddle-shaft bearings shall be turned and polished.

Submerged bearings shall be split cast-iron housing, grease-lubricated type and shall be lined and bushed with underwater bearing material in both the upper and lower halves. Drive shaft bearings installed in the dry well shall be split cast-iron housing type, fully bronze bushed. Shaft bearings shall have a length-to-diameter ratio of at least 1.0. Bearing pressures shall not exceed 10 kg/cm^2 of the projected bearing area. All bearings shall be equipped with pressure fittings for grease lubrication.

All bearings for horizontal shafts shall be provided with adequate means for anchoring to existing concrete piers and for shaft alignment and alignment maintenance. Bearing base castings shall be provided with slotted holes for lateral adjustment of the shaft alignment. Bottoms of base castings shall be carefully machined to provide an accurate and uniform bearing area. The bearing bases for horizontal shafts shall be supported on fabricated steel sub-bases not less than 16 mm in thickness, arranged so that bearing bases may be locked in position by adjusting screws and locknuts. Each sub-base shall be anchored to its supporting pier by four anchor bolts set in galvanized pipe sleeves.

The drive end of each paddle shaft shall extend through the wall of the dry well and shall be made watertight by means of a cast-iron stuffing box with a bronze split gland. Each stuffing box shall be furnished and installed complete with cast-iron wall pipe (with waterstop collar) through the wall and with all necessary packing material.

Each shafting assembly shall be provided with a slinger ring and rubber boot between the stuffing-box gland in the basin wall and the roller-chain enclosure. The slinger ring shall be installed as close to the stuffing box as is permitted by the equipment. Leakage from the stuffing boxes shall be discharged through a gutter and floor-drain system.

Each drive unit shall be designed and arranged to give infinitely variable rotational speeds from 10 rpm to 2 rpm for the first-stage, from 6 rpm to 1 rpm for the second-stage, and from 4 rpm to 0.5 rpm for the third-stage flocculators, respectively. Controls for adjusting the output speed of the drive units shall be readily accessible.

Each drive unit shall be installed on a concrete foundation pad on top of the dry well cover slab and shall have a sprocket on its output shaft. A roller-chain drive shall be provided in the dry well between the drive sprocket and a sprocket on the paddle shaft. The drive chain shall be nonmetallic, pintle type, with nonmetallic pins, 15 cm pitch. All pins shall be snap-lock type, with T-head feature to prevent pin rotation, and shall be held in place without the need for spring clips, cotter pins, or similar devices.

A chain tightener assembly of idler-sprocket type, designed to control tightening of the drive chain, shall be provided for each drive chain. Each chain tightener shall be furnished and installed complete with necessary accessories and shall have pressure fittings for lubrication of the idler sprocket.

All sprockets shall be semi-steel with case-hardened teeth and rims. Sprocket rims and teeth shall have an average Brinell hardness of 450 and a chill depth of not less than 5 mm. All sprockets shall be split for machining. All driven sprockets on paddle shafts shall be split for convenience in removal, and sprocket teeth shall be ground to fit the chain used.

8.10.3 Parshall Flumes

Four Parshall flumes are to be provided, each located upstream of the flocculation basin. Each flume shall have a 30-cm throat section and conform to U.S. Bureau of Reclamation dimension standards. The flume shall consist of a fiberglass liner embedded in a concrete basin. Each shall have a discharge of 0.328 m^3/s.

Problems and Discussion Topics

8.1 Visit a local water treatment plant. Determine the type and size of rapid-mix, chemical-feed, and flocculation facilities. Prepare layout sketches of the facilities. Determine the types and dosage rates of the coagulation and/or precipitation chemicals used. Discuss with the operator (and list) any operational problems that are experienced with these facilities.

8.2 Calculate the dimension of a rapid-mix single basin to treat a flow of 18,900 m^3/d (5 mgd) entering into four basins. Design for a detention time of 20 s. The basin shall be square with a depth-to-width ratio of 1.6 to 1 and a length-to-width ratio of 1 to 1.

8.3 Calculate the energy requirements and rotation speed of the rapid-mix basin designed in Problem 8.2. Provide a velocity gradient of 1000/s. Use a four-blade turbine mixer with an impeller diameter of 0.7 m. The gear efficiency is 0.9 and the ratio of blade width to blade diameter is 0.2.

8.4 Calculate the dimensions of a flocculation basin to treat a flow of 18,900 m^3/d (5 mgd). The flocculator has a three-stage basin, with a 10 minute detention time in each stage. The basin has a width of 12 m, and total length is three times the basin depth.

8.5 Calculate the requirement and rotation speed of paddle wheels in three stages of the flocculation basins designed in Problem 8.4. The efficiencies of gear box and bearings are 90% and 70% respectively. Basic requirements of paddle wheels are summarized in the following table.

Stage	*G*	Paddle Width	Paddle Length	Number of Paddles on Each Segment	No. of Segments	Paddle-Blade Diameter
1	45/s	15.2 cm	1 m	8	10	4 at 2.5 m, 4 at 2.0 m
2	30/s	15.2 cm	1 m	6	10	2 at 2.5 m, 2 at 2.0 m, 2 at 1.5 m
3	15/s	15.2 cm	1 m	4	10	2 at 2.5 m, 2 at 2.0 m

Assume that the absolute velocity exceeds the desired velocity relative to water by 33 percent.

8.6 In the Design Example, the rapid-mix basins were designed for a maximum flow rate of 113,500 m^3/d. Determine the impact of increasing the peak flow rate to 170,300 m^3/d. Discuss alternatives available to implement this increase.

8.7 Design a softening process to treat water having the following analysis.

CO^2	9 mg/L as CO_2	HCO_3^- = 120 mg/L as $CaCO_3$
Ca^{2+}	60 mg/L	SO_4^{2-} = 100 mg/L
Mg^{2+}	12 mg/L	Cl^- = 11 mg/L
Na^+	9 mg/L	

Calculate the amount of lime $Ca(OH)_2$ and soda ash required to soften the water to the solubility product of Ca^{2+} and Mg^{2+}. Express your answer as mg/L residual hardness.

8.8 Design a baffle wall for the outlet of the flocculation basin designed in Problem 8.4. Use a port diameter of 20 cm. Determine the head loss through the port and the number of ports required.

Determine the power requirement and the paddle area required to achieve G = 60/s in a tank that has volume = 4000 m^3, water temperature = 20°C, and C_D = 1.8. The paddle-tip velocity is 0.6 m/s.

8.9 Design a rapid mix, flocculation process to meet the following requirements.

Flow rate	20,000 m^3/d
Rapid-mix detention time	30 s
Rapid-mix velocity gradient	800/s
Flocculation detention time	3 stages at 12 min each
Flocculation velocity gradient	
1st stage	50/s
2nd stage	35/s
3rd stage	20/s

Select the number of basins, type of mixers, and basin geometry. Sketch the layout of your design. Assume D/W = 1.5 for the rapid-mix square basin. Efficiency of gear box is 90%. Assume four-blade turbine mixer with width/diameter = 0.18.

Width of flocculator basin = 9 m

Gearbox efficiency = 90%

Bearing efficiency = 70%

8.10 Determine the motor power required for a blending tank using redwood paddles. There are 12 paddles, six on each side of the central shaft. Each paddle is 0.10 m wide × 2.75 m high. The distances to the middle of each paddle from the center of the shaft are 0.5, 1.0, 1.5, 2.0, 2.5, and 0.3 m. Assume rotational speed of the shaft = 0.06 rps, C_D = 1.8, motor efficiency = 75 percent, and specific gravity of blended sludge = 1.01.

References

1. Tchobanoglous, D. and Schroeder, E. *Water Quality*, Addison Wesley Publishing Co., Reading, Mass, 1985.
2. Reynolds, T. and Richards, P. A. *Unit Operations and Processes in Environmental Engineering*, 2d ed., PWS Publishing Co., Boston, MA, 1996.
3. Viessman, W. and Hammer, M. *Water Supply and Pollution Control*, Harper & Row Publishers, New York, 1985.
4. Ravina, L. A. *Everything You Want to Know About Coagulation and Flocculation*, Zeta-Meter, Inc., Long Island

City, New York, 1988.

5. Montgomery, James M. Inc. Water *Treatment Principles and Design*, John Wiley & Sons, New York, 1985
6. Tekippe, R. and Ham, R. "Velocity-Gradient Paths in Coagulation," *Jour. AWWA*, vol. 63, no. 7, pp 439–448, July 1971.
7. Committee on Coagulation. "State of the Art of Coagulation: Mechanism and Stoichiometry," A Report Presented at the Annual Conference on June 23, 1970, *Jour. AWWA*, vol. 63, no. 2, pp 99–108,February 1971.
8. Amirtharajah, A. and O'Melia, C. R. *"Coagulation Process: Destabilization, Mixing, and Flocculation," Water Quality and Treatment*, 4th ed., (F.W. Pontius, editor), McGraw-Hill, Inc., New York, 1990.
9. Hiemenz, P. C. *Principles of Colloid and Surface Chemistry.* Marcel Dekker, Inc., New York, 1977.
10. Hunter, R. J. *Foundations of Colloid Science (Vol I)*, Clarendon Press, Oxford, 1987.
11. Packham, R. F. *Some Studies of the Coagulation of Dispersed Clays with Hydrolyzing Salts*, Journal Colloid Sci., 20 81–92, 1965.
12. Singley, J. *Theory and Mechanism of Polyelectrolytes as Coagulant Aids*, Proceedings of the American Water Works Association Seminar, *Polyelectrolytes—Aids to Better Water Quality*, Chicago, IL, June 4, 1972.
13. Sinsabaugh, R. H. III, Hoehn, R. C., Knocke, W. R., and Linkins, A. E. "Removal of Dissolved Organic Carbon by Coagulation with Iron Sulfate," *Jour. AWWA*, vol. 78, no. 5, pp. 74–82, May 1986.
14. Hering, J.G. and Elimelech, M. *Arsenic Removal by Enhanced Coagulation and Membrane Processes*, Final Report. Dept. of Civ. and Environ. Engrg., UCLA, Los Angeles, CA, 1995.
15. Zhu, G. *Arsenic Removal from Drinking Water by Enhanced Coagulation* (Ph.D. Dissertation, The University of Texas at Arlington), DAO 72699, UMI Co., Ann Arbor, MI, June 1997.
16. Stumm, W. and Morgan, C. J. *Aquatic Chemistry* 3d ed., John Wiley & Sons, New York, 1996.
17. Owen, J. and Cooper, I. L. *Adsorption*, John Wiley & Sons, Inc., New York, 1982.
18. Zhu, G., et. al. "Coagulation Diagrams for Arsenic Removal by Enhanced Coagulation," *Proceedings of the 1997 Fall Conference of Texas Section of American Society of Civil Engineers*, Arlington, Texas, October 2–3, 1997.
19. Crozes, G., White, P., and Marshall, M. "Enhanced Coagulation: Its Effects on NOM Removal and Chemical Costs," *Jour. AWWA*, vol. 87, no. 1, pp. 78–89, January 1995.
20. Jacangelo, J. G., DeMarco, J., Owen, D. M., and Randtke, S. J. "Selected Process for Removing NOM: An Overview," *Jour. AWWA*, vol. 87, no. 1, pp. 64–77, January 1995.
21. Randtke, S. J. et al. "A Comparative Assessment of DBP Precursor Removal by Enhanced Coagulation and Softening," *Proceedings of 1994 AWWA Annual Conference*, New York, 1994.
22. Qasim, S. R., Hashsham, S., and Ansari, N. I. "TOC Removal by Coagulation and Softening," *Technical Note in Journal of Environmental Engineering Division, American Society of Civil Engineers*, vol. 118, no. 3, pp. 432–437, June 1992.
23. Cheng, R. L., Liang, S., Wang, H. C., and Buehler, M. D. "Enhanced Coagulation for Arsenic Removal," *Jour. AWWA*, vol. 86, no. 9, pp. 79–90, September 1994.
24. Masterton, W. and Slowinski, E. *Chemical Principals*, W.B. Saunders Co., Philadelphia, PA., 1969.
25. Sawyer, C. N., McCarty, P. L., and Parkin, G. F. *Chemistry for Environmental Engineers*, New York, McGraw-Hill, Inc., New York, 1994.
26. Qasim, S.R. *Wastewater Treatment Plants: Planning, Design and Operation*, 2d ed., Technomic Publishing Co., Lancaster PA, 1999.
27. AWWA. *Water Quality and Treatment*, 5th ed., McGraw-Hill Book Co., New York 1999.
28. McGhee, T. *Water Supply and Sewerage*, 6th ed., McGraw-Hill Book Co., New York, 1991.
29. Metcalf and Eddy, Inc. *Wastewater Engineering: Treatment, Disposal and Reuse*, 3d ed., McGraw-Hill Book Co., New York, 1991.
30. Malcolm Pirnie, Inc., and HDR Engineering, Inc. *Guidance Manual for Compliance with the Filtration and Disinfection Requirements for Public Water System Using Surface Water*, U.S. Environmental Protection Agency, Office of Drinking Water, Washington, D.C., October 1990.
31. Hudson, E. *Water Clarification Process: Practical Design and Evaluation*, Van Nostrand Reinhold Co., New

York, 1981.
32. Letterman, R. J., Quar, J. E., and Gemmell, R. S. "Influence of Rapid Mix Parameters on Flocculation," *Jour. AWWA*, vol. 65, no. 11, pp. 716–722, November 1973.
33. McCabe, W. and Smith, J. *Unit Operations in Chemical Engineering*, McGraw-Hill Book Co., New York, 1976.
34. Hammer, M. J. and Hammer, M. J., Jr. *Water and Wastewater Technology*, Prentice-Hall, Inc., Englewood Cliffs, NJ, 1996.
35. Garrison, C. "How to Design and Scale Mixing Pilot Plants," *Chemical Engineering*, vol. 93, no. 3, pp. 63–70, Feb. 7, 1983.
36. Vrale, L. and Jorden, R. "Rapid Mixing in Water Treatment," *Jour. AWWA*, vol. 63, no. 1, p. 52, January 1971.
37. Tolman, S. *The Mechanics of Mixing and Flocculation*, Public Works, Vol. 93, No. 12, 1962.
38. JDV Equipment Corp. JDV-Carter Walking Beam Flocculator, JDV1098, JDV Equipment Corp., Fairfield, NJ, 1998.
39. EIMCO. *Flocculator and Above Water Horizontal Oscillating Flocculator for Water and Wastewater Treatment*, EIMCO Process Equipment Company, Salt Lake City, Utah, 1989.
40. ASCE and AWWA. *Water Treatment Plant Design*, 2d ed., McGraw-Hill Book Co., New York, 1990.
41. Great Lakes Upper Mississippi River Board of State Public Health & Environmental Managers. *Recommended Standards for Water Works,* Health Education Services, Albany, NY 1992.
42. Texas Natural Resource Conservation Commission. *Subchapter D: Rules and Regulations for Public Water Systems*, Section 290.38–290.47, Texas Register, Austin, TX, 1999.
43. Davis, C.V. and Sorensen, K.E. (Editors). *Handbook of Applied Hydraulics*, 3d ed. McGraw-Hill Book Co., New York, 1969.
44. Chow, V. T. *Open Channel Hydraulics*, McGraw-Hill Book Co., New York, 1959.
45. U. S. Department of the Interior, Bureau of Reclamation. *Water Measurement Manual*, U. S. Government Printing Office, Washington, D.C., 1967.
46. Parshall, R.L. *Measuring Water in Irrigation Channels with Parshall Flumes and Small Weirs*, U. S. Soil Conservation Service, Circular 843, May 1950.
47. Benefield, L.D., Judkins, J.F., and Parr, A.D. *Treatment Plant Hydraulics for Environmental Engineers*, Prentice-Hall, Inc., Englewood Cliffs, N J, 1984.

CHAPTER 9

Sedimentation

9.1 Introduction

Sedimentation is a physical treatment process that utilizes gravity to separate suspended solids from water. This process is widely used as the first stage in surface water treatment to remove turbidity-causing particles after coagulation and flocculation. Sedimentation is also used (1) to recover water in filter backwash water recovery systems and (2) to increase sludge solids concentration in sludge thickening. Presedimentation is also used in some cases to remove settleable solids such as gravel, grit, and sand from river water before it is pumped to the treatment plant.

In this chapter, the theory and design procedures for sedimentation basins are discussed. Example design calculations, operation and maintenance practices, and equipment specifications for sedimentation facilities are provided in the Design Example.

9.2 Theory of Sedimentation

The design of a sedimentation basin is dependent upon the concentration, size, and behavior of the solid suspension. In general, there are four types or classes of sedimentation. *Type I sedimentation*, known as *discrete settling*, describes the sedimentation of low concentrations of particles that settle as individual entities. Examples of *Type I* settling in water treatment plants are the settling of silt from river water before coagulation, the settling of water softening precipitates, and the settling of sand in filters after backwash. *Type II sedimentation*, known as *flocculant settling*, describes sedimentation of larger concentrations of solids that agglomerate as they settle. Sedimentation of coagulated surface waters is an example of flocculant settling. *Type III sedimentation*, known as *hindered settling* or *zone settling*, describes sedimentation of a suspension with solids concentration sufficiently high to cause the particles to settle as a mass. An example of hindered settling is the upper portion of the sludge blanket in sludge thickeners. *Type IV sedimentation*, known as *compression settling*, describes sedimentation of suspensions with solids concentration so high that the particles are in contact with one another and further sedimentation can occur only by compression of the mass. The lower portion of a gravity sludge thickener is an example of compression settling.[1]

The theories used in the design of a sedimentation basin depend upon the type of settling encountered in the basin. The design equations and procedures used to develop design parameters in each type of settling are therefore also different from those in the others.

9.2.1 Discrete Settling, Type I

Settling Behavior The theory of *Type I* settling is based on the physics that applies to a particle settling unhindered at a constant velocity through water. The forces acting on such a particle are limited to the gravitational and drag forces.[2] In such suspensions, a particle will begin to settle and will accelerate until a constant velocity is reached. At this time the drag force and the gravitational force are equal and cancel each other. This terminal settling velocity is calculated from Eq. (9.1), which is known as *Newton's Law.*[3]

$$v_s = \sqrt{\frac{4}{3} \times \frac{dg}{C_D} \times \frac{(\rho_s - \rho)}{\rho}} \text{ or } v_s = \sqrt{\frac{4}{3} \times \frac{dg}{C_D} \times (S_g - 1)} \tag{9.1}$$

where

v_s = settling velocity, m/s (ft/s)
d = particle diameter, m (ft)
g = acceleration due to gravity, m/s^2 (ft/s^2)
C_D = drag coefficient, dependent on Reynolds number
S_g = specific gravity of the particle
ρ_s = density of solids, kg/m^3 (lb/ft^3)
ρ = density of water, kg/m^3 (lb/ft^3)

This equation was originally developed for a spherical particle falling through a fluid. If a particle has a non-spherical shape, C_D will increase. Consequently, the settling velocity will also decrease. A simple method of correcting Eq. (9.1) for a non-spherical particle is to utilize a shape factor as given by Eq. (9.2).[3]

$$C_D = \frac{24\phi}{N_R} \tag{9.2}$$

where

ϕ = shape factor
N_R = Reynolds number, which can be calculated from (Eq. (9.3))

$$N_R = \frac{v_s d}{\nu} = \frac{v_s \rho d}{\mu} \tag{9.3}$$

where

ν = kinematic viscosity, m^2/s (ft^2/s)
μ = dynamic viscosity of water, $N{\cdot}s/m^2$ ($lb{\cdot}s/ft^2$)

Typical values of ϕ are 2.0 for sand and 2.25 for coal. Other ϕ values are documented in various textbooks.[4]

The value of C_D is dependent on the *Reynolds number* (Eq. (9.2)). In the laminar-flow range (Reynolds number below 1), the value of C_D for a spherical particle is obtained from Eq. (9.4). The value of C_D in the transition flow range (Reynolds number 1 to 10^4) is given by Eq. (9.5). The value of C_D remains constant around 0.4 for spherical particles in the turbulent flow range (Reynolds numbers greater than 10^4).

$$C_D = \frac{24}{N_R} \quad \text{(Laminar range)} \tag{9.4}$$

$$C_D = \frac{24}{N_R} + \frac{3}{\sqrt{N_R}} + 0.34 \quad \text{(Transition range)} \tag{9.5}$$

Settling of very small spherical particles in the laminar range is obtained by substituting the value of C_D from Eq. (9.4) in Eq. (9.1). A simplified equation (Eq. (9.6)) is obtained and is known as *Stokes Law.*[3,4]

$$v_s = \frac{gd^2(\rho_s - \rho)}{18\mu} = \frac{gd^2(S_g - 1)}{18\nu} \tag{9.6}$$

Ideal Sedimentation Basin An ideal horizontal-flow sedimentation basin exhibits the following characteristics, which are commonly used to describe the settling behavior of discrete particles: (1) the flow through the basin is evenly distributed across the cross section of the basin; (2) the particles are evenly dispersed in water; and (3) the settling of the particles is predominantly *Type I.*

An ideal sedimentation basin is divided into four distinct zones: the inlet, settling, sludge, and outlet zones (Figure 9-1).[1]

Inlet Zone. In this zone, the flow is uniformly distributed across the basin cross section; flow leaving the inlet zone is strictly horizontal and in the direction toward the basin outlet.

Settling Zone. In this zone, water is quiescent and gradually flows horizontally toward the basin outlet. In this zone, sedimentation occurs.

Sludge Zone. In this zone, the settled sludge accumulates. Once the sludge enters this zone, it remains there.

Outlet Zone. In this zone, the clarified water is collected evenly across the cross section of the basin.

Fraction of Particles Removed In the design of a settling basin, the settling velocity v_t (also called the *terminal velocity*) of the smallest particle is selected from Eq. (9.7) or Eq. (9.8). The basin is designed so that all particles having a settling velocity greater than the terminal velocity are fully removed. Particles having a settling velocity (v_i) smaller than the terminal velocity are partially removed (Figure 9-1); the velocity v_i is expressed by Eq. (9.9).

$$v_t = \frac{Q}{WH_0} \tag{9.7}$$

$$v_t = \frac{H_0}{t_0} \tag{9.8}$$

$$v_i = \frac{H_i}{t_0} \tag{9.9}$$

where

v_t = settling velocity of smallest particle that is fully removed, m/s (ft/s)
Q = flow rate through the basin, m^3/day (ft^3/day)
W = width of the basin, m (ft)
H_0 = side water depth in the basin, m (ft)
t_0 = theoretical detention time, ($t_o = V/Q$ (Eq. (8.22))), s
v_i = settling velocity of particles, (v_i less than v_t means they are not fully removed), m/s (fps)
H_i = falling depth of particles with v_i in time t_o, m (ft)

In a typical river water suspension sample, a large gradation of particle size occurs. To determine the removal efficiency at a given settling time, it is necessary to consider the entire range of particle settling velocities and the fractions that are removed. Therefore, the total removal efficiency of a settling basin can be determined by (1) a batch settling test or (2) a sieve analysis.

The batch settling test employs a settling column similar to that shown in Figure 9-2. The test sample is placed in the column, and samples are taken at timed intervals, usually each 30 to 60 seconds for 5 minutes and then every 1 to 2 minutes for the remainder of the test. The test is continued for from 30 minutes to 2 hours, depending on the size of particles in the sample.[5] The weight of suspended solids is measured for each sample and expressed as a fraction of solids remaining.

The sieve analysis technique requires a sample of dry solids that constitute the suspension to be removed in the basin. Standard sieve analysis is performed on dried samples to determine the fraction of particles in different diameter ranges. The settling velocity of each fraction is calculated from Eqs. (9-1) or (9-6).

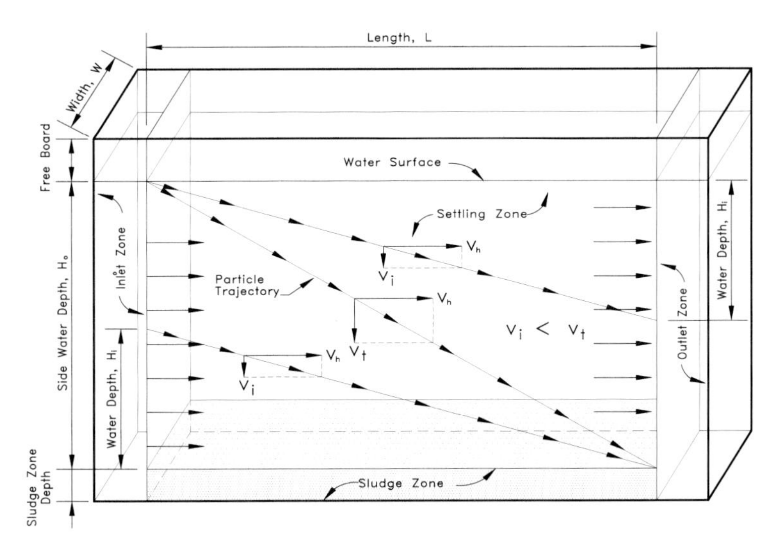

Figure 9-1 Settling behavior of discrete particles.

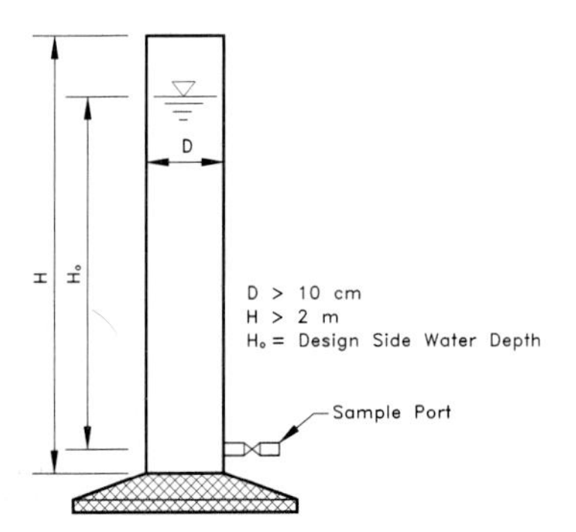

Figure 9-2 Standard settling column used for a discrete settling test.

With either method, a settling velocity analysis curve is developed, as shown in Figure 9-3. The total solids removal efficiency is then determined by graphically integrating the area under the curve.[1,5–7]

The fraction of particles removed in a settling basin having a terminal velocity of v_i is given by Eq. (9.10).

$$F = (1 - X_C) + \int_0^{X_C} \frac{v_i}{v_t} dx \tag{9.10}$$

where

F = fraction removed, dimensionless

X_C = fraction of particles with velocity v_i less than v_t, dimensionless

$(1 - X_C)$ = fraction of particles removed with settling velocity greater than v_t, dimensionless

$\int_0^{X_C} \frac{v_i}{v_t} dx$ = fraction of particles removed with velocity v_i less than v_t, dimensionless

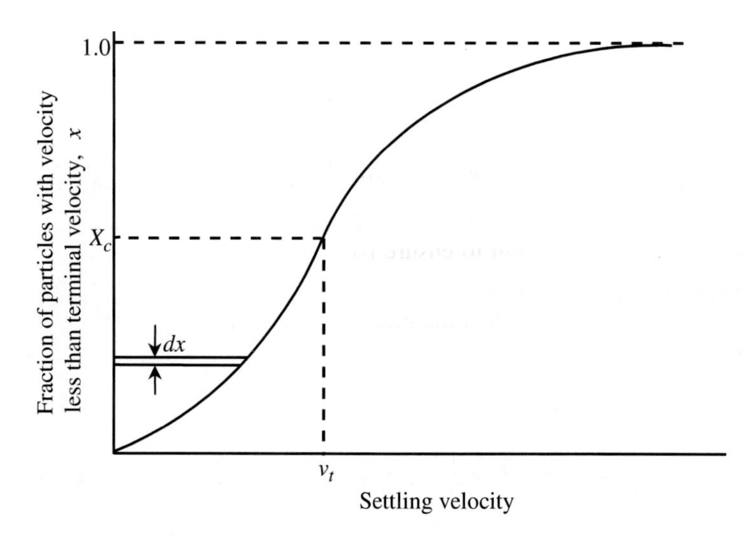

Figure 9-3 Discrete particles settling curve and fraction removal.

Overflow Rate The terminal velocity has great significance in the design of setting basins. It is also called *design overflow rate*, *surface loading rate*, or *hydraulic loading* ($m^3/m^2 \cdot d$). It is expressed numerically by Eq. (9.11).

$$v_t = \frac{Q}{A} \tag{9.11}$$

A = surface or plan area of setting basin, m^2 (ft^2)

In sedimentation basin design, the principal parameter affecting particle removal efficiency is the surface loading rate. When the surface loading rate is selected for an acceptable settled water quality, the required side water depth is often considered as a proxy for the detention time, which is another important design parameter. In fact, in an ideal basin, a shallower depth yields higher removal efficiencies at a given detention time, because the particles have a shorter distance to travel to reach the sludge zone, and the surface loading rate is typically lower. Unfortunately, the particle settling behaviors in an actual sedimentation basin are far from those achieved under ideal conditions. Factors such as density currents, thermal currents, wind action, and uneven flow distribution may cause a sedimentation basin to perform at an efficiency less than that under ideal conditions.

The most common use for discrete settling in water treatment is presedimentation basins. These basins are used to remove sand, gravel, and other discrete particles from raw water sources that have high solids loadıngs, such as river supplies. In the absence of any other design data, a surface loading rate of 200 to 400 $m^3/m^2 \cdot d$ and water depths of 3 to 5 m are typically used for presedimentation basin design.[6]

9.2.2 Flocculant Settling, Type II

Flocculant settling differs from discrete settling in that agglomeration (flocculation) of particles occurs.[5–7] The solid particles in the suspension begin to settle as discrete particles, but their concentration is sufficiently high to ensure particle collisions. The physical properties of the particles cause them to coalesce when collisions occur. The resulting combined particle is heavier, and so settles more rapidly, than the individual particles. As a result, particles have an increasing settling velocity that yields a settling path somewhat like that shown in Figure 9-4(a).[5,7] Due to such changes in particle size and settling behavior, flocculant settling is much complicated than discrete settling. No simple mathematical basis for flocculant settling has yet been developed for design purposes. Also, unlike in discrete settling, the side water depth is the most important parameter affecting the particles removal efficiency in flocculant settling. For a selected side water depth, other design parameters, including both overflow rate and detention time, must then be determined, either by batch settling tests or from experience with existing plants treating similar water. Typical design values for *Type II* sedimentation basins are discussed in Section 9.3.

Batch settling tests are performed in laboratory settling columns similar to that shown in Figure 9-4(b).[7–9] The suspension is thoroughly mixed and then placed in the column to a desired depth. At timed intervals (usually every 5 to 10 min), samples are withdrawn simultaneously from different ports. The suspended solids concentration is measured for each sample. A test duration of 1 to 3 hours should yield sufficient data for developing design parameters. Usually, the test should be repeated one or two times to ensure repeatability of the results. For illustration purposes, an example of a sedimentation basin follows. This example provides the procedure for obtaining the design overflow rate and detention time.

Example A sedimentation basin is designed to settle the flocculant suspension from a surface water source. It is desired to remove 80% TSS at a side water depth of 3.5 m. A batch column test analysis was performed in the laboratory. Determine the design values of overflow rate and detention time.

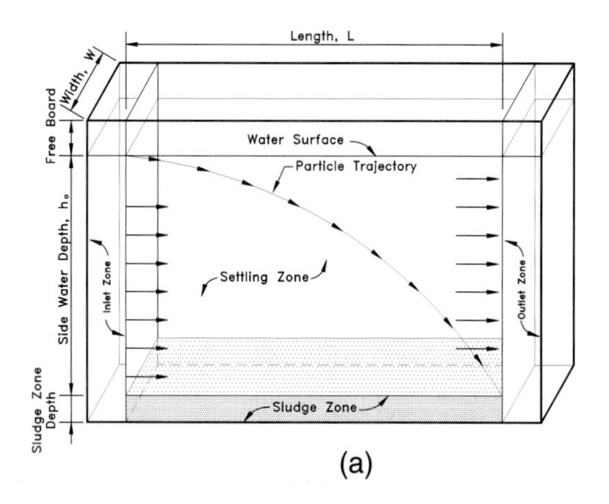

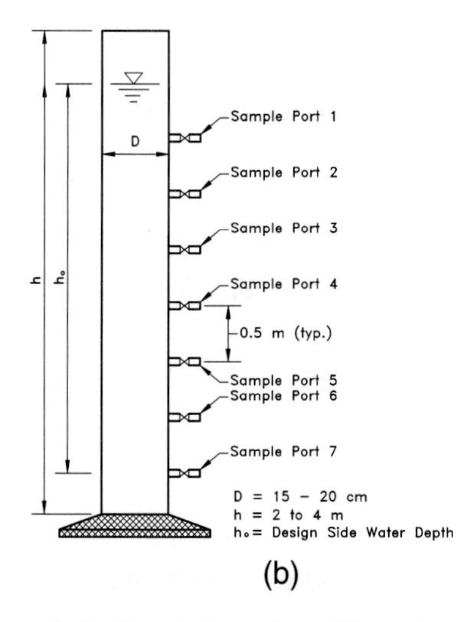

Figure 9-4 Flocculant settling. (a) Settling trajectories of flocculant particles in settling basin. (b) Standard settling column used for flocculant settling test.

Solution The following procedure is utilized in the batch column test analysis:

1. Fill the settling column with the flocculant suspension.
2. Draw samples from each port at timed intervals.
3. Determine TSS in each sample.
4. Reduce the results for each port, as illustrated in Table 9-1.
5. Prepare a summary table with reduced results for each port (Table 9-2).

Table 9-1 Example Batch Settling Test Results Reduction Analysis for Sample Port No. 1 (Figure 9-4(b))

Time, min	Measured TSS, mg/L	TSS Removed, mg/L	Removal Efficiency, percent
0	200	0	0
10	134	66[a]	33[b]
20	75	125	62
30	51	149	74
40	20	180	90

a 200 mg/L − 134 mg/L = 66 mg/L.

b $\frac{66 \text{ mg/L}}{200 \text{ mg/L}} \times 100\% = 33\%$.

Table 9-2 Reduced Batch Settling Test Results for Various Ports

Port No.	Depth, m	Sampling Time, min								
		10	20	30	40	50	60	70	80	90
1	0.5	33[a]	62	74	90					
2	1.0	21	41	65	71	80	89	90		
3	1.5	16	36	59	67	74	81	86	91	
4	2.0	17	33	56	64	71	78	82	88	91
5	2.9	14	32	54	64	70	78	82	85	88
6	3.0	14	30	52	63	69	75	81	83	85
7	3.5	12	30	51	60	69	74	80	83	84

a 33 percent of initial TSS is removed at sampling time of 10 minutes since start (Table 9-1).

6. Plot a grid showing percent TSS removal at each port at different time intervals (Figure 9-5).
7. Draw lines of equal percentage removal (*isoremoval*). These lines are drawn similarly to contour lines (Figure 9-5).
8. Draw a vertical line at each point an isoremoval line intersects the x-axis (3.5 m depth). For example, the $R = 60\%$ isoremoval curve intercepts the x-axis at 38 minutes. The 60-percent settling time t is therefore 38 minutes. The following observations can also be made:

    a. 90% of the particles have settled 0.51 m or more.

    b. 80% of the particles have settled 0.72 m or more.

    c. Likewise, 70% and 60% of the particles have settled 1.01 m, and 3.50 m or more, respectively.

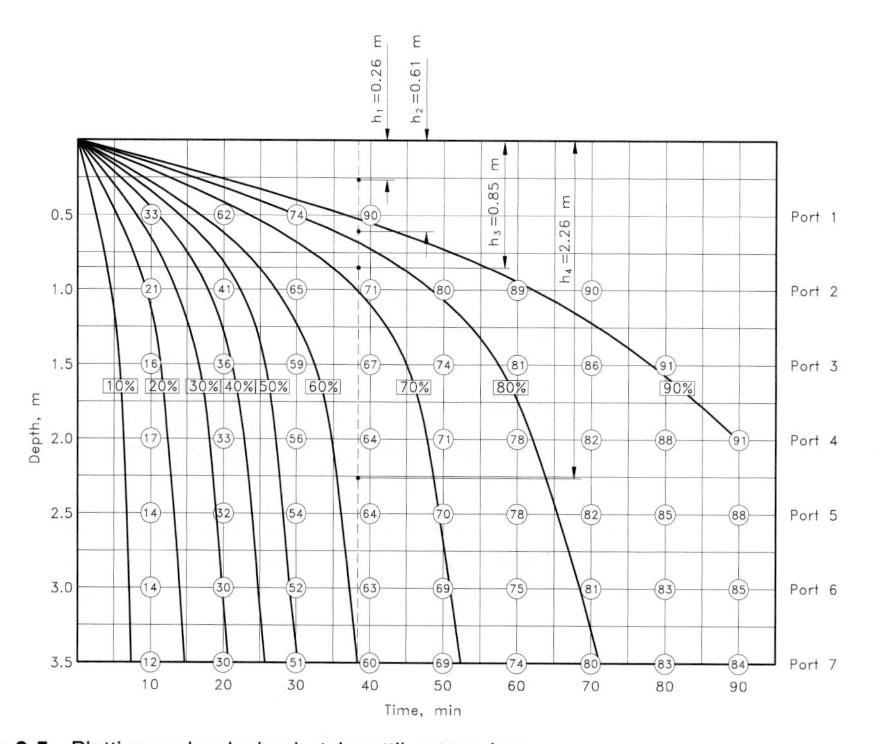

Figure 9-5 Plotting and reducing batch settling test data.

The total percent of TSS removal at a given detention time and at a given water depth h_0 is calculated from Eq. (9.12).

$$\text{Percent removal} = \frac{h_1}{h_0}(100 - R_1) + \frac{h_2}{h_0}(R_1 - R_2) + \ldots + \frac{h_{n-1}}{h_0}(R_{n-1} - R_n) + R_n \quad (9.12)$$

where

$h_1, h_2, ..., h_{n-1}$ = vertical distance from the top of the settling column to the midpoint between two consecutive lines of iso-removal at desired detention time (Figure 9-5), m (ft)

h_0 = desired design side water depth that is less than or equal to the depth of settling column (Figure 9-5), m (ft)

$R_1, R_2, ..., R_{n-1}$= consecutive iso-removal curves, percent removal

TSS removal efficiency of 71.3 percent at detention time t = 38 minutes and water depth h_0= 3.5 m is obtained as shown in Table 9-3.

Table 9-3 Calculation of Total Percent Removal at h_0 = 3.5 m and t = 38 minutes

n	h_0	$\frac{h_n}{h_0}$	R_{n-1}	R_n	$R_{n-1} - R_n$	$\frac{h_n}{h_0}(R_{n-1} - R_n)$
1	0.26	0.07	100	90[a]	10	0.7
2	0.61	0.17	90	80	10	1.7
3	0.85	0.24	80	70	10	2.4
4	2.26	0.65	70	60	10	6.5
5	-	-	-	60	-	60
				Total Percent Removal		71.3

a 90 percent TSS isoremoval contour.

9. Determine total percent of TSS removal at several detention times. Also, calculate the settling velocity of this suspension and overflow rate. The results are summarized in Table 9-4.

Table 9-4 Total Percent Removals at Percent Settling Time

Detention Time, min	Velocity, m/min	Overflow Rate, $m^3/m^2 \cdot day$	Percent Removal
7	0.50	720	20.9
15	0.23	336	34.9
21	0.17	240	45.8
26	0.13	194	55.3
30	0.12	168	62.5
38	0.092[a]	133[b]	71.3[c]
52	0.067	96.9	80.4
71	0.049	71.0	88.6

a 3.5 m/38 min = 0.092 m/min.

b 0.095 m/min ∞ m^2/m^2 ∞ 1440 min/d = 133 $m^3/m^2 \cdot d$.

c See Table 9-3 for calculation of percent TSS removed at h_0 = 3.5 m and t = 38 minutes.

10. Draw curves representing TSS removal with respect to overflow rate and detention time. These curves are shown in Figure 9-6.

11. Read theoretical values of overflow rate and detention time for a desired TSS removal efficiency from Figure 9-6.

In order to achieve a total TSS removal efficiency of eighty percent (80%) in a sedimentation basin with a side water depth of 3.5 m, it is required (1) that the overflow rate should not be

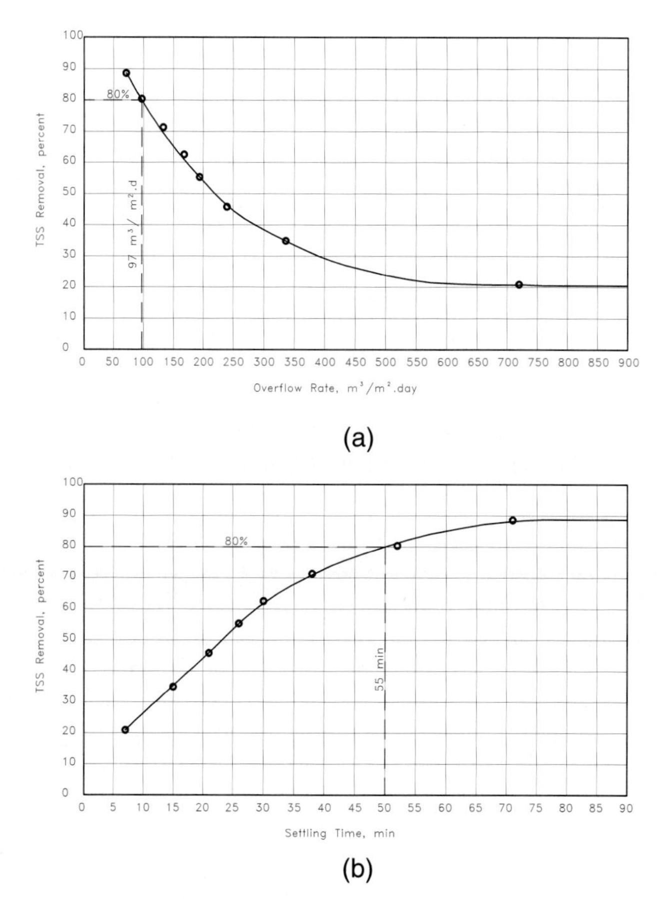

Figure 9-6 Percent TSS removal efficiency curves. (a) With respect to overflow rate. (b) With respect to settling time.

higher than 97 $m^3/m^2 \cdot day$ (Figures 9-6(a)), and (2) that the detention time should be at least 55 minutes (Figures 9-6(b)).

The batch settling test represents ideal settling conditions, which are rarely present in a full-scale continuous flow sedimentation basin. To account for such effects as density currents, temperature currents, and uneven flow distributions, correction factors can be applied to the experimental results. The test results are multiplied by the factors listed in Table 9-5 to yield the design parameters.[5,7,9] Using the factors of 0.7 and 1.6, the following design values are obtained:

Overflow rate = 68 $m^3/m^2 \cdot d$

Detention time = 88 min

Table 9-5 Correction Factors for Flocculant Settling Design Parameters in Full Scale Sedimentation Basin

Design Parameter	Correction Factor
Overflow rate	0.60 to 0.80
Detention time	1.25 to 1.75

9.2.3 Hindered Settling, Type III, and Compression Settling, Type IV

Hindered settling (*Type III*) and Compression settling, (*Type IV*) apply to gravity sludge thickeners in water treatment. The upper portion of the sludge blanket behaves like *Type III* settling; the lower portion behaves like *Type IV.* As with flocculant settling, no simple mathematical models of these settling conditions have been developed for design purposes. Design parameters need, therefore, to be developed from laboratory tests or from experience with existing plants handling similar sludges. Zone settling analysis or solids flux analysis are the typical methods used to develop these parameters. References 1, 3, 7 and 9 offer an excellent discussion of the procedures for these methods. Design of gravity thickeners is presented in Chapter 14.

9.3 Sedimentation Basin Design

The important considerations in sedimentation basin design are basin geometry, surface loading rate, detention time, inlet and outlet zone, weir loading rates, and the sludge collection and removal system. Table 9-6 lists typical values of such design parameters as overflow rate, detention time, and effluent weir loading rates. Other important design considerations are presented in later sections.

Table 9-6 Typical Water Treatment Sedimentation Design Parameters

	Detention Time, h	Surface Loading Rate, $m^3/m^2 \cdot d$ (gpd/ft^2)	Weir Loading Rate, $m^3/m \cdot d$ (gpd/ft)
Rectangular basins			
Coagulation	4–8	20–40 (50–1000)	250 (20,000)
Softening	2–6	40–60 (1000–1500)	250 (20,000)
Solids contact units			
Coagulation	2	40–60 (1000–1500)	170 (14,000)
Softening	1	60–100 (1500–2500)	350 (28,000)
Upflow basins			
Coagulation	2	40–60 (1000–1500)	170 (14,000)
Softening	1	60–1000 (1500–2500)	350 (28,000)

Source: Adapted in part from References 3, 6, 7 and 9.

Sedimentation basins often perform poorly in the field. The common reason is that actual detention time is significantly less than the designed detention time. This phenomenon is known as *short-circuiting*. It produces nonuniform time of passage and causes clarification efficiencies below expectations. The major causes of short-circuiting are currents induced by influent velocity, density and thermal gradients, wind action, and the effluent weir. Even in well-designed sedimentation basins, some short circuiting is expected. In a water treatment plant, poor performance of sedimentation basins normally results from breakup of floc during transport from flocculation to sedimentation basins.[1,10,11]

Actual detention time in a sedimentation basin can be determined from a tracer study. This study involves (1) slug or continuous injection of a tracer chemical, such as fluoride, at the basin influent and (2) measurement of the concentration of the tracer at the effluent. Common slug tracer feed responses for various geometric designs of sedimentation basin are shown in Figure 9-7. In most basins, depending upon the short-circuiting, the actual detention time is less than the theoretical detention time calculated from (Eq. (8.22)). This topic is presented in great detail in Chapter 12.

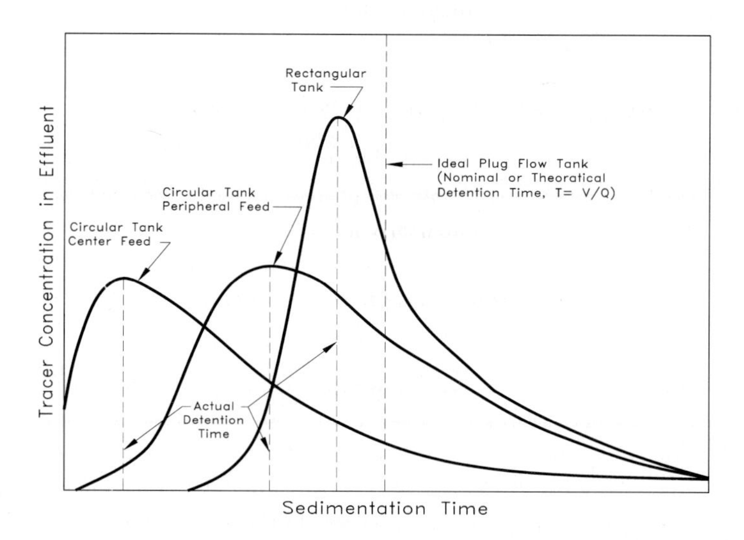

Figure 9-7 Typical results of a slug-dose tracer study for different geometric designs of sedimentation basins.

9.3.1 Geometries of Sedimentation Basin

There are three basic geometric shapes commonly used for sedimentation basins in water treatment: rectangular, circular, and square. A brief discussion on all these types of basins follows.

Rectangular Basins Rectangular basins typically are constructed with a length-to-width ratio ranging from 2:1 to 3:1. Monk has proposed that length-to-width ratios as high as 6:1 or 7:1 provide better control of short-circuiting.[11] With properly designed inlet and outlet facilities, the most nearly ideal flow pattern can be achieved in rectangular basins (Figure 9-7).[1] Rectangular basins also offer the advantage of lower construction cost for multiple units (as compared to circular basins) if common-wall construction can be employed.

Circular Basins Circular basins may be of either the center-feed or the peripheral-feed type. Tracer studies have demonstrated that peripheral-feed sedimentation basins experience less short-circuiting than do center-feed basins, yet center-feed basins are much more common. Generally, the flow pattern in circular basins is less nearly ideal than that in rectangular basins; however, circular sedimentation basins are more common, partly because they employ mechanically simple circular sludge collection equipment.

Square Basins Square basins are typically of the center-feed type. Square basins were developed to combine the advantages of common-wall construction with those of simple circular sludge-collection equipment. Because of the geometry of the effluent launders located on the basin perimeter, flows are not evenly distributed across the basin area.[6] Additionally, corner-sweep mechanisms utilized to remove sludge from basin corners are often potential sources of mechanical difficulties.

9.3.2 Enhancement of Sedimentation Basin Performance

Variations in basic sedimentation basin design have been employed to enhance the performance of the sedimentation process. The common variations include (1) laminar-flow devices that enhance performance by creating more nearly ideal flow conditions, (2) solids-contact units that attempt to enhance particle removal by promoting flocculation and enmeshment of incoming particles, and (3) a number of available proprietary systems. A brief discussion of some of these variations follows.

Laminar-Flow Devices One well-known modification of conventional sedimentation basins used in water treatment is the application of laminar-flow devices. These devices consist of banks of small square-shaped tubes (commonly called *tube settlers*) or plates (commonly called *plate settlers*), inclined at 45° to 60° angles from horizontal. The devices are installed near the outlet of the basin and are positioned so that the water must pass through the tubes or channels. These devices provide enhanced solid removal because: (1) the settling distance that a particle falls to enter the sludge zone is reduced (thus, the surface loading rate is reduced in the basin); (2) laminar-flow is achieved through the tubes (thus, more nearly ideal settling conditions are encountered); and (3) density currents, temperature currents, and wave action do not hinder the sedimentation process as they do in a conventional basin.[12–14]

The use of laminar-flow devices have been successful in reducing the settled water turbidity in some installations.[14] These devices are often used to enhance the performance of existing but overloaded sedimentation basins. Some manufacturers claim that the required detention time in new sedimentation basins can be greatly reduced by the use of laminar-flow devices. Designers are cautioned, however, against significantly reducing design parameters without thorough evaluation and testing. The design details of laminar-flow settlers are provided in Figure 9-8(a).

Solids-Contact Devices Another successful modification to conventional sedimentation basins is the solids-contact sedimentation basin. As illustrated in Figure 9-8(b), such a unit employs a sludge blanket at the bottom of the basin; the water stream must pass through the blanket. The high solids concentration in the sludge blanket promotes flocculation and enmeshment of incoming solids. As a result, the same solids removal efficiency as in conventional sedimentation process can be achieved at higher loading rates.[13] Solids-contact units have been used quite successfully in softening processes but they have been somewhat less successful in coagulation processes. Solids-contact clarifiers are usually designed to accommodate or utilize the equipment of a specific manufacturer. Normally, the manufacturer's representative will work closely with the design engineer in developing design parameters for such installations.[14,15]

Two-Tray System Monk proposed a two-tray sedimentation basin, as shown in Figure 9-9.[11] This configuration offers a large sedimentation area in a relatively compact space by effectively stacking one basin on top of another. In this system, flocculated water is fed through a perforated wall into a bottom chamber. The water flows horizontally to the end of the basin, then upwards into the upper chamber. The water then flows horizontally to the outlet zone in the end of the upper chamber. A chain-and-flight sludge collection system sweeps the sludge along the floor of the top chamber to the influent end. There the sludge falls through an opening in the floor into the lower chamber, where it settles and is then raked to the sludge pit to be pumped out of the basin.

Proprietary Sedimentation Systems A number of proprietary systems that claim to enhance sedimentation are on the market. These systems are mostly of the modifications of the solids-contact clarifier to prevent the formation of channels in the sludge blanket that generally reduce the settling effectiveness of a sedimentation basin. Three such systems are U.S. Filter's *Vortex Flow Clarifier*, Infilco Degremont, Inc.'s *Pulsation Sludge Blanket Clarifier*, and Krüger, Inc.'s *Micro-Sand Ballasted Flocculation*.[16–18]

The *Vortex Clarifier* is a conical tank, as illustrated in Figure 9-10(a). Water is introduced at the bottom of the tank so that a circular flow pattern is developed in the tank. Theoretically, this vortex keeps the sludge blanket in motion, prevents the development of channels in the sludge blanket, and produces thicker sludge.[16]

The *Pulsation Sludge Blanket Clarifier* is illustrated in Figure 9-10(b).[17] In this system, the water is fed through the bottom laterals. The feed rate is not constant. Instead, a specially designed vacuum chamber produces a pulsating flow. Theoretically, this pulsation stirs and mixes the sludge blanket and prevents the development of clearwater channels.

A more recent development in enhanced sedimentation is *Actiflo*® clarification, which is a combination of *Micro-Sand Ballasted Flocculation* and high-rate settling.[18] The technique consists of mixing the flocs (suspended solids) on micro-sand with the help of a polymer. The

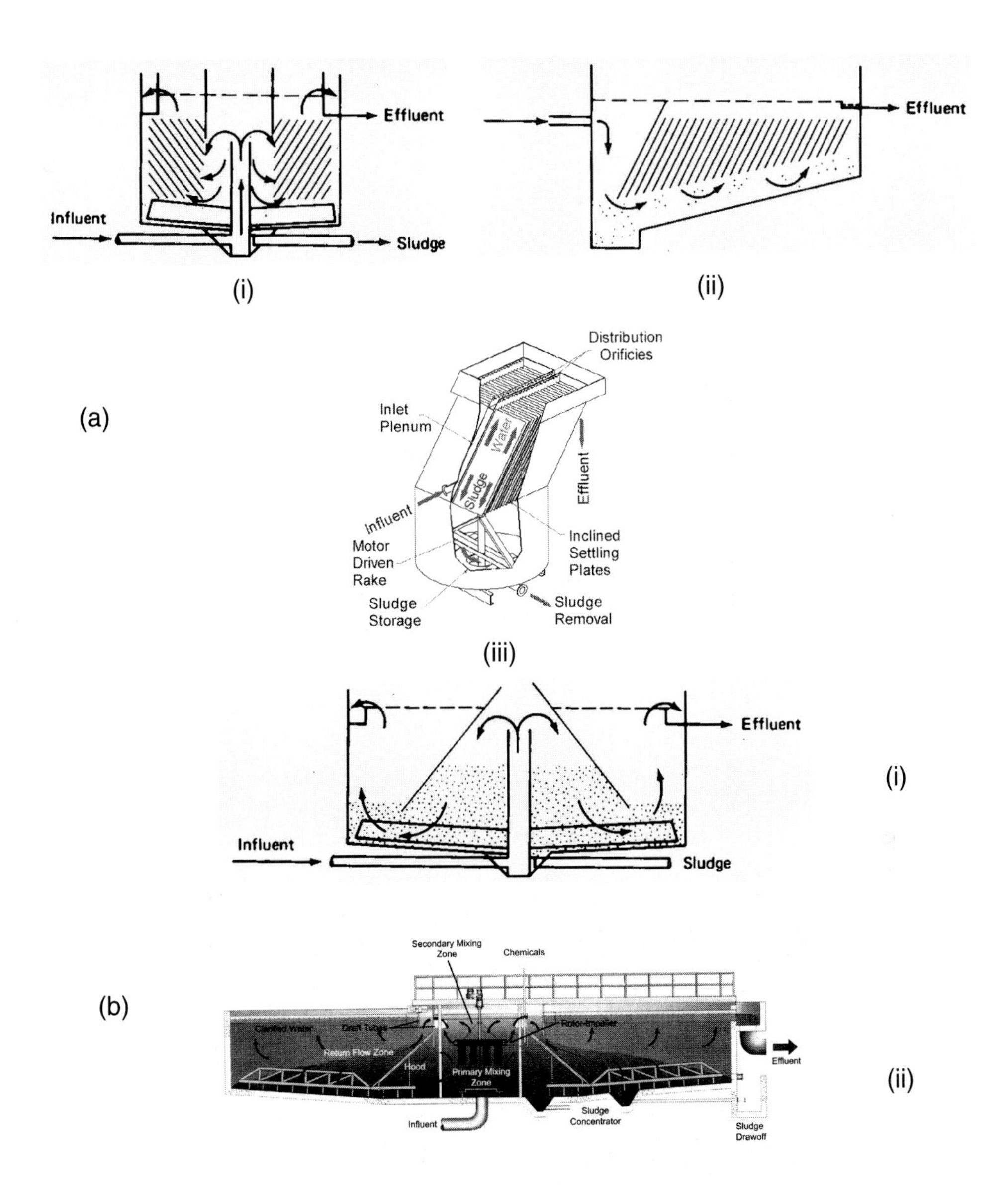

Figure 9-8 Types of sedimentation basins. (a) Laminar-flow settlers: (i) flow regeme in circular basin *(Reprinted from Reference 12, with permission from Technomic Publishing Co., Inc., copyright 1999.)*; (ii) flow regeme in rectangular basin *(Reprinted from Reference 12, with permission from Technomic Publishing Co., Inc., copyright 1999.)*; and (iii) equipment details. *(Courtesy of Hoffland Environmental, Inc.)* (b) Solids-contact clarifiers: (i) flow regeme *(Reprinted from Reference 12, with permission from Technomic Publishing Co., Inc., copyright 1999.)* and (ii) equipment details. *(Courtesy of Infilco Degremont, Inc.)*

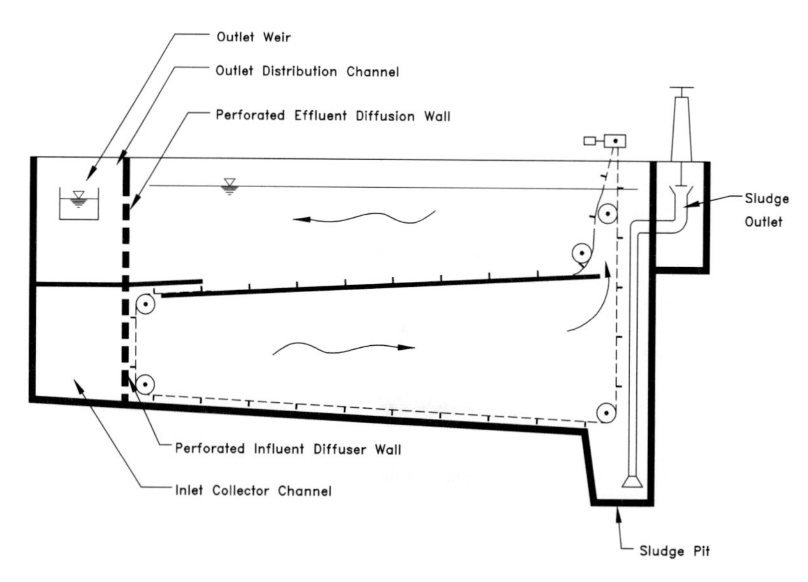

Figure 9-9 Two-tray sedimentation basin.

micro-sand is 50-100 μm in size, and the polymer is a flocculation aid. Reported benefits of the process in water treatment are increased removals of turbidity, color, TOC, and algae, with reduction in mixing, flocculation, and sedimentation times. The hardware, supplied by Krüger, Inc., utilizes an integral unit with a rapid mixing zone, a gentle mixing zone, and Lamella modules. The hydrocyclone is used for micro-sand recovery. The system configuration is shown in Figure 9-10(c).

In addition to these three specific proprietary systems, there are many more systems available. Some systems are effective; others are not. Engineers should carefully evaluate the claims and effectiveness of these proprietary systems through testing and through discussions with operators of other full-scale installations. Readers are referred to manufacturers for details of these and other proprietary systems. A list of manufacturers is provided in Appendix C.

9.3.3 Design Considerations

Surface Loading Rate Surface loading rates are used to calculate surface area requirements of a sedimentation basin (Eq. (9.9)).[6,9] A discussion of how to develop design values of surface loading rate by the batch settling test procedure for a flocculant settling suspension is presented in Section 9.2.2. Such a procedure is recommended prior to the design of the major facilities. Typical design parameters for the various water treatment processes in Table 9-6 can, however, be used in lieu of laboratory testing.

Detention Time Detention time is used in conjunction with the surface loading rate to calculate the volume and side water depth of the sedimentation basin. For a known detention

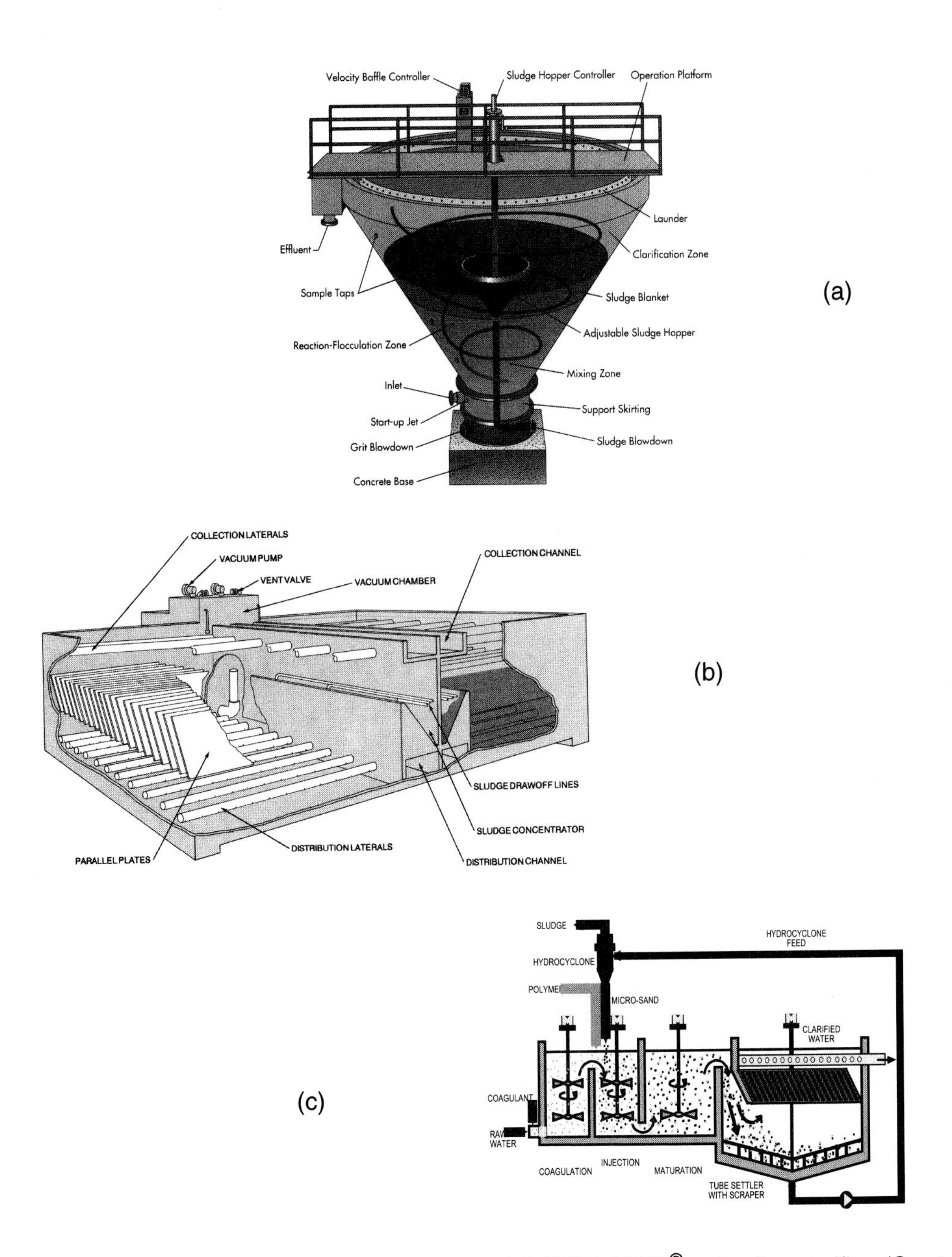

Figure 9-10 Proprietary sedimentation basins. (a) SPIRACONE® vortex flow clarifier. *(Courtesy of U.S. Filter.)* (b) Superpulsator® pulsation sludge blanket clarifier. *(Courtesy of Infilco Degremont, Inc.)* (c) Actiflo® micro-sand ballasted flocculation system. *(Courtesy of Krüger, Inc.)*

time, Eq. (9.8) is used to determine the basin volume. Often, batch settling tests (Section 9.2.2) are needed to obtain the design value of the detention time. Typical detention times of sedimentation basins are summarized in Table 9-6.[6,9]

Inlet Zone or Influent Structure The inlet zone evenly distributes the flow across the sedimentation basin and dissipates incoming velocity. If these two functions are achieved, the hydraulic flow characteristics of the basin will more closely simulate those of an ideal basin and make possible more efficient performance. Influent zones are designed differently for rectangular and circular basins.

1. Integration with Flocculation Basin

   Typically in water treatment, rectangular sedimentation basins are constructed to be integral with the flocculation basins. A baffle or diffusion wall separates the two basins and serves as the sedimentation basin inlet. The design of the diffusion wall is important, because performance of the sedimentation basin depends upon the floc quality. A detailed design of the diffusion wall is presented in Section 8.8.2, Step B. Readers should also review several references.[19–22]

2. Arrangements in Circular Basins

   The influent structure for a center-feed circular sedimentation basin typically consists of a metal skirt surrounding the inlet pipe. The influent structures for this type of basin are typically provided as a part of the sludge collection equipment and generally are designed by the equipment supplier. It is rear to use peripheral-feed circular sedimentation basin in water treatment.

Outlet Zone or Effluent Structure Like the inlet zone, the *outlet zone* or *effluent structure* has a great deal of influence on the flow pattern and settling behavior of floc in a sedimentation basin. Traditionally, overflow weirs and launder troughs have been used for outlet control in sedimentation basins. Either V-notch or submerged orifice weir plates are commonly used. Of these two, submerged orifice plates are preferred, because their use tends to result in less floc break-up between sedimentation and filtration units and also because they reduce ice problems in colder climates.[3,6,19,20] The length of weir required is determined by the weir overflow rate or weir loading rate. The typical design values of weir loading rates are provided in Table 9-6.

The weir plates typically discharge flow into launder troughs. Typical launder trough layouts for rectangular, square, and circular sedimentation basins are illustrated in Figure 9-11.[12] The flow in the launder troughs is described as spatially varied flow. Chow proposed an equation to estimate the water surface profile for such flow conditions.[23]

$$y_1 - y_2 = \frac{q_1(v_1 + v_2)}{g\,(q_1 + q_2)}\left[(v_2 - v_1) + \frac{v_2}{q_1}(q_2 - q_1)\right] + S_0 x - S_f x \tag{9.13}$$

where

y_1 and y_2 = the upstream and downstream water surface elevations, respectively, m. The difference $(y_1 - y_2)$ = change in water surface elevation.

q_1 = flow rate at the upstream point, m^3/s

q_2 = flow rate at the downstream point, m^3/s

v_1 = velocity at the upstream point, m/s

v_2 = velocity at the downstream point, m/s

g = acceleration gravity constant, m/s^2

x = incremental distance between upstream and downstream points, m

S_0 = slope of channel bottom

S_f = slope of energy gradient given by Eq. (9.14)

$$S_f = \frac{n^2 (v_{ave})^2}{(R_{ave})^{4/3}} \tag{9.14}$$

where

n = Manning's friction factor

v_{ave} = average velocity between the upstream and downstream sections, m/s

R_{ave} = average hydraulic radius between the upstream and downstream sections, m

Eq. (9.13) is valid only for launder troughs with a free discharge. Readers are referred to Reference 17 for other flow conditions.

The procedure for utilizing Eq. (9.13) is a trial-and-error procedure that is long and tedious. The calculation procedure is given in Reference 12. An alternative, approximate solution is obtained from Eq. (9.15). This equation provides a valid approximate solution for flat-bottom flumes with level inverts and parallel sides.[1] For small launder length, the term with friction factor is ignored, and the resulting equations reduce to Eq. (8.30). An allowance for friction and turbulence loses is made to the final value of y_1. The procedure is presented in Section 8.8.2, Step A.5b.iii.

$$y_1 = \sqrt{y_2^2 + \frac{2(Q^2)}{gb^2 y_2} + \frac{fLQ^2}{12gb^2 rd}} \tag{9.15}$$

where

Q = total discharge over length of launder trough, m^3/s

b = width of launder trough, m

f = Darcy friction factor, (0.03 to 0.12 for concrete)

L = length of channel, m

r = mean hydraulic radius, m

d = mean depth of channel, m

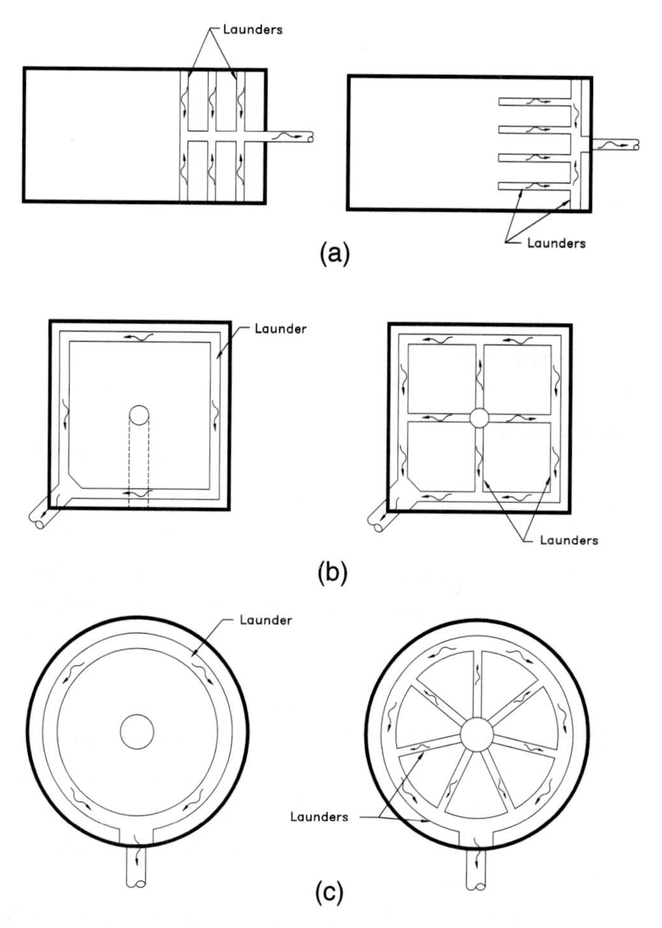

Figure 9-11 Typical launder trough layouts. (a) Rectangular basins. (b) Square basins. (c) Circular basins.

The weir loading rates shown in Table 9-6 typically result in a long weir-length requirement. In rectangular basins, these weir lengths require long launder troughs. As shown in Figure 9-12(a), the long launder length effectively lengthens the outlet zone and shortens the settling zone. Also, this configuration imparts a vertical velocity to the water near the outlet zone. As a result, the outlet zone configuration shown in Figure 9-12(a) offers less than ideal settling conditions. Short-circuiting due to density currents is promoted, and vertical velocities tend to resuspend some particles.[11] To alleviate these outlet zone problems of rectangular basins, Monk proposed an inlet and outlet configuration as shown in Figure 9-12(b).[11] This configuration employs a baffle-wall outlet in lieu of weirs. With this configuration, only a single launder trough is required behind the baffle wall to control the level in the sedimentation basin. This design will result in extremely high weir loadings, but the baffle walls promote parallel flow through the basin. This arrangement is considered better than a launder trough and weir arrangement, because more nearly ideal settling conditions are achieved.[11]

Another innovative design for effluent arrangement is a series of perforated effluent pipes provided at the effluent end of the rectangular basin. The supernatant enters through numerous circular openings provided into the effluent collection pipes. The number and size of the openings is such that comparable weir loading is achieved per unit length of the perforated pipe. The authors' experience with this design is very favorable. This arrangement is shown in Figure 9-12(c).

9.3.4 Sludge Collection

Quantity of Solids Produced The purpose of sludge collection is to withdraw residuals from the sedimentation basin for processing and disposal. The quantity of solids produced in coagulation is dependent upon the total suspended solids in the water, the type and dosage of coagulant, and the efficiency of the sedimentation basin. Typically in water treatment, turbidity values are used as an indicator of suspended solids loading; however, there is no absolute correlation between turbidity values (NTU) and total suspended solids (TSS). A ratio of TSS to NTU normally varies from 0.5 to 2.[3,6,24–26]

The sludge from a coagulation process also contains the metal hydroxide produced. Aluminum sulfate produces a hydroxide that has bound water.[3,6,24–26] Iron coagulants are ferric sulfate, ferrous sulfate, and ferric chloride. In all three cases, the iron precipitate formed is ferric hydroxide. The insoluble metal hydroxide is a bulky, gelatinous floc.[25,26] Theoretical quantities of sludge produced by various coagulant chemicals are listed in Table 9-7.

Because of bound water, the actual reported quantities of sludge from alum, ferric sulfate, and ferric chloride are 0.33, 0.59, and 0.48, respectively.[3,6] Therefore, it is advisable to increase the theoretical quantities of sludge when designing sludge-handling equipment and processes. In general, the solids content of coagulation sludge is 0.5 to 2 percent.

The quantity of sludge produced from lime-softening processes varies greatly, depending on the hardness of the water, the raw water chemistry, the chemical dosages used, and the desired finished-water hardness. Typical constants for estimating sludge produced by lime softening are also listed in Table 9-7. Reported quantities for various plants have been 43–2230 kg/1000 m^3 (360–18,7000 lb per million gallons), with an average of about 230 kg/1000 m^3 (2000 lb per million gallons) of water treated.[6] The solids content of softening sludge typically ranges between 2 and 15 percent.[3,6]

Sludge Collection. Generally, 80 to 95 percent of the total solids are removed in the sedimentation process.[6] This mass of solids must be removed from the sedimentation basin before accumulation becomes excessive. Although some plants drain basins and remove sludge manually, most use mechanical sludge collection equipment to move the sludge to a hopper for removal. The mechanical equipment employed in this process is generally classified as either chain-and-flight type, traveling-bridge type, or circular type. An example of each type is illustrated in

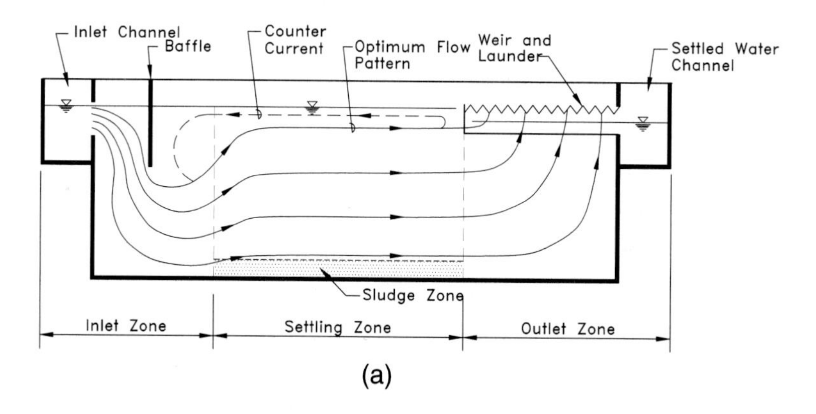

(a)

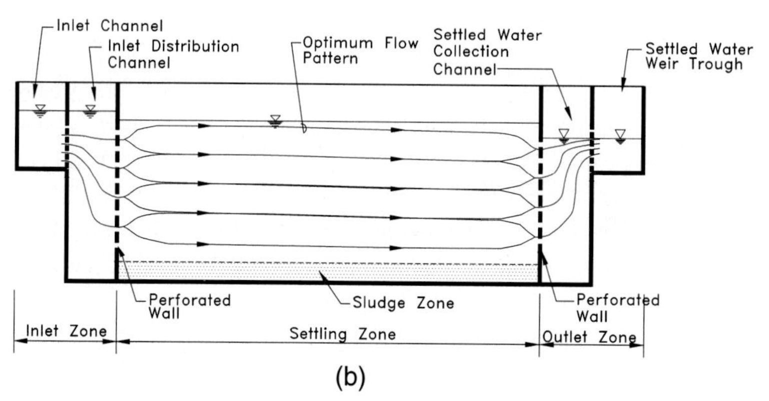

(b)

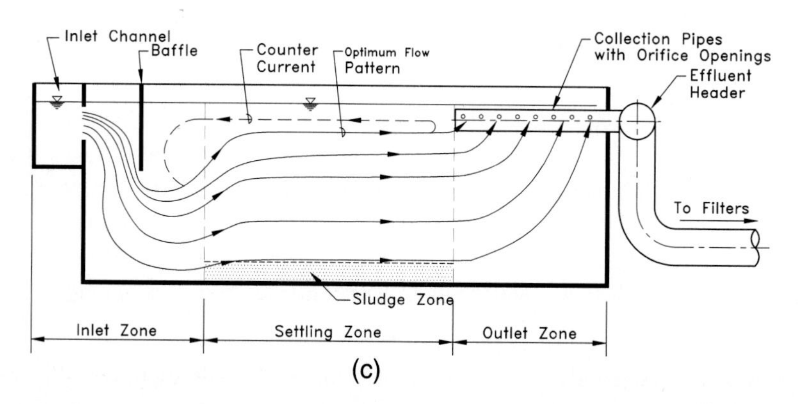

(c)

Figure 9-12 Effects of inlet and outlet structures on flow patterns through sedimentation basins. (a) Conventional horizontal flow sedimentation basin. (b) Sedimentation basin with inlet and outlet baffle walls. (c) Sedimentation basin with inlet baffle walls and effluent pipe in lieu of effluent launders.

Table 9-7 Typical Constants for Estimating Sludge Production

Source	Sludge Quantity Constant
Turbidity removal	0.5 to 2 mg TSS/L per NTU removed[a]
Coagulation	
Aluminum sulfate (alum), $Al_2(SO_4)_3.14H_2O$	0.26 kg per kg of coagulant[b]
Ferric sulfate, $Fe_2(SO_4)_3$	0.54 kg per kg of coagulant[c]
Ferrous sulfate, $FeSO_4.7H_2O$	0.39 kg per kg of coagulant[d]
Ferric chloride, $FeCl_3$	0.66 kg per kg of coagulant[e]
Polymers	1 kg per kg of coagulant
Powdered activated carbon	1 kg per kg of carbon
Softening	
Calcium	2.5 kg per kg of calcium (as Ca) removed
Magnesium	2.4 kg per kg of magnesium (as Mg) removed

a This value is very site-specific, but 1 is a common occurrence.

b $Al_2(SO_4)_3 \cdot 14H_2O + 3Ca(HCO_3)_2 \Leftrightarrow 2Al(OH)_3 \downarrow + 3CaSO_4 + 6CO_2 + 14H_2O$.

c $Fe_2(SO_4)_3 + 3Ca(HCO_3)_2 \Leftrightarrow 2Fe(OH)_3 \downarrow + 3CaSO_4 + 6CO_2$.

d $2FeSO_4 \cdot 7H_2O + 2Ca(HCO_3)_2 + (\int)(O_2) \Leftrightarrow 2Fe(OH)_3 \downarrow + 2CaSO_4 + 4CO_2 + 6H_2O$.

e $2FeCl_3 + 3Ca(HCO_3)_2 \Leftrightarrow 2Fe(OH)_3 \downarrow + 3CaCl_2 + 6CO_2$.

Source: Adapted in part from References 6, 23, and 24.

Figure 9-13. The operational characteristics of various types of sludge collection mechanisms are compared in Table 9-8.

Detailed design of sludge collection equipment is typically the responsibility of the equipment manufacturer. The responsibility of the design engineer includes the selection of the type of equipment to be used and the preparation of specifications for procurement of that equipment. It is essential that the mechanical collection and removal system should sweep and move all settled sludge, including that in the corners, and deposit it into a central hopper for removal from the basin by pumping or by gravity flow. Corner-sweep arrangements are presented in the Design Example.

9.4 Manufacturers of Sedimentation Equipment and Systems

A list of manufacturers of sedimentation basins and sludge collection equipment is provided in Appendix C. Information such as plant capacity, geometry of the treatment units, and basic design parameters for the process and equipment are required for making proper selection of sedimentation and sludge collection equipment. The equipment manufacturers and suppliers will aid the design engineer in selection and specification of the equipment for the proposed application. The reader is referred to the equipment manufacturers in Appendix C to obtain more nearly complete information.

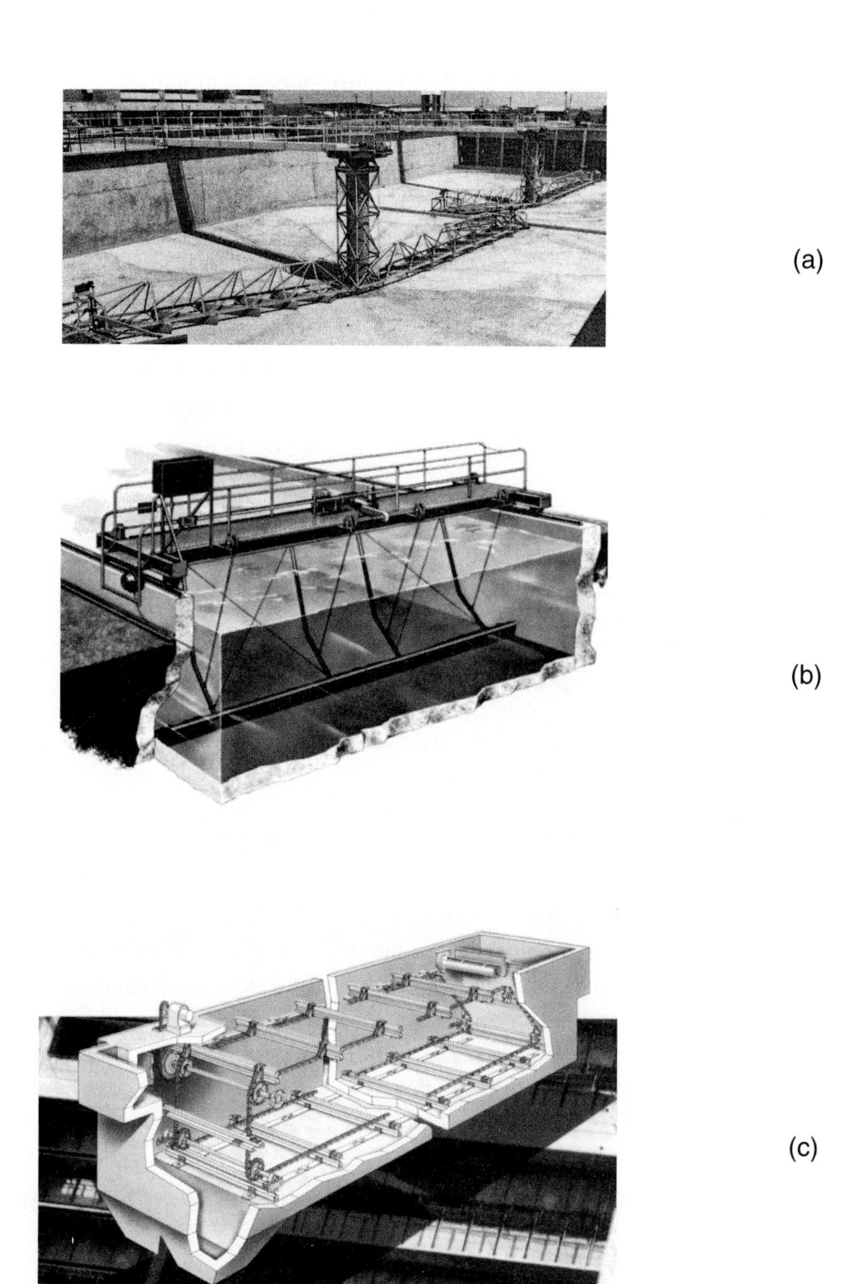

Figure 9-13 Typical sludge collection equipment. (a) Circular *(Courtesy of U.S. Filter/Link-Belt.)* (b) Traveling-bridge *(Courtesy of U.S. Filter/Envirex.)* (c) Chain-and-flight *(Courtesy of U.S. Filter/Envirex.)*

Table 9-8 Descriptions, Advantages, and Disadvantages of Sludge Collector Mechanisms

Type	Description	Advantages	Disadvantages
Chain and flight	1. Endless chain is connected to a shaft and drive unit. 2. Cross wood (flights) are attached to the chain. These flights scrape the sludge to a hopper at the end of the basin.	1. Simple to install. 2. Low power consumption. 3. Suitable for heavy lime sludge.	1. Chain and flights require intensive maintenance. 2.Tank must be dewatered to repair mechanism. 3. Light sludges may resuspend.
Traveling bridge	1. Bridge spans across the width of the basin. 2. Scrapper mechanism is suspended from bridge. 3. As the bridge travels towards the hopper, the scrapers are lowered and sweep sludge to the hopper at one end of the basin. 4. After the bridge reaches the sludge hopper, the scrapers are raised and the bridge returns to the other end of the basin. Cycle is repeated.	1. Most moving parts are above water. 2. Most maintenance can be achieved without dewatering tank. 3. Low maintenance cost for short span bridges. 4. Longer operating life.	1. High power requirement to move bridge. 2. Will not operate in ice-covered basins. 3. Wheels may derail frequently, particularly on long-span bridges. 4. Scraper may "ride up" on heavy sludges and cause accumulation of sludges.
Circular clarifiers, with or without corner sweep	1. Scrapers are attached to a truss, which rotates around the basin. Scrapers sweep sludge to center of the basin. 2. Truss arm is attached to a center cage. 3. The center cage is attached to a turntable and gear mechanism, which rotates the sludge collector. 4. Corner-sweep arms are used to collect sludge from corners of square or rectangular basins.	1. No submerged moving parts in circular basins. 2. Most maintenance can be achieved without dewatering tank. 3. Low maintenance cost. 4. Low operating cost.	1. Corner-sweep mechanisms are often a maintenance problem.

9.5 Information Checklist for Design of Sedimentation Facilities

The following information must be obtained and necessary decisions made before the design engineer can begin the design of a sedimentation facility.

1. Information to be obtained from predesign studies:

   a. design average flows and peak;
   b. detention time and surface loading rate;
   c. effluent weir configuration and overflow rate.

2. Information to be obtained from the preliminary design:

   a. preliminary plant layout, showing existing facilities and the general size, location and geometry of the proposed facilities;
   b. number of proposed units;
   c. preliminary hydraulic profile of the proposed units illustrating head loss parameters for the units being designed;
   d. type of sludge collection equipment.

3. Information to be obtained from the plant owner and operator:

   a. any particular preference with regard to equipment types and manufacturers;
   b. any particular preference with regard to the design details from other plants in operation;
   c. design criteria and any available guidelines.

4. Information to be developed by the design engineer:

   a. minimum design parameters established by regulatory authorities;
   b. types, sizes, and limitations of available equipment.

9.6 Design Example

9.6.1 Design Criteria Used

The following design criteria are used in the Design Example.

1. Flow rates

   a. Maximum flow rate = 113,500 m^3/d (30 mgd)
   b. Average flow rate = 57,900 m^3/d (15.3 mgd)

2. Design parameters

   a. Number of basins: four
   b. Detention time: 4 hrs
   c. Length-to-width ratio: 2 to 3
   d. Surface loading rate: 35 $m^3/m^2 \cdot day$ (860 gpd/ft^2)
   e. Weir loading rate: 250 $m^3/m \cdot day$ (20,000 gpd/ft)

3. Unit arrangement and general layout

   Provide four rectangular sedimentation basins. Each basin will be designed for 25 percent of the maximum design flow. The general layout of the sedimentation basins is provided in Figure 9-14. In this layout, the sedimentation basin is an integral part of the flocculation/sedimentation basin complex. (See Chapter 8 for design details of flocculation basins.) The width of each sedimentation basin is therefore the same as that of the flocculation basin. A perforated baffle wall or diffusion wall divides the two unit processes. This layout minimizes shearing and destruction of floc between the third stage of the flocculation chamber and the inlet zone of the sedimentation basin. It also provides excellent flow distribution at the inlet zone of the sedimentation basin. In each sedimentation basin, the settled water is collected by an effluent collection system that consists of launder troughs with V-notch weirs and a center effluent collection channel. The collected water flows into an effluent box and then is transferred to the filter complex through pipes. Two circular sludge collection mechanisms with corner-sweep arrangements are provided in each sedimentation basin.

9.6.2 Design Calculations

Step A: Sedimentation Basin Design

1. Unit dimensions

   a. Calculate the required area of each basin for one-fourth the maximum design flow.

   $$\text{Maximum design flow} = 113.5 \text{ MLD} = 113{,}500 \text{ m}^3/\text{day}$$

   $$\text{Design flow for each basin} = \frac{113{,}500 \text{ m}^3/\text{day}}{4} = 28{,}375 \text{ m}^3/\text{day}$$

   $$= 0.328 \text{ m}^3/\text{s}$$

   Using a surface loading rate = 35 $m^3/m^2 \cdot d$. (See Design Criteria.) The required basin area is calculated as follows.

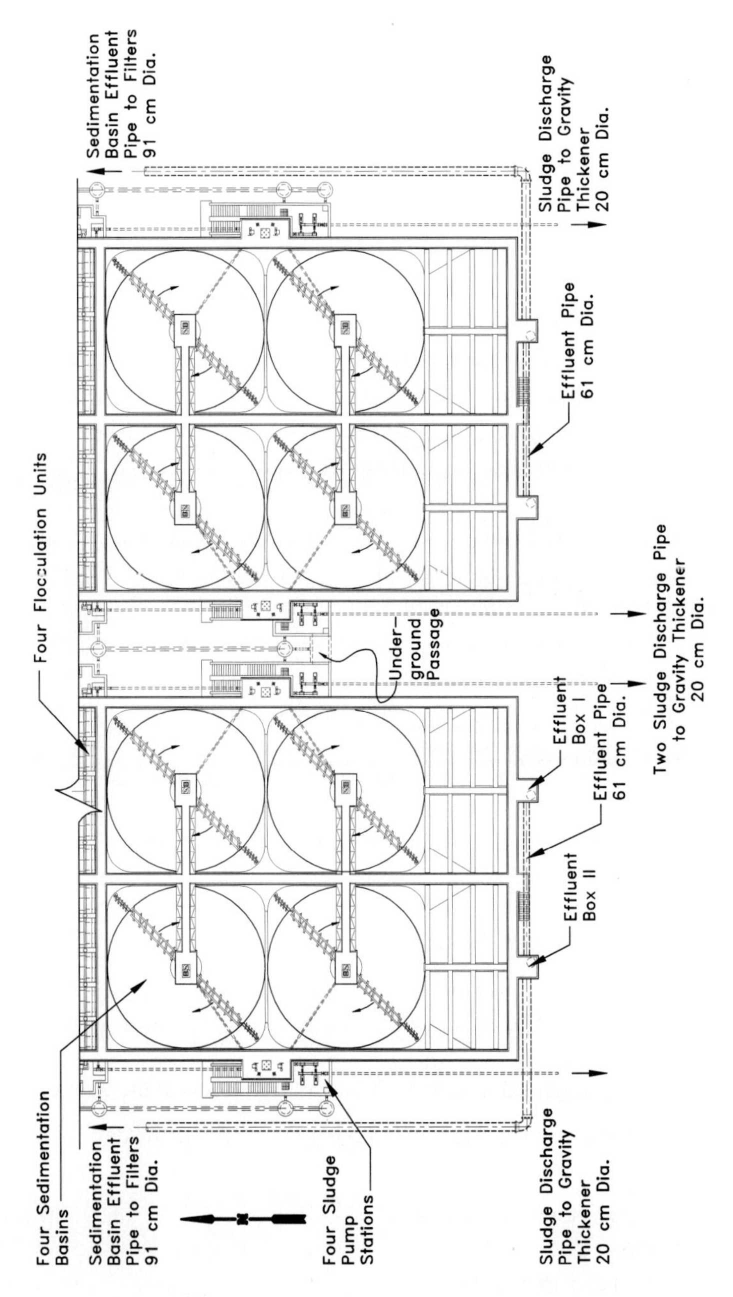

Figure 9-14 General layout of sedimentation basins for the Design Example.

$$A_{req} = \frac{28,375 \text{ m}^3/\text{d}}{35 \text{ m}^3/\text{m}^2\cdot\text{d}} = 811 \text{ m}^2$$

b. Calculate the required length of basin based on $W = 18.4$ m.

$$L_{req} = \frac{810.7 \text{ m}^2}{18.4 \text{ m}} = 44.1 \text{ m}$$

A design basin length $L = 45$ m is provided. The actual length-to-width ratio is calculated as follows:

$$L/W \text{ ratio} = \frac{45 \text{ m}}{18.4 \text{ m}} = 2.4 \text{ s}$$

This ratio is within the range of the design criteria.

c. Calculate required basin depth.

With a surface loading rate = 35 $m^3/m^2\cdot d$ and detention time of four hours, the required side water depth is calculated as follows:

$$H_{req} = \frac{35 \text{ m}^3/\text{m}^2\cdot\text{d} \times 4 \text{ h}}{24\text{h/d}} = 5.8 \text{ m}$$

Provide four basins with dimensions of 45 m long ∞ 18.4 m wide ∞ 5.8 m side water depth. The actual depth of the basins is 6.45 m, with a freeboard of 0.65 m above the water surface. The major dimensions of the sedimentation basin are shown in Figures 9-15 and 9-16.

2. Influent structure

The detailed influent structure design is provided in Chapter 8 (Section 8.8.2, Step B.5.d and Figure 8-21).

3. Effluent structure

The effluent structure in each basin consists of 90° V-notch weirs on both sides of each effluent launder trough, eight 0.5-m-wide launder troughs, and a 0.65-m-wide center collection channel that conveys the settled water into a final effluent box. A baffle board is provided between the settling and outlet zones, to direct flow beneath the outlet zone and to reduce surface currents and wave effects. This arrangement is also shown in Figure 9-16.

4. Equipment design

Sludge collection equipment is the only major equipment installed inside the sedimentation basin. Two circular collector mechanisms 17.5 m in diameter are provided in each basin. The diameter of a collector is slightly shorter than the basin width, to provide a space of 0.45 m at each side to install tracks for the corner-sweep guide. Each collector has corner-sweep mechanisms, to remove sludge from the corners. Motor contacts will monitor the status of sludge collection equipment. The sludge collection equipment is also

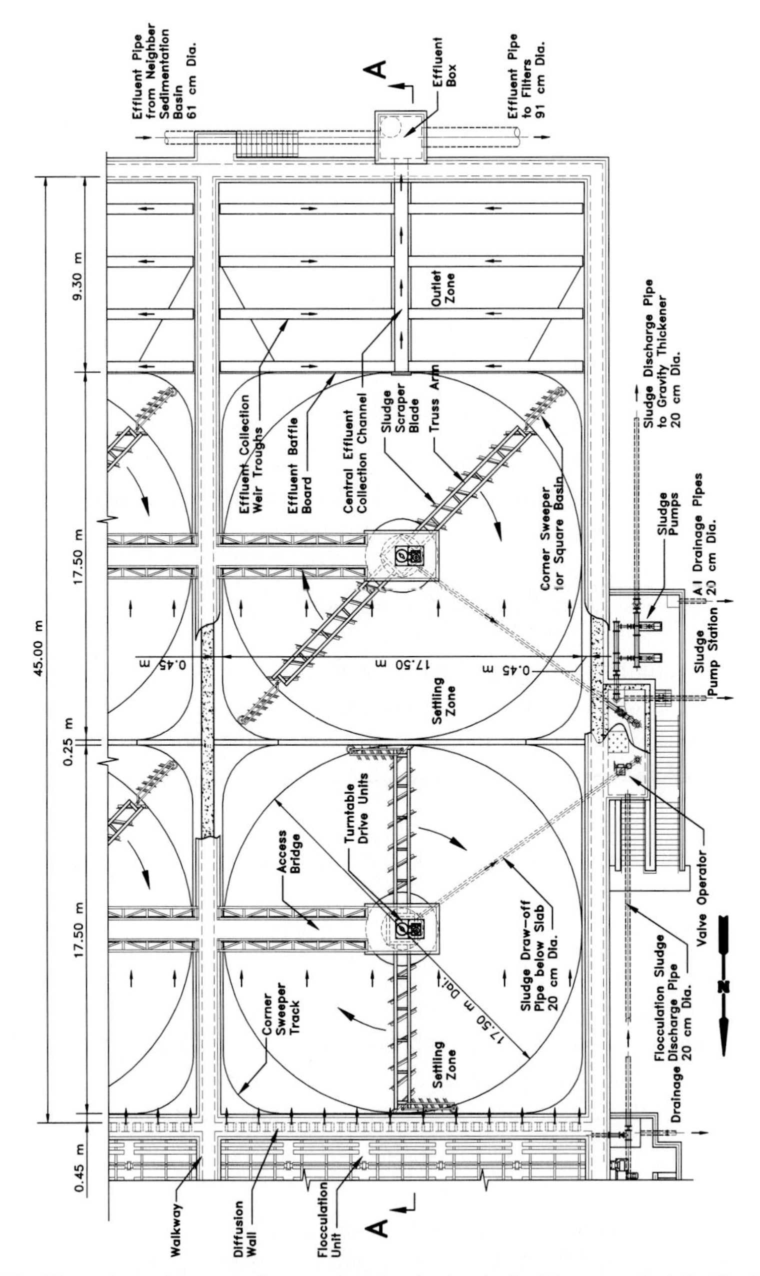

Figure 9-15 Plan view of one sedimentation basin (typical of four basins) in Design Example. (Section view of A-A is shown in Figure 9-16.)

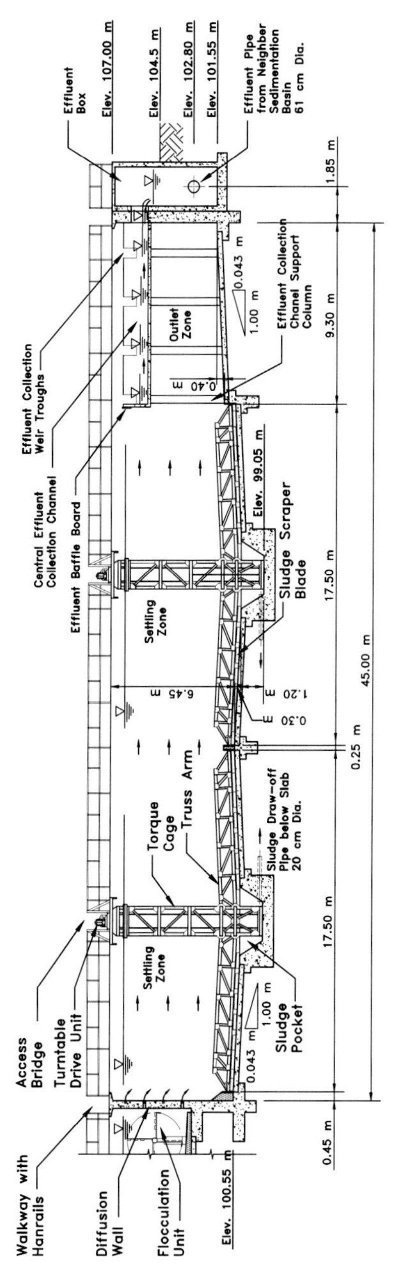

Figure 9-16 Section view of the sedimentation basin in Design Example. (Location of section A-A is shown in Figure 9-15.)

shown in Figures 9-15 and 9-16. The selection of telescope valves and sludge pumps is discussed with the sludge withdraw system (Section 9.6.2, Step B.2).

The effluent from the sedimentation basins will be monitored for turbidity, pH, and chlorine residual. The instrumentation is indicated in Table 16-3.

5. Head loss calculations and hydraulic profile through the sedimentation basin

In this section, the hydraulic calculations are conducted only for the head losses through the effluent structures. The head loss at the diffusion wall between the flocculation and sedimentation basins is provided in Chapter 8 (Section 8-8-2, Step B.5.d). The head loss through the sedimentation basin is small and so ignored.

a. Head loss at the V-notch weirs.

i. Calculate required weir length.

Weir loading = 250 m^3/m·d

$$\text{Weir length required} = \frac{28{,}375\ \text{m}^3/\text{day}}{250\ \text{m}^3/\text{m·d}} = 113.5\ \text{m}$$

Provide a total of 8 effluent launder troughs with a center effluent collection channel. Each trough is 8.5 m long (Figure 9-17).

Provide the first effluent launder troughs (nearest the basin inlet) with weir plates on one side; the remaining three troughs will have weirs on both sides. Actual weir length = 2 troughs ∞ 8.5 m ∞ 1 side per trough + 6 troughs ∞ 8.5 m ∞ 2 sides per trough = 119 m > 113.5 m (required length).

ii. Calculate total number of V-notches.

Use standard 90° V-notch weirs. Provide 9 notches per 2-m-long weir plate and 4 plates on each side of each 8.5-m long trough. The notches shall be 7.5 cm deep and spaced at 20 cm from center to center. Total number of notches in each basin = 2 troughs ∞ 1 side per trough ∞ 4 plates per side ∞ 9 notches per plate + 6 troughs ∞ 2 side per trough ∞ 4 plates per side ∞ 9 notches per plate = 504 notches. (See Figure 9-17(b)).

iii. Calculate head over the V-notch weirs.

The flow per notch is calculated as follows:

$$q = \frac{0.328\ \text{m}^3/\text{s}}{504\ \text{notch}} = 6.52 \times 10^{-4}/\text{m}^3/\text{s per notch}$$

The head over each notch can be calculated from the V-notch equation (Eq. (9.16)).[12]

$$q = \frac{8}{15} C_d (2g)^{1/2} \tan\frac{\theta}{2} H^{5/2} \tag{9.16}$$

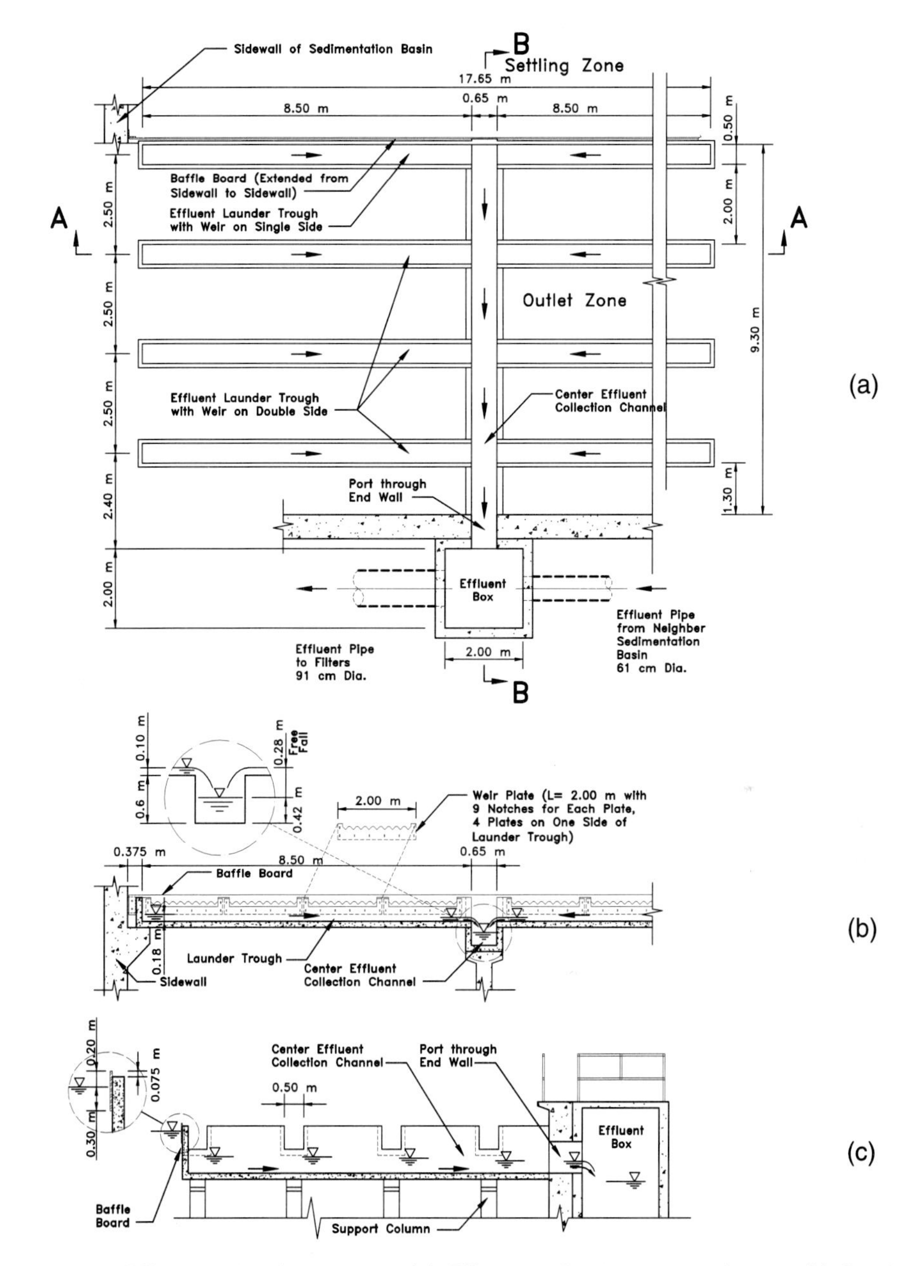

Figure 9-17 Effluent collection system. (a) Effluent collection system layout. (b) Section B-B. (Showing effluent launder trough with weir.) (c) Section A-A. (Showing center effluent collection channel.)

where

q = discharge per V-notch, m^3/s

C_d = coefficient of discharge = 0.6

H = head over notch, m

θ = angle of V-notch = 90°

Therefore,

$$6.52 \times 10^{-4}\ m^3/s = \frac{8}{15} \times 0.6 \times (2 \times 9.81\ m/s^2)^{1/2} \times \tan\left(\frac{90^\circ}{2}\right)(H)^{5/2}$$

Solving for H yields the following:

$H = 0.046$ m at maximum design flow (see Figure 9-18(a)).

This value is acceptable.[a] Weir freeboard is 7.5 cm − 4.6 cm = 2.9 cm. (See Figure 9-18(b).)

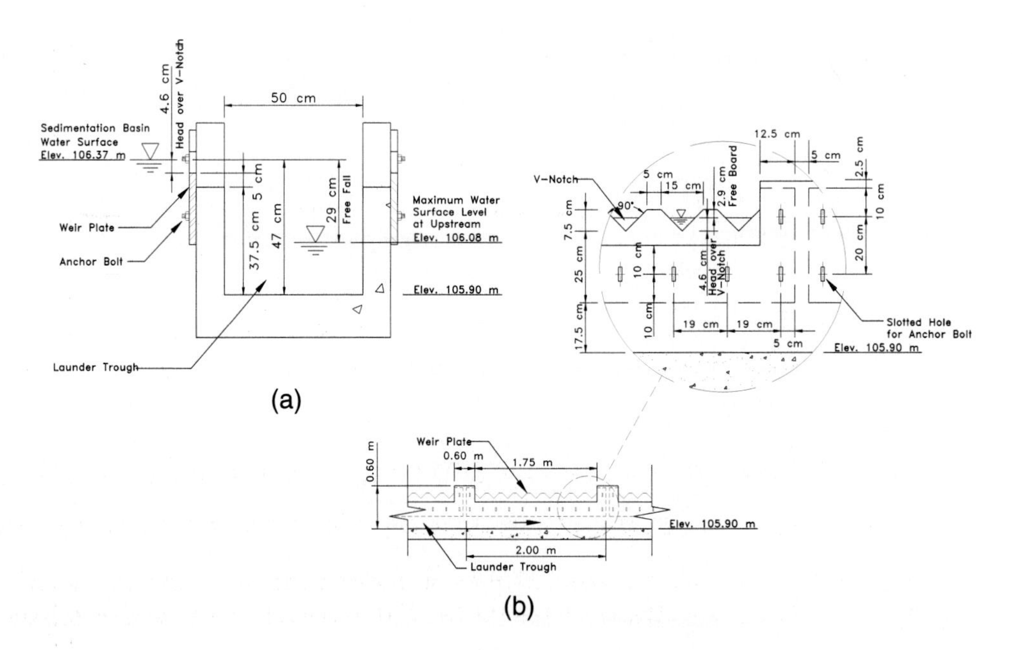

Figure 9-18 Launder trough and V-notch weir plate design. (a) Launder trough, detail. (b) Weir plate, detail.

a. Many designers use alternative procedures by selecting the head over the V-notch. Generally, the head over the V-notch is 3-8 cm. If a higher head over the weir notch is allowed, the discharge per notch will be high, and floc may be lifted up and lost in the effluent. From the selected head, the discharge per notch is calculated. Total number of notches is obtained from total flow divided by the discharge per notch.

b. Head loss through the effluent launder trough.

The head loss calculation is based on the flow in a launder trough with V-notch weirs plates on both sides.

i. Calculate flow in the launder trough.

The flow rate at the exit point of each effluent launder trough into the effluent launder is calculated as follows:

$Q' = q \infty 2$ sides per trough ∞ 4 plates per side ∞ 9 notches per plate

$= 6.52 \infty 10^{-4}$ m^3/s per notch ∞ 72 notches per trough

$= 0.0469$ m^3/s

ii. Calculate water depth at the downstream end of the launder trough.

Provide free fall at the exit point of the trough. Assume that the critical depth is reached at the exit point and is calculated from Eq. (9.17).[12]

$$y_c = \left(\frac{Q'^2}{gb^2}\right)^{1/3} \tag{9.17}$$

where

y_c = critical depth, m

Q' = flow rate per launder, m^3/s

b = launder width

Therefore,

$$y_c = \left(\frac{(0.047\ \text{m}^3/\text{s})^2}{9.81\ \text{m/s}^2 \times (0.5\ \text{m})^2}\right)^{1/3} = 0.096\ \text{m}$$

iii. Calculate the water depth at the upstream end of the launder trough.

Water depth at the upstream end of the trough (y_1) is calculated from Eq. (9.15). Assume the average depth in the effluent launder trough (d) = 0.14 m. The average hydraulic radius r_{avg} is calculated as follows:

$$r_{avg} = \frac{0.14\ \text{m} \times 0.5\ \text{m}}{2 \times 0.14\ \text{m} + 0.5\ \text{m}} = 0.090\ \text{m}$$

Assume Darcy friction factor $f = 0.10$ and $y_2 = y_c = 0.096$ m, from Eq. (9.16):

$$y_1 = \left((0.096\ \text{m})^2 + \frac{2 \times (0.0469\ \text{m}^3/\text{s})^2}{9.81\ \text{m/s}^2 \times (0.5\ \text{m})^2 \times 0.096\ \text{m}}\right.$$

$$+\frac{0.10\times 8.5\text{ m}\times(0.0469\text{ m}^3/\text{s})^2}{12\times 9.81\text{ m/s}^2\times(0.5\text{m})^2\times 0.090\text{ m}\times(0.14\text{ m})}\Bigg)^{1/2}$$

$$=(0.0092\text{ m}^2+0.0187\text{ m}^2+0.0050\text{ m}^2)^{1/2}$$

$$=0.18\text{ m}\quad(\text{See Figure 9-18(b)})^{\text{b}}$$

Check assumptions:

$$d=\frac{0.18\text{ m}+0.098\text{ m}}{2}=0.14\text{ m}$$

The head loss through the trough is therefore $y_1 - y_2 = 0.18\text{ m} - 0.098\text{ m} = 0.082\text{ m} \approx 0.08\text{ m}$.

Provide depth of effluent launder trough equal to 37.5 cm, and arrange V-notch 5 cm above the top of the effluent launder trough. The water surface in the sedimentation basin is 37.5 cm + 5 cm + 4.6 cm = 47.1 cm ≈ 0.47 m above the bottom of the launder trough. (See Figure 9-18(a).)

iv. Calculate free fall at the V-notch weirs.

The free fall at the V-notch weirs = water surface in sedimentation basin above the bottom of the launder trough – the upstream water depth in the laundertrough = 0.47 m – 0.18 m = 0.29 m. (See Figure 9-19(a).)

It is essential to provide adequate free fall to ensure that the flow is distributed equally to each weir. If the free fall is not sufficient, the water surface in the launder trough will rise and partially submerge the V-notch, and true weir flow will not occur. As a result, the flow along the weir plate will be unevenly distributed. Also, the free fall aerates the water and helps strip undesirable gases. The amount of free fall provided is based on the engineer's judgement and on the level of comfort with the available information. A larger free fall adds a safety factor to account for any unexpected hydraulic constraints and offers additional flexibility to increase the capacity of the basin in the future without requiring extensive improvements. On the other hand, an excessive free fall wastes hydraulic energy and costs the client needless energy for pumping. These conflicting considerations must be weighed when one is selecting appropriate freeboard and free-fall values and an overall design of the effluent structure.

b. This depth can also be obtained by using the simplified equation (Eq.(8.30) and adding an 8% allowance for friction losses.

c. Head loss through the center effluent collection channel.

The center effluent collection channel is divided into four segments, as shown in Figure 9-19. The head losses in each segment are calculated respectively. The total head loss in the channel is the sum of the losses in each segment.

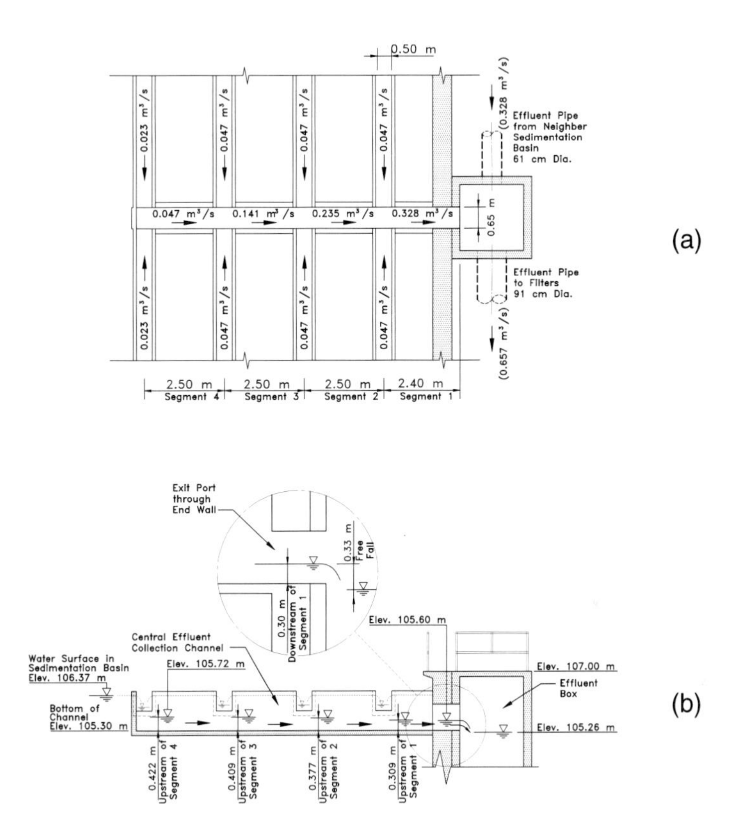

Figure 9-19 Flow and hydraulic profile through the center effluent collection channel. (a) Plan view. (b) Hydraulic Profile.

i. Calculate flow at each segment of the channel.

The center effluent collection channel receives flow from eight launder troughs and conveys the flow to the basin outlet. The flow rate at each segment of the center effluent collection channel is illustrated in Figure 9-19(a).

ii. Calculate the water depth at the exit point of the channel.

Provide free fall at the exit point of the channel. Assume that the critical depth is reached at the exit point. From Eq. (9.17), the water depth at the exit point $y_c = 0.296 \text{ m} \approx 0.30 \text{ m}$ (Figure 9-19(b)).

iii. Calculate the head losses in each segment of the channel.

The head losses in each segment include the friction loss due to channel roughness and an additional loss at each junction due to the turbulence induced by water flowing into the effluent channel from the launder troughs on both sides.

The friction losses are obtained from Manning's equation (Eq. (7.1)). Trial-and-error procedure is be utilized to solve the equation, because the water depth decreases along the flow direction. The head loss calculation begins from Segment 1, because the downstream water depth is known ($H_2 = y_c =$ 0.296 m). An initial upstream water depth in this segment is assumed first. The average water depth (H_{ave}), hydraulic mean radius (r_{ave}), and velocity (v_{ave}) are then calculated, respectively. Various terms in Eq. (7.1) are rearranged to solve for head loss ($h_f = L \infty S$). The upstream water depth (H_1) is then adjusted on the basis of this head loss. Repeat the above calculations; the water depth is calibrated each time. The final upstream water depth is determined when an acceptable difference between two consecutive depths is achieved.

Additional loss (h_L) at each junction is estimated from Eq. (7-39). In this calculation, v_2 and v_1 are velocity after and before the junction, respectively. The head loss constant K is assumed to be 1.

The results from the head loss calculations for each segment are summarized in Table 9-9. The water depth at the exit of the channel is 0.296 m ≈ 0.30 m. The water depth upstream of Segment 4, before the junction, is 0.423 m ≈ 0.42 m. The total head loss through the center effluent collection channel is, therefore, 0.42 m – 0.10 m = 0.12 m. The changes in water surface through the center effluent collection channel are shown in Figure 9-19(b).

iv. Calculate the free fall at the exit point of the launder trough.

Arrange the bottom of the center effluent collection channel 0.6 m (> 0.42 m) below the bottom of the effluent launder trough, to ensure the free fall condition at the exit point of the trough. (See Figure 9-17(b).)

The free fall at the exit point of the launder trough

= critical water depth at the exit of the trough + channel depth below the bottom of the trough – the upstream water depth in the channel

= 0.10 m + 0.6 m – 0.42 m = 0.28 m

v. Calculate the free fall at the exit of the center effluent collection channel.

A free fall of 0.34 m is provided at the exit point of the channel into Effluent Box I. (See Figures 9-14 and 9-19(b).)

Table 9-9 Calculation of Head Losses in the Center Effluent Collection Channel

Segment	Flow at Downstream (m^3/s)	Water Depth (m)			Average Hydraulic Mean Radius r_{ave} (m)	Length of Segment L (m)	Average Velocity v_{ave} (m/s)	Friction Head Loss through Channel h_f (m)	Velocity at Upstream Junction		Additional Head Loss at Junction h_L (m)	Subtotal Head Loss $h_{subtotal}$ (m)
		Downstream H_2	Upstream H_1	Average H_{ave}					After Junction v_2	Before Junction v_1		
1	0.328[a]	0.296[b]	0.309[c]	0.303	0.157	2.4[a]	1.67	0.013	1.63[d]	1.17[e]	0.066	0.079[f]
2	0.235	0.375[g]	0.379	0.377	0.175	2.5	0.96	0.004	0.95	0.57	0.029	0.033
3	0.141	0.408	0.409	0.409	0.181	2.5	0.53	0.001	0.53	0.18	0.013	0.014
4	0.047	0.422	0.422	0.422	0.184	2.5	0.17	0.000	0.17	0	0.001	0.001

a See Figure 9-20(a) for flow distribution in the channel.

b The water depth at the exit point in Segment 1, $H_2 = y_c = 0.296$ m (see Figure 9-20(b)).

c The upstream water depth is assumed initially. After performing a trial-and-error procedure, the final value from the last iteration is: $H_1 = H_2 + h_f = 0.296$ m + 0.013 m = 0.309 m.

d $v_2 = \dfrac{0.328 \text{ m}^3/\text{s}}{0.65 \text{ m} \times 0.309 \text{ m}} = 1.63$ m/s.

e $v_1 = \dfrac{0.235 \text{ m}^3/\text{s}}{0.65 \text{ m} \times 0.309 \text{ m}} = 1.17$ m/s.

f $h_{subtotal} = h_f + h_L$=0.013 m + 0.066 m = 0.079 m.

g The downstream water depth in Segment 2 is: $H_2 = H_2 + h_{subtotal}$ = 0.296 m + 0.079 m = 0.375 m.

d. Total head losses through the effluent structure.

• Free fall provided at the V-notch weirs	0.29 m
• Head loss through the effluent launder trough	0.08 m
• Free fall provided at the exit of the launder trough	0.28 m
• Head loss through the center effluent collection channel	0.12 m
• Free fall provided at the exit of the center effluent collection channel	0.33 m
Total =	1.10 m

e. Prepare a hydraulic profile through the sedimentation basin.

The water surface elevations in the effluent box of the sedimentation basin are controlled by the water level in the filter units and the head losses in the piping between the effluent box and the filter complex. The water surface elevation in Effluent Box I (Figure 9-16) is 105.26 m. This value may be found in Chapter 10, where the hydraulic profile through the filter system is developed. (See Section 10.13.2, Step A.5, and Figure 10-19.) Starting from Effluent Box I, the water surface elevation (WSEL) and major design elevations are calculated, and the hydraulic profile through the sedimentation basin is prepared. The results are shown in Figure 9-20.

The downstream WSEL in the center effluent collection channel = WSEL in the Effluent Box I + free fall provided at the exit of the channel = 105.26 m + 0.34 m = 105.60 m.

The upstream WSEL in the center effluent collection channel = downstream WSEL in the channel + head losses through the channel = 105.60 m + 0.12 m = 105.72 m.

The bottom elevation of the center effluent collection channel = the downstream WSEL in the channel - the downstream water depth in the channel = 105.60 m – 0.30 m = 105.30 m.

The downstream WSEL in the effluent launder trough = the upstream WSEL in the center effluent collection channel + free fall provided at the exit of the trough = 105.72 m + 0.28 m = 106.00 m.

The upstream WSEL in the effluent launder trough = the downstream WSEL in the trough + head losses through the trough = 106.00 m + 0.08 m = 106.08 m.

The bottom elevation of the effluent launder trough = the downstream WSEL in the trough – the downstream water depth in the trough = 106.00 m – 0.10 m = 105.90 m.

The WSEL in the sedimentation basin = the upstream WSEL in the effluent launder trough + free fall provided at the V-notch weirs = 106.08 m + 0.29 m = 106.37 m.

Step B: Sludge Withdraw System

1. Sludge quantity

   The residuals from the sedimentation basin consist basically of the solids carried by the raw water and the precipitates formed by adding the chemicals. The quantity of sludge is therefore estimated by totalling (1) the suspended solids and other constituents removed, (2) the amount of metal hydroxide, (3) the calcium carbonate produced, (4) the amount of polymer added, and (5) the powered activated carbon added. The sludge quantity is calculated from the optimum chemical dosages and constituents removed under both the maximum day and average day design flow conditions.

   a. Summary of the chemical dosages and raw water constituents producing sludge solids.

      The following design parameters are adapted from Section 8.8.1. These values are then used to calculate the residuals from the sedimentation process.

Design maximum day flow	113,500 m^3/d
Design average day flow	57,900 m^3/d
Maximum turbidity	17 NTU
Maximum seasonal iron concentration	0.7 mg/L
Maximum seasonal manganese concentration	0.4 mg/L
Optimum coagulant, ferric sulfate [$Fe_2(SO_4)_3$] dosage	25 mg/L
Optimum coagulant aid, cationic polymer	0.05 mg/L
Hydrated lime [$Ca(OH)_2$] for pH adjustment	15 mg/L
Seasonal potassium permanganate ($KMnO_4$) dosage	4 mg/L
Seasonal powdered activated carbon (PAC), for taste and odor control	5 mg/L

   b. Production of solids under the maximum day flow conditions.

      i. Calculate the solids from raw water (assuming a ratio of 1.0 mg-TSS/L/NTU).

$$q_{raw\ solids,\ max} = 17\text{ NTU} \times 1.0\text{ mg-TSS/NTU} \times 10^{-6}\text{ kg/mg}$$

$$\times 10^3\text{L/m}^3 \times 113{,}500\text{ m}^3\text{/d} = 1{,}930\text{ kg/d}$$

      ii. Calculate the solids due to precipitation of iron content.

$$q_{iron,\ max} = \frac{106.9\text{ g/mole Fe(OH)}_3}{55.9\text{ g/mole Fe}} \times 0.7\text{ mg-Fe/L} \times 10^{-6}\text{ kg/mg}$$

$$\times 10^3\text{ L/m}^3 \times 113{,}500\text{ m}^3\text{/d} = 152\text{ kg/d}$$

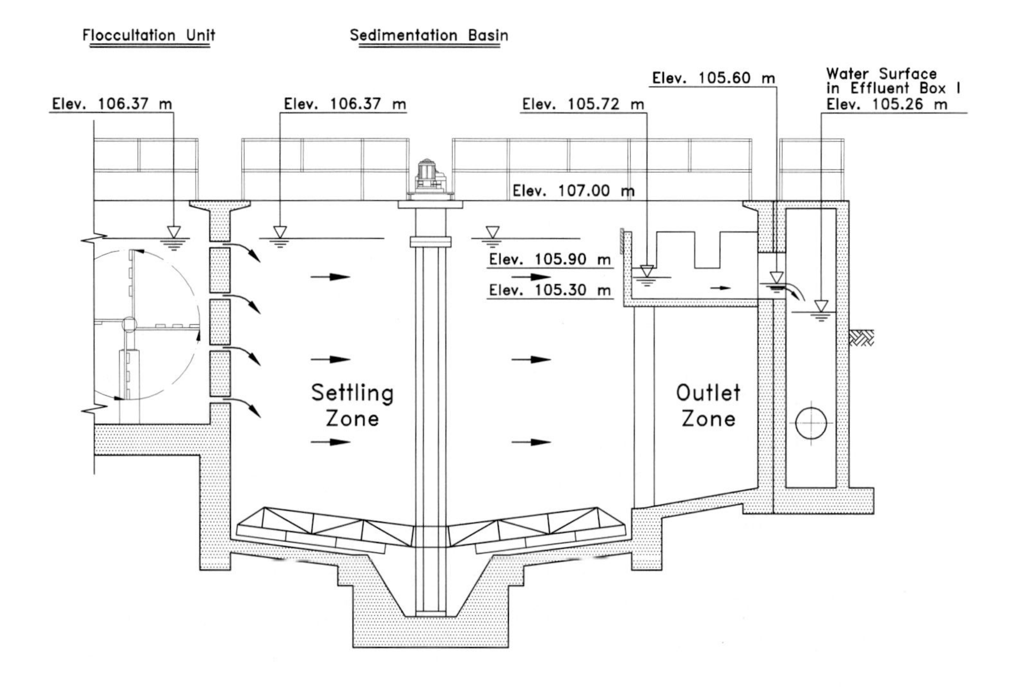

Figure 9-20 Hydraulic profile through the sedimentation basin.

iii. Calculate the solids due to precipitation of manganese content.

$$q_{manganese,\ max} = \frac{87.9 \text{ g/mole MnOOH}}{54.9 \text{ g/mole Mn}} \times 0.4 \text{ mg-Mn/L} \times 10^{-6} \text{ kg/mg}$$

$$\times 10^{3} \text{L/m}^{3} \times 113{,}500 \text{ m}^{3}\text{/d} = 73 \text{ kg/d}$$

iv. Calculate the solids due to precipitation of ferric sulfate. (See Table 9-7.)

$$q_{coagulant,\ max} = 0.54 \text{ kg-Fe(OH)}_3\text{/kg-Fe}_2\text{(SO}_4\text{)}_3 \times 25 \text{ g/m}^3 \text{ Fe}_2\text{(SO}_4\text{)}_3$$

$$\times 10^{-3} \text{ kg/g} \times 113{,}500 \text{ m}^{3}\text{/d} = 1532 \text{ kg/d}$$

v. Calculate the solids from the polymer.

$$q_{polymer,\ max} = 0.05 \text{ mg-polymer/L} \times 10^{-6} \text{ kg/mg} \times 10^{3} \text{ L/m}^{3}$$

$$\times 113{,}500 \text{ m}^{3}\text{/d} = 5.7 \text{ kg/d}$$

vi. Calculate the lime solids during pH adjustment. (Assume 20% of $Ca(OH)_2$ precipitates as $CaCO_3$.)

$$q_{lime,\ max} = 0.2 \times \frac{100 \text{ g/mole CaCO}_3}{74 \text{ g/mole Ca(OH)}_2} \times 15 \text{ mg-Ca(OH)}_2\text{/L} \times 10^{-6} \text{ kg/mg}$$

$$\times 10^3 \text{ L/m}^3 \times 113{,}500 \text{ m}^3\text{/d} = 460 \text{ kg/d}$$

vii. Calculate the solids due to precipitation of potassium permanganate.

$$q_{permanganate,\ max} = \frac{87.9 \text{ g/mole MnOOH}}{158 \text{ g/mole KMnO}_4} \times 4 \text{ mg-KMnO}_4\text{/L} \times 10^{-6} \text{ kg/mg}$$

$$\times 10^3 \text{ L/m}^3 \times 113{,}500 \text{ m}^3\text{/d} = 253 \text{ kg/d}$$

viii. Calculate the solids from PAC.

$$q_{PAC,\ max} = 5 \text{ mg-PAC/L} \times 10^{-6} \text{ kg/mg} \times 10^3 \text{ L/m}^3$$

$$\times 113{,}500 \text{ m}^3\text{/d} = 568 \text{ kg/d}$$

ix. Calculate the total solids produced.

$$q_{produced,\ max} = 1930 \text{ kg/d} + 152 \text{ kg/d} + 73 \text{ kg/d} + 1532 \text{ kg/d} + 5.7 \text{ kg/d}$$

$$+ 460 \text{ kg/d} + 253 \text{ kg/d} + 568 \text{ kg/d} = 4974 \text{ kg/d}$$

Because of many uncertainties associated with raw water quality, chemical dosages, and the amount of bound water, it is a common practice to increase this value by 20–60 percent. By increasing this quantity by 53 percent, total sludge solids and production is assumed to be 7600 kg/d.

c. Quantity of residuals removed under the maximum day flow conditions.

It is assumed that 90% of solids produced are removed from the sedimentation basins.

$$q_{removed,\ max} = 7600 \text{ kg/d} \times 0.90 = 6840 \text{ kg/d}$$

d. Specific gravity of wet sludge.

The specific gravity of the wet sludge is calculated from Eq. (9.18).

$$M_{ws}/S_{g,ws} = M_s/S_{g,s} + M_w/S_{g,w} \tag{9.18}$$

where

M_{ws} = mass of wet sludge, kg

$S_{g,ws}$ = specific gravity of wet sludge

M_s = mass of dry sludge solids, kg

$S_{g,s}$ = specific gravity of sludge solids

M_w = mass of water, kg

$S_{g,w}$ = specific gravity of water

The solids content of coagulation residuals range from 0.1 to 4 percent by weight (w/w). An average underflow solids concentration of 2 percent in sludge withdrawn is expected from the sedimentation basin provided in this Design Example. Assume the specific gravity of dry solids and water is 2.4 and 1.0, respectively. For one kg of wet sludge mass with 2 percent solids, the following result is obtained (Eq. (9.18)):

$$1\ \text{kg}/S_{g,\,ws} = 0.02\ \text{kg}/2.4 + 0.98\ \text{kg}/1.0$$

Solve for $S_{g,ws}$:

$$S_{g,\,ws} = 1.012$$

e. Volume of sludge withdrawn under the maximum day flow conditions.

For the sludge with 2 percent solids and a specific gravity of 1.012 at a production rate of 6840 kg/d, calculate the volume of wet residuals per day:

$$V_{ws,\,max} = \frac{6840\ \text{kg/d}}{0.02\ \text{g/g} \times 1012\ \text{kg/m}^3} = 338\ \text{m}^3/\text{d}$$

Calculate the quantity of sludge produced per 1,000 m^3 of raw water treated:

$$(M/Q)_{max} = \frac{6840\ \text{kg/d} \times 1000\ \text{m}^3}{113{,}500\ \text{m}^3/\text{d}} = 60.3\ \text{kg}/1000\ \text{m}^3$$

It has been reported in the literature that total solids from an alum coagulation facility may be 8–210 kg per 1000 m^3 of raw water treated (67–1800 lbs per million gallons).[6]

f. Production of solids under the average day flow conditions.

The solids production at average day design flow is estimated based on the ratio of two flow rates.

$$q_{produced,\,ave} = 5938\ \text{kg/d} \times \frac{57{,}900\ \text{m}^3/\text{d}}{113{,}500\ \text{m}^3/\text{d}} = 3029\ \text{kg/d}$$

The uncertainties associated with these calculations suggest that a higher quantity, 3860 kg/d, be used for the average day flow condition.

g. Quantity of residuals removed under the average day flow conditions.

It is assumed that same solids removal rate of 90% is achieved. Calculate the solids removed from the sedimentation facility:

$$q_{removed,\,max} = 3860\ \text{kg/d} \times 0.90 = 3474\ \text{kg/d}$$

h. Volume of sludge withdrawn under the average day flow conditions.

For the sludge with 2 percent solids and a specific gravity of 1.012 at a production rate of 3474 kg/d, calculate the volume of wet residuals per day:

$$v_{ws,\,ave} = \frac{3474 \text{ kg/d}}{0.02 \text{ g/g} \times 1012 \text{ kg/m}^3} = 172 \text{ m}^3\text{/d}$$

2. Sludge withdrawal

   The sludge is collected by two circular sludge collector mechanisms into two sludge pockets in each sedimentation basin and then removed from each pocket through a sludge pipeline that extends from the sludge pocket to a wet well of the sludge pump station located at the basin perimeter (Figure 9-15). The sludge withdrawal rate from each pocket is regulated by the use of a telescoping valve. A typical telescoping valve is illustrated in Figure 9-21. By adjusting the valve tube up or down, the differential head between the sludge pump station and the sedimentation basin is decreased or increased, and thus the flow rate is decreased or increased. A limit switch is utilized to monitor the position of the slip tube of the valve. The size of the valve is equal to that of sludge pipeline. The valve can be operated automatically, by timer, or manually, by operator, to release the sludge on a regular time interval or for a predetermined time duration. The application of telescoping valves is gaining popularity in water treatment plants.

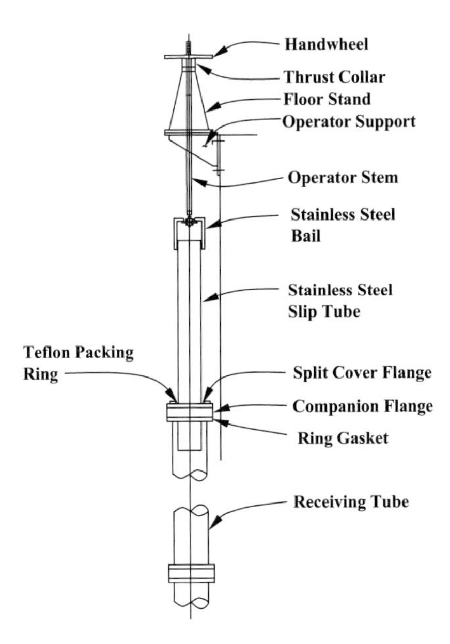

Figure 9-21 Telescoping valve.

a. Sludge withdraw system layout.

The sludge withdraw pipeline and pump station are schematically illustrated in Figure 9-22.

b. Sludge pipeline design.

The size of the sludge pipeline is determined basically by the velocity in the pipeline. The velocity must be high enough to prevent solids deposition. The velocities are typically maintained above 0.6 m/s (2 ft/s). The sludge pipeline is also sized as large as possible, to prevent clogging and to facilitate cleaning. Regardless of the velocity, a minimum pipe diameter of 20 cm (8 in) is therefore typically required. A cleanout is also provided for periodic cleaning of the sludge pipeline.

Under the maximum day flow conditions, the total sludge flow from four sedimentation basins over a 24-hour period is 338 m^3/d. There are a total of eight sludge withdraw pipelines. The flow through each pipeline is therefore 42.3 m^3/d. At a velocity of 0.6 m/s, the required diameter of sludge pipeline is 0.032 m (1∫ in).

A sludge pipe of this size will be too small. The minimum size of 20 cm is recommended. To achieve a velocity of 0.6 m/s in a 20 cm diameter pipe, calculate the required flow rate:

$$Q = 0.6 \text{ m/s} \times \frac{\pi}{4} \times (0.20)^2 = 0.019 \text{ m}^3/\text{s} = 1629 \text{ m}^3/\text{d}$$

Under the maximum day flow conditions, calculate the required opening time period for each valve:

$$T_{max} = \frac{42.3 \text{ m}^3/\text{d}}{1629 \text{ m}^3/\text{d}} = 2.6\% \text{ of the time, or 37 min per day}$$

Under the average day flow conditions, calculate the required opening time period for each valve:

$$T_{avg} = \frac{172 \text{ m}^3/\text{d}}{338 \text{ m}^3/\text{d}} \times 2.6\% = 1.3\% \text{ of the time, or 19 min per day}$$

To maintain the required minimum velocity of 0.6 m/s through the pipeline, the control system must therefore be capable of opening each telescoping valve for some minutes per day. An optimum sludge withdraw schedule shall be developed by considering such factors as (1) the capacity of the sludge pocket of the sedimentation basin, (2) the storage capacity of the wet well, and (3) the limits on the number of start/stop operations per hour for both pumps and telescoping valves.

c. Sludge pump station design.

There is one pump station for each sedimentation basin. Each pump station has two pumps including a backup pump in case of equipment failure. As in sludge pipeline

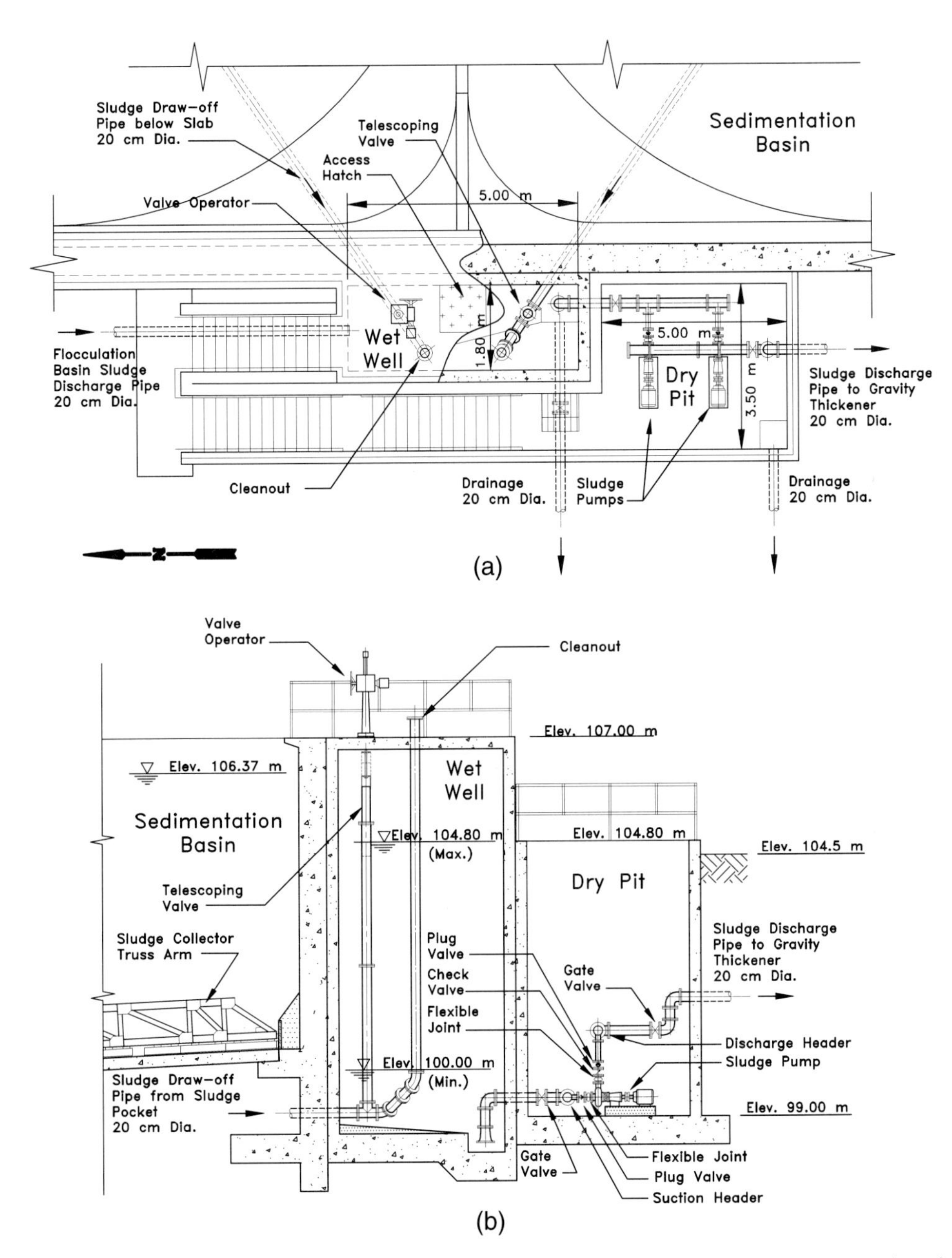

Figure 9-22 Sludge pump station design. (a) Pump station layout. (b) Schematic section view.

design, the required capacity of the sludge pump is also determined basically by the velocity in the pump discharge force main. A minimum pump capacity of 1629 m^3/d is required to achieve a velocity of 0.6 m/s in a force main with the minimum diameter of 20 cm. A pump capacity of 1630 m^3/d is, therefore, used to select the pump for the sludge withdrawal system. The pumps are started and stopped automatically in response to the outputs of the level sensors in the sludge wet well. The lead pump is alternated at each pump station to ensure equal wear on both pumps.

9.7 Operation, Maintenance, and Troubleshooting for Sedimentation Facilities

Sedimentation facilities can remove a large amount of raw water turbidity. Total suspended solids removal efficiencies of 80 to 95 percent are generally achievable.[14,27] Efficient removal in the sedimentation basin will result in better filter performance and longer filter runs. Performance of a sedimentation basin, however, is dependent upon the success of the coagulation and flocculation processes. Minimizing short-circuiting will also enhance solids removal.

9.7.1 Common Operational Problems and Troubleshooting

Operational problems associated with sedimentation basins typically relate to ineffective sludge removal or short-circuiting. Ineffective sludge removal commonly is associated with equipment problems or inadequate sludge removal practices. Short-circuiting is typically the result of improper inlet or outlet design. Short-circuiting can also be the result of wave action, density currents, or temperature currents. Common operational and maintenance problems and troubleshooting guides are the following:

1. Operational problems with sludge collection equipment may include the shear-pins or motor overloads, both generally due to improper sludge removal. Remove sludge more rapidly. Check for proper shear-pin installation and motor overload setting. Also, check for large debris in the basin.
2. Sludge withdrawal with low solids concentrations may result from an excessively rapid removal rate or from a sludge collection mechanism that is not operating properly. Decrease the removal rate, and check the operation of sludge collection equipment.
3. Clogged sludge withdrawal piping can be the result of insufficient sludge withdrawal. Increase the removal rate.
4. High effluent turbidity or floc carryover may result from an improper coagulation process. See Chapter 8 for trouble shooting. High turbidity or floc carryover may also result from short-circuiting in the sedimentation basin. Perform tracer studies, and make corrections. Possible corrective measures include inlet and outlet baffles.
5. Algae build up on basin walls or weirs may create taste and odor problems. Clean basin walls on a regular basis and, if possible, maintain a disinfectant residual in the basin.

6. Sludge with a high organic content may impart taste and odor problems to the finished water. Increase the sludge removal rate.

9.7.2 Operation and Maintenance

The following preventive maintenance procedures are necessary for satisfactory operation of the sedimentation facility.

1. Clean basins annually to remove any accumulated sludge and algae growth.
2. Lubricate sludge collection equipment as recommended by the manufacturer.
3. Test the sludge collection overload devices annually.
4. Test the solids content in the sludge withdrawal line daily.
5. Test turbidity of effluent on a regular basis and whenever the water quality or flow rate changes.

9.8 Specifications

The specifications for the sedimentation units designed in this chapter are briefly presented below. The purpose of these specifications is to describe more fully many components of the sedimentation system that could not be properly illustrated in the design example. These specifications are not intended to be complete, nor are they intended to be a guideline for a complete specification. The designer should consult with equipment suppliers and develop detailed specifications with the aid of information provided by the manufacturers.

Four sedimentation basins shall be provided each measuring 18.4 m ∞ 45 m. The basin walls shall be 6.45 m deep with a side water depth of 5.8 m leaving a 0.65 m freeboard. The basin shall be separated from the flocculation basin by a perforated diffusion wall made of concrete. The wall shall have a total of 178 openings 12.5 cm in diameter (Section 8.8.2, Step B.5).

Each basin shall be equipped with two pier supported circular sludge collection mechanisms, each equipped with corner sweeps. The sludge collection mechanisms shall consist of a drive motor, gear reducer, rake arms, corner sweeps, access bridge, and motor control unit. The gear reducer unit shall be driven by a 480-volt, 3-phase electric motor and equipped with torque switches to shut down the equipment in case of overload. The gear reduction unit shall be designed for continuous operation. Gears shall be designed for 100,000 hours of operation. Bearings shall have a B 10 life of 100,000 hours.[c] All underwater bearings shall be bronze, grease lubricated. All above water bearings shall be antifriction ball or roller type, grease lubricated. The rake arm access bridge and center cage shall be constructed of a ASTM A 36 structural steel truss.[d] The minimum thickness of the truss members shall be 8 mm.

c. B 10 Life - number of hours of normal operation at which 10 percent of a particular bearing will fail.

d. ASTM - American Society of Testing Materials.

Fiberglass V-notch weir plates shall be provided for the outlet structure of each sedimentation basin. Each plate shall be 2 m long, with 9 V-notches per plate. Each notch shall be at 90 degrees.

All nonsubmerged ferrous material shall be sandblasted to a commercial blast finish; submerged ferrous material shall be sandblasted to white metal. Blasted surfaces shall be primed and coated with an epoxy paint suitable for potable water service. All field welds and other damaged areas shall be sandblasted after welding and repainted with compatible paints.

Problems and Discussion Topics

9.1 Visit a local water treatment plant. Determine the type and size of the sedimentation facilities. Prepare sketches of the facility layout. Ask the operator to explain any operational problems experienced with these units.

9.2 Calculate the discrete settling velocity of 0.05-mm-diameter particles with a specific gravity of 2.65 at 20°C ($\nu = 1.004 \infty 10^{-6}$ m²/s).

9.3 Calculate the discrete settling velocity of 0.3-mm-diameter particles with a specific gravity of 2.65 ($\nu = 1.1306 \infty 10^{-6}$ m²/s).

9.4 Design a sedimentation basin to treat a flow of 18,900 m³/d. Design for an overflow rate of 30 m³/m²·d, 4-hour detention time, and 250 m³/m·d weir-loading rate. Use a rectangular basin with a length-to-width ratio of 4-to-1.

9.5 A sedimentation channel is designed for a hydraulic loading of 1,630 m³/m²·d (40,000 gpd/ft²) to remove sandy suspension. A sieve analysis was performed on a sample of dried solids. The average settling velocity of each fraction in the column test was experimentally determined; the results of the sieve analysis and column test are given below. Draw the fraction removal curve, and calculate the theoretical removal efficiency of the channel.

Fraction retained on sieve	0.45	0.15	0.10	0.15	0.10	0.05
Average velocity from column test, m/min	3.05	1.53	0.61	0.31	0.23	0.15

9.6 A settling column analysis was run on a discrete suspension. Water samples were withdrawn at different time intervals from a port 2 m below the original water level. The total suspended solids (TSS) analysis was performed on each sample withdrawn. The results of the TSS analysis are provided below. Calculate the theoretical removal efficiency of the suspension in a sedimentation basin that has an overflow rate of 30 m³/m²·d.

Time, min	0	60	90	120	180	260	400
Concentration of TSS, mg/L	250	153	148	138	120	70	24

9.7 A 30-m-diameter sedimentation basin has a side water depth of 3.0 m. It is treating a raw water flow of 0.3 m³/s. Compute overflow rate and detention time.

9.8 A sedimentation basin has an overflow rate of 80 m³/m²·d. What fraction of the particles that have a velocity of 0.02 m/min will be removed in this tank?

9.9 Design a circular clarifier to treat coagulated water having a flow of 0.22 m³/s. The design overflow rate is 36 m³/m²·d. The effluent launder is 0.5 m wide and is installed around the circumference of the basin, 1 m away from the concrete wall. Weir notches are provided on both sides of the launder. The effluent box is 1 m ∞ 1 m, and the depth of flow in the effluent box at peak design flow is 1 m. The invert of the effluent launder is 0.46 m above the invert of the effluent box. Also, calculate weir loading and the depth of the effluent launder. Provide 15 percent loss for friction and turbulence and 15 cm to allow for free fall.

9.10 A weir trough is 10 m long and 1 m wide. The weir crest is one side of the trough and covers the entire length of the trough. Calculate the depth of the trough if discharge through the basin is 0.3 m^3/s. Depth of flow at the lower end of the trough is 0.9 m. Assume friction loss is 15 percent of the depth of water at the upper end and free-fall allowance is 6 cm.

9.11 Determine the total head loss through a 4-m-wide sedimentation chamber. The details of the chamber are as follows:

(a) The influent channel is 1.5 m wide and has one submerged orifice 1.5 m ∞ 1.5 m. The invert of the influent channel is 0.5 m above the floor of the chamber. The depth of floor in the chamber is 3.0 m.

(b) The head loss in the chamber is small and can be ignored.

(c) The influent and effluent baffles occupy (respectively) 65 and 60 percent of the cross-sectional area of the chamber, and head losses due to obstruction are 0.001 m at each.

(d) The flow through the chamber is 1.6 m^3/s.

(e) There is an 0.4 m head loss into the effluent structure. This head loss is the difference between the water surface elevations in the chamber at the effluent weir and in the outlet box of the chamber.

9.12 A sedimentation facility was designed to treat an average flow of 0.6 m^3/s. The design overflow rate and the detention time are 45 $m^3/m^2{\cdot}d$ and 2.5 h, respectively. The length-to-width ratio of the rectangular basin is 4.3-to-1. Calculate the dimensions of each basin if two, three, or four basins are provided. Also, compute the weir loading rate in each case, if one weir trough and single weir plate is provided along the width in each basin. The outlet channel is 1 m wide.

9.13 What are the major differences between the following three types of sedimentation basin: horizontal flow, solids-contact, and inclined surface? Write the advantages and major applications of each type.

9.14 A conventional water treatment plant is treating 100,000 m^3/d surface water. Maximum turbidity is 16 NTU and optimum Al^{3+} dosage is 3.8 mg/L. PAC dosage for taste and odor control is 5.2 mg/L. No lime is added. Calculate the quantity of solids produced per day. Assume that TSS produced is 0.7 times NTU and that the solids removal efficiency of the plant is 95 percent.

References

1. Reynolds, T. P. and Richards, P. A. *Unit Operations and Processes in Environmental Engineering*, 2d ed., PWS Publishing Co., Boston, MA, 1995.
2. Daughterly, R. L. and Franzini, J. B. *Fluid Mechanics with Engineering Applications*, 8th ed., McGraw-Hill Book Co., New York, 1985.
3. AWWA. *Water Quality and Treatment: A Handbook of Public Water Supply*, 4th ed., McGraw-Hill, Inc., New York, 1990.
4. Coulson, J. and Richardson, J. *Chemical Engineering*, vol. 2, 3d ed., Pergamon Press, Oxford, United Kingdom, 1978.
5. Schroeder, E. *Water and Wastewater Treatment*, McGraw-Hill Book Co., New York, 1971.
6. James M. Montgomery, Inc., *Water Treatment Principles and Design*, John Wiley & Sons, New York, 1985.
7. Metcalf and Eddy, Inc., *Wastewater Engineering: Treatment, Disposal and Reuse*, 3d ed., McGraw-Hill Inc., New York, 1991.
8. Hammer, M. J. and Hammer, M. J. Jr. *Water and Wastewater Technology*, 3d ed., Prentice-Hall, Inc., Englewood Cliffs, NJ, 1996.

9. Tchobanoglous, G. and Schroeder, E. *Water Quality: Characteristics, Modeling, Modification*, Addison-Wesley Publishing Co., Reading, MA, 1985.
10. Viessman, W. and Hammer, M. J. *Water Supply and Pollution Control*, 6th ed., Adison-Wesley Longman, Inc., New York, 1998.
11. Monk, R. and Willis, J. F. "Designing Water Treatment Facilities," *Jour. AWWA*, vol. 79, no. 2, pp. 45–57, February 1987.
12. Qasim, S. R. *Wastewater Treatment Plants: Planning, Design and Operation*, 2d ed., Technomic Publishing Co., Lancaster, PA, 1999.
13. Nalco Chemical Co. *The Nalco Water Handbook*, McGraw-Hill Book Co., New York, 1979.
14. Degremont. *Water Treatment Handbook*, vol. 1, 6th ed., Lavoisier Publishing, Rueid-Malmaison, France, 1991.
15. Infilco Degremont, Inc. *Accelator® Treating Plant*, DB 530/540, Infilco Degremont, Inc., Richmond, VA, 1976.
16. U.S. Filter/General Filter. *Spiracone, Sludge Blanket Clarifier*, Bulletin No. 9604-496-50, General Filters Products, Ames, IA, undated.
17. Infilco Degremont, Inc. *Superpulsator® Type V Clarifier*, DB586, Infilco Degremont, Inc., Richmond, VA, 1997.
18. Krüger, Inc. *Actiflo® Process*, Krüger, Inc., Cary, NC, undated.
19. Hudson, E. *Water Clarification Process: Practical Design and Evolution*, Van Nostrand Reinhold Co., New York, 1981.
20. Kawamura, S. "Hydraulic Scale-Model Simulation of the Sedimentation Process," *Jour. AWWA*, vol. 73, no. 7, pp. 372–379, July 1981.
21. Kawamura, S. *Integrated Design of Water Treatment Facilities*, John Wiley and Sons, Inc., New York, 1991.
22. Great Lakes Upper Mississippi River Board of State Public Health and Environmental Managers. *Recommended Standard for Water Works*, Health Education Service, Albany, NY, 1992.
23. Chow, V. *Open Channel Hydraulics*, McGraw-Hill Book Co., New York, 1959.
24. ASCE and AWWA, *Water Treatment Plant Design*, 2d ed., McGraw-Hill Publishing Company, 1990.
25. Bishop, M. M., Rolan, A. T., Bailey, T. L., and Cornwell, D. A. "Test of Alum Recovery for Solids Reduction and Reuse," *Jour. AWWA*, vol. 79, no. 6, pp. 76–83, June 1987.
26. Cornwell, D. A. and Westerhoff, G. P. *Management of Water Treatment Sludge, Chapter 3 in "Sludge and Its Ultimate Disposal"*, J. A. Brochardt et. al. (Editors), Ann Arbor Science Publishers Inc./ The Butterworth Group, Ann Arbor, MI, 1981.
27. Tillman, G. M. *Water Treatment Plant: Troubleshooting and Problem Solving*, Ann Arbor Press, Inc., Chelsea, MI, 1996.

CHAPTER 10

Filtration

10.1 Introduction

The coagulation, flocculation, and sedimentation processes remove much of the colloidal material that causes turbidity. Further removal of colloidal particles is required to meet more stringent public health standards promulgated after the 1986 Safe Drinking Water Act Amendments. The most common process used for further removal of colloidal matter is filtration.

Filtration is one of the oldest forms of water treatment. In fact, filtration is nature's own water treatment mechanism. Water flowing slowly through porous sand and rock formations within the earth is cleaned and purified. The filtration process used in water treatment involves flow through a bed of granular media, such as sand, anthracite, garnet, or activated carbon. As the water passes through the media, the suspended particles are entrapped in the pore spaces of the media and thus removed from the water stream.

In this chapter, the theory of the filtration process is discussed. Typical calculations used in developing the design of filtration systems are presented in the Design Example. Operation and maintenance recommendations, as well as specifications for the design and construction of filtration facilities, are also presented in the Design Example.

10.2 Theory of Filtration

The mechanisms by which granular filtration media remove solids from water are complex and are not fully understood. Common theories suggest a number of mechanisms that act simultaneously in the solids removal process. These mechanisms are (1) straining, (2) sedimentation, (3) impaction, and (4) interception. A brief discussion of each mechanism follows.[1]

Some colloidal particles are too large to pass through pore spaces in the filter media bed. These particles become trapped and are removed. This removal mechanism is called *straining*.

This mechanism plays an important role in the direct filtration process, where flocculated water is fed directly into the filters. Most colloidal particles in settled water after a sedimentation process are, however, small enough to pass through the pore spaces in the media bed; therefore, additional mechanisms must be acting in the removal of colloidal matter. Water flowing through the filter bed is usually laminar, with the velocity and direction constantly changing because of the obstruction by the media grains. In low-velocity zones, some particles are removed by *sedimentation*. Other particles have too large a mass to follow sharp turns in the flow streamlines. The inertia of these particles carries them out of the flow stream. These particles strike the medium and are held there because inertia is greater than the hydrodynamic force. Thus, the removal of particles is by *impaction*. In some cases, the flow streamlines pass very close to a media grain. At times, a particle following these stream lines will touch a media grain and become lodged. These particles are removed by *interception*.

The physical and chemical properties of the filter media and colloidal particles further contribute to the removal mechanisms. These physical and chemical properties include: chemical bonding between the media and the colloidal particles; physical attraction between the media grains and the colloidal particles; and physical or chemical attraction between the colloidal particles, resulting in flocculation of the individual colloids.[1] As the colloidal particles pass through the media, *orthokinetic flocculation* occurs. Fluid motion through the annular spaces between media grains increases the number of interparticle contacts. The resulting particles become larger than the original individual particles, so removal by one of the four basic removal mechanisms is enhanced.[1] Also, as the colloidal particles come in contact with the media, some force of attraction must hold them there, otherwise the shear force of the water flowing through the media bed will dislodge the trapped colloidal particles. The principal attraction forces are (1) *chemical bonding* and (2) *physical attraction* (because of *electrostatic* or *van der Waals forces*).

As filtration continues, colloidal particles become trapped in the media. The pore spaces between the media grains become smaller, causing (1) increased straining action, (2) increased velocity of the water through the pore spaces, (3) increased shear forces of the water flowing through the media bed, and (4) increased head loss of the water flowing through the media. The net result is a decrease in filtration rate, filter performance, and efficiency. At this time, the filter bed is considered clogged. The shear forces of the water flowing through the media can be so great that the particles are dislodged as rapidly as they are deposited, to cause an unacceptable water quality (known as turbidity breakthrough), or the head loss of the water flowing through the media bed can exceed the available head. Ideally, both of these conditions must be met simultaneously. In actual practice, however, one of these conditions is usually met before the other. The filter should be cleaned when either of these conditions (or the preset filtration cycle) is reached.

10.3 Types of Filters

Filter types commonly utilized in water treatment are classified on the basis of (1) filtration rate, (2) driving force, and (3) direction of flow. All filter types are discussed next.

10.3.1 Filters Classified by Filtration Rate

Filters can be classified as *slow sand filters*, *rapid filters*, or *high-rate filters*. Slow sand filters have a hydraulic application rate less than 10 m^3/m^2·day (0.17 gpm/ft^2).[2] This type of filter is utilized extensively in Europe, where natural sand beds along river banks are used as a filter medium. Slow sand filters are also used almost exclusively in developing countries. An underdrain system is installed under the sand bed to collect the filtered water. When the medium becomes clogged, the bed is dewatered, and the upper layer of the sand is removed, washed, and replaced. This type of filter often does not utilize chemical coagulation in the water purification process. Details of a slow sand filter are illustrated in Figure 10-1.

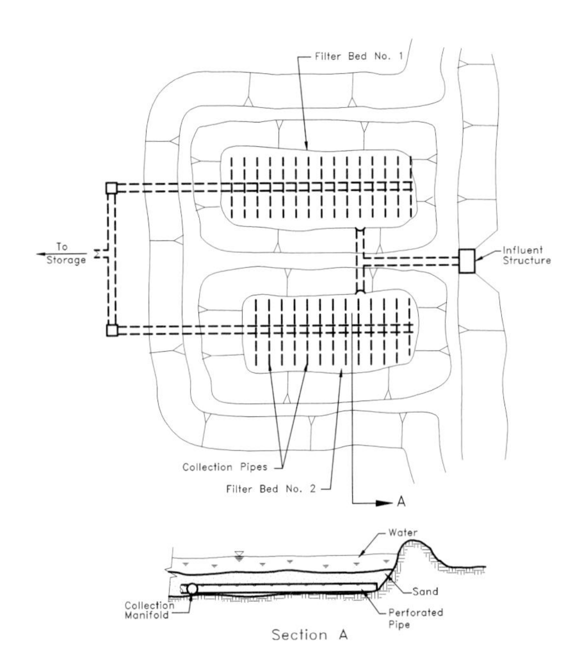

Figure 10-1 Slow sand filters.

Rapid sand filters have a hydraulic application rate of approximately 120 m^3/m^2·day (2 gpm/ft^2), and high-rate filters have a hydraulic application rate greater than 240 m^3/m^2·day (4 gpm/ft^2).[2,3] Both rapid and high-rate filters are used extensively in the United States. Construction of these systems is quite similar. As is illustrated in Figure 10-2(a), rapid and high-rate filters utilize concrete or steel basins filled with suitable filter media. The filter media are supported by a gravel bed and an underdrain system, which both collects the filtered water and distributes the backwash water used to clean the filter bed. Several types of proprietary filter underdrains are shown in Figure 10-3.

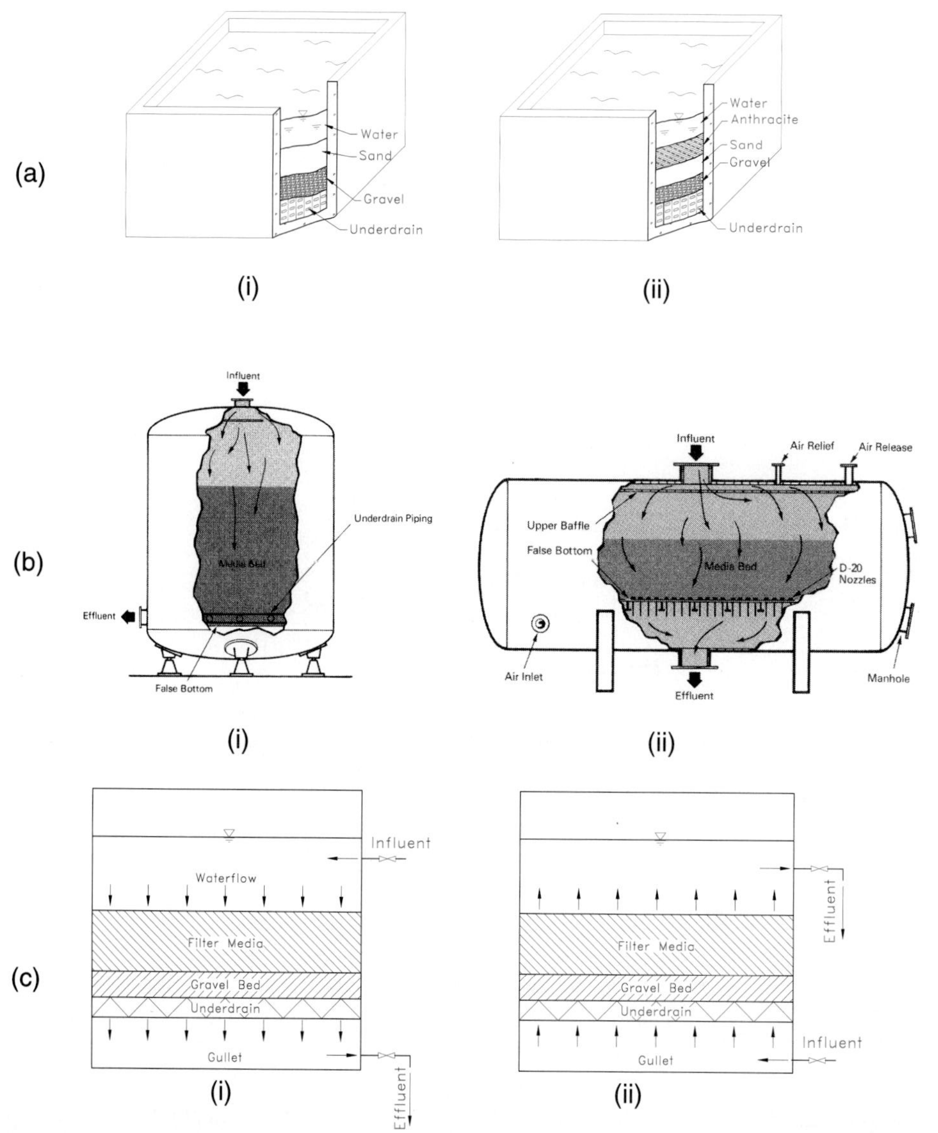

Figure 10-2 Filter types and components based on flow rate, driving force, and direction of flow. (a) Cross section of rapid and high-rate filters *(Reprinted from Reference 4, with permission from Technomic Publishing Co., Inc., copyright 1999):* (i) typical rapid sand filter, and (ii) typical high-rate dual-media filter. (b) Pressure filters *(Courtesy of Infilco Degremont, Inc.):* (i) vertical filter, and (ii) horizontal filter. (c) Direction of flow *(Reprinted from Reference 4, with permission from Technomic Publishing Co., Inc., copyright 1999):* (i) downflow filters, and (ii) upflow filter.

10.3.2 Filters Classified by Driving Force

Filters utilized in water treatment are also classified as gravity or pressure filters. The major differences between gravity and pressure filters are the head required to force the water through the media bed and the type of vessel used to contain the filter unit. Gravity filters usually require two to three meters of head and are housed in open concrete or steel tanks.[2] Pressure filters usually require a higher head and are contained in enclosed steel pressure vessels (Figure 10-2(b)). Because of the cost of constructing large pressure vessels, pressure filters typically are used only on small water purification plants; gravity filters are used on both large and small systems.

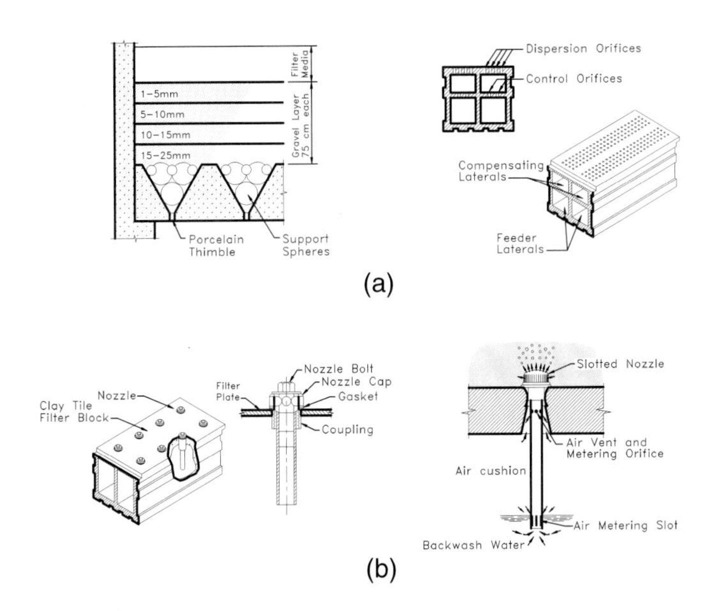

Figure 10-3 Various types of filter underdrain system. *(Reprinted from Reference 4, with permission from Technomic Publishing Co., Inc., copyright 1999.)* (a) With gravel support. (b) Without gravel support.

10.3.3 Filters Classified by Direction of Flow

Filter systems are classified as *downflow* or *upflow*. Downflow filters are the type most commonly used in water treatment. In this type of system, the flow through the media bed is downward, as is illustrated in Figure 10-2(c)(i). In upflow filtration, the water flows upward through the media bed, as is illustrated in Figure 10-2(c)(ii). This type of system is rarely used in granular filters, but it is sometimes used in granular activated carbon beds.

10.4 Filter Media

Selection of proper filter media is important in filter design as the media control (1) the solids-holding capacity of the filter bed, (2) the hydraulic loading rate of the filters, and (3) the finished water quality. Filters are classified as *single- (or mono-) medium*, *dual-media*, or *mixed- (or multi-) media*, depending on the number of different filter media used. In addition to the number of media materials used, other important media design considerations are media grain size, uniformity coefficient, and filter bed depth. The media design parameters are presented next.

10.4.1 Media Design Parameters

Filter media commonly are specified by *effective size* and *uniformity coefficient*. The effective size (d_{10}) is the size of the standard sieve opening that will pass ten percent by weight of the media. The uniformity coefficient (d_{60}/d_{10}) is the ratio of the standard sieve opening that will pass sixty percent by weight of the media to its effective size. Graphical representation of a standard sieve analysis and determination of d_{10}, d_{60}, and uniformity coefficient are illustrated in Figure 10-4. Other parameters often specified are media hardness and its resistance to acid attack (acid solubility). Standard B-100 of the American Water Works Association provides a basis for most filter media specifications.[a] Engineers are encouraged to obtain a copy of this standard and become familiar with its contents prior to utilizing it as a basis for filter media specifications.

10.4.2 Single-Medium Filters

Single-medium filters utilize a single material, most commonly well-graded sand. Typical effective size, uniformity coefficients, and bed depths for single-medium filters are listed in Table 10-1.[3,5,6] In a single-medium filter, after backwashing, larger grains settle faster than smaller grains, in a phenomenon called *stratification* or *reverse gradation*. This phenomena is illustrated in Figure 10-5(a). Reverse gradation is the major disadvantage of the single-medium filter. The smaller media grains at the top of the filter trap most of the particles; therefore, only the top 4 or 5 cm of the filter bed is utilized for filtration. Particles that pass through this layer are less likely to be removed in the larger media grains below. Additionally, because only the top 4 or 5 cm of the bed are used, the solids-holding capacity of the bed is small, and so filter runs are shortened. This problem can be minimized by specifying a larger-diameter medium with a uniformity coefficient close to unity (less than 1.4) and by using a deep medium bed. Such filters have been employed in California and Texas to successfully treat water with high filtration rates [in excess of 550 $m^3/m^2 \cdot day$ (9.5 gpm/ft^2)]. These filters are generally called *deep-bed* monomedium filters and utilize anthracite coal media.

a. American Water Works Association, "AWWA Standard for Filtering Material," ANSI/AWWA B100-89, Revised Edition.

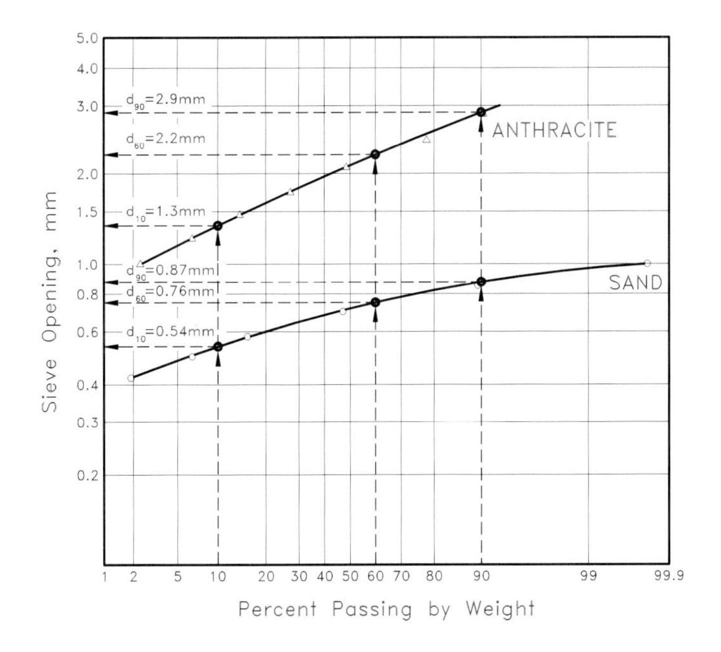

Figure 10-4 Typical sieve analysis of two filter media.

Table 10-1 Typical Media Design Values for Various Filters

Parameter	Single-medium Filters	Dual-media Filters	Mixed-media Filters
Anthracite layer			
Effective size, mm	0.50–1.5	0.70–2.0	1.0–2.0
Uniformity coefficient	1.2–1.7	1.3–1.8	1.4–1.8
Depth, cm	50–150	30–60	50–130
Sand layer			
Effective size, mm	0.45–1.0	0.45–0.60	0.40–0.80
Uniformity coefficient	1.2–1.7	1.2–1.7	1.2–1.7
Depth, cm	50–150	20–40	20–40
Garnet layer			
Effective size, mm			0.20–0.80
Uniformity coefficient			1.5–1.8
Depth, cm			5–15

Source: Adapted in part from References 3, 5 and 6.

10.4.3 Dual-Media Filters

Another solution to the problem of reverse gradation is the *dual-media* filter. Typical dual-media filters utilize anthracite coal and quartz sand as filter media. The anthracite with a specific gravity of 1.55, is lighter than the sand, which has a specific gravity of 2.65. Therefore, a larger anthracite grain has the same settling velocity as a much smaller sand grain. This characteristic allows coal grains to be placed on top of smaller sand grains to create a gradation, as shown in Figure 10-5(b). Typical design values for dual-media filters are listed in Table 10-1.[3,5,6] Dual-media filters behave very much like two single-medium filters in series, each with a different grain size. The larger anthracite grains remove the larger particles as well as some smaller ones; the sand, in its turn, captures the smaller particles. As a result, more depth of the filter is utilized for solids removal than in the stratified single-medium filter.

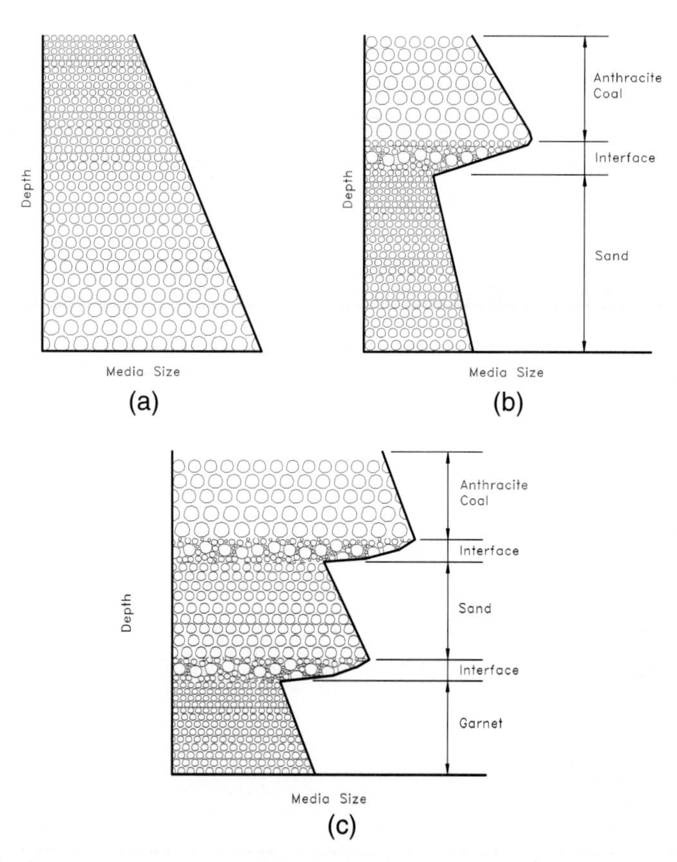

Figure 10-5 Gradation versus depth. *(Reprinted from Reference 4, with permission from Technomic Publishing Co., Inc., copyright 1999.)* (a) Single-medium filter. (b) Dual-media filter. (c) Mixed-media filter.

10.4.4 Mixed-Media Filters

Mixed-media filters are similar to dual-media filters, except that several materials are used. Typically, three materials are used: anthracite with a specific gravity of 1.55, sand with a specific gravity of 2.65, and garnet with a specific gravity of 4.05. The media are placed in the filter bed with the specific gravity of the media increasing and the size of the media decreasing with depth in the filter bed, to create a grain size distribution as shown on Figure 10-5(c). The typical design values for mixed-media filters are provided in Table 10-1.[7,8] Mixed-media filters are basically improved dual-media filters, with increased filter run times and water quality. The cost of mixed-media filters is much higher than that of dual-media filters, because garnet is quite expensive. The improved performance of mixed-media filters often cannot justify the additional cost.

10.4.5 Filter Media Selection

Filter efficiency is a function of certain physical characteristics of the filter bed that include the porosity of the media and the ratio of the media depth to media grain diameter. The determination of the appropriate media depth and size of media particles is the basic task of the design engineer. Currently, there are two ways to determine the depth and size of filter media: (1) pilot testing, and (2) experience from past studies. Pilot studies require significant time and capital expenditures. There are many combinations of filter media depth and size. Kawamura developed a relationship between depth of the filter media (l) and the effective size of the media (d_e) from many filter designs in operation.[2,9,10] This relationship for many filter designs is summarized in Table 10-2 and shown in Figure 10-6. The ratio of l/d_e is nearly constant. It may be noted that the average l/d_e ratio in Table 10-2 is around 1020 and is based on the actual value of l and d_e; whereas that in Figure 10-6 is 980 and is based on total combined depth and weighted average of d_e for dual- or multi-media bed.

10.5 Filter Components

Filter systems used in water treatment plants vary from simple to complex in design. Most designs utilize multiple filter units. A typical filter system employs a minimum of two, as many as thirty or more, individual filter cells operated in parallel. A system of filter boxes, pipe manifolds, and valves is utilized to distribute water to and from the individual filter units. Figure 10-7(a) illustrates a typical filter system using four filter cells. Following is a brief description of each part of the system.

10.5.1 Filter Box

A typical filter box is shown in Figure 10-2(a) and Figure 10-7(b). Water typically enters the filter on top and flows down through the filter media, the gravel support, and the underdrain. The underdrain system conveys the water to a central flume referred to as the *gullet*. Water flows out of the gullet into the effluent pipe.

Table 10-2 The Ratio of the Depth of the Media (l) to the Effective Size of the Media (d_e)

Filter Type	Material	Effective Size (d_e), mm	Media Depth (l), cm	Uniformity Coefficient	l/d_e
Small dual-media	Anthracite	1.00	50.8	1.5	1016
	Sand	0.50	25.4	1.3	
Intermediate dual-media	Anthracite	1.48	76.2	1.5	1023
	Sand	0.75	38.1	1.2	
Large dual-media	Anthracite	2.00	101.6	1.5	1016
	Sand	1.00	50.8	1.3	
Mixed-media	Anthracite	1.00	45.7	1.5	1306
	Sand	0.42	22.9	1.5	
	Garnet	0.25	7.6	1.3	
Mono-medium	Anthracite	1.00	101.6	1.4	1016

After running for a period of time, the filter media will become clogged with accumulated solids removed from the water. The filter must then be cleaned or *backwashed* to remove the accumulated solids. This operation involves closing the influent and effluent valves and opening the washwater and waste washwater valves (Figure 10-6(b)). This results in water flowing into the gullet and through the underdrain system toward the gravel support layer. Water flows upward through the gravel bed and into the filter media. As it flows through the filter media, the bed is expanded and the water picks up the deposited solids and carries them into the washwater trough and waste piping. The dirty washwater is then treated separately to recover water for recycling. The residuals are handled by one of a number of methods, as discussed in Chapter 14.

10.5.2 Influent Piping

The influent piping conveys water from the sedimentation units to the individual filter boxes. Each filter box has an influent valve that is used to stop the flow of water into the filter. The influent pipe typically is connected at the top of the filter unit, just below the water surface.

10.5.3 Effluent Piping

The effluent piping conveys filtered water from the individual filter boxes to the next treatment process. Each filter box has an individual *effluent valve* that is used to stop the flow of water out of the filter. The effluent pipe for an individual filter is usually connected to the gullet at the bottom of the filter.

Immediately after a filter is backwashed and set into operation, the turbidity of filtered water is generally quite high. Several methods can be used to control the turbidity. One com-

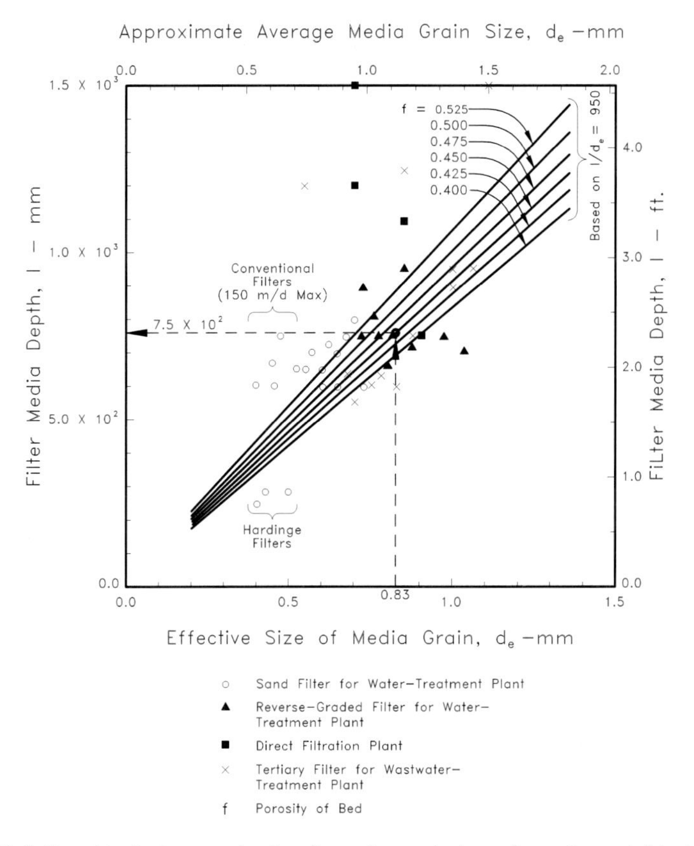

Figure 10-6 Relationship between depth of media and size of media weighted average to be used for dual- or multi-media bed. *(Reprinted from* Journal AWWA, *vol. 67, no. 10 (October 1975), by permission. Copyright® 1975, American Water Works Association.)*

monly used method is to install a waste-to-drain connection; the initially filtered water is wasted. This is called a *rinse cycle*. Methods to control initial turbidity spikes are further discussed in Section 10.8.7.

10.5.4 Washwater Piping

The washwater piping conveys clean water to the bottom of the filter for the backwash operation. Water flows from the washwater pipe into the gullet and through the underdrain system and the media. A valve is installed properly to stop flow into the filter box.

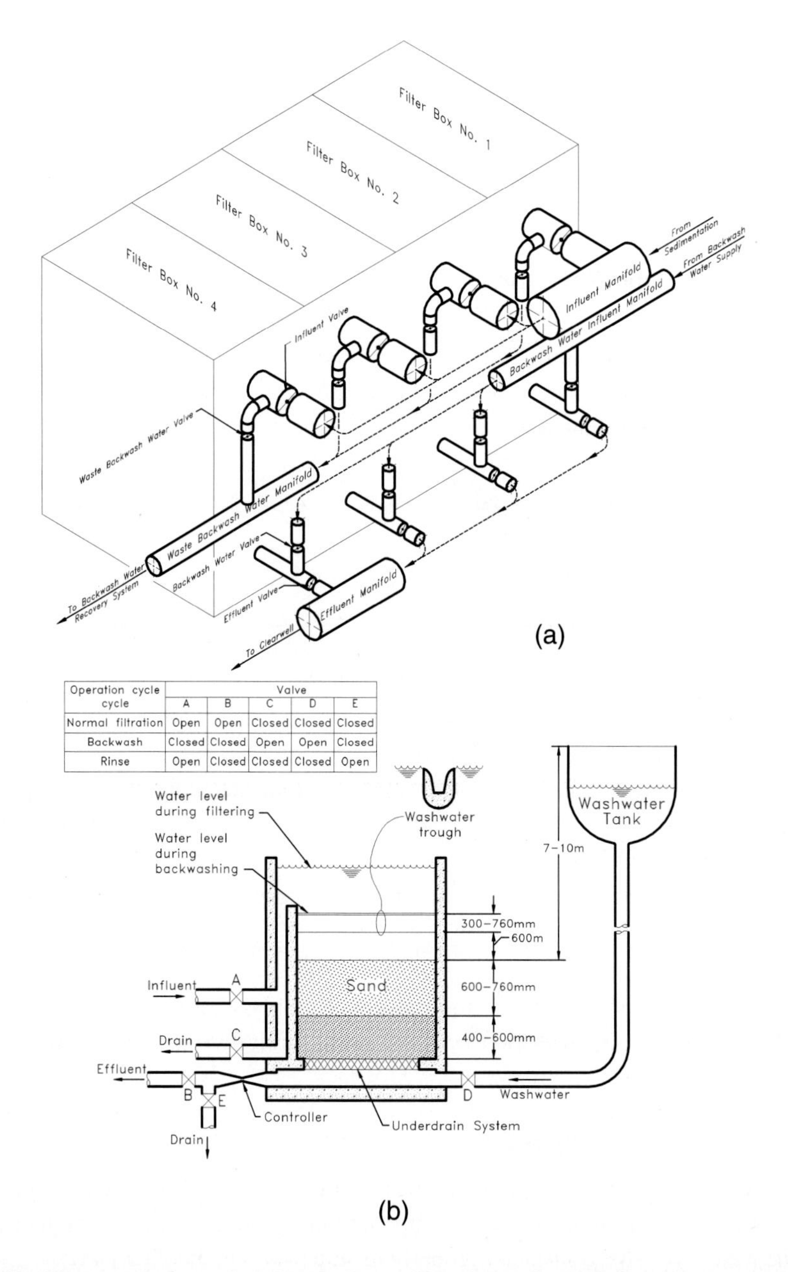

Operation cycle cycle	Valve				
	A	B	C	D	E
Normal filtration	Open	Open	Closed	Closed	Closed
Backwash	Closed	Closed	Open	Open	Closed
Rinse	Open	Closed	Closed	Closed	Open

Figure 10-7 Filter controls and piping system. *(Reprinted from Reference 4, with permission from Technomic Publishing Co., Inc., copyright 1999.)* (a) Typical arrangement of four filters and piping. (b) Filter operation.

10.5.5 Waste Backwash Water Piping

The waste backwash water is collected on top of the filter and directed to the waste backwash water pipe. The waste backwash water pipe also has a valve to stop flow into the waste drain during normal filtration operation.

10.6 Filter System Operation

10.6.1 Declining-Rate and Constant-Rate Filters

The flow of water through the filter system typically remains constant, although the flow through the individual filter units (boxes) may or may not be constant. Some filters are designed so that the flow rate through the individual filter units decreases (declines) as solids accumulate in the media bed. Such a filter is referred to as a *declining-rate* filter. Figures 10-8 (a) and (b) illustrate the relationship of a typical four-unit declining-rate filter system.

Other filter units use an external control called a *flow controller* to maintain a constant flow through the individual filters regardless of the amount of solids accumulated. Such filters are called *constant-rate filters*. Figure 10-8 (c) compares the flow rate and time relationship of declining-rate and constant-rate filter systems.

10.6.2 Variable-Level and Constant-Level

Some filter systems are designed to allow the water level in the filter influent channel or box to vary considerably as the filter media become clogged. Such systems are termed *variable-level* or *influent-controlled* filters. Other filter units are designed to maintain a relatively constant influent level regardless of the amount of solid deposition in the media. The difference in head loss is compensated for by an effluent rate-of-flow control valve. Such systems are called *constant-level* or *effluent-controlled* filters. Figure 10-9 illustrates the head loss versus filtration time relationships for influent- and effluent-controlled filter operations.

10.6.3 Comparison of Filter Operation Systems

Declining-rate and constant-rate filter systems can both be either influent-controlled or effluent-controlled. Table 10-3 compares the four basic control systems, and each is discussed further in the following paragraphs.[11] Figures 10-9 and 10-10 schematically illustrate each system.

The Effluent-Controlled Constant-Rate Filter In an effluent-controlled constant-rate filter, the total flow is divided equally among the filter units, and each unit has an equivalent influent head that remains relatively constant. The flow rate is controlled by an effluent rate-of-flow controller installed at each filter unit. These flow controllers are composed of a flow meter and a modulating valve. These devices adjust the head loss through the individual filter to maintain a constant flow rate. The head loss and filter run-time relationship is shown in Figure 10-9(a).

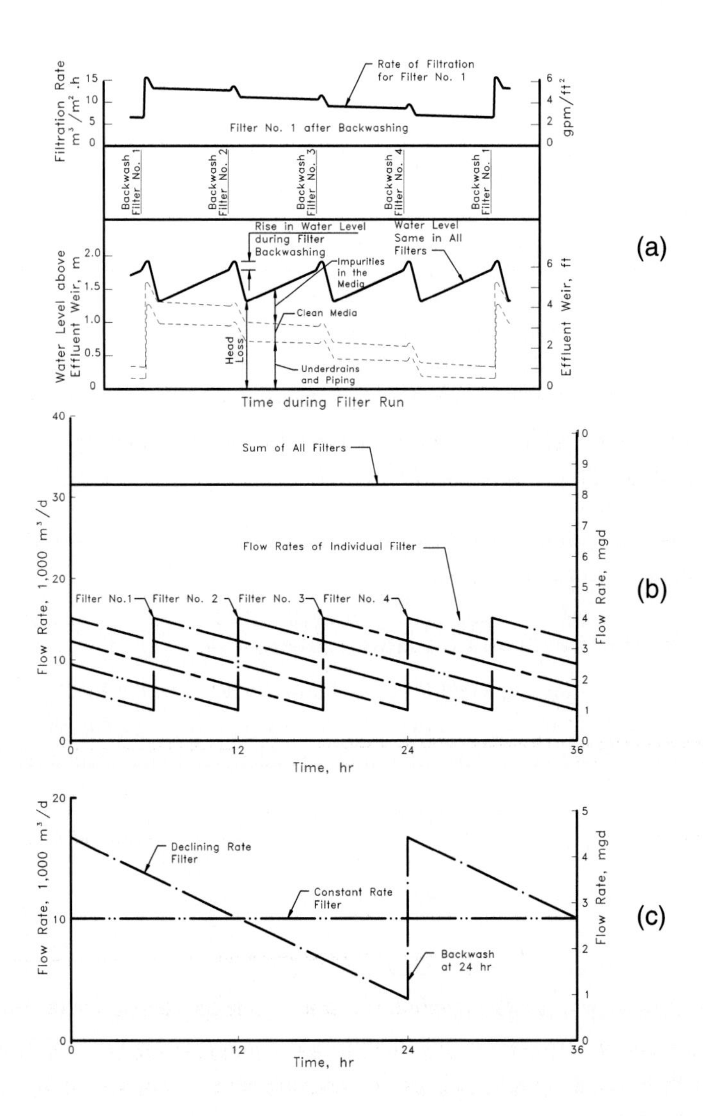

Figure 10-8 Declining-rate filtration and comparison of filter systems. (a) Declining-rate filtration. (Showing filtration rate, water level, and head losses for one filter run in a plant having four filters.) (b) Relationship between filter flow rate and filter run time for a bank of four filter cells. (c) Relationship between filter flow rate and filter run time for an individual filter cell.

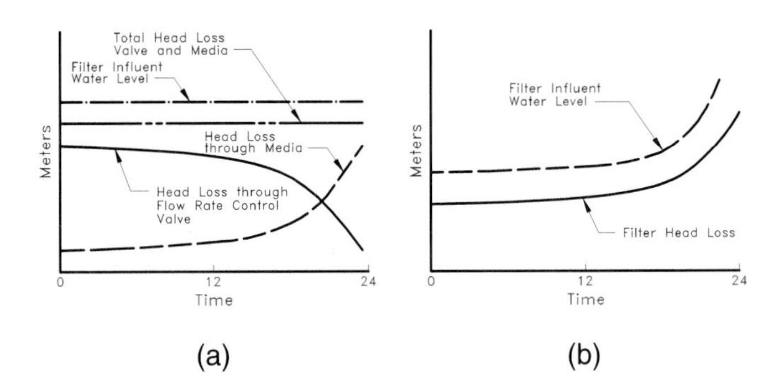

Figure 10-9 Comparison of influent- and effluent-controlled constant-rate filter operations. (a) Effluent-controlled filter. (b) Influent-controlled filter. The Influent-Controlled Declining-Rate Filter

Table 10-3 Characteristics of Filter Operation Systems

Design Parameter	Constant-Rate		Declining-Rate	
	Effluent-Controlled	Influent-Controlled	Effluent-Controlled	Influent-Controlled
Relative cost	High	Medium	Medium	Low
Simplicity of operation	Complex	Simple	Moderate	Simple
Length of filter run	Short	Short	Long	Long
Depth of filter bed used	Small	Small	Large	Large
Consistency in effluent quality	Good	Good	Good	Good

The control systems required for effluent-controlled constant-rate systems are rather complex (Figure 10-10(a)). The master controller senses changes in the influent hydraulic gradient and adjusts the effluent rate-of-flow controllers to maintain a constant influent water level. If the head loss through the filter increases, all valves are opened slightly. Likewise, all valves are closed slightly if the head loss decreases. Simultaneously, the flow through each filter is compared with that of the other filters. If one filter has a higher (or lower) flow than the other filters, its valve closes (or opens) slightly to equalize the filter output. Despite their complexity, effluent-controlled constant-rate filters are common in medium to large treatment plants. Their widespread use is attributed to their operational flexibility and to the general familiarity of operations personnel with the operating techniques of these systems.

The Influent-Controlled Constant-Rate Filter Influent-controlled constant-rate filters do not require such complex control systems as effluent-controlled filters do. The flow is divided into filters by a flow-splitting device, usually an overflow weir. The water surface in the individual filter units varies, depending on head loss. As a filter unit becomes clogged, its water surface rises to offset the head loss encountered. A disadvantage of the influent-controlled filter is the tendency for floc breakup at the influent weir; this often results in poorer finished water quality. Influent-controlled constant-rate filters are common in smaller treatment facilities. Their simplicity and relatively low construction cost make them well suited for such application. The head loss and filter run-time relationship is shown in Figure 10-9(b), and filter control systems are shown in Figure 10-10(b).

The Influent-Controlled Declining-Rate Filter The influent-controlled declining-rate filter systems are the simplest of all filter systems. No rate-of-flow controllers or flow-splitting devices are required. The hydraulic gradient is allowed to vary throughout the filter run to adjust for the increase in head loss through the filter system. When the hydraulic gradient becomes too high, one filter unit is backwashed. An overflow weir is usually placed in the filter effluent structure to maintain positive pressure in the filter bed at all times (Figure 10-10 (c)).

The influent-controlled declining-rate filter systems are able to tolerate large hydraulic gradient variation in the influent of the filter system. For this reason, this type of filter system usually cannot be used in converting a constant-rate system to a declining-rate system. The simplicity of an influent-controlled declining-rate system, however, makes it ideal for small treatment plants, where construction and operation budgets are limited.

The Effluent-Controlled Declining-Rate Filter The effluent-controlled declining-rate filter systems are similar to effluent-controlled constant-rate systems in that a constant hydraulic gradient is applied to all filters; however, in declining-rate systems, the rate-of-flow controller for each individual filter unit is not required. Instead, a single rate-of-flow controller is installed in the effluent piping. The flow rate through each individual filter unit is determined by the head loss through that unit. Therefore, the dirtiest filter has the lowest flow rate, and the cleanest filter has the highest flow rate. In some cases, a simple flow-limiting device, such as an orifice plate, is installed in the effluent piping of the individual filter units, to prevent excessive flow rates through a recently cleaned filter. The filter control systems are shown in Figure 10-10(d).

The declining-rate filter systems utilize the solids-holding capacity of the filter bed more efficiently than do constant-rate systems. Studies have shown that declining-rate filters produce a more constant and better quality effluent than constant-rate filters, and that filter runs for declining-rate filters are significantly longer than those for constant-rate systems.[7] However, unless the initial filtration rates are limited, excessive turbidities may be experienced immediately after backwash.

10.7 Filter Hydraulics

The hydraulic design of filters is complex and not fully understood. Several empirical equations are available to predict the head loss through a clean filter bed. Likewise, numerous methods of estimating the head loss through clogged filters are also available. The most depend-

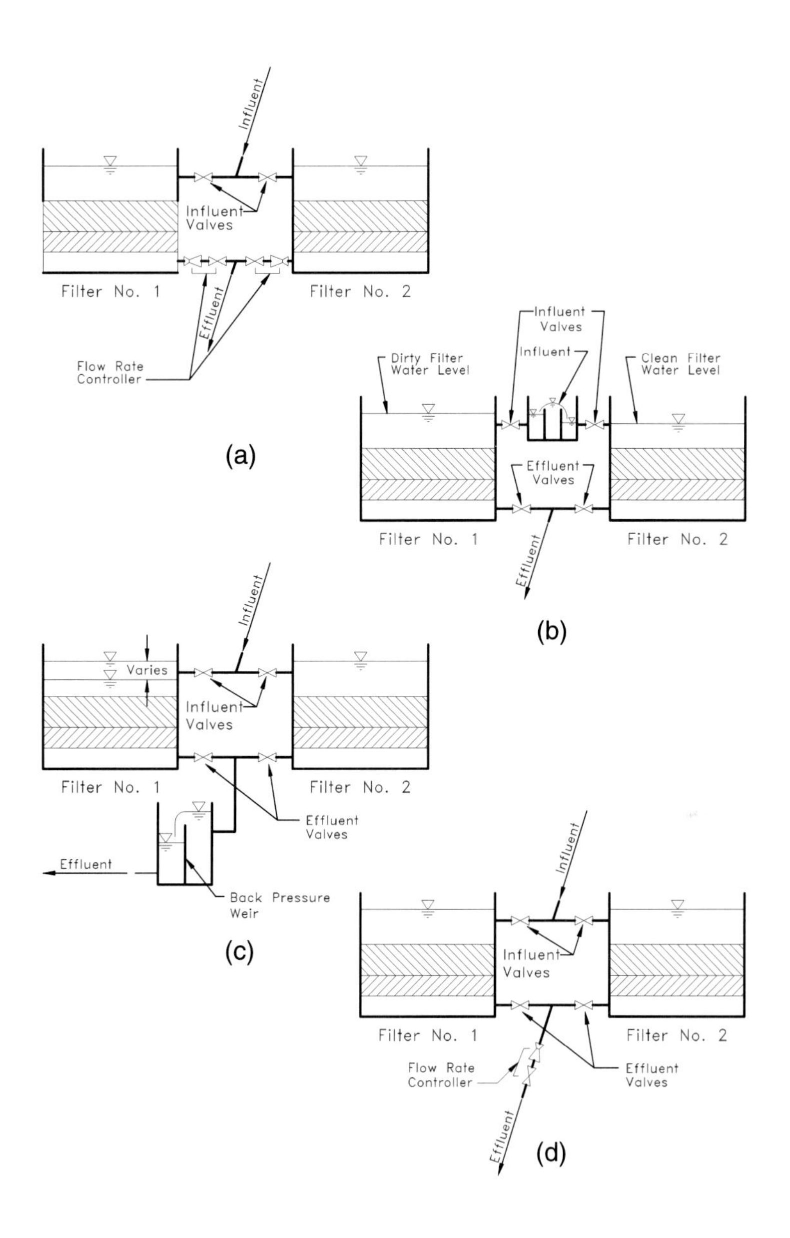

Figure 10-10 Filter control systems. (a) Effluent-controlled constant-rate filter. (b) Influent-controlled constant-rate filter. (c) Influent-controlled declining-rate filter. (d) Effluent-controlled declining-rate, filter.

able method of estimating the hydraulic characteristics of a filter system is a pilot study utilizing the water to be treated.

10.7.1 Head Loss in Clean Filter

Over the years, much research has been done on the hydraulics of sand filters. Numerous empirical and semi-empirical equations have been developed, which are used to predict clean filter performance. Some of the more commonly used equations are provided in Table 10-4.[1,6,11] Each of these equations requires an estimate of either the porosity of the bed or the relative compactness of the media. To estimate these parameters, the size, shape, and uniformity coefficient of the media must be known. Typical values of some parameters for commonly used media are provided in Table 10-1. The application of some of these equations in filter head loss calculations is shown in the Design Examples.

10.7.2 Head loss in Clogged Filter

Throughout a filter run, solid particles are deposited in the pores of the filter bed and cause a decrease in the porosity and a corresponding increase in the head loss through the filter bed. The design engineer must predict the head loss through the filter bed at the end of the filter run. The terminal head loss through a filter bed can be calculated by use of any of the equations listed in Table 10-4, if the porosity at the end of the filter run can be determined. The average change in porosity can be estimated by assuming that the volume of the particles removed equals the reduction in pore volume.[12] This method, however, is extremely inaccurate in predicting the change in porosity of various layers of filter media. Typically, filters have produced satisfactory turbidity levels during their entire run until the maximum head loss of 2 to 3 m is reached. More stringent water quality requirements in the future may require backwashing of the filters before these head loss values are reached or turbidity breakthrough occurs. More accurate estimates of maximum head loss values can be determine from pilot plant studies. With pilot plant studies, the variables in the Carmen–Kozeny equation (Eq. (10.1)) can be calibrated to yield more accurate solutions.

Another method of predicting clogged filter head loss is proposed by Tchobanoglous.[8] This method considers the total filter bed as being composed of several individual filter beds (of small depth) in series. The increase in head loss for each layer is related to the amount of solids deposited in that layer. By considering each layer of the filter bed separately, a more realistic hydraulic model of the filter is obtained. Extensive pilot testing data must be developed to calibrate the model to yield reliable results.

10.7.3 Miscellaneous Head losses

The hydraulic design of a filter system must consider not only the energy losses through the media but also the energy losses through the appurtenant systems, including (1) piping systems, (2) gravel beds, (3) underdrain system, and (4) wash water collection launders.

Piping Systems The piping systems include the influent, effluent, and wash water piping valves and flow rate controllers. These piping systems often include complex piping manifolds with numerous fittings and specials. The head losses through such systems are best calculated by use of the Darcy–Weisbach equation (Eq. (7.13)) or the Hazen–Williams equation (Eq. (7.14)). Losses through fittings can be expressed by Eq. (7.13). The application of these equations to typical filter piping is illustrated in the Design Example.

Gravel Beds The head loss through the supporting gravel bed is calculated from equations listed in Table 10-4. The procedure is the same as that for clean filter media. A gravel bed must also be considered as made up of several layers arranged in a series, to account for the variability in gravel media.

Underdrain System The underdrain system usually has numerous orifices. The hydraulic characteristics of these proprietary systems vary greatly. The designer should consult the manufacturers of the specified system to determine its hydraulic characteristics.

Table 10-4 Empirical Equations Used to Calculate Head Loss Through Clean Filter Beds

	Equation	Equation Number
Carmen–Kozeny	$h_L = \frac{f}{\phi}\frac{1-e}{e^3}\frac{L}{d}\frac{v^2}{g}$	(10.1)
Fair–Hatch	$h_L = k^2 \nu S^2 \frac{(1-e)^2}{e^3}\frac{L}{d^2}\frac{v}{g}$	(10.2)
Rose	$h_L = \frac{1.067}{\phi} C_d \frac{1}{e^4}\frac{L}{d}\frac{v^2}{g}$	(10.3)
Hazen	$h_L = \frac{1}{C}\frac{5.2\times 10^6}{T+10}\frac{L}{d_{10}^2}v$	(10.4)
Friction factor	$f = 150\frac{1-e}{N_R} + 1.75$	(10.5)
Reynolds number	$N_R = \frac{dv}{\mu}\rho_w$	(10.6)
Coefficient of drag	$C_d = \frac{24}{N_R} + \frac{3}{\sqrt{N_R}} + 0.34$	(10.7)

Source: Adapted in part from References 1, 6 and 11.

where

C = coefficient of compactness (600–1200)
C_d = coefficient of drag
d = media grain diameter, m
d_{10} = media effective size, mm
e = porosity ratio (usually 0.4 to 0.5)
f = friction factor (See Eq. (10.5).)
g = acceleration due to gravity (9.81 m/s^2)
h_L = head loss, m (ft)
k = filtration constant
L = media depth, m (ft)
N_R = Reynolds number
S = shape factor (6.0 to 8.5)
T = temperature ^oF
v = filtration velocity, m/s
μ = absolute viscosity, N-s/m^2 (kg/m·s)
ν = kinematic viscosity, m^2/s
ρ_w = density of water, kg/m^3
ϕ = particle shape factor (usually 0.85 to 1.0)

Washwater Collection Launders Washwater collection launders carry waste backwash water to the drain. These launders function like open channels with spatially-varied flow conditions. The design of these launders is similar to that of the effluent launder troughs of a sedimentation basin. The design procedure for these troughs was presented in Chapter 9.

10.8 Filter Cleaning or Backwashing

A filter cell must be cleaned when either (1) the head loss through the filter exceeds the design value, (2) turbidity breakthrough causes the effluent quality to be less than a minimum acceptable level, or (3) a pre-selected maximum filter run time has passed since it was last cleaned. Filter units are cleaned by backwashing. This involves passing water upward through the filter media at a velocity sufficient to expand the bed and wash out the accumulated solids.

During backwash, the filter media is expanded or fluidized. The particles become separated and the space between them becomes greater, increasing the porosity. In this state, the vertical velocity of the water is approximately equal to the settling velocity of the media and causes the media to be partially supported by the water. This allows the smaller floc particles that have a lower settling velocity to be removed and washed out of the bed. To ensure proper operation of the filter, the design engineer must consider (1) the settling velocity of the media, (2) the backwash flow rate, (3) the head loss during backwash, (4) the duration of the backwash cycle, and (5) the quantity of water required to backwash each filter.

10.8.1 Settling Velocity of Filter Media

The settling velocity of the filter media can be calculated from Newton's or Stoke's equations (Eqs. (9.1) or (9.6)). These equations assume spherical particles settling discretely, a valid assumption for most sand and garnet media. Anthracite media tend to be more irregular in shape and may have an actual settling velocity slightly lower than that calculated by these equations. This error, however, is small and insignificant.

In order for the backwash system to operate without washing out of the filter media, it is important, particularly for dual- and mixed-media filters, that the system be designed so that the entire filter bed has approximately the same settling velocity. Therefore, the size of the various layers of the filter media must be carefully selected. The following equation can be used to calculate the size of the media grains with different specific gravities and equal settling velocities.[2]

$$d_2 = d_1\left(\frac{S_{g1}-1}{S_{g2}-1}\right)^{2/3} \tag{10.8}$$

where

d_2 = effective size of the media with a specific gravity of S_{g2}, mm (in)
d_1 = effective size of the media with a specific gravity of S_{g1}, mm (in)

The normal design procedure is to select the sand medium that would yield acceptable finished water effluent quality. Then, by use of Eq. (10.8), select the effective size of the anthracite and/or garnet media.

10.8.2 Backwash Flow Rate

The purpose of backwashing the filter bed is to remove deposited floc without washing out the media; therefore, the vertical velocity of the water during the backwash must be greater than the settling velocity of the floc, but less than the settling velocity of the media. The settling velocity of the floc has been found to range between 2.5 and 25 cm/min. (1 and 10 inch/min.), and a minimum backwash rise rate should be 30 cm/min (12 inch/min.).[9,13] The settling velocity of the filter media can be calculated by Newton's equation (Eq. (9.1)).[9] For water at 20°C (68°F), Eq. (9.1) reduces to Eq. (10.9) for sand with specific gravity of 2.61, and Eq. (10.10) for anthracite with specific gravity of 1.5, respectively.[9,10]

$$v_s = 10d_{60} \qquad \text{(sand)}^{b} \tag{10.9}$$

$$v_s = 4.7d_{60} \qquad \text{(anthracite)} \tag{10.10}$$

where

v_s = terminal settling velocity of the media m/min (ft/min)
d_{60} = size of the standard sieve opening that will pass 60 percent by weight of the media, mm. ($d_{60} = d_{10} \infty$ uniformity coefficient (Section 10.4.1).)

Studies show that the most effective backwash rates equal approximately 10 percent of the media terminal settling velocities. Therefore, for sand and anthracite coal, respectively, the backwash rates at 20°C are calculated from Eqs. (10.11) and (10.12).

$$U_b = d_{60} \quad \text{(sand)} \tag{10.11}$$

$$U_b = 0.47 d_{60} \quad \text{(anthracite)} \tag{10.12}$$

U_b = backwash rate, m/min

The backwash rate should be adjusted for colder or warmer water by using Eq. (10.13).

$$U_{bT} = U_{b20} \times \mu_T^{1/3} \tag{10.13}$$

where

U_{bT} = backwash rate at temperature T°C, m/min (ft/min)

U_{b20} = backwash rate at 20°C, m/min (ft/min)

μ_T = absolute viscosity at temperature T°C, (N-s/m^2 or kg/m·s)

The proper combination of common filter media for a different backwash rate at 20°C are provided in Figure 10-11.[9,10,14]

10.8.3 Head Loss during Backwash

The head loss due to water flowing through a fluidized bed is equal to the weight of the supported media. It is generally expressed by Eq. (10.14).

$$h_L = \text{volume of media grains} \times (\text{density of media grains} - \text{density of water}) \tag{10.14}$$

Eq. (10.14) can also be mathematically expressed as Eq. (10.15).[9,10]

$$h_L = L(1 - e)(S_g - 1) \tag{10.15}$$

b. Kawamura used the modified equation proposed by Allen:

$$v_s = \left\{\frac{0.018(\rho_s - \rho_w)^2 g^2}{\rho_w \mu}\right\}^{1/3} d$$

Substituting $\rho_s = 2.61$, $\rho_w = 1.0$, $g = 9.81$ m/s^2, and $\mu = 10^{-3}$ kg/m·s (20°C).

$v'_s = 165\, d'$, where d' is in m and v'_s is in m/s. From this relationship, $v_s = 10\, d$, where v_s is in m/min and d is in mm.

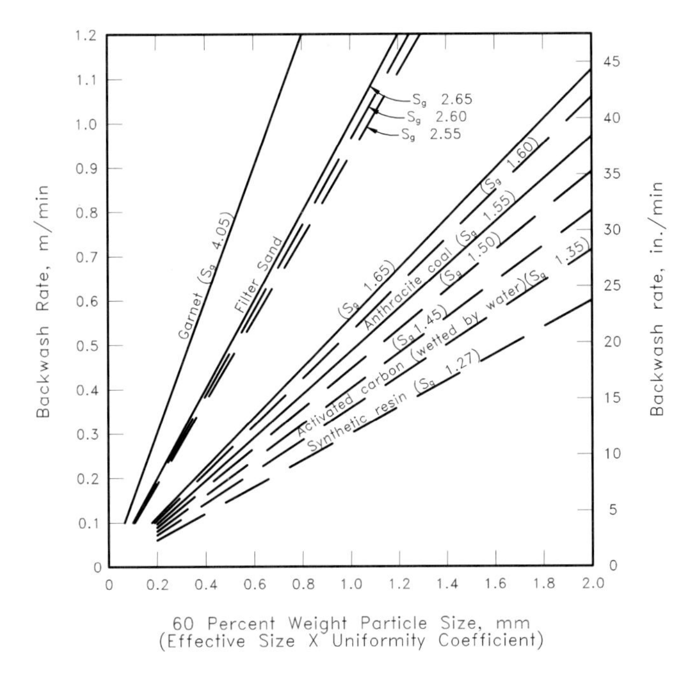

Figure 10-11 Proper combinations of common filter media for different backwash rates at 20°C. *(Reprinted from* Journal AWWA, *vol. 67, no. 10 (October 1975), by permission. Copyright 1975, American Water Works Association.)*

where

h_L = head loss through the media bed during backwash, m (ft)

e = porosity of the clean stratified bed at rest (not fluidized)

L = depth of the stratified bed at rest, m (ft)

S_g = specific gravity of the media

The gravel bed does not fluidize, so the head loss through it is computed in the same manner as for head loss calculations through clean filter media. Many of the equations listed in Table 10-4 are used for this calculation.

10.8.4 Duration of the Backwash Cycle

The backwash cycle should extend long enough to clean the media bed. The time required to accomplish this depends on (1) the nature of the floc, (2) whether air-scour (or surface wash) is used, and (3) the geometry of the filter bed and the washwater troughs. Usually the entire backwash cycle lasts from 8 to 15 minutes.[11,15,16]

10.8.5 Quantity of Water Required for Backwashing

The quantity of washwater depends on the washwater flow rate and the duration of the backwash cycle. If the washwater flow rate has cyclic variation, then the total quantity of washwater required is the area under the curve representing the flow rate versus time. The quantity of water required to operate the surface wash systems must also be included if these devices are utilized.

10.8.6 Auxiliary Cleaning Systems

Backwash alone does not always thoroughly clean clogged filter beds. Additional agitation of the filter media is often necessary to remove accumulated floc material. *Surface wash* and *air-scour* are two systems that have proven to be successful in providing supplemental cleaning of clogged filters.

Surface Wash Surface wash systems have become widely used throughout the United States. Many proprietary systems are available. The purpose of the surface wash system is to provide maximum agitation at the top of the filter bed, where the greatest accumulation of floc may be expected. This is done by spraying water through nozzles located on either fixed piping manifolds or rotating arms. Examples of both types of systems are given in Figure 10-12. The high-pressure spray from the nozzles provides added turbulence without increasing the vertical velocity of the backwash water, thereby preventing media from being washed out during backwash. The turbulence created by surface wash systems can be determined by Eq. (10.16).[9]

$$G_{sw} = \left(\frac{\Delta h g v_s \rho_w}{\mu \alpha L_e}\right)^{1/2} \quad (10.16)$$

where

G_{sw} = velocity gradient given by water-jet-type surface wash, 1/s
v_s = surface backwash rate, m/s
Δh = head applied to the media by surface wash system, m
α = coefficient 0.25 for surface wash, 0.5 for dual-arm surface or subsurface wash.
L_e = depth of expanded bed, m
μ = absolute viscosity of water, $N \cdot s/m^2$ (or kg/m·s)

Surface wash systems typically are proprietary units. They usually consist of 2.4–3.1-mm-diameter openings spaced along a fixed or rotating pipe at from 5 to 8.6 cm apart. Rotating pipe systems discharge from 1.2 to 2.4 $m^3/m^2 \cdot h$; fixed arm systems discharge from 5 to 10 $m^3/m^2 \cdot h$.[c] Typical operating heads are 350 to 520 kPa. The nozzles are placed at from 2.5 to 5 cm below the media bed at rest. They may also be placed just above the sand/anthracite interface in dual-media filters. They are started 1 to 3 min before the start of the backwash cycle and stopped 5 to

c. 1 $m^3/m^2 \cdot h$ = 3.28 $ft^3/ft^2 \cdot h$ or 0.409 gpm/ft^2.

10 min before the end of the backwash cycle.[15] Camp found that the velocity gradient at the top of a filter bed at the end of a filter backwash cycle is around 1000/s. Typically, surface wash velocity gradients are in the range of 1100 to 1300/s to remove tough compacted floc.[9] If dual-media filters are used, the heaviest concentration of floc often occurs at the interface between the sand and the anthracite. In these instances, a subsurface wash system located at the interface is utilized to apply the maximum turbulence at the interface. Surface wash systems work best in filters where there is little penetration of floc into the media bed or on media beds with relatively shallow depths.

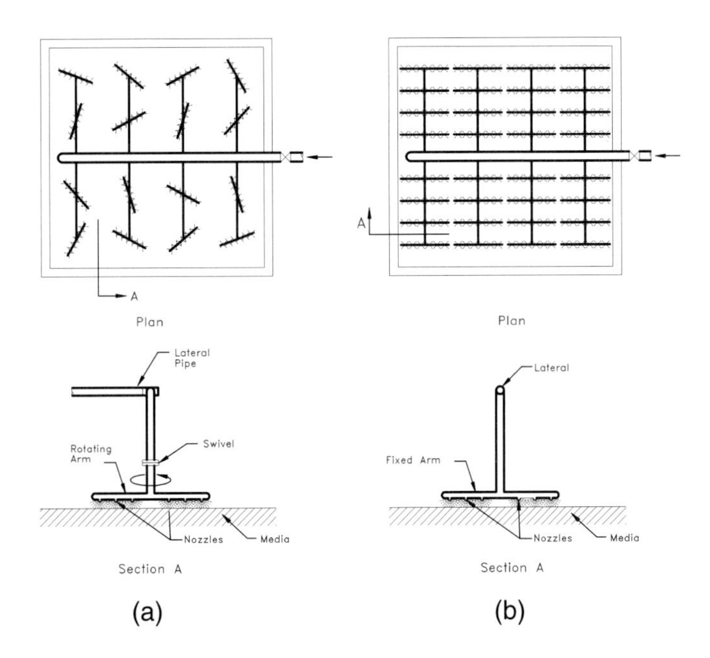

Figure 10-12 Surface wash systems. (a) Rotating arm. (b) Fixed manifold.

Air-Scour Air-scour systems have been used widely in Europe; and their use has become more common in the United States in recent years. In these systems, air is released from a manifold system located in specially designed underdrain systems. The rising air bubbles increase turbulence throughout the bed and clean the media grains by means of a scouring action.[11,16,17]

The typical air flow values for single-medium filters with small media sizes (0.5 mm effective size) are from 18 to 36 $m^3/m^2 \cdot h$, followed by water alone at from12 to 20 $m^3/m^2 \cdot h$. For dual-media filters in the U.S. (effective size sand 0.5 mm and anthracite 1.0 mm), a high air flow rate of from 55 to 91 $m^3/m^2 \cdot h$ is commonly used, followed by water backwash at from 37 to 49

$m^3/m^2{\cdot}h$. If air-scour and backwash are used simultaneously to clean larger sands (1.0 mm effective size), air flow rates of from 37 to 73 $m^3/m^2{\cdot}h$ are commonly used with a water flow rate of 15 $m^3/m^2{\cdot}h$.[14] For proper media bed stratification, it is essential in all cases for the air supply to be turned off several minutes prior to the stopping of the water backwash; this will allow stratification of the media.

Air-scour systems apply the turbulence throughout the bed; therefore, they are of greatest advantage in filter beds with an even accumulation of floc throughout the bed. If air-scour is used, specially designed underdrain systems and launder trough arrangements are necessary. Designers should consult manufacturers of proprietary systems before proceeding with detailed designs.

10.8.7 Controlling Turbidity Spikes After Backwashing

Immediately after backwashing, filter effluent will experience high turbidities. The authors have observed that, in some cases, this phenomenon is quite detrimental to the finished water quality of the plant. A typical turbidity spike after filter backwash is illustrated in Figure 10-13. These turbidity spikes must be controlled to meet the more stringent turbidity requirements of the Surface Water Treatment Rules.[18,19] There are several strategies commonly used to control these spikes. Among these are (1) to install a waste-to-drain connection, (2) to ripen filtration rate or to use a reduced filtration rate for the first few minutes after the filter is placed back in operation, and (3) to precondition the filter by introducing a coagulant in the backwash water during the last few minutes of backwash.[13]

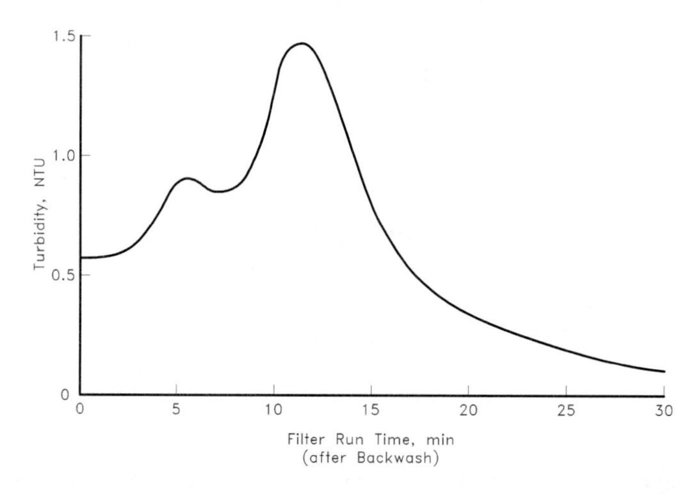

Figure 10-13 Example of a turbidity spike after backwashing.

Waste-To-Drain Waste-to-drain strategy is the least desirable alternative for controlling turbidity spikes. This concept involves rinsing the filters by connecting the filter effluent piping to the waste backwash drain system. Immediately after backwashing, the valve on this connection is opened and the high-turbidity rinse water from the filter effluent is wasted to the drain system. After the turbidity has been reduced, the valve is closed, and the filter is placed back into normal operation. This results in a large quantity of relatively clean water going to waste. There are other more desirable alternatives, so its use is becoming less common.

Ripening Filtration Rates This concept involves placing the filter on line at reduced filtration rates immediately after backwash. This allows the filter media and any remaining floc to settle and limits the amount of suspended material in the effluent. This process is known as *filter ripening***.** This is a simple process to control, it can be implemented at little or no cost, it usually is quite effective. It involves either changing to a manual backwash procedure or modifying the control logic of the automatic backwash system.

Preconditioning Filters Filter preconditioning involves introducing a coagulant, either a metal salt (such as ferric sulfate, ferrous sulfate, ferric chloride, or aluminum sulfate (alum)), or a polymer to the backwash water, just before the end of the backwash cycle.[13] The objective is to destabilize any restabilized suspended material remaining in the filter bed at the end of the backwash cycle. These destabilized solids are typically the source of post-backwash turbidity spikes. Adding a coagulant will restabilize this material so that it can be captured in the filter media instead of being released through the filter. To achieve this objective, a coagulant is added to the backwash water feed near the end of the backwash cycle and then stopped. The timing is such that the coagulant just begins to flow out of the washwater troughs while no coagulant remains in the gravel support, underdrain system, and bottom five cm of the lower media. The entire preconditioning process must carefully control the timing of the coagulant feed. If the timing is incorrect, several problems may develop: (1) coagulant may be wasted, (2) the treatment may be ineffective, or (3) effluent turbidities may increase because of coagulant in the underdrain systems.

10.8.8 Washwater Collection

To ensure that the filter beds are effectively cleaned, the flow of the backwash water (washwater) should be as nearly vertical as possible. In most filter installations, washwater collection troughs are utilized to ensure vertical flow. Typical arrangements of washwater collection troughs are illustrated in Figures 10-14(a) and 10-14(b). These troughs should be designed hydraulically to ensure free weir flow over the lip of the trough. The flow through the trough is considered spatially varied flow. Hydraulic calculations for such flow conditions are discussed in Chapter 9.

The spacing and the elevation of washwater troughs is an important consideration. If the troughs are spaced too far apart, the suspended floc will not be effectively removed during backwash. If the troughs are too low, filter media may be washed out. In general, it is better for the troughs to be closely spaced and set high above the expanded media. Eq. (10.17) can be utilized to determine a ratio of spacing to height.[10]

$$\frac{v'_s \pi}{U_b} < \frac{S}{D} < \pi \tag{10.17}$$

where

v'_s = settling velocity of the floc, m/s (ft/s)
U_b = backwash rise rate including surface wash rate, $m^3/m^2 \cdot s$ ($ft^3/ft^2 \cdot s$)
S = center-to-center spacing of the troughs, m (ft)
D = distance between top of the fluidized bed and water surface, m (ft)

Operating experience has indicated that the spacing of washwater trough (S) greater than 1.5 and less than 2 times H_0 yields satisfactory results in filter washing. The term H_0 is the vertical distance from the lip of the launder trough to the top of the media bed at rest. Typical values for H_0 range from 0.66 to 1.0 m (2 to 3 ft). The distance from the bottom of the trough to the top of the media should also be at least 0.2 m (8 in). Washwater troughs that are shallow, wide, and spaced closely together can create overly-high rise velocities that wash out anthracite coal during backwash. If the washwater trough covers more than half the area of the filter box, narrower and deeper troughs covering less surface area may be needed.

The cross section of the troughs is typically as shown in Figures 10-14(c) or (d). Both cross sections work effectively, but the section shown in Figure 10-14(d) is easier to construct, therefore is often slightly less expensive. The cross section shown in Figure 10-14(c) is often used for manufactured troughs constructed of fiberglass or other material.

10.8.9 Backwash Water Storage

A sufficient quantity of treated water must be stored at the treatment plant site to provide water for backwashing of the filters. The quantity to be stored should equal the maximum amount needed to backwash each filter once during a 24-hour period. In larger plants, where sufficient operating experience is available, the backwash can be reduced to the maximum amount of water required for backwashing two or three filters.

10.8.10 Backwash Water Return Systems

The amount of backwash water is quite large. This water must be disposed of by some means. One efficient means of disposing of waste backwash water is to return it to the head of the plant. Such a system has the following advantages: (1) the majority of the backwash water is reused, to reduce raw water cost, (2) the floc contained in the waste backwash water can provide nucleous in the flocculation process, thereby reducing the quantity of chemicals required, and (3) the amount of water to be disposed of is reduced. The major concern with backwash return water, however, is return and concentration of protozoan cysts in the system. Suggestions have been made to install a cyst-destruction system, such as ozonation or UV radiation, for recovered backwash water. Design considerations for backwash recovery systems are discussed in Chapter 14.

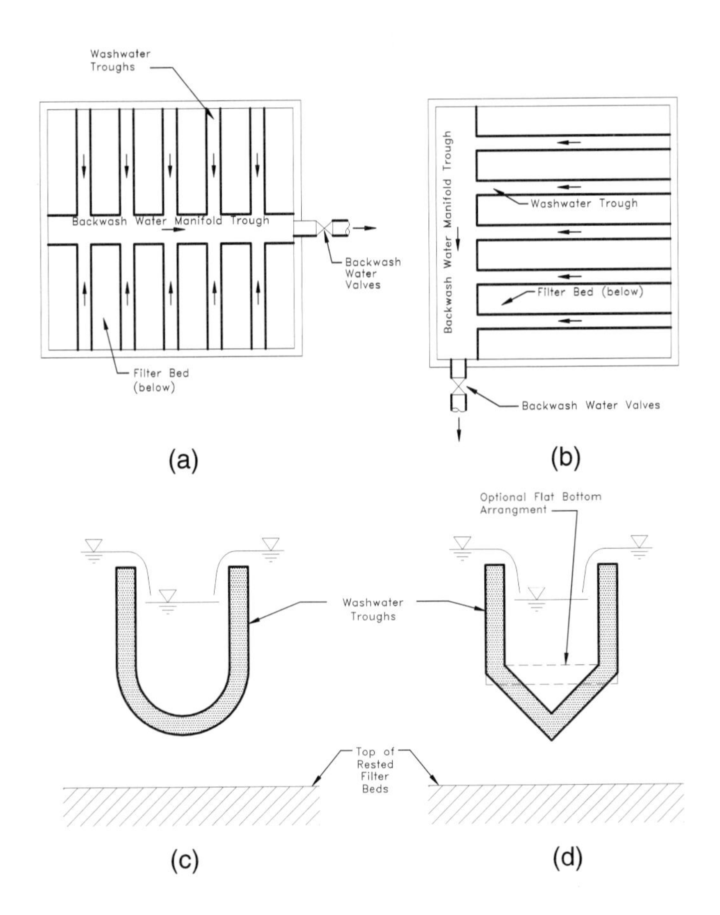

Figure 10-14 Typical backwash water (washwater) trough arrangements. (a) Center manifold. (b) Side manifold. (c) Typical U-shaped cross section. (d) Typical V-shaped cross section.

10.8.11 Backwash Water Delivery System

In most treatment plants, the clear water storage tanks are located at a hydraulic grade that is below the filter system. Therefore, the backwash water must be pumped to provide the hydraulic gradient required for backwash. A number of different types of systems are used to provide this gradient. Among these are (1) a high-capacity pump to deliver the entire backwash flow directly to the manifold, (2) an elevated storage tank, and lower capacity pumps to fill the tank, (3) pressure from the high-service pumps, and (4) the hydraulic gradient of the filtered water flume.

The first type of system usually has a lower capital cost, but, because of higher power requirements, may have a higher operating cost. The second system usually requires a higher capital cost due to the cost of the elevated tank, but it provides a more nearly constant power demand and, in many cases, a lower power cost. For the third system to be efficient, the head requirements of the backwash system and the high-service pumps should be compatible. The fourth type of system is employed by some proprietary systems. They require careful consideration of the head losses during backwash and proper selection of the elevation of the washwater collection trough. Typical schematic diagrams of these four types of backwash water system are shown in Figure 10-15.

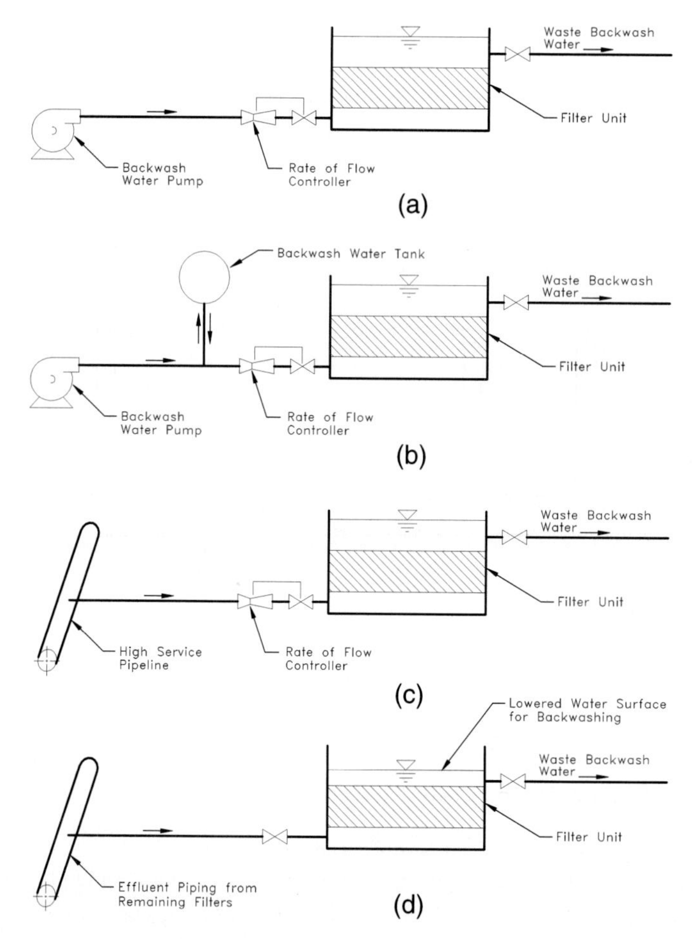

Figure 10-15 Typical backwash water systems. (a) Direct pumping. (b) Elevated storage.(c) High-service connection. (d) Filter system effluent connection.

10.9 Filter Media Support System

A porous support system for the filter media must be provided. In addition to providing structural support, this system must evenly distribute backwash water and collect filtered water across the entire filter cross section. Typically, the support system comprises two elements: (1) the *gravel layer* and (2) the *underdrain system*.

10.9.1 Gravel Layer

The gravel layer is located immediately below the filter media. Its purpose is to separate the filter media from the underdrain system, to prevent media particles from clogging the underdrain orifices, and to dissipate the backwash water jets from the orifices of the underdrain system. Gravel layers are typically 400–600 cm (150–240 in) deep and are made up of several layers of different-sized gravel grains. The gradation and depth of the gravel layer should be designed to match the hydraulic characteristics of the filter underdrain system. Design engineers should consult potential underdrain suppliers to select the most appropriate gravel support system.

Some manufacturers have developed a thin (2 to 3 cm) proprietary porous plate that is a substitute for gravel support systems. The plates are designed to support the media. They take up much less depth than gravel systems and distribute backwash flows as evenly as gravel. Therefore, they are particularly beneficial in filter renovations where filter cell depth is limited. Designers should consult with the manufacturers and their representatives for information on porous-plate systems.

10.9.2 Underdrain System

The underdrain system provides structural support for the entire filter system. It must have the structural strength to support the weight of the gravel and media layers and to resist the hydraulic pressures of the backwash water flow. The latter is an upward pressure requiring strong anchors to resist uplift. Several proprietary underdrain systems are available. Readers are referred to Appendix C for a list of manufacturers of proprietary systems.

10.10 Filter Instrumentation Systems

Instrumentation systems for water treatment systems are discussed in detail in Chapter 16. Because of their special needs, some instrumentation requirements for filter units are also discussed in this section. To meet the more stringent turbidity requirements specified under the Surface Water Treatment Rules, filter operation will be the heart of the treatment process. As a result, many important process control parameters such as head loss, effluent turbidity, and flow rates must be closely monitored and strictly controlled. These parameters are briefly discussed next.

10.10.1 Head Loss Instruments

Many existing filter units have loss-of-head-indicating instrumentation consisting of pressure taps at the filter influent and effluent. These taps are connected to pressure gauges, a manometer, or differential pressure instruments. Monk proposed a more elaborate loss-of-head instrumentation system, which is composed of several pressure taps spaced throughout the filter media bed.[13] These taps are connected to pressure instruments that can plot the pressure profile through the filter on a computer monitor. The shape of this plot will provide the operator a tool to closely monitor the filter operation and performance of the coagulation process. Figure 10-16 provides a visual indication of typical filter pressure monitors in a filter. If the head loss is occurring toward the top of the filter, either the floc is too large or too tough or both, the filter media are too fine (or need scalping), or the rate of filtration is too slow (Figure 10-16(a)). If the head loss curve is steeper, the floc is being carried well into the sand layer and may be the cause of premature turbidity breakthrough (Figure 10-16(b)). If the head loss gradually increases in both media, as shown in Figure 10-16(c), the floc is satisfactory.

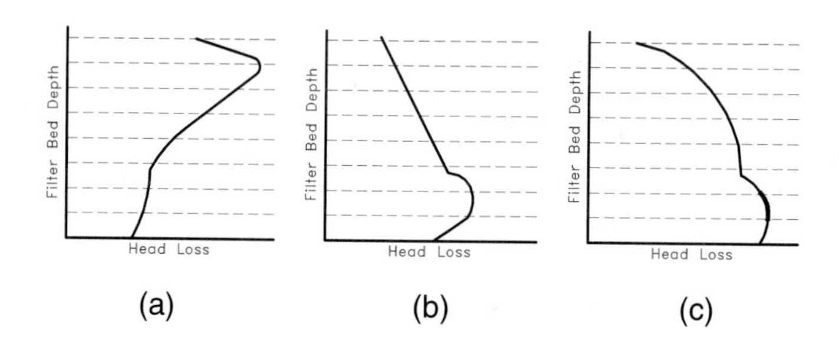

Figure 10-16 Typical indications from filter pressure monitors. (a) Floc too large or strong. (b) Floc too weak. (c) Satisfactory floc.

10.10.2 Filter Turbidity Monitoring

In the past, most of the filter operations focused on head loss and filter cycle time as indicators for backwashing. The more stringent turbidity requirements of the *Surface Water Treatment Rules* will necessitate the use of individual turbidity monitors or *turbidimeters* on the effluent of each filter unit.[17] The new rules leave operators with less margin to operate; therefore, turbidity breakthrough will also become an important consideration in deciding when to backwash a filter. Placing turbidimeters on each filter effluent will give the operator a tool to monitor the performance of each filter and to make informed decisions regarding filter backwash.

10.10.3 Flow Measurement

It is desirable to have some form of flow measurement device installed at each filter unit. Some types of control schemes, such as effluent-controlled constant-rate filters, require flow measurement. Most other control schemes do not necessarily have an individual flow meter for individual filter, although individual meters are desirable. Individual flow meters provide the operators the capability to measure the flow distribution between the filters and ensure that a single filter is not overloaded.

10.11 Manufacturers of Filtration Equipment

A list of manufacturers and suppliers of filter media, surface wash systems, filter underdrains, and filter washwater troughs is provided in Appendix C. Information such as the characteristics and volume of the water to be treated, the size and geometry of the proposed units, and the basic design parameters for the equipment are required in order to make equipment selections. The manufacturers can assist the designer in selecting the proper equipment for the proposed application. Readers are referred to Appendix C for a more complete discussion of selecting and specifying equipment for specific water treatment processes.

10.12 Information Checklist for Design of Filtration System

The following information must be obtained and necessary decisions made before the design engineer can begin to develop the design of a filtration facility.

1. Information to be obtained from pre-design studies:

   a. design flows;
   b. filtration rate;
   c. backwash rate;
   d. filter cycle time;
   e. surface wash rates and durations;
   f. influent and effluent water qualities.

2. Information to be obtained from the preliminary design report.

   As part of the facility plan, the design engineer develops design information that may generally include the following:

   a. number of proposed units;
   b. preliminary hydraulic profile, illustrating the head loss parameters proposed for the units being designed;
   c. type of filter media;
   d. type of underdrain system;
   e. type of surface wash system;
   f. type of washwater pressure system;
   g. method of handling and disposing of waste streams.

3. Information to be obtained from the plant owner and operator:

   a. any particular preference with regard to equipment type and manufacturers;
   b. any particular preference with regard to the design of other plants in operation.

4. Information to be obtained from research by the design engineer:

   a. minimum design parameters established by the regulatory authorities;
   b. types, sizes, and limitations of available equipment.

10.13 Design Example

10.13.1 Design Criteria Used

The following criteria are used in the design example.

1. Flow rates

   a. Maximum day flow = 113,500 m^3/d (30 mgd)
   b. Average day flow = 57,900 m^3/d (15.3 mgd)

2. Design parameters

   a. Filtration rate (hydraulic loading): 10 $m^3/m^2 \cdot h$ (4.1 gpm/ft^2)
   b. Backwash rate: 10 percent of the settling velocity of the media or a minimum of 37 $m^3/m^2 \cdot h$ (15 gpm/ft^2), which is higher
   c. Surface wash rate: 0.061 $m^3/m^2 \cdot min$ (1.52 gpm/ft^2)
   d. Minimum filtration cycle: 24 h
   e. Provision for applying filter aid at the designed capacity of the treatment plant.

3. General design guidelines

   a. Filters: A total of eight filter cells are provided. The capacity of each filter cell is one-seventh of the total plant capacity. This allows one unit to be out of service for backwashing or under maintenance, while maintaining the design filtration rate at full plant capacity.
   b. Filter media: The filters are of dual-media type, utilizing anthracite coal and quartz sand. The *effective size* of the sand is 0.50 mm. The *effective size* of the anthracite coal is such as to yield a settling velocity equal to the average settling velocity of the sand. The uniformity coefficient, specific gravity, and porosity ratio for the anthracite coal and sand are 1.6 and 1.4, 1.55 and 2.65, and 0.48 and 0.40, respectively.
   c. Underdrain: A perforated clay tile underdrain system is used.

d. Surface wash: A rotating-arm type surface wash system utilizing 0.061 $m^3/m^2{\cdot}min$ (1.52 gpm/ft^2) of water at a minimum pressure of 690 kPa (100 psi) is installed.

e. Backwash water system: Backwash water is supplied by an elevated tank having the capacity to hold the quantity of water sufficient to backwash two filters. Three backwash water pumps are provided to fill the elevated tank.

f. Disposal of backwash water: The backwash water will be discharged into a surge or equalization basin. The backwash water will then be released at a constant-rate into a recovery basin. The supernatant from the recovery basin will be returned to the rapid mix, and the sludge will be pumped into the sludge thickening system. The design details for the backwash water disposal system are presented in Chapter 14.

g. Filter control system: Each filter has an effluent-controlled constant-rate system.

4. Unit arrangement and layout

a. Unit arrangement:

Eight individual filter units are provided. Selection of the number of filter units is based on reducing the impact of backwashing a filter unit while also attempting to limit cost and the work load of the operators. Increasing the number of filter units has less effect upon the hydraulic loading on remaining filter units when one is taken out of service. Therefore, the size of the individual units is reduced. As an example, if only two units are provided, the loading on the other filter would increase 100 percent if one filter is taken out of service. This will require twice the area of each filter. In the Design Example, with eight units, the loading rate and filter area are increased by only 14 percent, a more acceptable value. Increasing the number of filter units results in less total filter area to offset the loss of capacity during backwashing, but also increases construction cost and operator work load. More filter units, while occupying slightly less area, require more walls, piping, valves, and equipment. Furthermore, each additional filter unit must be cleaned and monitored by the operator. Considerable experience in design operation and economics is needed to properly select the number of filters for a specific application.

In the judgement of the authors, eight filters is a good number for the size of treatment plant used in the Design Example. Smaller treatment plants would certainly require fewer units, larger plants more; however, a minimum of four units should be provided, even for small-or medium-sized treatment plants.

b. Unit layout:

The filter units are laid out as illustrated in Figure 10-17. The overall arrangement provides two banks of four filter units, with each bank divided by the administration and control building. Each bank of filters is further divided into two banks of two filter units with, a pipe gallery in the middle. Provision for eight additional filters, for future expansion, is also made. This arrangement provides the operator easy access to the filter units to monitor backwash operations. It also allows efficient layout of the filter piping.

The filter-complex plan is illustrated in Figure 10-18. This layout features an influent header that extends the length of the filter unit and the administration building complex. Water from the sedimentation basins is fed into the header at each end of the pipe gallery. Each filter output is measured by a flow meter in the effluent pipe. An effluent header also extends along the length of the pipe gallery and connects to a filtered water pipeline in the basement of the administration building. The pipeline carries the filtered water to a filter back pressure weir common bay before the chlorine contact channels. The third header is the backwash water header. This pipe carries backwash water to each individual filter unit. The waste backwash water piping is located around the perimeter and extends to a surge storage tank of the backwash water recovery system.

10.13.2 Design Calculations

Step A: Filter Unit Design

1. Unit dimensions

The filter units are designed at an average loading rate of 10 $m^3/m^2{\cdot}h$ (4.1 gpm/ft^2). Under maximum day flow conditions, the flow is distributed among seven filter units assuming that one is out of service for backwashing.

a. Calculate the required area for each filter unit.

Maximum design flow = 113,500 m^3/d

$$\text{Design flow for each basin} = \frac{113{,}500\ m^3/d}{7} = 16{,}214\ m^3/d$$

$$= 676\ m^3/h = 0.188\ m^3/s$$

At a filtration rate of 10 $m^3/m^2{\cdot}h$ (per Design Criteria), calculate the required unit area as follows:

$$A_{req} = \frac{676\ m^3/h}{10\ m^3/m^2{\cdot}d} = 67.6\ m^2$$

b. Calculate dimensional requirements.

Using square filter units, the required dimensions are:

Length (L) = Width (W)= $(67.6\ m^2)^{1/2}$ = 8.22 m

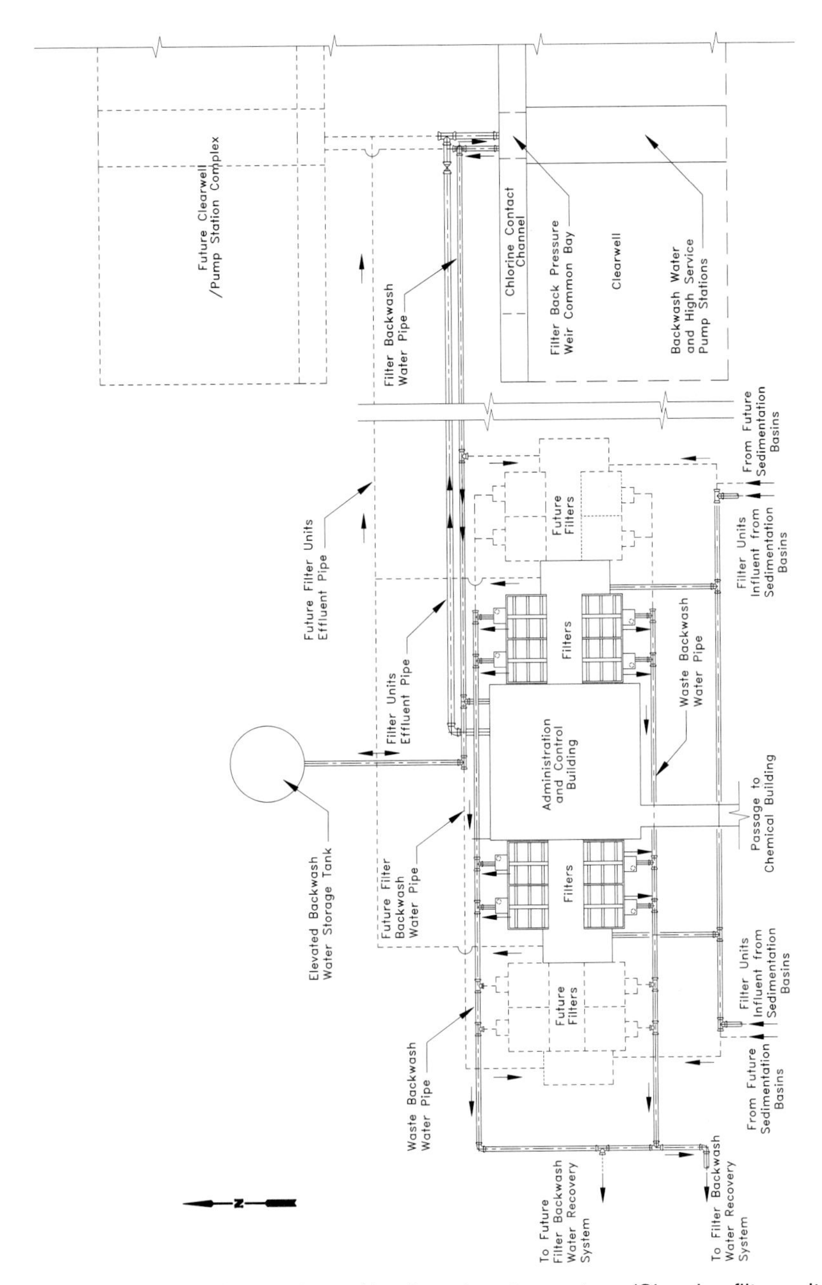

Figure 10-17 Layout of filter units and backwash water system. (Showing filter units, control and administration building, and backwash water system.

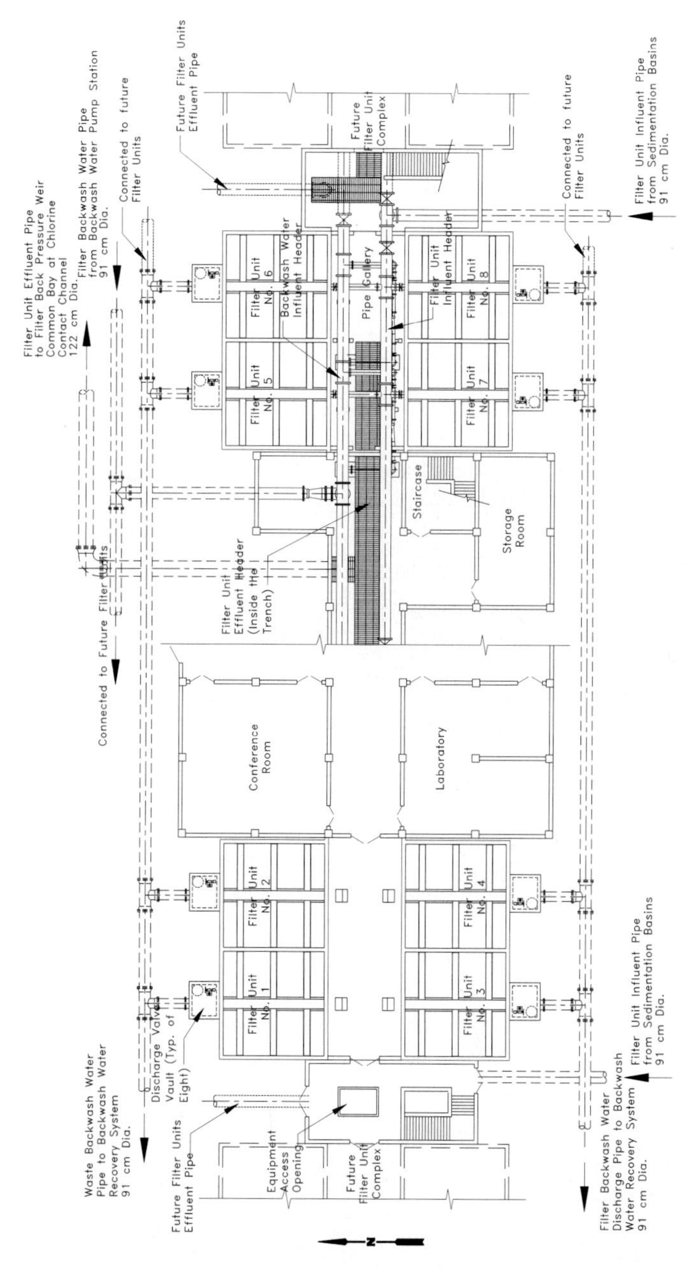

Figure 10-18 Filter-complex plan. (Left side showing ground floor and right side showing lower level.)

Provide dimensions of $L \infty W = 8.25\ \text{m} \infty 8.25\ \text{m}$. The surface area of each filter unit is therefore $A_f = 68.1\ \text{m}^2$. Detailed design dimensions of filter units are shown in Figure 10-19.

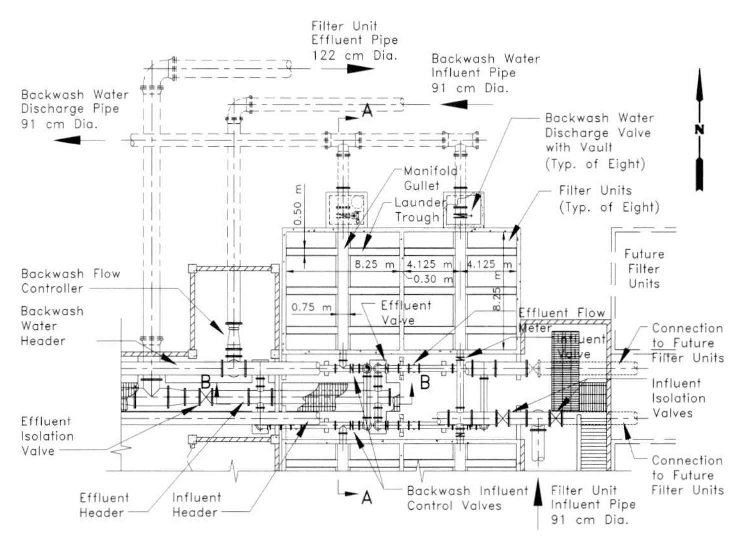

Figure 10-19 Plan view of filter units No. 5 and 6 in the Design Example. (Section views of A-A and B-B are shown in Figure 9-20.)

2. Process design and media selection

   a. Select media size.

   The procedure here used for media design and selection is one proposed by Kawamura.[2] This procedure is generally used to estimate the initial values for use in pilot plant tests. In many larger plants, media selection is based on extensive pilot testing to identify the optimum media design. Such testing is not always justified for small or medium-size plants, so experience and judgement must be used. In this design example, the media selection procedure is based on the authors' judgement, and the procedure is typical of many dual-media filters.

   The top media layer is anthracite coal of an undetermined size. The second media layer is quartz sand with an effective size $d_2 = 0.5$ mm. The upper anthracite layer must have a settling velocity compatible with that of the sand layer, to ensure effective backwashing. Eq. (10. 8) is used to select the media on the basis of the values given as the design criteria. From $S_{g1} = 1.55$ (anthracite), $S_{g2} = 2.65$ (sand), and $d_2 = 0.5$ mm (sand), the effective size of anthracite grains with the same settling velocity as the sand layer is calculated from Eq. (10.8):

$$d_1 = 0.5 \text{ mm} \times \left(\frac{2.65 - 1.0}{1.55 - 1.0}\right)^{2/3} = 1.04 \text{ mm or } 1.0 \text{ mm}$$

b. Determine the depth of filter bed.

Assume that the anthracite is 2/3 of the total bed depth and sand is the remaining 1/3. This ratio is also based on the judgement of the authors, and is typical for dual-media filters.

i. Calculate the weighted average of the media size.

$$\text{Weighted average media size} = 1.0 \times \frac{2}{3} + 0.5 \times \frac{1}{3} = 0.83 \text{ mm}$$

$$\text{Weighted average porosity ratio} = 0.48 \times \frac{2}{3} + 0.40 \times \frac{1}{3} = 0.45$$

ii. Determine the total bed depth.

The total bed depth is calculated from Figure 10-6 by using the average media size, 0.83 mm, and the average porosity ratio, 0.45.

$$l = 7.5 \times 10^2 \text{ mm or } 75 \text{ cm (29.5 in)}$$

iii. Calculate the depth of each media layer.

$$l_1 = 75 \text{ cm} \times \frac{2}{3} = 50 \text{ cm (19.7 in)}$$

$$\text{Sand depth } l_2 = 75 \text{ cm} \times \frac{1}{3} = 25 \text{ cm (9.8 in)}$$

The media details are summarized in Table 10-5. The total filter media depth is 0.75 m, as shown in Figure 10-20(a).

3. Head losses and hydraulic profile.

In this section, the hydraulic calculations are conducted for the head losses from the beginning of the filter influent header inside the filter piping gallery (point A in Figure 10-21) to the chlorine contact channels. These channels are an integral part of the clearwell/pump station complex. (See Figure 10-17.) The head losses through the 91-cm diameter pipe between the sedimentation basin box and the filter influent header are calculated in Chapter 15, where a hydraulic profile through the plant is prepared.

a. Calculate the head losses through the influent piping system.

Although the filter system is composed of eight filters configured in two symmetric banks of four filters each, the head loss calculations are based on one filter's being out of service under maximum day flow condition. The worst-case scenario is when one of four outside filters (i.e., filter units No. 1, 3, 6 or 8) is taken out of service. It

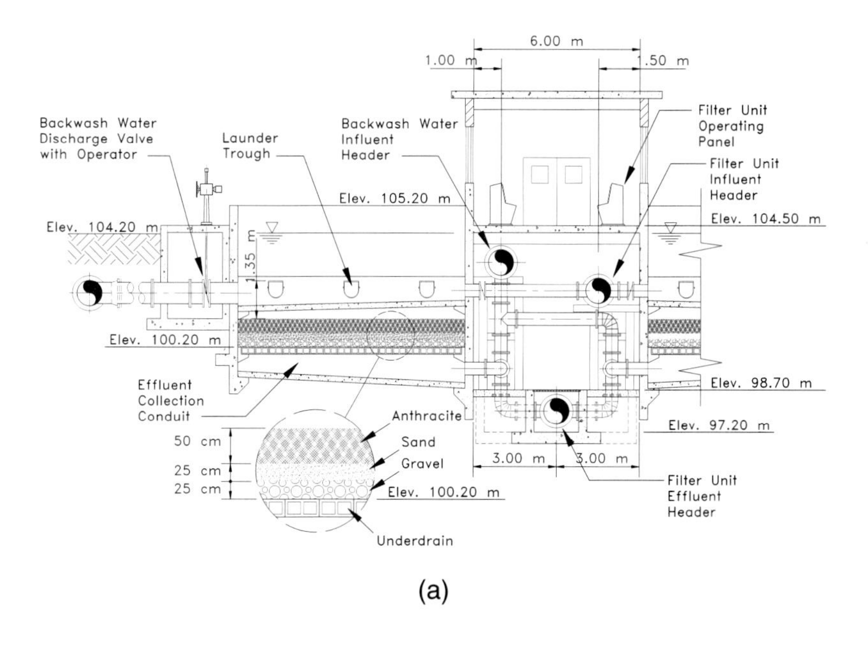

(a)

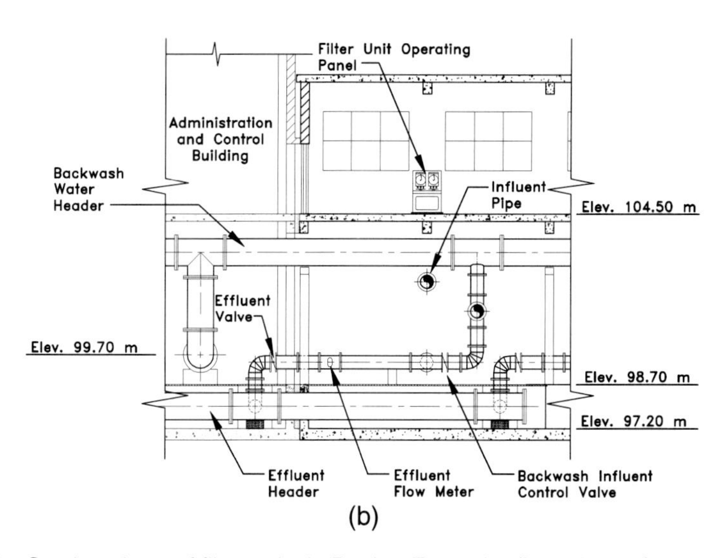

(b)

Figure 10-20 Section views of filter units in Design Example. (Locations of section A-A and B-B are shown in Figure 9-19.) (a) Section A-A. (Showing filter media, gravel support, underdrain, and backwash water collection launders and gullet.) (b) Section B-B. (Showing piping arrangement in piping gallery.)

Table 10-5 Summary of Filter Media Design

Layer	Material	Effective Size, mm	Depth, cm	Uniformity Coefficient	Specific Gravity	Porosity	L/d_e Ratio[a]
Top	Anthracite	1.0	50	1.6	1.55	0.48	500
Bottom	Sand	0.5	25	1.4	2.65	0.40	500
Total depth of filter media layer			75 cm or 0.75 m		Total L/d_e ratio		1000

a Average L/d_e ratio $= \dfrac{750 \text{ mm}}{0.83 \text{ mm}} = 904.$

is assumed that the flows in two pipes from the sedimentation basins are equal, and Filter No. 8 is closed. Under this condition, the flow distribution in the influent piping system at maximum day flow is shown in Figure 10-21(a). A transfer of flow (0.094 m^3/s) from the right bank to the left bank of filters will occur.[d] The control route for head loss calculations is therefore from Point A (the entering point of right influent header) to Point B (the entering of filter unit No. 2). The piping details and fittings are shown in Figure 10-21(b). The calculation procedures and results are shown in Table 10-6. The total head loss through the influent piping is 0.41 m.

b. Calculate head losses through the filter media.

i. Calculate head losses through the clean filter media.

The Carmen–Kozeny equation (Eq. (10.1)) is used to estimate the head loss through the clean filter media that consist of two layers: 0.5 m anthracite layer on top of 0.25 m sand layer. (See Section 10.13.2, Step A, 2.a and Figure 10-20(a) for filter media design details.) The head loss through each layer is calculated from Eqs. (10.1), (10.5) and (10.6) by using the following data:

Velocity $v = 10$ m/h $= 0.0028$ m/s (Section 10.13.1)

Density of water at 5° C (ρ_w) = 1000 kg/m^3

Viscosity of water at 5°C (μ) = 1.518 ∞ 10^{-3} N·s/m^2 (or kg/m·s)

Shape factor $\phi = 1.0$

The head loss calculations are listed in Table 10-7. Total head loss through the clean filter media is 0.46 m.

ii. Determine media terminal head loss and effluent monitoring.

d. It should be noted that this calculation procedure may not represent the actual head loss in the influent piping. The influent pipes from two sedimentation basins are interconnected outside the filter units, as shown in Figure 10-17. When one filter is closed, the transfer of flow from one bank to another bank of filters occurs not only through the pipe header Segment III, but also through the outer connecting pipe; however, the procedure presented above does offer simplified calculations and gives a slightly conservative design.

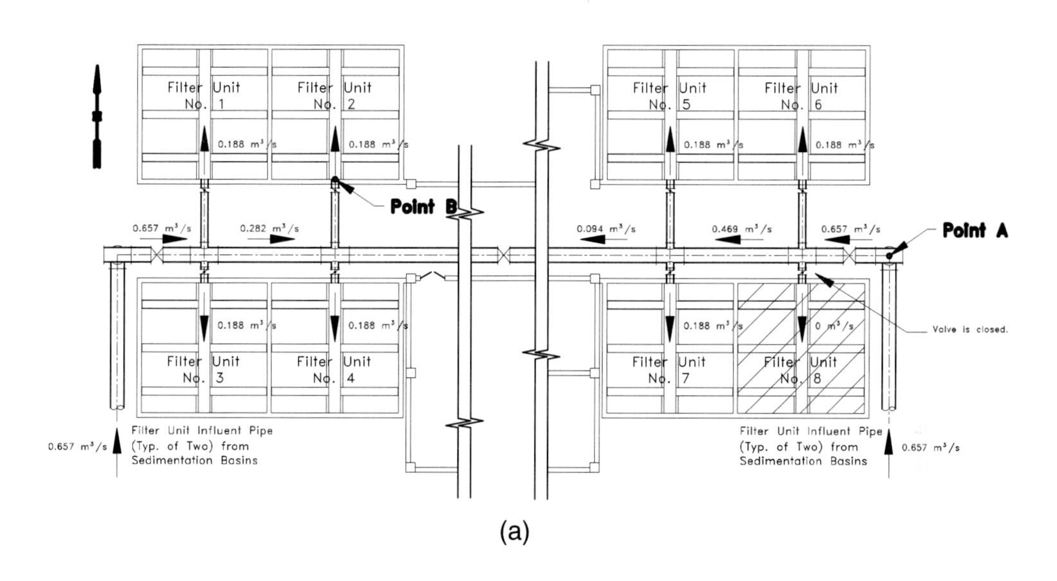

(a)

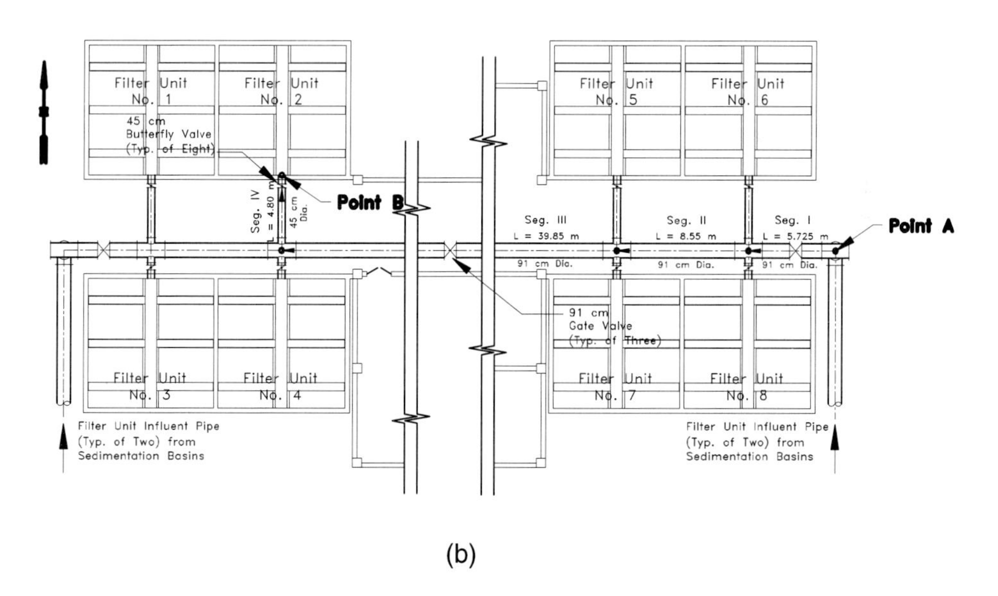

(b)

Figure 10-21 Filter influent piping layout for hydraulic calculations. (a) Flow distribution. (b) Segment assignment and dimensions.

Table 10-6 Calculations of Head Losses through Filter Influent Piping System

Appurtenance[a]	Equation Used	Flow Rate, m^3/s	Size, m	Length, m	Velocity, m/s	Head Loss, m
Segment I	Eq. (7.14), C = 120	0.657[b]	0.91	5.725	1.01	0.006[c]
Isolation valve in Segment I	Eq. (7.12), K = 1.0	0.657	0.91	N/A	1.01	0.052[d]
Cross (run-to-run) between Segments I and II	Eq. (7.12), K = 0.60	0.657	0.91	N/A	1.01	0.031
Segment II	Eq. (7.14), C = 120	0.469[e]	0.91	8.55	0.721	0.005
Cross (run-to-run) between Segments II and III	Eq. (7.12), K = 0.60	0.469	0.91	N/A	0.721	0.016
Segment III	Eq. (7.14), C = 120	0.094[f]	0.91	39.85	0.144	0.001
Isolation valve in Segment III	Eq. (7.12), K = 1.0	0.094	0.91	N/A	0.144	0.001
Cross (run-to-branch) between Segments III and IV	Eq. (7.12), K = 1.8	0.188[g]	0.91 ∞ 0.45	N/A	1.18	0.128
Segment IV	Eq. (7.14), C = 120	0.188	0.45	4.80	1.18	0.016
Butterfly valve in Segment IV	Eq. (7.12), K = 1.2	0.188	0.45	N/A	1.18	0.085
Exit	Eq. (7.12), K = 1.0	0.188	0.45	N/A	1.18	0.071
Total head loss through filter influent piping system						0.41

a See Figure 10-21 for segment numbering (Segments I-IV) and flow distribution.

b Flow in each pipe from sedimentation basins $= \dfrac{113{,}500\ m^3/d}{2 \times 86{,}400\ s/d} = 0.657\ m^3/s.$

c $h_f = 6.81 \times \left(\dfrac{1.01\ m/s}{120}\right)^{1.85} \times \dfrac{5.725}{0.91^{1.167}} = 0.006\ m.$

d $h_m = \dfrac{1.0(1,01)^2}{2 \times 9.81} = 0.052\ m.$

e Flow in Segment II = Flow in Segment I – Flow to Filter No. 6 = $0.657\ m^3/s - 0.188\ m^3/s = 0.469\ m^3/s$.

f Flow in Segment III = Flow in Segment II – Flow to Filters No. 5 & 7 = $0.469\ m^3/s - 2 \infty 0.188\ m^3/s = 0.094\ m^3/s$.

g Flow in Segment IV = Flow to Filter No. 2 = $0.188\ m^3/s$.

Table 10-7 Calculations of Head Loss through Clean Filter Media

Layer	Size, mm	Depth, cm	Porosity	Reynolds Number (N_R)	Friction Coefficient (f)	Head Loss (h_L), m
Anthracite layer	1.0	50	0.48	1.84[a]	44.1[b]	0.083[c]
Sand layer	0.5	25	0.40	0.922	99.4	0.372
Total depth of the media layers		75 cm or 0.75 m	Total head loss through clean filter media layers			0.455 or 0.46

a The Reynolds number is calculated from Eq. (10.6).

$$N_R = \frac{0.001\text{ m} \times 0.0028\text{ m/s} \times 1000\text{ kg/m}^3}{1.518 \times 10^{-3}\text{ N·s/m}^2} = 1.84.$$

b The friction coefficient f is calculated from Eq. (10.5).

$$f = 150 \times \frac{[1 - 0.48]}{1.84} + 1.75 = 44.1.$$

c The head loss h_L is calculated from Eq. (10.1).

$$h_L = \frac{44.1 \times (1 - 0.48) \times 0.50\text{ m} \times (0.0028\text{ m/s})^2}{1.0 \times (0.48)^3 \times 0.001\text{ m} \times 9.81\text{ m/s}^2} = 0.0829\text{ m}.$$

The terminal head loss of a clogged filter can be determined experimentally via pilot plant testing. In a pilot plant test, various filtration rates and terminal head loss values are tested to determine design values that optimize (1) finished water quality and (2) total water volume filtered between backwash cycles. A differential pressure gauge in each filter is provided to monitor, totalize, and record the head loss through each filter medium. In this Design Example, a terminal head loss of 2.5 m is assumed. This value is a typical value for high-rate filters. It will be used to determine head losses and the hydraulic profile through the filter system. A time clock and pH meter will record the filter run and monitor the pH of the filtered water. A turbidimeter in each filter effluent pipe will be provided to monitor the effluent turbidity. If justified, one on-line particle counter may be provided and sample piping routed in such a way that the same particle counter will monitor effluent quality from each filter. Valve limit switches will be provided in each filter to display effluent valve position (percent open).

iii. Determine minimum water depth in the filter.

The minimum water level in the filter unit is controlled by a downstream overflow weir. A weir structure containing a filter back-pressure weir common bay is provided outside the filter building, upstream of the chlorine contact channel. This weir controls the water level in the filter units (Figure 10-17). The minimum water level must be selected to provide sufficient head over the filter media to prevent negative pressures in the media. Such a phenomenon, called *air binding*, is detrimental to proper filter operation. To ensure positive pressure, the water depth over the media should exceed the maximum head loss through the media bed. The terminal head loss is 2.5 m, so a water depth of 3.01 m (> 3 m, required) above the filter media is provided in the filter unit. An ultrasonic level transducer will be provided in each filter to monitor, display, and record the water level over each filter.

c. Calculate head losses through the gravel support.

The calculations for head losses through the gravel support layers are similar to those for clean filter media. The gravel support is made up of four layers of gravel, as summarized in Table 10-8. This gravel support system is typical of rapid sand filters. The total depth of the gravel support layers is 0.25 m (see Figure 10-20(a)). The head loss calculations are also listed in Table 10-8. Total head loss through the gravel layers is 0.0028 m. This head loss is relatively small in comparison with that through the filter media and is assumed to be zero.

d. Calculate head loss through the underdrain system.

In this Design Example, a *clay tile* underdrain system is used. The hydraulic characteristics of such systems can be determined from data published in the manufacturing catalog or can be estimated by Eq. (10.19).[12]

The constant k_1 in Eq. (10.18), will vary with the type of underdrain system.

$$H_L = k_1 v^2 \tag{10.18}$$

where

H_L = head loss through the underdrain, m
v = filtration rate (hydraulic loading), m/h
k_1 = head loss constant that varies with the type of underdrain system (generally is given by the manufacturer; for the tile underdrain selected, $k_1 = 0.0005$)
$H_L = 0.0005 \infty (10.0 \text{ m/h})^2 = 0.05$ m

Table 10-8 Gravel Support Details and Head Loss Calculations[a]

Layer	Size, mm	Depth, cm	Porosity	Reynolds Number (N_R)	Friction Coefficient (f)	Head Loss (h_L), m
Top layer	2.5	5	0.45	4.61	19.6	0.0019
Second layer	4.5	5	0.45	8.30	11.7	0.0006
Third layer	9.5	7.5	0.50	17.5	6.04	0.0002
Bottom layer	15.5	7.5	0.50	28.6	4.37	0.0001
Total depth of the gravel layers		25 cm or 0.25 m	Total head loss through the gravel layers			0.0028 or 0.00

a See footnotes in Table 10-7 for detailed explanation of calculation procedures.

e. Calculate head losses through the effluent piping.

The head losses through the filter effluent piping system are calculated in a manner similar to that used for the influent piping. As with the influent piping, the flow distribution in the effluent piping system under maximum day flow condition is prepared as shown in Figure 10-22(a). The head losses through the effluent piping system are calculated only for the bank with all four filters in service, where the higher head loss is encountered. Although this approach is not totally accurate, it offers simplified calculations and a slightly conservative result. The calculations are performed from Point C (the exit of Filter No. 3) to Point D (the connecting point of the effluent headers with the filtered water conduit). The calculation steps are summarized in Table 10-9. The piping details and fittings are shown in Figure 10-22(b). The calculation procedures and results are shown in Table 10-7. The total head loss through the effluent piping is 0.80 m.

f. Calculate head loss through the filtered water conduit.

The filtered effluent is conveyed to the filter back-pressure weir common bay through a 122-cm diameter pipe. (See Figure 10-17.) The head losses include friction loss ($C = 120$) and minor losses that will occur at one 90° elbow ($K = 0.6$), one isolation gate valve ($K = 1.0$), one Tee (branch-to-run, $K = 1.8$), and one exit ($K = 1.0$) in the common bay. The head loss calculations show a total head loss of 0.52 m through the filtered water pipe.

g. Calculate the head over the common bay weir.

In this Design Example, the filtered water from all filters is conveyed to a filter back-pressure weir common bay. Readers should refer to Chapter 12 for design details of the back-pressure weir common bay. The flow is split at the common bay over two weirs and drops into two chlorine contact channels. The length of weir is flexible and can be selected by an engineer to meet the specific needs of the design. A greater length will cause lesser head variations between high and low flows but will increase the structure cost. In this example, a weir length of 2.5 meters is satisfactory.

The head over the weir is calculated from Eq. (8.24) by using $Q = \dfrac{1.314\ \text{m}^3/\text{s}}{2}$ $= 0.657\ \text{m}^3/\text{s}$, $C_d = 0.6$, and two end contractions ($n = 2$ and $L = 2.5$ m). By trial-and-error, the final results are as follows:

$$L' = 2.5\ \text{m} - 0.1 \times 0.28\ \text{m} \times 2 = 2.44\ \text{m}$$

$$H = \left(\frac{3 \times 0.657\ \text{m}^3/\text{s}}{2 \times 0.6 \times 2.44\ \text{m} \times \sqrt{(2 \times 9.81\ \text{m/s}^2)}}\right)^{2/3} = 0.28\ \text{m}$$

A free-fall of 0.39 m is provided at the weir.

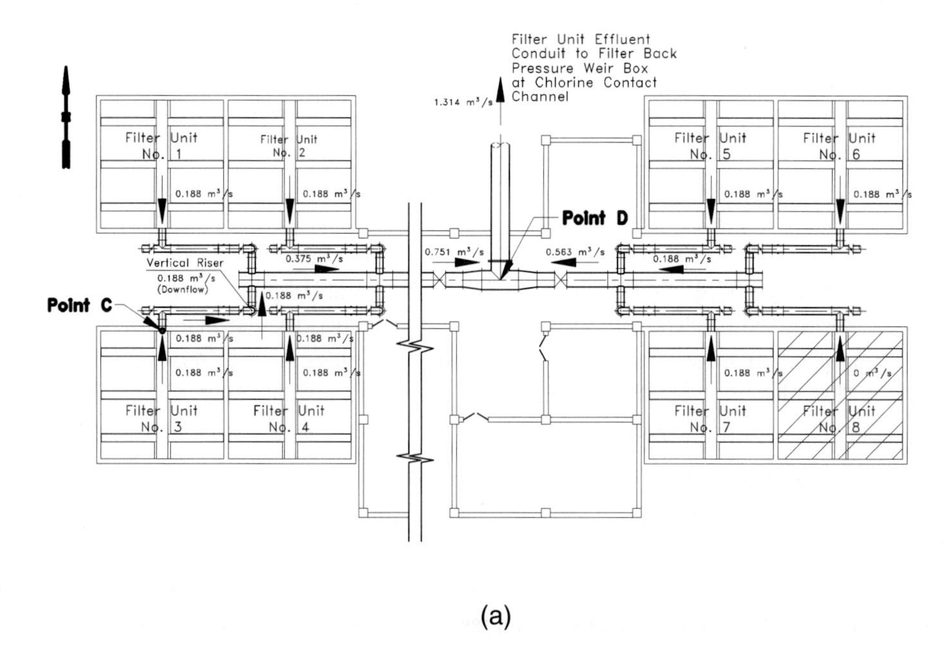

(a)

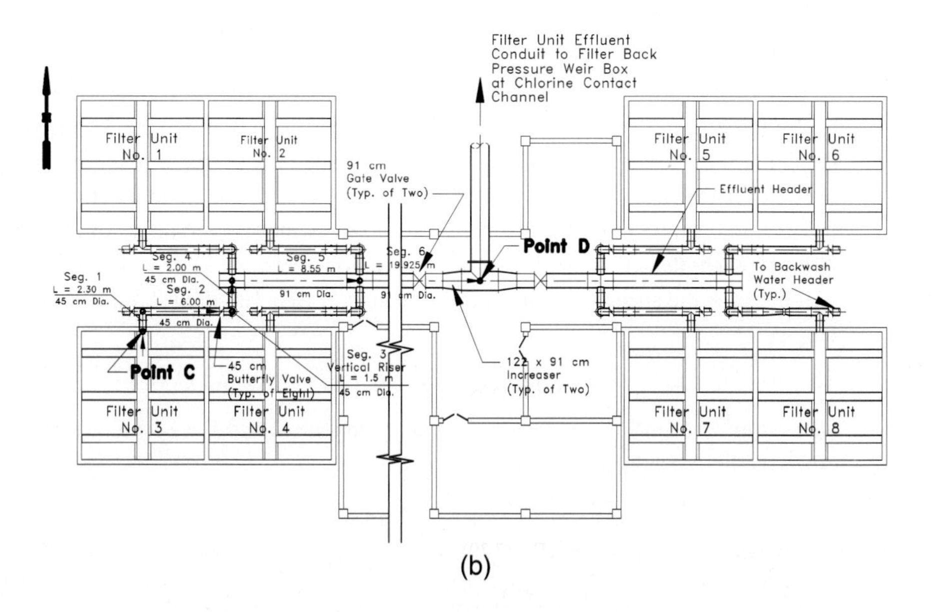

(b)

Figure 10-22 Filter effluent piping layout for hydraulic calculations. (a) Flow distribution. (b) Segment assignment and dimensions.

Table 10-9 Calculations of Head Losses through Filter Effluent Piping System

Appurtenance	Equation Used	Flow Rate, m^3/s	Size, m	Length, m	Velocity, m/s	Head Loss, m
Entrance	Eq. (7.12), $K = 0.5$	0.188[a]	0.45	N/A	1.18	0.036
Segment 1	Eq. (7.14), $C = 120$	0.188	0.45	1.3	1.18	0.004
Tee (branch-to-run) between Segments 1 and 2	Eq. (7.12), $K = 1.8$	0.188	0.45	N/A	1.18	0.128
Segment 2	Eq. (7.14), $C = 120$	0.188	0.45	6.0	1.18	0.020
Butterfly valve in Segment 2	Eq. (7.12), $K = 1.2$	0.188	0.45	N/A	1.18	0.085
90° Elbow (horizontal-to-vertical down) between Segments 2 and 3	Eq. (7.12), $K = 0.3$	0.188	0.45	N/A	1.18	0.021
Segment 3	Eq. (7.14), $C = 120$	0.188	0.45	1.5	1.18	0.005
90° Elbow (vertical down-to-horizontal) between Segments 3 and 4	Eq. (7.12), $K = 0.3$	0.188	0.45	N/A	1.18	0.021
Segment 4	Eq. (7.14), $C = 120$	0.188	0.45	2.0	1.18	0.007
Tee (branch-to-run) between Segments 4 and 5	Eq. (7.12), $K = 1.8$	0.188	0.45	N/A	1.18	0.128
Segment 5	Eq. (7.14), $C = 120$	0.375[b]	0.91	8.55	0.577	0.003
Cross (run-to-run) between Segments 5 and 6	Eq. (7.12), $K = 0.60$	0.751[c]	0.91	N/A	1.15	0.041
Segment 6	Eq. (7.14), $C = 120$	0.751	0.91	19.925	1.15	0.028
Isolation valve in Segment 6	Eq. (7.12), $K = 1.0$	0.751	0.91	N/A	1.15	0.068
Increaser after Segment 6	Eq. (7.39), $K = 1.0$	0.751	0.91 ∞ 1.22	N/A	before $v_1 = 1.15$ after $v_2 = 0.642$	0.047[d]
Cross (run-to-branch) after increaser	Eq. (7.12), $K = 1.8$	1.314[e]	1.22	N/A	1.12	0.116
Total head loss through filter effluent piping system						0.76

a Flow in Seg. 1 = Flow from Filter No. 3 = 0.188 m^3/s.

b Flow in Segment 5 = Flow from Filters No. 1 & 3 = $\frac{2}{7} \times 0.188\ m^3/s = 0.375\ m^3/s$.

c Flow in Segment 6 = Flow from Filters No. 1, 2, 3 & 4 = $\frac{4}{7} \times 1.314\ m^3/s = 0.751\ m^3/s$.

d The head loss in increase is calculated from Eq. (7.39):

$$h_L = \frac{K(v_1^2 - v_2^2)}{2g} = \frac{1 \times [(1.15\ m/s)^2 - (0.642\ m/s)^2]}{2 \times 9.81\ m/s^2} = 0.047\ m.$$

e Flow after the Tee at Point D = Maximum day flow = 1.314 m^3/s.

h. Calculate the total head loss through the filter system.

The head losses through the filter system are as follows:

• Head loss through the filter influent piping	0.41 m
• Filter media terminal head losses	2.50 m
• Head loss through the gravel support	0.00 m
• Head loss through the underdrain system	0.76 m
• Head loss through the filter effluent piping	0.05 m
• Head loss through the filtered water conduit	0.52 m
• Free fall provided at the filter common bay weir	0.39 m
Total =	4.63 m

The maximum head loss through the filter system is therefore 4.63 m when the filter media are dirty before the backwash. After backwash, the total head loss is reduced to 2.59 m, the minimum head loss that can be achieved through the system under the maximum day flow condition.

i. Prepare hydraulic profile through the filter system.

The water surface elevations in the filter units are controlled by the water level in the filter back-pressure weir common bay and the head losses in the piping between the filter units and the common bay. The water surface level in the common bay is, further, affected by the weir elevation. (See Figure 12-25 for design details of common bay.) A top elevation of 100.1 m is provided at the weir. Starting from the weir elevation, the water surface elevation (WSEL) and major design elevations are calculated, and the hydraulic profile through the sedimentation basin is prepared. The results are shown in Figure 10-23.

The WSEL in the common bay = weir elevation + head over the weir = 100.10 m + 0.28 m = 100.38 m. The WSEL in the chlorine contact chamber = WSEL in the common bay - free fall provided at the weir = 100.38 m – 0.39 m = 99.99 m.

The maximum WSEL in a filter unit when its filter media are dirty = WSEL in the common bay + head losses through the filtered water conduit + head losses through the filter effluent piping + head loss through the underdrain system + head losses through the gravel support + filter media terminal head losses = 100.38 m + 0.52 m + 0.76 m + 0.05 m + 0.00 m + 2.50 m = 104.21 m.

The elevation of the top of the filter media = the maximum WSEL in a filter unit when the filter media are dirty – water depth above the media = 104.21 m – 3.01 m = 101.20 m. The elevation of the bottom of the gravel support = elevation of the top of the filter media – total depth of filter media – total depth of the gravel layer = 101.20 m – 0.75 m – 0.25 m = 100.20 m.

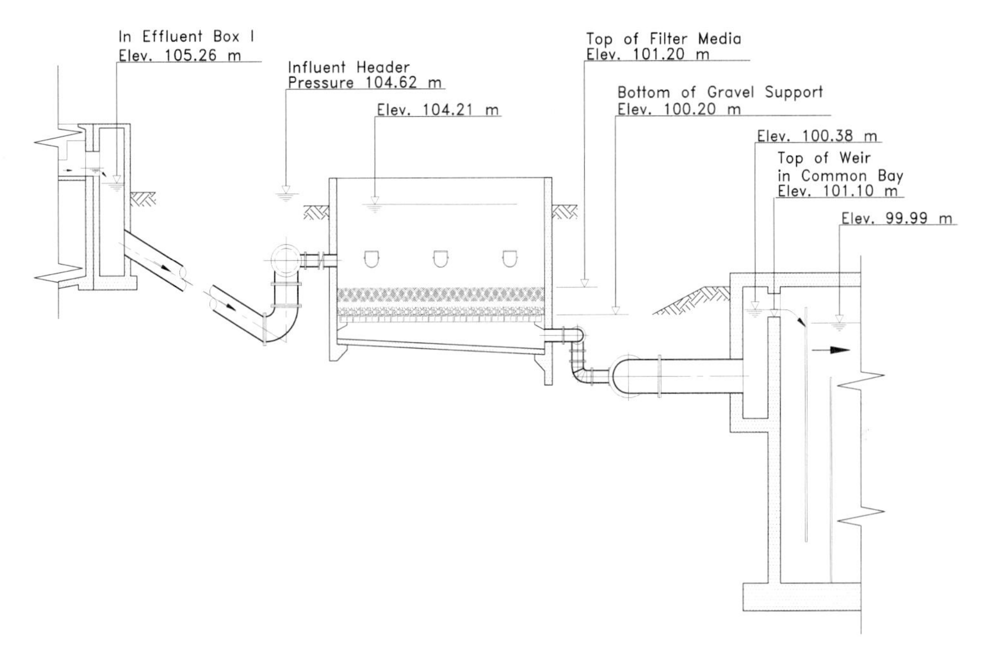

Figure 10-23 Hydraulic profile through the filter unit.

The WSEL (or pressure) at the beginning point of the filter influent header (Point A in Figure 10-21(a)) = the maximum WSEL in a filter unit when its filter media are dirty + head losses through the filter influent piping = 104.21 + 0.41 m = 104.62 m.

The WSEL in the Effluent Box I of the sedimentation basins (Figure 9-14) = WSEL (or pressure) at the beginning point of filter influent header + head losses through the settled water conduit between Effluent Box I and the filter influent header (Table 15-4) = 104.62 m + 0.64 m = 105.26 m. This value is used to prepare the hydraulic profile through the sedimentation basin in Chapter 9.

Step B: Filter Backwash System

1. Estimate backwash flow rate.

The backwash flow rate should achieve 40 to 50 percent expansion of both the sand and anthracite media layers. A backwash flow rate equal to 10 percent of the settling velocity of the media is used for backwashing. Several limit switches will be provided to monitor the position of influent, backwash, and waste valve positions and to monitor backwash flow rate by displaying percent opening of flow controller position. Flow meters will measure the filter backwash flow rate.

a. Calculate the media settling velocity at 20°C.

The approximate settling velocity (v_s) of the media is calculated from Eqs. (10-9) and (10-10) for water temperature of 20°C.

Sand:

$$v_{s,20} = 10d_{60} = 10d_{10} \times \text{uniformity coefficient}$$
$$= 10 \times 0.5 \times 1.4 = 7.0 \text{ m/min} = 0.117 \text{ m/s}$$

Anthracite:

$$v_{s,20} = 4.7d_{60} = 4.7d_{10} \times \text{uniform coefficient}$$
$$= 4.7 \times 1.0 \times 1.6 = 7.52 \text{ m/min} = 0.125 \text{ m/s}$$

b. Calculate the actual media settling velocity at 5°C.

Trial-and-error method is used to calculate the actual media settling velocity at 5°C.

i. Calculate the Reynolds number, N_R.

The Reynolds number at 5°C is first calculated from Eq. (9.3) by using the media settling velocity at 20°C ($v_{s,20}$).

Sand:

$$N_R = \frac{0.117 \text{ m/s} \times (0.0005 \text{ m} \times 1.4) \times 1000 \text{ kg/m}^3}{1.518 \times 10^{-3} \text{ m}^2\text{/s}} = 54.0$$

Anthracite:

$$N_R = \frac{0.125 \text{ m/s} \times (0.001 \text{ m} \times 1.6) \times 1000 \text{ kg/m}^3}{1.518 \times 10^{-3} \text{m}^2\text{/s}} = 132$$

Both numbers are in the transitional range.

ii. Calculate the drag coefficient, C_D.

The drag coefficient is calculated from Eq. (9-5) by using Reynolds number, N_R, calculated in last step.

Sand:

$$C_D = \frac{24}{64.0} + \frac{3}{\sqrt{64.0}} + 0.34 = 1.09$$

Anthracite:

$$C_D = \frac{24}{132} + \frac{3}{\sqrt{132}} + 0.34 = 0.783$$

iii. Calculate the terminal settling velocity.

The terminal settling velocity is calculated from Eq. (9.1).

Sand:

$$v_{s,5} = \frac{4}{3} \times \frac{0.0005 \text{ m} \times 1.4 \times 9.81 \text{ m/s}^2}{1.09} \times \sqrt{2.65 - 1}$$

$$= 0.118 \text{ m/s}$$

Anthracite:

$$v_{s,5} = \frac{4}{3} \times \frac{0.001 \text{ m} \times 1.6 \times 9.81 \text{ m/s}^2}{1.09} \times \sqrt{1.55 \times 1}$$

$$= 0.121 \text{ m/s}$$

These calculated settling velocities at 5°C are close to those at 20°C. It is safe to assume 0.12 m/s as the average settling velocity of the media (sand and anthracite).

c. Calculate the backwash flow rate.

At 10 percent settling velocity of the media (sand and anthracite), the average backwash rate $U_b = 0.012$ m/s.

$$\text{Backwash flow rate } Q_b = U_b A_f$$
$$= 0.012 \text{ m/s} \times 68.1 \text{ m}^2 \times 60 \text{ s/min} = 49 \text{ m}^3\text{/min}$$

d. Calculate the surface wash flow rate.

The surface wash flow rate is calculated from the surface wash rate given in the Design Criteria. (See Section 10.13.1.)

$$\text{Surface wash flow rate } Q_s = U_s A_f$$
$$= 0.061 \text{ m}^3\text{/m}^2\text{·min} \times 68.1 \text{ m}^2 = 4.2 \text{ m}^3\text{/min}$$

2. Calculate bed expansion.

a. Calculate the porosity of the expanded bed.

The quantity e_{eb} is a function of the settling velocity of the particles and the backwash velocity. An increase in backwash velocity will result in a greater expansion of the bed. The expression commonly used to relate the bed expansion to backwash velocity and settling velocity has been extensively reported in the literature. This relationship is given by Eq. (10.19).[1,9,11,19]

$$e_{eb} = \left(\frac{U_b}{v_s}\right)^{0.22} \qquad (10.19)$$

where

e_{eb} = porosity of the expanded bed

v_s = media settling velocity, m/s (ft/s)

Effective backwash is achieved at 10 percent of the settling velocity of the media (ratio U_b to v_s =0.1).

$$e_{eb} = (0.1)^{0.22} = 0.60$$

b. Calculate expanded bed depth.

The expanded bed depth is calculated from Eq. (10.20).[20]

$$L_{fb} = \frac{(1-e)L}{1-e_{eb}} \tag{10.20}$$

where

L_{fb} = expanded bed depth, m (ft)
L = bed depth at rest, m (ft)

Sand: $$L_{fb} = \frac{(1-0.40)\times 0.25\text{ m}}{1-0.60} = 0.38\text{ m}$$

Anthracite: $$L_{fb} = \frac{(1-048)\times 0.5\text{ m}}{1-0.60} = 0.65\text{ m}$$

The total expanded bed depth = 0.38 m + 0.65 m = 1.03 m

$$\text{Bed expansion ratio} = \frac{1.03\text{ m}}{0.75\text{ m}} = 1.37$$

The bed expansion is designed for sustained water temperature of 5°C during the winter months. If the water temperature is lower than 5°C, the settling velocity of the media will decrease. Therefore, the backwash rate should be reduced for colder water temperatures.

c. Determine the vertical dimensions of the filter launder troughs.

Place the washwater troughs 1.02 m above the top of the expanded bed. This will make the height of the washwater trough 1.30 m above the media bed at rest. Provide a launder height of 0.6 m, and leave 0.42 m between the inside bottom of the launder troughs and the top of the expanded media bed. Figure 10-24 illustrates the design vertical dimensions of the launder troughs.

3. Determine the backwash cycle.

The backwash cycle will be as follows:

a. Initially, the surface wash will operate at Q_s = 4.2 m^3/min (or U_s =0.061 m^3/m^2·min) during the operation time from 0 to 4 minutes.

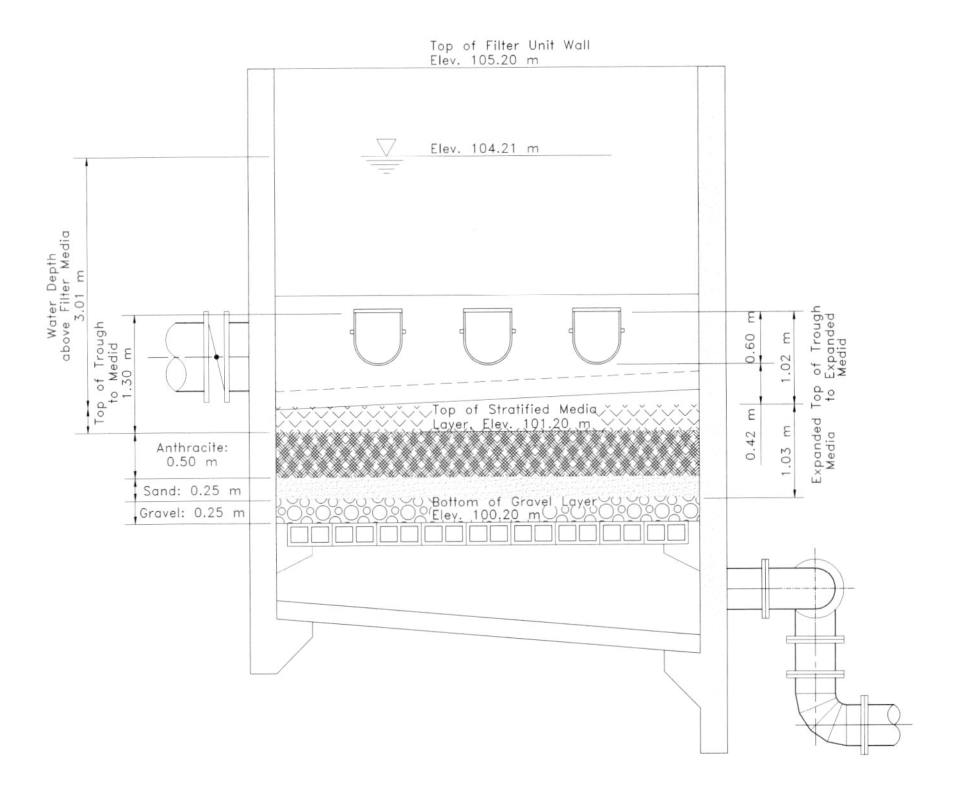

Figure 10-24 Design elevations of washwater launder trough.

b. The backwash will start at 3 min and will increase from zero to $Q_b = 49$ m^3/min over a time period of 1.3 minutes.

c. The backwash flow rate of 49 m^3/min will be maintained for an operation time period of 6 minutes.

d. The backwash flow rate will be reduced from $Q_b = 49$ m^3/min to zero over the 1.3 minutes after the 6-minutes backwash.

e. Total backwash cycle will take an operation time period of 11.6 minutes.

The backwash cycle is graphically illustrated in Figure 10-25. It was developed through the experience and judgment of the authors. Ideally, the most effective backwash cycle should be determined by pilot plant testing. The actual operating experience of the full-scale plant may indicate the need for adjustments in the backwash cycle. Design engineers should provide sufficient capacities in the backwash system to make possible for adjustments for changes in the operating conditions.

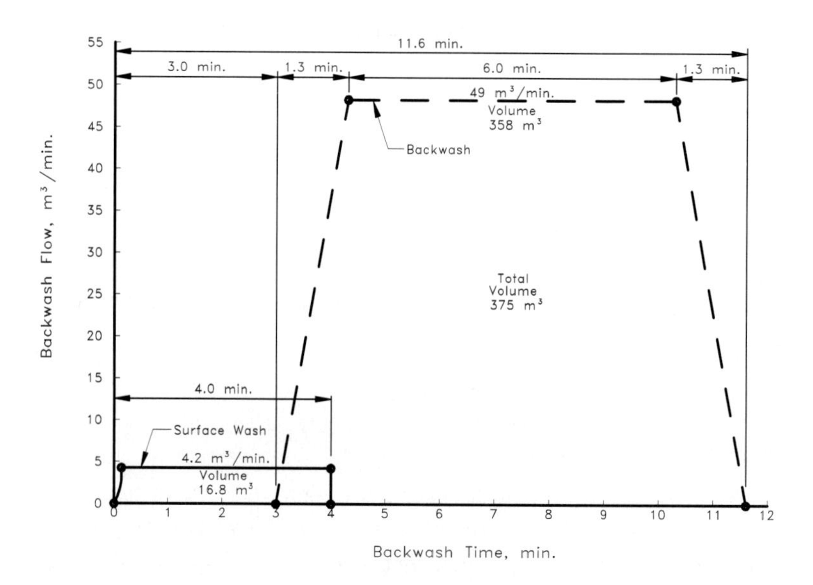

Figure 10-25 Filter backwash cycle design.

4. Determine the capacity of the backwash water tank.

   a. Calculate the volume of backwash water.

      i. Calculate surface wash water volume.

$$V_s = 4.2 \text{ m}^3\text{/min} \times 4 \text{ min} = 16.8 \text{ m}^3$$

      ii. Calculate backwash water volume.

$$V_b = 49 \text{ m}^3\text{/min} \times 6 \text{ min} + 2 \times \frac{49 \text{ m}^3\text{/min} \times 1.3 \text{ m}^3\text{/min}}{2} = 358 \text{ m}^3$$

      iii. Calculate total volume for each filter backwash cycle.

$$V_f = V_s + V_b = 16.8 \text{ m}^3 + 358 \text{ m}^3 = 375 \text{ m}^3$$

      iv. Calculate total volume of backwash water per day.

      Assume each filter has a minimum run of 24 hours. Calculate the total quantity of filter backwash as follows:

$$V_{daily} = 375 \text{ m}^3\text{/backwash cycle} \times 1 \text{ backwash cycle/filter·d} \times 8 \text{ filters}$$
$$= 3{,}000 \text{ m}^3\text{/d}$$

b. Determine the volume of the elevated storage tank for the backwash.

The storage volume should be sufficient to provide water for two backwash cycles.

$$V_{tank} = 375 \text{ m}^3\text{/backwash cycle} \times 2 \text{ cycles/tank}$$
$$= 750 \text{ m}^3\text{/tank (198,000 gal)}$$

In order to provide operational flexibility, increase storage by 10 percent.

$$V_{tank} = 750 \text{ m}^3\text{/tank} \times 1.10 = 825 \text{ m}^3 \text{ (218,000 gal)}$$

For economy in construction, provide one tank with the closest available standard size.

$$V_{tank} = 946 \text{ m}^3\text{/tank (250,000 gallons/tank)}$$

5. Size the backwash water pumps.

Select backwash water pumps to transfer a sufficient quantity of water from clearwells into the elevated tank within 30 minutes after a filter backwash operation. In order to ensure system reliability, provide three equal-size pumps, each having sufficient capacity that two pumps operating in parallel will provide the required flow rate.

a. Calculate the design flow rate.

$$Q = \frac{375 \text{ m}^3}{30 \text{ min}} = 12.5 \text{ m}^3\text{/min}$$

b. Calculate the capacity of each pump.

$$Q = \frac{12.5 \text{ m}^3\text{/min}}{2} = 6.25 \text{ m}^3\text{/min}$$

c. Add 10 percent capacity in order to allow flexibility in operation.

$$Q = 6.25 \text{ m}^3\text{/min} \times 1.10 = 6.9 \text{ m}^3\text{/min (18,20 gpm)}$$

The required pump head is determined by the system TDH (i.e., the static head plus the head losses between the clearwell and the elevated backwash water storage tank). Detailed procedures for designing a pumping system are presented in Chapter 7. (See Section 7.8.2.)

d. Determine the operational sequence of the pumps.

Provide three identical pumps, each with a capacity of 6.9 m^3/min (1,820 gpm). The pumps shall be controlled by a device that alternates the lead pump each day. The pumps shall be turned on and off by level switches in the elevated storage tank. The level control design of the pumps is summarized in Table 10-10.

Table 10-10 Control Level Design for Backwash Pumps

Tank Capacity	Level in Tank	First Pump	Second Pump	Third Pump
Full	maximum water depth	Off	Off	Off
90% full	90% water depth	On	Off	Off
60% full	60% water depth	On	On	Off
30% full	30% water depth	On	On	On
Empty	minimum water depth	On	On	On

6. Determine the height requirement of the backwash water storage tank.

a. Estimate head loss through the expanded media bed during backwash by using Eq. (10.15).

Sand: $h_L = 0.25 \times (1 - 0.4) \times (2.65 - 1) = 0.25 \text{ m}$

Anthracite: $h_L = 0.50 \times (1 - 0.48) \times (1.55 - 1) = 0.14 \text{ m}$

Total loss through expanded media = 0.14 m + 0.25 m = 0.39 m

b. Estimate head loss through gravel layer.

The head losses through the gravel layer are calculated from the Carmen–Kozeny Equation (Eq. (10.1)). The procedure for using this equation is shown in Section 10.13.2, Step A.5.c. Using the average backwash rate, $U_b = 0.012$ m/s, the head losses are calculated. These values are summarized in Table 10-11. The total head loss through the gravel layer is 0.02 m.

c. Estimate underdrain system losses.

The underdrain system losses are calculated from Eq. (10.19) by using $v = 0.012$ m/s ∞ 3,600 s/h = 43.2 m/h.

$$H_L = 0.0005 \times (43.2 \text{ m/h})^2 = 0.93 \text{ m}$$

d. Estimate backwash water piping losses.

The losses through the backwash water piping between the elevated storage tank and the filter underdrain entry are computed in a manner similar to the procedures used for the head losses in either influent or effluent heads. The results of this calculation indicate a head loss of 0.55 m.

e. Calculate the required minimum water surface level in the storage tank.

The head losses through the backwash system are as follows:

- Head loss through the expanded media bed 0.39 m
- Head loss through the gravel support 0.93 m
- Head loss through the underdrain system 0.02 m
- Head losses through the backwash water piping 0.55 m

Total = 1.89 m

Add 50% of head loss for additional flexibility. The minimum water surface level in the storage tank should be at least 2.85 m above the lip of the launder troughs. A flow-modulating valve is used to maintain a constant backwash rate of $U_b = 0.012$ m/s.

Table 10-11 Backwash Head Losses Through Gravel Layer[a]

Layer	Size, mm	Depth, mm	Porosity	Reynolds Number (N_R)	friction coefficient (f')	Head loss (h_L), m
Top layer	2.5	50	0.45	19.8	5.92	0.0105
Second layer	4.5	50	0.45	35.6	4.07	0.0040
Third layer	9.5	75	0.50	75.1	2.75	0.0013
Bottom layer	15.5	75	0.50	122	2.367	0.0007
Total depth of the gravel layers		250 mm or 0.25 m	Total head loss through the gravel layers			0.0165 or 0.02

a See footnotes in Table 10-7 for detailed explanation of calculation procedures.

7. Determine the characteristics of filter backwash.

a. Quantity of backwash

i. At maximum day flow

The total volume for each backwash cycle is 375 m^3, and the total volume of backwash water used in a day is 3000 m^3/d (Section 10.11.2, Step B.4.a). It has been assumed that each filter will be backwashed at most once during any 24-hour period. The maximum number of backwash cycles per day is therefore eight. The actual number of backwash cycles may be smaller if the filter loading rate is reduced and/or filter run times are longer.

ii. Quantity at average day flow

Total volume of filter backwash water during design average day flow is obtained in proportion to maximum day flow.

$$\text{Filter backwash quantity at average day flow} = \frac{3000\ \text{m}^3 \times 57{,}900\ \text{m}^3/\text{d}}{113.500\ \text{m}^3/\text{d}}$$

$$= 1530\ \text{m}^3/\text{d}$$

b. Quality of backwash

The solids concentration in the filter backwash ranges from 0.01 to 0.1 percent, depending on the efficiency of the filter and the degree of pretreatment provided. The total solids production for this design example is determined in Chapter 9. This quantity was 7600 kg/d under maximum day flow conditions. It has been assumed that 90 percent of these solids are removed in the sedimentation basins. The quantity of solids remaining in the sedimentation basin effluent is 760 kg/d. The solids production due to the addition of nonionic polymer as filter aid is 2.3 kg/d at a dosage of 0.02 mg/L under maximum day flow conditions. This quantity of solids from polymer is relatively small and is assumed to be zero in subsequent calculations. It is further assumed that 100 percent of incoming solids are captured by the filter and then cleaned out during filter backwash. The quantity of solids released into the backwash water (3000 m^3/d) is therefore 760 kg/d. This will result in a daily average TSS concentration of 254 mg/L (approximately 0.025 percent solids). In a similar way, the quantity of solids in the backwash water is 386 kg/d under the average day flow condition. The filter backwash characteristics under the maximum and the average day flow conditions are provided for mass balance calculations in Chapter 14.

10.14 Operation, Maintenance, and Troubleshooting for Filtration Systems

Filtration units should be operated and maintained only by personnel trained in operation of the filters. This training should include a review of the theory of filtration, filter hydraulics, and filter controls, as well as of the proper water quality testing procedures required to monitor filter performance. Maintenance personnel should be knowledgeable in basic electrical power systems and plumbing.

10.14.1 Operation of Filter Systems

Improper operation of filtration units can result in poor quality of finished water and in damage to the filter bed. In order to ensure proper operation, operators must continually monitor the operation of the filter units. The filtered water turbidity and the head loss through each filter unit are of particular interest.

The filters must be backwashed as soon as either the filtered water turbidity or the head loss through a filter unit reaches a preset maximum value. Also, if a filter unit has been idle for a period of time, it should be thoroughly backwashed prior to its being put back into service.

Improper filter backwashing may cause inadequate filter cleaning and possible damage to the unit. If the backwash water is introduced too rapidly, the filter bed can be disturbed, or, in extreme cases, the filter bottom can be damaged. In order to reduce the chance of damage to the filter beds from improper backwashing techniques, most filter systems utilize automatic backwash controls.

10.14.2 Preventive Maintenance on Filtration Systems

The preventive maintenance program for filtration systems should be based upon daily inspections of all pumps, valves, piping, and other mechanical equipment. Pumps should be inspected for proper lubrication of bearings, leaking seals, overheated motors, and noisy operation. Valves should be inspected for leaking seals. All valves and gates in water treatment plants should be operated at least once each month to prevent seizure. Piping should be inspected for corrosion and leaks. Other routine maintenance procedures called for in the equipment operation and maintenance manuals should be performed at the recommended intervals.

In order to prevent the buildup of algae and of growths of other microorganism in the filter beds, each filter unit should be treated at least *annually* with a highly concentrated chlorine solution or some other strong oxidizing agent. This procedure involves taking the filter out of service and backwashing it. Then, the chlorine solution is fed into the filter until the entire filter bed is submerged in the solution. After being allowed to set for several hours, the bed is backwashed again and put into service.

Each filter unit should be drained and inspected annually. The condition of the media is of prime concern. Determine if it shows signs of mud ball formation or other indications of poor cleaning. Measure media depth to determine whether any has been lost through backwashing. Have a sample analyzed for media size distribution and tested for acid solubility to determine whether the media are still within specifications.

Maintain accurate records on each filter operation. The record should include items such as head loss, filter run time, and turbidity of finished water. Compare the records with those of other filters and for previous years. These data will offer an indication of poor media condition, inappropriate chemical pretreatment, or other problems that may otherwise go undetected.

10.14.3 Troubleshooting Filter Systems

Following is a brief troubleshooting guide for filtration systems. Several actions are recommended in this section, but the operator should test each action individually to determine the proper corrective action.[21,22]

Conditions	Possible Causes
1. High head loss through a filter unit, or filter runs too short	Filter bed in need of backwashing
	Air binding
	Mud balls in the filter bed
	Improper rate-of-flow controller operation
	Clogged underdrains
	Improper media design: too small, or too deep
	Floc strength too strong—will not penetrate media
2. High effluent turbidity	Filter bed in need of backwashing
	Rate of flow too high
	Improper rate-of-flow controller operation
	Disturbed filter bed
	Mud balls in the filter bed
	Air binding
	Inappropriate media size or depth
	Low media depth (caused by loss during backwash)
	Floc too small or too weak caused by improper chemical pretreatment

The two most common problems encountered in filter operation are *mud ball formation* and *air binding*. Mud ball formation is usually the effect of improper backwashing techniques, but improper media selection can also be the cause. Single-medium filters historically show a greater tendency to form mud balls than do properly designed dual-media and mixed-media filters. Surface wash, sub-surface washing, or air-scouring of filters before and during backwash also reduces the tendency to form mud balls in the filter bed.

Once mud balls have formed in a filter bed, the most effective means of removing them is to remove the filter media and either replace them or thoroughly clean the media before placing them back into the bed. Once mud balls have begun to form in a filter bed, they will usually grow larger.

Air binding of filter beds is usually caused by improper hydraulic design of the filter system. This phenomenon is discussed more fully in Section 10.13.2, Step A.3.b.iii. Possible solutions to air binding include the following:

- replacing the filter media with one with a different gradation;
- reducing the maximum flow rate through the filter;
- inducing additional hydraulic head in the filter effluent, to raise the hydraulic gradient in the filter bed.

10.15 Specifications for Filtration Systems

The specifications for the filtration systems designed in this chapter are briefly presented in this section. The purpose of these specifications is to describe more fully the components of the filtration system that could not be properly illustrated in the design example. These specifications are not intended to be complete construction specifications, nor are they intended to be a guideline for construction specifications. The designer should consult equipment suppliers and develop detailed specifications with the aid of information provided by the manufactures. A more detailed discussion of specifying equipment for water treatment plants is provided in Appendix C.

10.15.1 General

The filtration system consists of eight filter beds and associated influent, effluent, backwash, surface wash, and control systems. The components shall be designed to treat effluent from a coagulation, flocculation, and sedimentation process. The influent water to the filtration system shall have the following characteristics:

	Maximum	Minimum
Flow Rate	113,500 m^3/d	28,300 m^3/d
Temperature	32°C	4°C
pH	7.2	6.8

The filter media shall be a dual-media system consisting of anthracite coal and quartz sand. The media shall conform to American Water Works Association Standard B100 and have the following physical characteristics:

1. Anthracite coal

Effective size, mm	1.0
Uniformity coefficient	1.6
Depth of bed, cm	50
Specific gravity	1.55

2. Quartz Sand

Effective size, mm	0.5
Uniformity coefficient	1.4
Depth of bed, cm	25
Specific gravity	2.65

10.15.2 Filter System Piping

The filter system piping consists of filter influent piping, filter effluent piping, washwater supply piping, surface washwater supply piping, and waste washwater piping. Also included in the filter system piping are all valves, fittings, control elements, and other pipe specials associated with the various piping systems. The materials and types of valves and appurtenances provided for each piping system are listed here:

1. Filter influent system

Piping	Steel pipe
Valves	Motor-operated butterfly valves
Fittings	Steel pipe
Control elements	None

2. Filter effluent system

Piping	Buried-prestressed concrete steel cylinder pipe
Fittings	Steel pipe
Control elements	Flow-controlled butterfly valves
Valves	Motor-operated butterfly valves

3. Washwater supply system

Piping	Steel pipe
Valves	Motor-operated butterfly valves
Fittings	Steel pipe
Control elements	Automatic rate-of-flow controller, electrically operated

4. Surface wash system

Piping	Steel pipe
Valves	Motor-operated butterfly valves
Fittings	Steel pipe
Control elements	None

5. Waste Washwater System

Piping	Buried-reinforced concrete pipe Interior-Steel pipe
Valves	Motor-operated butterfly valves
Fittings	Buried-reinforced concrete pipe Interior-steel pipe
Control elements	None

10.15.3 Filter Structure

The structure containing the filter system components shall be of reinforced concrete with a minimum 28-day compressive strength of 27,600 kPa.

The piping gallery shall be ventilated in order to reduce condensation. The ventilation system shall be designed to provide a minimum of one complete air change in the pipe gallery each minute.

The control room for the filter system shall be heated, air-conditioned, and humidity-controlled. All temperature and humidity sensitive instruments shall be located in the control room. The control room shall have sufficient room for all control and instrumentation panels, shall have two desks and chairs for the operators, and shall have a view of all filter beds.

Problems and Discussion Topics

10.1 Visit a local water treatment plant. Determine the number and size of filtration facilities. Determine the media type and size, design flow rates, backwash rates, backwash intervals, and filter performance parameters. Prepare sketches of the facility layouts. Ask the operator to explain any operational problems he/she experiences with these facilities.

10.2 In the Design Example, the filtration system was designed for a maximum flow rate of 113,500 m^3/d (30 mgd). Determine the head loss through the clean bed if flow rate is 170,300 m^3/d (45 mgd).

10.3 Calculate the head loss through a clean bed of uniform sand. The sand grains are 0.5 mm in diameter. The depth of bed is 0.3 m. The porosity ratio is 0.4. The shape factor is 0.85, and the specific gravity of sand is 2.65. The water temperature is 20°C. The filter has a hydraulic loading rate of 4.0 $m^3/m^2 \cdot h$. Use the Carmen–Kozeny equation (Eq. (10.1)).

10.4 A dual-media filter bed is being designed. The media used are sand and anthracite. The information about the media is as follows:

The effective size and specific gravity of sand are 0.60 mm and 2.65. The specific gravity of anthracite is 1.5. The anthracite layer is 2/3 of the total depth of the bed.

Calculate the following:

(a) the effective size of anthracite that has the same settling velocity as the sand, and

(b) the depth of the sand and the anthracite layers.

10.5 Calculate the head loss through a sand filter media, given the following:

Filtration rate	9.0 $m^3/m^2 \cdot h$
Media size	0.6 mm
Specific gravity	2.65
Media depth	28 cm
Porosity	0.45
Uniformity coefficient	1.6
Shape factor	1.0
Viscosity of water	$1.3 \infty 10^{-3}$ $N \cdot s/m^2$

10.6 Select an anthracite coal filter medium size and depth that is compatible with that given in Problem 10-5. The specific gravity of anthracite is 1.55, and media depth is 2/3 of total depth. Calculate the head loss through the dual-media at a filtration rate of 9.0 $m^3/m^2 \cdot h$.

10.7 A sieve analysis of filter sand yielded the following information. Compute the head loss across the clean stratified bed. The bed is 70 cm deep, and the porosity ratio is 0.41. The temperature of the water is 5°C, and the filtration rate is 150 $m^3/m^2 \cdot d$. Use the Carmen–Kozeny equation (Eq. (10.1)). Assume shape factor = 1.0.

Sieve Number (U.S. Sieve Series)	Weight Fraction of Particles Retained, ∞ 10^2	Geometric Mean Diameter, mm
14—20	0.8	1.09
20—25	4.25	0.77
25—30	15.02	0.65
30—35	16.65	0.54
35—40	18.01	0.46
40—50	15.25	0.35
50—60	15.65	0.27
60—70	9.30	0.23
70—100	2.07	0.18

10.8 In Problem 10-7, the sieve analysis is given. Determine the effective size and uniformity coefficient of the sand. Using the effective size value, calculate the head loss through a clean stratified bed. Use the necessary information in Problem 10-7.

10.9 A filter sand bed is 0.75 m deep. It is composed of a uniform-size sand medium 0.5 mm in diameter, with specific gravity of 2.64 and shape factor of 0.9. The porosity of the bed is 0.45. Plot the relationship between head loss and the filtration velocity of 2 to 7 $m^3/m^2 \cdot h$ at an increment of 1 $m^3/m^2 \cdot h$. The water temperature is 13°C.

10.10 A filter bed is composed of uniform sand of diameter 0.4 mm, with specific gravity of 2.65 and shape factor of 0.85. The porosity of the sand is 0.4, and the water temperature is 15°C. Calculate the filtration velocity in $m^3/m^2 \cdot h$, if a head loss of 2 m is maintained over the top of the media.

10.11 The filter bed in Problem 10-9 is backwashed. Calculate the backwash velocity at which the media is just fluidized.

10.12 Calculate the required backwash rate ($m^3/m^2 \cdot h$) if a bed expansion of 50 percent of the original bed depth is required during backwashing. Use the filter bed information given in Problem 10-9

10.13 The filter bed in Problem 10-7 must be fully fluidized during backwashing, to achieve filter cleaning. Calculate (a) the backwash velocity that will just fluidize the largest particle in the bed and (b) the expanded bed depth during backwashing.

10.14 The filter bed in Problem 10-3 is expanded to a porosity of 0.7 by backwashing. Calculate the required backwashing velocity and the resulting expanded bed depth.

10.15 The filter bed given in Problem 10-7 is backwashed at a velocity of 1.2 ∞10^{-2} m/s. Determine the depth of the expanded bed.

10.16 The effective size and the uniformity coefficient of the filter bed media in Problem 10-6 are calculated in Problem 10-8. Using this information, calculate the depth of the expanded bed if the filter backwash velocity is 1.2 ∞10^{-2} m/s.

10.17 A sand filter medium and a compatible anthracite coal filter medium are given in Problems 10-5 and 10-6. Calculate the appropriate backwash velocity for the dual-media filter and the head loss during backwash.

10.18 Prepare the hydraulic profile through the filtration system designed in the Design Example, if plant capacity is 170,300 m^3/d (45 mgd). The water surface elevation in Effluent Box I and the weir elevation in the filter back-pressure weir common bay are same as in Figure 10-20.

10.19 Design a filter system to treat 50,000 m^3/day of water. Select the number of filter units, dual-media sizes and type, head loss through media and gravel bed, backwash rate, backwash volume, and backwash head requirements. Make head loss calculations based on the filter unit only. Do not calculate piping losses.

10.20 Calculate the turbulence (velocity gradient) of a surface wash system having the following parameters.

$d = 0.25$

$v_s = 0.1\ m^3/m^2 \cdot min$

$P = 50\ m$

$d_e = 50\ cm$

$\mu = 1.3 \infty 10^{-3}\ N \cdot s/m^2$

10.21 Calculate the turbulence (velocity gradient) of an air-scour system, given the following:

Tank Volume	256.0 m^3
Water Depth	4.0 m
Air Flow	50.0 m^3/sec
Viscosity	$1.3 \infty 10^{-3}\ N \cdot s/m^2$

References

1. Tchobanoglous, G. and Schroeder, E. *Water Quality: Characteristics, Modeling, and Modification*, Addison-Wesley Publishing Co., Reading, MI, 1985.
2. Kawamura, S. "Design and Operation of High-Rate Filters - Part I," *Jour. AWWA*, vol 67, no. 10, pp. 535–544, October 1975.
3. Kawamura, S. "Design and Operation of High-Rate Filters," *Jour. AWWA*, vol. 91, no.12, pp. 77–90, December 1999.
4. Qasim, S. R. *Wastewater Treatment Plants: Planning, Design and Operation*, 2d ed., Technomic Publishing Co., Lancaster, PA, 1999.
5. McGhee, T. J. *Water Supply and Sewerage*, 6th ed., McGraw-Hill Book Co., New York, 1991.
6. Metcalf and Eddy, Inc. *Wastewater Engineering: Treatment Disposal and Reuse*, 3d ed., McGraw-Hill, Inc., New York, 1991.
7. Bernardo, L. and Cleasby, J. "Declining-Rate vs. Constant-Rate Filtration," *Journal Environmental Engineering Division, ASCE*, vol 106, no. EE6, pp. 1073–1041, December 1980.
8. Tchobanoglous, G. "Filtration Techniques in Tertiary Treatment," *Jour. AWWA*, vol. 42, no. 4, pp. 604–623, April 1970.
9. Kawamura, S. "Design and Operation of High-Rate Filtration - Part 2," *Jour. AWWA*, vol 67, no. 11, pp. 653–662, November 1975.
10. Kawamura, S. "Design and Operation of High-Rate Filtration - Part 3," *Jour. AWWA*, vol 67, no. 12, pp. 705–708, December 1975.
11. James M. Montgomery, Inc. *Water Treatment Principles and Design*, John Wiley & Sons, New York, 1985.
12. Hudson, H. *Water Clarification Processes—Practical Design and Evaluation*, Van Nostrand Reinhold Co., New York, 1981.
13. Monk, R. and Willis, J. F. "Designing Water Treatment Facilities," *Jour. AWWA*, vol. 79, no. 2, pp. 45–57, February 1987.
14. Kawamura, S. *Integrated Design of Water Treatment Facilities*, John Wiley and Sons, Inc., New York, 1991.
15. AWWA. *Water Quality and Treatment*, 5th ed., McGraw-Hill, Inc., New York, 1999.

16. ASCE and AWWA. *Water Treatment Plant Design*, 2d ed., McGraw-Hill Publishing Co., 1990.
17. Camp, J. R. "Backwashing Granular Filters", *Proceedings of the American Society Civil Engineers Sanitary Division*, p. 903, December 1971.
18. USEPA. *Technology for Upgrading Existing or Designing New Drinking Water Facilities*, Office of Drinking Water Center for Environmental Research Information, Cincinnati, OH, March 1990.
19. Malcolm Prinle Inc. and HDR, Inc. *Guidance Manual for Compliance with The Filtration and Disinfection Requirements for Public Water Systems Using Surface Water*, AWWA, Denver, CO, 1991.
20. Peavy, H. S., Rowe, D. R., and Tchobanoglous, G. *Environmental Engineering*, McGraw-Hill, Inc., New York, 1985.
21. Tillman, G. M. *Water Treatment: Troubleshooting and Problem Solving*, Ann Arbor Press, Inc., Chelsea, MI, 1996.
22. Gulp, G. L. and Heim, N. F. *Field Manual for Performance Evaluation and Troubleshooting to Municipal Wastewater Treatment Facilities*, EPA-430/9-78-001, USEPA, Washington, D. C., January 1978.

CHAPTER 11

Color, Taste, and Odor Control

11.1 Introduction

The principal objective of a water treatment plant is to produce water that conforms to the health and safety standards set in the *primary drinking water regulations*; however, aesthetic considerations are also important to maintaining public confidence in potable water supplies. Accordingly, *secondary drinking water standards* have been set to ensure potable water that is low in color, taste, odor, and other aesthetic parameters.

In this chapter, the sources of color, taste, and odor, the methods of measuring each, and the methods of removal are discussed. The step-by-step procedure for the design of color, taste, and odor removal systems for a specific situation is given in the Design Example.

11.2 Color

Color is a frequent characteristic of surface waters. The presence of noticeable color does not mean that water is necessarily unsafe to drink; however, color may decrease public confidence in the water supply, and its presence may degrade the quality of foods, beverages, paper, and textiles produced by local industry. Color is also of concern because it raises chlorine demand. Its presence may suggest a high trihalomethane formation potential (THMFP), and its root components may precipitate and later resuspend in the water distribution system.

11.2.1 Sources of Color

Color in natural waters is derived from organic and inorganic compounds of both natural and synthetic origin; these compounds may be suspended solids or dissolved materials. Suspended solids are removed by other treatment processes, so this discussion will deal primarily with *true color*, or that portion of *apparent color* that is composed of substances in the aqueous phase.

Inorganic color is usually owing to naturally-occurring metallic ions, such as iron and manganese, but it may also have industrial origin; however, the most common sources of color are organic compounds released by the decomposition of natural organic matter, by microbial metabolism, and in industrial discharges. Most organic compounds introduced into the water by microbial activity are called *humic substances*. They are amorphous, acidic, hydrophilic, chemically complex, and often aromatic anionic polyelectrolytes, ranging in molecular weight from several hundred to tens of thousands of grams per mole.[1] Humic materials are classified as humic, fulvic, or hyatomelanic acids. These acids have similar structures, though fulvic acid, with the lowest molecular weight, has the highest charge density.[2] Other sources include chlorophyll from algae and numerous synthetic organic chemicals that may impart color to the water.

11.2.2 Color Measurement

As explained previously, *apparent color* includes turbidity or suspended solids, while true color is owing to that portion of dissolved compounds that imparts color. Apparent color is determined on the original sample; true color is measured on samples taken after filtering or centrifuging. The color value of water is also pH dependent, increasing with increased pH; therefore, pH should be recorded when one is performing color measurements.

Color is measured by visual comparison of the water sample with known concentrations of colored solutions, by visual comparison with special color disks, or by spectrophotometric methods that correlate concentrations of given substances at specific wavelengths. *Standard Methods* specifies the platinum-cobalt method of measuring color.[3] The unit of color in this method is that produced by 1 mg/L of platinum as the chloroplatinate ion.

The National Secondary Drinking Water Regulations set a limit of 15 color units as measured on the platinum-cobalt scale. This limit was based on aesthetic considerations, because color by itself is not known to be physiologically harmful to humans.[4,5] Most municipal plants strive to maintain finished waters with color values of 5 units or less. Studies have found some correlation between residual color concentration and zeta potential that is frequently expressed as electrophoretic mobility. This, however, provides only a general guide in specific instances and cannot be used as a precise measure for water quality standards.

11.2.3 Color Removal

Conventional water treatment processes such as oxidation, coagulation, and filtration are effective in removing color. Adsorption and ion exchange can also be employed. In practice, color is usually removed to acceptable levels by normal water treatment operations. Each of these color removal processes is briefly discussed as follows.

Oxidation Color can be removed by such oxidizing chemicals as potassium permanganate, chlorine and chlorine compounds, and ozone. Color removal is most effective when followed by coagulation and other conventional treatment processes. Potassium permanganate can oxidize color-causing compounds but is normally used for disinfection or for taste and odor control.[6]

Chlorine and chlorine compounds partially oxidize color-causing compounds and act as a coagulant aid. This effect can result in better color reduction and decreased coagulant dosage.[6] Efficient color removal has also been reported with prechlorination followed by alum coagulation.[7,8] When coagulation and filtration could not remove color, a free chlorine residual has been known to bleach true color, an effect well-known in swimming pool operations.[6] However, unacceptable levels of trihalomethanes (THMs) may be formed by the use of free chlorine, and the use of chloramines (to prevent formation of THMs) is not an effective measure for color removal. Chlorine removes color most effectively in the pH range from 4.0 to 6.8. Unfortunately, this is usually below the normal pH range for water treatment.

Ozone is a more powerful oxidant than chlorine, and removes color better than chlorine. It has the added benefit that it does not depress the pH level as much-nor does it produce THMs. The cost of ozone is usually not justified merely for color removal, but ozone is frequently employed for taste and odor removal; it will be discussed more fully in Section 11.3.6.

Coagulation Conventional coagulation, as part of the overall water treatment process, effectively removes some color in addition to turbidity; however, the mechanism of color removal appears to be different from that for turbidity removal.[8] With enhanced coagulation, many dissolved solids also precipitate out of solution, like the coagulant, and may adsorb color-causing compounds on the growing floc. The quality of raw water, charge difference, and degree of ionization of the suspended and dissolved solids appear to account for the distinct responses during coagulation. Because of the varied nature and origins of color, jar tests and pilot testing can help to identify the best color removal strategy.

Filtration Filters remove suspended solids that cause turbidity and color. In a conventional treatment process of coagulation and flocculation followed by sedimentation, the solids removed in the filter consist of the lightest portion of floc which did not settle in the sedimentation basin. Thus, the filters remove most of the apparent color due to turbidity and some of the true color captured during the prior treatment processes. Filtered water is low in turbidity; most color remaining in the water is attributed to the dissolved solids and is measured as true color.

Adsorption As explained previously, some adsorption of color-causing compounds occurs during coagulation and flocculation. However, adsorption itself can also be an effective treatment process. The use of powdered activated carbon (PAC) or of filters composed of granular activated carbon (GAC) may effectively remove color.

Ion Exchange Ion exchange processes remove ionic species by employing resins and electrical charge methods.[6] The expense of ion exchange, however, inhibits its widespread application to remove color or even taste and odor. The subject of ion exchange is covered more fully in Chapter 18.

11.3 Taste and Odor

Taste and odor are frequent constituents in potable water supplies. Like color, the presence of taste and odor compounds in potable water does not imply that the water is unsafe. However, the public may find the water quality objectionable, and taste and odor can affect food processing and other industries. In ancient times, odors were believed to transmit disease, and this is still the public's first impression when confronted with water that has taste and odor problems.

Most organic and some inorganic chemicals contribute to taste or odor. These chemicals may originate in natural chemical or biological processes or from man-made sources. Frequently, there is no clear-cut distinction between whether a source is natural or synthetic, and often it is not important.[9] Often, the taste and odor problem may also develop within the distribution system. Some of the natural and synthetic sources are described below.

11.3.1 Natural Sources of Taste and Odor

Natural sources of taste and odor include the dissolution of salts and minerals. Groundwater, especially, may contain high salt and mineral concentrations. The presence of iron, manganese, and salts imparts a salty taste to the water and can render it unpalatable.[6] More commonly, microbial activity introduces taste and odor to both surface and ground waters. This activity is evident in decaying vegetation; in algae blooms in lakes and streams; in eutrophication, stratification, and turnover in reservoirs; and also in offensive tastes and odors in groundwater supplies. Each of these sources is presented below.

Decaying Vegetation Decaying vegetation, such as trees, brush, grasses, and aquatic plants, can cause taste and odor in surface water supplies. This is especially noted following periods of heavy rains when this material is swept into large streams, ponds, and reservoirs. Bacterial decomposition generally releases the taste- and odor-causing compounds that are associated with musty, earthy, woody, moldy, swampy, and fishy odors. Nutrient influx also spurs algal blooms, exacerbating the problem.[9]

Algae Algae are the most frequent cause of taste and odor in surface streams and reservoirs. In the 1880s and 1890s, research established algae as a major source of taste and odor.[10] It was not until the 1950s, however, that the taste and odor-causing chemicals produced by algae were actually isolated and identified.[11] Called *metabolites*, these chemicals originate either from the normal metabolic activity of the algae or from bacterial degradation of algal waste products. Information on five of these metabolites is given in Table 11-1. MIB and geosmin have been traced to actinomycetes and blue-green algae, and IBMP is associated with actinomycetes. These compounds are of particular concern to water utilities because of their sensitivity to detection by the human nose in concentrations down to the parts per trillion range.[5] Further, the blue-green algae can also produce mercaptans (odorous sulphur compounds).[9,12]

Reservoir Eutrophication Reservoir eutrophication is a problem caused by the continuous high influx of nutrients into the lake. These nutrients cause excessive algal and other aquatic growth, which gradually degrades water quality. This results in low dissolved oxygen

Table 11-1 Summary of Five Metabolites Originating from Microorganisms

Symbol	Name	Molecular Formula	Molecular Weight	Odor Characteristics	Threshold Odor Concentration, ng/L[a]
Geosmin	Trans-1,10-dimethyl-trans-9-decalol	$C_{12}H_{22}O$	182	Earthy, musty	10
TCA	2,3,6-Trichloroanisole	$C_7H_5OCl_3$	212	Musty	7
IPMP	2-Isopropyl-3-meth-oxy-pyrazine	$C_8H_{12}ON_2$	152	Earthy, musty, potato bin	2
IBMP	2-isobutyl-3-methoxy-pyrazine	$C_9H_{14}ON_2$	166	Earthy, musty, bell pepper	2
MIB	2-Methyl-isoborneol	$C_{11}H_{20}O$	168	Earthy, musty, bell pepper	29

a ng/L = nanogram /L = 10^{-9}g/L (parts per trillion).
Source: Adapted in part from References 5, 6, 9, and 12.

levels, high dissolved solids, increased turbidity, and an abundance of taste-and odor-causing compounds.

Reservoir Stratification Reservoir stratification is a seasonal source of taste and odor problems. Lakes normally yield a consistent water quality; however, due to seasonal temperature variations, many lakes become thermally stratified during the summer. Summer temperatures warm the *epilimnion* (the upper level of the lake) and effectively separate it from the cooler levels in the *hypolimnion* (lower level). A layer of sharp temperature change called the *thermocline* separates the two levels. This stratification allows low dissolved oxygen levels to prevail in the hypolimnion and cause anaerobic conditions. Under these conditions, anaerobic bacterial decomposition releases offensive odorous compounds such as hydrogen sulfide and reduces some metals to their soluble state; however, the quality of water withdrawn from the epilimnion may not be particularly affected by conditions in the hypolimnion.[11]

Reservoir Turnover Reservoir turnover is also a seasonal source problem. As fall weather cools, the epilimnion water temperature becomes lower than that in the hypolimnion below. The now-denser upper layer begins to settle. This creates strong vertical density currents, coupled with wind action, that stir up the lake bottom sediments and reintroduce organics and taste- and odor-causing compounds throughout the body of water. Similar action also causes spring turnover.

Bacterial Reduction in Groundwater Groundwater supplies are generally high in dissolved solids. In the case of sulfur-bearing waters under anaerobic conditions, bacteria present in the groundwater reduces sulfate (SO_4^{2-}) to sulfide (S^{2-}), which then becomes aqueous hydrogen sulfide (H_2S).[13] Offensive in odor, hydrogen sulfide can be detected by the human nose at 1 μg/L concentration.[5]

11.3.2 Synthetic Sources of Taste and Odor

Sources of taste and odor due to civilization include landfills and dumps, industrial waste discharges, urban and agriculture runoff, and wastewater discharges. These can introduce taste- and odor-causing substances directly into the ground or surface waters and can encourage bacterial or algal growth, which produces taste-and odor-causing compounds. A combination of natural and synthetic sources can also cause tastes and odors in potable water supplies.

Landfills and Dumps Landfills and dumps have received substantial attention in recent years as point sources of groundwater and surface water pollution. They often release a plume of leachate that may be high in chlorides, metal ions, and various organic chemicals and toxics.[14] These contaminants, even if at levels below those considered harmful, may still impart unpleasant taste and odor to the groundwater and render it unacceptable for consumption.[5]

Industry Industrial discharges to streams include innumerable organic and inorganic compounds, many of which cause serious taste and odor problems even in small concentrations. Notable examples include the chlorinated hydrocarbons, pesticides, and herbicides. Many synthetic and petrochemical compounds include naphthalene, tetraline, ethylene, benzene, styrene, and acetophenone. Some of the worst sources are waste from chemical dye, pharmaceutical coke, ammonia recovery, wood oil, phenol, cresol, petroleum products, textiles, and paper industries.[15] Frequently, these discharges not only directly introduce taste and odor but also cause increased algal or other microbial activity resulting in additional taste and odor.

Runoff Runoffs from agriculture lands and urban areas carry high concentrations of fertilizers and pesticides. Urban runoff may also contain petroleum products and other chemicals, which contaminate the raw water supply. Salts used to de-ice roads are responsible for contamination of groundwater supplies in the snowbelt states.[14] These nonpoint sources of pollution have been reported to cause taste and odor incidents in many water supplies.[16]

Wastewater Domestic wastewater may also be a source of taste and odor. Jenkins reported that an earthy-musty odor, similar to that from microorganisms, and also sulfurous odors are often associated with wastewater.[15] The wastewater source may be a wastewater treatment plant, leaking sludge lagoons, or drainage from private septic tanks. These can contaminate both groundwater and surface water, although groundwater is the more sensitive recipient. The groundwater plume from such sources is characterized by high nitrate-nitrite and chloride levels.[14]

11.3.3 Regrowth in the Distribution System

Potable water supplies frequently suffer taste and odor problems that develop in the distribution system. Stagnant water in isolated areas of the distribution system can exhaust the chlorine residual and so allow bacterial regrowth in the system. Iron- and sulfate-reducing bacteria can then propagate in colonies, corroding the piping and leading to rusty red or black water with accompanying taste.[9] These colonies can provide an environment for fecal coliform and associated pathogens to reside in, and so high bacterial counts can persist despite subsequent purging of the stale water. Another common source of taste and odor is the reaction of chlorine disinfectant with residual organics in the distribution system.[16] The end product may actually be more offensive than the original compound.

11.3.4 Taste and Odor Measurement

An appreciation for the complexity of taste and odor measurement will be facilitated by an understanding of the mechanisms of taste and odor detection before we proceed to actual methods of measurement.

The sense of taste is the least sensitive, the best understood, and yet a very complex sense. Taste is a combination of sensory responses from olfactory receptors located in the upper part of the nasal cleft, gustatory receptors on the tongue, and sundry receptors located on membranes in the nasal cavity.[6] The gustatory receptors (*papillae*), commonly called taste buds, respond to sweet, sour, salty, and bitter. Research suggests that the salty and sour papillae detect taste through an ion exchange process and that the sweet and bitter papillae utilize an enzyme-inhibition mechanism.[6] As previously stated, the interpretation of taste actually includes detection of taste by the tongue, combined with stimulation of olfactory receptors in the nose. Thus, the sense of taste is modified by the sense of smell.

The sense of smell is far more sensitive and much worse understood. Olfactory detection of some compounds can occur in the parts per trillion range. Attempts to explain odor response revolve around two theories: the stereochemical theory, and the chromatographic theory. Amoore avers in the stereochemical theory that there are only seven primary odors and that compounds with similar odors are similar in shape.[17,18] Olfactory receptor sites have corresponding slots where these compounds can attach and produce a sensory response.[6]

Roderick, on the other hand, suggests that the large surface area of the olfactory epithelium enables it to function as a gas chromatograph, able to isolate and distinguish between numerous compounds.[19] It has also been postulated that odor is a combination of olfactory and trigeminal sensations.[7]

Threshold Odor Measurement No satisfactory instrument has been developed for measuring odors. Perceived odors are often a complex of numerous compounds, many of which cannot be isolated and measured in ranges that the nose can detect without great difficulty. Mechanical techniques require a closed-loop stripping analysis (CLSA) to isolate the substances from water, followed by identification on a gas chromatograph-mass spectrometer (GC-MS). This method, which is quite involved and expensive, can identify and detect concentrations at or just below human detection limits.[5]

The easiest and most readily available method for measuring odors is also the most subjective. The human nose is extremely sensitive-but responses, descriptions, and threshold detection vary from person to person and over time. Olfactory fatigue may further prejudice response. Nonetheless, a test has been devised to provide a rough qualitative measure of odor. Called the Threshold Odor Test, this method requires a series of dilutions with odor-free water. Standard

Methods defines the resulting *threshold odor number* (TON) as the greatest dilution of sample with odor-free water still yielding a definitely perceptible odor.[3] The procedure for determining TON is explained below.

Normally a panel of five testers is convened, as suggested in Standard Methods.[3] An appropriate series of dilutions is prepared in 200-ml stoppered flasks. The flasks are heated to the specified temperature (generally 60°C) and brought before the panel. The flasks are then shaken, unstoppered, and sniffed in succession, starting with most dilute sample. Some blanks are inserted to offset the panelists. The first dilution in which odor is detected determines the TON. Eq. (11.1) is used to calculate TON.

$$\text{TON} = \frac{V_s + V_d}{V_s} \tag{11.1}$$

where

V_s = Volume of sample, mL
V_d = Volume of dilution water, mL
$V_s + V_d$ = 200 mL

The subjective approach of the TON method has led to dissatisfaction with the results.[3] Other approaches have been developed, such as ASTM Method D-1292, for industrial water and wastewater.[20] This method uses a different dilution methodology, in which the *Odor Intensity Index* is used as the scale. This index relates to TON but is defined as the number of times the sample must be diluted in half to reach the odor threshold.[21] The food and beverage industries and some water treatment systems are now using the *flavor profile* method for taste and odor testing.[7]

Threshold Taste Measurement Threshold taste measurement, like most threshold odor methods, provides a qualitative measure of taste intensity. A sample with strong odor may preclude taste tests, because odors can obscure individual taste responses. The standard taste test is similar to the TON test and, often, is performed concurrently. This test is described below.

Water deemed safe for ingestion is run through a series of dilutions and raised to 40°C. Panel members taste the sample by holding each dilution in the mouth for several seconds and then discharging it without swallowing. Beginning with the lowest concentration, this process continues until each member identifies the threshold taste and characterizes it according to primary descriptions.[3]

11.3.5 Taste and Odor Control at the Source

Taste and odor can be controlled at either the source or the treatment plant. A well-planned program of prevention or pretreatment at the source is often cost-effective and provides side benefits as well; however, the very complexity of the phenomena and the variety of origins may frustrate easy solutions. Some ways of controlling taste and odor at the source are discussed in the following paragraphs.[22]

Aquifer Protection In groundwaters, taste and odor control at the source is not normally possible; however, some recharge areas, such as the Edwards Aquifer in central Texas, are protected, by local and state laws, from agricultural, industrial, and urban contamination. This legislation constitutes a source protection strategy that, over an extended period of time, has a possible side benefit: preventing most synthetic organic sources of taste and odor in the groundwater.

Watershed Management For surface waters, a comprehensive watershed management program can help minimize both natural and man-made sources of pollution. Programs sponsored by the Soil Conservation Service[a] and the watershed protection efforts employed by Seattle[10] are examples of efforts that can regulate pollution and development upstream, and reduce erosion due to unsound farming or development practices. Ideally, a watershed or stormwater program would exercise jurisdiction over the complete watershed.[20] Unfortunately, this integration is not always practical. Accomplishing a significant reduction of nutrient loadings may be difficult for watersheds exceeding ten times the area of the reservoir, and could prove counterproductive.[23]

Stormwater Management Standards can also be adopted and enforced regarding agricultural and urban runoff, which constitute major nonpoint sources of pollution. Currently, under sections of the Clean Water Act, cities are being urged to develop stormwater utility districts that comply with new standards for urban runoff. These programs reduce nutrient loadings in streams and reservoirs, thus preventing eutrophication and algae blooms in raw water sources. This reduction, in turn can reduce many taste and odor problems.

The Use of Algicides Source control can include controlling algae directly. As mentioned before, algae are the primary source of taste and odor in surface waters. Algicides, usually modifications of copper sulfate, must be applied to the reservoir prior to the production of the odor-causing compounds.[23] The water quality expert must evaluate the potential impact of copper dosing for judicious application. Not all algae are readily controlled, and copper sulfate can be toxic to other aquatic organisms, including fish. Citric acid or other chemicals may be necessary to chelate the copper, and overdosing could have a detrimental effect on treated water quality. Copper sulfate is typically applied at 6 kg/ha (5.4 lbs/acre) when alkalinity is above 40 mg/L, and 1 kg/ha (0.9 lbs/acre) when it is less. A copper residual of 0.1 to 0.2 mg/L should be achieved. Application may be by the dragging of permeable bags of dry copper behind a boat or by the blowing of the dust.[23,24] Other methods include screened hoppers and solutions fed via spray nozzles from a boat, helicopter, or airplane.[20] Southern California Metropolitan Water

a. The Soil Conservation Service (SCS) of U.S. Department of Agriculture is now the Soil Service Division of the U.S. Department of Agriculture.

District reports effective utilization of chunk copper sulfate along selected shorelines where blue-green algae proliferate.

Biological Methods In addition to chemical methods, biological methods have been researched to degrade the odorous metabolites produced by algae. Silvey et. al. reported that some forms of bacteria reduced geosmin levels in water but these results have not been confirmed by other researchers.[25,26]

Reservoir Destratification Another approach to source quality control for reservoirs is destratification. As discussed previously, reservoirs may undergo thermal and/or dissolved oxygen stratification, followed by lake turnover in the fall. Each of these events is frequently the cause of pronounced seasonal taste and odor episodes in the treated water supply. Destratification, accomplished by the circulation of water between the epilimnion and hypolimnion, can minimize anaerobic conditions in the hypolimnion. The result is the elimination of dissolved iron, manganese, and hydrogen sulfide, as well as the reduction of many organic taste and odor compounds. Destratification also appears to limit the growth of algae, probably by diluting the warm water in the epilimnion with cooler water from the deeper hypolimnion.[6] Destratification does not resolve all taste and odor problems, but it maintains more nearly uniform water quality conditions in the reservoir and so reduces many sources of taste and odor.

The destratification technique suggested by Symons et. al. is to pump cooler water from the bottom of the lake to the top. This causes water to circulate from top to the bottom as well and so creates more nearly uniform water quality conditions throughout the full depth of the reservoir.[27] Providing oxygen is not necessary, because the epilimnion is high in dissolved oxygen and also because the increased oxygen gradient from circulating layers with low dissolved oxygen to the surface encourages additional diffusion.[11] By contrast, the most common method of destratification uses diffused air as the mechanism to achieve circulation.[20,28] In this method, piping conveys air from a blower system to diffusers at the bottom of the reservoir. The diffused air produces currents that circulate the stratified layers in the reservoir. A discussion on aeration devices can be found in Section 6.4.

11.3.6 Taste and Odor Control at the Plant

Source control of taste and odor is not always either possible or economical. In these cases, the water treatment plant must utilize in-plant treatment techniques for control of taste and odor. The sources of taste and odor are diverse; often there is not a single approach for their control. Ideally, plant design will be flexible enough to allow several methods for control of taste and odor. Control and removal methods include aeration, oxidation, and carbon adsorption, which are discussed in detail in the following paragraphs. Ion exchange, as mentioned previously, is normally utilized for removal of specific ionic species, including chlorides and other dissolved minerals. Many of these ionic species impart taste or odors. Therefore, ion exchange is discussed further in Chapter 18 (dealing with nonconventional treatment processes).

Aeration Aeration was one of the earliest methods for removal of taste and odor from water supplies. Also known as air stripping, aeration is effective in removing volatiles such as dissolved gases (hydrogen sulfide, carbon dioxide, etc.) and volatile organics, including several aromatics. Aeration also helps oxidize iron, manganese, and some organics as it raises the dissolved oxygen level in the water. Various types of aeration devices, Henry's Law, and the kinetics of gas-liquid equilibrium are presented in Section 6-4. Readers are referred to this section for additional details.

Oxidation Oxidation techniques include the use of several oxidizing compounds, such as potassium permanganate, chlorine, chlorine dioxide, ozone, and their combinations. Chemical oxidation by each of these compounds is discussed below. A more detailed discussion on several oxidizing compounds may be found in Chapter 12.

Potassium Permanganate ($KMnO_4$). Reportedly, it was first used by the City of Houston, Texas in treating their potable water. By the 1960s, the use of potassium permanganate had become widespread.[29] Potassium permanganate can be used as a disinfectant and algicide. It can remove taste and odor caused by decaying vegetation and can eliminate taste and odor caused by the halogen group of chemicals (chlorine, bromine and iodine).[20] It also precipitates color-causing iron and manganese. The physical and chemical properties of $KMnO_4$ are summarized in Table 11-2.[29] The general reaction of potassium permanganate with several constituents of water is given by Eq. (11.2).

$$\begin{bmatrix} \text{Organic compounds} \\ \text{Soluble iron} \\ \text{Soluble manganese} \\ \text{Hydrogen sulfide} \end{bmatrix} + KMnO_4 + H_2O \rightarrow \begin{bmatrix} \text{Carbon dioxide + water} \\ \text{Ferric hydroxide precipitate} \\ \text{Manganese dioxide} \\ \text{Soluble sulfate} \end{bmatrix} + MnO_2 \qquad (11.2)$$

Potassium permanganate can be delivered in drums, in liquid or dry form, or pneumatically from hopper trucks, in the dry state. A dry feeder is recommended when a plant is using more than 12 kg per day.[29] Figure 11-1 shows a typical delivery and feed system for a free-flowing grade of dry $KMnO_4$. Typically, the dry feeder discharges into a solution tank, for mixing prior to dosing the raw water, although some systems may feed dry potassium permanganate directly into the raw water stream. Liquid feed systems normally supply a 4-percent solution, fed into the stream by proportioning pumps.[29]

Potassium permanganate is usually applied at the raw water pump station or early in the water treatment process ahead of or in the rapid mix chamber, to allow sufficient time for oxidation to occur. Figure 11-2 provides a typical feed schematic for potassium permanganate feed at the rapid mix chamber. The potassium permanganate dosage should be adjusted so that the purple or pink color it imparts does not extend past the flocculation basin.[30] It is preferable to feed potassium permanganate well ahead of chlorine meant to oxidize organics. This can reduce the chlorine demand and lower the production of such halogenated organics as trihalomethanes. In

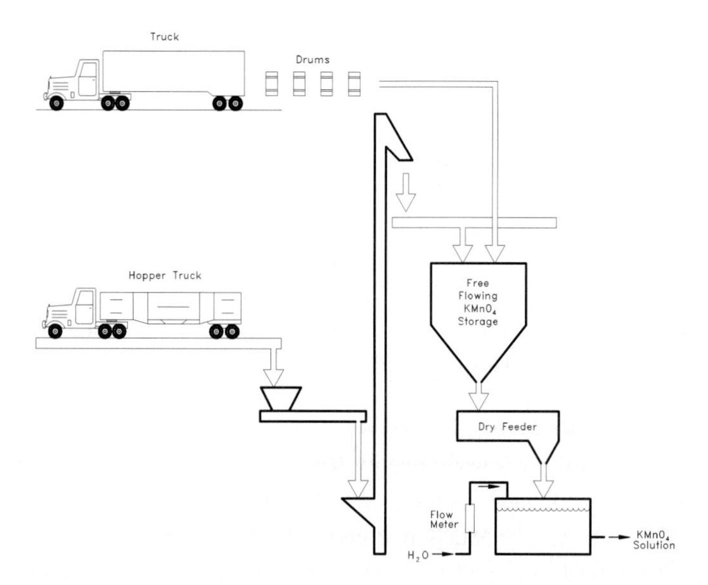

Figure 11-1 Potassium permanganate handling and storage system that uses drums or hopper trucks.

many instances, $KMnO_4$ is added at the raw water pump station to utilize the contact time in the pipeline.

Optimization of potassium permanganate dosing to control taste and odor requires jar tests. Jar tests must simulate actual conditions, including contact time, concentration, and pH. Potassium permanganate is more effective at alkaline pH. Typical dosages for taste and odor removal range from 0.5 to 2.5 mg/L, although dosages as high as 10 mg/L have been used occasionally.[29, 31] Overdosing should be avoided, because it can pass through the filters and enter the distribution system. In the distribution system, insoluble manganese dioxide (MnO_2) precipitates and blackens the water. It can also raise the manganese concentration above the limit set in the drinking water secondary standards.

When ozone is the primary disinfectant, potassium permanganate can interfere with ozone oxidation. Also, activated carbon consumes potassium permanganate; for this reason, ozone should be used only after dissipation of the permanganate residual.[29]

Chlorine (Cl_2). An ancient notion was that diseases were transmitted by the odors that emanated from swamps, lakes, and streams. This perception was not corrected until the mid-to-late 1800s, when Pasteur developed the germ theory of disease.[5] Prior to a clear understanding of disease transmission, chlorine compounds had been found effective against odors and were used as early as the 1830s to combat odors in drinking water.[10] Later, when chlorine was proven to kill pathogens, the practice of chlorination spread rapidly.

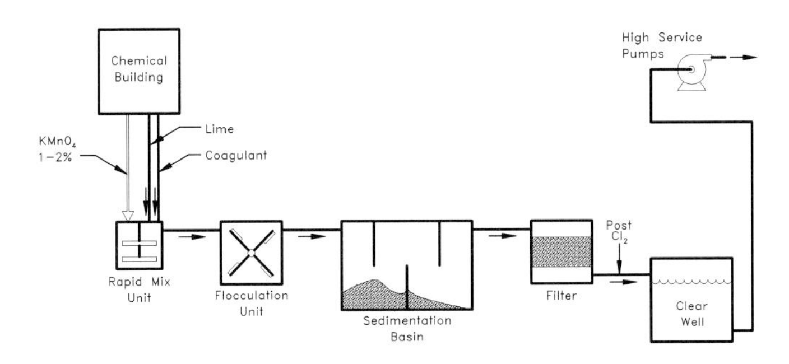

Figure 11-2 Potassium permanganate feed at the rapid mix unit.

The use of chlorine for taste and odor control has also been continued in order to meet aesthetic standards adopted for potable water quality. The major advantage of using chlorine is that it is readily available, is inexpensive, is easily applied, and maintains a residual. Some disadvantages are hazards associated with production, transportation, storage, and application. Additionally, the chlorination by-products are suspected carcinogens. Odorous by-products are produced by the interaction of chlorine with chemicals such as phenols. The new regulations on chlorinated organics make the chlorination alternative less attractive than in the past.[6] Chlorine controls taste and odor in two ways: it oxidizes the taste and odor compounds to less offensive products, and it controls the growth of microorganisms that often produce these compounds. Taste and odor control is a function of the chlorine residual, pH, temperature, and contact time. A detailed discussion of chlorine chemistry is given in Chapter 12. The effectiveness of chlorine for taste and odor control depends upon the type of compounds involved, which are oxidized variously by chlorine. Sulfides, disulfides, and mercaptans, for example, are readily removed by chlorination,13 but low levels of chlorine may accentuate odors from complex algal products such as MIB or geosmin.

Chlorine demand greatly depends upon the raw water quality, the initial dose, the contact time, and the temperature. The typical demand may vary from 2 to 10 mg/L, and the typical range of free chlorine residual is from 0.2 to 2.5 mg/L.[30]

Chloramines. Chloramines are formed by the reaction of free chlorine with ammonia naturally present in (or added to) the water. This combined chlorine consists of three species: monochloramine (NH_2Cl), dichloramine ($NHCl_2$), and trichloramine (NCl_3). The chemistry involved is explained more fully in Chapter 12. Chloramines require a longer contact time to oxidize and disinfect but produce no THMs. They are not especially effective at controlling taste and odor and can actually introduce or accentuate odors in the potable water. A chloramine disinfectant residual is maintained in potable water, primarily because of its longevity over free chlorine.

Chlorine dioxide (ClO_2). Chlorine dioxide reportedly was first used in drinking water at Niagara Falls, New York. Currently, between 300 and 400 facilities use chlorine dioxide in the United States, and several thousand employ it in Europe.[32–34] Like chlorine, it is primarily used for disinfection, with a side benefit of reducing taste and odor. Chlorine dioxide is actually classified as a form of chlorination. However, its generation and chemical reactions are much different. Chlorine dioxide is very unstable and must be generated on-site by utilizing a chlorine-chlorite solution. Current technology allows up to a 95 percent production of chlorine dioxide from the raw chemicals. Generation and chemistry of chlorine dioxide is presented in Chapter 12.

Oxidation by chlorine dioxide has all the advantages of chlorination previously mentioned. In addition, earthy-musty odors not easily controlled by chlorine are generally reduced by chlorine dioxide.[34] It also does not create the odorous organic compounds or the levels of THMs that chlorine produces. Other by-products, however, have come under close scrutiny to assess their potential effects. Also, the equipment for and cost of chlorine dioxide generation may be quite expensive. Chlorine dioxide demand is determined similar to chlorine demand. Samples and jar tests can optimize and confirm dosages. Typically, 0.1-to-5-mg/L concentrations of chlorine dioxide are used.

Ozone (O_3). Ozone reportedly was first used in France in the early 1900s.[21] Though widespread throughout Europe, the use of ozone in the United States has been limited. Ozone is a powerful oxidant and disinfectant that reduces taste and odor and can improve other treatment processes.

This highly reactive allotrope of oxygen is formed by electrical charge in the presence of oxygen. Ozone reacts in one of two ways: direct oxidation, or an indirect oxidation path via the formation of hydroxyl radicals. The first pathway is slower and is extremely selective in terms of the compounds that are oxidized. In the second path, ozone autodecomposes to the hydroxyl radical, a process is catalyzed by the presence of organics or other hydroxyl radicals. The hydroxyl radical is extremely fast and nonselective in oxidizing organic compounds.[6] For this reason, ozone disinfection is considered quite effective against virus.[10]

Ozone readily oxidizes many taste- and odor-causing compounds of natural origin. It is less effective at treating industrial contaminants.[6] The primary attraction of ozone is that it can reduce many taste and odor compounds more effectively than free chlorine, yet it does not produce THMs. It is also believed to deactivate *Giardia lamblia* cysts. Ozone is highly unstable and reactive, so it must be produced on site. For this reason, ozone had been less popular in the past, but today concern over THMs from chlorine has made this alternative more attractive. Ozone generation technology is covered in Chapter 12. The typical dosage ranges from 0.5 to 5 mg/L for most water treatment applications.

A variation of ozonation is the combined use of hydrogen peroxide and ozone, called *perozonation* or *peroxidation*. Known as an *advanced oxidative process* (AOP), the addition of hydrogen peroxide into the water prior to ozonation accelerates production of hydroxyl radicals. The effect of acceleration of oxidation in reducing taste and odor will require further research, and it may hold promise for oxidizing industrial contaminants.

Activated Carbon Adsorption Activated carbon comes in two forms: *powdered* activated carbon (PAC), and *granular* activated carbon (GAC). Both forms effect removal by physically adsorbing compounds. In the past, PAC was employed the more frequently, primarily for taste and odor control. PAC reportedly was first used in the United States in 1929.[20] Now, both PAC and GAC are generally viewed as a reliable last recourse for removal of taste and odors, particularly from industrial sources.[6]

In order to understand how activated carbon accomplishes adsorption, an explanation of how it is produced would be appropriate. Activated carbon is manufactured from wood, coal, bone char, coconut shells, or some other carbon source. The carbon source is heated in the absence of air to carbonize it, then activated by oxidation to remove noncarbon impurities. This activation burns away some carbon layers and produces a pore structure. Activation is accomplished by heating to a high temperature (between 200 and 1000°C) in the presence of steam, carbon dioxide, or air or by wet chemical treatment at lower temperatures, using such agents as concentrated phosphoric acid, potassium hydroxide, or zinc chloride. The result is a highly porous carbon surface with specific adsorptive properties. Pore-size distribution and adsorption characteristics are strongly influenced by the carbon source and the activation method employed.[6]

It is the high porosity and large surface area that give activated carbon its unique adsorptive properties. Surface areas in excess of 2000 m^2/g and a dry weight density of 500 kg/m^3 can be achieved. However, economics generally limits the use of carbon with a surface area between 500 and 1500 m^2/g in water treatment applications.

The large surface area available enhances the physical adsorptive properties of the carbon. In this process, water-borne solute molecules (*adsorbates*), pass through the maze of rough surfaces and irregular fissures of the activated carbon, or adsorbent. The molecules are removed by van der Waals forces (physical adsorption) and by very weak bonding between adsorbent and adsorbate. For this reason, the adsorbate is approximately at equilibrium with the solute concentration in solution. Any change in solute concentration, therefore, affects the amount of chemicals that can remain adsorbed.

In water treatment, the major considerations in selection of activated carbon are (1) the form of activated carbon (PAC or GAC), (2) the adsorptive capacity, and (3) the rate of adsorption.[35] PAC is normally more economical for treatment of seasonal or minor taste and odor problems; GAC may be desirable for more severe and persistent conditions. Both PAC and GAC are marketed with specific adsorptive properties, and virtually any sieve-size distribution is available.

Powdered activated carbon (PAC). The general feeling among water treatment plant operators is that, when everything else fails, it's time to use PAC. It is quite reliable, but its effectiveness depends on the type of carbon, the dosage, and the point of application. PAC is normally smaller than 300 mesh (0.074 mm), and the usual carbon dosage ranges from 0.5 mg/L to 5.0 mg/L.[35]

Normally, PAC dose is based on operator experience, although laboratory tests may provide more precise dosage. In these tests, a series of carbon dosages is added to the odorous water and the residual odor levels are determined. Where a single odorous chemical is present, the initial and residual concentrations can be measured by using advanced laboratory identification techniques. Where such equipment is not available, the odor source is unknown, or numerous compounds are involved, the TON method may suffice (Section 11.3.4). These data can then be plotted in one of several ways to clarify the removal response, to produce a predictive trend, and to determine the carbon dosages necessary to achieve acceptable water quality.
A simple plot of carbon dose versus TON (or the residual concentration) might be sufficient to highlight the relationship; however, the preferred method is to identify constants in an equation for removal and then apply these to dosage calculations. The *Freundlich* and *Langmuir isotherm* equations are commonly used. Both are applicable to both PAC and GAC situations. The primary difference is that the Freundlich isotherm equation is strictly an empirical approach, accepted as the standard, while the Langmuir isotherm has a theoretical basis and has wider applicability.[5,36,37]

The Freundlich isotherm is expressed by Eq. (11.3) and its corollary by Eq. (11.4).

$$\frac{X}{M} = kC^{1/n} \tag{11.3}$$

$$\ln\left(\frac{X}{M}\right) = \ln(k) + \frac{1}{n}\ln C \tag{11.4}$$

where

X = adsorbate actually adsorbed by carbon, mg/L ($X = C_0 - C$)
C_0 = initial concentration of adsorbate, mg/L
C = final or equilibrium concentration of the adsorbate, mg/L
M = carbon dose, mg/L
k = empirical constant (y-intercept)
n = slope-inverse constant

Similarly, the Langmuir isotherm is defined by Eqs. (11.5), (11.6) and (11.7)

$$\frac{X}{M} = \frac{abC}{1 + bC} \tag{11.5}$$

$$\frac{C}{XM} = \frac{1}{ab} + \left(\frac{1}{a}\right)C \tag{11.6}$$

$$\left(\frac{1}{X/M}\right) = \left(\frac{1}{ab}\right)\left(\frac{1}{C}\right) + \frac{1}{a} \tag{11.7}$$

where

a = maximum number of moles adsorbed per mass of adsorbent at monolayer saturation

b = empirical constant, L/mg

The Langmuir isotherm can be plotted in terms of $C/(X/M)$ versus C or of $1/(X/M)$ versus $1/C$. The plot that produces a clear trend provides the more accurate constants. Example 11-1 demonstrates the procedure for using both the Freundlich and Langmuir isotherms to develop predictive relationships.

Example 11-1

A water supply district has recognized watershed development as the cause of increasing taste and odor problems from algae blooms. Blue-green algae and actinomycetes have been identified as the most common source, with geosmin a representative odorous organic compound they produce. The district has no power to regulate development and expects decreased water quality. Accordingly, it has tested a commercial grade of PAC to determine if it removes geosmin effectively.

In the tests, raw water was placed in 1-L bottles and spiked with geosmin to reach a constant concentration of 40 μg/L (geosmin concentration in the raw water prior to spiking was less than 1 μg/L). A different dosage of PAC was applied to each bottle. After three hours, the samples were analyzed to determine the equilibrium concentration of geosmin. The results are given below.

Carbon Dosage (M), mg/L	Initial Geosmin (C_0), μg/L	Final Geosmin (C), μg/L
0	40	40.00
0.05	40	35.10
0.5	40	15.85
1.0	40	9.50
2.0	40	4.90
4.0	40	2.35
10.0	40	0.90
20.0	40	0.42
30.0	40	0.25

Given this information, produce the Freundlich isotherm to describe the removal of geosmin, then produce adsorption isotherm plots based on the Langmuir approach.

Solution

The data provided above are reduced to the variables used in plotting the isotherms. These results are tabulated below.

M, mg/L	C_0, µg/L	C, µg/L	X, µg/L	$\frac{X}{M}$, mg/g	$\frac{C}{X/M}$, mg/L	$\frac{1}{X/M}$, g/mg	$\frac{1}{C}$, L/µg
0.05	40	35.10	4.90	98.0	0.358	0.0102	0.0285
0.5	40	15.85	24.15	48.3	0.328	0.0207	0.0631
1.0	40	9.50	30.50	30.5	0.312	0.0328	0.105
2.0	40	4.90	35.10	17.5	0.280	0.0571	0.204
4.0	40	2.35	37.65	9.4	0.250	0.106	0.426
10.0	40	0.90	39.10	3.91	0.230	0.256	1.11
20.0	40	0.42	39.58	1.98	0.212	0.505	2.38
30.0	40	0.25	39.75	1.32	0.189	0.758	4.00

µg/mg = mg/g

Figure 11-3 shows the Freundlich isotherm plot. A clear linear relationship is apparent. The y-intercept, or k, taken from the plot is 4.4, and the slope, or $1/n$, is approximately 0.86, giving $n = 1.16$.

Figures 11-4 and 11-5 show the Langmuir adsorption isotherms. A linear trend is difficult to produce in Figure 11-4. A rough approximation is drawn. The y-intercept gives the term $1/ab = 0.22$ and the slope $1/a = 0.0046$. This calculation results in $a = 220$ and $b = 0.0207$. In the plot shown in Figure 11-5, it is much easier to establish the relationship. The y-intercept gives $1/a = 0.005$ and the slope $1/ab = 0.213$. This yields $a = 200$ and $b = 0.0234$. Once these constants are determined, the carbon dosage can be calculated for any level of geosmin removal.

This example is an idealized case. Not all plots will so readily reduce to a clear trend, on account of inaccuracies in measurement, uncontrolled variables, and other causes. Sometimes the plot may show more than one trend, because there are several compounds present, with distinctive adsorption removals. The isotherm approach is an attempt to set all variables, except for the carbon dosages and removal rates, constant and results must be interpreted in light of the assumptions that best represent the situation.

Granular activated carbon (GAC). The use of GAC is increasing in the United States, because GAC can be used as a filter medium in addition to providing adsorptive properties. Extensively used in the industry, GAC now is popular in potable water treatment. As the demand for high quality water increases, the cost of GAC regeneration declines, and its reputation for reliability and effectiveness is becoming established.

GAC is produced in the same manner as PAC and has essentially the same adsorptive characteristics. GAC often is installed instead of the conventional sand filter, or it can be used in a post-filtration column. Carbon capacity and rate of adsorption must be determined for either

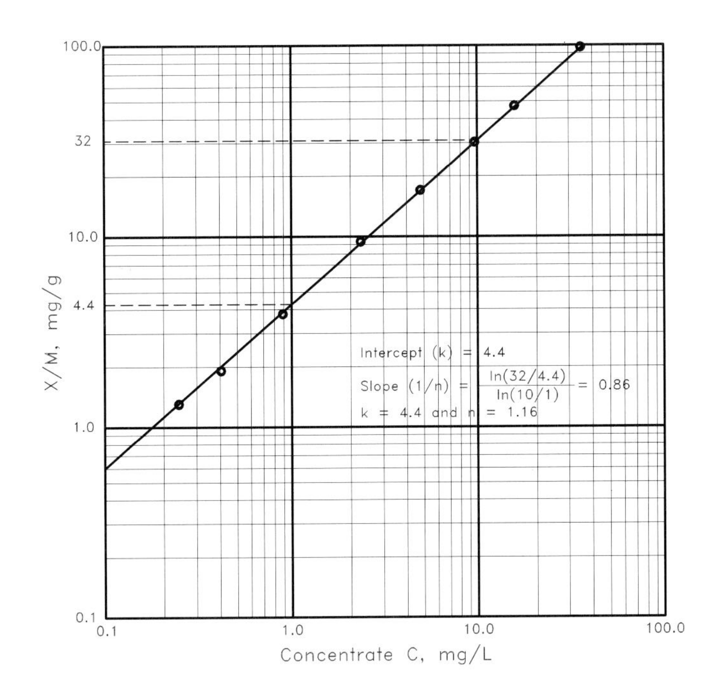

Figure 11-3 Freundlich isotherm plot for Example 11-1, PAC adsorption of geosmin.

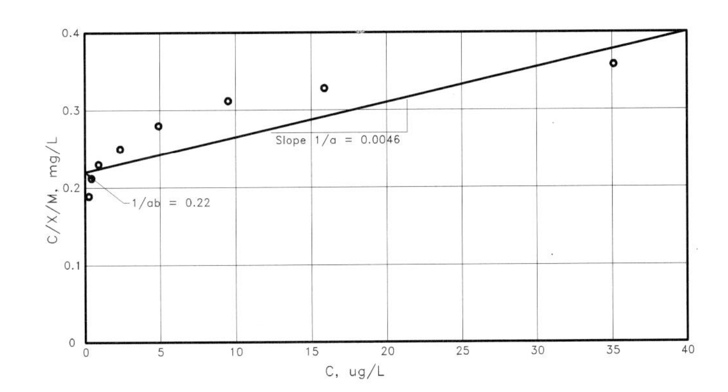

Figure 11-4 One type of Langmuir adsorption isotherm for Example 11-1, PAC adsorption of geosmin.

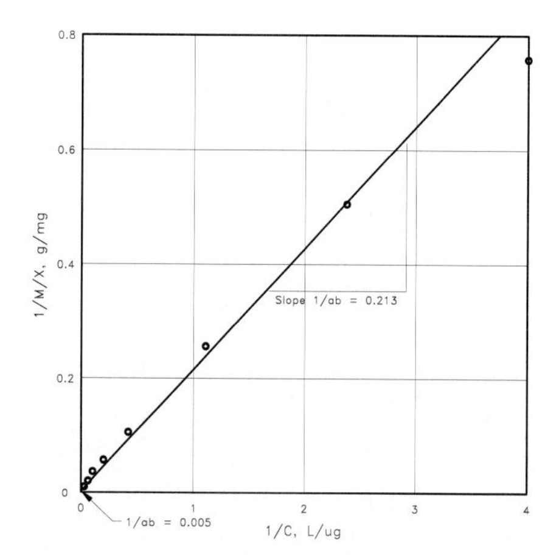

Figure 11-5 Another type of Langmuir adsorption isotherm for Example 11-1, PAC adsorption of geosmin.

location, but, for the filter option, an evaluation of GAC as a filter medium must also be performed.

Selection of a GAC treatment method should follow a laboratory-based study evaluating the feasibility and economics of this approach and establishing design parameters. Parameters include the adsorption rate, area and depth of the carbon bed or column, flow rate, run time, head loss, particle size, and GAC service life in terms of fraction lost and regeneration frequency.

The effectiveness of GAC in removing taste and odor or other compounds can be expressed by the Freundlich and Langmuir isotherms discussed previously. In addition, Bohart and Adams have proposed an approach that incorporates reaction rate theory to predict the performance of a continuous carbon column, as expressed by Eq. (11.8).

$$\ln\left[\frac{C_0}{C_e} - 1\right] = \ln[e^{KN_0D/V} - 1] - KC_0t \tag{11.8}$$

where

C_0 = influent concentration, mg/L or μg/L
C_e = allowable effluent concentration, mg/L or μg/L
K = rate constant, m^3/kg·h, (ft^3/lb carbon·h)
N_0 = adsorptive capacity, kg solute/m^3 carbon (1b solute/ft^3 carbon)
D = depth of bed, m (ft)
V = linear flow rate, m/h (ft/h)
t = service time, h

Usually, the term $e^{KN \cdot D/V}$ is much greater than unity, and Eq. (11.8) can be simplified as follows:

$$t = \frac{N_0 D}{C_0 V} - \frac{1}{KC_0} - \ln\left(\frac{C_0}{C_e} - 1\right) \tag{11.9}$$

A plot of t versus D in Eq. (11.9) presents a linear relationship. The slope S and intercept I are expressed by Eqs. (11.10) and (11.11).

$$S = \frac{N_0}{C_0 V} \quad \text{or} \quad (N) = C_0 VS \tag{11.10}$$

$$I = \ln\left(\frac{C_0}{C_e} - 1\right)\frac{1}{KC_0} \quad \text{or } K = \ln\left(\frac{C_0}{C_e} - 1\right)\left(\frac{1}{IC_0}\right) \tag{11.11}$$

The critical depth of the bed is obtained from Eq. (11.12) when $t = 0$.

$$D_0 = \left(\frac{V}{KN_0}\right)\ln\left[\frac{C_0}{C_e} - 1\right] \tag{11.12}$$

The rate constant K and the adsorptive capacity N_0 are determined from bench scale column tests. Once the relationship is developed, this equation can be used to predict the service time and critical bed depth. Typical laboratory equipment recommended by Eckenfelder and Ramalho for this test is shown in Figure 11-6.[37,38] The laboratory procedure is explained subsequently.

The laboratory procedure involves passing water through a series of carbon columns at a constant flow rate and measuring the desired compound in the effluent. As the effluent from the first column approaches some defined breakthrough concentration, the time of breakthrough is recorded, together with the column depth. The second column is added in series and the experiment is continued. Finally, when the last column reaches breakthrough, the test is concluded.

Upon completion of this procedure, the columns are replaced with fresh GAC, and the experiment is repeated at a different flow rate. The resulting data gives C_0, C_e, and V, with time and depth. A plot of time versus depth [Eq. (11.9)] produce straight lines for each flow rate. The slope and the intercept of each line are obtained, and from them the constants K, N_0, and D_0 are calculated. Plotting these constants against the flow rate produces curves used for full-scale design of the carbon filter facility. Examples 11-2 and 11-3 illustrate GAC design procedure when using the Bohart-Adams method.

Example 11-2

An industrial water treatment plant is withdrawing water downstream from an industrial region. The water is high in color, taste, and odor, primarily from organics. A specific synthetic organic chemical (SOC) has been identified as a representative contaminant and is to be tested for removal by GAC column. The results of this study will be useful in selecting a critical depth and flow rate to be used for design of a post-filtration adsorption column.

The Bohart-Adams method was used to develop the design data. A continuous bench scale carbon adsorption study was accordingly performed on filtered water from the plant. The selected SOC was at a constant concentration of 25 mg/L throughout the study. The maximum allowed concentration of SOC in the effluent was 1 mg /L. Temperature, pH, and other factors were relatively constant during the tests. The experiment used GAC columns 2.5 cm (one inch) in diameter, tested at four different flow rates and having various bed depths. Results from the experiment follow.

From the information given, prepare the necessary plots of time versus depth and of K, N_0, and D_0 versus flow rate.

Experiment number	Flow Rate, gpm/ft^2	Flow Rate, m^3/m^2·min	Bed Depth, m	Bed Depth, ft	Time, h	Through Volume m^3	Through Volume gallons
1	2	0.081	$D_1 = 0.61$	2	600	1.48	393
			$D_2 = 1.22$	4	1520	3.77	995
			$D_3 = 1.83$	6	2450	6.07	1604
2	4	0.163	$D_1 = 1.22$	4	430	2.13	563
			$D_2 = 2.44$	8	1110	5.50	1453
			$D_3 = 3.66$	12	1780	8.82	2330
3	8	0.326	$D_1 = 1.52$	5	180	1.78	471
			$D_2 = 3.05$	10	530	5.25	1388
			$D_3 = 4.57$	15	900	8.92	2356
4	16	0.652	$D_1 = 1.52$	5	70	1.39	367
			$D_2 = 4.57$	15	430	8.52	2251
			$D_3 = 7.62$	25	800	15.85	4189

From the information given, prepare the necessary plots of time versus depth and of K, N_0, and D_0 versus flow rate.

Solution

The plot of time versus depth is shown in Figure 11-7. Next, the velocity can be calculated, and the slope and absolute value of the y-intercept can be drawn from the figure. The result of these steps are shown here:

Experiment number	Flow Rate, gpm/ft^2	Flow Rate, m^3/m^3·min	Velocity ft/h	Velocity m/h	*S*-Slope h/ft	*S*-Slope h/m	*I*-Intercept h
1	2	0.081	16	4.88	463	1519	350
2	4	0.163	32	9.75	169	555	250
3	8	0.326	64	19.5	72	236	180
4	16	0.652	128	39.0	36.8	121	120

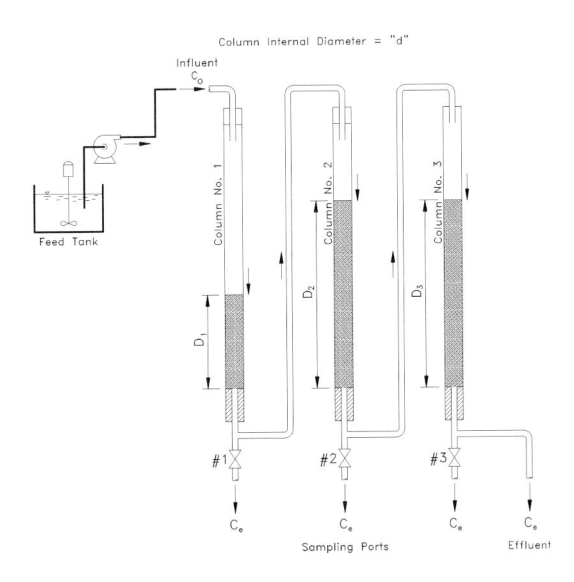

Figure 11-6 Laboratory equipment for granular activated carbon study.

This information is then used to calculate K, N_0 and D_0. Recall that C_0 = 25 mg/L = 25 ∞ 10^{-6} lb solute/lb solution, which, multiplied by 62.4 lb/ft^3, gives 1.56 ∞ 10^{-3} lb solute/ft^3 solution. The calculated values for these constants follow.

Experiment	Flow Rate,		N_o,		K,		D_o,	
Number	gpm/ft^2	m^3/m^2·min	lb/ft^3	kg/m^3	ft^3/lb·h	m^3/kg·h	ft	m
1	2	0.081	11.6	185	5.82	0.363	0.76	0.23
2	4	0.163	8.44	135	8.15	0.508	1.5	0.45
3	8	0.326	7.19	115	11.3	0.706	2.5	0.76
4	16	0.652	7.35	118	17.0	1.06	3.3	0.99

These data are plotted in Figure 11-8.

Example 11-3

The laboratory data on the GAC study developed in Example 11-2 is used to evaluate the performance of an activated carbon bed 1 m in diameter and 1.5 m deep. The concentration of specific synthetic organic chemical (SOC) in raw water is 25 mg/L, and the residual SOC concentration in the effluent is 1 mg/L. The flow through the carbon column is 95 m^3/d.

Calculate the following:

1. Service time (hours per cycle);
2. Number of carbon changes required per year and annual carbon volume (m^3).

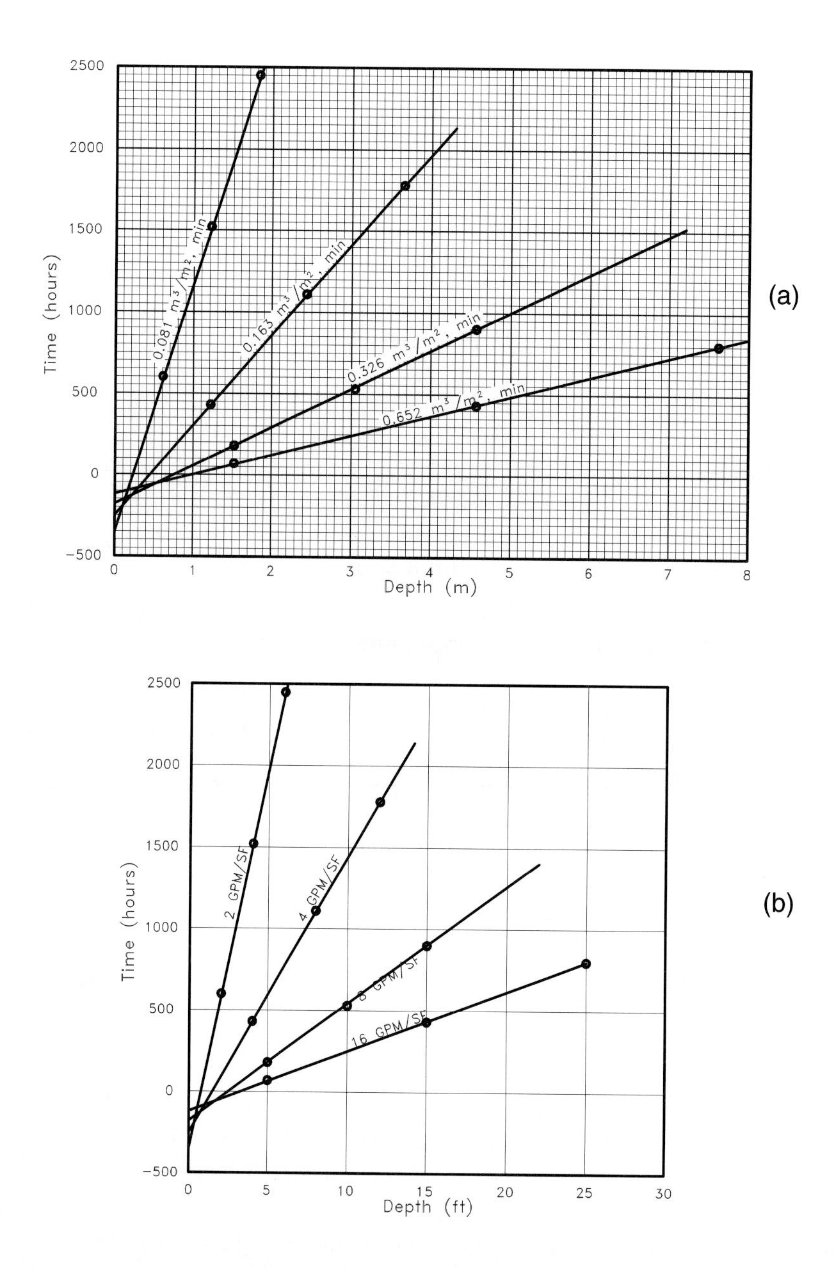

Figure 11-7 Bohart-Adams plot of time vs. depth for Example 11-2, GAC adsorption of target SOX. (a) Si units. (b) US customary units.

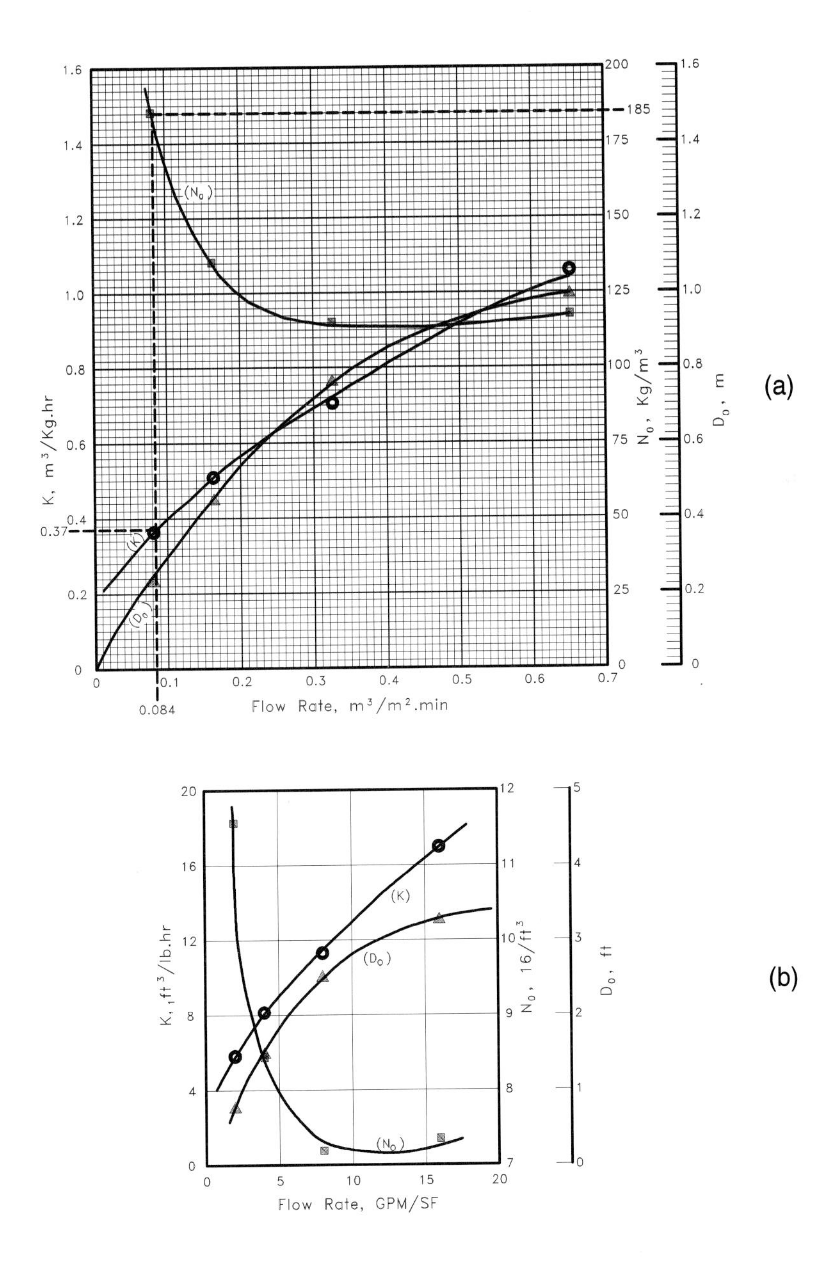

Figure 11-8 Bohart-Adams plots of N_0, K, and D_0 vs. flow rate for Example 11-2, GAC adsorption of target SOC. (a) SI units. (b) US customary units.

Solution:

1. Calculate service time.

$$\text{Flow rate to the bed} = 95.0\ \text{m}^3/\text{d} = 0.066\ \text{m}^3/\text{min}$$

$$\text{Cross-sectional area of the bed} = \frac{\pi}{4}(1\ \text{m})^2 = 0.785\ \text{m}^2$$

$$\text{Flow rate through bed} = 0.0663\ \text{m}^3/\text{min} \times \frac{1}{0.785\ \text{m}^2} = 0.084\ \text{m}^3/\text{m}^2\cdot\text{min}$$

$$\text{Flow velocity } V = 0.084\ \text{m/min} \times 60\ \text{min/h} = 5.04\ \text{m/h}$$

From Figure 11-8, for a flow rate of 0.084 $\text{m}^3/\text{m}^2\cdot\text{min}$, the following values are obtained:

$K = 0.37\ \text{m}^3/\text{kg}\cdot\text{h}$

$N_0 = 185\ \text{kg/m}^3$

Service time is calculated from Eq. (11-9). Use $D = 1.5$ m.

$$t = \left(\frac{185\ \text{kg/m}^3 \times 1.5\ \text{m}}{0.025\ \text{kg/m}^3 \times 5.04\ \text{m/h}}\right) - \left(\frac{1}{0.37\ \text{m}^3/\text{kg}\cdot\text{h} \times 0.025\ \text{kg/m}^3}\right)\ln\left(\frac{0.025\text{kg/m}^3}{0.001\ \text{kg/m}^3} - 1\right)$$

$$= 1849\ \text{h/cycle}$$

2. Calculate the number of carbon changes per year and the annual carbon volume.

$$\text{Number cycles/year} = \frac{365\ \text{d/year} \times 24\ \text{h/d}}{1849\ \text{h/cycle}} = 4.74\ \text{cycle}$$

$$\text{Annual carbon volume} = 0.785\ \text{m}^2 \times 1.5\ \text{m/cycle} \times 4.74\ \text{cycle/year}$$
$$= 5.58\ \text{m}^3/\text{year}$$

A more empirical approach to the use of GAC for taste and odor removal is the *empty bed contact time* (EBCT) method. This method can be very inexpensive yet still provide enough information to size and select carbon filters or columns. The design parameters are bed depth and filtration velocity. Other parameters can influence results. For a short EBCT, a deep bed at high velocity removes taste and odor better than a shallow bed. They also noted that, for a long EBCT, the opposite was true.[39] The EBCT is calculated from Eq. (11.13).

$$EBCT = \frac{H}{V} \tag{11.13}$$

where

$EBCT$ = empty bed contact time, s
H = filter bed depth, m
v = filtration velocity, m/s

This equation is often used as the basis for determining contaminant removal and GAC regeneration rates. Example 11-4 shows a simple application of the EBCT method. Langmuir equations may also be applied to the EBCT method to produce relationships.

Example 11-4

A newly-hired water treatment plant supervisor has two large mounds of GAC stored in one of the buildings. One was purchased recently, and the other was spent GAC needing regeneration by the manufacturer. Unfortunately, none of the employees remembers which GAC was the new shipment, and a visual inspection proves inconclusive. The supervisor, therefore, sent labeled samples of each to a laboratory to run the EBCT test and see whether this provides a clue.

The laboratory technicians set up parallel columns with 48 inches of GAC and spike the influent with methylisoborneol (MIB) at 15 μg/L. Tests are run in parallel at six different filtration rates with the following results.

Filtration Rate,		Velocity,		Depth,		*EBCT*,	C_0,	C_e, μg/L	
gpm/ft^2	m^3/m^2·min	ft/min	m/min	ft	m	min	μg/L	Test Column 1	Test Column 2
3	0.12	0.40	0.12	4	1.22	9.97	15	2.9	8.3
4	0.16	0.53	0.16	4	1.22	7.48	15	3.2	8.1
6	0.22	0.80	0.24	4	1.22	4.99	15	4.2	9.7
8	0.33	1.07	0.33	4	1.22	3.74	15	6.0	11.1
10	0.41	1.34	0.41	4	1.22	2.99	15	7.8	11.6
14	0.57	1.87	0.57	4	1.22	2.14	15	10.0	12.9
20	0.81	2.67	0.81	4	1.22	1.50	15	12.5	13.7

Figure 11-9 shows the effluent MIB concentration versus EBCT. Obviously, the carbon in test column 1 is the unused shipment.

11.4 Equipment Manufacturers of Color-, Taste-, and Odor-control Systems

Oxidative chemicals for color, taste, and odor control can be obtained from numerous local and national suppliers. Several sources specialize in producing different grades of activated carbon. Chemical feed systems are offered by a number of manufacturers, with several specializing in ozonation systems. A list of suppliers of chemicals and feed systems is provided in Appendix C.

The design engineer should work closely with the local representatives of the chemicals and equipment suppliers. Chemical selection should be based on the performance of the chemi-

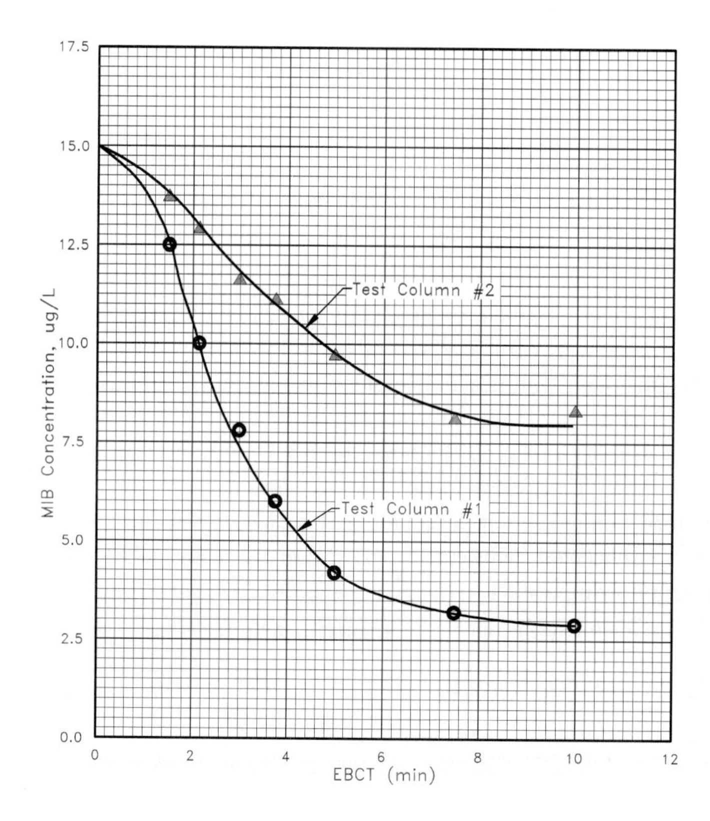

Figure 11-9 MIB concentrations plotted against *EBCT* for Example 11-3, GAC adsorption of MIB.

cals at water treatment plants with similar water quality and applications. Where possible, a bench-scale or pilot-scale study may be appropriate in choosing the chemicals for use at the plant

Equipment selection should also be based on performance history at similar installations. Complex systems, such as ozone and chlorine dioxide generation and dosing, might require a special consideration, because subcontractors may be involved. The design engineer should also perform a present worth analysis, to compare the chemical costs and the system operation and maintenance costs over time of the alternatives. Section 4.10 further explains responsibilities of the design engineer in proper equipment selection.

11.5 Information Checklist for Design of Color-, Taste-, and Odor-control Facilities

The following information is necessary for designing water treatment facilities for color, taste, and odor removal:

1. Information to be obtained from the predesign report and treatability studies:

   a. design flow: peak, average, and minimum;
   b. any color problem in raw water and suggested treatment methods, including coagulation, oxidation, combination of oxidation and coagulation, adsorption, and ion exchange;
   c. the chemical dosage and optimum in pH control in the flocculation units for color removal;
   d. need for oxidizing chemical, its application point, and dosage requirement for color removal;
   e. frequency and extent of taste and odor episodes in the raw water;
   f. proposed treatment methods, including aeration, oxidation, and carbon adsorption;
   g. chemical dosages and point of application.

2. Information to be obtained from the plant owner and operator:

   a. any particular preferences with regard to chemicals equipment types, and manufacturers;
   b. any preferences with regard to location of chemical addition and mixing facilities.

3. Information to be developed by the design engineer:

   a. design criteria established by the regulatory authorities;
   b. manufacturers' catalogs and design guides;
   c. types, sizes, and limitations of available equipment.

11.6 Design Example

11.6.1 Design Criteria Used

Design of a water treatment facility must address the need for removal of color, taste, and odor from the raw water. The following criteria are used to develop the design of these removal systems.

Flow Rates The flow rates are those cited previously. Removal systems must be designed to accommodate the maximum flow through the plant, which is 113,500 m^3/d at the maximum day during drought. Feed equipment must be capable of feeding at the lowest and highest desired dosage rates at the maximum and minimum plant flows.

Iron and Manganese Removal Seasonal problems with unacceptable concentrations of iron and manganese are expected. These cause color, taste, and odor problems. Iron concentrations of 0.2 to 0.7 mg/L have been reported, and 0.05 to 0.4 mg/L of manganese has been detected during these periods. The SMCLs[b] for iron and manganese are 0.3 and 0.05 mg/L, respectively.

Economical removal can be obtained by adding potassium permanganate in the raw water at the plant boundary. A maximum dosage of 4 mg/L (see Section 5.4.9) is sufficient to oxidize and precipitate iron and manganese to acceptable levels. This dosage of potassium permanganate will also reduce taste and odor to some degree and will reduce chlorine demand and THM formation potential.

Some color from organic sources is present in the raw water, but it is anticipated that the normal coagulation and flocculation systems will be sufficient to remove excess color not oxidized by the disinfectants. This expectation is confirmed by experience at another plant nearby with a water source of comparable quality.

Taste and Odor Control Seasonal taste and odor problems due to algae and actinomycetes affect source water quality. Potassium permanganate alone at 4 mg/L cannot reduce these problems. The client desired more flexibility in handling different types of taste- and odor-causing compounds. Bench-scale tests demonstrated the effectiveness of PAC in taste and odor control. The bench-scale tests determined that a maximum dosage of 8 mg/L, with a contact time of 25 to 30 minutes, would be sufficient to remove taste and odor below detectable level. Accordingly, the PAC feed system will be sized to feed up to 8 mg/L at the maximum flow rate. PAC will be added at the raw water pump station. The needed contact time will be provided by the travel time through the raw water pipeline. The spent PAC will be wasted along with other settleable floc in the sedimentation basins.

11.6.2 Design Calculations

Step A: Potassium Permanganate Feed System Potassium permanganate is usually fed in dry form, although some installations use a liquid feed system. The free-flowing potassium permanganate feed systems are used in this Design Example.

1. System geometry

   The potassium permanganate storage and feed systems will be located in the chemical building, as shown in Figure 8-23. The system will include delivery, bulk storage, feeder, solution tank, and feed pump and piping. (See Figure 8-22.) The $KMnO_4$ solution will be fed at Junction Box A inside the south boundary of the plant property. (See Figure 5-4 for

b. Secondary maximum concentration level (under current Secondary Drinking Water Regulations).

$KMnO_4$ feed point.) The solution will be fully mixed through free fall provided in Junction Box A, and the oxidation reactions will be completed in the pipeline before it reaches the rapid mix units. The schematic flow diagram and the potassium permanganate feed system for the design example are shown in Figure 11-10.

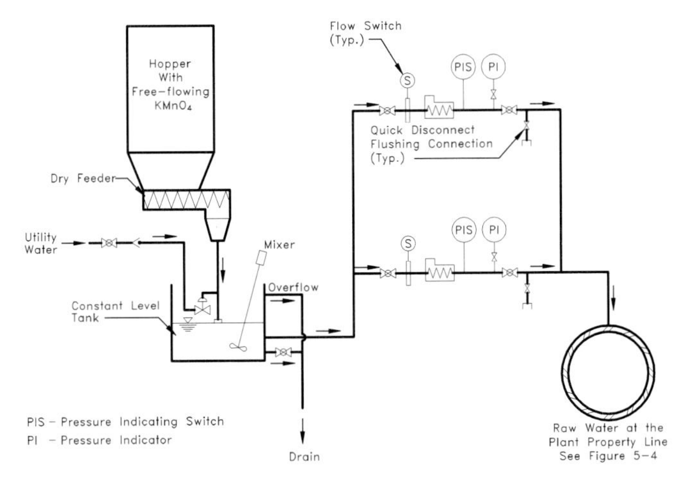

Figure 11-10 Schematic of potassium permanganate feed system.

2. Potassium permanganate delivery

   Delivery will be by hopper truck and by blowing into the silos. A dust filter will be provided, to prevent the fine particles from escaping during delivery. This filter will be a removable cartridge-type collector capable of cleaning an air flow of 11 standard m^3 per minute with less than 7.5 cm pressure loss in water column (W.C.)

3. Potassium permanganate storage

   The silo will be sized for an average dose of 4.0 mg/L from the potassium permanganate feed system. For a 30-day storage at the maximum day flow, 113,500 m^3/d, and 90 percent purity, the storage requirement are as follows:

Total storage requirement

$$= 113{,}500\ \text{m}^3/\text{d} \times 10^3\ \text{L/m}^3 \times 4.0\ \text{mg/L} \times (10^{-6}\text{kg/mg}) \times 30 \times \frac{1}{0.9}$$

$$= 15{,}100\ \text{kg/month}\ (33{,}300\ \text{lb/month})$$

There are two feed systems, so storage requirement per silo = 7550 kg.

Assume that the bulk density of dry potassium permanganate is 1500 kg/m^3 (Table 8-7); then the silo volume is 7550 kg/1500 kg/m^3 = 5.0 m^3. Rounding up and providing liberal freeboard allowance requires a total volume for each silo of approximately 8 m^3. Provide two silos, each 2 m in diameter and 2.5 m high.

4. Volumetric dry feeder

   The silo will be set above a hopper which discharges directly into the volumetric dry feeder. Several feeder arrangements are available from various manufacturers. Dry chemical is drawn from the hopper by means of a helical feed screw driven by an induction motor. The feed screw operating range is 20:1, complete with four pulley positions at the anticipated maximum dose of 4 mg/L and the maximum day flow of 113,500 m^3/d. Details of the volumetric dry feeder are shown in Figure 11-11.

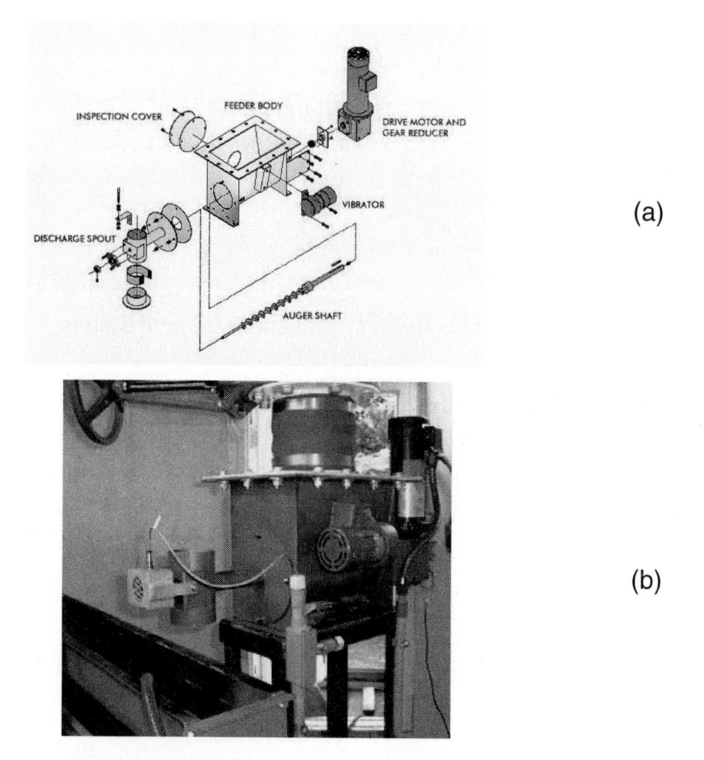

Figure 11-11 Volumetric screw dry feeder for delivery of $KmnO_4$. *(Courtesy of CHEMCO Equipment Co.)* (a) Schematic. (b) Equipment assembly.

The maximum capacity of each feeder

$$= \frac{1}{2} \times 113{,}500\ \text{m}^3/\text{d} \times 10^3\ \text{L/m}^3 \times 4.0\ \text{mg/L} \times (10^{-6}\text{kg/mg}) \times \frac{1}{24\ \text{h/d}}$$

$$= 9.5\ \text{kg/h}$$

$$\text{The volumetric capacity} = \frac{9.5\ \text{kg/h} \times 10^3\ \text{L/m}^3}{1500\ \text{kg/m}^3} = 6.3\ \text{L/h}$$

The capacities of available low-speed gearbox screw feeders can be between 1.0 L/h and 1500 L/h, so the above size is easily obtained. The feeder operation will be actuated in conjunction the operation of the solution-feed pump.

5. Solution tank

The solution tank will provide detention time for mechanical agitation prior to dosing. Dry potassium permanganate will drop into the tank from one side and will be mixed by a mounted mechanical mixer. The solution will flow via pipe to the pumps. A-200 liter tank will be sufficient. Dilution water will be fed by means of a float valve.

6. Feed pumps and piping

Two positive-displacement progressing cavity feed pumps in each feed system will convey the solution into the raw water pipe at the dosing point. One pump will normally be in service while the other is a standby. Each pump will be capable of pumping 0 to 10 L/min at sufficient pressure for delivery of this solution. The flow rate will be paced in proportion to the plant flow. The pumps will be rated for permanganate solutions up to 10 percent. All pipes and fittings must also be corrosion-resistant, able to handle solutions at this concentration. The dosing point will include a stinger for rapidly dispersing the solution into the raw water line.

Step B: PAC Feed System There are various approaches to design of PAC feed systems. Generally, simpler systems are less prone to problems and should be preferred over complex designs. In general, complex designs confuse operators, are difficult to repair, and have more components subject to failure. The system proposed here has flexibility and will provide the desired level of automation needed at a medium-size water treatment plant.

1. System geometry

The PAC storage and delivery system is provided at the raw water pump station. The location of the PAC storage and feed building was shown in Figure 7-18. The PAC system contains the following: two PAC-slurry bulk storage tanks with mixers, carbon delivery, and dust collection; two recirculation pumps; a volumetric liquid feeder; a constant-level tank; and three feed pumps, complete with piping and other necessary equipment and appurtenances. A schematic flow diagram is presented in Figure 11-12. Design of these components is presented below.

2. PAC-slurry bulk storage tanks

The anticipated dose of PAC is 5 to 8 mg/L. The system must be designed for maximum dose of 8 mg/L. For a 30-day storage at the maximum day flow of 113,500 m^3/d, the PAC quantity is calculated as follows:

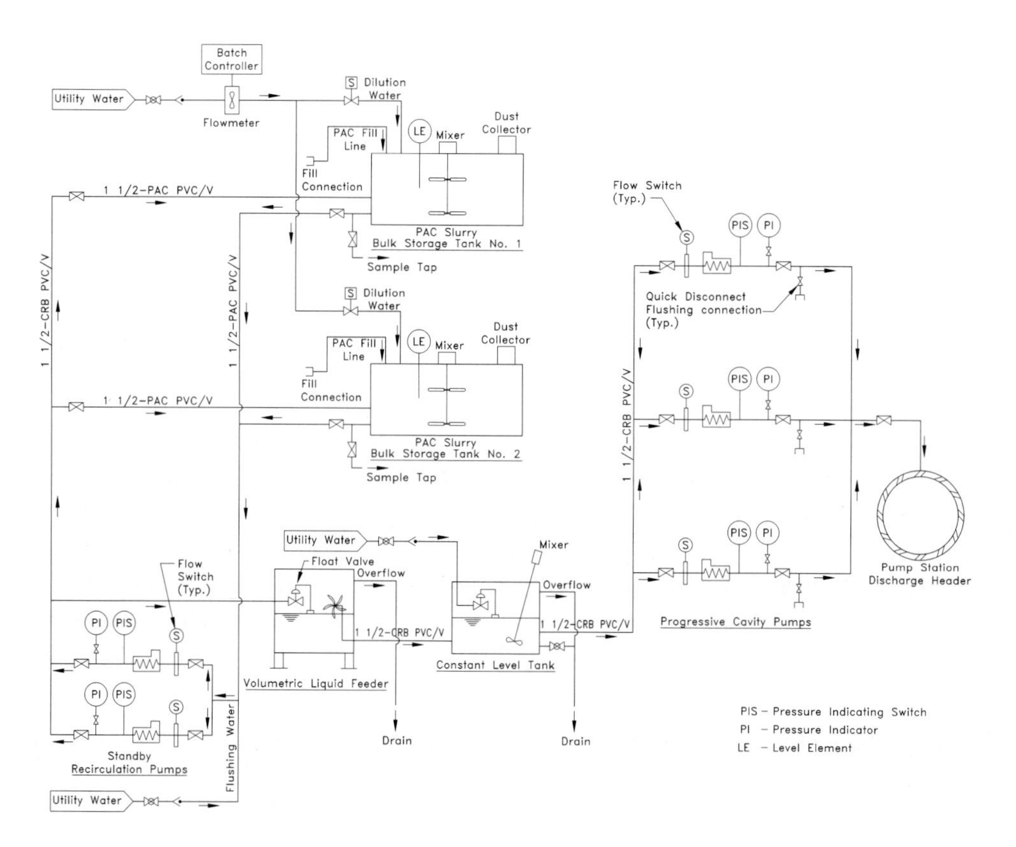

Figure 11-12 Schematic of powdered activated carbon feed system.

Total storage requirement

$$= 113{,}500 \text{ m}^3/\text{d} \times 10^3 \text{ L/m}^3 \times 8.0 \text{ mg/L} \times (10^{-6} \text{kg/mg}) \times 30 \text{ d}$$

$$= 27{,}200 \text{ kg/month } (60{,}100 \text{ lb/month})$$

Each slurry tank will be sized to store a full month's supply, so that one can be out of service or emptied for new PAC delivery while the other is maintained in service. The ultrasonic transducer will monitor the slurry level in the tank. The slurry will be maintained at a recommended concentration of 120 kg PAC per m^3 of fresh water. PAC expands the volume at a ratio of 1:1.2; therefore, the PAC concentration in the slurry will be 100 kg per m^3.

$$\text{The storage capacity required} = \frac{27{,}200 \text{ kg}}{120 \text{ kg/m}^3} \times 1.2 = 272 \text{ m}^3$$

Two structures, each 6.5 m square and 8.5 m deep, will each provide 359 m^3 capacity. Part of the surplus capacity will be dead storage above the overflow outlet and below the withdrawal piping. The structures will be located at the raw water pump station and adjacent to the access road, to permit easy unloading by the delivery trucks. The location of PAC facility is shown in Figure 7-18. Fresh (potable) water will be required to fill the storage tanks to the desired level prior to carbon delivery. A level sensor will be provided at each tank so that volume determination can be made. High- and low-level alarms will also be provided that can override system operation. From the quantity of fresh water added to the slurry bulk storage tank, total carbon quantity is calculated and delivered into the tank.

3. Carbon slurry mixer

   Before the PAC is delivered, the mixer will first be actuated in the receiving tank. A two-speed vertical shaft mixer will be employed, with one blade at 0.5 m above the floor and the second blade at 3 m above the floor. The mixer will be operated at high speed (35 to 50 rpm) when PAC is being delivered, to create a vortex into which the carbon can be mixed rapidly. Normal mixing will require a mixer operating at 25 to 35 rpm. From the manufacturer's catalog, a 15 kW (20 hp) motor will be required for this application. The motor contacts will monitor the status of the slurry mixers.

4. Carbon delivery

   Carbon will be delivered pneumatically from a pressure-differential (PD) truck. The PD truck will use an air compressor to fluidize the PAC and blow it into the slurry tank at a fixed rate. Typically, the carbon delivery rate is 12.2 m^3/min (430 cfm).[40] Connection will be made by use of a quick-disconnect flexible hose compatible with that used by the delivery vessel. Piping will direct the fluidized PAC into the vortex created by the slurry mixer.

5. Provisions for dust collection

   Pneumatic unloading requires design for air-volume displacement. Simple air vents would allow escape of the powdered carbon. Therefore, air scrubbers must be provided and placed above each tank. Both dry and wet scrubbers are available in the market. The design provided here will incorporate a dry scrubber for dust removal. The dry scrubber will be a self-cleaning cartridge-type collector capable of cleaning the required air flow at a specified pressure loss. Pressure indicators will monitor suction pressure, pressure loss across the filter, the contact monitor, and the fan motor. The high-pressure reverse-flow air jets will automatically clean the filter fabric when ever allowable head loss is exceeded. After the PAC has been mixed into the slurry, the mixer speed will be reduced to the low-level mixer setting.

6. Resuspension of the PAC

   When the tank has been out of service, but still contains slurry, the wetted PAC will settle in a thick layer at the bottom. Most agitator mixers will not provide the torque necessary to break this layer. Therefore, an air-sparge system shall be constructed on the bottom of each tank. When a tank with settled PAC is to be returned to service, high-pressure air must first be introduced into the sparge. This air will loosen and break up the PAC. To protect the agitator, a timer shall be provided to prevent the mixer from being activated prior to several minutes of air mixing. Once a portion of the PAC has been loosened, mixers will gradually erode and resuspend the rest of the PAC.

7. Recirculation pumps

   Two recirculation pumps should be provided for recirculation, transfer, and feeding of the carbon slurry. This will help maintain suspension in the feed piping and tanks. Centrifugal pumps are often provided for this purpose, but this design will utilize positive-displacement single-stage progressive cavity pumps, which are generally easier to maintain and offer better performance with abrasive slurries such as PAC. The pumps shall be rated for 0-40 L per min at 276 kN/m^2 (40 psi). Flushing water connections will be available for cleaning the lines and the pumps.

8. Volumetric liquid feeder

   The volumetric liquid feeder will be of the rotating dipper type, with an adjustable feed rate supplied via a float valve, rated for 0 to 20 L per min. A speed controller will adjust and monitor the slurry feed rate. The slurry will be conducted by gravity from the volumetric feeder to the adjacent constant-level tank. The type of volumetric feeder used in this design is shown in Figure 11-13.

(a)

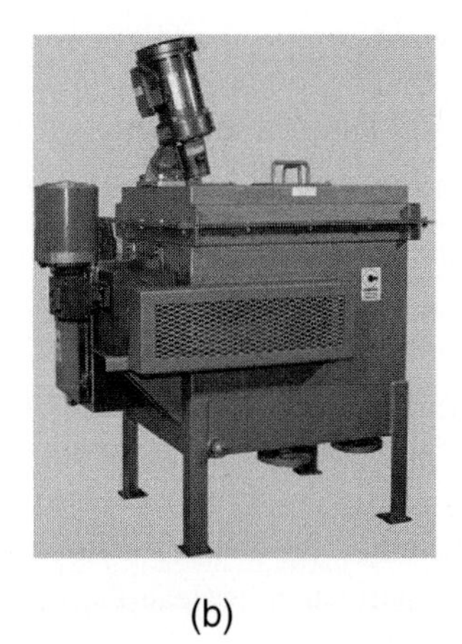

(b)

Figure 11-13 Volumetric rotating dipper type liquid feeder for PAC. *(Courtesy of CHEMCO Equipment Co.)* (a) Tubular dipping wheel. (b) Equipment assembly.

9. Constant-level tank.

   The constant-level tank will hold 200 L of carbon slurry and be equipped with an electric mixer to agitate the slurry. A float valve shall maintain liquid level in the tank.

10. PAC feed pumps.

    Three PAC slurry pumps will be provided, each capable of feeding 8 mg/L of PAC at the maximum plant flow. One pump will normally be in service, the second pump will be added for peak dosing, and the third is provided as a standby. These pumps will be of the positive-displacement single-stage progressive cavity type, paced proportionally to the plant flow. Motor contacts will monitor the status of the pumps.

11. Air and slurry piping.

    The air sparge and piping will conduct air at 690 kN/m^2 (100 psig) from an air compressor located in the PAC feed room. The air will be used to clean the filter and activate the air-sparge system. The slurry piping will conduct slurry to a dosing point located at the raw water pump station discharge header. The PAC will have a contact time of over 30 minutes at maximum day demand.

11.7 Operation, Maintenance, and Troubleshooting for Color, Taste, and Odor Control Facilities

Proper operation and regularly scheduled maintenance of color-, taste-, and odor-removal systems is essential to providing uninterrupted service and long life of the equipment; however, even the best-designed system requires a proper operation and maintenance schedule.

11.7.1 Common Operational Problems and Troubleshooting

Problems are frequently a function of the complexity of the system. Automated systems can be designed to avoid costly operator errors and reduce required O & M costs. On the other hand, such systems often utilize more sophisticated equipment requiring highly skilled personnel. The following are common problems encountered with PAC and potassium permanganate feed systems.

Potassium Permanganate Feed Systems

1. Potassium permanganate delivery from the hopper truck is aborted by high pressures. This condition implies that the air filter is clogged. The filter cartridge must be replaced with a clean cartridge.
2. Potassium permanganate is not reaching the dosing location. A pipe, a fitting, or the pump might have failed, or the feeder is turned off or has failed. Impurities in dry potassium permanganate can absorb moisture and cake, thus causing arching in the hopper above the feeder. This does not normally occur with the free-flowing variety of $KMnO_4$. Another possibility is that the feeder has been set too low.
3. The pink color due to potassium permanganate should disappear in the flocculator; otherwise, a faint pink color can remain in the water as it enters the sedimentation basin. Reduce the potassium permanganate dose if a pink color is clearly visible in the sedimentation basin.

PAC Feed Systems

1. Carbon unloading from the PD truck is aborted because of high pressures. This implies that the air scrubber is not functioning because of a plugged filter cartridge. The high-pressure air supply for cleaning the filter may have failed, or the filter cartridge needs replacement.
2. The agitator mixer sends a torque overload alarm. Such a false alarm may be due to bearing or other mechanical failure. If the air-sparge system is not operational or has not been activated prior to mixing, the lower set of mixer blades may be restrained by the settled PAC. Loosening the settled carbon with high-pressure air through the sparge should alleviate this condition.
3. There is blockage in the air-sparge system. A positive pressure must be maintained in the sparge piping, to reduce the influx of carbon into the sparge piping. Should check valves fail on the air line, PAC may enter and block the piping. If the air line does not have a loop above the liquid surface level, slurry could backflow into the air compressor when it is out of service.
4. PAC is not reaching the dosing location. One or more pumps may have failed, or the pipeline may be plugged or have failed. Automatic flushing can reduce the likelihood of plugged or broken lines. Other possibilities are: (1) open suction lines from the empty tank, (2) stuck float valve, (3) closed diaphragm valve, and (4) failed rotating dipper.

11.7.2 Operation and Maintenance

Operation and maintenance of the color-, taste-, and odor-removal systems is a function of the raw water quality. During periods of poor quality, these systems will require additional operator attention. Early detection of problems can minimize customer complaints. On the other hand, maintenance should not be neglected during periods when they are not in use. Routine maintenance should be scheduled to assure that equipment is not allowed to deteriorate. The following procedures are recommended for satisfactory operation of the facilities.

1. Perform daily sampling of finished water to determine the prevalence of color, taste, and odor. Adjust chemical dosage rates accordingly, or bring the PAC and/or potassium permanganate system into service when water quality goals are not met.
2. Verify dosage settings and record feed rates daily. Maintain a log of feed rates, together with raw and treated water quality parameters, for comparison. This log should be reviewed periodically to detect trends and relationships that might guide in future operator decisions.
3. Calibrate chemical feeders monthly. Verify proper operation of pumps, hoppers, feeders, mixers, scrubbers, piping, valves, and instrumentation.
4. Lubricate, clean, and service equipment as specified by the manufacturer. Manufacturers' O & M manuals on equipment usually are supplied, which specify O & M service required

on a daily, weekly, monthly, or longer basis. The manufacturer warranty may be voided by neglecting this recommended schedule.

5. Flush the PAC lines weekly when they are in service, monthly when they are out of service. Verify and make sure that the air lines and sparge for the PAC systems are not clogged. Start up the air compressor once each week when it is not in use; verify that the moisture blow-off is functional.
6. Monitor PAC and potassium permanganate levels when recording feed rates. Schedule delivery ahead of time. Periodically verify rated characteristics of samples.
7. Inspect bearings and motors on PAC mixers every six months. Also, inspect the air scrubber system, from compressor to filter.
8. Empty and clean out each PAC tank once each year. Inspect and repair sparge, piping, impeller, and other equipment.

11.8 Specifications

These specifications are intended not to be comprehensive but merely to be an initial guide towards producing complete design for use by the engineers and equipment manufacturer.

11.8.1 Potassium Permanganate Feed System

The contractor shall construct, assemble, test and make ready for operation the potassium permanganate system described in the drawings and the specifications. This system shall consist of one storage silo complete with hopper, fill opening, and air filter; one volumetric dry feeder and solution tank; and two feed pumps and associated piping and fittings.

Storage Silo and Hopper Two storage silos shall be sized so as to provide 8 m^3 of storage in each silo. The dimensions of each silo shall be 2 m in diameter and 2.5 m in height, having attached at the base a hopper with 60° slopes from the vertical. The silo and hopper shall be stainless steel. The silo shall include a port for truck delivery of dry potassium permanganate and a disposable cloth filter rated at 11 standard m^3 per min for this application. The filter chimney shall not allow moisture or air to enter the soil.

Volumetric Dry Feeder and Solution Tank The feeder and tank shall be the product of a single manufacturer and warranted against defects of compatibility or craftsmanship. All parts exposed to the feed chemicals shall be stainless steel. The feeder shall be mounted besides the tank. An induction motor will drive the helical feed screw. The motor shall have a four-step pulley. The system shall have a 20:1 operating range, with a timer set manually. All moving parts shall be housed, to prevent operator injuries during normal operation. Controls shall be mounted in a NEMA-1 enclosure adjacent to the feeder. The tank shall contain 200 L of liquid. It shall include a 0.04 kW (1/20 hp) electro-mechanical mixer mounted with a universal clamp. An adjustable float valve shall regulate the liquid level in the tank. An overflow shall be provided.

Feed Pumps, Piping, and Fittings The feed pumps shall be positive-displacement progressive cavity single-stage pumps. Two feed pumps, piping, and fittings shall be provided. They shall draw from the base of the solution tank via 2.5 cm (1") threaded PVC piping and a pump to the diffuser located in the raw water pipe at the plant property line. The pumps shall handle 0 to 400 L/min at 207 kN/m^2. Valves shall be standard PVC ball valves. All pumps, piping, and fittings shall be rated for handling potassium permanganate solutions without damage.

11.8.2 PAC Feed System

The contractor shall construct, assemble, test, and make ready for operation the PAC feed system described in the drawings and these specifications. This system shall consist of two PAC slurry storage tanks, each complete with fill connection, air scrubber system, sparge, two-speed mixer, level detector, and accessories; two positive-displacement progressive cavity recirculation pumps, complete with instrumentation, piping, valves, and fittings; one volumetric rotating-dipper liquid feeder; one constant-level tank with mixer, complete with instrumentation and piping; and three positive-displacement progressive cavity pumps, complete with piping, instrumentation, and stub-out into the raw water pump station discharge header.

PAC Slurry Storage Tanks The PAC slurry storage tanks shall be of reinforced concrete design, 6.5 m square by 8.5 m deep. The tanks shall be below grade and have one common wall. Following one month of curing, interior concrete surfaces shall be ground smooth, abrasive-blasted, and cleaned. A continuous lining shall cover the floor, walls, and ceiling, one produced by applying protective coatings appropriate to this application.

The PAC fill connection shall be 10-cm diameter (4-in), made of stainless steel grooved pipe, having a quick-disconnect coupling at the truck delivery point, and discharging immediately above the high-level mixer vortex. A 5-cm PVC pipe shall convey utility water to each tank. This line will be manually operated, and dilution water will be added up to the volume indicated by the level sensor and a calibration chart for this purpose.

The air scrubber will be a self-cleaning cartridge-type collector, certified for PAC application at 13 standard m^3/min with less than 7.6 cm of pressure loss W.C. The 690 kN/m^2 (100 psi) reverse-flow pressure jets will automatically clean the filter fabric when allowable head loss is exceeded. One scrubber will be placed above each tank.

The air-sparge system shall consist of a threaded black steel doubled cross set at 5 cm above the floor of each tank. The eight arms shall extend one foot past the reach of the mixer impeller. Each arm shall be perforated on the underside with ports alternating 45° from the vertical, the holes being spaced 10 cm apart. Arms (and the sparge double cross header) shall be restrained by type 316 stainless steel clamps, bolts, and nuts.

High-pressure air for the sparge (and also for the scrubber cleaner) shall be provided by a single-stage air compressor capable of 57 standard L/min at 690 kN/m^2. The air receiver shall hold 100 L (26 gallons) and be rated for 1400 kN/m^2, with safety valve, pressure gauge, and automatic condensate valve.

The vertical mixer for each tank shall include a 15-kW (20-hp) explosion-proof two-speed motor, of 1200 rpm maximum. Shaft and impeller shall be of stainless steel and designed to produce balanced loads. The upper impeller shall be a 1.5-m (60-in) diameter pitched-blade turbine with key position adjust; the lower impeller shall be a straight-blade turbine with key adjust, 1.3-m (50-in) in diameter. The lower impeller shall mount down to 0.3 m above the floor of the tank. The motor for each mixer shall be 15-kW (20-hp), explosion-proof, 1200 maximum rpm, with a reducer output of 35 to 50 rpm at high speed and 25 to 35 rpm at low setting.

Torque-overload protection shall be provided for the mixer unit, complete with alarms. Ultrasonic level sensors will be mounted on each tank. Malfunction alarms and shutdowns shall be provided as necessary for the mixer, scrubber, sparge, and level sensing systems.

Recirculation Pumps Two positive-displacement progressive cavity pumps shall recirculate PAC slurry when the system is in use. These pumps shall be single-stage, rated for abrasive slurry use, and able to pump 3 L/min at 138kN/m^2 of differential head. A flow switch at each pump will signal pipe or pump failure.

PAC slurry piping shall be 4 cm in diameter, withdrawn from each tank via an inverted expanded opening. Diaphragm valves with protective lining for abrasive service will be manually operated. Cleanouts shall be provided for all long runs, with unions for short segments.

Liquid Feeder and Constant-Level Tank One volumetric liquid feeder of the rotating-dipper type shall be provided to meter PAC slurry drawn from the circulation loop. The feeder shall be capable of a range from 0 to 20 L/min. The feeder shall be calibrated to deliver the desired feed rate to within one percent; the totalizer shall be accurate to within two percent.

The liquid feeder tank shall be of stainless steel, and so will the dipper shaft. The dipper shall be of cast iron. The tank shall include a stainless steel mechanical agitator, as per the manufacturer, to maintain the slurry in suspension. A float valve shall adjust to the proper liquid level. Dipper torque overload, motor failure, and tank high level shall be alarmed. When the feeder is taken out of service, an adjustable timer shall shut down the constant-level tank mixer and the three feeder pumps after the lines have been flushed by utility water.

The constant-level tank shall be gravity-fed from an adjacent volumetric feeder. It shall be constructed of stainless steel and hold 200 L of carbon slurry. A 0.2-kW (0.25-hp) electric mixer, capable of 1800-rpm severe duty, with impeller, shall be mounted on the tank by means of a universal clamp. A float valve shall maintain liquid level and permit flushing by utility water when the volumetric feeder is taken out of service.

Feed Pumps Three feed pumps shall be of the positive-displacement progressive cavity type. One shall normally be in service, the second shall be automatically brought into service during peak demand, the third will be a standby. They shall be sized the same as the recirculation pumps. Pumps shall include safety relief valves for bypass and also appropriate alarms. Piping will convey the slurry to the pump station discharge header, where a removable stinger pierces the header. The pump station will automatically shut down the PAC slurry feed system, beginning with the volumetric feeder and following with timed shutdown of the mixer and feed pumps.

Problems and Discussion Topics

11.1 Discuss the sources and impacts of color in surface water sources. Describe the various physical and chemical treatment process used for removal of color.

11.2 Distinguish between true color and apparent color. Describe the methods used for color measurement.

11.3 A water sample has a true color of 20 units. This sample has 48-percent transmittance. After conventional treatment, the transmission is 92 percent. Determine the color remaining in the treated water and the percent of color removal. Assume that Beer's law applies:

$$\log\left(\frac{I}{I_0}\right) = -kC$$

where

I_0 = intensity of light entering

I = intensity of light leaving

k = constant

C = concentration of absorbing solution

11.4 True color of a water sample is generally determined by centrifugation. Why is centrifugation preferred to filtration to remove turbidity?

11.5 What are the possible effects of chlorination on natural waters containing color-causing compounds?

11.6 A 10-mL original water sample was diluted with distilled water to a volume of 100 mL for color measurement. The result gave a color of 5 units. What is the color of the original sample?

11.7 List the various sources of taste and odor in raw water supply sources. Discuss methods of controlling taste and odor in the source of raw water.

11.8 Define TON. Describe the method of odor testing commonly used at water treatment plants.

11.9 In a threshold-odor test, a panel of five individuals was used. Two panel members detected odors at a lowest sample concentration of 12 mL diluted to 200 mL with odor-free water. The other three panel members detected odors at 8.3 mL sample concentration. Calculate TON.

11.10 What is the purpose of the blanks used in between the series of dilutions for threshold taste or odor test?

11.11 In a threshold-odor test, the following results were obtained with a five-member panel. Calculate mean TON.

Dilution (mL sample to 200 mL)	0	4	5.7	0	8.3	12	0	17	25
Response of member A	–	–	–	–	+	+	–	+	+
Response of member B	–	–	+	–	+	+	–	+	+
Response of member C	–	–	+	–	+	+	–	+	+
Response of member D	–	–	–	–	–	–	–	+	+
Response of member E	–	–	+	–	+	+	–	+	25

11.12 A raw water supply source has MIB concentration of 50 μg/L. A carbon adsorption study was conducted by using locally available PAC. The results of this study are summarized below. Calculate the value of constants *a* and *b* in the Langmuir isotherm. What will be the carbon dosage if initial concentration of MIB is 140 μg/L and treated water has MIB concentration at threshold odor concentration of 29 μg/L?

C_0, μg/L	*M*, μg/L	*C*, μg/L
50	301	10
50	161	20
50	91	30

11.13 A batch water treatment study was conducted by using PAC for removal of an organic contaminant. Determine the various constants in the Freundlich and Langmuir isotherm equations. The results of a 1-L batch test are given below.

Mass of Carbon (*M*), μg/L	Equilibrium Contaminant Concentration (*C*), μg/L
0	20
0.9	13
1.7	10
4.0	6
7.0	4
10.0	3

11.14 A GAC column study was conducted that used a 2.54-cm diameter column to remove a chemical dye. The initial concentration of the dye was 10 μg/L. The results of this experiment are summarized below. Determine (1) the coefficients in the Bohart-Adam equation and (2) the quantity of carbon required per year to treat a flow of 37.7 L/min that has an initial dye concentration of 10 μg/L and a residual dye concentration of 0.5 μg/L. Also, calculate the adsorption efficiency of the carbon. The dimensions of the carbon column are 0.6 m in diameter and 1.5 m in total.

Flow, L/min·m^2	Column Depth, m	Throughput Volume, m^3
0.102	0.76	1.38
	1.52	4.62
	2.28	8.16
0.204	0.76	0.54
	1.52	2.77
	3.04	8.32
0.407	1.52	1.26
	3.04	5.24
	4.57	10.49

11.15 Describe the application of PAC and GAC in water treatment practice.

11.16 TON of a water sample is 3. Calculate the volume of sample diluted to 200 mL.

11.17 In a threshold-taste test, 15 mL of sample was diluted with 185 mL of taste-free water. What will be the threshold taste number at this dilution?

11.18 Discuss various treatment processes for controlling taste and odor in water supply systems.

11.19 A small lake has a surface area of 45 acres. The total alkalinity of the lake water is 80 mg/L as $CaCO_3$. Copper sulfate is applied for algae control. The average application rate of copper sulfate, 0.2 mg/L as Cu, is maintained. Calculate the total $CuSO_4$ dosage in a depth of 3 ft.

References

1. Schnitzer, M. and Khan, S. U. *Humic Substances in the Environment*, Marcel and Dekker, New York, 1972.
2. Shaw, D. J. *Introduction to Colloid and Surface Chemistry*, Bultersworth, Boston, MA 1970.
3. APHA, AWWA, and WEF. *Standard Methods for the Examination of Water and Wastewater*, 18th ed., American Public Health Association, New York, 1992.
4. AWWA. *Water Quality and Treatment, A Handbook of Public Water Supplies*, 3d ed., McGraw-Hill Book Company, New York, 1971.
5. AWWA. *Water Quality and Treatment*, 5th ed., McGraw-Hill, Inc., New York, 1999.
6. James M. Montgomery, Inc. *Water Treatment Principles and Design*, John Wiley & Sons, New York, 1985.
7. Krasner, W. et al. "Tastes and Odors: The Flavor Profile Method," *Jour. AWWA*, vol. 77, no. 3, pp. 34–39, March 1985.
8. Sanks, R. L. *Water Treatment Plant Design for Practicing Engineers*, Ann Arbor Science Publishers, Ann Arbor, MI, 1978.
9. Lin, S. D. *Sources of Taste and Odors in Water-Part I*, Water and Sewage Works, June 1976.
10. American Public Works Association. *History of Public Works in the United States—1776–1976*, American Public Works Association, Chicago, IL, 1976.
11. Silvey, J. K. G. and Biederman, W. J. *Tastes and Odors - Known Organisms and Chemicals Responsible and Control Measures*, Manual of Water Utility Operations, Texas Water Utilities Association, Austin, TX, 1976.
12. Palmer, C. M. *Algae in Water Supplies*, U.S. Department of Health, Education, and Welfare, Public Health Service, Division of Water Supply and Pollution Control, Washington D.C., 1962.
13. McFarland, W. E. "Groundwater Treatment Alternatives for Industry—Part III, Sulfur and Other Taste and Odor Problems," *Plant Engineering*, vol. no. 3, pp. 162–165, March 13, 1986.
14. Qasim, S. R. and Chiang, W. W. *Sanitary Landfill Leachate: Generation, Control, and Treatment*, Technomic Publishing Co., Inc., Lancaster, PA, 1994.
15. Jenkins, D. "Effects of Organic Compounds-Taste, Odor, Color, and Chelation, Organic Matter in Water Supplies: Occurrences, Significance, and Control," *Proceedings of the 15th Annual Public Water Supply Engineers' Conference*, Champaign, IL, pp. 15–21, 1973.
16. USEPA, *Control of Biofilm Growth in Drinking Water Distribution Systems*, Seminary Publication, EPA/625/R-92/001, USEPA, Office of Research and Development, Washington, D.C., June 1992.
17. Amoore, J. E. "The Chemistry and Physiology of Odor Sensitivity," *Jour. AWWA*, vol. 78, no. 3, pp. 70–76, March 1986.
18. Amoore, J. E., Johnston, J. W., and Rubin, M. "The Stereochemical Theory of Odor," *Scientific American*, 42, February 1964.
19. Roderick, W. "Current Ideas on the Chemical Basis of Olfaction," *Jour. Chemistry Ed.*, vol. 43, p. 510, 1969.
20. Lin, S. D. "Tastes and Odors in Water Supplies: A Review," *Water and Sewage Works*, pp. R-141–R-163, April 1977.
21. Baker, R. A. "Tastes and Odors - Examination of Present Knowledge," *Jour. AWWA*, vol. 58, no. 6, pp. 695–699, June 1966.
22. Meadows, M. "Reservoir Management," *Jour. AWWA*, vol. 79, no. 8, pp. 26–31, August 1987.
23. Raman, K. "Controlling Algae in Water Supply Impoundments," *Jour. AWWA*, vol. 77, no. 8, pp. 41–43, August 1985.
24. McGuire, M. J., Jones, R. M., Means, E. G., Izaguirre, E. G., and Preston, A. E. "Controlling Attached Blue-Green Algae With Copper Sulfate," *Jour. AWWA*, vol. 76, no. 5, pp. 60–65, May 1984.

25. Silvey, J. K. G., Henley, D. E., Hache, B., and Nunz, W. J. "Musty-Earthy Odors and Their Biological Control," *Proceedings of the AWWA Conference on Water Quality Control*, Atlanta, GA, 1975.
26. Danglot, C., Amar, G., and Viligines, R. "Ability of Bacillus to Degrade Geosmin, Water Science and Technology," vol.15, *In Taste and Odor in Waters and Aquatic Organisms*, by P. E. Persson (Editor), Pergamon Press Ltd., Oxford, United Kingdom, pp. 291–299, 1983.
27. Symons, J. M., Irwin, W. H., and Boebeck, G. G. "Impoundment Water Quality Changes Caused by Mixing," *Journal of the Sanitary Engineering Division—Proceedings of the American Society of Civil Engineers*, vol. 93, no. SA2, April 1967.
28. Scott, A. R. *Aeration in Water Quality and Treatment*, 3d ed., McGraw-Hill Book Co., New York, 1971.
29. Ficek, K. J. "Potassium Permanganate for Iron and Manganese Removal, and Taste and Odor Control," *Water Treatment Plant Design for the Practicing Engineer*, R. L. Sanks (Editor), Ann Arbor Science Publishers, Ann Arbor, MI, 1978.
30. Westvaco, Chemical Division-West Virginia Pulp and Paper Company. *Taste and Odor Control in Water Purification*, Githens-Sohl Corporation, New York, 1986.
31. Qasim, S. R., Hashshaw, S., and Middleton, G. D. "Study Aimed at Odor Reduction in Water from Reservoir," *Water Engineering and Management*, pp. 58–61, April 1991.
32. Aieta, E. M. and, Berg, J. D. "A Review of Chlorine Dioxide in Drinking Water Treatment," *Jour. AWWA*, vol. 78, no. 6, pp. 62–72, June 1986.
33. Walker, G. S., Lee, F. P., and Aieta, E. M. "Chlorine Dioxide for Taste and Odor Control," *Jour. AWWA*, vol. 78, no. 3, pp. 84–93, March 1986.
34. Lalezary, S., Pirbazari, M., and McGuire, M. J. "Oxidation of Five Earthy-Musty Taste and Odor Compounds," *Jour. AWWA*, vol. 78, no. 3, pp. 62–69, March 1986.
35. Kornegay, B. K. "Activated Carbon for Taste and Odor Control," *Water Treatment Plant Design for the Practicing Engineer*, R. L. Sanks (Editor), Ann Arbor Science, Ann Arbor, MI, 1978.
36. Perrich, Jerry R. *Activated Carbon Adsorption for Wastewater Treatment*, CRC Press, Inc., Boca Raton, FL, 1981.
37. Eckenfelder, W. W. *Industrial Water Pollution Control*, McGraw-Hill Book Co., New York, 1989.
38. Ramalho, R. S. *Introduction to Wastewater Treatment Processes*, 2d ed., Academic Press, New York, 1983.
39. McGuire, J. M., Suffet, H. I., and Irwin, H. *Activated Carbon Adsorption of Organics from the Aqueous Phase*, Vol.2, Ann Arbor Science Publishers Inc., Ann Arbor, MI, 1980.
40. American Norit Company, Inc. *Powdered Carbon Handling,* Bulletin AN 89-2, American Norit Company, Inc., Jacksonville, FL. Undated.

CHAPTER 12

Disinfection and Fluoridation

12.1 Introduction

Disinfection of potable water is employed to inactivate and/or remove pathogens in order to meet primary drinking water standards. Disinfection satisfies part of the primary objective of water treatment, which is to provide water that is free from disease-causing organisms. Fluoridation, on the other hand, is employed as a general public health measure to prevent dental cavities. Public sensitivity to fluoridation and recent discoveries of disinfection by-products have made both subjects controversial issues.

This chapter provides discussion on some of the controversies surrounding disinfection. More particularly, a survey of various methods of potable water disinfection and of current fluoridation practices is presented. In addition, types, design, operation and maintenance considerations, and specifications for disinfection and fluoridation systems are discussed. Step-by-step design calculations are covered in the Design Example.

12.2 Disinfection

Disinfection of drinking water has been practiced for centuries. Over the past century, chlorination became the accepted means of disinfection, and it is the single most important discovery in potable water treatment. Recently, however, the concern over disinfection by-products (DBPs) produced by chlorine has given new impetus to investigating alternative disinfectants. The *Safe Drinking Water Act* of 1974 and its *amendments* in 1986 have further spurred interest in providing safe and aesthetically pleasing drinking water. Accordingly, in addition to a survey on chlorination, in this chapter, chloramines and other alternative disinfectants such as ozone, chlorine dioxide, potassium permanganate, and ultraviolet irradiation are discussed. Advantages and disadvantages of each alternative are also presented.

Disinfection of potable water is the specialized treatment for destruction or removal of organisms capable of causing disease; it should not be confused with sterilization, which is the destruction or removal of all life. Although disinfection of water has been practiced for centuries, it was not until the 1880's and the advent of the germ theory of disease that its importance in the treatment of potable water was understood.[1]

Pathogens (disease-producing organisms) are present in both groundwater and surface water supplies. These organisms, under certain conditions, are capable of surviving in water supplies- for weeks at temperatures near 21°C, and for months at colder temperatures. Destruction or removal of these organisms is essential in providing a safe potable water supply. Some bacteria, viruses, protozoa, and larger organisms ingested from contaminated water cause diseases varying from mild illnesses to life-threatening. Many water-borne disease and their associated pathogens have been presented in greater detail in Chapter 2. Disinfectants widely used in water treatment are oxidizing agents (halogens, halogen compounds, ozone) and physical agents (ultraviolet (UV) and radiation).

While the exact effect of disinfection agents on microorganisms is not clearly understood, some factors that affect the efficiency of disinfection are the following:

- type and concentration of microorganisms to be destroyed;
- type and concentration of disinfectant;
- contact time provided;
- chemical character and temperature of the water being treated. [2-5]

Oxidizing agents, by their different oxidation potentials, vary in their abilities to kill or inactivate pathogens. The oxidation potentials of several disinfectants are compared in Table 12-1. Ozone by far has the highest oxidation potential.

Table 12-1 Oxidation Potential of Selected Disinfectants in Water at 25°C

Disinfectant	Molecular Weight	Oxidation Potential (V)
Hydroxyl free radical, OH^-	17.0	-2.80
Ozone, O_3	48.00	-2.07
Hydrobromous acid, HOBr	96.91	-1.59
Hydrochlorous acid, HOCl	52.46	-1.49
Chlorine, Cl_2	70.90	-1.36
Bromine, Br_2	159.81	-1.07
Chlorine dioxide, ClO_2 (aq.)	67.50	-0.95
Mono chloramine, NH_2Cl	51.47	-0.75
Dichloramine, $NHCl_2$	95.93	-0.74

Source: Adapted in part from Reference 6.

12.2.1 Objectives of Disinfection

To assure compliance with all applicable regulations (both current and anticipated), the specific objectives of disinfection for all public water supply systems are as follows:

- assure 99.9 percent (3-log) and 99.99 percent (4-log) inactivation of Giardia lamblia cysts and viruses, respectively;
- do not impart toxicity and taste and odor to the disinfected water;
- minimize the formation of undesired disinfection by-products;
- meet the maximum contaminant levels (MCLs) for the disinfectants used and the by-products that can form.

Conventional treatment that includes coagulation, flocculation, sedimentation, and filtration, along with disinfection, can achieve the minimum mandatory inactivation of *Giardia lamblia* and enteric viruses, when properly designed and operated. Primary disinfection systems that use free-chlorine residual, ozone, and chlorine dioxide can achieve better than the above stated inactivation of both organisms. This is not the case, however, when chloramination is used as a primary disinfectant. Pilot-scale tests are needed to determine compliance against disinfection by-products when free chlorine and chloramine are used for primary disinfection.

12.2.2 Disinfection Requirement Under Surface Water Treatment Rule (SWTR)

Among the 1986 Amendments to the Safe Drinking Water Act of 1974 (PL-93-523) was the requirement for disinfection of all public water supplies. This requirement was addressed in a regulation that was promulgated on June 24, 1989 and has become known as the *Surface Water Treatment Rule (SWTR).*[6,7] This Rule established a treatment requirement that includes achievement of 99.9 percent (3-log) and 99.99 percent (4-log) removal or inactivation of *Giardia lamblia* cysts and *enteric viruses*, respectively. In order to supplement the disinfection and other surface water treatment requirements, in 1989, the USEPA published a manual that provides guidance in determining the levels of disinfection to be provided and the effectiveness of several disinfection chemicals.[6]

A parameter known as the "*CT value*" was developed to quantify disinfection efficiency and to provide a way to gauge the removal or inactivation of *Giardia lamblia* and viruses. A *CT* value is the product of the residual disinfectant concentration, C, in mg/L, determined at the end of the process, with the corresponding "disinfectant contact time", T, in minutes. The disinfectant contact time is based on the time that water with 10 percent of an approximate tracer concentration (T_{10}) takes to appear at the sampling point at peak hourly flow.

Determination of T_{10} The determination of T_{10} is based on two methods: (1) dye tracer, and (2) theoretical methods. The dye tracer method uses application of a suitable dye tracer to simulate the actual flow conditions. The theoretical method involves the use of a rule-of-thumb to establish T_{10} value. This method provides an approximate value of T_{10}. Both methods are presented below.

Determination of T_{10} by Dye Trace. The procedure for determining T_{10} is based on a profile of flow through a basin over time, generated by tracer studies. A suitable tracer material, such as calcium fluoride (CaF_2), sodium fluoride (NaF), sodium silica fluoride (Na_2SiF_6), or hydrofluosilic acid (H_2SiF_6), is applied at the influent section of the reactor, and time series samples are collected at the outlet. The tracer concentration is determined and results are plotted to obtain the T_{10} value. Two methods of tracer additions are employed: step-dose and slug-dose. The step-dose method uses introduction of a tracer chemical at a constant dose until the concentration at the desired end point reaches a steady-state level. The slug-dose method uses instantaneous dose of a concentrated tracer solution. Tracer profile at the exit point is developed and used to manipulate the T_{10} value. Both methods are described in detail in the following examples.[6]

Example 12-1 Two tracer studies employed the step-dose and slug-dose methods of tracer addition in a clearwell that has a volume of 197.9 m^3. A theoretical detention time T of 30 minutes[a] was obtained at an average flow rate of 9500 m^3/d. The procedures used for step-dose and slug-dose dye-tracer studies are given below. Determine T_{10} by both methods.

Step-dose Testing Method. Used a constant feed of sodium fluoride solution. Necessary feed equipment was installed, and 1% fluoride solution was fed at a rate of 1319 mL/min. This feed rate provided 2 mg/L[b] fluoride concentration in the influent. Prior to the start of testing, a fluoride baseline concentration of 0.2 mg/L was established for the water exiting the clearwell. The fluoride levels in the clearwell effluent were monitored and recorded every three minutes. The raw tracer-study data, along with the results of further data analyses, are shown in Table 12-2.

Slug-dose Testing Method. utilized a concentrated fluoride slug dose at the influent structure. This method is generally used where constant-feed equipment is not available. The duration of tracer injection should be less than 2 percent of the theoretical detention time of the basin, so a dosing time of less than 0.6 minute (maximum dosing time) was used. The fluoride-injection apparatus consisted of a funnel and a length of copper tubing. Four liters of dye-tracer solution was injected at a rate of approximately 7.5 L/min. This gave a dosing time of 0.53 min. The dye concentration in the feed solution was 109 g/L, or approximately 10.9 percent by weight. The effluent samples were collected at three-minute intervals, and dye concentration was determined in each sample. The result of slug-dose testing is provided in Table 12-3.

a. $$\frac{197.9\ \text{m}^3 \times 1440\ \text{min/d}}{9500\ \text{m}^3\text{/d}} = 30\ \text{min}$$

b. $$\frac{10{,}000\ \text{mg/L}\ (1\%) \times 1319\ \text{mL/min} \times 10^{-6}\ \text{m}^3\text{/mL} \times 1440\ \text{min/d}}{9500\ \text{m}^3\text{/d}} = 2\ \text{mg/L}$$

Table 12-2 Evaluation of Clearwell Data from Step-Dose Tracer Test[a]

Time (t), min	Fluoride Concentration, mg/L		C/C_0	t/T	$1 - C/C_0$	Log $(1 - C/C_0)$
	Measured	Tracer				
0	0.20	0	0	0	1	0
3	0.20	0	0	0.1	1	0
6	0.20	0	0	0.2	1	0
9	0.20	0	0	0.3	1	0
12	0.29	0.09[b]	0.045[c]	0.4[d]	0.955[e]	–0.020[f]
15	0.67	0.47	0.24	0.5	0.76	–0.119
18	0.94	0.74	0.37	0.6	0.63	–0.201
21	1.04	0.84	0.42	0.7	0.58	–0.237
24	1.44	1.24	0.62	0.8	0.38	–0.420
27	1.55	1.35	0.68	0.9	0.32	–0.495
30	1.52	1.32	0.66	1.0	0.34	–0.468
33	1.73	1.53	0.76	1.1	0.24	–0.620
36	1.93	1.73	0.86	1.2	0.14	–0.854
39	1.85	1.65	0.82	1.3	0.18	–0.744
42	1.92	1.72	0.86	1.4	0.14	–0.854
45	2.02	1.82	0.91	1.5	0.09	–1.046
48	1.97	1.77	0.88	1.6	0.12	–0.921
51	1.84	1.64	0.82	1.7	0.18	–0.745
54	2.06	1.86	0.93	1.8	0.07	–1.155
57	2.05	1.85	0.92	1.9	0.08	–1.097
60	2.10	1.90	0.95	2.0	0.05	–1.301
63	2.14	1.94	0.96	2.1	0.04	–1.395

a Measured concentration = tracer concentration + baseline concentration; baseline concentrate = 0.2 mg/L; fluoride dose C_0= 2.0 mg/L; and T = 30 min.

b Tracer concentration C = measured concentration - baseline concentration = 0.29 mg/L – 0.20 mg/L = 0.09 mg/L.

c C/C_0 = 0.09 mg/L ÷ 2.0 mg/L = 0.045.

d t/T = 3 min ÷ 30 min = 0.4.

e $(1 - C/C_0)$ = (1 – 0.045) = 0.955.

f Log (0.955) = –0.020.

Source: Data adapted from Reference 6.

Solution

1. *Step-Dose Testing*[6]

a. Process test data.

The concentration of fluoride added in the feed solution is 2 mg/L. The evaluation results for all data points are provided in Table 12-2.

b. Determine T_{10} using graphical analysis.

In order to determine T_{10} by the graphical method, a plot of C/C_0 versus t is drawn. A smooth curve is drawn through the data points, and T_{10} value of 12 min. is read from the plot. The procedure is shown in Figure 12-1.

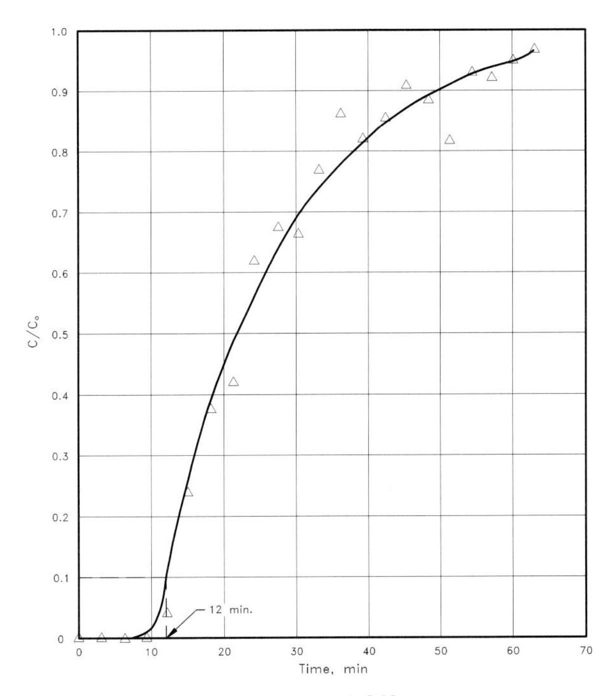

Figure 12-1 Graphical analysis for T_{10} from plot of C/C_0 vs. t.

c. Determine T_{10} using numerical analysis.

The dye-tracer curve represents the classical first-order reaction rate with a lag time. Several numerical methods can be used to transform the data into a straight line. Among these are the Thomas, Fujimoto, moments, least-squares, daily-difference, and rapid-ratio methods.[8] One numerical method uses Eq. (12.1).

$$\log(1 - C/C_0) = m\left(\frac{t}{T}\right) + b \qquad (12.1)$$

where

t = elapsed time, min

T = theoretical detention time, min

m = slope of the line

b = intercept

The results of step-dose tests are presented in the form log $(1 - C/C_0)$ and t/T and then plotted to obtain a straight line. The calculations and procedure are shown in Table 12-2 and Figure 12-2. The slope m and intercept b are –0.752 and 0.261, respectively. The correlation coefficient is 0.94. T_{10} is calculated from Eq. (12.1) by substituting $C/C_0 = 0.1$ and $T = 30$ min. A T_{10} value of 12 min is also obtained.

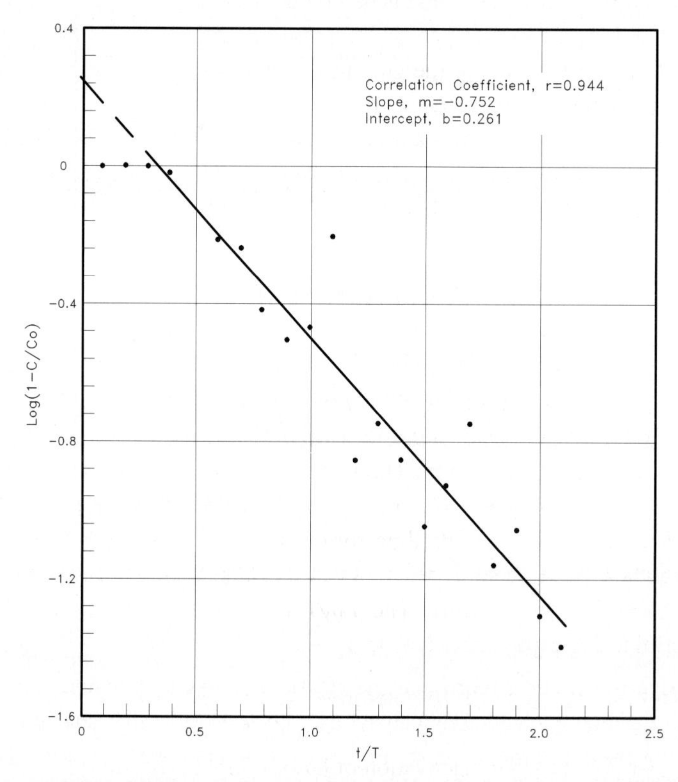

Figure 12-2 Numerical analysis for T_{10} from plot of log $(1 - C/C_0)$ vs. t/T, using Eq. (12.1).

2. Slug-Dose Testing

a. Evaluate test data.

The theoretical concentration of fluoride added in the feed solution is 2.2 mg/L. The results of raw slug-dose tracer-test data are provided in Table 12-3. It is observed that the fluoride concentration rises from 0 mg/L at $t = 0$ min to the peak concentration of 3.6 mg/L at $t = 18$ min. The fluoride concentration then gradually recedes to near zero at $t = 63$ min. It should be noted that the theoretical fluoride dose of 2.2 mg/L is obtained by assuming a complete mix of the tracer added throughout the entire clearwell volume; however, as shown in Table 12-3, the fluoride concentration in the clearwell effluent exceeded 2.2 mg/L for about 6 min between 14 and 20 min. The reason for a peak concentration exceeding the theoretical concentration is that the tracer is dispersed in only a portion of the clearwell volume.

To determine the T_{10} value from the slug-dose tracer-test data the dimensionless ratios C/C_0 are plotted as a function of t, as shown in Figure 12-3.[6] The required step-dose data format can be developed by either the graphical or the numerical method. The graphical method is based on a physical measurement of the area using a planimeter. In numerical technique, the area is determined by numerical integration. The results by either the graphical or the numerical method would be similar. The numerical method is therefore used in this example to develop the results. The necessary calculating steps are also provided in Table 12-3. The equivalent step-dose C/C_0 data are finally obtained from the testing results. These equivalent C/C_0 values are also plotted in Figure 12-3. T_{10} value of 12 minutes is thus obtained directly from this graph.

Determination of T_{10} by the Theoretical Method. Under such conditions as lack of funds, manpower, and equipment, or facilities under planning stages, the USEPA suggests that states may allow theoretical methods for estimation of T_{10}. In this procedure, T_{10} value is calculated from the theoretical detention time using a "rule of thumb" parameter T_{10}/T. The ratio of T_{10}/T for a clearwell is estimated from, the length-to-width ratio, the degree of baffling, and the effect of inlet baffling and outlet configuration. The baffling is generally utilized to reduce short-circuiting by reducing dead space areas, the effect of wind, temperature, and density currents, and influent and effluent turbulence. The baffling for clearwell is shown in Figures 12-4 and 12-5. These figures illustrate poor, average, and superior baffling arrangements for rectangular and circular clearwater basins, respectively. The T_{10}/T fractions for poor, average, and superior degrees of baffling are provided in Table 12-4.[6]

Inactivation Ratio In order to determine the total percent inactivation, the *CT* value may be calculated at each point where "*C*" is measured. Each calculated *CT* value is divided by the tabular value of *CT* for desired inactivation of *Giardia lamblia* cysts and viruses found in the SWTR tables. These tabular values are for different disinfectants at specified conditions of pH,

Table 12-3 Evaluation of Clearwell Data from Slug-Dose Tracer Test[a]

Time (t), min	Fluoride Concentration, mg/L Measured	Fluoride Concentration, mg/L Tracer	C/C_0	Incremental Area, mg-min/L	Cumulative Area, mg-min/L	Equivalent Step-Dose C/C_0 Data
0	0.2	0	0	0	0	0
3	0.2	0	0	0	0	0
6	0.2	0	0	0	0	0
9	0.2	0	0	0	0	0
12	1.2	1.0	0.45	3.0	3.0	0.05
15	3.6	3.4[b]	1.55[c]	10.2[d]	13.2[e]	0.22[f]
18	3.8	3.6	1.64	10.8	24.0	0.40
21	2.0	1.8	0.82	5.4	29.4	0.49
24	2.1	1.9	0.86	5.7	35.1	0.59
27	1.4	1.2	0.55	3.6	38.7	0.65
30	1.3	1.1	0.50	3.3	42.0	0.71
33	1.5	1.3	0.59	3.9	45.9	0.77
36	1.0	0.8	0.36	2.4	48.3	0.81
39	0.6	0.4	0.18	1.2	49.5	0.83
42	1.0	0.8	0.36	2.4	51.9	0.87
45	0.6	0.4	0.18	1.2	53.1	0.89
48	0.8	0.6	0.27	1.8	54.9	0.92
51	0.6	0.4	0.18	1.2	56.1	0.94
54	0.4	0.2	0.09	0.6	56.7	0.95
57	0.5	0.3	0.14	0.9	57.6	0.97
60	0.6	0.4	0.18	1.2	58.8	0.99
63	0.4	0.2	0.09	0.6	59.4	1.00
				Total Area = 59.4		

a Measured concentration = tracer concentration + baseline concentration; baseline concentrate = 0.2 mg/L; feed solution concentration = 109 g/L, theoretical fluoride dose.

$$C_0 = \frac{109 \text{ g/L} \times 4 \text{ L} \times 1000 \text{ mg/g}}{197.9 \text{ m}^3 \times 1000 \text{ L/m}^3} = 2.2 \text{ mg/L}.$$

b Tracer concentration = measured concentration – baseline concentration= 3.6 mg/L – 0.20 mg/L = 3.4 mg/L.

c C/C_0 = 3.4 mg/L ÷ 2.2 mg/L = 1.55.

d Incremental area = time increment from previous sampling time ∞ tracer concentration = (15 min – 12 min) ∞ 3.4 mg/L = 10.2 mg-min/L.

e Cumulative area = previous cumulative area + incremental area = 3 mg-min/L + 10.2 mg-min/L = 13.2 mg-min/L.

f Equivalent slug-dose data = cumulative area/ total area = 13.2 mg-min/L ÷ 59.4 mg-min/L = 0.22.

Adapted from Reference 6.

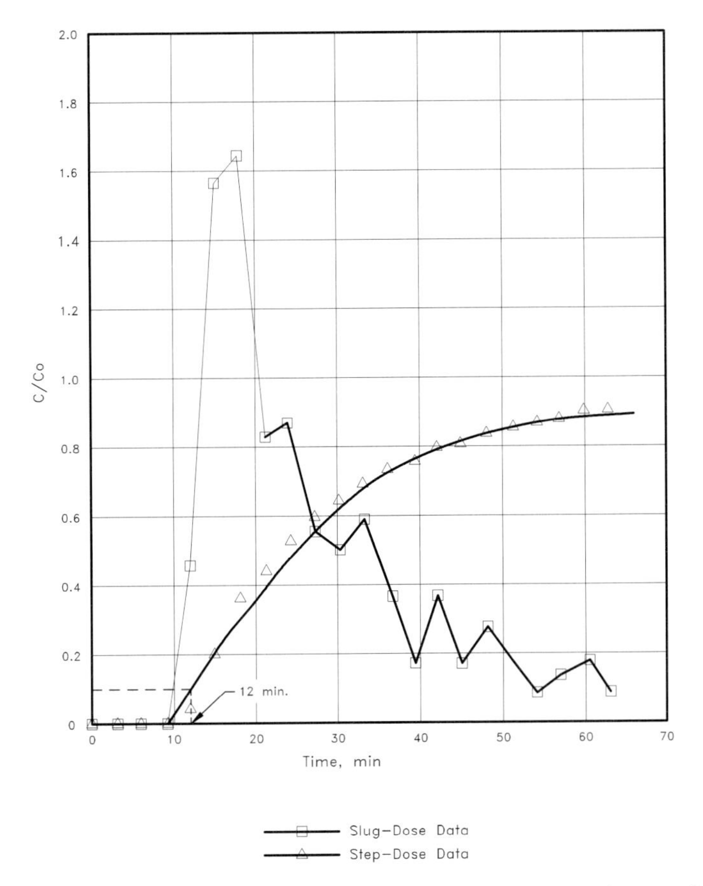

Figure 12-3 Development of equivalent step-dose C/C_0 data from slug-dose testing results used to plot C/C_0 vs. t.

temperature, and residual disinfectant concentrations.[6] The sum of the inactivation ratios for all treatment units up to the first customer should be equal to or greater than 1.0 (Eq. 12.2).

$$\sum \frac{CT_{Cal}}{CT_{Tab}} \geq 1.0 \tag{12.2}$$

where

$CT_{Cal} = CT$ calculated (C in mg/L, and $T = T_{10}$ in min)
$CT_{Tab} = CT$ obtained from SWTR tables

Table 12-4 Baffling Classifications

Baffling Condition	T_{10}/T	Baffling Description
Unbaffled (mixed flow)	0.1	None, agitated basin, very low length-to-width ratio, high inlet- and outlet-flow velocities.
Poor	0.3	Single or multiple unbaffled inlets and outlets, no intrabasin baffles (Figures 12-4(a) and 12-5(a)).
Average	0.5	Baffled inlet or outlet, with some intrabasin baffles (Figures 12-4(b) and 12-5(b)).
Superior	0.7	Perforated inlet baffle, serpentine or perforated intrabasin baffles, outlet weir or perforated launders (Figures 12-4(c) and 12-5(c)).
Perfect (plug flow)	1.0	Very high length-to-width ratio (pipeline flow), perforated inlet, outlet, and intrabasin baffles.

CT_{Tab} values for inactivation of *Giardia lamblia* cysts and viruses for disinfection by free chlorine, chlorine dioxide, ozone, chloramine, and UV are provided in Tables 12-5, 12-6, and 12-7.[6]

For the designing of a disinfection facility, the SWTR requires that disinfection credit for the treatment processes used in the process train should also be given. For example, well-operated conventional treatment plants which have been optimized for turbidity removal can be expected to achieve at least a 2.5-log removal of *Giardia lamblia* cysts.[6] Likewise, well-operated diatomaceous earth, slow-sand filtration, and direct-filtration plants can be expected to achieve at least 2-log removal of *Giardia lamblia* cysts. Table 12-8 provides the required primary disinfection needed for the entire treatment process to meet the overall treatment requirement of 3-log *Giardia lamblia* and 4-log virus removal/inactivation.[6] The use of contact time available in the pipeline up to the first external customer has been of some controversy. The authors believe that the contact time in the pipeline should not be used for *CT* calculations. The treatment plant will use water for potable purposes and, therefore, is generally the true first customer.

The following example provides guidelines for determining the *CT* compliance of SWTR.

Example 12-2 A water utility provides water treatment by a conventional filtration water treatment plant. Chlorine is applied at the filter effluent. A free chlorine residual of 1 mg/L is maintained through the first compartment of the clearwell. Ammonia is added and mixed ahead of the second compartment to convert free chlorine into the combined chlorine residual of 1 mg/L. The pH values of treated water in the summer and winter are 8 and 7, respectively. The average summer and winter water temperatures are 20° and 5°C, respectively. A dye-tracer study through each compartment of the clearwell at peak demand condition was conducted. The T_{10} values from dye-tracer response curves for summer and winter conditions are as follows: first compartment, 8 and 13 minutes; second compartment, 38 and 56 minutes, respectively. Also, the length of the force main from the high-service pumps to the first customer is 2.3 km.

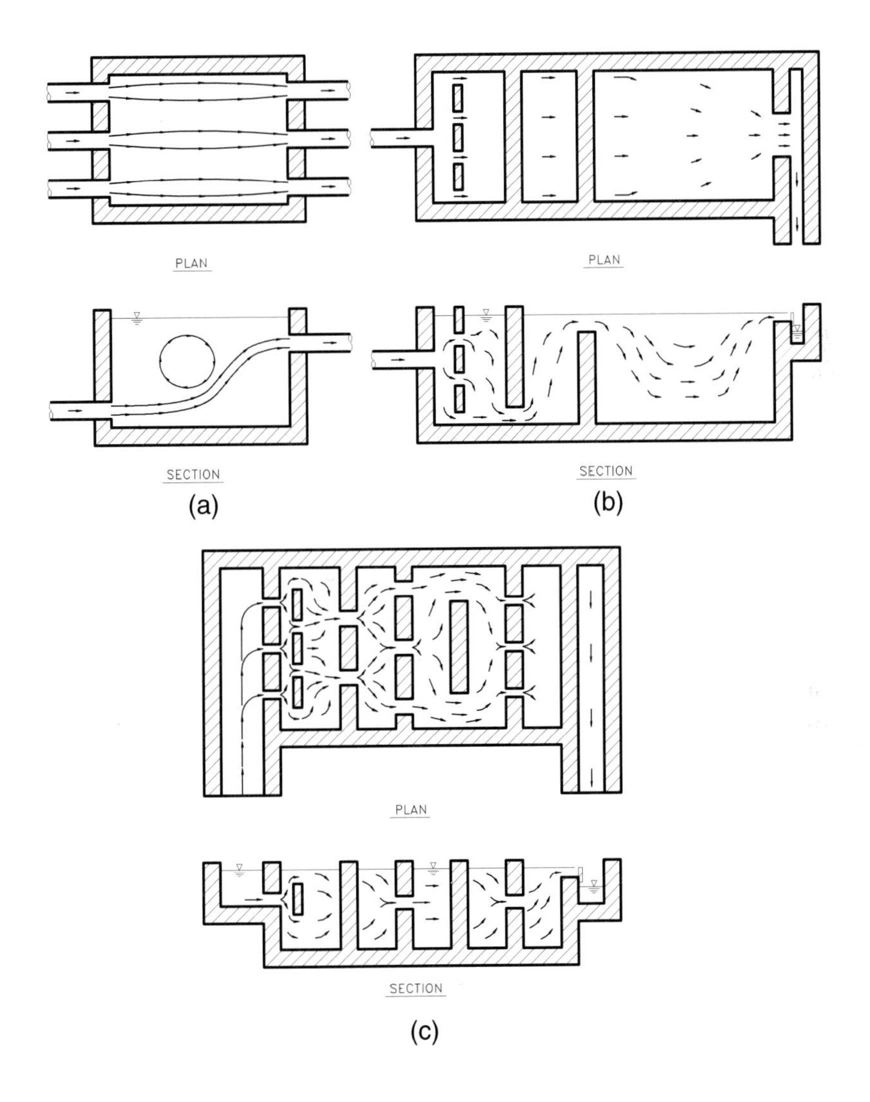

Figure 12-4 Baffling conditions for rectangular clearwater basins. (a) Poor. (b) Average. (c) Superior.

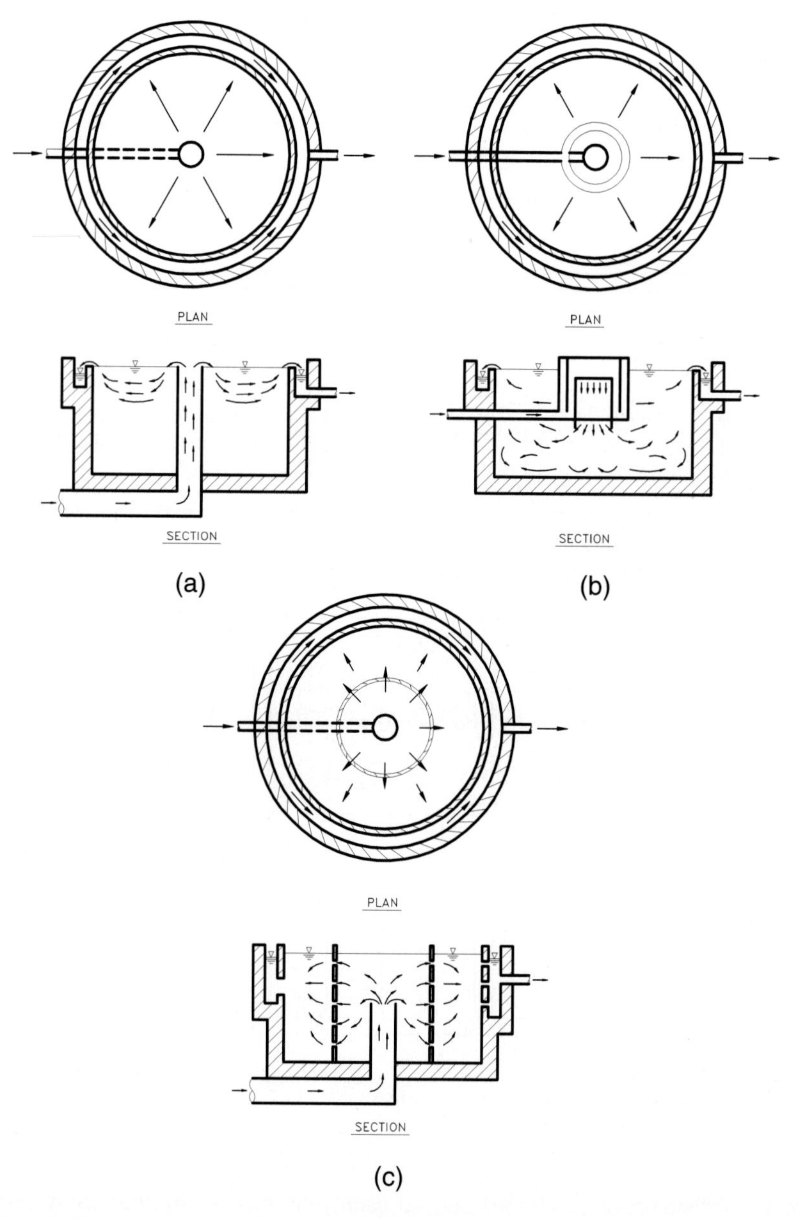

Figure 12-5 Baffling conditions for circular clearwater basins. (a) Poor. (b) Average. (c) Superior.

Table 12-5 *CT_{Tab}* Values for Achieving Inactivation of *Giardia lamblia* Cysts by Free Chlorine Residual

pH and Temperature	Free Chlorine, 1 mg/L			Free Chlorine, 2 mg/L			Free Chlorine, 3 mg/L		
	0.5-log	1-log	2-log	0.5-log	1-log	2-log	0.5-log	1-log	2-log
pH ≤ 6									
Temperature, °C									
<1	25	49	99	28	55	110	30	60	121
5	18	35	70	19	39	77	21	42	84
10	13	26	53	15	29	58	16	32	63
20	7	13	26	7	15	29	8	16	31
25	4	9	17	5	10	19	5	11	21
pH 7									
Temperature, °C									
<1	35	70	140	39	79	157	44	87	174
5	25	50	99	28	55	110	30	61	121
10	19	37	75	21	41	83	23	46	91
20	9	19	37	10	21	41	11	23	45
25	6	12	25	7	14	27	8	15	31
pH 8									
Temperature, °C									
<1	51	101	203	58	115	231	64	127	255
5	36	72	144	41	81	162	45	89	179
10	27	54	108	30	61	121	34	67	134
20	14	27	54	15	30	61	17	34	67
25	9	18	36	10	20	41	11	22	45
pH 9									
Temperature, °C									
<1	73	146	291	83	167	333	92	184	368
5	52	104	208	59	118	235	65	130	259
10	39	78	156	44	88	177	49	97	195
20	20	39	78	22	44	88	24	49	97
25	13	26	52	15	29	59	16	32	65

Source: Adapted from Reference 6.

Table 12-6 *CT_{Tab}* Values for Inactivation of Viruses by Free Chlorine

Temperature, °C	pH = 6–9			pH = 10		
	2-log	3-log	4-log	2-log	3-log	4-log
0.5	6	9	12	45	66	90
5	4	6	8	30	44	60
10	3	4	6	22	33	45
20	1	2	3	11	16	22
25	1	1	2	7	11	15

Source: Adapted from Reference 6.

The velocities in the main in summer and winter are 1.3 and 0.3 m/s, respectively. Check on whether the *CT* requirements for SWTR are met.

Table 12-7 CT_{Tab} Values for Inactivation of *Giardia lamblia* Cysts and Viruses by Chlorine, Dioxide, Chloramine, and UV

Disinfectant	Log Inactivation of *Giardia lamblia* Cysts			Log Inactivation of Virus		
	0.5-log	1-log	2-log	2-log	3-log	4-log
Chlorine Dioxide[a]						
Temperature, °C <1	10	21	42	8.4	25.6	50.1
5	4.3	8.7	17	5.6	17.1	33.4
10	4.0	7.7	15	4.2	12.8	25.1
20	2.5	5	10	2.1	6.4	12.5
25	2	3.7	7.3	1.4	4.3	8.4
Ozone						
Temperature, °C <1	0.48	0.97	1.9	0.9	1.4	1.8
5	0.32	0.63	1.3	0.6	0.9	1.2
10	0.23	0.48	0.95	0.5	0.8	1.0
20	0.12	0.24	0.48	0.25	0.4	0.5
25	0.08	0.16	0.32	0.15	0.25	0.3
Chloramine[b]						
Temperature, °C <1	635	1270	2535	1243	2063	2883
5	365	735	1470	857	1423	1988
10	310	615	1230	643	1067	1491
20	185	370	735	321	534	746
25	125	250	500	214	356	497
UV				21	36	

a Inactivation of viruses by chlorine dioxide at pH 6-9.
b Inactivation of *Giardia lamblia* cysts by chloramine at pH 6-9.
Source: Adapted from Reference 6.

Solution The conventional filtration water treatment plant provides 2.5-log removal of *Giardia* cysts and 2-log removal of viruses. Inactivation of 0.5-log for *Giardia* and 2-log for viruses through primary disinfectants is required. The *CT* values and CT_{Cal}/CT_{Tab} ratios are provided in Table 12-9. The contact time available in the pipeline is ignored.

The calculated ratios of $\Sigma CT_{Cal}/CT_{Tab}$ clearly indicate that the treatment plant is in full compliance for inactivation of viruses but not for that of *Giardia lamblia* cysts. The ratio of $\Sigma CT_{Cal}/CT_{Tab}$ for *Giardia lamblia* cysts during summer is 0.78, which is less than the required sum of 1.0 or greater. This value can be increased by (1) providing longer T_{10} in the first compartment of clearwell, or (2) increasing free chlorine residual. T_{10} is generally increased by installing baffles in the clearwell. Some discussion on this subject can be found in Chapter 13. Increase in free chlorine residual and contact time may increase THM formation potential. Therefore, such modifications to meet *CT* requirements must be carefully evaluated by bench-scale and pilot plant investigations. A brief discussion on the Milwaukee *Cryptosporidium* incident and DBP formation, and regulations regarding them, is provided below.

Table 12-8 Recommended Disinfection Credits for Entire Treatment Process to Meet SWTR Requirements

Process Train	Expected Log Removal		Recommended Disinfection (Log Inactivation) by Primary Disinfectant	
	Giardia lamblia	Viruses	*Giardia lamblia*	Viruses
Conventional filtration plant	2.5	2	0.5	2
Direct filtration plant	2	1	1	3
Slow-sand filtration	2	2	1	2
Diatomaceous-earth filtration	2	1	1	3
Treatment plant without filtration	–	–	3	4

Source: Adapted from Reference 6.

12.2.3 Cryptosporidium Incident in Milwaukee

The 1986 amendments to the Safe Drinking Water Act emphasize the disinfection requirements for municipal water supply; however, the *Cryptosporidium* incident in Milwaukee, Wisconsin, has led to a re-evaluation of standards relative to the guidelines on disinfection.[9]

Incident In March and April of 1993, approximately 400,000 residents of Milwaukee became ill from a water-borne disease. It was caused primarily by the protozoan *Cryptosporidium parium*. Of those individuals, approximately 4000 required hospitalization, and there were several reported deaths. In the months following the outbreak, it became apparent that the infection could severely affect people with weakened immune systems (i.e., the immunocompromised or immunosuppressed). It is now believed that approximately 100 people may have died as a result of this contamination incident.

The Milwaukee water treatment system utilizes two conventional water treatment plants receiving raw water from Lake Michigan. It is believed that high concentrations of *Cryptosporidium* organisms passed through one of these plants. The following conditions may have contributed to this event: (1) the rivers were swollen by spring rains and snow runoff, (2) the plants' watershed included cattle, slaughterhouses, and a sewage treatment plant, (3) the treatment chemicals may not have been properly applied because chemical feed rate instrumentation (a

Table 12-9 *CT* Calculation Procedure for the Example Problem

Condition	Summer	Winter
General conditions		
pH	8	7
Temperature, °C	20	5
T_{10} of first compartment, min	8	13
T_{10} of second compartment, min	38	56
Free chlorine residual (C_{free}), mg/L	1	1
Combined chlorine residual ($C_{combined}$), mg/L	1	1
CT_{Cal} of first compartment	8[a]	13
CT_{Cal} of second compartment	38[b]	56
Inactivation of *Giardia lamblia* cysts		
CT_{Tab} of first-compartment 0.5-log inactivation of *Giardia lamblia* cysts	14[c]	18
CT_{Tab} of second-compartment 0.5-log inactivation *of Giardia lamblia* cysts	185[d]	365
CT_{Cal}/CT_{Tab} of first compartment	0.57[e]	0.72
CT_{Cal}/CT_{Tab} of second compartment	0.21[f]	0.15
$\mathbf{\Sigma CT_{Cal}/CT_{Tab}}$	**0.78**[g]	**1.00**
Inactivation of viruses		
CT_{Tab} of first-compartment 2-log inactivation of viruses	1[aa]	4
CT_{Tab} of second-compartment 2-log inactivation of viruses	321[bb]	857
CT_{Cal}/CT_{Tab} of first compartment	8	3.25
CT_{Cal}/CT_{Tab} of second compartment	0.12	0.07
$\mathbf{\Sigma CT_{Cal}/CT_{Tab}}$	**8.12**	**3.32**

a CT_{Cal} of first compartment = free chlorine residual (C_{free}) ∞ T_{10} of first compartment = 1 mg/L ∞ 8 min = 8 mg/L-min.

b CT_{Cal} of second compartment = Combined chlorine residual ($C_{combined}$) ∞ T_{10} of second compartment = 1 mg/L ∞ 38 min = 38 mg/L-min.

c CT_{Tab} of first compartment is obtained from Table 12-5.

d CT_{Tab} of second compartment is obtained from Table 12-7.

e CT_{Cal}/CT_{Tab} of first compartment $= \dfrac{CT_{Cal} \text{ of first compartment}}{CT_{Tab} \text{ of first compartment}} = \dfrac{8 \text{ mg/L-min}}{14 \text{ mg/L-min}} = 0.57.$

f CT_{Cal}/CT_{Tab} of second compartment $= \dfrac{CT_{Cal} \text{ of second compartment}}{CT_{Tab} \text{ of second compartment}} = \dfrac{38 \text{ mg/L-min}}{185 \text{ mg/L-min}} = 0.21.$

g $\Sigma CT_{Cal}/CT_{Tab}$ for *Giardia lamblia* cysts inactivation (from Eq. (12.2) = CT_{Cal}/CT_{Tab} of first compartment + CT_{Cal}/CT_{Tab} of second compartment = 0.57 + 0.21 = 0.78.

aa CT_{Tab} of first compartment is obtained from Table 12-6.

bb CT_{Tab} of second compartment is obtained from Table 12-7.

streaming current meter) was not operating, and (4) the monitors designed for continuous measurement of the turbidity of filtered water were not in operation. Nevertheless, it is significant to note that, before and during this outbreak, all drinking water quality standards were met by the water utility.

The Milwaukee incident was not the first *Cryptosporidium* outbreak in this country. Since 1984, outbreaks in Oregon, Texas, and Georgia have affected thousands of people. All this had led many to believe that standards relative to the operation of water treatment plants must be re-evaluated.

Public Health Issues Human infection with *Cryptosporidium* was first documented in 1976. The illness is called *Cryptosporidiosis*. Symptoms include diarrhea, abdominal cramps, fever, nausea, and vomiting. The disease is generally not distinguishable from other short-term flu-like diarrhea infections caused by other common pathogens. The only way to confirm *Cryptosporidiosis* is by stool examination. Currently, no medication can successfully treat the illness. In general, healthy adults who contract the disease recover within 48 hours or a few weeks. Some patients experience a severe protracted illness, requiring hospitalization and rehydration therapy. Patients who are immunocompromised or immunosuppressed as a result of acquired immune deficiency syndrome (AIDS), patients receiving treatment for certain cancers, and organ-transplant recipients are at greater risk, and the disease is certainly life-threatening to them.

There are six known species of *Cryptosporidium*. Of these, *Cryptosporidium parium* is thought to pose the greatest threat to human infection. *Cryptosporidium* is relatively widespread in the environment. The principal source of this organism is believed to be livestock (cattle, sheep, or pigs), some wild animals, and municipal wastewater treatment plants. A national survey done in 1991 determined that it was present in 87 percent of the surface water sources studied.

Enhanced Surface Water Treatment Rule (ESWTR) The USEPA has been developing the Enhanced Surface Water Treatment Rule (ESWTR) that regulates the public health risks associated with such microbial contaminants as *Cryptosporidium parium* in drinking water. This rule will amend the existing Surface Water Treatment Rule (SWTR) and will include new requirements for improved removal of particles from drinking water. The SWTR of 1989 established the goals of microbial integrity and focused on reducing risks from *Giardia lamblia* cysts and viruses in surface water treatment. The USEPA has promulgated the Interim ESWTR (IESWTR) in December 1998.[10] In addition to *Giardia* and virus removal/inactivation, the IESWTR adds requirements for control of *Cryptosporidium*.

Currently, water supplies are not routinely required to test for *Cryptosporidium*; however, the Information Collection Rule (ICR) does require that water systems treating surface water and serving over 10,000 people conduct testing for the presence of the organism. The Long-Term 2 ESWTR will incorporate ICR data and may include some site-specific treatment requirements to give greater assurance that the incidence similar to the Milwaukee outbreak will not be repeated elsewhere.

Strategies for Removing *Cryptosporidium* The chlorine dose needed to kill *Cryptosporidium* is 8000 to 16,000 mg/L, a level much too high for human consumption. Traditionally, in municipal water supply, chlorine is applied at levels no higher than 2 mg/L. Other disinfectants have been tested for effectiveness against *Cryptosporidium*. Chlorine dioxide has shown limited effectiveness. Ozone is more effective and may be a good candidate for disinfection. The major defense against the organism, therefore, is the filter. Generally, *Cryptosporidium* is 4 to 6 microns in size. The *Giardia* organism is larger than 6 to 10 microns.

Similar to the 1989 SWTR, the IESWTR emphasizes the treatment techniques as the condition for compliance, rather than establishing MCL for *Cryptosporidium* in finished water. The IESWTR requires that (1) combined filtered water effluent turbidity must be less than or equal to 0.3 NTU in at least 95 percent of the measurements taken each month, with measurements taken every four hours of operation and (2) combined filtered water effluent turbidity must not exceed 1.0 NTU at any time with measurements taken in four-hour intervals. The public water systems using conventional rapid granular filtration and meet these turbidity requirements will be considered to be achieving at least 2-log removal of *Cryptosporidium*. They will automatically comply with the IESWTR.[10]

A properly designed and operated conventional water treatment plant will remove a high percentage of *Cryptosporidium* organisms; however, it is possible that some organisms will occasionally pass through a plant, even though it meets current standards for drinking water quality. The resistance of *Cryptosporidium* to chlorination places renewed emphasis on the traditional "multiple barrier" approach to drinking water treatment. Some of these approaches are the following.

- Protect the watershed against contamination.
- Optimize turbidity level in finished water by maintaining stable flow rates through filters. (If changes in flow rate are necessary, make them slowly.) In addition to measurement of turbidity in system's combined filtered-water in the plant, provide continuous monitoring of effluent turbidity for each individual filter. Avoid turbidity breakthrough occurrences; after backwashing, rinse or purge the initial filter flow to waste until the turbidity spike is reduced.
- Make sure that the filter backwash recovery system and other recycled side-streams do not concentrate the population of such pathogens as *Cryptosporidium* in the system.
- Monitor the distribution system so that water quality does not degrade via stagnation without a chlorine residual.

12.2.4 Disinfection By-Products

The oxidation of natural organic material (NOM), e.g. humic substances, produce aldehydes, ketones, alcohols, and carboxylic acids upon the addition of ozone, chlorine, chlorine dioxide, or potassium permanganate. Halogenation of organic materials can occur in the presence of free chlorine to produce THMs and other halogenated organic or organic halides (TOX).

Chlorine can also form chloramines by reacting with nitrogen-containing organic compounds (amino acids and proteins). In addition, monochloramines can produce organonitrogen compounds. If bromide ion is present, it may be oxidized by ozone or chlorine to form hypobromous acid, which in turn can brominate organic materials. Haloacetic acids (HAAs), identified as another major class of halogenated DBPs, are also produced by chlorine disinfection of water.[2,11]

Public Health Issues Many halogenated DBPs have been identified in chlorinated drinking water. Some of these DBPs are listed in Table 12-10. The current knowledge of the health effects of many selected chlorinated DBPs is provided in Reference 2. The USEPA is currently studying the by-products associated with ozonation.[2,11]

Disinfectants and Disinfection By-products Rule (D/DBPR) Total trihalomethanes (TTHMs) were first regulated in finished drinking water in 1979.[2,11] In November 1992, USEPA started a negotiated process to prepare a Disinfectants and Disinfection Byproducts Rule (D/DBPR). The D/DBPR regulates the public health risks associated with DBPs and disinfectant chemicals in drinking water. This rule, being promulgated in two phases, will limit the amount of DBPs and disinfectant residual allowed in the distribution system. After many years of intense study, the USEPA promulgated the Stage 1 D/DBPR in December 1998.[12] This rule is applicable to all community water systems (CWSs). The intent of the D/DBPR and national impacts are covered in several References 10-16.

The Stage 1 D/DBPR uses two approaches to limit the amount of DBPs and residual disinfectants in drinking water. First, the rule requires the removal of normal organic matter (NOM) that is measured in the form of TOC. Second, the rule establishes (1) the Maximum Contaminant Levels (MCLs) and the Maximum Contaminant Level Goals (MCLGs) for DBPs, and (2) the Maximum Residual Disinfectant Levels (MRDLs) and the Maximum Residual Disinfectant Level Goals (MRDLGs) for disinfectants in drinking water.

TOC Removal. The precursor-removal requirement based on total organic carbon (TOC) as a substitute or surrogate measurement will apply to conventional water treatment plants and to softening plants. A specified percentage of the TOC in the source water will need to be removed before addition of the disinfectant. TOC removal will be based on the alkalinity of source water, as summarized in Table 12-11.[12]

MCLGs for DBPs. The Stage 1 D/DBPR regulates MCLs and MCLGs of total THMs, HAAs, bromate, chlorite, free and combined chlorine, and chlorine dioxide. These values are based on

Table 12-10 Common Halogenated By-Products Formed by Chlorination

HOCl + Br + NOM →THMs and other halogenated DBPs	
Trihalomethanes (THMs)	**Haloacetic acids (HAAs)**
Chloroform	Monochloroacetic acid
Bromodichloromethane	Dichloroacetic acid
Dibromochloromethane	Trichloroacetic acid
Bromoform	Monobromoacetic acid
	Dibromoacetic acid
	Bromochloroacetic acid
Cyanogen halides	
Cyanogen chloride	**Haloacetonitriles (HANs)**
Cyanogen bromide	Dichloroacetonitrile
	Trichloroacetonitrile
Halopicrins	Dibromoacetonitrile
Chloropicrin	Tribromoacetonitrile
Bromopicrin	Bromochloroacetonitrile
Chloral hydrate	**Haloketones, Haloaldehydes, Halophenols**

Source: Adapted in part from References 2 and 11.

Table 12-11 Percent TOC Removal by Enhanced Coagulation Required by D/DBPR

Source Water TOC, mg/L	Source Water Alkalinity, mg/L as $CaCO_3$		
	0–60	60–120	>120
>2.0–4.0	35.0	25.0	15.0
4.0–8.0	45.0	35.0	25.0
>8	50.0	40.0	30.0

Source: Adapted in part from Reference 12.

current knowledge, additional monitoring requirements under ICR, health effects, and treatment studies. The MCLs and MCLGs for the DBPs in Stage 1 DBPR are the following.[12]

	MCL, mg/L	MCLG, mg/L
• TTHMs	0.080	N/A
• HAA5	0.060	N/A
• Bromate	0.010	Zero
• Chlorite	1.0	0.8

MRDLs and MRDLGs for Disinfectants. The Stage 1 D/DBPR establishes MCLs and MCLGs for three disinfectants as the follows.[12]

Disinfectant	MRDL, mg/L	MRDLG, mg/L
Chlorine	4, as Cl_2	4, as Cl_2
Chloramine	4, as Cl_2	4, as Cl_2
Chlorine dioxide	0.8, as ClO_2	0.8, as ClO_2

Strategies for Controlling DBPs Formation of DBPs depends upon a number of factors, including (1) concentration and types of organic material present when chlorine is added, (2) dose of chlorine, (3) temperature and pH of the water, (4) bromide concentration and (5) reaction time. Several strategies for controlling DBPs can be used. Among these are the following:

- source control by effective watershed management—this involves both nutrient input and nutrient cycling in reservoirs, and algae control;
- precursor removal by enhanced coagulation, carbon adsorption, and membrane processes;
- alternative oxidants and disinfectants that do not produce undesired by-products;
- removal of by-products after they are formed, by processes including stripping, carbon adsorption, and membrane filtration.

The removal of by-products after they are formed is difficult and costly. Reducing the concentrations of organic precursors before adding chlorine or other oxidants will provide the highest-quality finished water.

12.3 Primary Disinfection Technologies

Primary disinfection is a key step in water treatment. Typically, in a conventional water treatment plant, this step occurs either just before filtration or just after filtration. Pre-disinfection by using chlorine dioxide and ozone can also be used by careful review of chlorite, chlorate, and bromate formation potential.

Several primary disinfection technologies are generally evaluated for conventional water treatment plants. These are chlorine and chloramines, ozone, chlorine dioxide, potassium permanganate, and UV radiation. These technologies are discussed in the following sections.

12.3.1 Chlorination

Chlorine is the chemical predominantly used in the disinfection of potable water supplies. The first application of chlorine in potable water treatment was for taste and odor control in the 1830s.[3,17] At that time, diseases were thought to be transmitted by odors. This false assumption led to chlorination even before disinfection was understood.

Currently, chlorine is used as a primary disinfectant in potable water treatment. Other uses include taste and odor control, algae control, filter-media conditioning, iron and manganese removal, hydrogen sulfide removal, and color removal.[3] Chlorine is available in a variety of forms, including elemental chlorine (liquid or gas), solid hypochlorite compounds of calcium or sodium, and gaseous chlorine dioxide.

Chlorine Chemistry

Free Chlorine Residual. Many properties of chlorine make it an ideal disinfectant. It is highly soluble in water, so it is easy to apply. It is easily measured and controlled; it forms a persistent residual; and, when compared with other disinfection chemicals, it is relatively inexpensive.

When chlorine gas is dissolved in water, it hydrolyzes, to form hydrochlorous acid (HOCl), according to Eq. (12.3). In this form, chlorine exists as *free chlorine residual.*

$$Cl_2 + H_2O \rightarrow HOCl + H^+ + Cl^- \quad (12.3)$$

HOCl is a weak acid that is ready to undergo the partial-dissociation reaction expressed by Eq. (12.4).

$$HOCl \Leftrightarrow H^+ + OCl^- \quad (12.4)$$

The relative concentrations of HOCl and OCl^- vary with pH, water temperature, and the concentration of chlorine in solution. HOCl is many times stronger an oxidant than OCl^-. The predominant concentrations of HOCl and OCl^- are below pH 6.0 and above pH 7.5, respectively. Therefore, the disinfecting power of chlorine decreases with increase in pH. Near a pH of 9, free chlorine residual has little disinfecting power, because of the high hypochlorite-ion concentration.[18,19]

Free chlorine is also added in water from hypochlorite salts (calcium and sodium hypochlorites). The dissociation reactions of calcium and sodium hypochlorites are given by Eqs. (12.5) and (12.6).

$$Ca(OCl)_2 + 2\ H_2O \Leftrightarrow 2\ HOCl + Ca^{2+} + 2\ OH^- \quad (12.5)$$

$$NaOCl + H_2O \Leftrightarrow HOCl + Na^+ + OH^- \quad (12.6)$$

As seen from these reactions, chlorine gas lowers the pH (Eqs. (12.3) and (12.4)), whereas hypochlorite in solution raises the pH (Eqs. (12-4) through (12-6)). Therefore, for equal amounts of chlorine added to poorly buffered water, a higher disinfection efficiency is achieved with chlorine gas than with hypochlorite solution. Factors that affect the disinfection efficiency of chlorine and other disinfectants include the following: (1) nature of the disinfectant; (2) concentration of the disinfectant; (3) length of contact time with the disinfectant; (4) temperature; (5) type and concentration of organisms; and (6) pH.

Reaction With Natural Organic Matter (NOM). Hypochlorous acid is a strong disinfectant because of its ability to diffuse easily through the cell walls of a microorganism and disrupt its life functions.[5] Both HOCl and OCl^- react indiscriminately with NOM. The chlorine reaction with NOM is provided by Eqs. (12.7) through (12.10).

$$HOCl + Precursors \rightarrow CHCl_3 + \text{other chlorinated DBPs} \quad (12.7)$$

$$HOCl + Br \rightarrow HOBr + Cl^- \quad (12.8)$$

$$HOBr + Precursors \rightarrow CHBr_3 + \text{other brominated DBPs} \quad (12.9)$$

$$HOCl + Br + Precursors \rightarrow CHCl_3 + CHBrCl_2 + CHBr_2Cl + CHBr_3 + \text{other halogenated DBPs} \quad (12.10)$$

Combined Chlorine Residual. Chlorine reacts readily with ammonia to form chloramines and exerts combined chlorine residual (Eqs. (12.11), (12.12) and (12.13). In this form, chlorine is a weak disinfectant, does not product THMs, and provides a stable residual in the distribution system.

$$NH_3 + HOCl \rightarrow NH_2Cl \text{ (monochloramine)} + H_2O \quad (12.11)$$

$$NH_2Cl + HOCl \rightarrow NHCl_2 \text{ (dichloramine)} + H_2O \quad (12.12)$$

$$NH_2Cl + HOCl \rightarrow NCl_3 \text{ (trichloramine)} + H_2O \quad (12.13)$$

Breakpoint Chlorination. When chlorine is added to water, it is consumed in oxidizing a wide variety of compounds present in water. No chlorine can be measured until the initial chlorine demand is satisfied. Then, chlorine reacts with ammonia to produce combined chlorine residual. The combined chlorine residual increases with additional dosage until a maximum combined residual is reached. Further addition of chlorine causes a decrease in combined residual. This is called "breakpoint chlorination." At this point, the chloramines are oxidized to oxides of nitrogen or other gases. After breakpoint chlorination is reached, free chlorine residual develops at the same rate as that of the applied dose. A typical breakpoint-chlorination curve and chemical reactions are shown in Figure 12-6.

Chlorine Dosages and Residuals The chlorine dosage is the amount of chlorine added to the water to produce a specified residual at the end of a fixed contact time. The residual is the dosage minus any chlorine demand exerted by compounds and organic material present in the water. Chlorine dosages required for any of the application points in the treatment process are best determined by bench scale or pilot plant testing.

Chlorine dosages may vary with the raw water quality, temperature, and other climatic conditions. Typically, the dosages may range from 0.2 to 4 mg/L.[2,3,6,19] Table 12-12 indicates representative chlorine dosages.

Chlorine residuals will vary according to operating experience and requirements of the governing regulatory agency. Typically, the combined chlorine residuals should be 0.5 to 1 mg/L at distant points in the distribution system. Higher residuals may be required, depending on pH, temperature, and other characteristics of water.

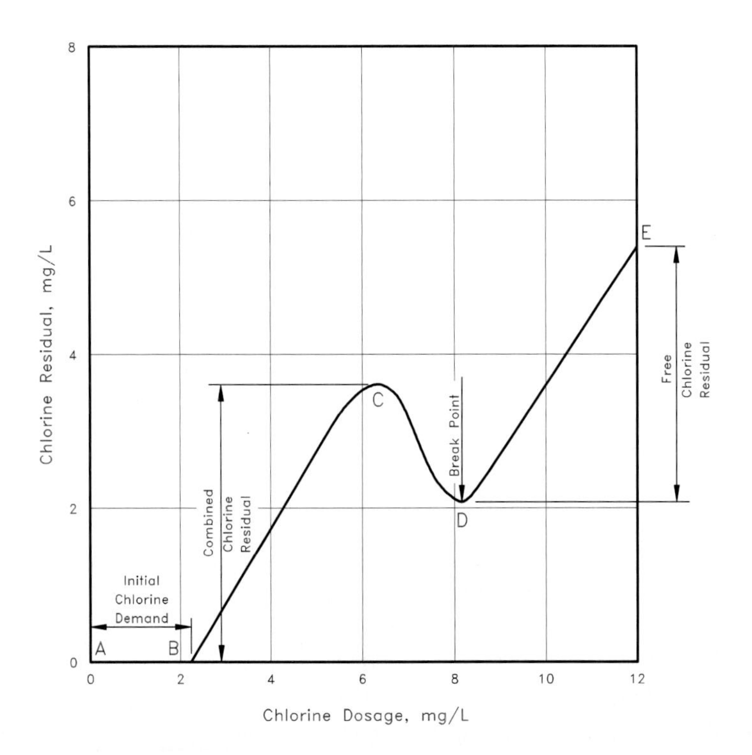

Figure 12-6 Chlorine residual curve. (Showing breakpoint and chemical reactions.)

Table 12-12 Representative Chlorine Dosage Required for Disinfection

Treatment Objective	Chlorine Dosage, mg/L	Contact Time, min	pH Range
Combined chlorine residual	1–5	[a]	7–8
Free chlorine residual	0.5–4	[a]	7–8
Breakpoint reaction	6–8 times NH_3[b]	30	6.5–8.5 (7.5 optimum)
Monochloramine formation (Some dichloramine will form if pH drops below 7.)	3–4 times NH_3[b]	20	7–8
Free chlorine residual formation	6–8 times NH_3[b]	20	6.5–8.5

a As needed for CT requirement.

b Expressed in mg/L as NH_3.

Source: Adapted in part from References 2, 3, 6, and 19.

The chlorine residual maintained for the specified contact time must demonstrate effective disinfection at the plant and protection against regrowth of undesired organisms.

Design of Chlorination System A chlorination system for disinfection of water supply consists of six separate subsystems: (1) chlorine supply; (2) storage and handling; (3) safety provisions; (4) chlorine feed and application; (5) diffusion, mixing and contact; and (6) the control system. Design considerations for each system are discussed below.

Chlorine Supply. Chlorine is usually supplied as a liquefied compressed gas under pressure. Chlorine can be supplied in containers or in bulk shipment.[3,20]

The containers typically used in chlorine supply are (1) cylinders (45.4-kg and 68-kg (100-lb and 150-lb) capacity cylinders predominate) and (2) container (only one size of 907-kg (2000-lb or 1-ton) container is available, called the *ton container*). Each ton container has a gross weight of 1587 kg (3500 lbs or 1.75 ton). The basic engineering features of the container system are provided in Reference 3 and 21.

The methods most commonly used for bulk-chlorine supply are (1) railroad tank cars (two common sizes of 49,900 kg and 81,600 kg (55 tons and 90 tons)) and (2) tank motor vehicles (usually in the range from 13,600 kg to 20,000 kg (15 tons to 22 tons)). Bulk chlorine may also be supplied by pipeline, portable tanks, and barge tanks.

Selection of the size of chlorine containers or the method of bulk shipment mainly depends on (1) the quantity of chlorine used, (2) the technology used in the chlorination system, (3) the space available for storage, (4) transportation and handling costs, and (5) the preference of the plant operator. The cylinders are most likely applied to small water supply systems. The use of 907-kg containers is generally desirable for moderate-size users. Bulk shipment may be the most cost-effective for large-scale water utilities.

Chlorine Storage and Handling. The chlorine storage and handling systems must be designed with full safety considerations; chlorine gas is very poisonous and corrosive.

The cylinders and ton-containers are usually housed in an enclosure or building. A designer's checklist for ton-container storage and handling facilities should include, but not be limited to, the following:[3] (1) appropriate auxiliary ton-container valves (captive Yoke type), flexible copper tubing, and a rigid black seamless steel manifold header with valves, fitting, and shut-off valves; (2) container-weighing scales or load cells; (3) trunnions for ton containers; (4) ton-container lifting bar; (5) overhead crane or monorail with 3600-kg (4-ton) capacity; (6) chlorine-gas filter; (7) external chlorine pressure-reducing valve as necessary; (8) pressure gauges; (9) drip legs (i.e., condensate traps) at inlet to chlorinators; (10) continuous chlorine-leak detector with sensors and alarms; and (11) emergency-repair kit for ton-container.[3,20–23]

The use of bulk-storage tanks for chlorine storage has increased in recent years. This increase is due in part to larger chlorine tank fleets and increasing bulk-chlorine deliveries by truck. The capacity of a bulk-storage tank will vary from 22,700 kg up to 81,600 kg (25 tons to 90 tons), depending on whether the delivery is by truck or by tank car. Tank design should be of steel or iron construction, and should follow recommendations included in the guidelines of the

Chlorine Institute, Inc.[20–22] The bulk-tank storage system must be specifically designed for each installation, but it should include the following essential items: (1) unloading station for truck or unloading platform for tank-car deliveries; (2) weighing device; (3) piping system with appropriate fittings, gauges, and safety relief and other valves; (4) liquid-chlorine pressure-relief system with expansion chambers, pressure switches, and rupture discs; (5) gas-chlorine pressure-relief system with pressure switches, rupture discs, and pressure-relief valves; (6) chlorine-gas filter or strainer; (7) continuous chlorine-leak detector with sensors and alarms; (8) pressure-reducing valves as necessary; (9) automatic shut-off valves as necessary; (10) sun shield as necessary; (11) air-padding system as necessary; (12) pressure switches and alarms as necessary; and (13) emergency-repair kit for the bulk tank.

Safety Considerations. USEPA had set forth the regulations that intended to minimize the risk of injury, death, or damage to the operation personnel and potential off-site impact on public and environmental receptors during an accidental release of chlorine.[24,25] The 40 CFR Part 68 Accidental Release Prevention Program Rule (ARPPR) applies to many water treatment facilities that have inventories of regulated substances (i.e., chlorine, ammonia, chlorine dioxide, etc.) in greater quantities than those minimum threshold quantities specified in the regulation. To comply with the requirements in the ARPPR, the Risk Management Program (RMProgram) had to be prepared for all regulated facilities by June 21, 1999 and be updated by every five-year anniversary and after any major changes in regulated processes. The shipment, storage, handling, and use of hazardous materials (i.e., chlorine and ammonia) are subject to regulation by DOT, OSHA, and state legislatures.[26–28]

The adequate design approaches of chlorine system will provide more solid basis for development of safer operation and maintenance procedures. The important safety considerations in chlorine system design include operation personnel exposure limits, ventilation, fire protection, delivery and unloading, storage and transfer, surrounding development, feed system, and monitoring.[26–31] The Chlorine Institute, Inc. provides standards for chlorine-handling equipment and safety procedures. All chlorine systems should conform to these standards. Many other design guides and safety considerations for chlorine-handling equipment are provided in References 3, 20 and 21.

Installation of an emergency chlorine-vapor scrubber system is another effective measure to reduce the risk of damage and injury to plant personnel and the potential impact to the off-site receptors. The most commonly used scrubbers are single-pass (or once-through) multi-stage absorption systems.[32–34] The scrubber system is designed to neutralize the chlorine-contaminated air from a building or enclosure where a chlorine leak occurs. A 20-percent caustic (NaOH) solution is typically used to clean up the contaminated air.

Chlorine Feed and Application. The chlorine feed and application system mainly include the following: [3,20]

- chlorine withdrawal (as gas or liquid chlorine);
- evaporator (necessary for liquid chlorine withdrawal only);

- automatic switchover;
- vacuum regulator;
- chlorinator;
- injector system (with utility water supply);
- defusion, mixing, and contact;
- control system.

The chlorine feed and application system may also include liquid and gas pressure-relief systems, gas pressure-reducing valves, gas pressure and vacuum gauges with high pressure and vacuum alarms, gas filters, and several vent-line systems. The design information on the major components of the systems is provided below.

Chlorine Withdrawal. The chlorine supply from storage to feed systems can be either gas withdrawal or liquid withdrawal. Each ton container is equipped with two identical outlet valves near the center of one end inside a removable steel valve protective housing. The top and bottom valves on a ton container are used for gas and liquid withdrawal, respectively.[3,20] If gas withdrawal from top valve is used, the maximum withdrawal rate for a ton container is 180 kg/d (400 lbs/d) at room temperature. If the withdrawal rate is 180 to 680 kg/d, two or more ton containers must be manifolded together. If chlorine is being used in this manner, the room temperature must be maintained above 18°C, to provide the heat required to replace the heat of evaporation. Withdrawal rates higher than 680 kg/d should use liquid withdrawal and employ evaporators. If liquid withdrawal is used, the bottom outlet valve of the ton container is connected to the feed piping. Liquid withdrawal from a ton container has certain advantages: (1) it is not affected by ambient temperature, so container storage is possible in an open structure with only a sun shield if the storage site is located in a remote area; (2) there is no danger of reliquification between the container and the chlorinator; and (3) fewer containers need to be connected at one time, because liquid-withdrawal rates are much higher than gas-withdrawal rates. The significant disadvantages of liquid withdrawal from ton container include the following: (1) evaporators are required; and (2) the risk of injury or off-site impact can be significantly higher if liquid chlorine is accidentally released from the liquid-chlorine line between the containers and the evaporators.

Evaporators.[3,20] Chlorine evaporators are generally used when the chlorine withdrawal exceeds 680 kg/d. Evaporator capacities range from 180 to 4500 kg/d (400-10,000 lb/d). The evaporator receives the liquid chlorine from the storage facility and vaporizes it to chlorine gas in a sealed pressurized chamber. The chamber is heated either by electric heater, by recirculated hot water, or by steam, to provide the heat for vaporization. The temperature control system will maintain the water temperature between 76 and 82°C (170 and 180°F) when a hot-water bath is used. All evaporators should be equipped with a pressure-reducing automatic shut-off valve on the discharge line from the evaporator, to prevent liquid chlorine from entering the gas-chlorine piping. A cathodic protection system should be provided to protect against corrosion, and the exterior of the water bath should be insulated. Some type of corrosion-inhibitor solution needs to be added into the water bath, to reduce the chance of scaling problems if cathodic protection is not utilized.[3,20] Sizing of the evaporator can be accomplished by using the equipment manufacturers' catalog data. A typical chlorine evaporator is shown in Figure 12-7.

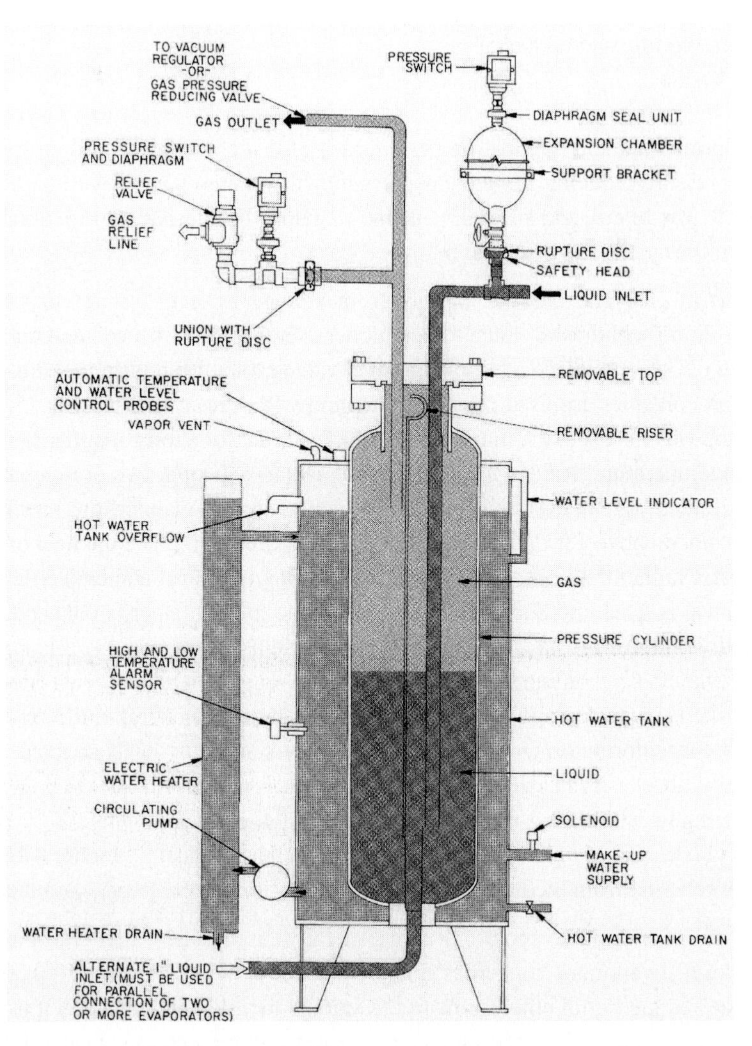

Figure 12-7 Typical chlorine evaporator with relief system. *(Courtesy of Wallace & Tiernan/U.S. Filter.)*

Automatic Switchover. Provision for automatic switchover from one container to another should be included, to increase system reliability. There are two types of automatic switchover systems: vacuum, and pressure. Both types switch the withdrawal of chlorine from one container to the other as soon as one container becomes empty.

Chlorinator. The chlorinator receives the chlorine gas from the storage container or evaporator and regulates the flow to the injector. Different types of chlorinators are (1) direct feed; (2) pressure type; (3) remote-vacuum type; and (4) sonic-flow type. A conventional chlorinator consists of the following units: an inlet pressure-reducing valve, a rotameter, a metering control orifice, and a vacuum-differential regulating valve. The driving force comes from the vacuum created by the chlorine injector. The feed rate varies from 30 to 5000 kg/d. The selection of any type of chlorinator should be based on flow rate and on type of application. Chlorinators are usually designed by the equipment manufacturers to limit the concentration of chlorine solution at the injector to 3500 mg/L. The control schemes for chlorination range from simple to complex and from manual to varying degree of automation.[3,20] Equipment manufacturers should be consulted for design and selection of chlorinators. A typical chlorinator assembly is shown in Figure 12-8.

Injector System.[3,20] The chlorine feed, ejector, or injector system is very essential because it provides the required dosage at the point of application. There are two types of chlorine injector systems: (1) pressure gas injection and (2) vacuum gas feed. The pressure injection of gas may pose risks of gas release into the atmosphere. It is normally used in small plants or, in large facilities, where safety precautions are rigidly followed. In vacuum-feed systems, a specified vacuum is created at the injector gas inlet. This vacuum is applied to suck chlorine gas from the chlorine-storage source to the chlorinator and then through a vacuum piping system to the injector. A vacuum regulator, which is typically an integral part of the chlorinator, is utilized to control the chlorine-gas withdrawal rate under the vacuum condition. The amount of water must be enough (1) to maintain chlorine concentration in the solution below 3500 mg/L and (2) to create the required amount of vacuum in the line to the chlorinator and in all of the components of the chlorinator system. Manufacturers of chlorination systems provide injector operating curves that specify the amount of water and pressure required for a given amount of chlorine to be applied against a given back pressure (Figure 12-9). From the injector, the chlorine solution (in the form of hypochlorous acid) flows to the point at which it is applied into the water. The injector system usually includes the following items: (1) water supply pump, piping to the injector, pressure gauge, and backflow prevention device; (2) injector; (3) vacuum piping from the chlorinator to the injector, including vacuum gauge (for remote injector installation), built-in vacuum switch and alarm for both high and low vacuum, and vacuum-line shutdown system (for remote injector location); and (4) chlorine-solution piping, including compound back-pressure gauge located immediately downstream from the injector (for variable-throat injector only), flow meter, and pressure switch and low-water-pressure alarm.[3,20]

Diffusion, Mixing, and Contact. Rapid mixing of chlorine solution into water, followed by a contact period, is essential for effective disinfection. The chlorine solution is provided through a diffuser system. It is then mixed rapidly by either (1) mechanical means, (2) a baffle arrangement, or (3) a hydraulic jump created downstream of a weir, Venturi flume, or Parshall flume. A diffuser is the device at the end of the solution piping that introduces the chlorine solution into the treated water at the application point. Various types of chlorine diffusers and mixing arrangements are illustrated in Figures 12-10 and 12-11.[3] A velocity gradient of above 400 per second is

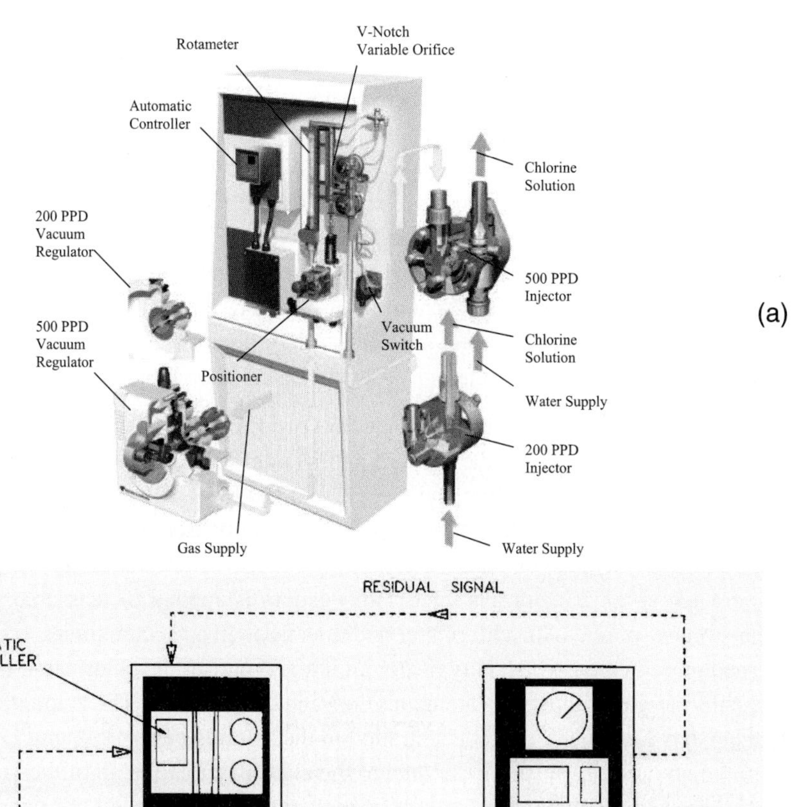

(a)

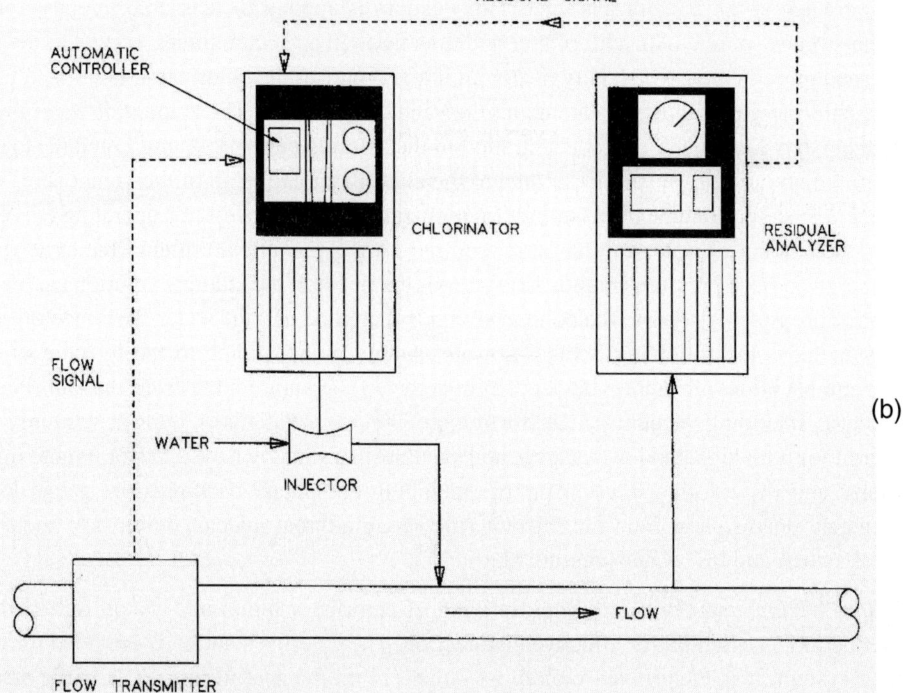

(b)

Figure 12-8 Typical chlorinator assembly. *(Courtesy of Wallace & Tiernan/U.S. Filter.)* (a) Chlorinator details. (b) Typical automatic compound-loop chlorinator assembly.

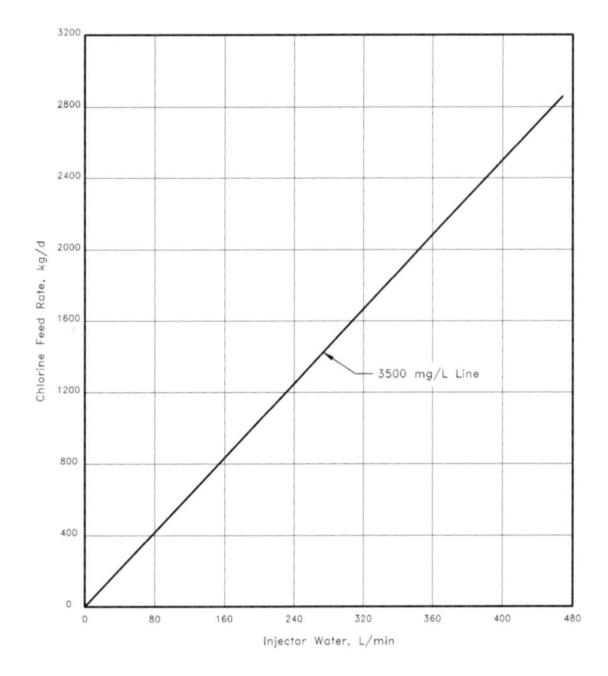

Figure 12-9 Injector-water flow rate versus chlorine feed rate, limiting chlorine concentration to 3500 mg/L. *(Reprinted from Reference 18, with permission from Technomic Publishing Co., Inc., copyright 1999.)*

sufficient to provide the desired mixing. The velocity gradient to satisfy a mixing requirement is calculated from Eq. 8.20.

The purpose of a chlorine contact chamber is to provide the contact time necessary for the disinfecting compound to provide the degree of inactivation desired. The procedure for calculating *CT* values has been provided earlier. Chlorine may be applied at several points in a water treatment plant. Possible application points are pre-chlorination (at the raw water pump station), pre-coagulation, pre-sedimentation, pre- and post-filter, and prior to clearwell. Applying chlorine toward the end of the water treatment process minimizes THM formation.

A strong chlorine solution is applied in water (3500 mg/L). Negative-pressure conditions should be avoided in chlorine solution piping downstream of the injectors. They could result in the undesired release of chlorine gas at the application point. An exit velocity of 7 to 9 m/s at the injection point will provide adequate mixing. A head loss of 2 to 4 m will help to maintain back-pressure at the injector discharge and develop a jetting effect. A check valve and an isolation valve are essential for diffusers in a pressurized pipe.

Control System. The chlorination system must maintain a given chlorine residual at the end of the specified contact time. Chlorine dosage must be adjusted frequently to maintain the required residual. At small installations, manual control is enough to provide the required chlorine dos-

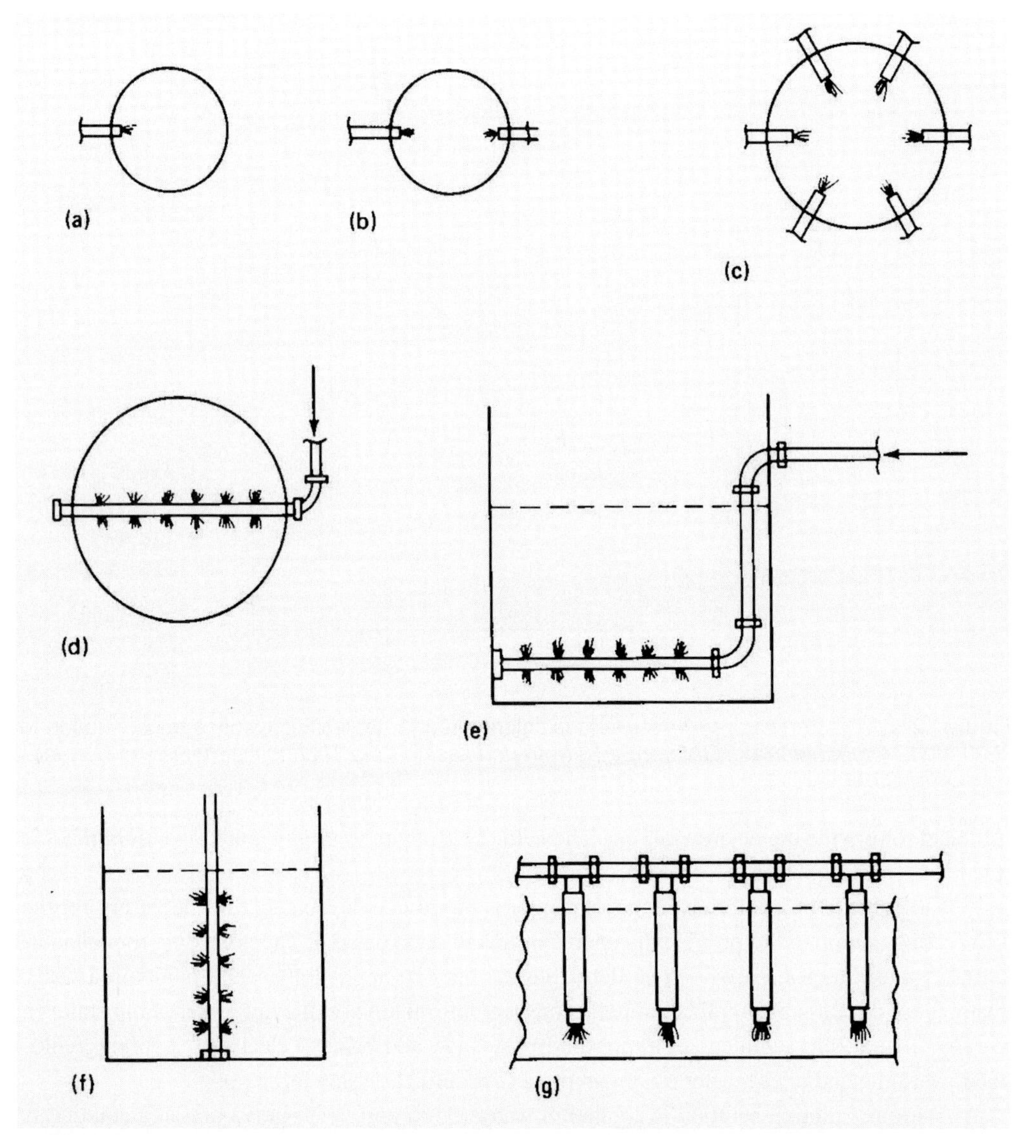

Figure 12-10 Typical chlorine diffusers. *(Reprinted from Reference 18, with permission from Technomic Publishing Co., Inc., copyright 1999.)* (a) Single injector for small pipes. (b) Dual injector for small pipes. (c) Multiple injectors for medium-size pipes. (d) Injector system for large pipes. (e) Single horizontal diffuser in open channel. (f) Single vertical diffuser for open channel; for wide channels, multiple diffusers along the width can be used. (g) Typical hanging-nozzle-multiple diffusers along the length of an open channel.

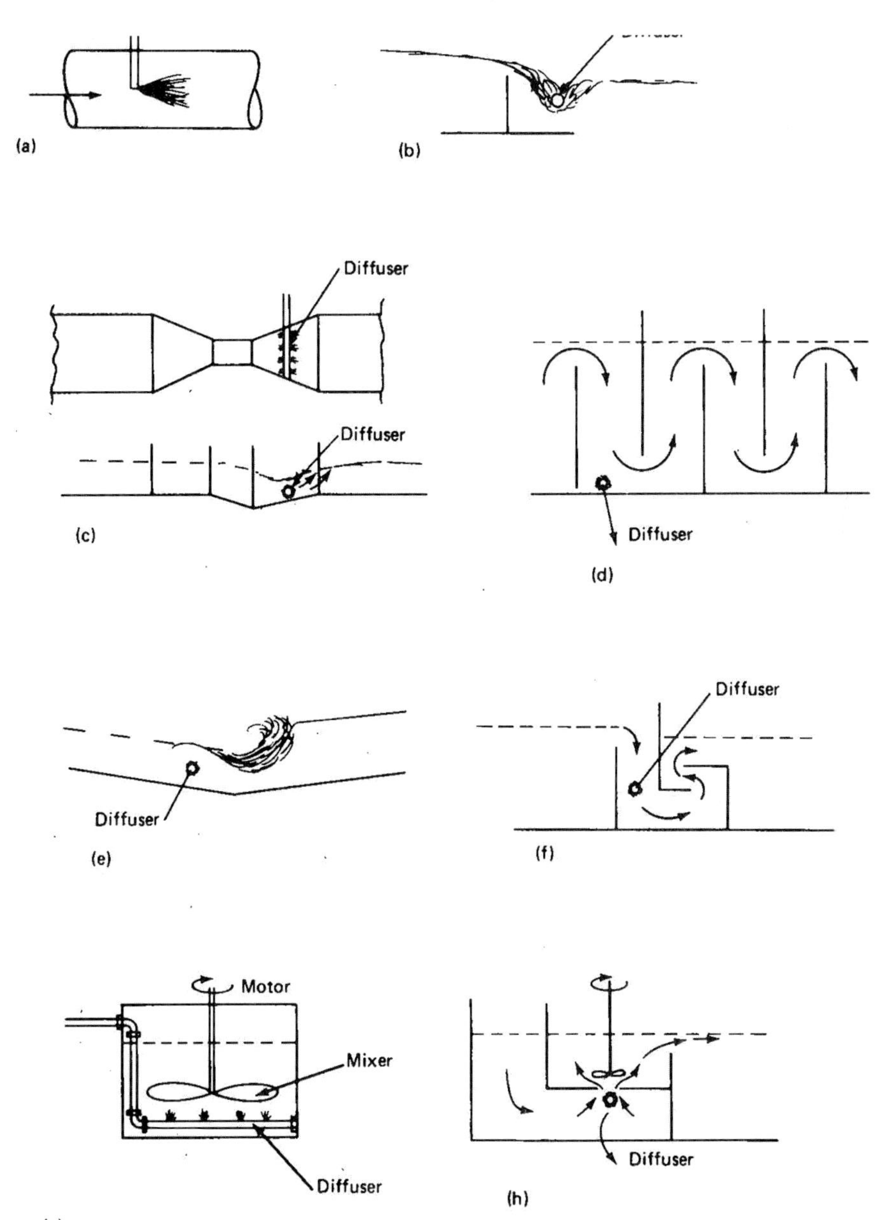

Figure 12-11 Different types of mixing arrangements used for dispersing chlorine solution in water. *(Reprinted from Reference 18, with permission from Technomic Publishing Co., Inc., copyright 1999.)* (a) Mixing methods by natural turbulence and diffusion: (i) jet nozzle in a closed conduit; (ii) downstream of a weir; (iii) Parshall flume; (iv) over-and-under baffles in a basin; (v) hydraulic jump in a channel; (vi) baffle arrangement. (b) Mechanical mixers: (i) in open channel; (ii) with baffles in open channel.

age. The operator determines the chlorine residual and then adjusts the feed rate of chlorine solution. A simple orifice-controlled constant-head arrangement or low-capacity proportioning pumps are used to feed the chlorine solution. Often, constant-speed feed pumps are programmed by time-clock arrangement to operate the pump at the desired intervals.

At large facilities, complex automatic proportional-control systems with recorders are used. Signals from a flow meter transmitter and chlorine residual analyzer are transmitted to the chlorinator to adjust the chlorine feed rate (Figure 12-9) and to maintain a constant chlorine residual that is preset in accordance with the design criteria and standard operating procedures (SOPs). The chlorine analyzers and automatic control-loop chlorinator systems are supplied by many manufacturers.

Several alarms are also considered an essential part of the control system. These include high and low pressures in storage vessels, liquid or gas chlorine lines, high and low injector vacuum lines, high and low temperatures for evaporator water bath, high and low chlorine residuals, and chlorine leaks.

Disinfection by Hypochlorite Compounds The potential hazard associated with transportation, storage, and handling of chlorine has resulted in the use of hypochlorite solution at many plants. Hypochlorite is somewhat more expensive, has less strength on storage, and may be difficult to feed; however, for safety reasons alone, and especially after the USEPA issued the Accidental Release Prevention Program Rule (ARPPR), many larger plants in urban areas have changed to hypochlorites.

Sodium hypochlorite solution is available in 1.5 to 15 percent strength, in 4.9 to 7.6 m^3 tanks and in tank cars. Stronger solutions decompose readily upon exposure to light and heat. High-grade calcium hypochlorite contains at least 70 percent available chlorine. It is available in 45- to 360-kg drums as powder, granules, or compressed tablets or pellets. Hypochlorites and solutions must be stored in cool, dry places for a better shelf life. A comparison of important properties of chlorine, chlorine compounds, ozone, and hydrogen peroxide is provided in Table 12-13.

12.3.2 Chloramination

The addition of ammonia to form a combined chlorine residual was first used in potable water treatment in the United States in 1917.[3] The practice continued for many years, to control the formation of certain taste and odor compounds, growth of algae, and bacteria on walls and weirs of the treatment units and filters. The ability of resultant chloramines to maintain a disinfectant residual for longer than free chlorine in the distribution system has also assured its continued use. Current interest in ammoniation, however, is owing to the recent concern over chlorination by-products, particularly the THMs discussed earlier. Combined chlorine (or chloramines) apparently do not form THMs, and many systems may eventually convert to chloramine. Therefore, a discussion on design aspects of ammoniation is provided as follows.

Ammonia Properties At normal pressures and temperatures, anhydrous (dry) ammonia is a gas. Table 12-14 lists the physical properties of ammonia.[3]

Ammonia Supply Ammonia used in water treatment is provided mainly in four commercial forms: anhydrous ammonia, aqua ammonia, ammonium sulfate, and ammonium chloride. Anhydrous ammonia is a dry gas, or liquefied gas. Aqua ammonia (ammonium hydroxide) is usually a 20-percent solution of ammonia gas in water ($NH_3 + H_2O \rightarrow NH_4OH$). Ammonium sulfate and ammonium chloride salts are dissolved in water to produce 25 to 30 percent solutions in plastic or fiberglass containers. These solutions under alkaline condition produce aqua ammonia.

Anhydrous ammonia. Because it can be easily liquefied under pressure, anhydrous ammonia is usually supplied as a liquefied compressed gas under pressure. It is supplied in containers or by bulk shipment in ways very similar to those used for chlorine. Typically, the containers used in ammonia supply are 45.4-kg and 68-kg (100-lb and 150-lb) cylinders. The most commonly used methods for bulk anhydrous ammonia supply are (1) railroad tank cars with a common size of 73,000 kg (80 tons), (2) tank motor vehicles with a maximum capacity of about 5400 kg (6 tons), and (3) portable tanks, generally in the range from 526 kg to 8200 kg (600 lb to 9 tons). The basic engineering features of the cylinder system are provided in References 3 and 36.

Aqua Ammonia. Aqua ammonia is supplied in 208-L (55-gal) drums. If bulk shipment is needed, unpressurized or low-pressure (180 kN/m^2) carbon steel tanks should be used. The ammonium sulfate and ammonium chloride are supplied in 45.4-kg (100-lb) bags and are generally 95 to 98 percent pure.[2]

Ammonia Storage and Handling Ammonia storage and handling systems must be designed with full safety considerations , because release of ammonia is considered a hazard. The cylinders are usually housed in an enclosure or building. A capacity between 3600 kg and 5400 kg (4 tons and 6 tons) is most commonly used for a bulk-ammonia tank receiving tank truck delivery. Bulk tanks are usually installed outdoors. Tanks must be designed and constructed in accordance with the code and standards for storage and handling of anhydrous ammonia.[36–38] Copper, silver, brass, bronze or galvanized piping should not be used for ammonia service, because moist ammonia reacts readily with these materials, to cause corrosion.

Ammonia Safety Ammonia is a colorless gas which has a very sharp, pungent odor that is detected at concentrations of 50 ppm. Ammonia gas at high concentration can irritate the mucous membranes, eyes, nose, throat, and lungs. It is not poisonous, but long exposure time can cause burns and damage. Both anhydrous ammonia and aqua ammonia (concentration equal to or higher than 20%) are regulated substances in the Accidental Release Prevention Program Rule (ARPPR). Special safety considerations that must be taken in storage, handling and use of ammonia are provided in References 3, 36 and 38.

Ammonia Dosage For disinfection, the addition of ammonia is primarily for the purpose of converting all of the free chlorine residual to combined chlorine residual (or chloram-

Table 12-13 Comparison of Important Properties and Common Applications of Chlorine, Chlorine Compounds, Ozone, and Hydrogen Peroxide

Characteristics	Chlorine	Sodium Hypochlorite	Calcium Hypochlorite	Chlorine Dioxide	Ozone	Hydrogen Peroxide
Chemical formula	Cl_2	$NaOCl$	$Ca(OCl)_2$	ClO_2	O_3	H_2O_2
Form	Liquid, gas	Solution	Powder, pellets, or 1 percent solution	Gas	Gas	Liquid
Container shipping	45.5-kg and 68-kg cylinder; and 907-kg container	N/A	45-kg to 360-kg, drums	On-site generation	On-site generation	114-L to 208- L fiberglass drums
Bulk shipping	49,900-kg and 81,600-kg tank cars; and 13,600-kg to 20,000 kg tank truck	4.9 to 7.6 m^3 tanks, and tank cars	N/A	N/A	N/A	Tank cars.
Commercial strength, percent	100	12–15	70	Up to 0.35	2	35–70
Stability	Stable	Light yellow liquid, unstable	Stable	Greenish yellow gas, explosive	Unstable	Stable
Toxicity to microorganisms	High	High	High	High	High	Medium
Hazards associated with handling and use	High	Medium	Medium	High	High	Medium
Corrosion	High	Medium	Medium	High	High	Low
Deodorizing	High	Medium	Medium	High	High	High
Cost	Low	Medium	Medium	Medium	High	High
Common applications	Control of slime growth, taste and odor, ammonia oxidation, disinfection	Control of slime growth, disinfection	Control of slime growth, disinfection	Control of slime growth, odor, disinfection	Taste and odor control, oxidation of precursors and refractory organics, disinfection	Control of slime growth, taste, and odors

Table 12-14 Physical Properties of Ammonia

Properties	Typical Values
Molecular formula	NH_3
Molecular weight	17.031
Boiling point at 1 atm[a]	–33.3° C (–28°F)
Freezing point at 1 atm	–77.7° C (–107.9°F)
Critical temperature	133° C (271.4°F)
Critical pressure	114.2 bars (1657 psia)
Latent heat at –33.3°C (–28°F) and 1 atm	13.71 ∞ 10^5 J/kg (589.3 Btu/lb)
Relative density of vapor[b]	0.5970
Vapor density at –33.3°C (–28° F) and 1 atm	0.8899 kg/m^3 (0.0555 lb/ft^3)
Specific gravity of liquid at –33.3°C (–28°F)[c]	0.6819
Liquid density at –33.3°C (–28°F) and 1 atm	681.9 kg/m^3 (42.57 lbs/ft^3
Specific volume of vapor at 0°C (32°F) and 1 atm	1.297 m^3/kg (20.78 ft^3/lb)
Heat of solution at 0% concentration by weight	8.081 ∞ 10^5 J/kg (347.4 Btu/lb)
Heat of solution at 28% concentration by weight	4.999 ∞ 10^5 J/kg (214.9 Btu/lb)

a 1 atm = 760 mm Hg (29.92 inch) or 1.01325 bar.
b compared to dry air at 0°C (32°F) and 1 atm atm.
c compared to that of water at 4°C (39.2°F).
Source: Adapted in part from References 2, 3, and 22.

ines).[2] The SWTR requires chlorine to be added before ammonia. Higher level of disinfection is achieved by addition of chlorine first, but lowest levels of THM formation are obtained by ammoniation prior to chlorination, or ammoniation that is concurrent with chlorination. Optimum pH of the water for the addition of ammonia and chlorine is between 7.0 and 8.3. Some basic design considerations of ammonia and chlorine doses are listed here.[2,3]

- Chlorine or ammonia, after being added individually, must be mixed quickly and thoroughly.
- Once the ammonia and chlorine have been added to the water, they should also be mixed as quickly as possible.
- The reaction between the two chemicals in solution to form chloramines is quick at a pH of 7.0 to 8.5. The reaction time is approximately two seconds. Lower pH and temperature (below 10°C) or poor mixing conditions will retard the formation of chloramines, either by slowing the reaction rate or by allowing side reactions to occur between the chlorine and organic material present in the water.
- The ammonia dose used in disinfection is generally measured as a chlorine-to-ammonia (NH3) ratio. The chlorine-ammonia weight ratio range commonly used is from 3:1 to 4:1.

A chlorine-ammonia weight ratio of 4:1 is approximately equivalent to one mole of chlorine to one mole of ammonia[c].

- In the weight ratio range up to 4:1, only monochloramine is formed, and disinfection is achieved by combined chlorine residual. No THM is formed.
- The breakpoint (Figure 12-6) is reached at chlorine-ammonia weight ratio of approximately 5:1.[2] If free chlorine is desired, weight ratios of 6:1 to 10:1 are usually required. At this dose ratio, breakpoint is passed (Figure 12-6).
- Pilot testing is essential to optimize the chlorine-ammonia weight ratio for taste, disinfection efficiency, and THM formation potential.
- Overdosing of ammonia can encourage the growth of nitrifying bacteria in the treatment processes and the distribution system. This affects the quality of water supplied to the consumers.[2]

Ammonia Feed and Application

Anhydrous Ammonia. Ammonia feed and application equipment is similar to chlorine equipment. The ammoniators are evaporator-type, direct-feed, or solution-feed.[22] A typical anhydrous ammonia feed system is illustrated in Figure 12-12.[13] The design details of various components are given here[2,3,22]:

- The ammoniators control the feed rate of the ammonia gas to the injector.
- The direct-feed ammoniators have a maximum capacity of 450 kg/d (1000 lbs/d) at the low discharge pressure of 120 kN/m^2 (15 psig).
- In designing the evaporator for ammonia, it should be noted that (1) the feed rate of ammonia is one-third to one-fifth of chlorine feed rate and (2) the heat of vaporization of ammonia is almost five times that of chlorine. Therefore, for the same feed rate, an ammonia evaporator may consume five times the amount of energy needed for a chlorine evaporator. If the ammonia feed rate is one-fifth of the chlorine feed rate, the ammonia evaporators will need 1 to 1.5 times the size of evaporators at the same treatment plant.
- The solution-type ammoniators have capacities up to 1800 kg/d (4000 lb/d) and can operate with discharge pressures of up to 1200 kN/m^2 (150 psig).
- The injector mixes the ammonia gas with injector water. Smaller amounts of injector water are required for ammoniation, because of ammonia's high solubility in water. The formation of ammonia hydroxide increases the pH and so cause a water-softening effect, which results in scaling and clogging of injector and solution lines. If the hardness of the injector water exceeds 35 mg/L as $CaCO_3$, provision for softening of the injector water must be made[2,3].

c. 4:1 Chlorine to ammonia weight ratio means 4 mg/L (or lb) Cl_2: 1 mg/L (or lb) NH_3.

- Ammonia-solution piping should be sized to minimize head losses.
- The diffuser openings should be designed to cause a head loss from 3 to 4 m and produce a G from 250 to 300 per second.[39] Eq. (8.20) is used to calculate velocity gradient.

Aqua Ammonia. The aqua ammonia feed system consists of a diaphragm metering pump, pulsation dampener, flow meter, back-pressure valve, and diffuser. Aqua ammonia solution is much more concentrated than anhydrous ammonia; therefore, it is used for systems requiring larger amounts of ammonia. For systems requiring lower ammonia, aqua ammonia is diluted with demineralized water for accurate delivery. The diffusers and piping are similar to those used for anhydrous ammonia solution. A typical aqua ammonia feed system is illustrated in Figure 12-13.[13]

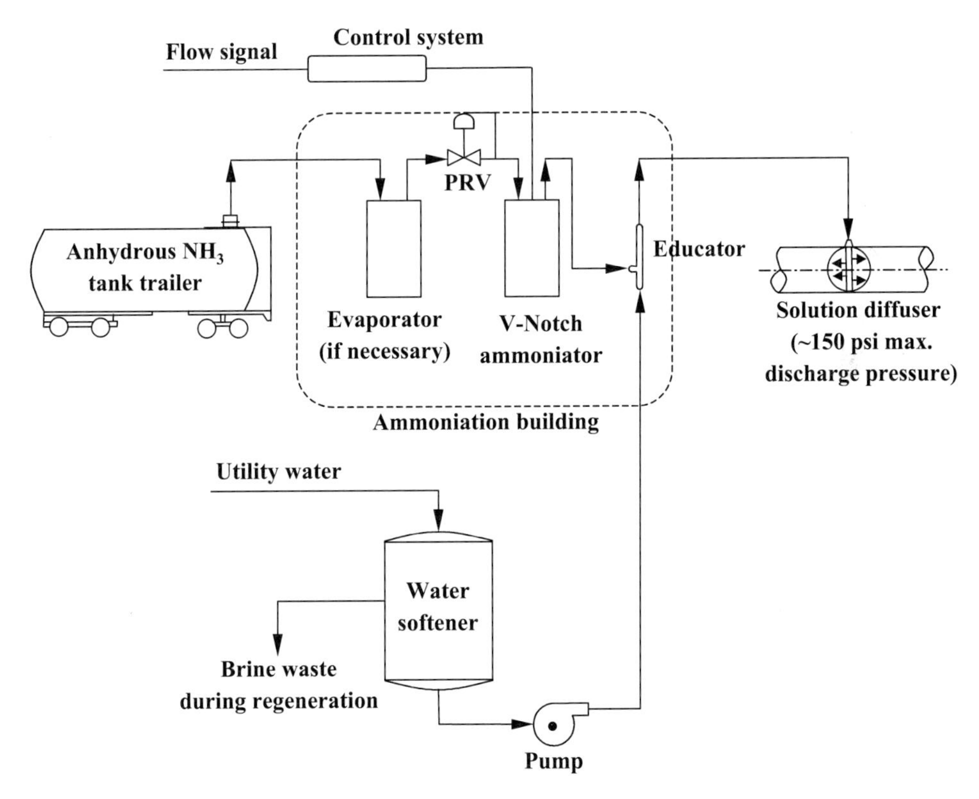

Figure 12-12 Typical solution-feed ammoniation system using anhydrous ammonia.

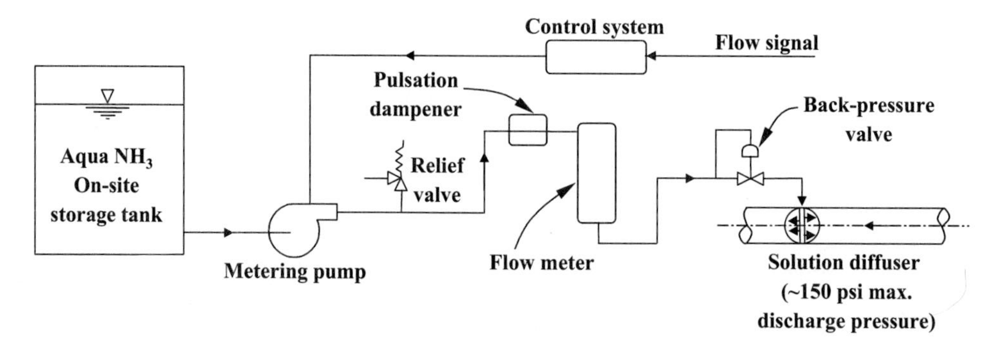

Figure 12-13 Typical aqua ammonia feed system.

12.3.3 Chlorine Dioxide

Chlorine dioxide has been used for many years to bleach flour, paper, and textiles. Its use in water treatment was limited mainly to taste and odor control (and to iron and manganese removal at smaller installations). Since the mid-1970s, its use has been considered more frequently for disinfection, on the premises that it does not react with organic compounds to form THMs and that it also reduces THM precursors.[2,3,22] A discussion of chemical properties, generation, and application of chlorine dioxide in water treatment is given below. Advantages and disadvantages of chlorine dioxide disinfection in a water supply system are presented in Table 12-15. A comparison of important properties of chlorine dioxide with those of other chlorine compounds, ozone, and hydrogen peroxide is provided in Table 12-13.

Table 12-15 Advantages and Disadvantages of Chlorine Dioxide as a Disinfectant[2,3]

Advantages	Disadvantages
1. Powerful bactericide and viricide	1. High chemical costs (approximately 5 times that for chlorine)
2. No reaction with ammonia nitrogen	2. Must be generated on site
3. No reaction with organic compounds to form THMs	3. Possibility of health hazards from by-products of production
4. Reactions to destroy 30 to 40 percent of THM precursors.	4. Pronounced metallic taste in treated water at high dosages
5. Efficiency of disinfection relatively unaffected by pH levels between 6 and 10	5. Lack of reliable and practical techniques for evaluation and differentiation of chlorine dioxide from other chlorine compounds

Properties of Chlorine Dioxide Chlorine dioxide is a greenish/yellow gas with a strong, disagreeable odor that is similar to but stronger than that of chlorine. It is toxic to humans when inhaled. Its odor is detectable above concentration of 0.1 ppm. It is an unstable gas, explosive in air in concentrations above 10 percent by volume (corresponding to solution concentration of about 12 g/L). Because of its instability, chlorine dioxide is always generated on site in a gaseous solution and used shortly after its preparation. Solutions up to 5 g/L can be stored for several days. Specific properties of chlorine dioxide are listed here.[2,3]

- It is a more powerful disinfectant than chlorine, although it has lower oxidation potential (Table 12-1).
- When prepared in the absence of excess free chlorine, it does not produce THMs or other chlorinated organic by-products of concern. Some chlorine dioxide generation methods do create conditions of excess free chlorine in which chlorination by-products are produced.
- Excess free chlorine, if present, produces hypobromous acid and brominated compounds.
- Excess free chlorine, if present, can slowly produce chlorate and chlorite ions by disproportionation under pH below 2 or above 11.
- In the Stage 1 D/DBPR, the USEPA establishes a Maximum Contaminant Level (MCL) of 0.8 mg/L as ClO_2 for chlorine dioxide.[12] Typical dosage of chlorine dioxide for disinfection ranges from 0.1 mg/L to 0.3 mg/L.[2,3]
- Chlorine dioxide does not react with ammonia. Therefore, it lasts longer than free chlorine residual.
- Chlorine dioxide can be used to preoxidize phenolic compounds and to separate iron and manganese from some organic complexes that are stable against chlorination. Reduction in THM precursors by 30 to 40 percent has been reported.[2,3]

Production of Chlorine Dioxide For drinking water treatment, chlorine dioxide is generated from solutions of sodium chlorite ($NaClO_2$). Sodium chlorite may be purchased as a 25-percent aqueous solution or as a solid of 80-percent purity. Chlorine dioxide is produced by reacting the aqueous solution of sodium chlorite with chlorine gas or hypochlorous acid. These reactions are expressed by Eqs. (12.14) and (12.15).

$$2NaClO_2 + Cl_2(g) \rightarrow 2ClO_2(g) + 2NaCl \tag{12.14}$$

$$2NaClO_2 + HOCl \rightarrow NaCl + NaOH + 2ClO_2(g) \tag{12.15}$$

The reaction is carried out in a chlorine dioxide reactor by metering appropriate strengths of reactant solutions. The resulting yellow solution is pumped out for use. Important considerations are the following[2,3]:

- If excess chlorine solution is added, free chlorine can oxidize chlorine dioxide to form chlorate ion (ClO_3^-), which is difficult to remove from solution.
- Chlorine dioxide added to water is converted to chlorite ion (ClO_2^-), which, along with chlorate ion, has been associated with undesired health effects. The conversion is from 70 to100 percent. Therefore, chlorine dioxide feed must also be controlled to meet the MCL of 1.0 mg/L for chlorite ion in the Stage 1 D/DBPR.[12]

- Excess chlorine in chlorine dioxide will produce THMs. If excess chlorine is controlled, the formation of chlorinated organic compounds is reduced significantly.

Several types of chlorine dioxide generation equipment are commercially available. Manufacturers provide details of their systems, methods of preparation, and excess chlorine expected. The schematic flow diagram of chlorine dioxide generation from sodium chlorite is illustrated in Figure 12-14.[2]

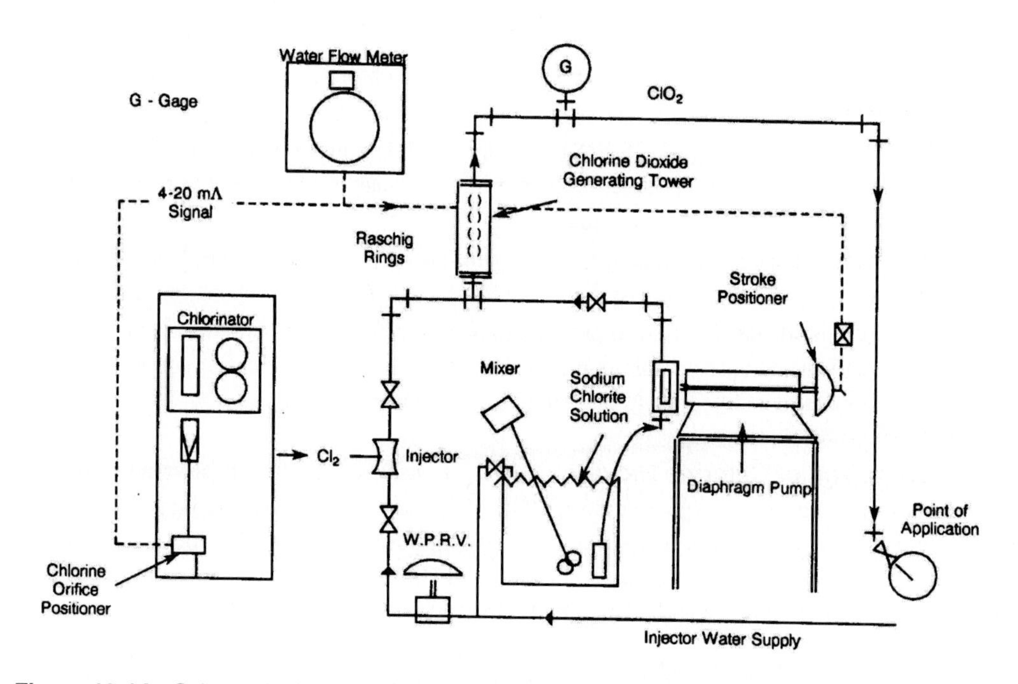

Figure 12-14 Schematic diagram of an automatic-feed, flow-proportional system for generating chlorine dioxide from sodium chlorite. *(Adapted from Reference 2.)*

Chlorine Dioxide System Design Considerations The system design for chlorine dioxide includes (1) the chlorine dioxide generator, (2) transport of metered chlorine dioxide feed solution to the application point, and (3) dispersion of the solution in water. Generators are commercially available. The design of these systems is similar to that for chlorine and chloramine, presented in earlier sections.

12.3.4 Ozonation

Ozone has been used extensively in Europe for disinfection and for taste and odor control in water supplies. Interest in the United States and Canada has increased in recent years because of a growing concern about THM formation during chlorination of drinking water. In addition to its use as a disinfectant, preozonation is also used for (1) removal of taste and odor, (2) removal of color, (3) removal of iron and manganese, (4) enhanced removal of NOM, and (5) oxidation and volatilization of organics.

Ozone is an unstable gas; therefore, it has to be generated on site. In addition, ozone cannot be used as a secondary disinfectant, because an adequate residual in water can be maintained for only a short period of time. Because of its high oxidation potential, ozone requires a short CT_{Cal} value. As a microflocculation aid, ozone is added during or before rapid mix followed by coagulation. Many studies have shown that preozonation enhances coagulation-flocculation and improves performance of sedimentation and filtration processes.[3,40-43] The advantages and disadvantages of ozonation in water treatment are provided in Table 12-16.

Table 12-16 Advantages and Disadvantages of Ozone Disinfection

Advantages	Disadvantages
1. Complex taste, odor, and color problems are effectively reduced or eliminated.	1. The residual does not last long.
2. Organic impurities are rapidly oxidized.	2. High electric energy input and high capital and operating costs (about 10 to 15 times higher than chlorine) are required.
3. Effective disinfection is achieved over a wide temperature and pH range.	3. High temperature and humidity may complicate ozone generation.
4. Bactericidal and sporicidal action is rapid (300 to 3000 times faster than chlorine); only short contact periods are required.	4. The process is less flexible than those for chlorine in adjusting for flow rate and water quality variations.
5. Odors are not created or intensified by formation of complexes.	5. Analytic techniques are not sufficiently specific or sensitive for efficient process control.
6. It reduces chlorine demand and in turn lowers chlorine dosage and so THM formation potential.	6. Waters of high organic and algae content may require pretreatment to reduce ozone demand.
7. It improves overall treatment efficiency.	

Source: Adapted in part from References 2, 3, and 5.

Ozone Formation Reactions Ozone, being an unstable gas, must be generated on site and used quickly. It is generated by applying energy to oxygen or dried air. A high-energy electrical field (corona) causes oxygen to dissociate (Eq. (12.16). Each dissociated oxygen molecule reacts with another oxygen molecule to form an ozone molecule (Eq. (12.17)). The overall ozone formation reaction is, therefore, expressed by Eq. (12.18).

$$O_2 + 2e^- \rightarrow 2O^- \tag{12.16}$$

$$O^- + O_2 \rightarrow O_3 \tag{12.17}$$

$$3O_2 + 2e^- \rightarrow 2O_3 \tag{12.18}$$

This reaction is reversible, and, once ozone is formed, it decomposes to oxygen. This reversible reaction occurs quite readily above 35°C (95°F). Therefore, an ozone generation system utilizes cooling components to dissipate heat produced during generation. The energy input per kg O_3 generated is 0.82 kW-h.[3]

Ozonation Process Descriptions Ozone is a strong oxidizing agent. When exposed to a neutral or alkaline environment (pH above 6), UV light, or hydrogen peroxide, it decomposes in water to produce more active hydroxyl free radicals (Eqs. 12.19 and 12.20).[3] This reaction is accelerated at pH above 8.

$$O_3 + H_2O \rightarrow 2\ HO_2 \tag{12.19}$$

$$O_3 + HO_2 \rightarrow 2\ O_2 + HO \tag{12.20}$$

The hydroxyl free radicals (HO_2 and HO), or a mixture of ozone and hydroxyl radicals, can be a powerful oxidizing agent. The free-radical species are more effective oxidizing agents than the molecular ozone, but they are extremely short-lived. The CT_{Tab} values for ozone (Table 12-7) are for the molecular ozone. Because of its high oxidizing capability, ozone is a powerful disinfectant. It is believed that ozone can kill microorganisms effectively by destroying cell-wall structures (in cell *lysis*) . Unlike chlorine, ozone produces little or no THM (the primary DBP of concern). Ozone does not produce dissolved and suspended solids. Therefore, ozonation does not cause either high-TDS or residuals problems.

Ozone reacts with such inorganic compounds as nitrites, ferrous, manganous, sulfides, and ammonium ions. The oxidation of these inorganic substances by ozonation process is very fast and complete. Ozone, being a powerful and effective oxidant, destroys many organic compounds that produce color, taste, or odor in potable water.[3,18] Therefore, it is widely used in taste and odor control, color removal, and iron and manganese removal.[3]

Ozone provides a significant contribution to breaking down organic compounds, although ozonation alone has a negligible effect on the overall concentration of TOC in raw water. Ozone also reacts with NOMs. Among these are aliphatic and aromatic compounds, humic acids, and pesticides. Ozone segments these organic compounds to lower molecular species, such as aldehydes and ketones. Some reactions of ozone with NOMs and bromide are presented in Figure 12-15. Ozone does not produce halogenated organic matter directly; however, in presence of bromide ion, hydrobromic acid is formed, which may encourage formation of brominated organics.[16]

Properties of Ozone and Safety Ozone is a faintly blue, pungent-smelling, and unstable gas. It is detectable even at low concentrations (0.01 to 0.02 ppm by volume). Higher concentrations may cause olfactory and other reactions, and much higher concentrations may be toxic in some instances. The longer the exposure to ozone, the less noticeable the odor. The maximum allowable ambient ozone concentration for an 8-hour working period is 0.1 ppm, which is much less than the ozone concentration of greater than 500 ppm in volume in the off-gas.[3] An effective off-gas ozone destruction system therefore must be included in the design of

O_3 + NOM ⟶ Oxidation By-products

- Aldehydes
  - Formaldehyde
  - Acetaldehyde
  - Glyoxal
  - Methyl glyoxal
- Aldo– and Ketoacids
  - Pyruvic acid
- Acids
  - Oxalic acid
  - Succinic acid
  - Formic acid
  - Acetic acid
- Hydrogen Peroxide

O_3 + Br^- + NOM ⟶ Brominated By-products

- Bromate
- Bromoform
- Brominated acetic acids
- Bromopicrin
- Brominated acetonitriles

Figure 12-15 Principal by-products of ozonation.

an ozonation process, in order to eliminate environmental risk and damage or injury to the plant personnel.

The solubility is governed by Henry's law. The solubility of ozone in water depends upon its temperature and its concentration in the feed gas as it enters the ozone contactor. The higher is the concentration of ozone in the feed gas, the more soluble it is in water. Increasing pressure in the ozone contactor also increases its solubility. A solubility of 4.7 mg O_3/L at 20°C has been reported.[3] Half-life of ozone in water ranges from 8 minutes to 14 hours, depending on the level of ozone-demanding contaminants in water and on the water temperature. The limitation of the solubility of ozone in water is one significant factor affecting the efficiency of ozonation processes.

Design Considerations The design of an ozonation system has six major elements: (1) feed-gas preparation, (2) electrical power supply, (3) ozone generator, (4) ozone contactor, (5) ozone destruction of exhaust gas, and (6) control and monitoring system. All of these systems are discussed below.

Feed Gas Preparation. Ozone may be generated from either air or oxygen. When air is utilized, the air fed to the ozone generator must be dried to a maximum dew point of –65°C. Moisture in air may reduce ozone production, cause fouling of the dielectric tubes, and increase corrosion in the ozone generator and downstream equipment.

The ozone generation systems using ambient-air feed can be operated at low, medium, or high operating pressure. Low-pressure systems operate at partial vacuum created by submerged turbine or other ejector device. Medium-pressure systems range from 0.7 to 1.05 kg/m^2 (10 to 15 psig). High-pressure systems operate at pressures from 4.9 to 7.03 kg/cm^2 (70 to 100 psig) and reduce the pressure prior to the ozone generator. Pressure desiccant dryers are also used in conjunction with compression and refrigerant dryers for generating large and moderate quantities of ozone. A low-pressure air-feed gas preparation schematic for ozone generation is shown in Figure 12-16. Very small systems may use two desiccant dryers in series (no compression or refrigerant drying). The desiccant dryers use silica gel, activated alumina, or molecular sieves to dry air to the necessary dew point.

Feed gas can also be pure oxygen. Pure-oxygen generating systems have many benefits over air-feed systems: (1) higher ozone-production density (more ozone produced per unit area of the dielectric), (2) high concentration of ozone in the product gas (almost double), (3) less energy requirement for creating corona, (4) smaller feed-gas volume for the same ozone output, and (5) less ancillary equipment. For small-to-medium-size systems, oxygen may be purchased from chemical suppliers in containers as a gas or as a liquid. The liquid-oxygen containers are under pressure. For large operations, oxygen generation on site may be necessary. There are two methods of producing oxygen on site for ozone generation: (1) pressure-swing adsorption of oxygen from air, and (2) cryogenic production (liquification of air, followed by fractional distillative separation of oxygen from nitrogen). Systems for production of oxygen on site contain many of the same elements as the air-preparation system, because the feed gas must be clean and dry irrespective of oxygen content.[2,4,18]

Another benefit of an oxygen-generation system is peak-power shedding. The oxygen-enriched feed gas can be generated during off-peak power demand. During the period of peak power demand, the system may be switched to oxygen-enriched feed, thus reducing the power requirement for ozone generation during peak power demands.

Electrical-Power Supply. The voltage or frequency of the power to the ozone generation must be varied to control the amount and rate of the ozone produced. Zone generators use high voltages (up to 710,000 V) or high-frequency electrical current (up to 2000 Hz). Therefore, specialized power-supply equipment and design considerations such as proper insulation or wiring and cooling of transformers are necessary.

Ozone Generators. Ozone can be generated by two methods: (1) UV light, and (2) cold-plasma or corona discharge. Ozone generation by UV light is much the way ozone is formed in the upper atmosphere. UV light (less than 200 nm) is produced by an arc-discharge lamp and passes through dry or oxygen-enriched air. Ozone is generated by photochemical reaction. Ozone generated by this method is much lower in concentration (0.25 percent) than that produced by corona discharge. This method is suited only for small scale systems, requires low capital investment, and is relatively easy to maintain.

The most common method of ozone generation for water treatment is the corona-discharge cell. The discharge cell consists of two electrodes separated by a discharge gap. High voltage potential is maintained across a dielectric material, and feed gas flows between the electrodes (Figure 12-17). Ozone concentration of from 1 to 3.5 percent by weight is generated from cool, dry feed air, from 2 to 7 percent from pure oxygen.

Commercially available ozone generators are typically horizontal or vertical tubes or plates with a water-, air-, or oil-cooled system. Operating conditions of these generators are as follows:

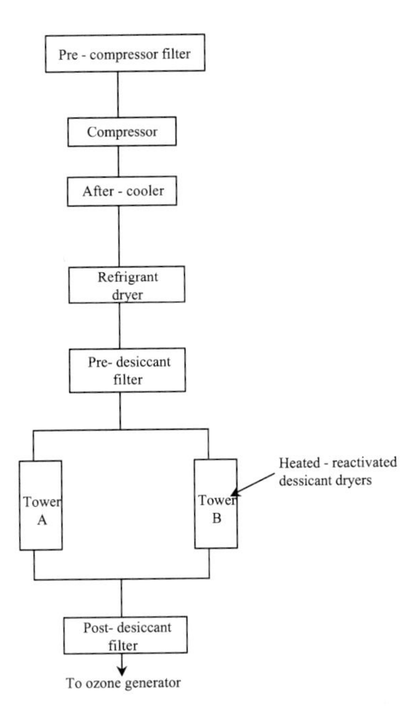

Figure 12-16 Low-pressure air-feed gas treatment schematic for ozone generation. *(Adapted from Reference 2.)*

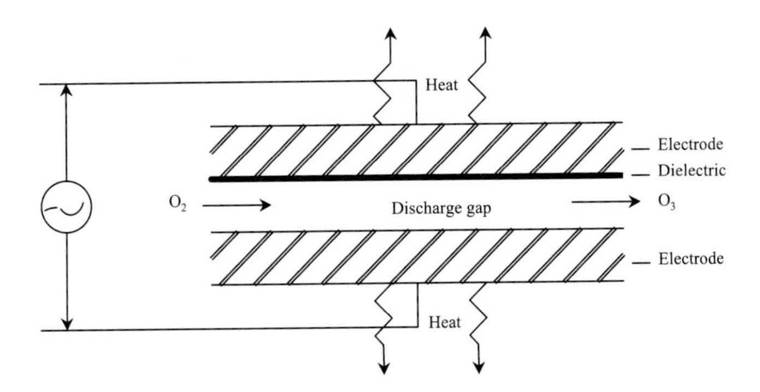

Figure 12-17 Typical ozone-generating configuration for a corona-discharge cell. *(Adapted from Reference 2.)*

- low frequency (60 Hz), high voltage (>20,000 V);
- medium frequency (60–1000 Hz), medium voltage (10,000 V–20,000 V);
- high frequency (>1000 Hz), low voltage (<10,000 V).

Currently, low-frequency high-voltage units are used most commonly, but higher-frequency, lower-voltage units are more desirable on account of recent improvements in electronic circuitry.[2,4,5] Ozone generation at from 60 to 70 percent of the maximum generation capacity is most cost effective. Multiple units, if selected properly, should satisfy average and peak demands, with a necessary standby unit for maintenance. The following example shows the selection process.

- Average ozone requirement is 40 kg/d (88 lb/d).
- Peak period ozone requirement is 60 kg/d (132 lb/d).
- Provide three ozone generators, each designed for 30 kg/d (66 lb/d).
- Two generators, each operating at 67 percent of maximum capacity, will provide average ozone requirement (30 kg/d ∞ 2 ∞ 0.67 = 40.2 kg/d).
- Three generators, each operating at 67 percent of maximum capacity, will provide maximum ozone requirement (30 kg/d ∞ 3 ∞ 0.67 = 60.3 kg/d).
- One standby unit will be available for maintenance during average demand.

Ozone Contactor. Ozone is dispersed into water in a contact chamber. The contact chamber could be a tall vertical water column, a two-compartment tank, an inclined packed column, a static mixer, or a high-speed agitator. The delivery of ozone-rich air can be made under pressure through a porous-plate diffuser at the base, or under negative pressure, through an injector device or a Venturi-type nozzle. The fine bubbles of ozone and air mixture cause mass transfer, and oxidation and disinfection take place in the contact chamber. When ozone is supplied under pressure, the supply pressure must be enough to overcome the static pressure, the exit pressure through the diffuser, and piping losses. If delivery is under negative pressure, the injector device must create vacuum action to draw the ozone mixture from the generator. The ozone transfer efficiency depends on the concentration of ozone in feed gas, pressure of feed gas at release point, ozone bubble size, water temperature, baffle conditions in the contact chamber and design capacity of the plant.[3]

Initially, ozone applied is consumed up to satisfaction of the ozone demand, then a residual is detected. Typical ozone residual concentrations range from 0.3 to 0.9 mg/L.[2,3] Typical ozone feed rate is 1 to 5 mg/L, depending upon the purposes (i.e., disinfection, taste and odor control, color removal, and control of THM precursors). Contact time T_{10} is difficult to determine, because of mixing and turbulence in the contact chamber. Several complete mixed chambers can be connected in a series to provide better plug flow conditions. The T_{10} value is safely assumed equal to 50 percent of the theoretical detention time in this type of arrangement. In a two-compartment system (Figure 12-18), for instance, the first chamber may receive two-thirds of the total ozone dose, to meet the ozone demand and have a negligible residual. In the second compartment, the remaining ozone dosage is added to achieve a desired and stable residual level. For *CT* calculations, the contact time of the second compartment and average ozone concentration at the inlet and exit points of the second compartment are used. The ozone dose is best determined by pilot studies. A typical schematic flow diagram of an ozone-generation system is illustrated in Figure 12-19.

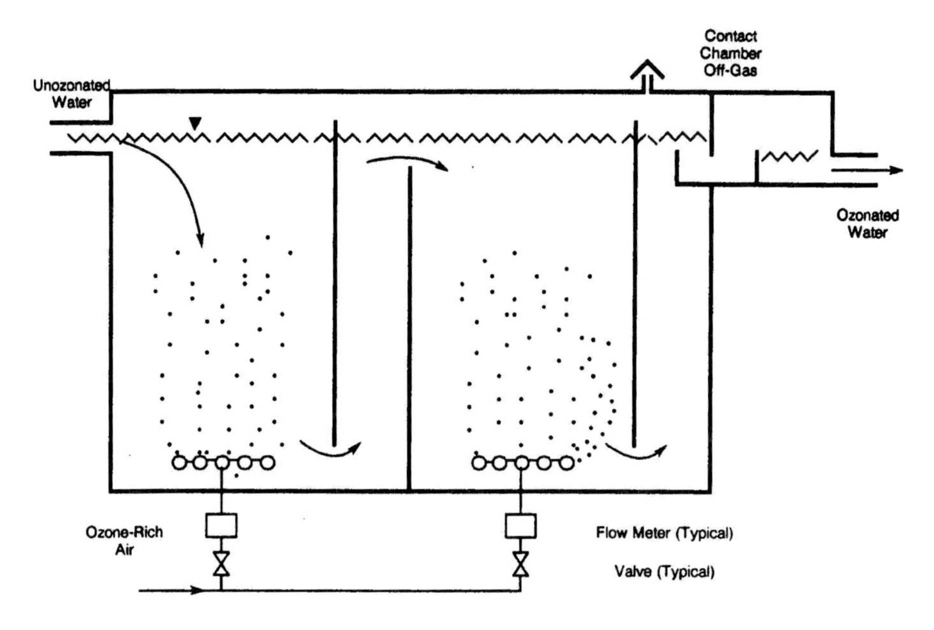

Figure 12-18 Two-compartment ozone contact chamber with porous diffusers. *(Adapted from Reference 2.)*

Ozone-Destruction System. The off-gas from a contactor generally contains ozone concentration greater than 1 g/m^3 (500 ppm by volume). The ozone must be destroyed in the final exhaust gas. Most ozonation processes, including those that use oxygen feed gas, include one certain type of ozone-destruction system, because it is more cost effective. An ozone-destruction system can use any one of the following four primary methods: (1) thermal (300 to 350°C for 3 seconds), (2) thermal/catalytic, (3) catalytic, and (4) moist granular activated carbon. The most common catalyst is metal oxide. A cartridge made of granular manganese oxide is an effective catalyst for ozone destruction.

Control and Monitoring System. The ozonation process can be operated simply by manually changing the applied ozone dosage with the flow rate. Automatic control systems can also utilized in the operation of an ozonation process. The purpose of automation is mainly to reduce ozonation energy cost rather than to improve process performance. The simplest automatic control system is to change the power input to the ozone generator in proportion to the flow rate to the plant. A microprocessor is usually used to determine the ozone feed rate from such information as plant flow rate, feed-gas flow rate, and concentration of ozone in the feed gas. This

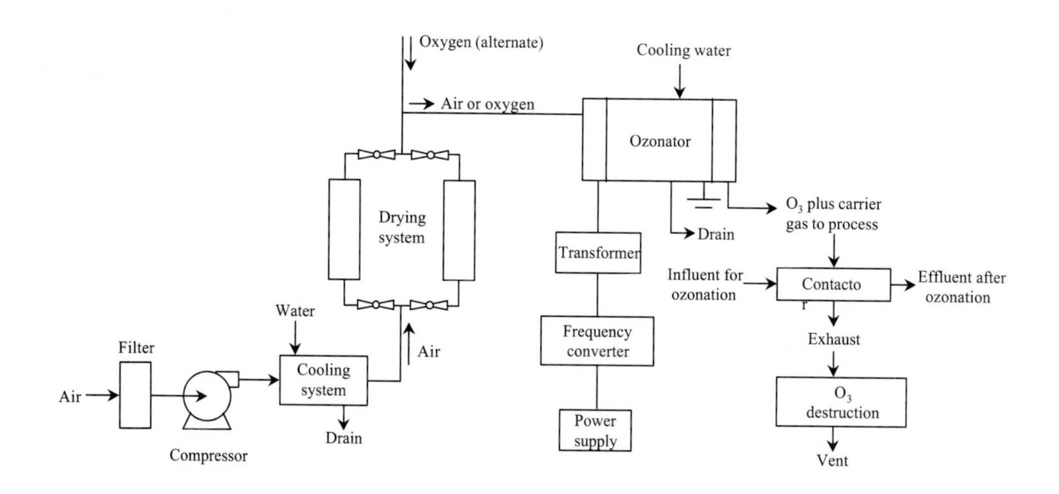

Figure 12-19 Ozonation system schematic.

method is believed to be adequate if accurate flow measurement is obtained and relatively constant water quality is encountered. Other complicated systems are also available for automation of ozonation process. These systems may utilize either an ozone-residual or an off-gas-concentration control strategy.[3]

No matter what type of control strategy is utilized for control of an ozonation process, a proper metering and monitoring system is required. Following is a list of some equipment that can be used for this purpose:

- gas pressure and temperature gauges or monitors at air preparation unit;
- continuous monitoring of dew point of feed gas to ozonator, alarm at high dew point, and generator shutdown;
- air temperature monitoring of feed and exit gases from the ozonator, temperature of coolant media (water, oil, or air), and an automatic shutdown of ozone generator, if coolant flow is interrupted or if temperature and discharge pressure exceed a set value;
- flow rate, ozone concentration, temperature, and pressure monitors in the ozone gas line;
- ozone monitor for plant ambient air (in case of leaks);
- ozone monitoring of exhaust gas;
- separate power-input monitors for ozonator and air preparation units.

Other Design Considerations Materials of construction for ozone generation and application equipment should be capable of resisting the strong oxidizing and corrosive effects of ozone. Recommended construction materials are (1) reinforced concrete for the contact chamber; (2) series 300 stainless steel for piping, fittings, and valves; and (3) Teflon for gaskets.

12.3.5 Potassium Permanganate

Potassium permanganate is a strong oxidant and is used widely for taste and odor control and for manganese removal. Chemical properties and use of potassium permanganate for taste and odor control have been presented in Chapter 11 (Section 11.3.6). Potassium permanganate is a weak disinfectant; therefore, the contact times and concentrations required for adequate primary disinfection are quite large. For example, at a 2-mg/L residual, a 24-hour contact time is required for satisfactory disinfection. As a result, it has limited use as a disinfectant, especially for potable surface water treatment.[2,22,44]

Potassium permanganate is a purple, crystalline salt of permanganic acid. It is produced by the action of chlorine in potassium manganate. When it is added to water, its presence can usually be detected visually by a pinkish tint in the water. Table 12-17 lists the advantages and disadvantages of the use of potassium permanganate as a disinfectant.

Table 12-17 Advantages and Disadvantages of Potassium Permanganate as a Disinfectant

Advantages	Disadvantages
1. It is easily detected in water by pink color.	1. Feeding can be messy; causes staining.
2. It is safe to handle and transported in solid form; and there is no explosion danger.	2. Feeding as solid is more labor-intensive than with liquid chemicals.
3. Risk of overdosing is reduced by solid-batch feeding and by seeing pink color in solution.	3. It requires long contact times to be effective.
	4. Filtration is required to remove the insoluble manganese oxide hydrates that are formed.

12.3.6 Ultraviolet (UV) Irradiation

The germicidal effects of UV have been known in some detail since the late 19th Century. Sunlight was used to purify water, even in ancient times. Direct application of the technology in water treatment came when the mercury-vapor lamp was introduced in 1901[2]. Currently, UV technology is used extensively in wastewater treatment because of improvements in modern lamp and system design, lack of a toxic chemical residual, excellent virucidal properties, improved reliability, and simplicity of operation.[3–5] In water treatment, the USEPA considers UV ineffective for inactivation of *Giardia* cysts. As the presence of *Giardia* is confirmed in more streams across the nation, the acceptability of UV disinfection becomes more narrow.[2] Nevertheless, future process improvements may enable UV to compete effectively against other primary disinfectants. The advantages and disadvantages of UV radiation as a disinfectant are provided in Table 12-18.

The primary mechanism for inactivation of microorganisms by *UV light* is direct damage to the cellular nucleic acids. When UV energy is absorbed by the genetic material (DNA) of microorganisms, structural changes are induced that prevent propagation of the organism. The peak absorption is between 250 and 265 nm, which is the optimum wavelength range for germi-

Table 12-18 Advantages and Disadvantages of UV Radiation as a Disinfectant[5]

Advantages	Disadvantages
1. No chemical is introduced into the water, so water quality is not significantly affected.	1. Spores, cysts, and viruses are less susceptible than bacteria.
2. Water constituents in solution, such as ammonia, do not exert any effect on disinfection capacity.	2. Thorough water treatment is required prior to UV, because the radiation is absorbed by many other constituents.
3. Tastes or odors are not produced (but UV does not remove taste, odor or color).	3. There is no residual; therefore, a secondary disinfectant is needed.
4. Exposure time is short.	4. Expensive equipment and large amounts of electrical energy are required.
5. Overdosing does not produce any detrimental effects.	5. Frequent, expensive maintenance of apparatus is necessary to ensure stable energy application and relatively uniform density throughout the effective irradiation area.
6. Treatment efficiency is not readily determinable (lack of rapid field test).	

cidal effectiveness. Radiation effectiveness is a direct function of the quantity of energy, or dosage, which was absorbed by the organism. This dosage is described by the product of the rate at which the energy is delivered (the *intensity*) with the time for which the organism is exposed to this intensity.[2,4,5,18]

The ideal inactivation of bacteria by UV radiation can be approximated by the first order expression (Eq. (12.21)).[4,18]

$$\ln \frac{N_t}{N_0} = -kIt \qquad (12.21)$$

where

N_t =bacterial density remaining after exposure to UV and often after photorepair, number of organisms per unit volume of water (organisms/cm^3 or organisms/100 mL)

N_0 = initial bacterial density before exposure to UV, number of organisms per unit volume of water (organisms/cm^3 or organisms/100 mL)

k = inactivation rate constant, exposure area of water per unit energy input per second or minute (cm^2/μW·s or cm^2/μW·min)

I= intensity of UV energy input, energy input per unit exposure area of water, μW/cm^2

t = exposure time, s or min

This equation represents idealized conditions. In practice, however, deviation from the ideal model can occur because of many factors that affect the efficiency of inactivation of bacteria by UV radiation. The procedure for determining the intensity of UV energy input from reactor UV density and other coefficient for different lamp arrangements is provided in References 18, 45, and 46. Readers may refer to Reference 18 for information about the application of a series of empirical equations and a detailed design example on UV disinfection. Many factors and considerations that affect design and performance of a UV disinfection facility are also discussed in References 2, 3, 18, and 46.

12.3.7 Advanced Oxidation Processes

Recent studies have shown that ozone, when used in combination with UV radiation or hydrogen peroxide, can adequately disinfect and, at the same time, oxidize many refractory organic compounds (such as halogenated organics) present in raw water. The advanced oxidation processes (AOP) involve (1) combining ozone with UV radiation, (2) combining ozone with hydrogen peroxide, or (3) simply conducting ozonation at elevated pH (between 8 and 10). Under these conditions, hydroxyl free radicals that have high oxidation potential (Table 12-1) are produced. Many taste- and odor-causing compounds and priority compounds such as trichloroethylene (TCE) and tetrachloroethylene (PEC), which normally are stable under direct reaction with the ozone molecule, can be rapidly oxidized by hydroxyl free radical.[47,48]

A number of AOP utilize electrotechnology that also destroy microorganisms as well as chlorinated hydrocarbons. Among these are electron-beam disinfection and photocatalytic oxidation. These are emerging technologies and will be used in disinfection, taste and odor control, and destruction of priority compounds.[49–51]

12.4 Secondary Disinfection Technologies

The secondary disinfectants provide an essential residual that prevents regrowth in the distribution system. Chlorine, chloramines, and chlorine dioxide are typically used as secondary disinfectants. Among these, however, chloramine is the most commonly selected secondary disinfectant. It is ineffective as a viricide and is only marginally effective against *Giardia* cysts. The most favorable properties of chloramine are its low reactivity, persistence, lack of THM formation, and the easy conversion of free chlorine into chloramine residual by addition of ammonia. As secondary disinfectants, chlorine, chloramine, and chlorine dioxide are handled in the same way as they are as primary disinfectants. These discussions may be found in Sections 12.3.1 through 12.3.3.

12.5 Fluoridation

Fluoride occurs naturally and can enter raw water sources by dissolution of fluoride-containing sedimentary or igneous rocks. It is seldom found in appreciable quantities in surface waters and appears in groundwater in only a few geographical regions. Fluoride in large quantities is toxic to humans and other animals, but small concentrations can be beneficial. Concentrations of approximately 1 mg/L in drinking water help to prevent dental cavities in children. During formation of permanent teeth, fluoride combines chemically with tooth enamel, resulting in harder and stronger teeth that are resistant to decay. As a result, the decayed, missing, and filled (DMF) tooth rates may thus be lowered. However, larger concentrations of fluoride (exceeding 2 mg/L) may cause discoloration (mottling) of teeth. Excessive dosages of fluoride (exceeding 5 mg/L in drinking waters) may cause brittle teeth, fluorosis, or skeletal abnormalities.[52]

Many water utilities use fluoridation to maintain fluoride concentration around optimum levels. On the assumption that people drink more water in a warmer climate, fluoride levels in water supplies are often adjusted in accordance with ambient temperatures.[15] Mean maximum temperature and corresponding recommended fluoride concentrations that have been used as a guide are given in Table 12-19.

Table 12-19 Mean Maximum Temperatures and Recommended Optimum Fluoride Concentrations

Mean Maximum Temperature, °C	Recommended Optimum Fluoride Concentration, mg/L
10.0–12.1	1.2
12.2–14.6	1.1
14.7–17.7	1.0
17.8–21.4	0.9
21.5–26.2	0.8
26.3–32.5	0.7

Note: $°F = °C \times \frac{9}{5} + 32$

12.5.1 Fluoridation Chemicals and Storage

Several compounds are currently used in the fluoridation of public water supplies: calcium fluoride, sodium fluoride, sodium silicofluoride, ammonium silicofluoride, and hydrofluosilicic acid. The characteristics of these compounds are given in Table 12-20.[3,17,22,52,53]

Dry Chemicals Dry forms of fluoridation chemicals are stored indoors to protect from moisture. Sodium fluoride dust is corrosive; therefore, dust-collection equipment should be included in facility design if conventional dry-feed equipment is provided. The storage area should be well ventilated. The storage area should have a capacity of 1.2 to 1.5 times the normal shipment of the chemical. In larger facilities, steel silos should be used. A conveyance system will be required to transfer the chemical from the bulk-storage silos to feed equipment.

Liquid Chemical Drums containing a solution of hydrofluosilicic acid should be stored in an enclosed well-ventilated area. For larger quantities, a bulk-storage tank should be used. Steel tanks lined with rubber (neoprene) or a suitable thermoplastic (such as cross-linked polyethylene) should be designed for bulk storage. The capacity should be sufficient to hold a full shipment load or at least one to two months' supply. An alternative is to provide two smaller bulk-storage tanks. The following items warrant consideration[53]: (1) Provision must be made for unloading tank, truck, or rail car; (2) A service water line should be provided for washing down leakage or spills; (3) Above ground storage should have a containment structure to hold the

Table 12-20 Characteristics of Fluoridation Compounds

Characteristics	Calcium Fluoride (CaF_2)	Sodium Fluoride (NaF)	Sodium Silicofluoride (Na_2SiF_6)	Ammonium Silicofluoride ($(NH_4)_2 SiF_6$)	Hydrofluosilicic Acid (H_2SiF_6)
Common or Trade Name	Fluorspar	Fluoride	Sodium Silicofluoride	Ammonium Silicate	Fluosilicic Acid
Form	Powder	Powder or crystal	Powder	Crystal	Liquid
Molecular Weight	78.08	42.00	188.05	178.14	144.08
Fluoride ion (100% pure material), %	48.84	45.3	60.7	63.9	79.2
Commercial purity, %	85–98	90–98	98.5	98	22–30
Weight, kg/m^3	1618	1041–1442	881–1153	1298	1258 (liquid)
Solubility, g/1000 g H_2O at 25°C	0.0016	4.05	0.762	18.5 (17.5 °C)	High
pH of saturated solution	6.7	7.6	3.5	3.5	1.2 (1% Solution)
Type of shipping container	Bag, drum, barrels, hopper car, truck	Bag, barrel, drum	Bag, barrel, drum	Bag	Rubber-lined drum, truck, railroad tank car
Material of shipping container	Steel, iron, lead	Steel, iron, lead	Steel, iron, lead	Steel, iron, lead	Rubber-lined steel, PVC

Source: Adapted in part from References 3, 17, 22, 52, and 53.

entire storage capacity in the event of tank or piping rupture; (4) The concrete containment structure must have a protective coating or an acid-resistant lining; (5) A below ground storage facility should have a leak-detection system to comply with current regulations concerning underground storage tanks; (6) All piping and fittings should be of suitable thermoplastic.

12.5.2 Fluoride-Feed Equipment

The fluoridation chemicals are added to the water in dry form, or as a solution. A fluoride solution can be prepared from dry chemicals prior to application. Dry chemicals can be dissolved with varying degrees of difficulty, depending upon the solubility of the compound. Ammonium silicofluoride and sodium fluoride are the most readily soluble chemicals available in dry form. The feeders are either dry-feed or solution-feed, although dry feeders also utilize a solution tank to prepare a solution for delivery at the application points.

Dry Feed A dry-feed system usually consists of (1) bulk-chemical transfer equipment (screw, belt, or pneumatic conveyors), (2) a dry feeder, (3) a one-day supply solution tank with stirrer, and (4) a dust collection system. Dry feeders are divided into two types: volumetric, and gravimetric. Volumetric feeders deliver a specific volume of the chemical (cubic centimeter, cubic meter, cubic foot) in a specified time. Gravimetric feeders deliver a certain weight (gram, kilogram, lb) in a specified time. All components of dry feeders are constructed of iron or steel. Operating features of volumetric and dry feeders are compared here.

- Volumetric feeders are used to deliver smaller quantities than dry feeders.[17,21,52]
- Accuracy of volumetric feeders is in the range of from 3 to 5 percent by weight, as compared to one percent for gravimetric feeders.
- Volumetric feeders are simpler and less expensive to install.
- Gravimetric feeders are readily adapted to automatic control.
- Gravimetric feeders are generally provided at large water treatment facilities.

Typical dry-fluoride feed system is illustrated in Figure 11-11.

Solution Feed A solution-feed system usually consists of the following components: (1) bulk storage and transfer pump, (2) saturator or day tank, and (3) solution feeder. The saturator or day tank is made of chemically-resistant thermoplastic and should be sized for approximately a one-day supply of fluoride solution. The solution can be saturated or of any desired strength. A corrosion-resistant mixer, totally enclosed and fan-cooled, is necessary. Solution feeders are devices for measuring and delivering a specific volume of liquid during a given time interval. General types of solution feeders include diaphragm pumps, piston pumps, rotary-displacement pumps, and rotating cups. Diaphragm pumps are the most widely used. The pump manufacturer should utilize the type of diaphragm material suitable for any particular fluoridation compound and chemical strength.[17,21,52] The pump must provide the required discharge pressures and be protected against back pressures. Typical rotating dipper-type liquid feeder is illustrated in Figure 11-13.

12.5.3 Fluoride Application Points

The fluoride application point should be located downstream of the coagulation and filtering processes. Fluoride losses due to precipitation or adsorption onto floc can be as high as 30 percent at alum doses of 100 ppm.[17,52] A recommended application point is in the filtered effluent header, prior to the clearwell. Whenever possible, the application point should be located where all of the water can be fluoridated with one feed system. Multiple application points will require multiple feed systems to control the dosage properly.[17]

The application point should be designed to prevent the flow of potable water back into the feed system. An air gap between the fluoride-feed piping and the receiving water is the most dependable back-flow protection system. Common types of back-flow prevention devices are vacuum breakers, and spring-loaded check valves.[21,52]

12.6 Equipment Manufacturers of Disinfection and Fluoridation Systems

Disinfection and fluoridation chemicals are available from local and national suppliers. Numerous manufacturers produce the various elements necessary for an effective disinfection or fluoridation feed system. These include manufacturers of chemical-storage vessels, feed pumps, evaporators, injectors, hoppers, on-site generation systems, instrumentation, and safety equipment. Appendix C provides a list of manufacturers and suppliers that can provide these services.

The design engineer should work closely with the local representatives of these chemical and equipment suppliers. Each system may have special requirements that should be considered during the preliminary design stages of the project. System selection should also be based on an understanding of current state and federal design criteria and of future anticipated restrictions regarding DBPs.

As always, system performance at other installations, combined with effective pilot or bench-scale studies, should be used to help direct design decisions.

12.7 Information Checklist for Design of Disinfection and Fluoridation Systems

The following information is necessary for designing disinfection and fluoridation treatment facilities.

1. Information to be obtained from the predesign report and treatability studies.

    a. design flow-peak, average, and minimum;
    b. the water quality parameters essential for disinfection systems selected: turbidity, pH, alkalinity, temperature, background coliform levels, algae, organic compounds, SOCs and VOCs (background fluoride levels should be ascertained before selection of ozone and fluoride systems);
    c. required chemical doses and optimum water quality parameters-some pretreatment studies may be necessary prior to designing disinfection facility;
    d. location of chemical application; this should also consider hydraulic-mixing and contact-time requirements.

2. Information to be provided by the plant owner and operator:

    a. preferences with regard to chemicals, systems, and manufacturers;
    b. preferences with regard to dosing locations and operation and maintenance considerations;

c. expectations regarding flexibility of design to comply with future treatment regulations, expectations regarding complexity of the system.

3. Information to be developed by the design engineer:

a. design criteria established by the regulatory authorities;
b. manufacturers' catalogs and design guides;
c. types, sizes, and limitations of available equipment.

12.8 Design Example

12.8.1 Design Criteria Used

The following design criteria shall be used for the design of the chlorination facility in the Design Example.

1. Flow rates

a. Maximum day flow = 113,500 m^3/d (30 mgd)
b. Average day flow = 57,900 m^3/d (15.3 mgd)

2. Design parameters and criteria

a. Chlorine and chloramines will be used as the primary disinfectant to satisfy the *CT* requirement of SWTR. Chloramines will be used as secondary disinfectant.
b. Sufficient chlorine dose will be applied to maintain an average free chlorine residual of 2 mg/L in two contact channels in front of clearwells. Free chlorine residual will be maintained in the channels. Chloramine residual will be maintained in the clearwells. pH meters and chlorine residual meters will monitor the pH and chlorine residuals in the chlorine contact channels and in the clearwells, respectively.
c. The anhydrous ammonia (stored in a bulk-storage tank) will be applied downstream of the effluent weir structure of the contact channel. A chlorine (Cl_2) to ammonia (NH_3) weight ratio of 5:1 will be used. The raw water does not have any appreciable naturally occurring ammonia. A chloramine residual of 2 mg/L will be maintained through the clearwell at maximum design flow of 113,500 m^3/d.
d. The detention time in the chlorine contact channel at maximum design flow of 113,500 m^3/d will be used to calculate T_{10} for the channel that will have superior baffling.
e. The clearwell is unbaffled; therefore, T_{10} in the clearwell at maximum design flow will be estimated from the theoretical detention time of an unbaffled basin.

f. The chlorination feed equipment shall be capable of delivering 5 mg/L chlorine at maximum design flow. The ammonia feed equipment shall be able to deliver ammonia in the weight ratio of 1:5.

g. A sodium fluoride solution prepared from powder or crystal will be used for fluoridation of water. The fluoride application point will be downstream of the effluent structure of the chlorine contact channel.

h. Fluoridation equipment shall be designed to deliver a maximum of 1.2 mg/L fluoride concentration at maximum design flow.

3. Unit arrangement and layout

a. Unit Arrangement:

The filtered water from all filter units reaches a filter common bay with two back pressure weirs. This arrangement is shown in Figure 10-18. Chlorine is applied in the 122-cm diameter filter effluent pipe ahead of the filter back pressure weir common bay. Free fall over the filter back pressure weir provides mixing of chlorine. The flow then divides into two chlorine contact channels in which desired free chlorine residual and contact time are maintained to meet the *CT* requirement. The effluent weir at the far end of each chlorine contact channel is provided to maintain desired water depth. A 12.5 m extra length of the influent channel of the clearwell is provided to feed ammonia, sodium fluoride, and sodium carbonate solutions through different diffuser systems. The free chlorine will be converted into chloramine within this channel. The addition of sodium carbonate is required for water stability control. (See Chapter 13.) The chlorinated, fluoridated, and stability-controlled water will flow through a 90 cm ∞ 90 cm isolation gate into the clearwell. The desired chloramine residual is maintained for a sufficient detention time in the clearwells. Finally, the finished quality water is delivered by five high-service pumps to the customers through the distribution system. (See Chapter 13.)

b. Unit layout:

Two chlorine contact channels are provided. One channel is located on each side of the filter back pressure weir common bay and in front of one of the two clearwells. The dimensions of each chlorine contact channel are 30-m long, 4.5-m wide, and 9.5-m deep (approximate water depth at maximum day flow). The influent channel of the clearwell is 12.5 m long, with the same cross section as the chlorine contact channel. The dimensions of each clearwell are 40 m ∞ 40 m ∞ 9 m deep (maximum water depth). The clearwell dimensions are computed in Chapter 13 (Section 13.8.2, Step B.3). A common wall is constructed between the chlorine contact channel and the clearwell. The high-service pump station is located in the middle between two clearwells. A total of six pumps are installed, with five pumps in service and one stand-by under the maximum day flow condition. The design of the high-service pump station is provided in Chapter 13. A design example for a raw water pump station design is also included in Chapter 7. Readers may refer to Sections 7.8 and 13.8 for detailed procedures and calculations. Three filter backwash water pumps are also installed in the same pump station and located at the end near the filter back pressure weir common bay. The backwash water system design is provided in Section 10.13.2, Step B. The general layout and basic dimensions of the facilities described above are shown in Figures 12-20 and 12-21.

12.8.2 Design Calculations

The step-by-step design calculations of the Design Example are as follows:

Step A: Chlorine Contact Channel and Clearwell Design The disinfection is achieved in the chlorine contact channel by free chlorine residual and in the clearwell by combined chlorine residual. The detention time in the influent channel of the clearwell can not be used to calculate the T_{10} for free chlorine disinfection, because ammonia is added in this section. For conservative design, this detention time is not used to calculate the T_{10} for combined chlorine disinfection, because it is relatively short compared to the detention time in the clearwell.

1. Calculate T_{10} for free chlorine disinfection in the chlorine contact channel.

$$\text{Volume of each chlorine contact channel} = 30\ \text{m} \times 4.5\ \text{m} \times 9\ \text{m} = 1215\ \text{m}^3$$

$$\text{Flow in each chlorine contact channel} = \frac{113{,}500\ \text{m}^3/\text{d}}{2} = 56{,}750\ \text{m}^3/\text{d}$$

$$\text{Theoretical detention time } T = \frac{1215\ \text{m}^3 \times 1440\ \text{min/d}}{56{,}750\ \text{m}^3/\text{d}} = 30.8\ \text{min}$$

For superior baffling in the chlorination channel, the ratio of T_{10}/T is assumed to be 0.7 (Table 12-4).

$$T_{10} = 0.7 \times 30.8\ \text{min} = 21.6\ \text{min}$$

Use a conservative value of 20 min for *CT* calculation.

2. Calculate T_{10} for combined chlorine disinfection in the clearwell.

$$\text{Volume of each clearwell} = 40\ \text{m} \times 40\ \text{m} \times 9\ \text{m} = 14400\ \text{m}^3$$

$$\text{Theoretical detention time } T = \frac{14400\ \text{m}^3 \times 1440\ \text{min/d}}{56{,}750\ \text{m}^3/\text{d}} = 365\ \text{min}$$

For unbaffled clearwell, the ratio of T_{10}/T is assumed to be 0.1 (Table 12-4).

$$T_{10} = 0.1 \times 365\ \text{min} = 36.5\ \text{min}$$

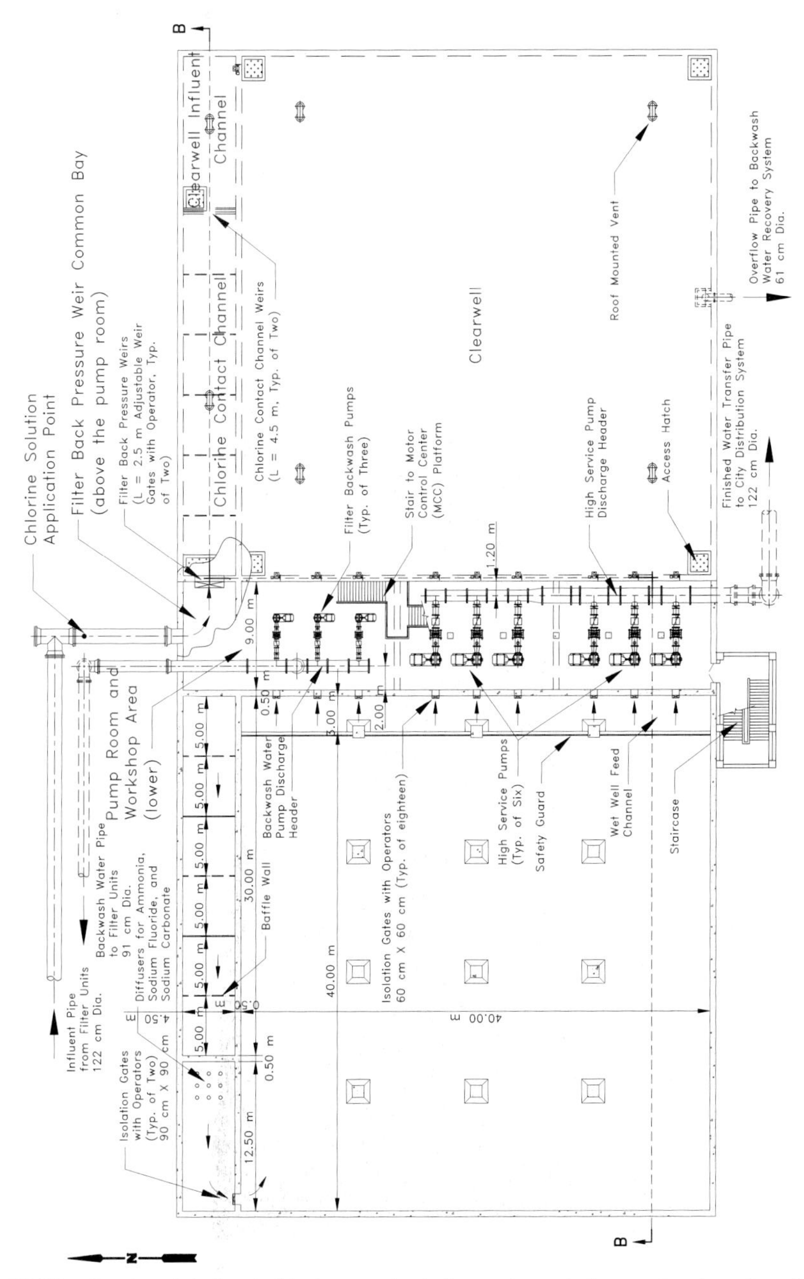

Figure 12-20 Design details of chlorine contact channels and clearwells-plan view A-A. (Section A-A location is shown in Figure 12-21.)

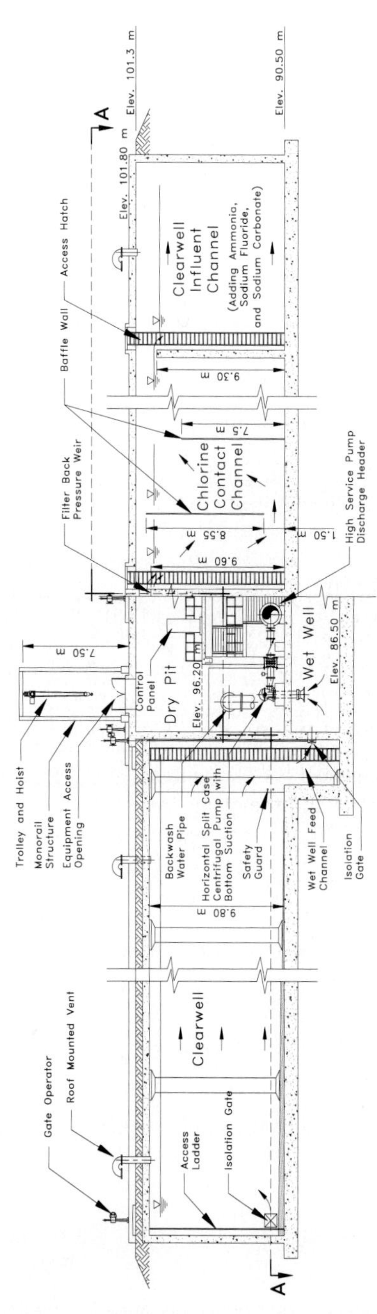

Figure 12-21 Design details of chlorine contact channels and clearwells-section B-B. (Section B-B location is shown in Figure 12-20.)

3. Check CT requirement

   The minimum disinfectant contact time (CT value) requirements are achieved by providing the free chlorine contact in the channel and the chloramine contact in the clearwell. The winter condition will be critical. The conventional filter water treatment plant will provide 2.5-log removal of *Giardia lamblia* cysts and 2-log removal of enteric viruses. Therefore, primary disinfectants must provide 0.5-log and 2-log deactivation of *Giardia lamblia* cysts and viruses. The CT calculations are summarized in Table 12-21.

   It can be noted from Table 12-21 that the values of inactivation ratio ($\Sigma CT_{Cal}/CT_{Tab}$) exceed 1.0. This means that 0.5-log inactivation of *Giardia lamblia* cysts and 2-log inactivation of viruses through disinfection is achieved by primary and secondary disinfectants. With 2.5-log removal credit for *Giardia lamblia* cysts and 2-log credit for viruses available through the conventional filtration plant, the total CT values are calculated as follows:

   Total CT available for *Giardia* = 0.5-log + 2.5-log
   = 3.0 log (99.9% Giardia inactivation)

   Total CT available for virus = 2.0 log + 2.0-log
   = 4.0-log (99.9% virus inactivation)

4. Calculate the hydraulic profile through the chlorine contact channel and the clearwell.

   In this section, the hydraulic calculations are performed to develop the hydraulic profile through the chlorine contact channel and the clearwell. The head losses in these two facilities are the results of (1) the free fall at the filter back pressure weir, (2) the free fall at the effluent weir of the chlorine contact channel, and (3) the minor head loss at the isolation gate between the influent channel and the clearwell. The friction losses through the channels and clearwell are expected to be small and are assumed to be zero.[d] The head loss at the filter back pressure weir common bay is calculated in the hydraulic design of the filter unit in Chapter 10. The head over the weir is 0.28 m, and a free fall of 0.39 m is provided. (See Section 10.13.2, Step A.5.g.) The head losses that occur at other two locations are calculated next.

   a. Calculate the head loss at the effluent weir of the chlorine contact channel.

      The effluent weir is across the 4.5 m width of the chlorine contact channel. The head over the effluent weir of the chlorine contact channel is calculated from Eq. (8.24) by using

      $Q = \dfrac{56{,}750\ \text{m}^3/\text{d}}{2} = 0.657\ \text{m}^3/\text{s}$, $(C_d) = 0.60$, and no end contractions ($n =$ 0 and $L = 4.5$ m).

d. The reader may refer to Section 7.8.2, Step D.2 and Table 7-9 for detailed equations and procedures to calculate the friction head loss through an open channel.

Table 12-21 *CT* Calculations for the Design Example (Winter Condition)

Condition	Value
pH	8.0
Temperature, °C	5
T_{10} for free chlorine, residual, min (conservative value)	20.0
Free chlorine residual concentration, mg/L	2.0
T_{10} for combined chlorine residual, min	36.5
Combined chlorine residual concentration, mg/L	2.0
0.5-log inactivation of *Giardia lamblia* cysts	
CT_{Cal} for free chlorine residual	40
CT_{Tab} for free chlorine residual (Table 12-5)	36
CT_{Cal} for combined chlorine residual	73
CT_{Tab} for free combined chlorine residual (Table 12-7)	365
CT_{Cal}/CT_{Tab} for free chlorine residual	1.1
CT_{Cal}/CT_{Tab} for combined chlorine residual	0.2
$\mathbf{\Sigma CT_{Cal}/CT_{Tab}}$	**1.3**
2.0-log inactivation of viruses	
CT_{Cal} for free chlorine residual	40
CT_{Tab} for free chlorine residual (Table 12-6)	4
CT_{Cal} for combined chlorine residual (Table 12-7)	73
CT_{Tab} for combined chlorine residual	857
CT_{Cal}/CT_{Tab} for free chlorine residual	10.0
CT_{Cal}/CT_{Tab} for combined chlorine residual	0.1
$\mathbf{\Sigma CT_{Cal}/CT_{Tab}}$	**10.1**

$$H = \left(\frac{3 \times 0.657 \text{ m}^3/\text{s}}{2 \times 0.6 \times 4.5 \text{ m} \times \sqrt{(2 \times 9.81 \text{ m/s}^2)}}\right)^{2/3} = 0.19 \text{ m}$$

A free fall of 0.40 m is provided at the weir. The water level in the clearwell will fluctuate, but 0.40 m is the minimum free fall expected in the clearwell. This free fall is considered sufficient to provide energy for mixing of ammonia, sodium fluoride, and sodium carbonate solutions.

b. Calculate the head loss through the isolation gate.

The dimensions of the gate are 90 cm ∞ 90 cm. The velocity through the gate is calculated as follows:

$$v = \frac{0.657 \text{ m}^3/\text{s}}{0.9 \text{ m} \times 0.9 \text{ m}} = 0.81 \text{ m/s}$$

The head loss through the gate is calculated from Eq. (6.4) by using $C_d = 0.60$:

$$h_L = \frac{1}{2 \times 9.81 \text{ m/s}^2}\left(\frac{0.81 \text{ m/s}}{0.6}\right)^2 = 0.09 \text{ m}$$

c. Calculate the total head losses through the chlorine contact channel and the clearwell.

• Free fall provided at the effluent weir	0.40 m
• Head loss through the isolation gate	0.09 m
Total =	0.49 m

d. Prepare the hydraulic profile through the chlorine contact channel and the clearwell.

The water surface elevation (WSEL) of 99.99 m in the chlorine contact channel is calculated in Chapter 10. (See Section 10.13.2, Step A.5.i.) Starting from this elevation, the water surface elevations and major design elevations are calculated below. The hydraulic profile is then prepared and is shown in Figure 12-22.

The WSEL in the influent channel of the clearwell = WSEL in the chlorine contact channel – free fall provided at the weir = 99.99 m – 0.40 m = 99.59 m. The top elevation of the weir = WSEL in the chlorine contact channel – head over the weir = 99.99 m – 0.19 m = 99.80 m.

The maximum WSEL in the clearwell = WSEL in the influent channel of the clearwell – head loss through the isolation gate = 99.59 m – 0.09 m = 99.50 m. The bottom of the clearwell = WSEL in the clearwell – design water depth in the clearwell = 99.50 m – 9.00 m = 90.50 m. Provide the same bottom elevation for the chlorine contact channel; then the minimum water depth in the chlorine contact channel = top elevation of the weir – bottom of the chlorine contact channel = 99.80 m – 90.50 m = 9.30 m. This minimum water depth is greater than the required water depth of 9 m to achieve sufficient *CT* for disinfection.

Step B: Chlorine Storage and Feed System Design

1. Chlorine storage

a. Calculate average chlorine required per day.

Assume average chlorine dosage is 3 mg/L; thus,

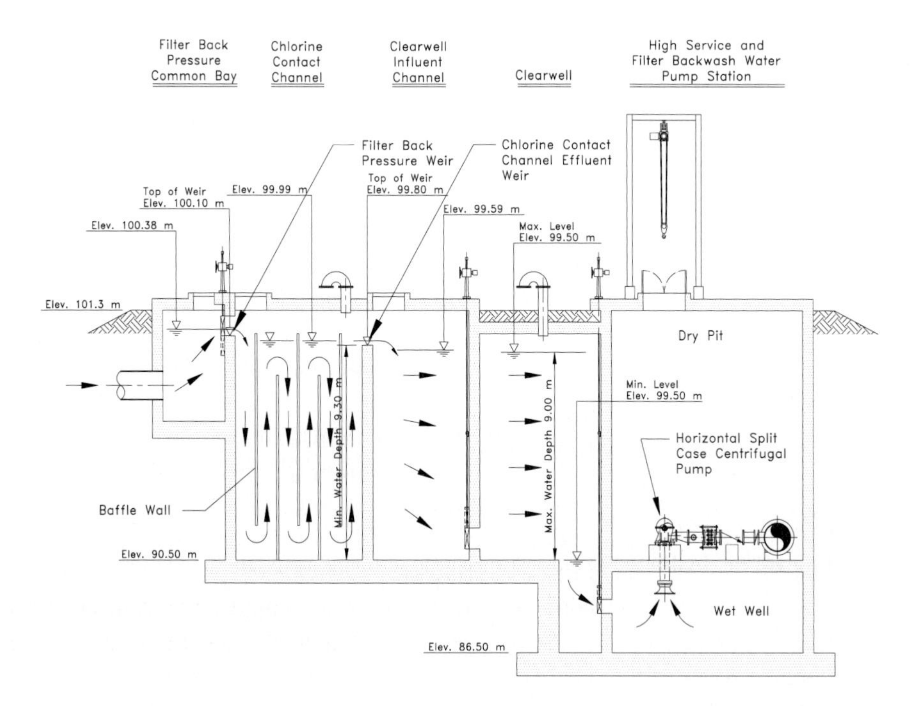

Figure 12-22 Hydraulic profile through the chlorine contact channel and the clearwell.

Average chlorine required

$$= 3 \text{ mg/L} \times 113{,}500 \text{ m}^3\text{/d} \times 10^{-6} \text{mg/kg} \times 10^3 \text{ L/m}^3$$
$$= 340.5 \text{ kg/d}$$

b. Calculate chlorine storage for 45 days.

$$\text{Chlorine storage} = 340.5 \text{ kg/d} \times 45 \text{ d} = 15322.5 \text{ kg}$$

c. Calculate the number of 907-kg (1-ton) containers required.

$$\text{Number of containers required} = \frac{15322.5 \text{ kg}}{907 \text{ kg/container}} \approx 17 \text{ containers}$$

Provide a chlorine storage facility for 17 ton-containers.

2. Chlorine withdrawal

a. Calculate the maximum chlorine withdrawal per day.

Maximum chlorine withdrawal rate

$$= 5 \text{ mg/L} \times 113{,}500 \text{ m}^3\text{/d} \times 10^{-6} \text{ mg/kg} \times 10^3 \text{ L/m}^3 = 567.5 \text{ kg/d}$$

b. Calculate the number of chlorinators required.

Select 450 kg/d chlorinators.

$$\text{Number of chlorinators} = \frac{567.5 \text{ kg/d}}{450 \text{ kg/d}} = 1.3 \text{ chlorinators}$$

Provide two chlorinators: one will be used as a standby unit under average demand.

c. Compute the number of chlorine containers manifolded.

Assume that the maximum chlorine withdrawal rate is 400 kg/d per container.

$$\text{Number of chlorine containers manifolded} = \frac{567.5 \text{ kg/d}}{400 \text{ kg/d}} = 1.4 \text{ containers}$$

Two banks of containers are provided. Two containers are manifolded in each bank. An automatic switchover device will automatically change operation from an empty bank to a full bank of containers. The maximum chlorine withdrawal rate per container is within the allowable range, so no evaporator will be needed.[e] Chlorine gas will be withdrawn from the top angle valves of the containers. Vacuum gauges will monitor vacuum condition in the chlorine gas piping between the vacuum regulators and the injectors. A chlorine-leak detector with multiple sensors will monitor the chlorine leakage around the facility. The chlorine storage and feed facility is shown in Figure 12-23.

3. Chlorine feed

The chlorine application point is in the filter effluent pipe, ahead of the filter back pressure weir common bay (Figure 12-20). The injector water is the treatment plant service water. The maximum chlorine feed rate is 567.5 kg/d. The injector water flow of 120 L/min is estimated from Figure 12-9. The diffuser design should consider the following issues: (1) the back pressure in the chlorine piping, (2) the head loss through the piping, (3) the orifice sizing, and (4) the velocity gradient for the desired mixing.[f] The velocity gradient should be greater than 480 per second to provide adequate mixing.

Step C: Ammonia Storage and Feed System Design

1. Ammonia storage.

a. Calculate average ammonia required per day

Provide for chlorine to ammonia weight ratio of 5:1.

e. In case the chlorine container storage is outside and the ambient air temperature drops below 5°C, chlorine evaporators should be provided to insure that the gaseous chlorine is reaching he chlorinator.

f. The velocity gradient is typically greater than 480 per second to provide adequate mixing.

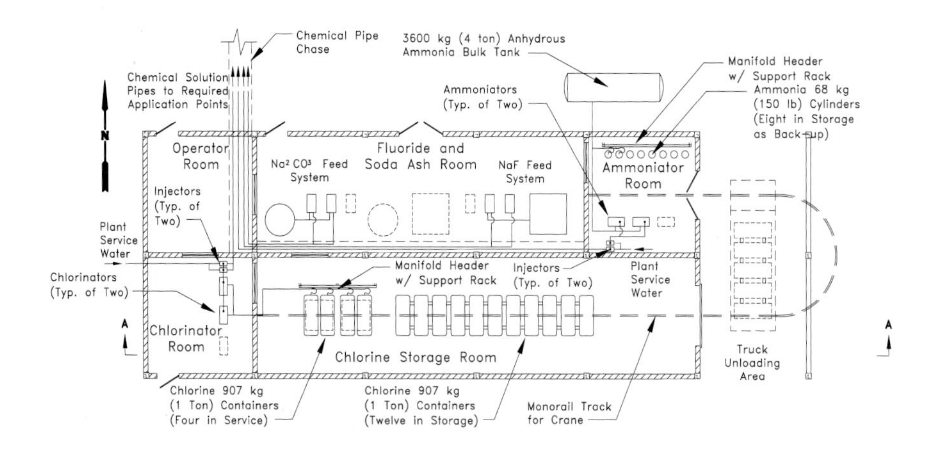

(a)

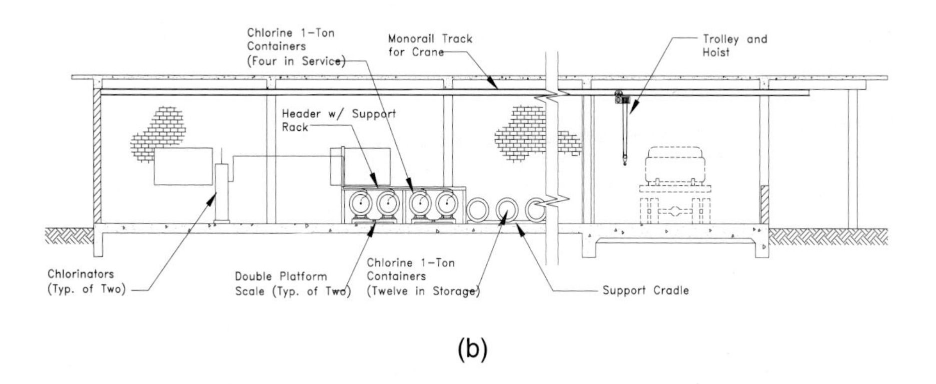

(b)

Figure 12-23 Chlorine and ammonia storage and feed facilities. (a) Plan. (b) Section A-A.

$$\text{Average ammonia required per day} = \frac{340.5 \text{ kg-chlorine/d}}{5 \text{ chlorine/ammonia}}$$

$$= 68.1 \text{ kg/-ammonia (anhydrous)/d}$$

b. Calculate anhydrous ammonia storage for 45 days.

$$\text{Chlorine storage} = 68.1 \text{ kg/d} \times 45 \text{ d} = 3064.5 \text{ kg}$$

Provide one 3600-kg (4-ton) bulk-storage tank. Eight 68-kg (150-lb) cylinders are also maintained, as a five-day backup supply under maximum day flow condition, in case the bulk tank needs to be taken out of service for routine maintenance or emergency repair.

2. Ammonia withdrawal

a. Calculate the maximum ammonia withdrawal per day.

The ammonia withdrawal rate is one-fifth of the maximum chlorine withdrawal rate.

$$\text{Maximum ammonia withdrawal rate} = \frac{567.5 \text{ kg/d}}{5} = 113.5 \text{ kg/d}$$

b. Calculate the number of ammoniators required.

Select a direct feed ammoniator of 100 kg/d capacity.

$$\text{Number of ammoniators} = \frac{113.5 \text{ kg/d}}{100 \text{ kg/d}} = 1.13 \text{ ammoniators}$$

Provide two ammoniators of 100 kg/d capacity. One ammoniator will be a standby unit under average conditions. Because the anhydrous ammonia withdrawal rate is small, no evaporator will be needed.[g] A manifold header is provided, for connection with two cylinders in service if necessary. The anhydrous ammonia storage and feed facility is also shown in Figure 12-23.

3. Ammonia feed

The anhydrous ammonia feed solution from the injectors is applied in the clearwell influent channel (Figure 12-20). A water softener will be needed to provide injector water. The quantity of injector water should be selected to deliver 113.5 kg/d of ammonia at the application point. It is assumed that the injector water flow for ammonia is one-fifth of the flow for chlorine. The injector water flow of 24 L/min is therefore obtained. The feed system design should include injector, booster pump, solution piping, and diffusers. The manufacturer's information should be used to select the equipment for ammonia feed system properly. The G value for ammonia diffusion shall be greater than 300 per sec.

Step D: Fluoride Storage and Feed System Design

1. Fluoride storage

a. Calculate the maximum fluoride required per day.

$$\text{Maximum flouride required} = 1.2 \text{ mg/L} \times 113{,}500 \text{ m}^3\text{/d} \times 10^{-6} \text{ mg/kg} \times 10^3 \text{ L/m}^3 = 136.2 \text{ kg/d}$$

g. In case ammonia evaporators are designed, it should be noted that the heat of evaporation of ammonia is almost five times greater than that of chlorine. If chlorine type evaporators are provided, the higher heat of evaporation of ammonia should be considered in the design. Most designs use same-size evaporators for chlorine and ammonia.

b. Calculate fluoride storage for 45 days.

$$\text{Fluoride storage} = 136.2 \text{ kg/d} \times 45 \text{ d} = 6129 \text{ kg}$$

Quantity of 98% sodium fluoride (NaF)

$$= \frac{6129 \text{ kg} \times 42 \text{ g/mole NaF}}{19 \text{ g/mole F} \times 0.98} = 13825 \text{ kg}$$

The sodium fluoride will be stored in a silo. The volume of the silo for a 45-day supply

$$= \frac{13525 \text{ kg}}{1041 \text{ kg/m}^3} = 13.0 \text{ m}^3 \text{ (Table 12-20).}$$

Provide two silos, each with a 6.5-m^3 capacity. Each silo will be 2.5 m in diameter and 2.6 m deep, with a liberal freeboard allowance. The delivery will be by hopper truck and blowing into the silo. A dust filter (cartridge type) will be provided to prevent the fine particles of NaF from escaping during delivery.

2. Fluoride solution preparation

Provide a dry-feed volumetric feeder. Each silo will be set above a hopper that will discharge directly into a volumetric feeder. Dry chemical will be drawn from the hopper by means of a helical feed screw driven by an induction motor. The feed screw operating range is 20:1, and both feeders will deliver the dry chemical directly into a one-day supply.

a. Calculate maximum NaF required per day.

Maximum quantity of NaF delivered by each feeder

$$= \frac{136.2 \text{ kg/d F} \times 42 \text{ g/mole NaF}}{19 \text{ g/mole F} \times 0.98} = 307 \text{ kg/d}$$

Maximum volume of NaF delivered by each feeder

$$= \frac{307 \text{ kg/d}}{1041 \text{ kg/m}^3} = 0.29 \text{ m}^3\text{/d}$$

b. Calculate volume of NaF solution tank for one-day supply.

Three percent (30 kg/m^3) solution is prepared for one-day supply.

$$\text{Volume of NaF solution tank} = \frac{307 \text{ kg/d} \times 1 \text{ d}}{90 \text{ kg/m}^3} = 10.2 \text{ m}^3$$

Provide a one-day supply tank, 2.5 m ∞ 2.5 m ∞ 2 m (deep). This arrangement will provide the daily need from a single one-day supply tank, while one feeder and silo will serve as standby units. The storage and feed facility for sodium fluoride is also shown in Figure 12-23.

3. Fluoride feed

The fluoride solution will be applied in the influent channel of the clearwell through diffusers (Figure 12-20). This channel is located downstream of the chlorine contact channel. Ammonia and sodium carbonate solution are also added in the same channel.

Maximum daily pumping rate for 3% NaF solution

$$= \frac{10.2\ \text{m}^3/\text{d} \times 1000\ \text{L/m}^3}{1440\ \text{min/d}} = 7.1\ \text{L/min}$$

The fluoride solution will be delivered by the variable flow rate positive-displacement progressive-cavity feed pumps. Provide two pumps, each capable of meeting the maximum requirement. The fluoride level in the clearwell is measured. The in-use pump is then adjusted for the required fluoride feed rate.

12.9 Operation, Maintenance, and Troubleshooting for the Disinfection Facility

Routine maintenance should be scheduled to assure that problems are corrected before unnecessary damage occurs to the equipment. In this way, unplanned chemical and labor costs can be reduced, treatment efficiency maintained, and many safety hazards prevented.

Routine operation and maintenance of the chlorine and ammonia feed systems includes the following:

1. Inspect the chlorinators, ammoniators, evaporators, and storage tanks each day to ensure proper operation. Low gas pressure or no feed may indicate flow restrictions, empty vessels, clogged injectors, or damaged equipment.
2. Inspect the diffusers. Diffusers may become plugged. The ammonia diffuser is especially likely to clog.
3. Monitor the combined and total chlorine residual daily. Excess variations may indicate equipment malfunction.
4. Monitor treated water quality daily. Perform a periodic review of treated water quality. This should include analysis of daily reports, as well as periodical assessment of DBPs and other parameters of concern.
5. Drain contact chambers annually, and repair structures and equipment as needed.
6. Test leak detectors and emergency equipment every six months, and verify operator training in emergency procedures.

12.10 Equipment Specifications

The specifications for the chlorination system designed in this chapter are briefly presented in this section. The purpose of these specifications is to describe more fully the various components of the chlorination system illustrated in the design example. These specifications are not intended to be complete construction specifications. The designer should consult equipment manufacturers for information to assist in the development of detailed specifications. A more detailed discussion of specifying equipment for water treatment plants is provided in Appendix C.

12.10.1 General

The Contractor shall furnish and install a complete and operating chlorination system. The system shall include cabinet-mounted chlorinators, remote chlorine injectors, and all necessary valves and appurtenances for a complete and working chlorination system.

The chlorination system shall be designed for the following operating conditions:

Water Flow Rate, m^3/day	113,500
Temperature, °C	32
pH	7.2

The chlorine gas shall be stored as a liquid under pressure. The chlorine shall be withdrawn from storage as a gas under pressure. This pressurized chlorine gas shall pass through pressure-reducing valves and vacuum regulators. The gas shall then be transported under vacuum through cabinet-mounted chlorinators and on to the various points of chlorine application. The chlorine under vacuum shall be fed to an injector where it is mixed with water to create a chlorine solution.

12.10.2 Chlorinator

The chlorinator shall be of vacuum operated, solution-feed type. Each chlorinator shall include a rotameter or gas-flow meter mounted in a corrosion-resistant cabinet. Flow meter capacities shall accommodate the maximum capacity of the chlorinator. Flow meter shall be manually adjustable and shall be capable of automatically adjusting the amount of chlorine fed through the chlorinator on the basis of an external signal.

12.10.3 Injector Assembly

The injector assembly shall be a chlorine-water type and shall create the vacuum necessary for operation. The injector assembly shall be equipped with a check valve to prevent water from entering the chlorine system and shall be mounted in a remote fiberglass enclosure. The injector shall be capable of feeding the required rate of chlorine gas at the application point operating conditions.

12.10.4 Chlorine System Piping and Valves

Piping for the chlorination system shall be supplied in accordance with the following schedule:

Service	Pipe Material
Pressurized Chlorine Gas	Schedule 80 Seamless Steel
Chlorine Liquid	Schedule 80 Seamless Steel
Chlorine Gas Under Vacuum	Schedule 80 PVC or PVDF
Chlorine Solution	Schedule 80 PVC
Designated Drain Lines	Schedule 80 PVC or PVDF

Valves for pressurized chlorine gas service and for chlorine liquid service shall be designated specifically for chlorine at 300 psi working pressure. Where ball valves are called out for pressurized liquid chlorine and chlorine gas service, ball valves shall have forged carbon-steel body, having a monel ball and stem in conjunction with teflon seats and seals. The valves shall be rated at 2000 psi working pressure. Valves seats shall be of special design and composition to ensure the automatic internal relief of pressure build-up within the ball cavity at the time of valve closure.

Problems and Discussion Topics

12.1 List the factors that influence the chlorine disinfection efficiency.

12.2 A disinfection process achieves 92 percent *Giardia* deactivation. Calculate log removal.

12.3 In a dye-tracer study, a dye is continuously fed into the influent to a basin. The concentration of dye tracer is 12 g/L and injection rate is 1.2 L/min. The influent flow to the basin is 16,000 m^3/d. Calculate the concentration of dye tracer in the flow discharging into the basin.

12.4 In a basin that has a volume of 600 m^3, a slug of dye tracer is injected. The volume and concentration of dye tracer are 3 liter and two percent. Calculate the concentration of the bye tracer in the basin if it is completely mixed.

12.5 A continuous-feed dye-tracer study was conducted in a sedimentation basin to determine T_{10}. The concentration of dye tracer in the effluent was measured every 5 minutes. The results are tabulated below. The background concentration of dye in the influent was 0.2 mg/L. Prepare the plots of C/C_0 versus t, and $\log(1 - C/C_0)$ versus t/T. Determine the value of T_{10}.

Concentration of tracer in the feed = 15,000 mg/L

Tracer feed rate = 1000 mL/min

Influent flow rate = 7500 m^3/d

Volume of the basin = 150 m^3

Results of the study:

t, min	Measured Concentration, *Ct*, mg/L
0	0.2
5	0.2
10	0.2
15	0.3
20	1.0
25	1.4
30	2.4
35	2.6
40	2.8
45	3.0
50	2.8
55	2.6
60	2.8

12.6 A storage basin was tested for T_{10} by using a slug of solution of sodium fluoride. The slug-test data is given below. Convert the slug-dose test data into step-dose data format by plotting C/C_0 versus t. Obtain the value of T_{10}.

t, min	Measured Concentration *F*, mg/L
0	0.1
5	0.1
10	0.1
15	0.8
20	2.5
25	3.8
30	2.2
35	1.5
40	0.6
45	0.4
50	0.3
55	0.3
60	0.2

Flow to the basin = 9000 m^3/d

Volume of the basin = 750 m^3

Concentration of F in NaF solution = 10 percent

Volume of slug pumped in 30 seconds = 15 Liters

12.7 Calculate the volume of the chlorine contact basin and the quantity of chlorine needed in kg/d. The average design flow is 3 m^3/s. T_{10} is 20 minutes and the CT_{10} required for 0.5 -log desired deactivation of *Giardia lamblia* cysts is 40. The pilot plant study shows that the following relationship holds:

$$\log(1 - C/C_0) = -0.723(T_{10}/T) + 0.174$$

12.8 A polluted lake is considered for recreational purposes. A bench scale study was conducted to determine the performance of chlorine at different dosages. The results of the study are provided below. Plot the data, and determine the value of the exponent n and the constant K in the equation $C^n t_p = K$ for two values of residual coliform counts: 200/100mL, and 1000/100 mL. In this equation, C = concentration of disinfectant in mg/L, and t_p = required time in minutes to achieve a certain percentage kill.

Chlorine Dosage, mg/L	Residual Coliform Count, No/100 mL		
	Contact time, min		
	15	30	60
1	10000	2000	500
2	3000	350	90
4	400	65	20
6	110	30	12
8	54	19	6
10	30	10	1

12.9 A perfect plug-flow basin is determined, with a very large length-to-width ratio and with perforated inlet, outlet, and interbasin baffles. If the volume of the channel is 30 m^3 and the influent flow rate is 1 m^3/min, calculate T_{10}. Use the theoretical approach.

12.10 In a conventional water treatment plant, preozonation is used. Free chlorine is maintained in the chlorine contact channel and combined chlorine is maintained in the clearwell. Determine whether 0.5-log *Giardia lamblia* cysts and 2-log viruses deactivation is achieved. The temperature and pH of the water are 20°C and 7.0, respectively. The disinfection information is summarized below:

Ozone concentration = 0.2 mg/L

T_{10} in the ozone contact chamber = 0.5 min

Free chlorine residual concentration = 1 mg/L

T_{10} in the chlorine contact channel = 0.5 min

Combined chlorine residual concentration = 2 mg/L

T_{10} in the clearwell = 27 min

12.11 List seven halogenated by-product categories formed by chlorination.

12.12 List the factors that may cause the formation of DBPs.

12.13 Disinfection by-products are associated with many toxicological effects. Match the terms in two lists:

List A	List B
Chloroform	a. Carcinogenic
2-Chlorophenol	b. Developmental
Chloroacetonitrile	c. Phytotoxic
Dichloroacetonitrile	d. Genotoxic
Bromoform	e. Hepatotoxic
	f. Metagenic
	g. Renal Toxic
	h. Tumor Promotor

12.14 Calculate the number of chlorinators, the number of chlorine ton containers attached to the header, and the number of containers required for a 4-week chlorine supply. Use the following data:

Maximum day flow = 2 m^3/s

Average day flow = 0.67 m^3/s

Maximum chlorine feed rate = 5 mg/L

Rated capacity of chlorinator = 450 kg/d

Gas withdraw rate per container = 180 kg/d

12.15 The residual chlorine and chlorination data are given below. Plot the chlorination curve. Obtain the break point chlorination dosage. What will be the initial demand and total chlorine demand (kg/d) to give a free chlorine residual of 1.2 mg/L? The flow is 180 L/s.

Chlorine dosage, mg/L	Chlorine Residual, mg/L
1	0
2	0.8
3	1.4
4	1.0
5	1.1
6	1.6
7	2.1

12.16 The reduction of organism in a chlorination process is expressed by the following equation: $\ln(N/N_0) = -kt$, where, N_0 and N are number of organisms present initially and after time t, and k is die-off coefficient. Using the midpoint chlorine dosage of 3 mg/L and 130 coliform organisms per 100 mL remaining in the raw water after 30 minutes of contact time, calculate k. $N_0 = 10^6$ coliform per 100 mL. Also, calculate the number of coliform organisms remaining after 20 minutes of contact time.

12.17 Calculate the volume of a chlorine contact basin and the quantity of chlorine needed in kilograms per day. The average design flow is 0.2 m^3/s, contact time is 18 minutes, total chlorine demand is 6 mg/L, and the chlorine residual maintained is 1.5 mg/L.

12.18 Define the following terms: free chlorine residual, combined chlorine residual, total chlorine residual, pasteurization, and disinfection.

12.19 The laboratory data for a chlorination study are given below. Using the equation in Problem 12-16, calculate the contact period to reduce the coliform count from 10^5 organisms per 100 mL to 50 organisms per 100 mL after disinfection. The chlorine dosage is the same as that used in the laboratory study.

Contact Time at Chlorine Dosage of 10 mg/L, min	Number of Coliform Organisms Remaining per 100 mL
0	10^5
5	5000
10	250
15	12

12.20 Write about the advantages and disadvantages of ozonation versus chlorination.

12.21 What is the significance of breakpoint chlorination in water disinfection? Why is ammonia added before water is pumped into the distribution system?

References

1. Prudden, T. M. *Drinking Water and Ice Supplies, and Their Relations to Health and Disease*, G. P. Putnam's Sons, The Knickerbocker Press, New York, 1891.
2. USEPA. *Technologies for Upgrading Existing or Designing New Drinking Water Treatment Facilities*, Office of Drinking Water, EPA/625/4-89/023, USEPA, Cincinnati, OH, March 1990.
3. White, C. G. *Handbook of Chlorination and Alternative Disinfectants*, 4th ed., John Wiley & Sons, Inc., New York, 1999.
4. Task Force on Wastewater Disinfection. *Wastewater Disinfection*, Manual of Practice FD-10, Facilities Development, Water Pollution Control Federation, Alexandria, VA, 1986.
5. WPCF Disinfection Committee. *Wastewater Disinfection: A State-of-the-Art Report*, Water Pollution Control Federation, Alexandria, VA, 1984.
6. Malcolm Pirnie, Inc. and HDR Engineering, Inc. *Guidance Manual for Compliance with Filtration and Disinfection Requirements for Public Water Supply Systems Using Surface Water Sources*, ISBN 0-89867-558-8, Science and Technology Branch, USEPA, Washington, D.C., October 1990.
7. Huben, H.V. *Surface Water Treatment: The New Rules*, AWWA, Denver, CO, 1991.
8. Metcalf and Eddy, Inc. *Wastewater Engineering: Treatment, Disposal and Reuse*, 3d ed., McGraw-Hill Book Co., Inc., New York, 1991.
9. Utah Division of Drinking Water. *Cryptosporidium*, Utah Guidance Document, Revised February 2, 1995.
10. USEPA. "National Primary Drinking Water Regulations: Interim Enhanced Surface Water Treatment; Final Rule," *Federal Register*, vol. 63, no. 241, pp. 69478–69521, December 16, 1998.
11. Regle, S. "Intent of the D/DBP Rule and the Regulatory Framework," *AWWA Satellite Teleconference on Disinfectants and Disinfection By-Products: Understanding the Proposed D/DBP Rule*, AWWA, 1993.
12. USEPA. "National Primary Drinking Water Regulations: Disinfectants and Disinfection Byproducts, Final Rule," *Federal Register*, vol. 63, no. 241, pp. 69390–69476, December 16, 1998.
13. McGuire, M. J. "Content of the D/DBP and Related Rules," *AWWA Satellite Teleconference on Disinfectants and Disinfection By-Products: Understanding the Proposed D/DBP Rule*, AWWA, 1993.
14. Cromwell, J. E. "National Impact of the D/DBP Rule," *AWWA Satellite Teleconference on Disinfectants and Disinfection By-Products: Understanding the Proposed D/DBP Rule*, AWWA, 1993.
15. Means, E. G. "Utility Impact of the D/DBP and Related Rules," *AWWA Satellite Teleconference on Disinfectants and Disinfection By-Products: Understanding the Proposed D/DBP Rule*, AWWA, 1993.
16. Pontius, F. W. "Disinfection By-Products - A Regulatory Balancing Act," *Opflow*, vol. 19, no. 12, American Water Works Association, December 1993.
17. AWWA. *Water Quality and Treatment*, 4th ed., McGraw-Hill Publishing Co., Inc., New York, 1990.
18. Qasim, S. R. *Wastewater Treatment Plants: Planning, Design, and Operation*, 2d ed., Technomic Publishing Co., Inc., Lancaster, PA 1999.
19. ASCE and AWWA. *Water Treatment Plant Design*, 2d ed., McGraw-Hill Publishing Co., Inc., New York, 1990.
20. Chlorine Institute, Inc. *The Chlorine Manual*, Pamphlet No. 1, 6th ed., Chlorine Institute, Inc., Washington, D.C., 1997.
21. Chlorine Institute, Inc. *Non-refrigerated Liquid Chlorine Storage*, Pamphlet No. 5, 5th ed., Chlorine Institute, Inc., Washington, D.C., October 1992.
22. James M. Montgomery, Inc. *Water Treatment Principles and Design*, John Wiley & Sons, New York, 1985.
23. The Chlorine Institute, Inc. *Recommended Practices for Handling Chlorine Tank Cars*, Pamphlet No. 66, 2th ed., the Chlorine Institute, Inc., Washington, D.C., November 1994.
24. USEPA. "List of Regulated Substances and Thresholds for Accidental Release Prevention and Risk Management Programs for Chemical Release Prevention; Final Rule and Notice," *Federal Register*, vol. 59, no. 20, pp. 4493–4499, January 31, 1994.

25. USEPA. "Accidental Release Prevention Requirements: Risk Management Programs Under the Clean Air Act, Section 112(r)(7); List of Regulated Substances and Thresholds for Accidental Release Prevention, Stay of Effectiveness; and Accidental Release Prevention Requirements: Risk Management Programs Under Section 112(r)(7) of the Clean Air Act as Amended, Guidelines; Final Rules and Notices," *Federal Register*, vol. 61, no. 120, pp. 31668–31730, June 20, 1996.
26. AWWA Research Foundation and USEPA Office of Research and Development. *Compliance Guidance and Model Risk Management Program for Water Treatment Plants*, AWWA Research Foundation and American Water Works Association, Denver, CO, 1998.
27. Chlorine Institute, Inc. *Chlorine Pipeline*, Pamphlet No. 60, 4th ed., Chlorine Institute, Inc., Washington, D.C., May 1997.
28. Chlorine Institute, Inc. *Atmospheric Monitoring Equipment for Chlorine*, Pamphlet No. 73, 5th ed., the Chlorine Institute, Inc., Washington, D.C., April 1991.
29. The National Institute for Occupational Safety and Health (NIOSH). *Pocket Guide to Chemical Hazards*, U.S. Government Printing Office, Washington, D.C., 1997.
30. American Conference of Governmental Industrial Hygienists (ACGIH). *Documentation of the Threshold Limit Values and Biological Exposure Indices*, American Conference of Governmental Industrial Hygienists, Cincinnati, OH, 1991.
31. National Fire Protection Association (NFPA). *The Fire Protection Guide to Hazardous Materials*, 11th ed., National Fire Protection Association, Quincy, MA, 1994.
32. RJ Environmental, Inc. *Chlorine and Sulfur Dioxide Vapor Scrubbing System (Equipment Brochure)*, RJ Environmental, Inc., San Diego, CA, Undated.
33. EST Corporation. *Caustic Management in EST Emergency Chlorine Scrubbing Systems*, Technical Bulletin 92, EST Corporation, Quakertown, PA, 1994.
34. Chlorine Institute, Inc. *Chlorine Scrubbing Systems*, Pamphlet No. 89, 1st ed., Chlorine Institute, Inc., Washington, D.C., December 1991.
35. Chlorine Institute, Inc. *Chlorine Vaporizing System*, Pamphlet No. 9, 5th ed., Chlorine Institute, Inc., Washington, D.C., April 1997.
36. Compressed Gas Association, Inc. *Anhydrous Ammonia*, CGA G-2, 8th ed., Compressed Gas Association, Inc., Arlington, VA, 1995.
37. American National Standards Institute and the American Society of Mechanical Engineers. *Boiler and Pressure Vessel Code, Section VIII-Unfired Pressure Vessels- Division 1*, CGA G-2, American Society of Mechanical Engineers, New York, NY, 1997.
38. American National Standards Institute and the Compressed Gas Association, Inc. *Safety Requirements for the Storage and Handling of anhydrous ammonia*, ANSI K 61.1, American National Standards Institute, New York, NY, 1989.
39. Reynolds, T. T. and Richards, P. A. *Unit Operations and Processes in Environmental Engineering*, 2nd ed., PWS Publishing ,Co., Boston, MA, 1996.
40. Saunier, B. M., Selleck, R. E., and Trussell, R. R. "Preozonation as a Coagulant Aid in Drinking Water Treatment," *Jour. AWWA*, vol. 75, no. 5, pp. 239–245, 1983.
41. Chang, S. D. and Singer, P. "The Impact of Ozonation on Particle Stability and the Removal of TOC and THM Precursors," *Jour. AWWA*, vol. 83, no. 3, pp. 71–78, 1991.
42. Grasso, D. and Weber, W. J. "Ozone Induced Particle Destabilization," *Jour. AWWA* , vol. 80, no. 8, pp. 73–80, 1988.
43. Qasim, S. R. and Hussain, I. *The Effect of Preozonation on Turbidity Removal*, Department of Civil Engineering, The University of Texas at Arlington, Arlington, TX, December 1992.
44. Ficek, K. J. "Potassium Permanganate for Iron and Manganese Removal, and Taste and Odor Control," *Water Treatment Plant Design for the Practicing Engineers*, R. L. Sanks (Editor), Ann Arbor Science, Ann Arbor, MI, pp. 461–479, 1978.

45. Scheible, O. K. and Bassell, C. D. *Ultraviolet Disinfection of a Secondary Wastewater Treatment Plant Effluent*, EPA 600/2-81-182, MERL, ORD, USEPA, Cincinnati, Ohio, 1981.
46. Johnson, J. D. and Qualls, R. G. *Ultraviolet Disinfection of a Secondary Effluent: Measurement of Dose and Effects of Filtration*, EPA 600/2-84-160, MERL, ORD, USEPA, Cincinnati, Ohio, 1984.
47. Glaze, W. H. and Kang, J. W. "Advanced Oxidation Processes for Treating Ground Water Contaminated with TCE and PCE: Laboratory Studies," *Journal AWWA*, vol. 80, no. 5, pp. 57–63, May 1988.
48. Aieta, E. M. et al. "Advanced Oxidation Processes for Treating Ground Water Contaminated with TCE and PCE: Pilot-Scale Evaluation," *Journal AWWA*, vol. 80, no. 5, pp. 64–72, May 1988.
49. "Electrotechnologies for Water and Wastewater Disinfection," Technical Commentary, *EPRI Journal*, vol 1, no. 4, pp. 1–5, 1993.
50. Douglas, J. "Electrotechnologies for Water Treatment," *EPRI Journal*, vol. 8, no. 2, pp. 1–13, March 1993.
51. USEPA. *Demonstration Bulletin on CAV-OX® Ultraviolet Oxidation Process, Magnum Water Technology*, EPA/540/MR-93/520, USEPA, Cincinnati, Ohio, 1993.
52. AWWA, *Water Fluoridation Principles and Practices*, No. M4, 2nd ed., AWWA, Denver, CO, 1984.
53. Texas State Technical Institute and Texas Department of Health. *Operator Training in Fluoridation Procedures*, Texas State Technical Institute and Texas Department of Health, Austin, TX, July 1980.

CHAPTER 13

Water Stability, Clearwell, High-service Pumps, and Distribution System

13.1 Introduction

Water is an excellent solvent. It dissolves many minerals, including those used in the construction of water pipelines and plumbing. Conversely, a water that is supersaturated with minerals tends to deposit these minerals on the piping and plumbing fixtures in the water supply system. The tendency of water to either dissolve or deposit minerals varies with its chemical make-up. This characteristic is known as the *stability* of water.

A *clearwell* (or *clearwater well*) is a storage tank commonly located at a water treatment plant. Treated water from the plant is stored in such tanks prior to transfer into the distribution system. This storage provides the required contact time for disinfection, many functional demands in the service water system at the plant (i.e., chemical dilution and feeding, filter backwash, pump lubrication, and domestic needs), and a buffer to offset differences between treatment plant flow rates and distribution system demands. In most cases, water must be pumped from a clearwell into the distribution system. The *high-service pump* provides the energy to lift water into the distribution system. The *water distribution system* serves as a network that consists of conduits to deliver water from the treatment facility to the customers. It is composed of pumps, pipelines, ground and elevated storage reservoirs, valves, and other appurtenances.

In this chapter, the theory and design considerations for water stability control, clearwells, high-service pumps, and distribution systems are presented. The Design Example provides typical calculations and details for these facilities. Operation and maintenance recommendations and equipment specifications are also presented.

13.2 Water Stability

Water can exhibit a tendency to either dissolve or deposit certain minerals in pipes, plumbing, and appliance surfaces. This tendency is known as *stability*. Water that tends to dissolve minerals is considered *corrosive*. Conversely, a water that tends to deposit minerals is considered *scaling*.

13.2.1 Importance of Water Stability

The tendency of a corrosive water to dissolve minerals can be detrimental to water quality and to the distribution system. In addition to dissolving calcium and magnesium, corrosive waters can also dissolve such harmful metals as lead and copper from plumbing. Recent regulations, promulgated by the USEPA in response to the 1986 Amendments to the Safe Drinking Water Act restrict the use of lead materials in water supply and plumbing systems.[1] To further reduce the chances of lead contamination in the water supplies, the regulations require utilities to test for the presence of dissolved lead or copper in tap water. If detected, these utilities are required to implement treatment techniques to reduce the corrosiveness of the water.

Scaling waters deposit a film of minerals on the pipe wall which reduces and, in some cases, prevents corrosion of metallic surfaces. If the scale deposition is too rapid, however, it can also be harmful. Excessive buildup of scale can damage appliances, such as water heaters, and increase pipe friction coefficients; in extreme cases, scale may clog pipes. Therefore, to prevent corrosion yet but limit scale deposition, the most desirable water is one that is just slightly scaling.

13.2.2 Common Methods Used to Measure Water Stability

The most common method of calculating the stability of a water is the *Langelier saturation index* (*LI*). This index compares the actual pH of the water to the pH of the water if it were saturated with calcium carbonate.[2] Eqs. (13.1) and (13.2) are used to determine the Langelier saturation index.[3]

$$LI = pH - pH_s \tag{13.1}$$

$$pH_s = (pK_2 - pK_s) + pCa^{2+} + pAlk \tag{13.2}$$

where

LI = Langelier saturation index

pH = measured pH of water

pH_s = pH at $CaCO_3$ saturation

pK_2 and pK_s = constants based on the ionic strength and the total dissolved solids (TDS) of the water (Table 13-1)

pCa^{2+} = negative logarithm of calcium ion concentration in moles per liter, $-\log[Ca^{2+}]$

$pAlk$ = negative logarithm of total alkalinity in equivalents per liter, $-\log[Alk]$

A negative Langelier index indicates that the water will have a corrosive tendency; a positive value indicates the water will tend to scale. The magnitude of the index, however, is not indicative of the severity of the corrosivity or scaling tendency of the water. To overcome this shortcoming, Ryznar modified the Langelier index by the Ryznar index (RI) as shown in Eq. (13.3).[2-4]

$$RI = 2pH_s - pH \tag{13.3}$$

Table 13-2 lists the scale formation or corrosive tendencies of waters with various Ryznar index values.

An analytical method of measuring water stability is the *marble test*. This test involves holding the water in contact with calcium carbonate marbles. After 24 hours of contact, the alkalinity is measured. If the alkalinity decreases, the water is supersaturated and will tend to scale. Conversely, if the alkalinity increases, the water is undersaturated and will tend to be corrosive.[5]

13.2.3 Treatment Options to Enhance Water Stability

Corrosive Waters One treatment option to enhance water stability involves adjustment of the water pH to yield the desired saturation index. In reviewing Eqs. (13.1) and (13.3), it is apparent that increasing the pH of the water will tend to make the water less corrosive. The addition of hydrated lime ($Ca(OH)_2$) is a common treatment strategy to achieve this; it increases the pH and neutralizes free carbon dioxide. If the water is soft, it also adds calcium ions. Other strong bases, such as soda ash (Na_2CO_3) or sodium hydroxide (NaOH) can also be used. Engineers should carefully evaluate alternative application points of chemicals for pH adjustment, to ensure that performance of treatment processes is not affected. Coagulation and disinfection are two processes that are sensitive to pH.

Scale Forming Waters If the Ryznar index is too low (indicating heavy scale formation), a typical treatment strategy involves lowering the pH. This can be achieved by adding an acid or by *recarbonation*. In recarbonation, carbon dioxide (CO_2) is added to the water to shift the pH downward. This chemical reaction is discussed in Chapter 8. Another strategy to treat scaling waters is to add sequestering chemicals to prevent precipitation of excess calcium carbonate.[6–8] Typically, metaphosphates or orthophosphates are used. These sequestering chemicals allow the use of a slightly scaling water without causing serious scale buildup. This is because the precipitates that form scale cannot crystallize. Another treatment strategy for scale-forming waters involves softening. Waters with high calcium and magnesium concentrations are softened by precipitation (Chapter 8) or ion exchange (Chapter 18).

13.3 Clearwells

The design of clearwells involves the selection of the appropriate size and geometry. Other design considerations include sanitary protection of the stored water, protection of the structure from excessive hydraulic loadings, and protection of workers who must maintain the tank throughout its service life.

Table 13-1 Values of $pK_2 - pK_s$ with Respect to Temperature and Total Dissolved Solids (TDS)

TDS, mg/L	$pK_2 - pK_s$ 0°C	10°C	20°C	30°C	40°C	50°C	80°C
	2.45	2.23	2.02	1.86	1.68	1.52	1.08
40	2.58	2.36	2.15	1.99	1.81	1.65	1.21
80	2.62	2.40	2.19	2.03	1.85	1.69	1.25
120	2.66	2.44	2.23	2.07	1.89	1.73	1.29
160	2.68	2.46	2.25	2.09	1.91	1.75	1.31
200	2.71	2.49	2.28	2.12	1.94	1.78	1.34
240	2.74	2.52	2.31	2.15	1.97	1.81	1.37
280	2.76	2.54	2.33	2.17	1.99	1.83	1.39
320	2.78	2.56	2.35	2.19	2.01	1.85	1.41
360	2.79	2.57	2.36	2.20	2.02	1.86	1.42
400	2.81	2.59	2.38	2.22	2.04	1.88	1.44
440	2.83	2.61	2.40	2.24	2.06	1.90	1.46
480	2.84	2.62	2.41	2.25	2.07	1.91	1.47
520	2.86	2.64	2.43	2.27	2.09	1.93	1.49
560	2.87	2.65	2.44	2.28	2.10	1.94	1.50
600	2.88	2.66	2.45	2.29	2.11	1.95	1.51
640	2.90	2.68	2.47	2.31	2.13	1.97	1.53
680	2.91	2.69	2.48	2.32	2.14	1.98	1.54
720	2.92	2.70	2.49	2.33	2.15	1.99	1.55
760	2.92	2.70	2.49	2.33	2.15	1.99	1.55
800	2.93	2.71	2.50	2.34	2.16	2.00	1.56

Source: Adapted in part from Reference 3.

13.3.1 Size of Clearwells

Clearwells serve several purposes. Their size, therefore, is determined from the volume requirement for various operational needs, including (1) water storage for plant operations, (2) a

Table 13-2 Scale and Corrosion Tendencies of Waters with Various Ryznar Index Values

RI Range	Indication
Less than 5.5	Heavy scale formation
5.5 to 6.2	Some scale will form
6.2 to 6.8	Non-scaling or corrosive
6.8 to 8.5	Corrosive water
More than 8.5	Very corrosive water

Source: Adapted from Reference 4.

storage reservoir to balance plant output with distribution system demands, and (3) disinfection contact time. These three volumes are additive, as is expressed mathematically by Eq. (13.4).

$$V = PW + B + C \tag{13.4}$$

where

V = clearwell storage volume
PW = volume required for plant water uses
B = volume required to balance plant output with distribution demand
C = volume required for disinfection contact

The plant water uses include the water required to operate the plant. Among these requirements are chemical dilution and feed, pump lubrication, wash down, in-plant domestic needs, and filter backwash requirements. Chemical dilution and feed requirements are estimated from chemical feed rates and raw chemical requirements. For example, gaseous chlorine and ammonia are dissolved in water to facilitate their transportation around the plant, their injection, and their mixing. The appropriate quantity of water required is determined from information provided by the suppliers of chemicals and chemical feed equipment. The calculation of the amount of water required for filter backwashing is discussed in Chapter 10. Generally, a sufficient quantity of water should be stored to satisfy at least one day's filter washing requirements. Water demand for wash-down is estimated from the number and size of the outlets provided. In-plant domestic needs can be estimated from the projected number of plant employees. Typical water demand factors for both of these cases are listed in Chapter 4. In the absence of more detailed information, the quantity of water required for domestic and wash-down uses typically falls in the range from 1 to 2 percent of the plant capacity. Water requirements for chemical dilution and feed and for filter backwash should be specifically calculated for each application.

The clearwells must also satisfy the difference between the treatment plant production rate and the peak outflow of treated water into the distribution system. Typically, the high-service pump station operates at higher flow rates than the treatment plant during the higher water demand periods of the day and at lower flow rates during the lower water demand periods, to reduce the required volume for water storage facilities of the distribution system. The clearwell volume requirement is determined from a summation of the difference between the hourly production rate and hourly transmission flow rate. This procedure is illustrated in the Design Example.

The Surface Water Treatment Rule dictates minimum disinfection contact times for every water treatment plant that serves as a public water supply.[9] These requirements are discussed in Chapter 12. For some water treatment plants, contact time in the clearwells is used to satisfy a portion of the disinfection contact time requirements. The volume necessary to provide this contact time dictates the minimum water level in the clearwells; therefore, this volume must be calculated and added to the total clearwell volume requirement.

13.3.2 Clearwell Geometry

The geometry of a clearwell is important, particularly if it is used as a disinfection contact chamber. The geometry should promote a flow condition that approaches plug flow, where the actual contact time is very near the theoretical detention time (Eq. (9.8)).

The effective disinfection detention time, T_{10}, is best determined from tracer studies for existing basins, however for proposed basins, or in cases where tracer studies are impractical, the effective disinfection detention time can be estimated from the basin geometry and baffling. Typically, the theoretical contact time should be calculated and checked for the minimum water surface level in the basin at the maximum flow rate. Discussion of tracer studies and the procedure for determining T_{10} values are provided in Chapter 12. In the cases where the requirement of $CT_{cal} >> CT_{tab}$ is achieved in the chlorine contact channel, the detention time in the clearwell may be ignored.[10]

Poor, average, and superior baffling arrangements for rectangular and circular clearwells are illustrated in Figures 12-4 and 12-5 in Chapter 12, respectively. Other important clearwell geometry considerations include inlet and outlet hydraulics. The maximum water surface in the clearwell should not cause a backwater effect on the filter or other upstream processes. Similarly, the minimum water surface elevation should be above the high-service pump elevation, to maintain prime on the pumps. The clearwell outlets should also be configured so as to promote smooth flow into the outlet pipe and to prevent vortices. This is achieved by properly placing outlet in the clearwell, and by limiting entrance velocities to from 0.15 to 3 m/s (from 0.5 to 1.0 fps). Many other wet well and pump suction requirements also apply to high-service pumps. Some of these requirements are shown in Figures 7-13 and 7-14.

13.3.3 Other Design Considerations for Clearwells

Security and Safety Clearwells contain potable water and must be protected from possible contamination. Therefore, the clearwells must be covered, to prevent intrusion of rainwater and atmospheric particulates. Access to the inside of clearwells is through roof hatches that are rainproof and locked. Ladders to the bottom of the tank must conform to the Occupational Safety and Health Administration (OSHA) standards. Often a few inches of water remain in a clearwell even after it is drained. Therefore, any bottom openings such as pipe penetrations or sumps should be protected by handrails or other similar guards, to prevent workers from slipping, falling, or tripping while doing routine maintenance work inside the clearwell.[11,12]

Ventilation Clearwells must be able to breathe or expel and take in air, as the water level rises and falls. This capability is achieved by installing vents designed to pass the quantity of air that equals the maximum water inflow and outflow rates. The vent design should include a rain cover and insect screen, to protect the potable water quality inside the clearwell.

Overflow Clearwells may inadvertently become filled beyond their capacity. To protect the structural integrity of the tank in such cases, an overflow system is incorporated in the design. The overflow system should also incorporate screens or flap valves, to protect the water quality from insects and animal intrusions. The discharge of the overflow should be routed in such a way that it will not flood adjacent areas or otherwise cause damage from ponding or flowing water. The designer should provide a sufficient length of weir so that, during the overflow condition, the water level does not exceed the top of the walls of the clearwell.[11]

13.4 High-service Pumping

The basic design considerations for high-service pumps and pump stations are the same as those for raw water pump stations, discussed in Chapter 7. The capacity of the high-service pumps is dependent on the capacity of the distribution, storage, and pumping systems. Some smaller systems utilize the high-service pumps for boosting the flow to the distribution pressures. Because these systems do not use distribution pumping, the high-service pumps must be designed to supply peak hour and fire demand flows. Other systems that utilize ground storage and pumping in the distribution network require lower capacities of high-service pumping, because peak demands are supplied by the distribution reservoirs. A reservoir capacity analysis, as discussed in Section 13.3.1, is required to determine the high-service pumping requirements in such cases. In more complex systems, a simulated distribution system computer model study conducted over extended periods is used to establish the design capacities. These models are discussed in Section 13.5.3.

13.5 Water Distribution Systems

13.5.1 Purpose of Water Distribution Systems

The primary purpose of a water distribution system is to transport treated water from the treatment facility to the consumer. It should have sufficient capacity to meet the water supply needs of the consumers under all demand conditions. In most communities, the water distribution system serves the secondary purpose of providing water supply for fire-fighting. This demand, known as *fire demand*, is often the most severe design condition for the water system.

13.5.2 Hydraulic Analysis of Distribution Systems

The hydraulic analysis of a water distribution system involves determining the flow rate and head loss in each pipe and the resulting pressure at critical points in the system under different demand conditions. A typical water distribution system consists of *pipes*, *nodes*, and *loops*.

Direct solution of even a simple network is not possible. Several methods have been used to analyze the distribution system. These are (1) sectioning, (2) the circle method, (3) pipe equivalence, (4) trial and error (or relaxation), (5) digital computer analysis, and (6) electrical analogy.[13] The *Hazen–Williams* equation (Eq. (7.19)) is rearranged to develop Eq. (13.5). This equation is more readily used for network analysis.

$$Q = KCD^{2.63}\left(\frac{h_f}{L}\right)^{0.54} \tag{13.5}$$

Q, C, D, h_f, and L are as defined for Eq. (7.19). These are respectively, pipe flow, coefficient of roughness, pipe diameter, friction head loss, and equivalent length of pipe. K is a proportionality constant

It is evident that the pipe size, pipe length, friction coefficient, and flow rate all must be known to determine head loss. Unfortunately, there is no direct method to calculate the flow rate through each pipe in a system such as that illustrated in Figure 13-2, so, a trial-and-error procedure is generally necessary.[14–18]

Trial-and-Error or Relaxation Solutions Before the wide-spread use of digital computers in civil engineering, the *Hardy Cross Method* of pipe network hydraulic analysis was the most widely used procedure. In this procedure, a trail flow rate (Q) is assumed for each pipe in the system. The resulting head loss in each pipe is then calculated. The head losses of all pipes in each loop are added with proper sign.[a] If the sum of all head losses is not zero, a correction is applied to all flows. The correction factor is calculated from Eq. (13.6).[14]

$$\Delta Q = -\frac{\Sigma h_f}{n\Sigma \frac{h_f}{Q}} \tag{13.6}$$

where

ΔQ = flow correction for each pipe in a loop

Σh_f = summation of head loss in each pipe of the loop

n = inverse of the exponent of the friction slope term in the head loss equation (1.85, for the Hazen–Williams equation)

$\Sigma \frac{h_f}{Q}$ = summation of the ratio of head loss and flow in each pipe Q. This ratio always has a positive sign.

a. A sign convention that designates clockwise direction as positive and counterclockwise direction as negative signs to flows and corresponding head losses has been adopted.

The Hardy Cross method, a procedure used for water distribution network analysis, may be found in several references.[3,5,14]

Computer Programs Solution of large and complex pipe networks by the trial-and-error method is tedious and time-consuming. Over the past twenty years, numerous computer programs have been developed to perform complex network analyses. Typically, the program input includes pipe sizes, friction coefficients, lengths, demands, and the ground elevation of each node. The output includes flow, head loss and velocity in each pipe, and the pressure at each node. Many computer programs include subroutines that simulate pumps, pressure-control valves, check valves, reservoirs, and other appurtenances in the distribution system that affect the hydraulic characteristics of the distribution system.[18–22]

The computer solution of water distribution systems is called a *model*. To accurately represent the characteristics of a distribution system, a model must simulate (1) the demand distribution throughout the system, and (2) the system configuration and layout (pipe size, friction coefficients, and length).

Demand Distribution Water demand distribution involves assigning the water demand to appropriate nodes. Most computer programs require that demands can be applied only at nodes. Water demands used in a computer model are typically the maximum hour demand and the fire demand. The maximum hour demands are a function of land use within each zone of the community. Typical water demand factors for various types of land uses are presented in Chapter 4. Utilizing land use or zoning maps, the designer can estimate the water demand for each type of land use. By overlaying the water distribution system map over the zoning map, the applicable demand at each node can be determined. This procedure is illustrated in Figure 13-1 and involves constructing a perpendicular bisector on each pipe segment. These bisectors will form polygons representing the area served by each node. The area of each polygon is calculated and multiplied by the appropriate demand factor to estimate the water demand applied to each node.

Major water users, such as heavy industries or large commercial users (any single user that exceeds approximately 1 percent of total system demand), should be reviewed separately and applied to the closest node. An engineer should review the results of this procedure and modify them to account for any unusual system characteristics that could contradict the results.

System Input The details of the distribution system should be accurately defined to represent the system configuration and characteristics. Information required to build the model includes pipe sizes, equivalent lengths, connection points, demands, and friction coefficients. The pump characteristics, the pressure- and flow-control valves, the location, size, and operating characteristics of the reservoir(s) are also required. Pump characteristics include horsepower, type of pump, head and capacity relationships, efficiency curves, name of pump manufacturer, pump series, and model number. The pressure-control valve information needed for input data include the size, the type of valve, and pressure-control settings.

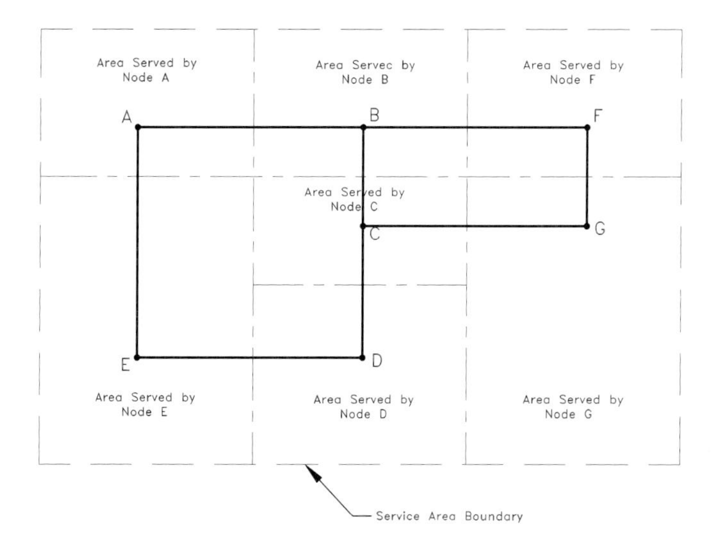

Figure 13-1 Example of water demand distribution to individual pipe nodes.

The detail of information and input data required for an accurate computer model typically is not available from a single source. Construction and maintenance records, zoning plans historical water consumption data, and critical water zones are excellent sources to obtain information necessary to develop the system characteristics. Senior members of the operations and maintenance staff of the water department can also provide helpful information.

Field testing is usually required to estimate friction coefficients. This testing, known as a *C factor test*, involves accurately measuring the head loss along isolated segments of the distribution pipes under known flow conditions. Using the Hazen–Williams equation, the friction coefficient can be computed if the flow rate, pipe length, and head loss are known. The test should involve measurements at a number of flow rates. The *C factor test* procedure, for conducting the C factor test, is provided in the Design Example.

Model Calibration A distribution system model should be compared with actual field monitoring data to confirm its validity. This procedure, known as *calibration*, involves taking field measurements of pressure and flow rates and comparing the measured values with results predicted by the model. Depending on the details of the study desired, the predicted values should fall within 5 to 15 percent of the measured values. If the desired accuracy is not attained, the model should be adjusted to provide satisfactory results.[23,24]

Quasi Dynamic Models Some computer programs have the capability to simulate water distribution systems over a full day's operations. This allows an engineer to model a distribution system over a set of diurnal water demand variations and test the storage capability of the system. The input required for these models includes all data necessary for a steady-state model, plus information on reservoir levels and capacities, set points to turn pumps on and off, and diurnal water demand variations. Output includes system pressures and flow rates, plus reservoir levels at various time intervals.

13.5.3 Distribution System Design Considerations

Pressure Planes If a water distribution system serves an area that has a great deal of topographic relief, wide variations in pressures can be experienced. Some locations at higher elevations may experience low pressures, while locations at low elevations may develop high pressures. To alleviate this situation, systems are often divided into *pressure planes* or *pressure zones*. With this concept, higher elevations are served at a pressure adequate to meet their needs, while lower elevations are served at a reduced pressure. The systems are usually interconnected, but separated by pressure control valves.

Operationally, these systems can be of two basic forms, as shown schematically in Figure 13-2. Figure 13-2(a) illustrates a type of system in which water is pumped to the higher pressure plane. Water flows back down into the lower pressure plane through pressure control valves. These systems are simple and they usually have a low capital expense but a high energy cost. This type of system is appropriate for small systems or for systems having small, isolated areas of low elevation.

Figure 13-2(b) illustrates a system in which water is pumped into the lower pressure system, then pumped again into the higher-pressure system. This system is more energy efficient, because only the water required for the higher-pressure plane is pumped to the higher gradient. This type of system is appropriate for systems with large lower-pressure plane areas or systems with isolated high-pressure areas.

Ground Storage Reservoirs Important design considerations for ground storage reservoirs are the capacity of the reservoir, its location, and often its aesthetics. Important design details for ground storage reservoirs include provisions for venting the tank, overflow, security, and corrosion control. Brief discussions on some of these design details have been provided in Section 13.3.

Capacity. The ground storage and elevated storage reservoirs collectively must provide sufficient capacity to offset the difference between distribution system demand and treated water transfer flow rates from the treatment facility. Ground storage reservoirs are the least costly form of treated water storage. From a cost viewpoint, it is desirable to use ground storage reservoirs for a large percentage of the storage volume. From an operational viewpoint, the elevated storage reservoirs offer more reliability. No pumping or mechanical equipment is needed to deliver water from elevated storage to the consumers; thus, the chances for equipment failure and interruptions of water supply to the consumers are essentially eliminated.

The designer must first estimate the total reservoir capacity required. This is accomplished either by a mass balance calculation, as outlined in Section 13.3.2, or by a suitable quasi

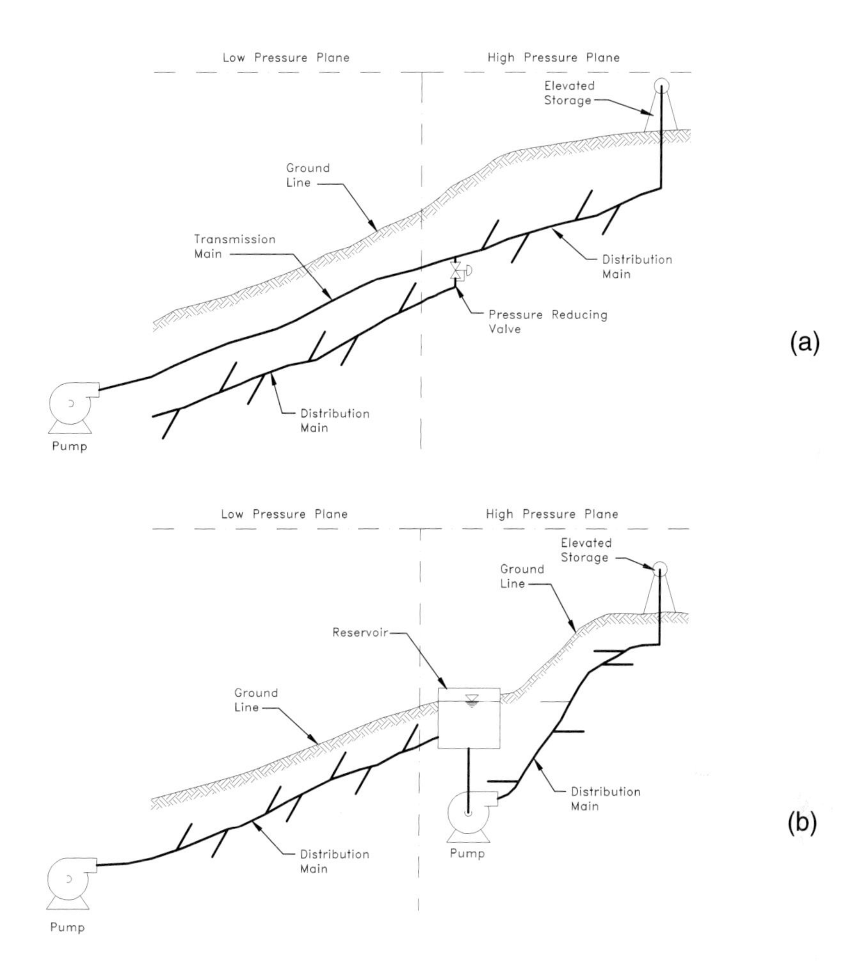

Figure 13-2 Alternative pressure plane systems. (a) Single-pump distribution system. (b) Multiple-pump system.

dynamic distribution system model. A determination of the volume of water that can economically be stored in elevated reservoirs is then made; the remaining volume must be stored in ground storage reservoirs. An evaluation of elevated storage capacity need is discussed later in this section.

Location. Ground storage facilities must be located near major transmission mains for needed pipeline capacity to move water in and out of the tank. If a reservoir is associated with a pump station, then it should be conveniently located adjacent to the pump station and its bottom should be higher than the pump centerline. The top of the reservoir should be set at an elevation close to that of the hydraulic gradient of the influent transmission mains.

Aesthetics. In many cases, the visual impact of a large storage reservoir must be considered. In some cases, architects are retained to enhance the visual aesthetically effects through architectural designs and landscaping. Figure 13-3 illustrates an example of architectural design for a ground storage reservoir.

Figure 13-3 Examples of architectural design for a ground storage reservoir. *(Courtesy of Caldwell Tank, Inc.)*

Ventilation. Like clearwells, ground storage reservoirs must be vented to allow air to flow into and out of the reservoir as the water level rises and falls. A fail-safe vent sized for a capacity equal to the maximum inflow and outflow rate must be installed. The vent should have an insect-proof screen and rain cover incorporated to protect water quality.

Overflow. Ground storage reservoirs must also have provisions to discharge excess water should the tank inadvertently be filled in excess of its capacity. The design considerations for the overflow system for the ground storage reservoir are the same as those discussed for clearwells in Section 13.3.

Security and Safety. Like any potable water facility, ground storage reservoirs should be located on a secured site and protected by locked anti-personnel fences and other security measures. Protection of the water inside the tank and provisions for access are the same as discussed for clearwells in Section 13.3.

Corrosion Control. Often, ground storage reservoirs are constructed of steel. Steel tanks must be protected from corrosion by high-performance epoxy painting systems designed for potable water service. The designers should consult suppliers of high-performance industrial coatings for guidance in specifying painting systems. Often, cathodic protection systems consisting of zinc anodes are immersed in the water and electrically connected to the tank. These systems protect the tank by promoting an electric current that tends to corrode the zinc in lieu of the steel. Special attention to installation details must be used in areas subject to freezing. Ice layers on the surface of the tank tend to pull apart the wires used for the cathodic protection system.

Pumps Distribution pump stations are subject to the same considerations as the raw water pumps discussed in Chapter 7. The capacity of distribution pump stations, however, should be sufficient to meet the maximum hour or fire demand of the system that is not stored in the elevated storage. The capacity of the distribution pump station should be equal to the sum of the combined capacities of the all pumps, except that the largest unit can be out of service. This will ensure that the pump station can meet the emergency, even in case of equipment failure.

Elevated Storage The important considerations for design of elevated storage reservoirs are similar to those for ground storage reservoirs. Considerations include capacity, location, and aesthetics. Important design details include ventilation, overflows, access, security, and corrosion control.

Capacity. As previously discussed, the storage volume requirement must be distributed between elevated and ground storage in an economical way. From an operational standpoint, elevated storage is more desirable than ground storage, but it is more expensive to construct. An economic model comparing the capital cost of elevated storage tanks to that of ground storage and pump stations can be used to evaluate alternatives in storage capacity distribution. In larger systems, or in systems with high peak hour electricity rates, the model should also include the present worth of future power supply costs. Regardless of the results of the economic evaluation, the capacity of the elevated storage must be sufficient to maintain system pressures for both peak hour demands and fire demands. This capacity requirement can be evaluated with a quasi-dynamic model. If a steady-state model is used, the maximum hour flow rate from the tank can be used to calculate the tank volume by extending the tank outflow to a four-hour period. Elevated storage reservoirs should also be checked for adequate emergency storage to meet the water demands for a short period if the pump station fails. In addition, the elevated storage volume should meet the minimum capacity requirements of the regulatory agencies. The designers should review the applicable regulatory authority requirements before recommending storage capacity.

Location. The effect of an elevated storage tank on the pressure of the distribution system is sensitive to the hydraulic location of the tank. The hydraulic location is the point in the system at which where the hydraulic effects of the reservoir are fully utilized. The hydraulic location does not necessarily relate to the physical location of the tank. This point is illustrated in Figure 13-4. In Figure 13-4(a), the physical location of the elevated tank and its hydraulic effect are both located at the end of the distribution system. This is achieved by connecting the transmission main to all distribution pipes. In the system illustrated in Figure 13-4(b), the hydraulic effects of the tank are realized near the pump station, even though it is physically located at the opposite end of the system. This is achieved by connecting the transmission main only to the distribution pipes near the pump station. In the system illustrated in Figure 13-4(c), the hydraulic effects of the tank are shifted to the center of the distribution system by connecting the transmission main to the distribution pipes in that area only. In most cases, the hydraulic effects of the tank are most desirable at a point slightly past the half-way point between the source and the most remote part of the system (Figure 13-4(c)). This scheme results in less pressure variations, throughout the day than do other examples. In systems covering larger areas, multiple reservoirs are necessary to distribute pressures more evenly.

Aesthetics. Aesthetics are of more concern with elevated storage tanks than with ground storage reservoirs. Because of their height, they often dominate the skyline of small communities. Figure 13-5 illustrates some examples of tank styles that are commonly available. Other styles can be custom designed, but construction of custom tanks is usually quite expensive. Elevated tanks often utilize special painting schemes. To reduce the visual impact of tanks, soft colors, such as earth tones or light pastels, are often used; light browns, light blues, and light greens are most common. Local high school or college colors are also used. Many communities desire to have their name, logo, or local high school mascot painted on the tank. For the most part, aesthetic treatment is a sensitive issue that engineers should leave to the architects or community leaders.

Design Details. Concerns relating to ventilation, security, overflow, and corrosion control all apply to elevated tanks in the same manner as they do to ground storage tanks. These issues have been discussed in earlier sections. The issue of access, however, is of particular concern with elevated tanks. Their height makes the safety issue of access ladders is an important consideration. Designers should consult and follow OSHA guidelines for ladder or stair design. Ladder or stairway access should also be designed to limit access to only authorized personnel. Elevated tanks tend to attract vandalism. In addition to the cost of repairing vandalized items, there

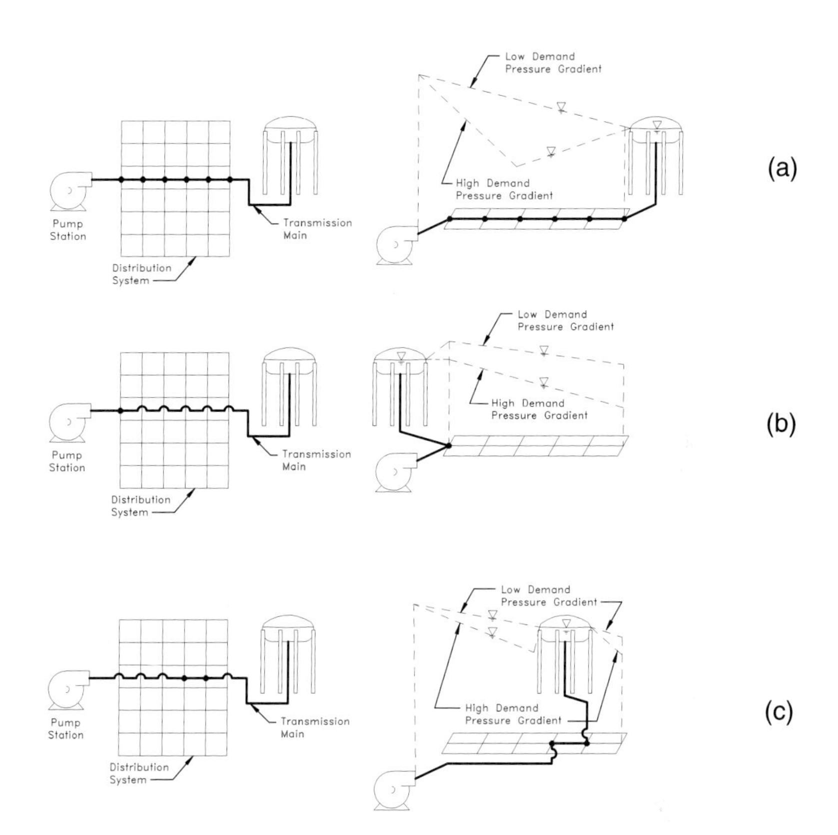

Figure 13-4 Examples of effects of piping and physical location on the hydraulic effect from an elevated storage tank. (a) Transmission main connected to all distribution pipes. (b) Transmission main connected distribution main near the pump station. (c) Transmission main connected to distribution main near the center of the demand.

is a potential liability for both the engineer and the tank owner should a vandal become injured. Ladders should not extend to the ground and should be accessible only by a portable ladder. Preferably, ladders should be enclosed inside the tank riser. The examples of elevated storage tanks are illustrated in Figure 13-5.

Pipelines The water distribution system should provide pipeline access to every potential user. This usually requires locating service pipelines in the right-of-way of all streets. To reduce the potential for water stagnation in the system, dead ends should be avoided. Instead, each line should be looped or connected at each end to another pipeline. If dead end lines are necessary, they should be sized no larger than necessary to meet the demands. Also, provisions should be made to flush dead-end lines at least once per month.

(a)

(b)

(c)

Figure 13-5 Examples of elevated storage tanks. (a) Traditional multi-column tank. (b) Pedesphere tank. (c) Composite tank. *(Courtesy of Caldwell Tanks, Inc.)*

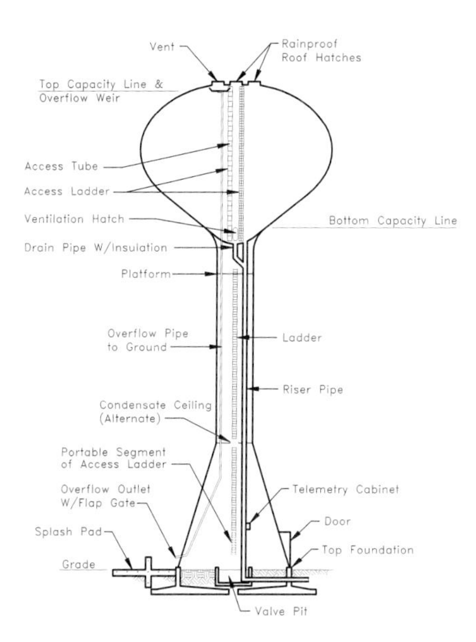

Figure 13-6 An example of an elevated storage tank.

The size of water distribution pipelines is dependent on maximum and minimum allowable velocities and minimum allowable pressures in the distribution system. The maximum and minimum velocities and pressures for several major water supply systems throughout the United States are listed in Table 13-3.[25]

Pipeline size and location both are typically determined from computer model studies, as discussed in Section 13.5.3. A computer run is made, with the existing network receiving the design flows. If the analysis indicates values outside the acceptable limits, pipe sizes are adjusted, or pipes are added or removed. The system is analyzed again, and the process is repeated until acceptable results are obtained. In general, 30-cm (12-in) mains should be placed on approximately 1.5-km grids in residential areas. Industrial areas and areas with high water or fire demand may require closer spacing of major supply mains.

Appurtenances

Fire Hydrants. Required spacing for fire hydrants varies somewhat from one area to the next. The spacing is dependent mostly on the length and characteristics of fire hose available to the local fire department. Typical spacings are 180 m (600 ft) for residential areas, and 90 m (300 ft) for industrial or commercial areas. Designers should, however, refer to the local officials for specific requirements. To maintain adequate pressure during fire demand, hydrants should be located on a looped pipe at least 15 cm (6 in) in diameter, if possible. If more than one hydrant is

Table 13-3 Typical Distribution System Design Criteria

City	Velocity, m/s (fps)		Pressure, Kpa (psi)	
	Maximum	Minimum	Maximum	Minimum
Austin, TX	1.5 (5)	None	827 (120)	138 (20)
Albuquerque, NM	3 (10)	None	690 (100)	138 (20)
Boca Raton, FL	None	None	None	241 (35)
Corpus Christi, TX	1.5 (5)	None	None	207 (30)
Dallas, TX	1.2	0.15 (0.5)	793 (115)	207 (30)
Denver, CO	1.5 (5)	0.9 (3)	758 (110)	276 (40)
Durham, NC	None	None	None	276 (40)
El Paso, TX	1.5 (5)	None	None	310 (45)
Fort Worth, TX	None	None	None	241 (35)
Hollywood, FL	None	None	None	138 (20)
Houston, TX	None	None	None	138 (20)
Las Vegas, NV	2.1 (7)	None	690 (100)	207 (30)
Orlando, FL	1.5 (5)	None	None	138 (20)
Phoenix, AZ	1.5(5)	None	690 (100)	414 (60)
Raleigh, NC	1.5 (5)	None	None	138 (20)
Sacramento, CA	1.8 (6)	None	None	138 (20)
St. Petersburg, FL	None	None	None	345 (50)
San Antonio, TX	1.8 (6)	None	1207 (175)	172 (35)
San Diego, CA	4.6 (15) (fire only)	None	None	138 (20)
San Jose, CA	None	None	862 (125)	276 (40)
Tucson, AZ	None	None	586 (85)	241 (35)
West Palm Beach, FL	None	None	None	276 (40)

Source: Adapted from Reference 25.

necessary between connecting pipes, the minimum pipe size should be increased to 20 cm (8 in). Figure 13-7 illustrates a typical fire hydrant installation.

Valves. In general, valves should be located in the water distribution system to minimize disruption of water service in case of pipe failure or of pipe maintenance requirements. A general practice is to place valves at any point where two pipes connect, so that no more than one fire hydrant will be shut off by isolation of a pipe segment.

Pressure Control Valves. Pressure-control valves are installed on pipes connecting two pressure planes. They may be either pressure-reducing or pressure-sustaining, depending on the

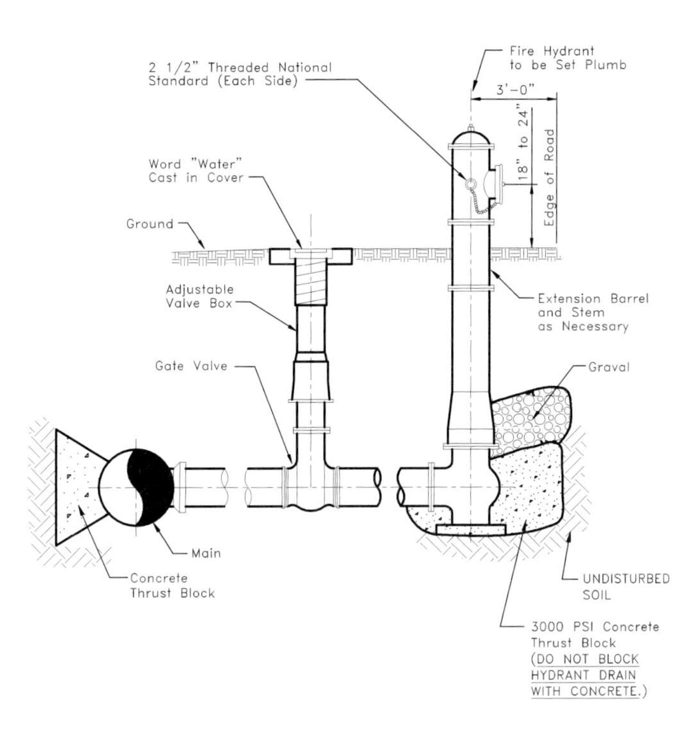

Figure 13-7 Typical fire hydrant installation details.

need. A pressure-reducing valve adjusts to maintain a preset downstream pressure and closes if the downstream pressure exceeds the pressure of setting the valve. The pressure-reducing valve is the most common type of pressure-control valve in water distribution systems and is typically used in feeder lines between a high-pressure plane and a low-pressure plane. A pressure-sustaining valve adjusts to maintain a preset upstream pressure and opens if that preset pressure value is exceeded.

An important design consideration for pressure-control valves is prevention of cavitation. High velocities in the valves may cause negative pressures within the valve body and result in cavitation damage. Manufacturer's data provides guidelines on maximum flow rates and pressure drops across these valves. If these values are not exceeded, the valve should provide satisfactory service.

Service Connections and Meters. Service connections include the plumbing lines from the water main to the water meter and then the user. The primary design consideration for service connections is the selection of a size appropriate for the customer's demand. Residential connections are typically 19 mm (3/4 in). Commercial and industrial connections are typically specified by the design engineer who determines the actual demand and flow conditions.

13.6 Equipment Manufacturers for Clearwells, High-service Pumps, and Distribution Systems

A list of manufacturers of equipment for clearwells, high-service pumps, and distribution systems is provided in Appendix C. General design guidelines for review and selection of equipment in the design are discussed in Appendix C. The equipment selected for water supply projects should meet the project requirements, especially for flexibility of operation, maintainability, and durability, while still meeting the primary design requirements.

13.7 Information Checklist for Design of Clearwells, High-service Pumps, and Distribution Systems

The following information should be obtained and necessary decisions made before the design engineer begins the design of clearwells, high-service pumps, and distribution systems.

1. Information to be used for stability calculations.

   a. finished water quality data:

      i. pH;
      ii. calcium concentration;
      iii. total alkalinity;
      iv. total dissolved solids.

   b. anticipated temperature range of the finished water;
   c. existing condition of the distribution system and stability data on the current water supply.

2. Information to be used for clearwell design.

   a. filter backwash water rate, duration, and frequency;
   b. plant water needs:

      i. chemical feed systems;
      ii. domestic and sanitary uses;
      iii. wash-down.

   c. Disinfection requirements:

      i. current and future disinfection chemicals used;
      ii. contact time available.

3. Information to be used for high-service pump station design.

    a. required capacity;
    b. treated water transmission pipeline length, size, and friction coefficient;
    c. hydraulic elevation of the point of discharge into the distribution system or distribution system pressure at the point of discharge into the distribution system;
    d. water surface elevation of the clearwells.

4. Information to be used for distribution system design.

    a. peak day and maximum hour demands;
    b. expected hourly diurnal flow pattern;
    c. fire demand;
    d. characteristics of the existing distribution system.

13.8 Design Example

This design example provides detailed design calculations on water stability, clearwells, and distribution systems. The design of high-service and distribution pump stations is not provided in this section. The design procedure for these facilities is similar to that for raw water pump stations. Readers are referred to Chapter 7 for design information on pump stations.

13.8.1 Design Criteria Used

1. Raw water quality data

    a. Total alkalinity = 80–110 mg/L as $CaCO_3$

        i. Bicarbonate alkalinity = 80–110 mg/L as $CaCO_3$
        ii. Carbonate alkalinity = 0
        iii. Hydroxide = 0

    b. Hardness = 100–120 mg/L as $CaCO_3$

        i. Calcium = 38–46 mg/L as Ca^{2+}
        ii. Magnesium = 1.2 mg/L as Mg^{2+}

    c. Total dissolved solids = 200–250 mg/L
    d. pH = 7.5 to 8.2
    e. Temperature = 5–30°C

2. Clearwell design data

   a. Plant water needs

      i. Chemical feed systems

         The chemicals that are diluted and applied include potassium permanganate, chlorine, ammonia, fluoride, lime, coagulants, and polymers.

      ii. Pump-seal flush = 0.5–2 L/d per pump

      iii. Plant domestic and sanitary needs

         Employees = 15 persons

         Per capita consumption = 190 L/d

   b. Filter backwash requirements

      i. Volume required to backwash each filter = 375 m^3 (Chapter 10)

      ii. Minimum filter backwash cycle time = 24 hrs (Chapter 10)

3. High-service pumping

   Capacity = 120 percent of plant capacity

4. Distribution system

   a. Plant capacity = maximum day flow = 113,500 m^3/d (30 mgd)

   b. Maximum hour demand = 181,600 m^3/d

   c. Maximum high-service pumping capacity = 120 percent of treatment plant capacity (136,200 m^3/d)

   d. Maximum distribution pump station capacity = 145 percent of treatment plant capacity (164,600 m^3/d)

   e. Minimum pressure at maximum hour demand = 276 kN/m^2 (40 psig)

   f. Minimum pressure at fire demand plus average day = 207 kN/m^2 (30 psig)

   g. Minimum pressure at maximum day demand plus fire demand =138 kN/m^2 (20 psig)

13.8.2 Design Calculations

Step A: Calculations of Water Stability The raw water quality analysis indicates the following:

Temperature = 5°C to 30°C

pH = 7.9 to 8.2

Total dissolved solids = 200 to 250 mg/L

Calcium ion concentration = 38 to 46 mg/L (as Ca^{2+})

Total alkalinity (bicarbonate only) = 80 to 110 mg/L as $CaCO_3$

In general, the raw water quality will be indicative of the mineral composition of the finished water; however, the pH and temperature can be expected to vary slightly. The pH will be adjusted in the coagulation process by the addition of hydrated lime ($Ca(OH)_2$) to yield optimum pH for coagulation (about 7.5). The temperature will also be affected by contact with air during sedimentation and filtration. Extreme values from 5°C to 30°C are expected in winter and summer. By using these assumptions, the stability is calculated for the temperature extremes. The Ryznar index will be used, because it gives an indication of the severity of the corrosion or scaling potentials.

1. Determine $(pK_2 - pK_s)$.

   The value of TDS ranges from 200 to 250 mg per liter. The temperature of the finished water ranges from 5°C to 30°C. By interpolating from Table 13-1, the following values of $(pK_2 - pK_s)$ are obtained.

	Value of $(pK_2 - pK_s)$	
TDS	5°C	30°C
200 mg/L	2.60	2.12
250 mg/L	2.63	2.15

   To check the stability of water at extreme conditions, use $(pK_2 - pK_s)$ values of 2.63 and 2.12.

2. Calculate pCa^{2+}.

   The calcium ion concentration ranges from 38 mg/L to 46 mg/L. The value of pCa^{2+} is calculated from the calcium concentration in moles per liter; therefore,

$$[Ca^{2+}] = \frac{38 \times 10^{-3} \text{ g/L}}{40.1 \text{ g/mole}} = 0.000948 \text{ mole/L}$$

   Similarly, for Ca^{2+} concentrations of 46 mg/L the molar concentration is 0.00115 mole/L.

   The value of pCa^{2+} is given by the negative logarithm of the calcium ion concentration; therefore, for Ca^{2+} concentration of 0.000948 mole/L as Ca^{2+},

$$pCa^{2+} = -\log[0.000948] = 3.02$$

   For Ca^{2+} concentration of 0.00115 mole/L as Ca^{2+},

$$pCa^{2+} = -\log[0.00115] = 2.94$$

3. Calculate pAlk.

   The negative logarithm of the total alkalinity (in equivalents per liter) is calculated.

$$[\text{Alk}] = \frac{80 \text{ mg/L as } CaCO_3 \times 10^{-3} \text{ g/mg}}{50 \text{ g/eq as } CaCO_3} = 0.0016 \text{ eq/L}$$

Similarly, total alkalinity of 110 mg/L as $CaCO_3$ is equal to 0.0022 eq/L.

For total alkalinity of 80 mg/L as $CaCO_3$,

$$pAlk = -\log[0.0016] = 2.80$$

For total alkalinity of 110 mg/L as $CaCO_3$,

$$pAlk = -\log[0.0022] = 2.66$$

4. Calculate *RI* for extreme cases.

   The maximum RI will occur with the highest values of $(pK_2 - pK_s)$, pCa^{2+}, and $pAlk$, and the minimum value will correspond to the lowest values of these parameters; therefore, the values of *RI* are calculated from Eq. (13.3):

$$RI_{max} = 2 \times (2.63 + 3.02 + 2.80) - 7.5 = 9.4$$

$$RI_{min} = 2 \times (2.12 + 2.94 + 2.66) - 8.2 = 7.2$$

5. Evaluate Ryznar indices.

   Comparing the calculated values of the Ryznar index with the values given in Table 13-2, the water has corrosive (RI_{min} = 7.2) to very corrosive tendency (RI_{max} = 9.4). This tendency is not acceptable; therefore, the finished water must have pH adjustment after filtration. The *RI* must be adjusted to a range of from 6.2 to 6.8 (non-scaling and non-corrosive).

   The options available for stability adjustment are the addition of lime ($CaOH_2$), soda ash (Na_2CO_3), or caustic soda (NaOH). Each of these chemicals adds dissolved solids and alkalinity and raises the pH of the water. These chemicals are quite effective in making the water less corrosive. In general, lime is the least expensive, but it has a low solubility, and controlling its feed rate is difficult because it causes scaling on equipment. Soda ash is less expensive than caustic soda; therefore, soda ash will be used, because of its solubility and cost advantages. The desired pH values for stability of water exceed the optimum coagulation pH for alum; thus, adjusting the pH prior to filtration might interfere with turbidity removal and adversely affect the finished water turbidity. Consequently, the chemical application point for pH adjustment is selected to be after filtration.

6. Soda ash feed.

   From the laboratory titration results, it was determined that the amount of soda ash required (at 100 percent purity) to maintain the Ryznar index at desired levels under the critical conditions of the finished water quality is 10 mg/L.

$$Na_2CO_3 \text{ dosage} = 10 \text{ mg/L} \times 113{,}500 \text{ m}^3\text{/d} \times 10^{-6} \text{ mg/kg} \times 10^3 \text{ L/m}^3$$
$$= 1135 \text{ kg/d}$$

Commercial soda ash powder or granules are 99.2% Na_2CO_3, so the feed rate is approximately 1135 kg/d ÷ 0.992 = 1145 kg/d. The soda ash injection point is together with the ammonia in the downstream portion of the chlorine contact channel.

The Na_2CO_3 feed system will be similar to those designed for several other chemicals in Chapters 8, 10, and 11. These details are not repeated here.[b] A workable solution of Na_2CO_3 is prepared and stored for delivery. Based on dissolution rate of commercial grade Na_2CO_3, the minimum contact time in the mixing and storage basins shall be 10 min and 20 min for dry feed rates of 30 g/L (0.25 lb/gal) and 60 g/L (0.50 lb/gal), respectively. For the higher dry feed rate, warm water and efficient mixing can reduce requirement on detention time. Positive-displacement pumps are used to deliver the solution at the point of application. Provide three pumps, each capable of pumping from zero to a maximum feed rate of 10 L/min, to deliver the Na_2CO_3 solution at the application point. Two will be normally be in service, while the other is a standby. The Na_2CO_3 feed system is located in the chlorine and ammonia building (Figure 12-24). The application point is downstream of the chlorine contact channel (Figure 12-21).

Step B: Design of Clearwells The volume required for the clearwell is dependent on: (1) plant water uses, (2) the balance of plant output with distribution system demand, and (3) disinfection contact time. In this example, a separate disinfection contact channel is provided; therefore, the clearwell volume is based only on plant water needs and on balancing pumping rates with plant production. Each component of the volume requirement is estimated here:

1. Plant water uses

The plant service water requirements include (a) chemical dilution and feed, (b) pump operation, (c) wash down, (d) in-plant domestic use, and (e) filter backwash.

a. Calculate water demand for chemical dilution and feed water.

Chemical dilution and feed flows are more or less at a constant rate throughout the day; therefore, some storage should be provided, to allow adjustments in feed rates necessary to account for changes in water quality or flow rate through the plant. In the authors' judgement, the volume of storage allocated to chemical dilution and feed should be equal to 50 percent of the *total daily* chemical dilution and feed demand at peak plant flow rate. The required in-plant water use is for chemical dilution of chlorine, ammonia, potassium permanganate, lime, sodium carbonate, coagulants, and polymers. The water needed for chemical dilution is provided in Table 13-4.

b. Na_2CO_3 feed system is similar to $KMnO_4$ system presented in Chapter 11 (Section 11.6.2, Step A). The specific gravity of Na_2CO_3 in bulk is: 1.04 (dense), 0.64 (medium), and 0.48 (light).

Clearwell volume for chemical feed and dilution

$= 50\%$ of one day demand $\times$ total dilution water demand

$= 0.5 \text{ d} \times 357.6 \text{ m}^3/\text{d}$

$= 178.8 \text{ m}^3$

Table 13-4 A Summary of Clearwell Volume for Chemical Dilution and Feed

Chemical System	In-Plant Dilution Flow, m^3/d	Sources
Lime	68[a]	Section 8.8.2, Step C.2 and Table 8-8
Potassium permanganate	28.8[b]	Section 11.6.2, Step A.6
Chlorine	172.8[c]	Section 12.8.2, Step B.3
Ammonia	34.6[d]	Section 12.8.2, Step C.3
Fluoride	10.2[e]	Section 12.8.2, Step D.3
Sodium carbonate (soda ash)	43.2[f]	Section 13.8.2, Step A.6
Total dilution water	357.6	

a Assume lime solution has a concentration of 25 kg/m^3 as pure $Ca(OH)_2$,

the dilution water quantity $= \dfrac{1700 \text{ kg/d}}{25 \text{ kg/m}^3} = 68 \text{ m}^3/\text{d}$.

b 2 pumps (total) ∞ 10 L/min·pump ∞ 1440 min/d ∞ 10^{-3} $m^3/L = 28.8$ m^3/d.

c 120 L/min ∞ 1440 min/d ∞ 10^{-3} $m^3/L = 172.8$ m^3/d.

d 24 L/min ∞ 1440 min/d ∞ 10^{-3} $m^3/L = 34.6$ m^3/d.

e 7.1 L/min ∞ 1440 min/d ∞ 10^{-3} $m^3/L = 10.2$ m^3/d.

f 3 pumps (total) ∞ 10 L/min·pump ∞ 1440 min/d ∞ 10^{-3} $m^3/L = 43.2$ m^3/d.

Provide a clearwell volume of 200 m^3 for chemical feed and dilution.

b. Calculate clearwell volume for pump operation water demand.

Pump operations require water for pump seal lubrication. Like chemical dilution and feed, this represents a small and consistent demand. The clearwell volume required for this demand is usually negligible, but, for this example, calculations are shown here. The clearwell volume allocated for pump operation is based on 50 percent of the peak daily water demand for pump lubrication.

A brief discussion about the design of high-service pump stations will be given in Section 13.8.2, Step C. The high-service pump station design includes a total of six high-service pumps. The filter backwash system requires three backwash water pumps (Section 10.13.2, Step B.5). In addition to these nine pumps, numerous other chemical feed pumps are provided, but the water demand for these pumps is included in the chemical dilution and feed system. Therefore, the pump operation demand is based on these nine pumps.

Centrifugal pumps have seals that prevent leakage of water between the shaft and the casing. Two common types of seals are the stuffing box and the mechanical. In both cases, water is injected to lubricate the seal. The amount of water required is quite small (about 0.5 to 2 L per hour per pump). For this example, the upper value is used.

Water demand for pump lubrication
$= 9 \text{ pumps (total)} \times 2 \text{ L/h-pump} \times 24 \text{ h/day} \times 10^{-3} \text{ m}^3/\text{L}$
$= 0.432 \text{ m}^3/\text{d}$

Required clearwell volume for pump lubrication
$= 50\% \text{ of one day demand} \times \text{total water demand}$
$= 0.5 \text{ d} \times 0.432 \text{ m}^3/\text{d}$
$= 0.216 \text{ m}^3$

This quantity is small and is ignored.

c. Calculate clearwell volume for in-plant domestic water demand.

Water must be provided for the domestic needs of the employees. A maximum staff of 15 employees is anticipated. Per Table 4-10, the daily demand factor for service station workers with showers is 190 L per employee per day. This classification best describes the employment conditions at water treatment plants; therefore, the domestic water demand is calculated as follows:

In-plant domestic water demand
$= \text{number of employees} \times \text{demand factor}$
$= 15 \times 190 \text{ L/employee·d} \times 10^{-3} \text{ m}^3/\text{L}$
$= 2.85 \text{ m}^3/\text{d}$

Domestic demand is not constant through the day; therefore, the full daily demand is allocated for clearwell storage. Total in-plant demand is small and can be ignored.

d. Calculate clearwell volume for filter backwash water demand.

The filter backwash calculations were presented in Chapter 10. Each filter requires a water volume of 375 m^3 per filter backwash cycle. To provide sufficient water in reserve for washing filters, a clearwell volume sufficient to backwash all eight filters will be allocated to the clearwells.

Clearwell volume allocated to filter backwash
$= 1 \text{ d} \times 375 \text{ m}^3/\text{backwash cycle} \times 1 \text{ filter backwash cycle/filter·d} \times 8 \text{ filters}$
$= 3000 \text{ m}^3$

Provide a clearwell volume of 3000 m^3 for filter backwash water demand.

2. Balance of plant output and high-service pumping rate

a. Calculate the diurnal flow and the design pumping rates.

The peak hourly demand and maximum day demand for the design example were established in Chapter 5. These flow rates are 181,600 m^3/d and 113,500 m^3/d, respectively (Table 5-3). The design capacity of the plant is equal only to the maximum day demand. The difference between the peak hour and the maximum day demands must be supplied by clearwell storage, high-service pumping, distribution storage, and distribution pumping. Figure 13-8 illustrates the diurnal flow variations in the water demand. As illustrated, the peak hour demand occurs at approximately 7:00 p.m. and is 160 percent of the maximum day demand, or 181,600 m^3/d. The minimum hour demand occurs from 12:00 midnight to 4:00 a.m., with a demand of 48.5 percent of the maximum day demand, or 55,050 m^3/d.

The requirements of balanced pumping rate and plant output are met as follows.

(1) The treatment plant produces a constant maximum day demand of 113,500 m^3/d. This flow reaches the clearwells.

(2) The high-service pumps supply water from the clearwells into the ground storage reservoirs in the distribution system. Water is not directly fed into the distribution piping by high-service pumps.

(3) The distribution pumping from ground storage provides the distribution system demand under low and average flow situations and also fills the elevated storage.

(4) The peak distribution system demand is provided by the distribution pumping from the ground storage and from the elevated storage.

It is desirable to provide distribution storage in both elevated and ground storage. Sufficient storage should be provided to meet the demand in emergency situations, including temporary plant shut-downs. This storage is necessary and is provided so it would be cost-effective to use this volume to meet some of the peak demand and reduce the required high-service pumping capacity. In this example, the high-service pumps are designed for a maximum of 120 percent of the plant output capacity; the distribution pumps are designed for a maximum of 145 percent of the plant output capacity. The remainder of the peak hour demand will be provided from elevated storage reservoirs. These values are based on the authors' experience and are consistent with those at numerous water treatment facilities currently in operation. These values, however, are not necessarily typical. Pump capacities may be as low as 50 percent of the plant output, or, in many situations, as high as the peak hourly flow. Figure 13-8 also illustrates a pumping scheme for both the high-service and the distribution pumps, to meet the diurnal demand. Other schemes could also be used, provided that in all cases the total volume pumped during a day is 113,500 m^3. A

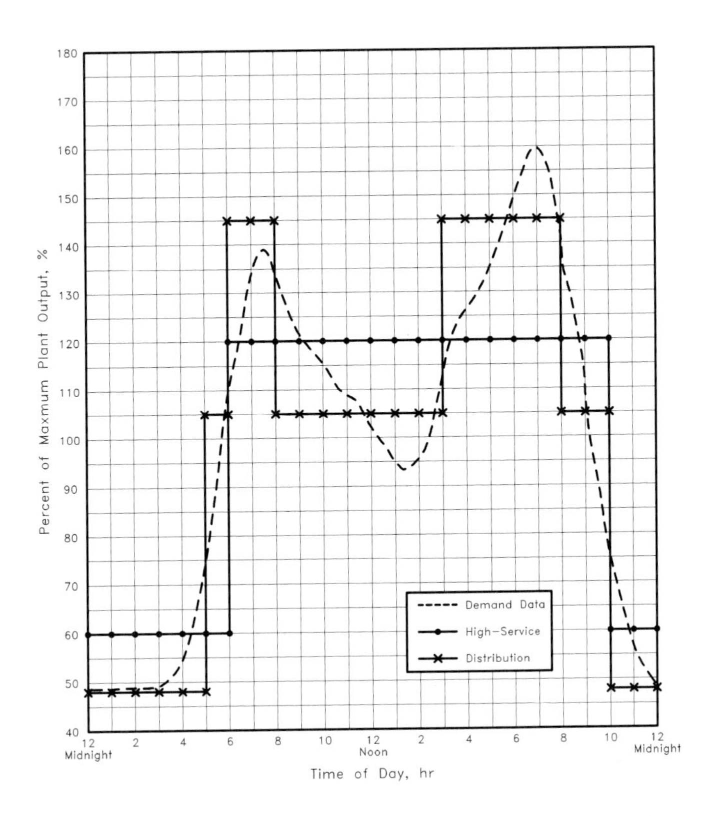

Figure 13-8 The diurnal flow and the high-service and distribution pumping rates. (Maximum plant output is 113,500 m^3/d.)

brief description of Figure 13-8 is necessary to provide an understanding of the pumping rates of the high-service pumps and of the distribution pumping.

(1) At low demand (10:00 p.m. to 5:00 a.m.), the diurnal hourly demand is at 48.5 percent of the plant capacity. The distribution pumps deliver 48 percent of the plant capacity into the distribution system to provide this demand. The high-service pumps operate at 60 percent of the plant output to fill the distribution storage.[c]

(2) As the hourly demand increases, the distribution pumping rate increases step-wise, to meet this demand. The high-service pumping is also stepped up, to fill the distribution ground storage. At 6:00 a.m., the high-service

c. From Figure 13-8, at midnight, the high-service pumping is 60% of plant output, or 113,500 $m^3/d \infty 0.6 = 68{,}100$ m^3/d.

pumping rate reaches 120 percent of the plant output. This flow is maintained until 10:00 p.m.; after that, the pumping rate is reduced in one step. During the low demands, the ground and elevated distribution storages are filled. It is possible to have more step-wise increases and decreases in the high-service pumping to match the distribution demand. This will reduce the volume of clearwell. In authors' judgment, many step-wise increases or decreases in high-service pumping will make pumping operation more complex and undesirable.

(3) The distribution pumping increases step-wise to keep pace with the hourly demand. At 6:00 a.m., the distribution pumping reaches 145 percent of the plant capacity, to supply the morning hourly peak. As the morning peak is satisfied and the demand decreases, the distribution pumping fills the elevated reservoirs, and the rate drops to 105 percent of plant capacity between 8:00 a.m. and 3:00 p.m. The pumping rate again steps up to meet the evening demand. At 3:00 p.m., the maximum distribution pumping capacity of 145 percent of the plant output is reached. This pumping rate is maintained for five hours, then it is reduced to 48 percent of the plant output. The peak hourly demand at 7:00 p.m. is met by the distribution pumping and elevated storage. As with high-service pumping, step-wise increases and decreases in distribution pumping to match the diurnal demand curve can reduce the ground and elevated storage requirements in the distribution system, but, again, the operation will become complex and difficult to manage.

b. Calculate the clearwell storage required to balance the pumping rates.

The clearwell volume is determined from the balance differences between the pumping rates and the water treatment plant production rate. The hourly high-service pumping rate and cumulative high-service pumping flow are provided in Table 13-5. The hourly plant production and cumulative plant production rates are provided in this table. The plant production rate is constant throughout the day, so the clearwell capacity can be calculated from (1) the mass diagram, or (2) the maximum-deficit and maximum-surplus values. These two methods are presented below.

i. The maximum-deficit and -surplus method

The maximum-deficit and -surplus method uses the hourly plant production rate and hourly high-service pumping rate. These flow rate values are provided in Columns 2 and 4 of Table 13-5. The cumulative plant output and high-service pumping volume are calculated in Columns 3 and 5, respectively. The difference between these two values (Column 6) is obtained by subtracting Column 4 from Column 2. In Column 6, a positive value represents net inflow into the clearwell—a surplus; a negative value represents a deficit, or net flow out of the clearwell. Column 7 is the cumulative change in storage. The total storage requirement is computed by adding the absolute value of the maximum deficit to the maximum surplus of change in storage from Column 7.

Clearwell storage required
= absolute value of maximum deficit + maximum surplus
= $|-3{,}800|\ m^3 + 11{,}400\ m^3$
= $15{,}200\ m^3$

A closer look at the data in Table 13-5 provides some important information about the operation of the clearwell. Around 6 a.m., the clearwell is at its full capacity (maximum level). Gradually, the water is withdrawn from the well and its level drops. Around 10 p.m. the usable capacity has been withdrawn, and the lowest water elevation is reached. After that, the filling operation begins and the cycle repeats. The hourly volume of usable water remaining in the clearwell is shown graphically in Figures 13-9 by plotting against time the values in column 7 of Table 13-5.

ii. The mass diagram method

Figure 13-9 illustrates a graphical representation of the clearwell volume required to offset pumping and plant output rates. This represents the mass diagram method. Curve A is plotted from the cumulative plant output (Column 3) in Table 13-5. This curve is also obtained by drawing a straight line from zero (at 12:00 midnight) to 113,500 m^3/d (at 12:00 midnight the following day) and represents the water treatment plant production. Curve B is obtained by plotting the cumulative high-service pumping rate (Column 5 of Table 13-5). The volume of storage required is equal to the vertical distance between two tangent lines drawn from its furthest distance above and below the Curve B and parallel to Line A. The required volume of the clearwell is approximately 15,200 m^3.

3. Clearwell capacity requirement

The total capacity requirement of the clearwells is summarized in Table 13-6. A total capacity of 27,600 m^3 is required for clearwells.

Two clearwells will provide the operational flexibility needed.

$$\text{Volume of each clearwell required} = \frac{27{,}600\ m^3}{2} = 13{,}800\ m^3$$

The dimensions of each clearwell are 40 m ∞ 40 m ∞ 9 m (water depth). (See Figure 12-21.)

$$\text{Volume of each clearwell provided} = 40\ m \times 40\ m \times 9\ m = 14{,}400\ m^3$$

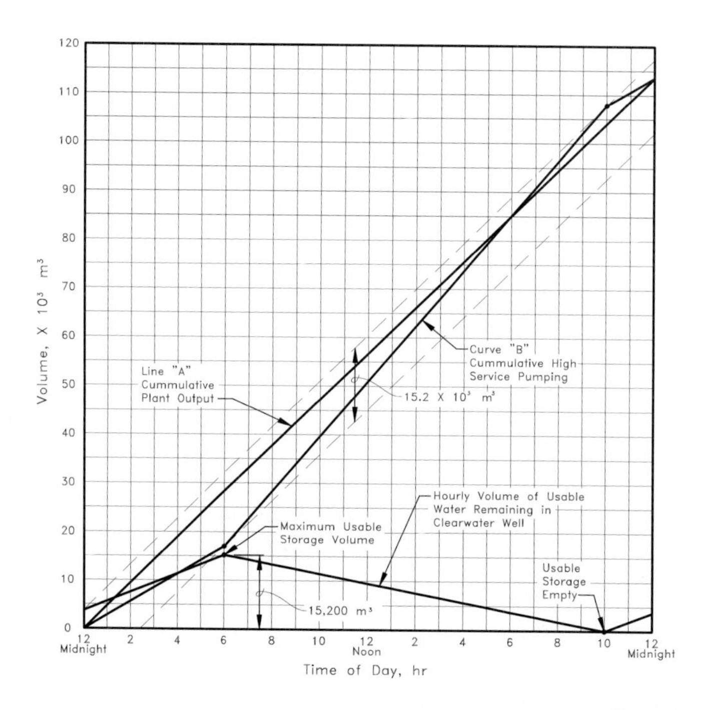

Figure 13-9 Mass diagram for determination of clearwell capacity. (Showing cumulative plant output, high-service pumping curves, and hourly change in usable water volume remaining in the clearwells.)

The clearwell volume provided is greater than that required. The actual volume will vary with the water level in the clearwell. Ultrasonic transducers are provided to monitor the clearwell elevation, which will vary, depending upon the pumping rate of high-service pumps.

4. Dimensions of clearwells

Often the clearwell volume is used to satisfy the disinfection contact time requirements of the Surface Water Treatment Rule. For this example, the disinfection contact time is achieved in separate chlorine contact channels. These channels are structurally attached to the clearwells, but the volume of the chlorine contact channels is not considered part of the total clearwell capacity. The disinfection requirements and contact time calculations were presented in Section 12.8.2, Step A. The layout of the filter back pressure weir common bay, the chlorine contact channels, the clearwells, the high-service pumps, and the filter backwash water pumps is also shown in Figures 12-20 and 12-21.

Table 13-5 Clearwell Storage Capacity Requirements

Time	Plant Output,[a] ∞10^3 m^3/h	Cumulative Plant Output, ∞10^3 m^3/h	High-Service Pumping Flow, ∞10^3 m^3/h	Cumulative High-Service Pumping Flow, ∞10^3 m^3/h	Difference between Plant Output and High-Service Pumping Flow,[b] ∞10^3 m^3	Surplus or Deficit in Storage, ∞10^3 m^3
(1)	(2)	(3)	(4)	(5)	(6)	(7)
1 a.m.	4.73	4.73	2.83[c]	2.83	+1.90[d]	+1.90
2	4.73	9.46	2.83	5.66	+1.90	+3.80
3	4.73	14.19	2.83	8.49	+1.90	+5.70
4	4.73	18.92	2.83	11.32	+1.90	+7.60
5	4.73	23.65	2.83	14.15	+1.90	+9.50
6	4.73	28.38	2.83	16.98	+1.90	+11.40[e]
7	4.73	33.11	5.68[f]	22.66	–0.95	+10.45
8	4.73	37.84	5.68	28.34	–0.95	+9.50
9	4.73	42.57	5.68	34.02	–0.95	+8.55
10	4.73	47.30	5.68	39.70	–0.95	+7.60
11	4.73	52.03	5.68	45.38	–0.95	+6.65
12 Noon	4.73	56.76	5.68	51.06	–0.95	+5.70
1	4.73	61.49	5.68	56.74	–0.95	+4.75
2	4.73	66.22	5.68	62.42	–0.95	+3.80
3	4.73	70.95	5.68	68.10	–0.95	+2.85
4	4.73	75.68	5.68	73.78	–0.95	+1.90
5	4.73	80.41	5.68	79.46	–0.95	+0.95
6	4.73	85.14	5.68	85.14	–0.95	±0
7	4.73	89.87	5.68	90.82	–0.95	–0.95
8	4.73	94.60	5.68	96.50	–0.95	–1.90
9	4.73	99.33	5.68	102.18	–0.95	–2.85
10	4.73	104.06	5.68	107.86	–0.95	–3.80[g]
11	4.73	108.79	2.83	110.69	+1.90	–1.90
12 Mid-night	4.71[aa]	113.50	2.81[bb]	113.50	+1.90	±0

a The hourly plant production rate $= \dfrac{113{,}500\ m^3/d}{24\ h/d} = 4{,}730\ m^3/h = 4.73 \times 10^3\ m^3/h.$

b Surplus (+) or deficit (–).

c High-service pumping rate is 60% of maximum day flow (Figure 13-8), so

$$\text{the flow} = \frac{0.60 \times 113{,}500\ m^3/d}{24\ h/d} = 2{,}830\ m^3/h = 2.83 \times 10^3\ m^3/h.$$

d Column (6) = Column (2) – Column (4) = ($4.73 \infty 10^3\ m^3/h$ - $2.83 \infty 10^3\ m^3/h$) ∞ 1 h = $1.90 \infty 10^3\ m^3$.

e The maximum surplus = $11.40 \infty 10^3\ m^3/h$.

f High-service pumping rate is 120% of maximum day flow (Figure 13-8), so

$$\text{the flow} = \frac{1.20 \times 113{,}500\ m^3/d}{24\ h/d} = 5{,}680\ m^3/h = 5.68 \times 10^3\ m^3/h.$$

g The maximum deficit = $-3.80 \infty 10^3\ m^3/h$.

aa Value adjusted to account for rounding error.

bb Value adjusted to account for rounding error.

Table 13-6 Clearwell Capacity Requirements

Purpose of In-Plant Service Water Supply	Volume Requirement, m^3
Chemical dilution and feed	200
Pump operation	small and assumed zero
In-plant domestic water	small and assumed zero
Filter backwash	3000
To balance pumping rates	15,200
Subtotal	18,400
Add 50 percent for reserve	9200
Total clearwell capacity required	27,600

Step C: Design of High-Service Pump Station

1. High-service pumping strategy

The treated water transmission pipeline connects the high-service pump station to the distribution ground storage reservoirs. This pipeline begins at the high-service pump station, and its size decreases as it passes by each ground storage reservoir site. The flow of water into each ground storage reservoir is regulated by a flow modulating valve located at each site. Each valve is controlled automatically by the level of water in the connected ground storage reservoir. As the water level in the reservoir drops, the valve opens gradually, thus increasing the flow into the tank. Likewise, as the water level rises, the valve closes gradually, thus reducing the flow into the ground storage reservoir. The plant operator at the plant control room has the ability to override the automatic control system and remotely set a flow rate, using the telemetry system.

2. High-service pumping capacity

   The total peak hour demand to be supplied by the elevated storage and the distribution pump station is 181,600 m^3/d. The maximum treatment plant capacity is 113,500 m^3/d. The clearwell volume calculations utilized are for balanced pumping and storage. (See Fig. 13-9.) In most plants, the high-service pump capacity is typically larger than the plant capacity, allow the operator to meet short-term peak demands from the clearwell storage volume. Consistent with this approach, the high-service pump capacity is kept at 120 percent of the plant capacity. Therefore, high-service pump capacity = 1.2 ∞ 113,500 m^3/d = 136,200 m^3/d.

   Ideally, the actual high-service pump capacity should be determined from complex model studies using *quasi-dynamic* computer models. Such models provide the flow rate from the high-service pumping needed to maintain the required levels in the ground storage reservoirs during peak demand and during filling periods. Sometimes, such computer models may not be required, and a simplified approach is sufficient. The simplified approach followed in this example is typical and is used quite often by designers.

3. Number of high-service pumps

   In this example, six pumps will be provided, although the total demand can be met by only five pumps. Again, the number of pumps is determined primarily from the authors' experience and judgement. Five pumps, each ranging from approximately 25 to 35 percent of the plant capacity, provide the operators with a broad range of pumping rates from zero to a full capacity of over 120 percent. The sixth pump provides a stand-by unit, to ensure full capacity in the event of equipment failure.

   Often to maintain the flexibility to meet different demand conditions, a large number of smaller constant speed pumps is provided. In other facilities, fewer but larger pumps, each with a variable speed drive unit, are used to achieve the same result. The decision to use variable speed units can be made from an economic analysis, but such an analysis must also consider the operator's ability to maintain variable speed equipment.

4. Pump selection and station layout

   The pump selection and station layout involves (1) identification of pump and manufacturer's information, (2) preparation of modified head-capacity curve, (3) preparation of system head-capacity curves and pump combination, and (4) pump arrangement. These steps have been presented in great detail in Chapter 7 for a raw water pump station. The procedure is similar for high-service pumps and will not be repeated here. Readers should refer to appropriate sections in Chapter 7 to develop background knowledge on this subject.

   The layout of the high-service pump station is shown in Figure 12-26. In this case, the layout is based on radial flow horizontal pumps arranged in a dry pit. These pumps are used for higher head applications. Refer to Chapter 7 for additional discussion on various types of pumps.

The necessary instrumentation for a high-service pumping station includes pump and motor contacts, temperature probes, current meters and pressure indicators, flow meters, and limit switches to monitor valve positions.

Step D: Improvement of Distribution Pumping Stations

1. Distribution pumping strategy

   The distribution pumping stations pump water from the ground storage reservoirs into the distribution system. The distribution pump operation is controlled by the levels in the *elevated tanks*. The telemetry system sends to the designated pump controls a signal proportioned to the tank level. As the level drops, more pumps are turned on; as the level rises, some of the pumps are turned off. This type of pump operation automatically balances the number of pumps operating with water demand and maintains the level in the elevated tanks.

2. Distribution pumping capacity

   As was previously discussed, the designed distribution pumping has been selected for 145 percent of the plant capacity. This provides a distribution pumping rate = 1.45 ∞ 113,500 m^3/d = 164,600 m^3/d, or approximately 91 percent of the peak hour demand (181,600 m^3/d). The remaining 17,000 m^3/d must be supplied from elevated tanks. The design of the elevated storage facilities necessary to provide this demand is discussed in subsequent sections.

   The existing ground storage and distribution pumping facilities and the distribution system of Modeltown were discussed in Chapter 5. Currently, there are three pump stations, with a combined pumping capacity of 80,000 m^3/d, available to serve the community. To meet future needs, an additional pumping capacity of approximately 85,000 m^3/d must be provided. To achieve this, a complete new pumping facility (the West Side Pump Complex), complete with ground storage reservoir, will be constructed on the west side of the community. This pump station will have a capacity of 47,000 m^3/d. The three existing pump stations will be expanded to increase their combined capacity from 80,000 m^3/d to 118,000 m^3/d. This modification will provide a total pumping capacity of 165,000 m^3/d (118,000 m^3/d + 47,000 m^3/d). The capacities of the pump stations, after modifications, are summarized here.

   West Side Pump Complex (new): Total Capacity 47,000 m^3/d (544 L/s)

   North Pump Station (existing): Total Capacity 62,500 m^3/d (723 L/s)

   Modeltown Pump Station (existing): Total Capacity 27,500 m^3/d (318 L/s)

   Northeast Pump Station (existing): Total Capacity 27,500 m^3/d (318 L/s)

Although the total combined pumping capacity of all four pump stations provided in this design is 165,000 m^3/d, it is desirable to provide 10 percent additional capacity to each distribution pumping station to meet any emergency situations.

The optimum location of distribution pump stations is best determined by the computer model studies of the distribution system. These studies generally pinpoint both areas where low pressure problems would occur and possible pumping station sites. The computer model can also be used to simulate alternative pump station, elevated storage, and piping system improvements and thus to help engineers develop an optimum design.

The site location of the proposed West Side Pump Complex is illustrated in Figure 5-1. This complex features flow meters to measure flow into and out of the complex, a ground storage reservoir, and a pump station. This pump station features vertical turbine pumps. The vertical pumps are selected because they offer construction cost savings. The pump suction must be below the bottom of the reservoir, in order to utilize the entire reservoir capacity. The site is essentially flat, so the vertical pumps must be set in a separate suction well, with their suction below the reservoir. If horizontal pumps were used, a structural pit would be necessary in the reservoir, resulting in an increased structure cost. These conditions are site-specific and should be included in developing the design alternatives. Chlorine and ammonia storage and feed systems must also be included. Engineers must evaluate different constraints and options before selecting a pump type for a specific situation. The design of distribution pumping equipment is similar to that for high-service or raw water pumping, so it is not necessary to repeat the details. Readers should utilize the concepts presented in Chapter 7 to develop a design specific to the site and the conditions of the problem.

Step E: Improvement of Ground Storage Reservoirs The ground storage reservoirs provide storage to offset the differences between the high-service pumping rates and the pumping rates of the distribution pump station. They also provide a hydraulic break between the treated water transmission system and the distribution system, thereby reducing hydraulic transient impacts. Finally, the ground storage reservoirs offer a storage volume of water to mitigate any emergencies such as fires or temporary loss of water supply from the treatment plant. The volume requirement for the ground storage reservoirs is based on the differences between the high-service pumping rate and the distribution system pumping rate. In addition, many state regulatory authorities require a minimum ground storage volume that must be provided in the distribution system.

1. Minimum ground storage requirement

   The hourly pumping rates for both the high-service and the distribution pump stations and the associated storage requirements can be determined from the diurnal demand curve and pumping scenario illustrated in Figure 13-8. The procedure for calculating the deficit and surplus flows to ground storage basins is similar to that for clearwells. This procedure is covered in Section 13.8.2, Step B.2. The cumulative high-service and distribution pumping curves are shown in Figure 13-10.

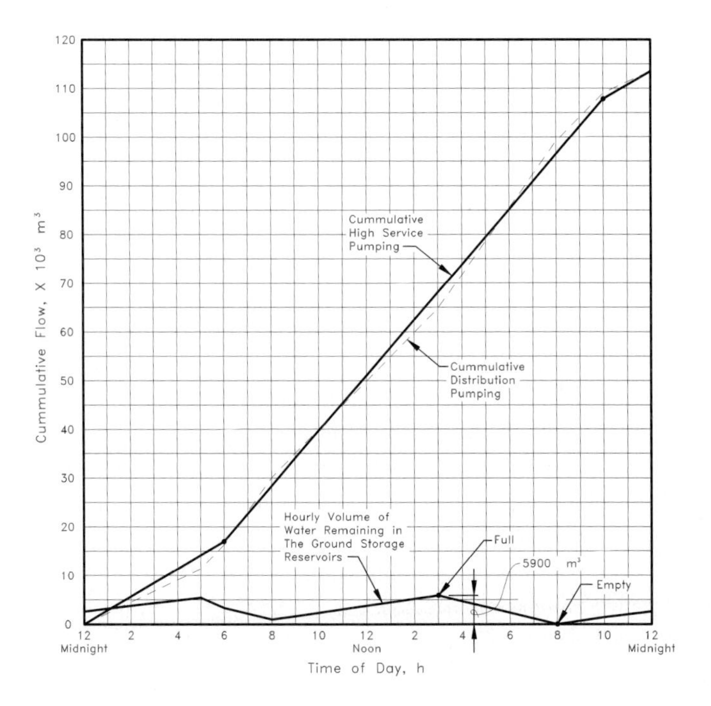

Figure 13-10 Mass diagram for determination of ground storage capacity. (Showing cumulative high-service and distribution pumping curves and hourly change in usable water volume remaining in the distribution ground storage reservoirs.)

2. Ground storage capacity requirement

   By adding the absolute values of the maximum surplus (3270 m^3) and deficit (–2,630 m^3), a total storage requirement of 5900 m^3 is obtained. By adding 50 percent extra capacity for emergency storage and operational flexibility, therefore, the total ground storage required is 1.5 ∞ 5900 m^3, or 8850 m^3. The by-hour volume of usable water remaining in the ground storage reservoirs is also illustrated in Figure 13-10. It is clearly indicated that all ground storage reservoirs will be full at 3 p.m. The total ground storage volume, 5900 m^3, will deplete by 8 p.m., when refilling starts. Full storage capacity of 5900 m^3 will be reached again at 3 p.m. next day.

3. Expansion of ground storage capacity

   The total existing ground storage volume, as discussed in Chapter 5, is 12,000 m^3 (Table 5-5). The existing capacity thus exceeds the minimum required storage of 8850 m^3;

therefore, no additional storage is needed. A new distribution pumping facility will be constructed at the West Side Pump Complex, however, so additional ground storage will be added at this pump station site. This ground storage is necessary to provide a hydraulic break between the water transmission and distribution systems and to provide for emergency operation of the pump station should the water supply to this location be temporarily interrupted. The selection of ground storage volume often must be made on the basis of economics as well as of an engineering evaluation. More volume will provide greater operational flexibility but would also add to the project cost.

The volume provided at the West Side Pump Complex will be equal to two hours' flow at the maximum pumping rate of the pump station (47,000 m^3/d). In the authors' judgment, two hour's storage will provide sufficient water supply for emergency operation and will offer flexibility for the operation of the high-service pump and other distribution pumping facilities.

Ground storage volume required at the West Side Pump Complex

$$= \frac{2 \text{ h} \times 47{,}000 \text{ m}^3/\text{d}}{24 \text{ h/day}} = 3920 \text{ m}^3$$

Provide ground storage capacity of 4000 m^3 at the West Side Pump Complex. Thus, total ground storage capacity provided = 12,000 m^3 + 4000 m^3 = 18,000 m^3.

Step F: Improvement of Elevated Storage Facilities The elevated storage reservoirs provide storage to offset the differences between the distribution system demand and the high-service pumping rates. Elevated storage reservoirs provide sufficient volume of water stored at distribution pressure to meet any emergency requirements, such as fire demands or temporary failure of the high-service pump station or the distribution pumping. Furthermore, the elevated storage reservoirs provide a hydraulic buffer to stabilize pressure in the distribution system.

1. Minimum elevated storage requirement

   Typically, the minimum volume requirement for elevated storage tanks is based on the differences between distribution pumping capacity and anticipated maximum hourly demand. In addition, many state regulatory authorities require a minimum volume of elevated storage.

   The associated elevated storage requirements can be determined from the anticipated hourly demand curve and pumping scenario for the distribution pump stations illustrated in Figure 13-8. The procedure for calculating the deficit and surplus flows at elevated storage basins is similar to that for clearwells. This procedure is covered in Section 13.8.2, Step B.2. A total storage requirement of 2640 m^3 is obtained by adding the maximum surplus, 460 m^3, and deficit, −2,180 m^3. It is necessary to provide an additional 100 percent of volume, for safety and to maintain operational flexibility. Therefore, the minimum elevated storage volume provided is 5280 m^3. The cumulative hourly demand, the cumulative distribution pumping rate, and the usable elevated reservoir volume remaining at different hours of the day are shown in Figure 13-11. The results indicate that the elevated reservoir capacity will be full at 10 p.m. and empty at 11 a.m.

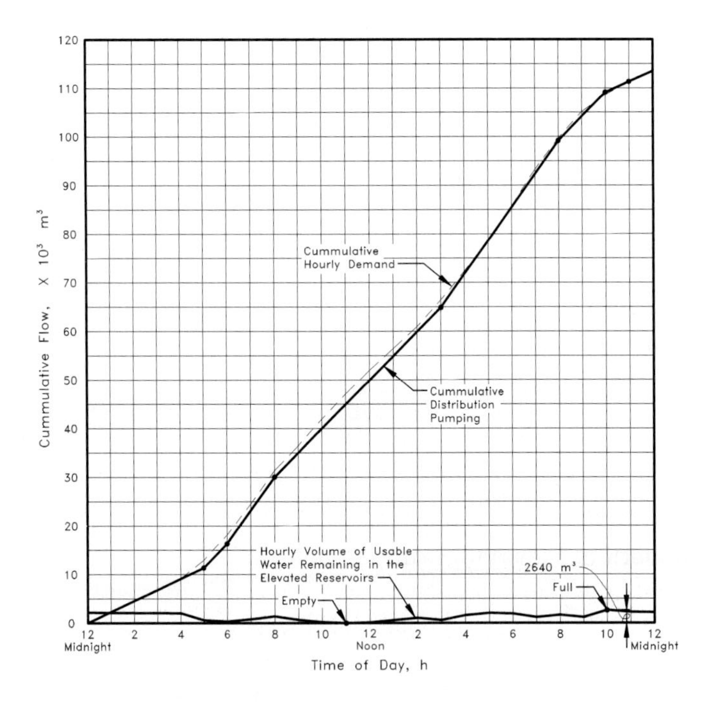

Figure 13-11 Mass diagram for determination of elevated storage capacity. (Showing cumulative distribution pumping and water demand curves and hourly change in usable water volume remaining in the elevated storage tanks.)

2. Emergency elevated storage

Emergency elevated storage is desired to meet the demand should temporary failure of distribution pumping occur. In situations where a single distribution pump station is provided, the emergency elevated storage volume should be sufficient to meet a minimum of two hours of the highest demand without resupply from pump stations. This offers the necessary emergency storage in case of pump station failure. There are four pump stations in this Design Example. The emergency volume should be sufficient to supply the maximum two-hour demand with the largest pump station out of service. The largest pump station in the Design Example is the North Pump Station, with a capacity of 62,500 m^3/d or 2600 m^3/h. Therefore, the additional emergency storage required to balance pumping failure is calculated as follows:

$$\text{Emergency elevated storage} = 2\ \text{h} \times 2600\ \text{m}^3/\text{h} = 5200\ \text{m}^3$$

There is another way to calculate the capacity for emergency storage. With multiple distribution pumping, the capacity of each pump station is calculated with its largest pump out of service. The largest individual pump station capacity thus calculated provides the system capacity for meeting two-hour storage emergency requirements. The authors prefer using the lower of these two values.

3. Total elevated storage required

The total amount of storage needed is equal to the emergency storage volume plus the storage volume required to balance pumping demand.

$$\text{Total storage needed} = 5200\ \text{m}^3 + 2640\ \text{m}^3 = 7840\ \text{m}^3$$

The available existing storage is 6,000 m^3. This amount was established in Table 5-5 in Section 5.4.7.

4. Total elevated storage provided

The existing emergency storage available is 3360 m^3 (6000 m^3 – 2640 m^3). An additional elevated storage capacity of 1840 m^3 (5200 m^3 – 3360 m^3) is therefore required in this capital improvement project. An elevated tank will be constructed by duplicating the smallest existing elevated tank. The smallest existing elevated tank is the Modeltown Tower, which has a capacity of 2000 m^3 (Table 5-5). This elevated storage tank will provide sufficient capacity for a significant time in the future. Total elevated storage volume for the city, therefore, will be 8000 m^3.

The authors have observed many existing water distribution systems that have emergency elevated storage significantly less than that required. In these cases, either the volume of emergency supply was compromised to cater to for limited capital resources, or there was sufficient backup capacity in the distribution pumping for emergency situations. In the Design Example, since there are four distribution pumping stations, so sufficient back up capacity is available for emergency supply, and the distribution system could function well without additional elevated storage.

The design of elevated tanks includes consideration of volume, height, and location. Ideally, the optimum height and location will be developed from the distribution model studies discussed in the following sections. Most engineers prefer to match the top overflow elevation of the new tank with that of the existing tanks, and locate the new tank in the lowest pressure area. Where suitable land is available, higher ground elevations are desirable to minimize construction cost. Lower ground elevations require extra tank height to provide the required hydraulic gradient. Height substantially increases the cost of the elevated storage tanks.

Another consideration for elevated tank design is the volume required to sustain system pressures. An elevated storage tank may have adequate volume for emergency storage and for balancing pumping with demands, but if the tank is not properly located in the distribution system, its total volume may not be usable to sustain the pressures. Therefore, location and hydraulic pressure are important design considerations; they are obtained from computer model studies. This topic is covered in subsequent sections.

Step G: Improvement of Distribution System Hydraulic analysis of the Modeltown water distribution system is conducted by using computer simulation technology. The results from the simulation runs provide important guideline information for improvement of the distribution system. The hydraulic analysis procedure consists of the following steps: (1) evaluate the existing distribution system, (2) develop improvement recommendations, (3) check system performance under critical conditions after improvements, (4) check system pressure under a fire condition, and (5) report final results with recommended improvements.

1. Performance of distribution system without improvement

   a. Development of distribution system network for computer simulation

      i. Create distribution system network.

         The distribution system is designed to deliver the peak hourly demand to each customer. The distribution system is composed of a network of pipes of various sizes, valves, reservoirs, and pumps. Generally, the pipelines are routed in street or alley right-of-way, to provide access to water supply from each parcel of land. The first step in evaluation of a water distribution system is to set up a network with pipe nodes and links representing the distribution system. The land use map, as well as existing and proposed water systems for Modeltown, are described in Chapter 5 and are shown in Figure 5-2. Figure 13-12 shows the pipe layout and the ground and elevated storage of the existing distribution system. A numbering system is developed to identify a particular pipe or node (pipe junction) throughout this distribution system. In this numbering system, each pipe and node is coded with a number. This step is illustrated in Figure 13-13, which represents the existing and expanded distribution system.

      ii. Determine water demand.

         The next step is to distribute the total system demand to each pipe node. This task involves determining the water demand for each area of the city, on the basis of land use. A great deal of effort and judgment is needed to develop the demand at various nodes. The area served by each node is determined by constructing perpendicular bisectors for each pipe in the system. The areas bounded by the resulting polygons are served by respective nodes (Figure 13-13). Some judgement, however, must be used in completing the polygons. For example, zoning or land use boundaries are often the limits of a service area (example node 101, Figure 13-13). Also, in some cases, the bisector may have to be skewed to logically complete a polygon (example node

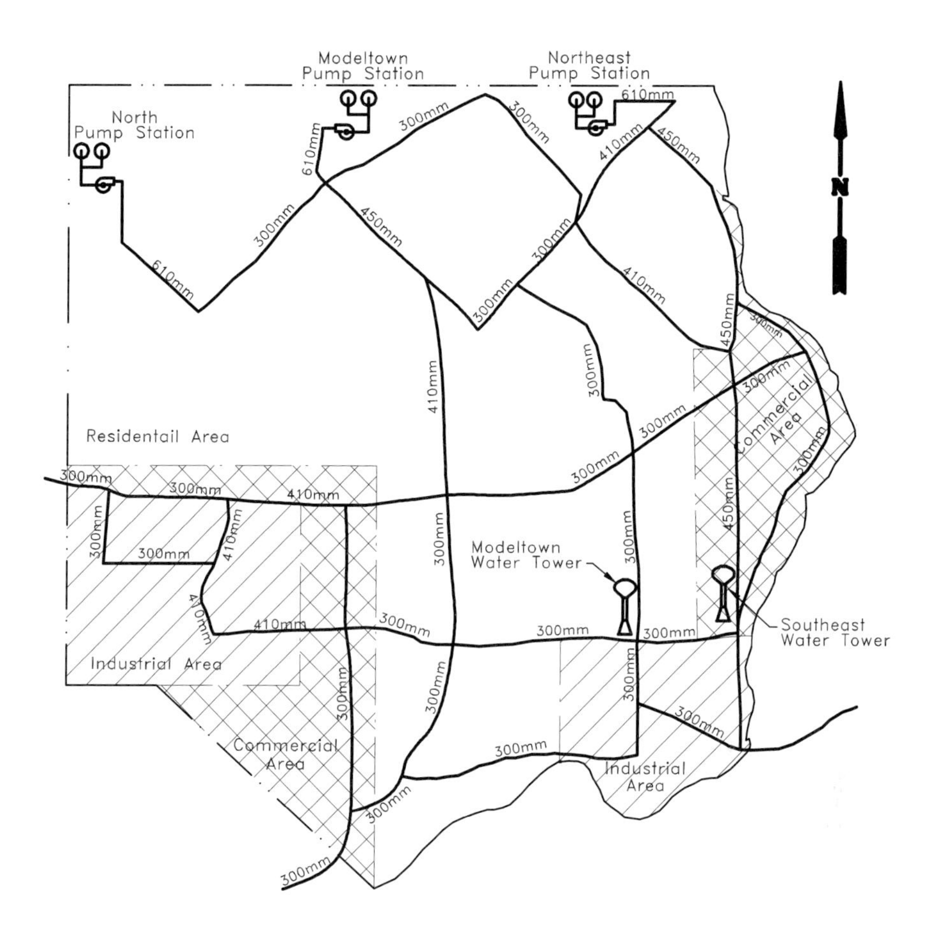

Figure 13-12 Modeltown-existing distribution system.

122, Figure 13-13). Engineers must keep in mind that neither this procedure nor any other is exact in distributing flows; therefore, judgement must be used to develop an accurate flow distribution model.

The area of each land use type is then multiplied by a demand factor, to estimate the node demand. The following demand factors were developed for the Modeltown area from city water use records and from field measurements. Field measurements included installation of recording meters on a number of user facilities, to develop actual average flow multipliers. The average per capita water demand may be verified from the demand at the node and the population served within that polygon. The population densities in residential,

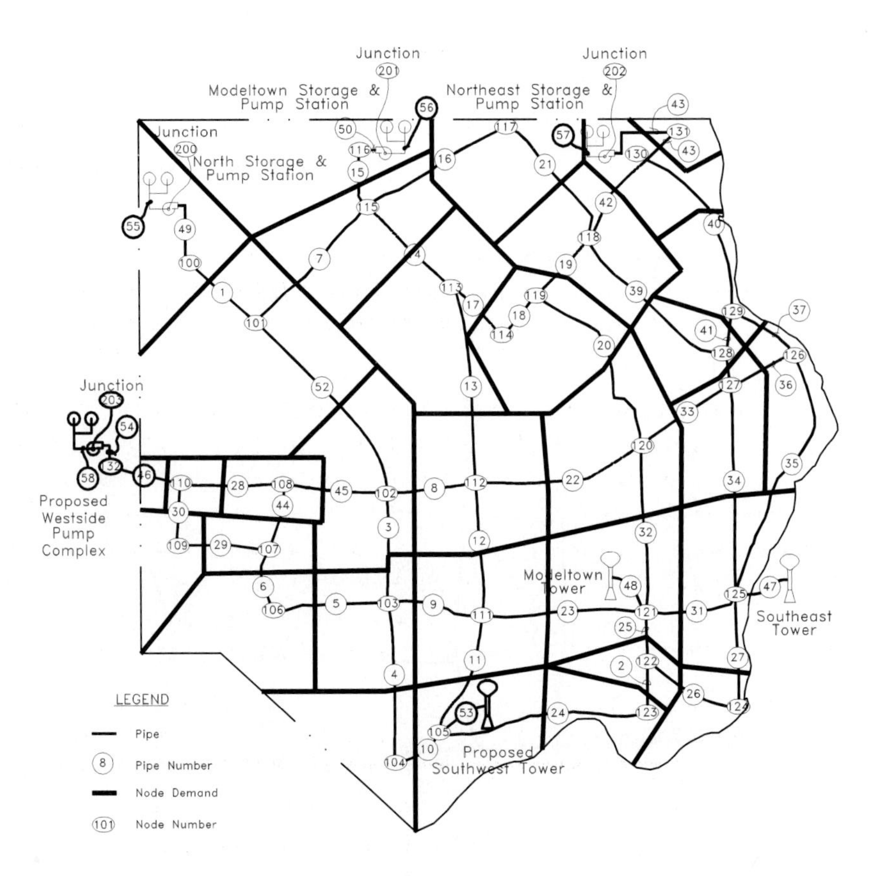

Figure 13-13 Modeltown-existing and improved distribution system. (Showing pipe and node coding system, node service polygon boundaries, and location of ground and elevated storage facilities.)

commercial, and industrial areas depend upon land use conditions, and are provided in Chapter 4 (Table 4-5). The following average flow multiplier values were obtained:

Residential	1300 $m^3/km^2{\cdot}d$
Commercial	2100 $m^3/km^2{\cdot}d$
Industrial	3000 $m^3/km^2{\cdot}d$

It may be noted that the above demand multipliers are for average demand conditions. The demand multipliers for maximum day and peak hour conditions are significantly larger. In Chapter 5, the ratio of maximum day to average day demand and the ratio of peak hour to maximum day demand were developed from the water demand analysis of the distribution system of

Modeltown. These ratios were 1.96 and 1.60, respectively (Section 5.4.4). On the basis of these ratios, the maximum day and peak hour demand multipliers are developed. These multipliers are listed here:

	Peak Hour, $m^3/km^2 \cdot d$	Maximum Day, $m^3/km^2 \cdot d$	Average Day, $m^3/km^2 \cdot d$
Residential	4200	2600	1300
Commercial	6500	4100	2100
Industrial	9400	5900	3000

The following example illustrates the peak hour flow calculation procedure for one node (node 102). Total node demand is calculated from Eq. (13.7).

$$\text{Total node demand} = A_R \times D_R + A_C \times D_C + A_I \times D_I \tag{13.7}$$

where

A_R, A_C, and A_I = area of service polygon for each type of land use: residential, commercial, and industrial, respectively. These are established from the respective areas within the polygon (Figure 13-13).

D_R, D_C, and D_I = unit demand for residential, commercial, and industrial areas, respectively. The unit demand factors may be derived from water use records or from field testing.

Therefore, the projected peak hour water demand at node 102 is calculated as follows:

$$\begin{aligned}
&\text{Total node demand at node 102} \\
&= 0.72\ \text{km}^2 \times 4200\ \text{m}^3/\text{km}^2\cdot\text{d} + 0.66\ \text{km}^2 \times 6500\ \text{m}^3/\text{km}^2\cdot\text{d} \\
&\quad + 0.18\ \text{km}^2 \times 9400\ \text{m}^3/\text{km}^2\cdot\text{d} \\
&= 9006\ \text{m}^3/\text{d} = 9.00 \times 10^3\ \text{m}^3/\text{d}
\end{aligned}$$

Maximum and average day demands can also be calculated. These values for node 102 are 5640 m^3/d ($5.64 \infty 10^3$ m^3/d) and 2860 m^3/d ($2.86 \infty 10^3$ m^3/d), respectively. These values should represent the measured demands based on the actual metered flows supplied to the customers. A study of Modeltown's pumping records and the city's water meter records indicated that the city typically pumps approximately 11.3 percent more water than is actually used by the customers (Section 5.4.4, Step A). The difference is due to unaccounted water losses, such as leaks, unauthorized uses, and unmetered municipal uses. Such losses are typical in water distribution systems and must be accounted for in the distribution system design. Each node demand is therefore multiplied by 1.113, to include this unmetered demand. Therefore, the actual peak hour demand at node 102 is calculated as follows:

Actual demand at node 102 $= 1.113 \times (9.00 \times 10^3\ m^3/d)$
$= 10.02 \times 10^3\ m^3/d$

Table 13-7 lists the measured and actual demand (including unaccounted water uses and losses) for each node. This table also lists the service polygon areas for each node.

The results presented in Table 13-7 are important for the analysis of the distribution system, and therefore, must be verified. Total service area of Modeltown obtained in Column 5A is 30.65 km^2. Total area of the town, obtained from the land use in the design year, is 30.5 km^2 (Section 5.4.2). The total peak hourly demand for the entire service area, calculated from the peaking ratios, is 118,660 m^3/d (181.66 ∞ 10^3 m^3/d). This flow checks with that developed in Table 5-3.

iii. Estimate friction factor "*C*".

The water distribution analysis requires use of a computer simulation model and input data. The complete analysis requires estimate of friction factors and other input data. The method for estimating a friction coefficient is presented in this section.

An important consideration in developing a computer model for a water distribution system is the estimate of the friction coefficients. The most accurate means of estimating this is field testing; however, rule-of-thumb values can be used in some cases. For this example, and in most cases, field testing is used. The standard procedure for field testing is illustrated next:

From the city records, the diameter, age, and material of each water main are obtained. Representative pipeline segments for each pipe material and age category are selected for field testing. Figure 13-14 illustrates the representative data from a field test on pipe 27. The test is performed by closing a valve at one end of a short segment of pipe 27, thus ensuring that the supply comes only from the opposite end of the pipe. Pressure gauges are installed at two or three locations along the pipeline, and the relative elevation of each pressure gauge is obtained. The customer's yard water hose connection pipes make convenient locations to install these pressure gauges. A flow meter is installed on the fire hydrant located near the end of the pipe, with the valve closed. The hydrant is opened and allowed to flow. The flow rate and pressure gauge readings are all recorded. The fire hydrant valve is adjusted to yield a series of flow rates, and the data is recorded for each condition. This procedure is repeated for at least three or four flow rates. It is important to request customers to refrain from using water during the period, which usually lasts for about one hour. The reduction of test data and the procedure to calculate the *C* value at each flow in the pipe section are summarized in Tables 13-8 and 13-9. The

Table 13-7 Estimate of Peak Hour Demand for Each Distribution System Node

Node	Residential @4.2 ∞10^3 m^3/m^2·d		Commercial @6.5 ∞10^3 m^3/m^2·d		Industrial @9.4 ∞10^3 m^3/m^2·d		Total		Total Demand Plus Unmetered Use, ∞10^3 m^3/d
	Area, km^2	Demand, ∞10^3 m^3/d	Area, km^2	Demand, ∞10^3 m^3/d	Area, km^2	Demand, ∞10^3 m^3/d	Area, km^2	Demand, ∞10^3 m^3/d	
(1)	(2A)	(2B)	(3A)	(3B)	(4A)	(4B)	(5A)	(5B)	(6)
100	1.22	5.12	–	–	–	–	1.22	5.12	5.70
101	2.96	12.43	–	–	–	–	2.96	12.43	13.83
102	0.72	3.02	0.66	4.29	0.18	1.69	1.56	9.00	10.02
103	–	–	0.68	4.42	0.17	1.60	0.85	6.02	6.70
104	–	–	0.77	5.00	–	–	0.77	5.00	5.57
105	1.06	4.45	–	–	–	–	1.06	4.45	4.95
106	–	–	0.23	1.49	0.78	7.33	1.01	8.82	9.82
107	–	–	–	–	0.44	4.14	0.44	4.14	4.61
108	–	–	0.35	2.27	0.16	1.50	0.51	3.77	4.20
109	–	–	–	–	0.69	6.49	0.69	6.49	7.22
110	–	–	0.15	0.97	0.12	1.13	0.27	2.10	2.34
111	1.22	5.12	–	–	–	–	1.22	5.12	5.70
112	1.73	7.27	–	–	–	–	1.73	7.27	8.09
113	1.36	5.56	–	–	–	–	1.36	5.56	6.19
114	0.47	1.98	–	–	–	–	0.47	1.98	2.20
115	1.24	5.21	–	–	–	–	1.24	5.21	5.80
116	1.47	6.18	–	–	–	–	1.47	6.18	6.88
117	0.96	4.03	–	–	–	–	0.96	4.03	4.49
118	1.20	5.04	–	–	–	–	1.20	5.04	5.61
119	0.64	2.69	–	–	–	–	0.64	2.69	2.99

Table 13-7 Estimate of Peak Hour Demand for Each Distribution System Node (continued)

Node	Residential @4.2 $\infty 10^3$ $m^3/m^2 \cdot d$		Commercial @6.5 $\infty 10^3$ $m^3/m^2 \cdot d$		Industrial @9.4 $\infty 10^3$ $m^3/m^2 \cdot d$		Total		Total Demand Plus Unmetered Use, $\infty 10^3$ m^3/d
	Area, km^2	Demand, $\infty 10^3$ m^3/d	Area, km^2	Demand, $\infty 10^3$ m^3/d	Area, km^2	Demand, $\infty 10^3$ m^3/d	Area, km^2	Demand, $\infty 10^3$ m^3/d	
(1)	(2A)	(2B)	(3A)	(3B)	(4A)	(4B)	(5A)	(5B)	(6)
120	1.64	6.89	–	–	–	–	1.64	6.89	7.67
121	0.88	3.70	–	–	0.26	2.63	1.16	6.33	7.05
122	0.02	0.08	–	–	0.36	3.38	0.38	3.46	3.85
123	0.18	0.76	–	–	0.42	3.95	0.60	4.71	5.24
124	–	–	–	–	0.42	3.95	0.42	3.95	4.40
125	0.12	0.50	0.63	4.10	0.22	2.07	0.97	6.67	7.42
126	–	–	0.84	5.46	–	–	0.84	5.46	6.08
127	0.17	0.71	0.61	3.97	–	–	0.78	4.68	5.21
128	0.48	2.02	0.11	0.72	–	–	0.59	2.74	3.05
129	0.32	1.34	0.21	1.37	–	–	0.53	2.71	3.02
130	0.81	3.40	–	–	–	–	0.81	3.40	3.78
131	0.20	0.84	–	–	–	–	0.20	0.84	0.93
132	–	–	0.05	0.32	0.05	0.47	0.10	0.79	0.88
Total	21.07	88.34	5.29	34.38	4.29	40.33	30.65	163.05	181.66

average C measured for this tested section of pipe 27 is 120. The authors have experienced more reliable results when an additional gauge in between the two end gauges is included in the test runs. Comparison of C with the intermediate gauge reading provides a check of the accuracy of the flow meter and the pressure gauges.

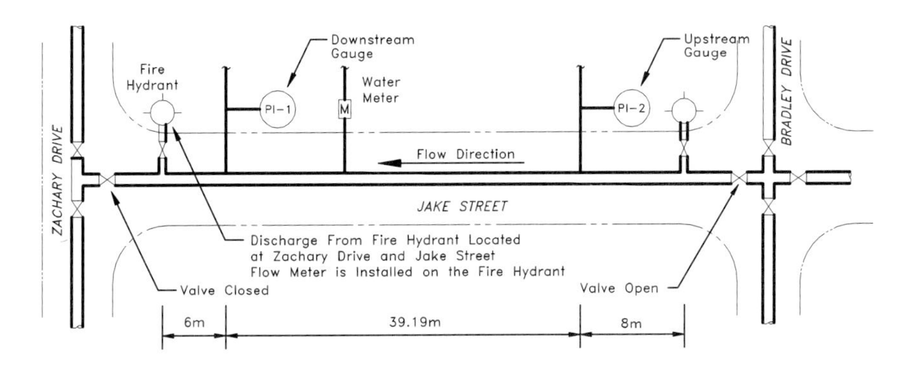

Figure 13-14 Example C-factor test data sheet.

Table 13-8 Field Test Data Reduction for Determination of C Value[a]

Time	Meter Reading, L	Flow Rate, L/s	PI-1 Pressure Gauge Reading, m	PI-2 Pressure Gauge Reading, m
10:01 a.m.	374,621,561[b]	55.6[c]	4.916[d]	6.075[e]
10:15 a.m.	374,668,265			
10:30 a.m.	374,688,135	100.5	4.739	6.090
10:45 a.m.	375,592,635			
10:59 a.m.	375,638,420	123.9	4.679	6.181
11:15 a.m.	375,757,364			
11:25 a.m.	375,856,310	166.8	4.154	5.970
11:40 a.m.	376,006,430			

a Pipe No. 27 data:
Material: ductile iron;
Diameter: 300 mm;
Length of section tested: 39.19 m.

b Meter readings at start and completion of test at one flow rate.

c $\text{Flow} = \dfrac{(374{,}668{,}265\text{ L} - 374{,}621{,}561\text{ L})}{14\text{ min} \times 60\text{ s/min}} = 55.6\text{ L/s}.$

d Downstream pressure gauge reading.

e Upstream pressure gauge reading.

Table 13-9 *C*-Factor Test Reduction

Flow, L/s	PI-1 Pressure, m	PI-1 Elevation, m	PI-1 Head, m	PI-2 Pressure, m	PI-2 Elevation, m	PI-2 Head, m	Head Loss (PI-2) - (PI-1), m	Length, m	Friction Head Loss Slope	*C* Value
(1)	(2)	(3)	(4)	(5)	(6)	(7)	(8)	(9)	(10)	(11)
55.6	4.916	104.311[a]	109.227	6.075	103.251	109.326	0.099[b]	39.19	0.00252[c]	120[d]
100.5	4.739	104.311	109.05	6.090	103.251	109.342	0.291	39.19	0.00743	121
123.9	4.679	104.311	108.99	6.181	103.251	109.432	0.442	39.19	0.01130	119
166.8	1.154	104.311	108.465	5.970	103.251	109.221	0.756	39.19	0.01930	120
					Average *C* value					120

a Elevation is obtained from the design plans or obtained from field survey.

b Head loss = PI-2 Head (Column 7) – PI-1 Head (Column 4) = 109.326 m – 109.227 m = 0.009 m.

c $\text{Friction slope} = \text{headloss/length} = \dfrac{0.099\ \text{m}}{39.19\ \text{m}} = 0.00252.$

d C value is calculated from Eq. (7.19).

$$C = \frac{55.6\ \text{L/s}}{0.278 \times 1000\ \text{L/m}^3 \times \left(\dfrac{300\ \text{mm}}{1000\ \text{mm/m}}\right)^{2.63} \times (0.00252)^{0.54}} = 120.$$

For this Design Example, eight additional locations were tested with pipes of different ages, sizes, and materials. The results obtained from these tests are presented in Table 13-10. The results of the nine tests are used to generalize the *C* values for all pipes in the distribution system. The generalized values are provided in Table 13-11. From the generalized values in Table 13-11, the *C* value of every pipe in the distribution system is estimated. These values, along with the length and diameter of existing pipes, are provided in Table 13-12 and are used later in the model analysis.

Table 13-10 *C*-Factor Test Results

Test No.	Pipe No.	Pipe Size, mm	Material	Age, years	Average *C*
1	5	410	Ductile iron	7	130
2	9	300	Ductile iron	9	130
3	15	610	Ductile iron	25	120
4	27	300	Ductile iron	16	120
5	35	300	Cast iron	35	100
6	38	450	Cast iron	31	100
7	51	610	Cast iron	35	100
8	1	610	Concrete pressure	17	120
9	48	450	Concrete pressure	17	120

Table 13-11 Generalized *C* Values for the Distribution System Modeling

	Age (years)	*C* Value
Ductile iron (300–610 mm)		
	1–15	130
	16–25	120
Cast iron (300–610 mm)		
	25–35	100
Concrete pressure pipe (410–610 mm)		
	15–35	120

b. Hydraulic analysis of distribution system without improvement

The common practice of water distribution system analysis is to input the data on the existing pipe network, distribution pumping, elevated tanks, and the projected design flow conditions. This simulation essentially provides a forecast of the deficiencies of the existing system under future flow conditions. From these results, the needed improvements are identified. Many simulations are necessary to fully understand and pinpoint the critical situation and develop several capital improvement plans. Finally, a solution is reached that provides the needed improvements for the least capital investment.

i. Initial model input data

Initially, the existing water distribution system without improvement, in the Design Example, was analyzed by using a computer network hydraulic analysis program. This program is one of many commercially available programs that are used to perform such an analysis. These input data incorporate only the major components of the existing water distribution system. Among these components are all pumps, reservoirs, and pipes of 300 mm (12 in) diameter and larger. The model does not include many smaller pipes in the distribution system. To reduce the time for and the expense of the analysis, skeleton systems such as these typically are analyzed. Trying to include smaller pipes and minor components is time consuming and does not significantly improve the accuracy. Internal loops can be analyzed separately by using the results of the skeleton system analysis. The flows imputed in the initial analysis are projected peak hour demands in the design year. To simplify this analysis, fixed flows totaling future distribution pumping capacity (145 percent of maximum plant capacity) are imputed at the pumping stations. This run has also the following conditions:

- Total peak distribution demand is the projected design peak hour flow, 2103 L/s or 181,660 m^3/d.
- It considers the existing 51 pipes of diameter 300 mm or larger.
- It considers the existing 36 nodes.
- It considers the three existing distribution pumping stations.
- It considers the two existing elevated towers with a design gradient of 140 m for each.

The information sources for this simulation study include the following: (1) the previous improvement and design report; (2) as-built street and utility plans; (3) aerial maps; (4) interviews with maintenance personnel; and (5) field surveys. Naturally, the accuracy of the results is highly dependent on the accuracy of the pipe and node data, but the cost of exhaustive research must be weighed against the expected accuracy gained. It may not be desirable to allocate unnecessary funds and effort, yet increase accuracy by only a small amount.

Table 13-12 Pipeline Inventory of Existing and Improved Water Distribution System[a]

Pipe No.	Pipe Size, mm	Material	Age, years	Length, m	Assigned *C* Value
1	610	Concrete pressure	17	1100	120
2	300	Ductile iron	10	800	130
3	410	Ductile iron	9	1000	130
4	300	Ductile iron	9	1250	130
5	410 (508)	Ductile iron	7 (1)	1100	130
6	410 (508)	Ductile iron	7 (1)	700	130
7	300	Ductile iron	17	1450	120
8	410	Ductile iron	9	850	130
9	300 (450)	Ductile iron	7 (1)	850	130
10	300	Ductile iron	9	700	130
11	300	Ductile iron	10	950	130
12	300	Ductile iron	12	1200	130
13	410	Concrete pressure	17	1800	130
14	450	Concrete pressure	17	1050	120
15	610	Ductile iron	25	650	120
16	300	Ductile iron	20	1500	120
17	300	Ductile iron	25	600	120
18	300	Ductile iron	25	450	120
19	300	Ductile iron	20	800	120
20	300	Ductile iron	25	1650	120
21	300	Ductile iron	20	1200	120
22	300	Ductile iron	17	1500	120
23	300 (410)	Ductile iron	12 (1)	1450	130
24	300	Ductile iron	10	1850	130
25	300	Ductile iron	16	800	120
26	300	Ductile iron	16	1000	120
27	300	Ductile iron	16	900	120
28	300	Ductile iron	5	800	130
29	300 (610)	Ductile iron	5 (1)	700	130

Table 13-12 Pipeline Inventory of Existing and Improved Water Distribution System[a] (continued)

Pipe No.	Pipe Size, mm	Material	Age, years	Length, m	Assigned *C* Value
30	300 (610)	Ductile iron	5 (1)	350	130
31	300 (410)	Ductile iron	25 (1)	750	120 (130)
32	300	Ductile iron	25	1400	120
33	300	Ductile iron	20	900	120
34	300	Cast iron	30	1900	100
35	300	Cast iron	35	2500	100
36	300	Cast iron	25	850	100
37	300	Cast iron	35	1000	100
38	450	Cast iron	31	400	100
39	410	Ductile iron	25	1550	120
40	450	Cast iron	35	1250	100
41	450	Cast iron	31	550	100
42	410	Ductile iron	20	800	120
43	610	Cast iron	35	500	100
44	410	Ductile iron	7	350	130
45	410	Ductile iron	7	1000	130
46	300 (610)	Ductile iron	5 (1)	400	130
47	450	Cast iron	35	100	100
48	450	Concrete pressure	17	100	120
49	610	Concrete pressure	17	10	120
50	610	Ductile iron	25	10	120
51	610	Cast iron	35	10	100
52	(610)	Ductile iron	(1)	(2000)	(130)
53-FG	(450)	Ductile iron	(1)	(100)	(130)
54	(610)	Ductile iron	(1)	(100)	(130)
55-FG PU	(760)	Ductile iron	(1)	(10)	(130)
56-FG PU	(610)	Ductile iron	(1)	(10)	(130)
57-FG PU	(610)	Ductile iron	(1)	(10)	(130)
58-FG PU	(610)	Ductile iron)	(1)	(10)	(130)

a All values in parentheses are that after proposed capital improvement.

Table 13-13 Output File for the Hydraulic Analysis of Existing Water Distribution System under Design Flow Condition[a]

Pipeline Results				
Pipe No.	Flow Rate, L/s	Head Loss,[b] m	Line Velocity, m/s	HL/1000, m/m
1	839.1	13.3	2.9	12.1
2	108.4	6.0	1.5	7.5
3	114.5	1.8	0.9	1.8
4	–3.3	0.01	0.1	0.0
5	145.4	3.1	1.1	2.8
6	31.7	0.1	0.2	0.2
7	679.0	375.6	9.6	259.1
8	–421.7	17.2	3.2	20.2
9	–105.2	6.0	1.5	7.1
10	–67.8	2.2	1.0	3.3
11	77.4	3.8	1.1	4.0
12	132.3	13.0	1.9	10.8
13	552.2	59.9	4.2	33.3
14	809.7	52.3	5.1	49.8
15	420.4	2.2	1.4	3.4
16	222.6	49.2	3.2	32.8
17	183.9	13.8	2.6	23.1
18	158.4	7.9	2.2	17.5
19	31.9	0.7	0.4	0.9
20	155.7	27.9	2.2	16.9
21	170.6	24.1	2.4	20.1
22	–95.3	10.2	1.4	6.8
23	–116.4	12.4	1.7	8.5
24	–47.7	3.0	0.7	1.6
25	110.4	7.2	1.6	9.0
26	–42.6	1.5	0.6	1.5
27	93.5	5.9	1.3	6.6
28	–60.0	2.0	0.9	2.5
29	60.9	1.8	0.9	2.6
30	22.7	0.1	0.3	0.4
31	–24.9	0.4	0.4	0.6
32	102.0	10.8	1.4	7.7
33	130.5	11.0	1.9	12.2
34	104.5	21.5	1.5	11.3
35	83.5	18.7	1.2	7.5

Table 13-13 Output File for the Hydraulic Analysis of Existing Water Distribution System under Design Flow Condition[a] (continued)

36	53.8	2.9	0.8	3.3
37	100.1	10.5	1.4	10.5
38	348.9	5.9	2.2	14.7
39	229.9	11.8	1.8	7.6
40	289.4	13.0	1.8	10.4
41	154.3	1.8	1.0	3.2
42	156.0	3.0	1.2	3.7
43	489.2	3.1	1.2	6.2
44	82.6	0.4	0.6	1.0
45	191.2	4.7	1.5	4.7
46	10.2	0	0.1	0.1
47[c]	−16.5	0	0.1	0.1
48[d]	−181.4	0.3	1.1	3.1
49	905.1	0.1	3.1	13.9
50	500.0	0.1	1.7	4.6
51	500.0	0.1	1.7	6.5

Junction Node Results

Junction No.	Hydraulic Grade, m	Junction Elevation, m	Pressure Head, m	Junction Pressure, m
100	641.4	111.0	530.4	5201.6
101	628.1	109.0	519.1	5090.8
102	123.1	106.0	17.1	168.0
103	121.3	105.0	16.3	160.1
104	121.3	102.0	19.3	189.7
105	123.5	102.0	21.5	211.2
106	118.2	107.0	11.2	110.2
107	108.1	108.0	10.1	99.2
108	118.5	109.0	9.5	92.8
109	116.3	109.0	7.3	71.8
110	116.5	109.0	7.5	73.2
111	127.3	104.0	23.3	228.8
112	140.3	106.0	34.3	336.3
113	200.2	107.0	93.2	913.8
114	186.3	106.0	80.3	787.9
115	252.5	107.0	145.5	1426.8
116	254.7	107.0	147.7	1448.3
117	203.3	105.0	98.3	963.5

Table 13-13 Output File for the Hydraulic Analysis of Existing Water Distribution System under Design Flow Condition[a] (continued)

118	179.2	103.0	76.2	747.1
119	178.5	106.0	72.5	710.7
120	150.5	104.0	46.5	456.3
121	139.7	103.0	36.7	359.8
122	132.5	101.0	31.5	309.2
123	126.6	100.0	26.6	260.4
124	134.1	99.0	35.1	343.8
125	134.0	100.0	40.0	392.2
126	158.7	99.0	65.4	585.4
127	161.5	102.0	59.7	583.6
128	167.4	102.0	59.5	641.2
129	169.2	99.0	70.2	688.1
130	182.2	102.0	80.2	786.1
131	185.3	101.0	84.3	826.5
132	116.4	111.09	5.4	53.2
200[e]	641.6	111.0	530.6	5203.0
201[f]	254.7	105.0	149.7	1468.3
202[g]	185.3	103.00	82.3	807.5

a The hydraulic analysis is conducted using the University of Kentucky Hydraulic Analysis Program, KYPIPE 2, 1000 Pipe, Version 1.10 (08/25/92).

b The minor loss coefficient is zero for all pipes of existing water distribution system.

c A fixed hydraulic gradient level of 140.0 m is assumed in the Southeast Tower, which is connected by pipe 47 to node 125.

d A fixed hydraulic gradient level of 140.0 m is assumed in the Modeltown Tower, which is connected by pipe 48 to node 121.

e The North Storage and Pump Station is located at node 200.

f The Modeltown Storage and Pump Station is located at node 201.

g The Northeast Storage and Pump Station is located at node 202.

ii. Simulation results

The results of the hydraulic analysis study are provided in Table 13-13. This table lists the operational parameters for each pipe and each node in the distribution system. The simulation results indicate the following issues:

- The sum of total node demand is 2103 L/s or 181,700 m^3/d.
- Total flow from the three pumping stations is 1905.1 L/s or 164,600 m^3/d.
- Total supply from elevated tanks is 197.9 L/s or 17,000 m^3/d.
- The total demand is, therefore, met by the distribution pumping and the elevated reservoirs.

- Pressures in excess of 600 kPa exist in the areas of nodes 100, 101, 113, 114, 115, 116, 117, 118, 119, 128, 129, 130, and 131.
- Low pressures (less than 275 kPa) are indicated in numerous joints throughout the system.
- The pipes with high head loss and velocity are 1, 7, 8, 13, 14, 16, 17, 18, 20, 21, 38, and 49.
- The pressure heads needed at the pump stations are in the range from 82 m to 531 m (from 117 to 758 psi).

2. Proposed improvements for the distribution system

From the results of the initial simulation, it is clear that many improvements are needed to supply the design flow throughout the distribution system. The presence of both excessive pressures and low pressures clearly indicates insufficient pipeline capacity to transport water from the Northern Pump Station to the system. All pipes with high head loss do not have to be replaced to improve the situation. First, adding a pipeline between nodes 101 and 102 will reduce the flows in pipes 7, 13, 14, and 20 to transport water south. Selecting a size for this pipe is based on judgement as well as on trial and error. One pipe size is added to the model. If the results are not satisfactory, the pipe size is increased until the desired results are obtained. For this example, a new pipe (pipe 52) is added with a 610-mm size. Also, pipes 1 and 49 are replaced by 760-mm-diameter pipelines to reduce a bottleneck situation arising at the North Pump Station. The location of the new pump station must also be established. The area near node 110 in the western industrial zone is ideal. This is a convenient location for a transmission pipeline route from the water treatment plant and is in a high demand area that has low pressures. To convey the flow from the new pump station to the remainder of the system, pipes 5, 6, 29, 30, and 46 must be replaced by larger lines. Pipes 5 and 6 are increased to 508 mm, and the remaining three lines are increased to 610 mm. (See Table 13-12.) These line sizes are based on judgment and are used in the subsequent trial runs.

Finally, a suitable location for the new elevated storage reservoir must be identified. During review of the hydraulic gradients of each node (Table 13-13), the lowest gradients are found at nodes 102 through 111, 123, and 132. Hydraulically, these are ideal locations for the tank, but, economically, higher ground is more desirable. Most nodes on higher ground have a gradient above 140 m, the design gradient for the existing tanks, so no benefit would be gained by locating a tank in these areas. The best location is probably near the extreme end of the system, in a commercial area where its aesthetic impact will be slight, and because no other tank is located in the western commercial/industrial area. Therefore, nodes 104 and 105 are possible candidates for a new tank location. At any of these locations, the elevated tank should stabilize the pressures in the western area of the town, as well as in the entire system. Node 105 is preferable because it is closer to the demand center. A design gradient of 140 m is selected initially for the new tower, because it matches that of the other two tanks. To achieve this gradient requirement, pipes connecting the new and old elevated towers are increased; thus, pipes 9, 23, and 31 are replaced by 45-, 41-, and 41-diameter pipe, respectively. (See Table 13-12.) This gradient is the mid-point of the tank depth and represents the minimum operating level that should be maintained in the tank at the end of the peak hour demand.

3. Performance of distribution system with improvement

   a. Check system improvement for critical conditions under static conditions.

      The improvements indicated above are inputted into the computer simulation models, and numerous runs are made to study the behavior of the distribution system under extreme conditions. These extreme conditions are simulated from static conditions, such as flow input at the pump stations, or by imputing low or high operating levels in the reservoirs. Readers must realize that modeling and checking a complex water distribution network is not a simple task. There is no set procedure for making checks for extreme conditions. The authors, on the basis of their experience gained from numerous water distribution network projects, have developed a few basic checks to analyze the critical components of the network under extreme situations. These basic checks are as follows:

      - Develop and check the neutral points of the reservoirs.
      - Check the operating storage volumes.
      - Check the minimum system pressure.
      - Check the reservoir filling time.
      - Check the maximum system pressure.

      The procedure and results of these basic checks are presented next.

      i. Develop and check the neutral points of the reservoir.

         The reservoir neutral point is the water surface elevation at which no flow from or into the reservoir occurs if the total pumping capacity (input to the system) equals the system demand. The reservoir neutral point must lie within the operating range of the reservoir. Ideally, the neutral point must be close to the mid-point elevation of the reservoir. The mid-point or mid-elevation of the reservoir has a great significance. Above this point lies the operating storage, and below this point is the emergency storage. Various operating levels of a reservoir are shown in Figure 13-15.

         The neutral points of the reservoirs in the Design Example are developed and checked under the following conditions.

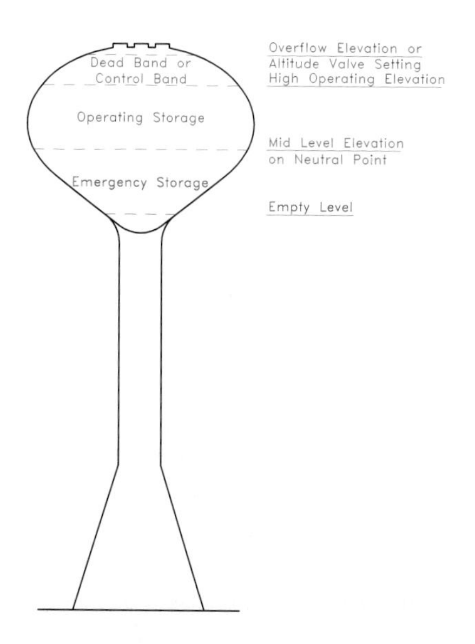

Figure 13-15 Reservoir operating levels.

- Maximum pumping capacity (145 percent of plant capacity) = 164,600 m^3/d (1905.1 L/s). This capacity is distributed at all four pumping stations in the system, in accordance with the size of the pumping stations.
- Distribution demand = 164,600 m^3/d (1905.1 L/s).
- The input design neutral point value, 140 m, is used at the most remote existing elevated tower (node 125). Disconnect the second existing elevated tower, and represent it as a node similar to node 105 (one of the proposed sites of the new elevated tower).

The simulation run is completed. The results of this simulation indicate operating conditions at the pumping stations and at reservoir locations. The major simulation results are summarized in Table 13-14. The results indicate that the pressure head developed at the node of the existing reservoir (node 121) and at the proposed site for new reservoir (node 105) are 140.4 m. These pressure heads are the static neutral points of the reservoirs at those sites. These levels are close to the design mid-point elevations of the reservoirs (140 m). The static head at each pump station is representative of the operating point, which is the intersection of the *system head-capacity curve* and the *pump head-capacity curve* when all pumps at different stations are operating in parallel. As a check, the same simulation is made again with all reservoirs in line and at neutral points. These final results are also provided in Table 13-14. The simulation test results show that there are very small flows in to or out of the reservoirs, which are essentially zero.

ii. Check the operating storage volumes.

The operating storage of the reservoirs should be sufficient to meet the duration of peak hour demand. This simulation is made under the following conditions.

- Maximum pumping capacity = 164,600 m^3/d (1905.1 L/s). Distribution pumping is the same as above.
- Maximum distribution demand (peak hours) = 181,700 m^3/d (2103.0 L/s).
- Reservoir elevations at maximum operating level. In this example, the designed value of the high operating level in all reservoirs is assumed 5 m above the mid-level elevation. The high operating level of the reservoirs is developed by the designer from the storage requirements and the structural constraints.

The results of this simulation are provided in Table 13-14. The maximum drawdown from each reservoir is obtained. These values and the available volumes are used to calculate the time needed to deplete this storage. For a well-designed system, the time to deplete the operating storage volume should be significantly longer than the duration of peak hour demand.

The times to deplete the operating storage volumes are summarized in Table 13-15.

It may be noted that the time to deplete the operating storage is sufficiently long. In the event that any elevated tower depletes before the end of the peak hour demand, the supply will continue to come from the emergency storage volume, which is almost as large as the operating storage volume.

iii. Check for low pressure in the network.

The lowest pressure in the network occurs when the water surface elevations in the reservoirs are at mid-point, distribution pumping is at maximum capacity, and system demand is under the peak hour condition. The selected results of this simulation are summarized in Table 13-14. The simulation results also show that the lowest pressure, 36.7 m (52.2 psi), occurs at node 108. This value is significantly higher than the recommended minimum pressure of 276 kN/m^2 or 28.1 m (40 psig).

iv. Check elevated tower filling time.

The elevated tower filling time should be checked under low demand conditions. The lowest system demand, 55,000 m^3/d (637.0 L/s), is at 48.5 percent of the treatment plant capacity. This demand lasts for approximately four

Table 13-14 Static Distribution System Simulation to Check System Improvements for Critical Conditions[a]

System Simulation	Flow, L/S				Hydraulic Grade, m			Remarks
	Existing Pumping Stations			Proposed Pumping Station	Existing Elevated Tower			Proposed Elevated Tower
	Node 200	Node 201	Node 202	Node 203	Node 125	Node 121	Node 105	
i. Determine static neutral point of reservoirs and check these points.	650.0[b]	400.0	400.0	455.4	140.0	Closed	Closed	High and low pressure heads at nodes 131 and 121 are 52.6 and 37.4 m. Total pumping and demands are 1905.4 L/s.
					(140.0)	(140.4)	(140.4)	
	650.0	400.0	400.0	455.4	140.0	140.4	140.4	
					(140.0)	(140.4)	(140.4)	
ii. Check for operating storage volumes.	650.0	400.0	400.0	455.4	145.0	145.0	145.0	Times to empty storage volume are given in Table 13-15. Total pumping and demands are 1905.4 and 2103.0 L/s.
					(144.9)	(145.0)	(145.0)	
iii. Check for minimum system pressures.	650.0	400.0	400.0	455.4	140.0	140.0	140.0	Minimum pressure head of 36.7 m is at node 108. Total pumping and demands are 1905.4 and 2103.0 L/s.
					(139.9)	(140.0)	(140.0)	
iv. Check for reservoir filling time. Three simulations are made to determine the filling rates of the reservoirs.	313.8	193.1	193.1	219.7	140.0	140.0	140.0	Times required to refill storage volumes are given in Table 13-16. Total pumping and demands are 919.7 and 637.0 L/s.
					(140.1)	(140.2)	(140.1)	
	313.8	193.1	193.1	219.7	140.0	Closed	140.0	
					(140.4)	(141.8)	(140.1)	
	313.8	193.1	193.1	219.7	140.0	Closed	Closed	
					(141.0)	(144.4)	(148.1)	
v. Check for high system pressures.	313.8	193.1	193.1	219.7	145.0	Closed	Closed	Maximum pressure head of 56.4 m is at node 131. Total pumping and demands are 919.7 and 637.0 L/s.
					(146.0)	(149.4)	(152.1)	

a The hydraulic analysis is conducted by using the University of Kentucky Hydraulic Analysis Program, KYPIPE 2, 1000 Pipe, Version 1.10 (08/25/92). All values in parentheses are obtained from computer simulations.

b Distribution of total flows at each pumping station is based on their capacities. Total input capacity = 1905.3 L/s.

Table 13-15 Summary of Drawdown Conditions and Depletion Times of Operating Storage Volumes under Static Conditions

Conditions	Existing Elevated Reservoir		Proposed Elevated Tower (Node 105)
	Southeast Tower (Node 125)	Modeltown Tower (Node 121)	
Total Storage Volume[a], m^3	4000	2000	2000
Usable Storage Volume[b], m^3	2000	1000	1000
Maximum Drawdown[c], L/s	84.9	53.1	59.6
Time to Deplete Operating Storage Volume, h	6.5[d]	5.2	4.7

a Existing elevated storage volumes are given in Table 5-5. Proposed elevated storage volume is determined in Section 13.8.2, Step F.4.

b Usable volume is half of the total volume.

c Values obtained from simulation rum (Table 13-14).

d Depletion time > $\dfrac{2000\ \text{m}^3 \times 10^3\ \text{L/m}^3}{84.9\ \text{L/s} \times 3600\ \text{s/h}} = 6.5\ \text{h}$.

hours: 12:00 midnight to 3:00 a.m. (Figure 13-8). The pumping rate should be sufficient to provide this demand and also fill the reservoirs for this example. In the authors' judgment, a pumping rate of 70 percent of the plant capacity, or 79,500 m^3/d (919.6 L/s), can be considered satisfactory. At this pumping rate, the average time to fill the elevated towers is obtained as follows:

$$\text{Lowest water demand} = 54{,}000\ \text{m}^3/\text{d}\ (630.9\ \text{L/s})$$

$$\text{Average reservoir-filling rate} = 24{,}500\ \text{m}^3/\text{d}\ (284\ \text{L/s})$$

$$\text{Total operating storage of all reservoirs} = 4000\ \text{m}^3/\text{d}$$

$$\text{Average time to fill all reservoirs} = \frac{4000\ \text{m}^3/\text{d} \times 24\ \text{h/d}}{24{,}500\ \text{m}^3/\text{d}} = 3.9\ \text{h}$$

Therefore, a total pumping rate of 919.6 L/s will satisfy the low-demand condition and fill all reservoirs within the low-demand duration of 4 hours. The total pumping rate is distributed to different stations in proportion to their capacities. The conditions of this simulation, therefore, are as follows:

$$\text{Total pumping capacity} = 79{,}500\ \text{m}^3/\text{d}\ (920.1\ \text{L/s})$$

$$\text{Distribution system demand} = 55{,}000\ \text{m}^3/\text{d}\ (637.0\ \text{L/s})$$

$$\text{Water surface elevation in towers at mid-level} = 140\ \text{m each}$$

The results of this simulation are provided in Table 13-14. It may be noted that the reservoir-filling rates are not uniform. As a result, the highest operating level in one reservoir is reached first, and the filling operation of this reservoir is stopped. The flow is diverted into the other two making their filling rate faster. Similarly, the second reservoir and then the final reservoir are filled. These conditions are also shown by computer simulations with the results provided in Table 13-14. The actual filling times for each reservoir are summarized in Table 13-16.

Table 13-16 Check for Filling Times of Elevated Reservoirs under Static Condition[a]

Condition	Existing Elevated Towers		Proposed Elevated Tower (Node 105)
	Southeast Tower (Node 125)	Modeltown Tower (Node 121)	
Storage volume to be filled[b], m^3	2000	1000	1000
Filling rates (three reservoirs in-line), L/s	(68.7)	(126.7)	(93.3)
Time to fill, h	–	2.2[c]	–
Filling rate (two reservoirs in-line), L/s	(165.6)	0	(123.1)
Time to fill, h	–	Full	0.6[d]
Filling rate (one reservoir in-line), L/s	(288.7)	0	0
Time to fill, h	1.1[e]	Full	Full
Total time to fill, h	2.2 + 0.6 + 1.1 = 3.9	2.2	2.2 + 0.6 = 2.8

a The hydraulic analysis is conducted using the University of Kentucky Hydraulic Analysis Program, KYPIPE 2, 1000 Pipe, Version 1.10 (08/25/92). All values in parentheses are obtained from the computer simulation.

b Storage volume date from Table 13-15.

c $\text{Time to fill} = \dfrac{1000\ \text{m}^3 \times 10^3\ \text{L/m}^3}{126.7\ \text{L/s} \times 3600\ \text{s/h}} = 2.2\ \text{h}.$

d $\text{Time to fill the remaining volume} = \dfrac{1000\ \text{m}^3 \times 10^3\ \text{L/m}^3 - 2.2\ \text{h} \times 93.3\ \text{L/s} \times 3600\ \text{s/h}}{93.3\ \text{L/s} \times 3600\ \text{s/h}} = 0.6\ \text{h}.$

e Time to fill the remaining volume

$$= \frac{2000\ \text{m}^3 \times 10^3\ \text{L/m}^3 - (2.2\ \text{h} \times 68.6\ \text{L/s} + 0.6\ \text{h} \times 165.6\ \text{L/s}) \times 3600\ \text{s/h}}{288.7\ \text{L/s} \times 3600\ \text{s/h}} = 1.1\ \text{h}.$$

v. Check for high pressures in the network.

High pressures in the system will develop when the demand is lowest, pumps are filling the elevated towers, and the water surface elevation in the towers is at the high operating level. A computer simulation of this condition is presented in Table 13-14. The highest pressure, 56.4 m (80.2 psi), is noted at node 131. This pressure is within the permissible limit.

b. Check system improvement for critical conditions under dynamic conditions.

Pump characteristic curves for all distribution pumping stations are developed from the results of the system simulations given in Table 13-14. Each head-capacity value represents the operating point of the pump station. Smooth curves are drawn that represent the combined head-capacity relationship when all pumps are operating in parallel. A thorough study of existing pumping equipment at the stations, modification requirements, and availability of new equipment is necessary before such curves can be drawn. Readers are referred to Chapter 7 for in-depth discussions on (1) manufacturers' characteristic curves, (2) preparation of modified head-capacity curve for one pump and pump parallel combination curves, (3) preparation of system head-capacity curve, (4) pump operating points, and (5) pump selection procedure and criteria. It is emphasized that a great deal of engineering judgment is needed to develop the combined head-capacity curves and pump selection at each pumping station. Consideration should also be given to changes in the head-capacity curves when fewer pumps are in operation.

From the results of previous simulations and in-depth engineering investigations, the characteristic curves for each distribution pumping station are developed. These curves represent the parallel combinations when all pumps at a station are in operation. These curves are expressed by three points. Table 13-17 provides this information, as well as number of pumps provided at each distribution pumping station.

Table 13-17 Combined Pump Characteristic Curve Input Data Points and Number of Pumps Arranged in Parallel

Distribution Pumping Station	Total Capacity, L/s	Total Number of Pumps	Pump Operating Points					
			Point 1 (Shutoff Head)		Point 2		Point 3	
			Flow, L/s	Head, m	Flow, L/s	Head, m	Flow, L/s	Head, m
Northern Pump Station (Node 200)	723.6	4	0	53.4	504.7	50.1	857.9	40.8
Modeltown Pump Station (Node 201)	318.5	3	0	55.5	397.4	51.0	643.4	42.3
Northeast Pump Station (Node 202)	318.5	3	0	55.5	397.4	51.0	643.4	42.3
West Side Pump Complex (Node 203)	544.4	4	0	49.8	378.5	46.4	567.8	40.4

Note: Each pump station has all identical pumps and one additional pump as a spare.

Several computer simulations were made with pump head-capacity curves as discussed above. All five static conditions simulated earlier were rerun with pumping curves expressed by three points. These simulations represent the dynamic conditions, with the results presented in Table 13-18. The operating points at the pumping stations, and the reservoir operating conditions thus developed by computer simulations, are compared against the results of the earlier runs. A comparison of the results with static and dynamic pumping simulations is presented next.

i. Check for reservoir neutral points.

   The static neutral points in all reservoirs under static and dynamic pumping simulations lie between 140.0 m and 140.4 m.

ii. Check the operating storage volumes.

   The operating storage of the reservoirs is sufficient to meet the duration of peak hour demand. The times to deplete the elevated reservoirs under maximum drawdown situations and dynamic pumping simulations are summarized in Table 13-19. The time to deplete the reservoirs under dynamic pumping compare very well with those under static pumping (Table 13-15).

iii. Check for low pressure in the network.

   The minimum pressure in the system under dynamic pumping condition will occur when all pumps are supplying the peak hour demand. This condition is critical, because maximum pumping corresponds to lowest operating head. Under this condition, the pressure heads at the reservoir nodes 121 and 103 are 135.6 m and 136.0 m, respectively. These pressure heads are significantly less than the reservoir neutral points. The lowest pressure, 51.2 m (72.8 psi), is developed at node 131.

iv. Check elevated tower filling time.

   The elevated reservoirs' filling time is checked under the dynamic pumping condition. The demand in the distribution system is lowest then and lasts for four hours; the reservoirs must be filled in this time. The reservoir operation is such that, as high operating level in one reservoir is reached, the difference in pumping and demand is diverted to the remaining reservoirs. The filling times of the reservoirs are summarized in Table 13-20. It may be noted that all reservoirs are filled within four hours of minimum demand duration.

v. Check for high pressures in the network.

   The system high pressures are developed under the low-demand condition, and this demand is met by the distribution pumping. This condition is simulated by closing all reservoirs. Highest pressure, 58.5 m (83.2 psi), is developed at node 129. This pressure is within the acceptable range.

Table 13-18 Dynamic Distribution System Simulation with Pump Head-Capacity Curve to Check System Improvements for Critical Conditions[a]

System Simulation	Flow, L/s				Hydraulic Grade, m			Remarks
	Existing Pumping Stations			Proposed Pumping Station	Existing Elevated Tower		Proposed Elevated Tower	
	Node 200	Node 201	Node 202	Node 203	Node 125	Node 121	Node 105	
i. Determine static neutral points of reservoirs.	(656.1)	(393.9)	(401.0)	(457.4)	Closed (140.0)	Closed (140.3)	Closed (140.4)	High and low pressure heads at nodes 131 and 121 are 52.6 m and 37.3 m. Total pumping and demand are 1908.4 and 1908.1 L/s.
ii. Check for operating storage volume.	(656.1)	(395.6)	(386.8)	(443.1)	145.0 (144.9)	145.0 (145.0)	145.0 (145.0)	Times to empty storage volume are given in Table 13-19. Total pumping and demands are 1881.6 and 2103.3 L/s.
iii. Check for minimum system pressures.	(708.4)	(439.3)	(447.8)	(507.5)	Closed (140.0)	Closed (135.6)	Closed (136.0)	Minimum pressure head of 51.2 m is at node 131. Total pumping and demands are 2103.0 and 2103.0 L/s.
iv. Check for reservoir filling time. Three simulations are made to determine the filling times of the reservoirs.	(457.4)	(200.8)	(236.9)	(163.0)	140.0 (140.3)	140.0 (140.5)	140.0 (140.2)	Times required to refill storage volumes are given in Table 13-20. The system demand is 630.9 L/s. The total pumping rates are 1158.1, 1117.1, and 994.4 L/s when three, two, and one reservoirs are filling.
	(449.0)	(192.9)	(226.1)	(249.1)	140.0 (141.0)	Closed (144.1)	140.0 (140.4)	
	(419.8)	(173.0)	(211.9)	(189.7)	140.0 (141.6)	Closed (146.5)	Closed (152.0)	
v. Check for high system pressures.	(374.6)	(113.8)	(75.9)	(66.5)	Closed (157.2)	Closed (157.3)	Closed (157.7)	Maximum pressure head of 58.5 m is at node 129. Total pumping rate and demands are 630.9 and 630.9 L/s.

a The hydraulic analysis is conducted by using the University of Kentucky Hydraulic Analysis Program, KYPIPE 2, 1000 Pipe, Version 1.10 (08/25/92). All values in parentheses are obtained from computer simulations.

Table 13-19 Summary of Drawdown Conditions and Depletion Times of Operating Storage Volumes under Dynamic Condition

Condition	Existing Elevated Reservoir		Proposed Elevated Tower (Node 105)
	Southeast Tower (Node 125)	Modeltown Tower (Node 121)	
Usable storage volume[a], m^3	2000	1000	1000
Maximum drawdown, L/s	92.2	63.9	65.6
Time to deplete operating storage volume, h	6.0	4.4	4.2

a Storage volume data from Table 13-15.

4. Pressures under fire flow

Low-pressure condition may develop under fire flows; therefore, the distribution system must be checked with fire flow coincident with maximum day demand. The total system demand is kept at 100 percent of the plant capacity (maximum day demand), and total fire flow is placed at a node that can cause a critical situation. Generally, the most remote nodes are checked, because they are associated with critical condition.

In the Design Example, the critical pressures under fire flow are checked by placing the fire flow at a remote node. The water surface elevation in the elevated reservoirs is kept at mid-elevation, so that the emergency storage is utilized and critical pressure condition is created. Computer simulation results are checked for low pressures. Again, the fire flow is moved to the next node and the results are checked by a computer run. The process is continued until the entire distribution system is thoroughly checked and verified. The critical pressures under the fire flow condition are checked for the following conditions as required by the State Board of Insurance.

System demand at maximum day flow = 113,500 m^3/d

Water surface elevation in reservoirs = mid-depth

Fire flow, residential area, high density = 47.5 L/s (750 gpm)

Fire flow, residential area, single family = 31.7 L/s (500 gpm)

Fire flow, commercial area = 95 L/s (1500 pm)

Fire flow industrial area = 190 L/s (3000 gpm)

The fire flow is applied on nodes 106 (industrial area), 102 (residential area), and 120 and 124 (commercial areas). The results are summarized in Table 13-21. The minimum required pressure, coincident with fire flow and maximum day demand, is 138 kPa (20

Table 13-20 Check Filling Time of Elevated Tanks under Dynamic Conditions[a]

Condition	Existing Elevated Towers: Southeast Tower (Node 125)	Existing Elevated Towers: Modeltown Tower (Node 121)	Proposed Elevated Tower (Node 105)
Storage volume to be filled[b], m^3	2000	1000	1000
Filling rates (three reservoirs in-line), L/s	(132.9)	(222.0)	(172.3)
Time to fill, h	–	1.3[c]	–
Filling rates (two reservoirs in-line), L/s	(277.0)	Full	(209.2)
Time to fill, h	–	0	0.3[d]
Filling rates (one reservoirs in-line), L/s	(363.5)	Full	Full
Time to fill, h	0.8[e]	0	0
Total Time to Fill, h	1.3 + 0.3 + 0.8 = 2.4	1.3	1.3 + 0.3 = 1.6

a The hydraulic analysis is conducted by using the University of Kentucky Hydraulic Analysis Program, KYPIPE 2, 1000 Pipe, Version 1.10 (08/25/92). All values in parentheses are obtained from the computer simulation.

b Storage volume data from Table 13-15.

c $$\text{Time to fill} = \frac{1000\ \text{m}^3 \times 10^3\ \text{L/m}^3}{222.0\ \text{L/s} \times 3600\ \text{s/h}} = 1.3\ \text{h.}$$

d $$\text{Time to fill remaining volume} = \frac{1000\ \text{m}^3 \times 10^3\ \text{L/m}^3 - 1.3\ \text{h} \times 172.3\ \text{L/s} \times 3600\ \text{s/h}}{209.2\ \text{L/s} \times 3600\ \text{s/h}} = 0.3\ \text{h.}$$

e Time to fill the remaining volume

$$= \frac{2000\ \text{m}^3 \times 10^3\ \text{L/m}^3 - (1.3\ \text{h} \times 132.9\ \text{L/s} + 0.3\ \text{h} \times 277.0\ \text{L/s}) \times 3600\ \text{s/h}}{363.5\ \text{L/s} \times 3600\ \text{s/h}} = 0.8\ \text{h.}$$

psi). The actual minimum pressure in the distribution system is significantly higher than the required value.

5. Final results of computer simulation, with recommended improvements

The improvements recommended in Step G.2 in the distribution system were checked for critical and fire flow components in Steps 3 and 4. Numerous computer runs were made, and the summary results of these component checks are provided in Tables 13-14 through 13-21. The distribution network with all improvements has checked out very well. For the sake of completeness, the results of one simulation with complete computer input and output data are presented in this section. This simulation includes (1) peak hour demand, (2) distribution pumping operating at full capacity, and (3) all reservoirs full and ready to provide the remainder demand. The pump data used in the simulation are provided in Table 13-22. The complete output results of a computer simulation are provided in Table 13-23. The results indicate a good overall improvement in the distribution system.

The system pressures range from 404.4 kPa to 519.0 kPa, well within the acceptable ranges. Pipeline velocities generally are around 2 m/s. The total system demand, 181,700 m^3/d (2103.3 L/s), is met by distribution pumping. The flows from the distribution pump stations and elevated reservoirs are given in Table 13-24. The readers should compare these results with those for the existing system receiving the design flows, (Table 13-13) and note the deficiencies of the existing system and how well the proposed capital improvements relieve these deficiencies.

Table 13-21 Check for Distribution System under the Fire Flow Condition[a]

Fire Flow Applied at Node	Total Fire Flow, L/s	Total Pumping Rate, L/s	Node and Minimum Pressure Head			
			Node	Pressure Head		
				m	kPa	psi
106 (Industrial)	189.3	(1701)	121	37.1	363.9	52.7
112 (Residential)	47.5	(1615)	121	37.2	364.4	52.8
120 (Commercial)	95	(1630)	121	37.1	364.0	52.7
124 (Commercial)	95	(1582)	124	36.5	358.1	51.9

a The hydraulic analysis is conducted using the University of Kentucky Hydraulic Analysis Program, KYPIPE 2, 1000 Pipe, Version 1.10 (08/25/92). All values in parentheses are obtained from the computer simulation.

Table 13-22 Input Pump Data for the Hydraulic Analysis of Improved Water Distribution System under Design Flow Conditions

Pump Identification		Characteristics	
Name	Location	Head, m	Flow Rate, L/s
Northern Storage and Pump Station	Pipe 55	53.4	0
		50.1	504.7
		40.8	857.9
Modeltown Storage and Pump Station	Pipe 56	55.0	0
		51.0	397.4
		42.3	643.4
Northeast Storage and Pump Station	Pipe 57	55.0	0
		51.0	397.4
		42.3	643.4
New Westside Pump Complex	Pipe 58	49.8	0
		46.4	378.5
		40.4	567.8

The model hydraulic analysis study of the distribution system by computer simulation has been presented briefly. Many other simulations that were conducted to arrive at the improved system design are not presented here. Other recommended simulations include (1) pump station failure for each pumping facility, (2) filling of elevated tanks under different demands, (3) other conditions under which the elevated tanks are filled and supply the demand, and (4) the overflow conditions of the elevated tanks.

13.9 Operation, Maintenance, and Troubleshooting for Water Stability, Clearwells, High-service Pumps, and Distribution Facilities

13.9.1 Common Operational Problems and Troubleshooting

1. Water Storage Reservoirs

Clearwells, ground storage reservoirs, and elevated storage tanks have few mechanical systems, and causing minimal operational problems are associated with them. Following are some common problems and associated troubleshooting guidelines.

a. Periodic musty or poor-tasting water withdrawn from tanks is an indication of (1) severe short-circuiting and stagnation, (2) accumulation of dirt, sludge, or solids in the reservoir bottom, and/or (3) lack of chlorine residual in the tank. Conduct chlorine residual test in the tank outflow as well as other locations in the tank. A chlorine residual in the outflow, but not in parts of the tank, indicates poor tank circulation. Baffling in the tank should be considered to improve circulation. Lack of chlorine residual in the outflow indicates insufficient chlorination (or perhaps a large chlorine demand in the tank). Check the tank for solids accumulation, unusual growth, or contamination. Clean the tank.
b. Noisy pump operation can be caused by vortexing in the clearwell or ground storage reservoirs. Check the outlet conditions of the reservoirs, and observe vortexing if possible. Install a vortex plate over the outlet pipe or place some floating plates in the tank as part of the corrective measure. (See Chapter 7.)

2. Water Conditioning and Stabilization Systems

Operational problems associated with stabilization systems are mainly those commonly associated with liquid- or dry-chemical feeders. Refer to Chapter 8 for troubleshooting on chemical system. Other stabilization system related operational problems and associated troubleshooting guides are listed next.

a. Red or brown water in the distribution system is an indication of a corrosive water. Other indicators of corrosion are higher lead, copper, and/or iron concentrations than found in the finished water. Stabilize water to reduce corrosive tendencies.

Table 13-23 Output File for the Hydraulic Analysis of Improved Water Distribution System under Design Flow Condition[a]

Pipeline Results				
Pipe No.	Flow Rate, L/s	Head Loss,[b] m	Line Velocity, m/s	HL/1000, m/m
1	589.9	2.1	1.3	1.9
2	17.9	0.2	0.3	0.3
3	142.4	2.7	1.1	2.7
4	55.5	2.7	0.8	2.2
5	–126.5	0.8	0.6	0.8
6	–240.2	1.8	1.2	2.5
7	30.2	1.2	0.4	0.8
8	121.1	1.7	0.9	2.0
9	135.9	1.3	0.9	1.6
10	–9.0	0.1	0.1	0.1
11	–43.5	1.3	0.6	1.4
12	52.5	2.3	0.7	2.0
13	74.8	1.5	0.6	0.8
14	210.9	4.3	1.3	4.1
15	316.0	1.3	1.1	2.0
16	68.2	5.5	1.0	3.7
17	62.5	1.9	1.0	3.1
18	37.0	0.5	0.5	1.2
19	36.6	0.9	0.5	1.2
20	39.0	2.2	0.6	1.3
21	16.2	0.3	0.2	0.3
22	49.8	3.1	0.7	2.1
23	78.9	1.3	0.6	0.9
24	42.8	2.5	0.6	1.3
25	59.1	2.3	0.8	2.8
26	–3.3	0.0	0.1	0.0
27	54.2	2.2	0.8	2.4
28	53.0	1.6	0.8	2.0
29	–269.2	0.9	0.9	1.3
30	352.8	0.7	1.2	2.1
31	22.9	0.1	0.2	0.1

Table 13-23 Output File for the Hydraulic Analysis of Improved Water Distribution System under Design Flow Condition[a] (continued)

32	20.9	0.6	0.3	0.4
33	20.9	0.4	0.3	0.4
34	20.2	1.0	0.3	0.5
35	4.9	0.1	0.1	0.0
36	29.5	0.9	0.4	1.1
37	45.8	2.5	0.7	2.5
38	130.8	1.0	0.8	2.4
39	82.2	1.8	0.6	1.1
40	164.7	4.6	1.0	3.7
41	83.9	0.6	0.5	1.1
42	167.5	3.4	1.3	4.2
43	376.0	1.9	1.3	3.8
44	24.5	0.0	0.2	0.1
45	20.1	0.1	0.2	0.1
46	–432.9	1.2	1.5	3.1
47[c]	–92.2	0.1	0.6	1.3
48[d]	–63.9	0.1	0.4	0.5
49	655.9	0.0	1.5	2.3
50	395.6	0.0	1.4	3.0
51	386.8	0.0	1.3	4.0
52	399.6	5.3	1.4	2.6
53[e]	–65.6	0.0	0.4	0.4
54	443.1	0.3	1.5	3.2
55[f]	655.9[g]	0.3	1.5	2.3
56[aa]	395.6[bb]	0.3	1.4	2.6
57[cc]	386.8[dd]	0.3	1.3	2.5
58[ee]	443.1[ff]	0.4	1.5	3.2

Junction Node Results

Junction No.	Hydraulic Grade, m	Junction Elevation, m	Pressure Head, m	Junction Pressure, m
100	157.6	111.0	46.6	457.4
101	155.6	109.0	46.6	456.9
102	150.3	106.0	44.3	434.6
103	147.6	105.0	42.6	417.8

Table 13-23 Output File for the Hydraulic Analysis of Improved Water Distribution System under Design Flow Condition[a] (continued)

104	144.9	102.0	42.9	420.8
105	145.0	102.0	43.0	421.3
106	148.5	107.0	41.5	406.5
107	150.2	108.0	42.2	413.9
108	150.2	109.0	41.2	404.4
109	151.1	109.0	42.1	412.8
110	151.8	109.0	42.8	420.0
111	146.3	104.0	42.3	414.5
112	148.6	106.0	42.6	417.9
113	150.1	107.0	43.1	422.6
114	148.2	106.0	42.2	414.0
115	154.4	107.0	47.4	465.0
116	155.7	107.0	48.7	477.7
117	148.9	105.0	43.9	430.7
118	148.6	103.0	45.6	447.3
119	147.7	106.0	41.7	408.8
120	145.5	104.0	41.5	407.3
121	145.0	103.0	42.0	411.4
122	142.7	101.0	41.7	408.9
123	142.5	100.0	42.5	416.7
124	142.7	99.0	43.7	428.7
125	144.9	100.0	44.9	440.1
126	145.0	99.0	46.0	450.8
127	145.9	102.0	43.9	430.5
128	146.9	102.0	44.9	439.9
129	147.4	99.0	48.4	475.0
130	152.0	102.0	50.0	490.4
131	153.9	101.0	42.9	519.0
132	153.1	111.0	42.1	412.4
200	157.7	111.0	46.7	457.6
201	155.7	105.0	50.7	497.6
202	154.0	103.0	51.0	499.7
203	153.4	109.0	44.4	435.1

a The hydraulic analysis is conducted using the University of Kentucky Hydraulic Analysis Program, KYPIPE 2, 1000 Pipe, Version 1.10 (08/25/92).

b The minor loss coefficient is zero for all pipes of improved water distribution system.

c A fixed hydraulic gradient level of 145.0 m is assumed in the Southeast Tower, which is connected by pipe 47 to node 125.

d A fixed hydraulic gradient level of 145.0 m is assumed in the Modeltown Tower, which is connected by pipe 48 to node 121.

e A fixed hydraulic gradient level of 145.0 m is assumed in the proposed Southwest Tower, which is connected by pipe 53 to node 105.

f A fixed hydraulic gradient level of 111.0 m is assumed in the ground storage reservoir of North Storage & Pump Station, which is located at node 200.

g At a TDH of 47.0 m.

aa A fixed hydraulic gradient level of 105.0 m is assumed in the ground storage reservoir of Modeltown Storage & Pump Station, which is located at node 201.

bb At a TDH of 51.0 m.

cc A fixed hydraulic gradient level of 103.0 m is assumed in the ground storage reservoir of Northeast Storage & Pump Station, which is located at node 202.

dd At a TDH of 51.3 m.

ee A fixed hydraulic gradient level of 109.0 m is assumed in the ground storage reservoir of proposed Westside Pump Complex, which is located at node 203.

ff At a TDH of 44.8 m.

Raise pH and alkalinity by adding lime ($Ca(OH)_2$), caustic soda ($NaOH$), or soda ash (Na_2CO_3). Check the stability of water. Be sure the water is slightly scaling before it leaves the clearwells.

b. Complaints regarding heavy scale on appliances and plumbing fixtures is an indication of a scaling water. The pH of water must be lowered by recarbonation or acid addition.

3. Pumping Stations

Typical operational problems and a troubleshooting guide for high-service pumps and distribution pumps are similar to those of raw water pumps. Raw water pumps were presented in Chapter 7.

4. Distribution Systems

Typical problems associated with the distribution systems include low water pressure, high water pressure, and stagnated water. Following are some troubleshooting guidelines for each condition.

a. Low water pressure in a section of the distribution system can be due to a number of causes. A common cause is a closed or partially closed valve. Check all valves in the area and valves on water supply mains supplying the area. Another cause for low water pressures is inadequate pipe size. Conduct a computer simulation model study of the system to identify high head loss areas and develop corrective measures. Low

Table 13-24 Flow from Distribution Pump Stations and Elevated Reservoirs

Inflow Sources	Flow, L/s
Distribution pumping stations	
Northern Pump Station (Node 200 and Pipe 55)	655.9
Modeltown Pump Station (Node 201 and Pipe 56)	395.6
Northeast Pump Station (Node 202 and Pipe57)	386.8
Proposed Westside Pump Complex (Node 203 and Pipe 58)	443.1
Total distribution pumping	1881.4
Elevated reservoirs	
Southeast Tower (Node 125 and Pipe 47)	92.2
Modeltown Tower (Node 121 and Pipe 48)	63.9
Proposed Elevated Tower (Node 105 and Pipe53)	65.6
Total flow from elevated reservoirs	221.7
Total flow to the distribution system	2103.1

pressures in a section of distribution may occur because of high ground level relative to the pressure plane gradient. This too can be confirmed with a computer model. Corrective measures include raising the gradient by booster pumps and creating another, higher pressure plane. Often, large leaks in the distribution system cause low pressure. Check for signs of leaks and repair the mains.

b. High water pressure may be caused by a closed or partially closed valve. Check all valves in the area, particularly valves leading to the elevated tanks. Inadequate pipe sizes can also cause high water pressures. Conduct a computer simulation study to identify the problem. Another cause might be ground elevations (relative to hydraulic gradient) that are too low. Confirm with a model study, and install pressure reducing valves. High pressures can also be caused by excessive pumping capacity. Reduce the number of pumping units on line, or change pump impellers or speed to reduce capacity.

c. Water stagnation is a result of dead-end pipelines or of closed valves that in effect create dead-end pipelines. Open the valves fully, and eliminate the dead ends of the grid. Connect with pipes to complete loops. If this is not possible, flush dead end pipes regularly (weekly). Low pipeline velocities because of low demands and/or large pipe sizes can also cause water stagnation. Correct the situation by flushing the pipes on a regular basis.

13.9.2 Operation and Maintenance Guide

The following operation and maintenance items are necessary to keep the water distribution system in satisfactory operating condition.

1. Water Storage Reservoirs

   The operation and maintenance requirements for clearwells, ground storage reservoirs, and elevated storage reservoirs are generally the same. These are listed next.

   a. Inspect tanks annually for corrosion, solids deposition, clogged inlets and outlets, and general condition of the structure.
   b. Repair any corroded or damaged sections of the steel tanks and repaint as necessary.
   c. Check vents and overflows monthly to make sure they are clean enough to operate.
   d. For security reasons, drive around remote sites at least daily. Check fences, gates, doors, and locks.

2. Water Conditioning and Stabilization System

   a. Check the stability of water daily. Use routine laboratory data to calculate Langelier and Ryznar indexes, or perform a marble test. In addition, check the stability of water when the quality of the raw water changes and when chemical doses or treatment processes are modified. Adjust chemical feed as necessary.
   b. Each month, determine lead and copper levels in samples taken from the test taps in the distribution system, to assure that the corrosion control measures are working.

3. Pumping Stations

   The operation and maintenance requirements for high-service and distribution pump stations are similar to those for the raw water pumping stations. This information is provided in Chapter 7.

4. Distribution Systems

   a. Patrol pipelines regularly, to look for leaks and unauthorized taps.
   b. Check system pressures during high- and low-demand periods. In particular, check pressure at high and low points in remote parts of the system. Take corrective measures immediately if unusual situations develop.
   c. Check the chlorine residual at remote locations in the distribution system. Be sure that minimum chlorine levels are maintained during the low demand periods.
   d. Flush out, at least monthly, the dead-end pipes and areas subject to stagnation. In extreme cases, a higher frequency of pipe flushing may be necessary.
   e. Use good sanitation practices when working on or repairing the distribution pipelines. Sanitize tools and fixtures that come in contact with water. Prevent contamination of the water supply. As a prudent course of action, use water with a high chlorine residual. Feed chlorine in to local lines to increase the residual temporarily in the affected areas.

13.10 Specifications

Brief specifications for stability control equipment, clearwells, high-service pumping, ground storage reservoirs, elevated storage reservoirs, distribution pumping, and distribution piping are presented in this section. These specifications are provided to describe the design more fully. Many details have been intentionally omitted from these specifications, to keep the contents brief and simple. Therefore, these specifications should be used only as a guide to preparing more detailed specifications. Manufacturer representatives should be consulted during the design and preparation of specifications for any equipment utilized in the design.

13.10.1 Stability Control Equipment

The Contractor shall furnish and install a soda ash (Na_2CO_3) storage and feed system. The system shall consist of the following components: (1) two dry soda ash (Na_2CO_3) storage bins, (2) two dry-chemical feeders, (3) two solution tanks, (4) three metering pumps, (5) all piping, fittings, valves, and miscellaneous items necessary to provide a complete system, and (6) a secondary containment facility around the solution tanks to restrain any spills. The design details and specifications of similar chemical systems are covered in Chapters 8 and 11. This information is not repeated here. Readers should review the related information in these chapters and develop the design and specifications of the required components.

13.10.2 Chlorine Contact Channel

The Contractor shall construct a concrete chlorine contact channel adjacent to the clearwells. Concrete shall have a minimum 28-day compressive strength of 27,600 kPa (4000 psi). The chamber shall be 9.0 m deep by 4.5 m wide by 40 m long, with the effective length of the channel being 30 m. Water shall flow the length of the channel before it flows over a weir into the clearwell. A chlorine diffuser shall be installed near the beginning of the channel, and diffusers for ammonia and sodium carbonate (Na_2CO_3) solution shall be installed near the end of the channel, leaving approximately 30 m effective length for chlorine disinfection.

13.10.3 Clearwell

The Contractor shall construct two clearwells, each of 14,400 m^3. Each well shall be constructed out of 27,600 kPa (4000 psi) concrete and shall be 40 m square by 9.0 m deep. The clearwells shall be connected to the high-service pump station and the chlorine contact channel. The floor of the wells shall be sloped toward the pump suction piping. Prior to being placed in service, the entire interior surface of the clearwells shall be disinfected with a strong hypochlorite solution. After disinfection, water samples shall be tested to ensure that no bacteria remain.

13.10.4 High-Service Pump Station

The Contractor shall furnish and install a high-service pumping station. The pumping station structure shall be integral with the clearwells and shall house six high-service pumps and three filter backwash water pumps. Plant service water will be down from the treated water transmission main. Each pump shall be capable of drawing suction from either or both of the clearwells.

1. Pumps

   All pumps shall be of horizontal split-case, radial-flow type. The design capacity and head requirements for high-service and backwash pumps shall meet the specified requirements. Six high-service pumps and three filter backwash pumps shall be required.

   Each pump shall have an alternating current electric induction motor with a horsepower in excess of the maximum brake horsepower requirement of the pump across the entire operating range. The motors shall be of 480-volt, 60-cycle, 3-phase type.

   Pumps shall be constructed of materials as follows:

Casing	Cast iron
Impeller	Stainless steel or bronze
Impeller shaft	Carbon steel
Shaft sleeve	Stainless steel or bronze
Base plate	Carbon steel

   Pumps shall be fitted with mechanical seals, sleeve-type radial bearings, and antifriction thrust bearings. Pumps and motors shall be mounted on a base or bed plate with a true drip lip along the entire perimeter.

2. Piping

   All piping shall be either (1) ductile iron conforming to American Water Works Standard C150 or (2) steel conforming to American Water Works Standard C200.

3. Valves

   Isolation valves shall be of butterfly type; check valves shall be of tilting-disc type.

4. Structure

   The pump station structure shall be constructed of concrete and have a 28-day compressive strength of 27,600 kPa (4000 psi).

13.10.5 Ground Storage Reservoir

1. New Ground Storage

   Construct a new ground storage reservoir at the new Westside Pump Station site. The tank shall be of either welded steel or prestressed concrete construction. Welded steel tanks shall conform to the reference codes, procedures, and standards of the American Water Works Association (AWWA), American Welding Society (AWS), and American Institute of Steel Construction (AISC) and shall be painted with a high-quality epoxy coating suitable for potable water use. Prestressed concrete tanks shall conform to reference codes, procedures, and standards of AWWA and American Concrete Institute (ACI).

Regardless of the type of construction, tanks shall be equipped with an access ladder, a roof hatch with rain curb and locking hasp, a fail-safe vent with stainless steel insect screen, a 76-cm (30-inch) diameter manway, and inlet and outlet pipe connections.

2. Existing Ground Storage

All existing ground storage basins shall be renovated as indicated in the design plans.

13.10.6 Distribution Pump Station

1. New Distribution Pump Station

Construct one new distribution pump station complex at the new Westside Pump Station site. The pump station shall have a total of three pumps, with the capacity to add two additional pumps. The pumps shall be of vertical turbine type, with alternating current electric induction motors. Pump column and head shall be of either steel or cast iron construction. Pump shaft shall be stainless steel, with product lubricated sleeve bearings. Impeller shall be bronze.

The pump station structure shall be of masonry block construction, with a built-up asphaltic roof. Piping shall be either ductile iron or steel. Isolation valves shall be of butterfly type; check valves shall be of tilting disc type.

2. Existing Distribution and Pump Station

All existing pumping equipment shall be renovated to meet the distribution pumping requirements as specified in the design plans.

13.10.7 Pipelines

Pipelines shall be constructed in locations shown on the plans. Size and type of pipe and appurtenances shall also be as indicated on the drawings. Where indicated, ductile iron pipe shall conform to American Water Works Standard C150, steel pipe shall conform to American Water Works Standard C200, prestressed concrete cylinder pipe shall conform to American Water Works Standard C301 or C303, and polyvinyl chloride pipe shall conform to American Water Works Standard C900.

Pipelines shall be laid in a trench with a width approximately one meter wider than the outside diameter of the pipe. The trench wall shall be vertical up to 30 cm above the top of the pipe. Above 30 cm the trench shall be sloped at a safe angle.

13.10.8 Elevated Storage Tanks

1. New Tower

   Construct one new 2-million-liter steel elevated tank of the height indicated on the drawings. The general shape and style of the tank shall be as selected by the Owner. The tank shall be furnished with a dry standpipe from the bottom of the tank to the top. The standpipe shall have a ladder permanently attached, with a suitable safety device. The roof shall have an access hatch to the standpipe, and the tank interior shall have a fail safe vent with insect screen. The interior of the tank shall also have a permanently attached ladder leading to the bottom of the tank. The tank shall be painted inside and outside with a higher-performance industrial coating suitable for potable water service.

2. Existing Towers

   Both existing towers shall be renovated to meet the requirements as specified in the design plans.

Problems and Discussion Topics

13.1 Calculate the Langelier Index and Ryznar Index of the following water sample. Is the water scaling or corrosive? Discuss any optional treatment needs.

Temperature	10°C
pH	8.5
Total dissolved solids	280 mg/L
Calcium ion concentration	60 mg/L as Ca^{2+}
Total alkalinity (bicarbonate only)	91 mg/L as $CaCO_3$

13.2 A marble test is conducted on a water sample. The beginning alkalinity is 96 mg/L as $CaCO_3$. The ending alkalinity is 105 mg/L as $CaCO_3$. Is the water scaling or corrosive? Discuss any optional treatment needs.

13.3 Given the following table of demands and pumping rates, calculate the minimum size of the reservoir needed to meet differences between demand and pumping. Use both the graphical and spreadsheet methods.

Time	**Average demand, m^3/day**	**Average pumping, m^3/day**
12 midnight	10,000	13,833
4	11,000	13,833
8	15,000	13,833
12 noon	20,000	13,833
4	15,000	13,833
8	12,000	13,833

13.4 Two elevated storage reservoirs are connected by a pipe that is 1 mile long. The *C* value of this 18-inch- diameter pipe is 110. The water surface elevations in two reservoirs are 244 and 214 feet. Calculate the flow in the pipe.

13.5 At a point A in a horizontal 12-inch pipe ($f = 0.02$), the pressure head is 200 ft. At a point B, 200 feet from A, the 12-inch-diameter pipe suddenly reduces to 6 inches in diameter ($f = 0.015$). At a further distance of 100 feet, the 6-inch-diameter pipe suddenly enlarges to a 12-inch-diameter pipe at point C. At a distance of 100 feet from this point of expansion, the velocity in the pipe is 8.05 ft/s. Draw the EGL and HGL.

$$\text{Contration head loss} = \frac{K_c V_6^2}{2g}$$

$$\text{Expansion head loss} = \frac{(V_6^2 - V_{12}^2)}{2g}$$

$K_c = 0.37$

13.6 Water is flowing from reservoir A to reservoir B. The water surface elevations in reservoirs A and B are 215 and 140 m, respectively. Two parallel pipes connecting from reservoir A to B are 60 cm and 50 cm in diameter and 400 m and 600 m long, respectively. Assume that entrance, exit, and contraction loss are small; calculate the flow and head loss in each pipe.

13.7 Calculate the demand at each node of the subdivision shown in Figure 13-1. The pipe lengths and service area information are given here:

Length of pipe AB, BF, CG, DE and AE = 400 m, BC and FG = 150 m, and CD = 250 m.

Areas served by nodes F and G are industrial zoning with demand factors of 9000 m^3/ha·d; the areas served by A, B, D, and E are residential with demand factors of 4000 m^3/ha·d; and the area served by node C is commercial with a demand factor of 6000 m^3/ha·d. Assume that the outer boundary of the subdivision is at a distance of half-pipe length from the exterior pipes.

13.8 The hourly time-demand data for a distribution system are given here. The water is pumped into an elevated storage reservoir at a constant rate of 281,400 gallons per hour. Determine the reservoir capacity by using the mass diagram and deficiency methods.

Time	Demand, gpm	Time	Demand, gpm
1 a.m.	1900	1 p.m.	6470
2	1875	2	6460
3	1840	3	6440
4	1795	4	6520
5	1810	5	6710
6	1920	6	7115
7	3220	7	9010
8	5000	8	8700
9	5550	9	5320
10	6020	10	2220
11	6220	11	2110
12 noon	6320	12 midnight	2000

13.9 Calculate the theoretical detention time for the basin shown in Figure 12-4(a). Estimate the T_{10} time. Make recommendations that will improve the T_{10} to within 70 percent of T_{100}.

13.10 A test was conducted to determine the C factor of a water main. The set up is shown in Figure 13-5. The following results were obtained:

Upstream gauge pressure and elevation = 8.20 m and 107.32 m

Downstream gauge pressure and elevation = 11.31 m and 103.81 m

Distance between pressure gauges = 40.5 m

Flow through the pipe = 118 L/s

Calculate the C factor. Also, calculate the pressure-gauge reading at the midpoint of the pipe. Assume that the elevation of this gauge is 105.57 m.

13.11 Obtain a copy of the water distribution system map of a local community. Study the map; identify the locations of elevated tanks, pump stations, ground storage tanks, and wells or water treatment plants. Visit the facilities. Determine the design concepts. Discuss the design of the facility with the operators and determine their thoughts about the strengths and weaknesses of the design.

13.12 What is a neutral point of a reservoir? Discuss how the neutral point of a reservoir is utilized in checking critical condition of the distribution system.

13.13 Proposed improvement needs of the distribution system in the Design Example are given in Section 13.8.2, Step G.2. The final computer simulation results are provided in Table 13-23 and 13-24. Do you think that all improvement objectives have been met in this design? Give your commentary.

13.14 Analyze the pipe network shown below. Use the Hardy-Cross Method. Calculate the flow and velocity in each pipe. Also, calculate pressure at each node if supply node has a pressure of 300 feet of water. Assume all nodes have a datum elevation of zero. Draw 10-ft-interval pressure contours. All flow are in gpm, pipe lengths in ft, and pipe diameters in inches. Use $\Delta Q \leq$ 10 gpm and $C = 100$.

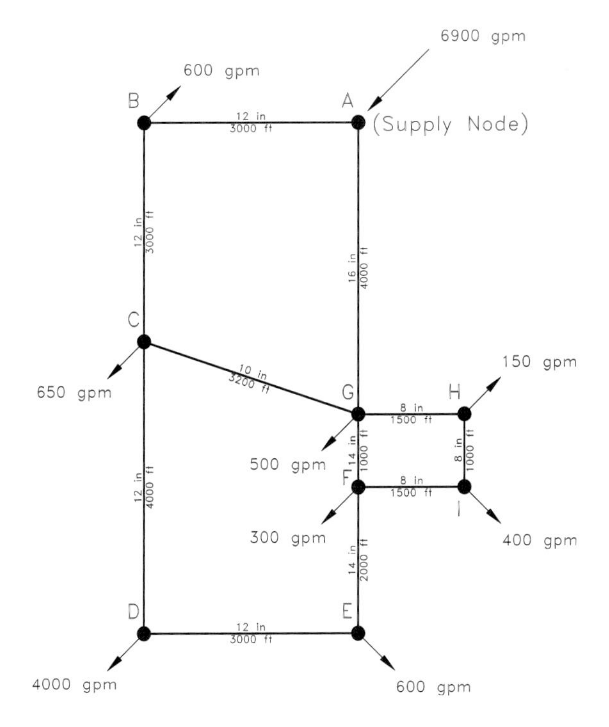

13.15 In Problem 13.14, the fire flow is given at node D. If fire flow of 4000 gpm is placed at node I, and normal demand of 400 gpm is placed at node D, what will be the flow and velocity in each pipe and the head at each node?

13.16 A simple pipe loop is shown below. Calculate flow, velocity, and head loss in each pipe. Also, calculate the pressure at node B if the supply pressure is 91 m. Use $C = 100$.

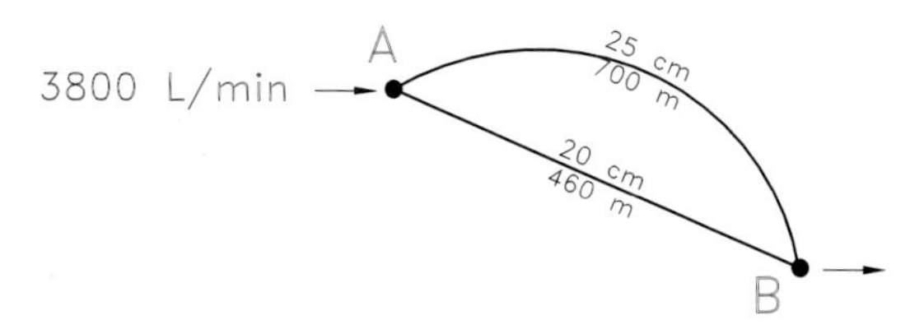

13.17 Calculate the velocity, head loss, and flow in each pipe of the network shown below. What will be the pressure at node B? Use $C = 100$.

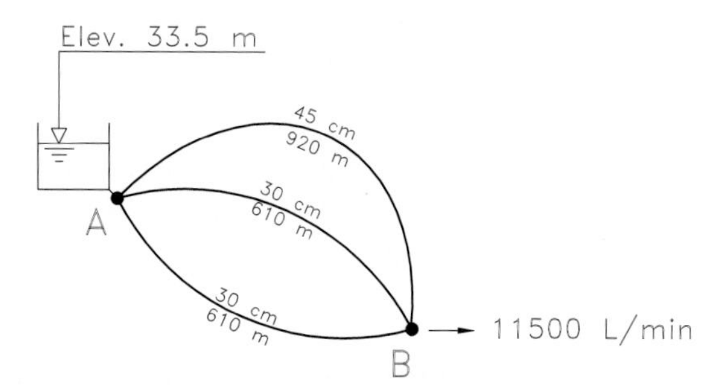

13.18 A new elevated reservoir is added at node B in the figure of Problem 13.17. The water surface elevation in the new reservoir is 25.0 m. Calculate the flow, velocity, and head loss in each of three pipes between nodes A and B. Use $C = 100$.

References

1. Pontius, F. W. "New Horizons in Federal Regulations," *Jour. AWWA*, vol. 90, no. 3, pp. 38–50, March 1998.
2. Tchobanoglous, G. and Schroeder, E. D. *Water Quality: Characteristics, Modeling, and Modification*, Addison-Wesley Publishing Co., Reading, MA, 1985.
3. Viessman, W. and Hammer, M. J. *Water Supply and Pollution Control*, 6th ed., Addison-Wesley, Reading, MA, 1998.
4. Snoeyink, V. L. and Jenkins, D. J. Water Chemistry, John Wiley & Sons, New York, 1980.
5. McGhee, T. J. *Water Supply and Sewerage*, 6th ed., McGraw-Hill, Inc., New York, 1991.
6. Dodrll, D. M. and Edwards, M. "Corrosion System on the Basis of Utility Experience," *Jour. AWWA*, vol. 87, no. 7, pp. 74–85, July 1995.
7. Holm, T. R. et. al. "Polyphosphate—Water Treatment Products: Their Effects on the Chemistry and Solubility of Leak in Potable Water System," *Proceedings of 1989 Water Quality-Technology Conference (WQTC)*, Philadelphia, PA, Nov. 12–16, 1989.
8. Boffardi, B. P. "Polyphosphate Debate," *Jour. AWWA*, vol. 83, no. 12, pp. 10–12, December 1991.
9. White, C. G. *Handbook of Chlorination and Alternative Disinfectants*, 4th ed., John Wiley & Sons, Inc., New York, 1999.

10. Malcolm Pirnie, Inc. and HDR Engineering, Inc. *Guidance Manual for Compliance with Filtration and Disinfection Requirements for Public Water Supply Systems Using Surface Water Sources*, ISBN O-89867-558-8, Science and Technology Branch, USEPA, Washington, D.C., October 1990.
11. AWWA. *Water Transmission and Distribution*, 2d ed., American Water Works Association, Denver, CO, 1996.
12. AWWA. *Safety Practices for Water Utilities*, Manual M3, American Water Works Association, Denver, CO, 1996.
13. Peary, H. S., Rowe, D. R., and Tchobanoglous, G. *Environmental Engineering*, McGraw-Hill Book Co., New York, 1983.
14. Gross, H. *Analysis of Flow in Network of Conduits or Conductors*, Bulletin No. 286, University of Illinois Engineering Experiment Station, Urbana, IL, 1936.
15. Jeppson, R. W. *Analysis of Flow in Pipe Network*, Ann Arbor Science, Ann Arbor, MI, 1976.
16. Walski, T. M. *Analysis of Water Distribution System*, Van Nostrand Reinhold, New York, 1984.
17. Alexander, S. M., Glenn, N. L., and Bird, D. W. "Advanced Technologies in the Mathematical Modeling of Water Distribution Systems," *Jour. AWWA*, vol. 67, no. 7, pp. 343–346, July 1975.
18. Wood, D. J. and Charles, C. O. A. "Hydraulics Network Analysis Using Linear Theory," *American Society of Civil Engineers, Journal of the Hydraulics Division*, vol. 98, no. HY7, pp. 1157–1170, July 1972.
19. Wood, D. J. *A Computer Program for the Analysis of Pressure and Flow in Pipe Distribution System*, College of Engineering, University of Kentucky, Lexington, KT, 1975.
20. Martin, D. W. and Peters, G. "The Application of Newton's Method to Network Analysis by Digital Computer," *Journal of the Institute of Water Engineers*, vol. 17, pp. 115–129, 1963.
21. Epp, R. and Fowler, A. G. "Efficient Code for Steady-State Flows in Networks," *American Society of Civil Engineers, Journal of the Hydraulics Division*, vol. 96, no. HY1, pp. 43–56, January 1970.
22. Cesario, L. *Modeling, Analysis, and Design of Water Distribution Systems*, American Water Works Association, Denver, CO, 1995.
23. Ormsbee, L. E. and Lingireddy, S. "Calibrating Hydraulic Network Models," *Jour. AWWA*, Vol. 89, No. 2, pp. 42–50, February 1997.
24. Walski, T. M. "Sherlock Holmes Meets Hardy Cross, or Model Calibration in Austin, Texas," *Jour. AWWA*, vol. 82, . no. 3, pp. 34–38, March 1990.
25. Rhoades, S. D. "Water Systems Standards Survey," *Jour. AWWA*, vol. 78, no. 11, pp. 30–34, November 1986.

CHAPTER 14

Residuals Processing and Disposal

14.1 Introduction

Water treatment plants produce residuals from various treatment processes. These residuals contain organic and inorganic turbidity-causing solids, including algae, bacteria, viruses, silt and clay, and precipitated chemicals that are produced during treatment. Historically, the water treatment residuals were discharged into the natural water system. Such discharges are now prohibited under the Water Pollution Control Act Amendments of 1972 and the Clean Water Act of 1977.[1,2] These residuals are considered industrial wastes, and their discharge into the natural waters is subject to a permit system under the National Pollutant Discharge Elimination System (NODES).

With restrictions on surface water discharges of residuals from water treatment plants, other alternatives available are disposal on land and disposal into the sanitary sewers. A summary of different residuals-management systems commonly used in the overall process train has been presented in Section 3.4. The purpose of this chapter is to present an overview of the sources of and types of residues produced during water treatment, of their characteristics, of the various methods of processing, of reclamation of useful by-products and disposal, and of process design and selection considerations. The step-by-step design procedure, operation and maintenance, and equipment specifications are provided in the Design Example.

14.2 Sources and Characteristics of Residual Streams

The major sources of residuals from a conventional water treatment plant are alum or iron coagulation sludges, softening process sludge, and filter backwash. To a lesser extent, the residuals generating processes are precipitation of iron and manganese, use of coagulant aids and activated carbon, disposal of diatomaceous earth filter washwater, and disposal of spent brines.

On the whole, these wastes are stable, or nonputrescible, and are, therefore, collected or discharged intermittently rather than continuously.

The characteristics of the residual streams generated in various treatment processes are primarily a function of treatment process, added chemicals, and quantity of raw water. An understanding of the quantity of sludge, solids content, and the nature of solids is essential to select and design the processing equipment properly. The basic characteristics of various sources of residual streams are discussed below.

14.2.1 Alum or Iron Coagulation Sludge

Aluminum and iron salts are the principal chemicals used for removal of colloidal particles. Many plants utilize lime in conjunction with alum or iron to achieve partial softening and to improve coagulation. Polymers are also used as coagulant and/or filter aids to improve removal of colloidal particles in coagulation and filtration processes. Often, powdered activated carbon is also used for taste and odor control. In each case, the characteristics of the sludge will differ, and such conditions should be considered in quantity and quality estimation.

Physical and Chemical Quality Coagulation sludges are produced by flocculating and settling natural turbidity. Aluminum and iron salts react with alkalinity and form precipitates of aluminum and ferric hydroxides. The sludge contains these hydroxide precipitates and turbidity-causing organic and inorganic compounds. Although the BOD and COD values may be high, the sludges are stable, because there is no organic matter to undergo active decomposition or promote an anaerobic condition. As a result, the sludge is often allowed to accumulate in sedimentation basins for days and months, and then removed intermittently.

Basic physical and chemical properties of coagulation sludges are listed below. Numerical values are summarized in Table 14-1.[3–10]

- The solids concentration, thickening, density, and dewaterability of the sludge produced are highly dependent upon the raw water quality and seasonal variations.
- High-turbidity waters usually result in sludges that are more concentrated and less difficult to dewater; low-turbidity waters present a more difficult sludge processing problem.[3,4]
- Coagulation sludge from water containing high algae concentration will result in sludges of light and low solids concentration.[4]
- Sludges with high proportions of metal hydroxides are not as easily dewatered as are sludges that have high proportions of other materials. This is because metal hydroxides have water molecules in their structure that separate the floc and other particles. Thus, direct contact of particles is more difficult, unless large amounts of lime and optimum polymer dosages are used to improve the bond between the particles.[3,4]
- In general, settled iron sludges have a higher solids concentration than alum sludges.[3]
- The addition of polymers, lime, or PAC increase the solids concentration of both alum and iron sludges.[3,4]

- Alum sludge is a voluminous gelatinous sludge with poor compressibility. It will generally concentrate 0.5 to 2 percent solids in the sedimentation basin. It can achieve 10 percent concentration in 2 days in sand beds.[5]
- Solids concentration in a sedimentation basin depends on the design and operation of the facility. Typical underflow concentration may vary from 0.1 to 1 percent solids. If, however, the sludge is allowed to accumulate for a month or longer, the sludges may thicken to 4 to 6 percent solids.[3]
- The density of sludge depends upon moisture content. Sludge with one percent solids has almost the same density as water. The density of dry sludge is 1200 to 1520 kg/m^3 (75 to 95 lb/ft^3).[4] The specific gravity of sludge solids is 1.5 to 2.5.
- Sludge viscosity and shear strength increase as solids concentration increases. Below solids concentration of about 2 percent, the specific resistance of sludge increases as solids concentration decreases. Above 2 percent solids concentration, the specific resistance remains almost constant.[4]
- The volatile fraction of the suspended solids does not reflect the organic fraction, because much of the loss on ignition is probably due to bound water in hydrated oxides.
- Ferric oxides are insoluble over a wider pH range than are aluminum oxides, so the pH of the water in contact with hydrated iron oxide may be found to vary over a wider range.[5]

Quantity The quantity of solids produced in coagulation is dependent upon total suspended solids in the water, on type and dose of coagulant, and on the efficiency of the sedimentation basin. Generally, 60–90 percent of total solids are removed in the sedimentation basin.[4] The remaining solids are removed in the filters. There is no absolute correlation between the turbidity units (TU) and total suspended solids (TSS). A ratio of TSS to TU normally varies from 0.5 to 2.0. Discussion on sludge quantity calculations are presented in Sections 9.3.4 and 9.6.2, Step B.1. It has been reported that total solids from an alum coagulation facility may vary from 8 to 210 kg/per 1000 m^3 of raw water treated (67 to 1800 lbs per million gallon).

14.2.2 Softening Sludge

Surface- and groundwaters containing high concentrations of calcium and magnesium are softened by lime and soda-ash processes. The residue most frequently consists of calcium carbonate, magnesium hydroxide, and unreacted lime (called lime grit). Because of the colloidal nature and poor settling characteristics of the lime-soda precipitate, softening normally is supplemented with coagulation. As a result, the softening sludge also contains hydrated iron and aluminum oxides. If combined coagulation and softening of surface water is used, the sludge also contains the turbidity-causing solids. Highly turbid waters are normally treated in two stages. The turbidity is removed by coagulation prior to softening.[4,5]

Physical and Chemical Quality The physical and chemical properties of softening sludge are important. It is generally stable, dense, inert, and relatively pure. For these reasons,

Table 14-1 Physical and Chemical Characteristics of Alum and Iron Coagulated Sludges

Characteristics	Alum Sludge	Iron Sludge	Reference
Physical:			
Quantity, $kg/1000m^3$	8–210, Typical 48	80	4,5
lb/million gal.	67–1750, Typical 400	668	4–6
Specific resistance, s^2/g	$1.05–5.4 \infty 10^{10}$	$4.1 \infty 10^8$	7–10
Viscosity, g/cm·sec	0.03 (non-Newtonian)	0.03	5–7
Initial settling velocity, cm/sec	0.06–0.15	–	7
Dry density, kg/m^3	1200–1520	1200–1800	8
Dewaterability	10% concentration in 2 days on sand beds	–	3
Chemical:			
BOD_5, mg/L	30–300	30-300	5
COD, mg/L	30–5000	30–5000	5
pH	6.0–8.0	7.4–8.5	5,9
Total Solids, %	0.1–4	0.25–3.5	7
Solids characteristics:			
$Al_2O_3\ 5{\cdot}5H_2O$, %	15–40	–	9
Fe, %	–	4.6–20.6	9
Silicate and inert material, %	35–70	–	5,6
Organics, %	15–25	–	5,6
Volatiles, %	–	5.1–14.1	5,6

Source: Adapted in part from References 3–10.

the recovery of lime by recalcination is a relatively attractive option at larger plants. Important physical and chemical properties of these sludges are summarized here:

- The pH of the sludge is typically high (10.5–11.5), and it is white, unless colored by turbidity or iron and manganese. The sludge is odorless, has little or no BOD or COD, and (because of its high pH) does not contain significant number of bacteria.[3,5]
- The calcium carbonate solid is more crystalline, dense, and discrete. Magnesium hydroxide sludge is gelatinous and similar in nature to aluminum and iron coagulant solids. The calcium carbonate solids have specific gravity of 2.71. The density calculation for wet sludge may be found in Section 9.6.2, Step B.1.
- The dewaterability of lime-softening sludge varies with the amount of magnesium hydroxide. $Mg(OH)_2$ may range from a few percent to as much as 30 percent. Sludges

with low $Mg(OH)_2$ can dewater to sludge cake having up to 60 percent solids but cake solids can be as low as 20–25 percent with higher magnesium hydroxide concentrations.[3,5] The softening sludge at 50–60 percent solids is sticky and is difficult to discharge cleanly from dump trucks.[3] The physical and chemical properties of softening sludge are summarized in Table 14-2.

Table 14-2 Physical and Chemical Properties of Softening Sludge

Characteristics	Value	Reference
Physical:		
Color	White	5,6
Odor	None, musty	5,6
Quantity, % of treated water	0.3–6	8
mg/L per mg/L hardness removed	2–4	4
Settleability, % of initial volume	50 in a week	5
cm/sec	0.001–0.1	4
Compaction achieved, % solids	2–15	5
Specific gravity of dry solids	2.7–2.75	4
Wet density of 25% solids, kg/m^3 (lb/ft^3)	1920 (120)	8
Specific resistance, s^2/g	$0.20–25.6 \infty 10^7$	10, 11
Plastic viscosity, g/cm·s	0.06	12
Sludge character, % solids	Liquid at 0–10	3
	Viscous liquid at 25–35	3
	Semisolid 40–50 (toothpaste-like)	3
	Crumbly cake 60–70	3
Chemical:		
BOD, mg/L	Very low	5
COD, mg/L	Low	5
pH	10.5–11.5	4
Composition, %		
$CaCO_3$	75	4,13
CaO	42	4,13
Silica as SiO_2	6	4,13
Total carbon	7	4,13
Aluminum as $A1_2O_3$	3	4,13
Magnesium as MgO	2	4,13

Source: Adapted in part from References 4–13.

Quantity The quantity of lime-softening sludge produced at water treatment plants varies greatly, depending on the hardness of the water, the raw water chemistry, the dosages of chemicals used, and the desired finished-water quality. As a general rule, for each mg of calcium (as Ca) removed, 2.5 mg of $CaCO_3$ sludge is produced[4]. Likewise, removal of 1 mg of magnesium (as Mg) produces 2.4 mg of $Mg(OH)_2$ in sludge.[4] Reported quantity for various plants has ranged from 43 to 2230 kg per 1000 m^3 (360 to 18,700 lbs per million gallons), with an average of about 230 kg per 1000 m^3 (2000 lb per million gallons) of water treated.[4] The solids content of softening sludge typically ranges between 2 and 15 percent.[5]

14.2.3 Filter Backwash

Filter backwash contains a large volume of water and a relatively small concentration of solids (50 to 1000 mg/L). It constitutes 1 to 5 percent (average of 2 percent) of the total water processed. The solids are difficult to separate. A filter backwash recovery system is needed to recover water for reprocessing and to concentrate solids.

Physical and Chemical Quality The physical and chemical quality of the filter backwash stream is summarized here:

- The solids in filter backwash are similar to those in the sedimentation basin. Typical backwash contains hydrous oxides of aluminum, iron, manganese, and magnesium, carbonates of calcium and iron, spent activated carbon, and silicate materials (which constitute the largest fraction of solids). It may also contain the algae, bacteria, protozoan cysts and, detached slime bacteria that are often supported on filter media.
- The BOD, COD, and bacterial count in filter backwash is high; therefore, the water recovery system should be carefully selected, because backwash has the potential for release of taste-and odor-causing compounds.
- The chemical analysis of filter backwash solids reveals that aluminum and silica oxides predominate, and carbon and organic matter compose roughly one-fifth of the residue.[5] Physical and chemical properties of filter backwash are provided in Table 14-3.
- Filter backwash solids typically are difficult to separate from the liquid.[3] Washwater recovery ponds achieve 80 percent concentration of solids in 24 hours with the use of polymers or other coagulant aids.[3]
- Where direct filtration is used, the solids loading is a function of the coagulant dosage and the raw water turbidity. The filter backwash contains more solids and behaves like coagulation sludge.

Quantity The quantity of solids in filter backwash depends on the efficiency of the filter and on the pretreatment provided. Where sedimentation precedes the filter, typical suspended solids concentration escaping the sedimentation basins range from 4 to 10 mg/L, which corresponds to turbidity units of 2 to 6.[3,4] At this loading, the backwash water will contain solids at

Table 14-3 Physical and Chemical Quality of Filter Backwash

Characteristics	Value
Physical:	
Volume, % of treated water	1–5, average 2
Solids, %	0.01–0.1
Color	Varies widely; grey–brown–black
Odor	None
Settleability, % of initial volume	80 in 2 to 24 hr with polymer
Specific resistance, s^2/g	1.5×10^{10}
Viscosity, g/cm·s	0.01
Dewaterability	Must be coagulated and settled
Chemical:	
BOD, mg/L	2–10
COD, mg/L	28–160
pH	7.2–7.8
Solids, %	
Al_2O_3	25–50
SiO_2	34–35
Carbon and organics	15–22

Source: Adapted in part from References 3–6.

from 4 to 10 kg/1000 m^3 (from 35 to 85 lbs per million gallons) of water treated. If direct filtration of flocculated water is practiced, the quantity will be considerably higher.

14.2.4 Iron and Manganese Precipitation Sludge

Iron and manganese are generally present in small concentrations in surface- and groundwaters. Therefore, their effect upon overall quantity of sludge processing is small. The sludge produced typically is inert and is red or black in color.

Soluble iron and manganese are precipitated by oxidation with potassium permanganate, chlorine, or ozone. Oxidation and precipitation is also achieved by simple aeration. The precipitate thus formed is ferric hydroxide, ferric carbonate, or manganese dioxide. The quantity of sludge is 1.5 to 2 mg/L for each mg/L of iron or manganese in solution.[3] The iron and manganese sludge is removed in sedimentation tanks or in filters.

14.2.5 Residues From Coagulant Aid

Clays such as bentonite, activated silica, and synthetic organic polyelectrolytes or polymers are often added to some waters to provide negatively charged particles to aid in coagulation and to increase sludge density. The use of these compounds improves floc settleability and in some cases lowers sludge volume.[5]

14.2.6 Residues From Filter Aid

Filter aids are generally synthetic organic polymers or metal salts similar to those used in coagulation processes. These compounds help to produce floc particles that are more resistant to the greater shear forces exhibited at higher filtration rates. Thus, filter performance is improved. A filter aid may also be added to the backwash water to reduce initial turbidity breakthrough.

14.2.7 Spent Powdered Activated Carbon (PAC)

Powdered activated carbon is often added seasonally to reduce odor problems. For cost reasons, the quantity of PAC added in raw water is relatively small (4 to 10 mg/L); therefore, its impact upon the sludge quantity is also small. PAC adds a grey or black color to coagulation sludge.

14.2.8 Diatomaceous-Earth Filter Washwater

Generally, small water treatment plants use diatomaceous-earth filters. The filter consists of leaves or discs that are assembled inside a housing. The filter operation is composed of three cycles: (1) precoating, (2) filtering, and (3) cleaning. In the precoating cycle, a slurry of filter aid (diatomaceous earth or other material) is added into the tank. Either under pressure or under suction from a filter pump, the slurry is recirculated through the tank until a uniform layer of precoat is deposited on the filter leaves or discs. In the filtration cycle, the water is pumped through the filter. Filtration continues until a thick cake of solids is formed on the precoated material. In many cases, small amounts of filter aid are injected into the influent (body feed) during the filtration cycle. During the cleaning cycle, drain valves are opened and the sludge cake is washed from the leaves or discs by water jets. Some filter arrangements allow for a dry-cake discharge. These types of filters have been widely applied in swimming pools and industries.[14]

The solids produced from filter wash are mainly pure silica (diatomaceous earth). It is easily concentrated and dewatered. The BOD and COD are around 106 and 340 mg/L, respectively. Total suspended solids and volatile suspended solids are 7600 mg/L and 260 mg/L, respectively. The density of diatomaceous earth is 160 kg/m^3 (10 lb/ft^3), and the specific gravity of the solids is 2.0.[4,8]

14.2.9 Spent Brine

The spent brines from the regeneration of ion-exchangers, or from a reverse osmosis unit, form a special class of water treatment residues. These streams contain high concentrations of soluble salts. Total dissolved solids may range from 15,000 to 35,000 mg/L. Major cations and anions are sodium, potassium, calcium, magnesium, chloride, and sulfate. Depending upon the raw water quality, the volume of brine may range from 3 to 10 percent of the treated water.[5]

14.3 Residuals Processing

Residuals-management processes include thickening, conditioning, dewatering, drying, chemical recovery, and disposal. It is important that solids processing systems be investigated in the planning phase of the project, because capital and operation and maintenance (O&M) costs of the facilities often represent a substantial portion of the overall costs of water treatment plants. Many factors are considered in the planning and design of solids-handling facilities. Among these factors are plant size, available land, sludge transport and disposal, and recovery and reuse of coagulants and lime.

A summary of 20 residuals-management systems that are used in development of process trains has been presented in Table 3-3. Readers should refer to this table for a quick review. An overview is provided in Figure 14-1. A complete residuals-management process train would be made up of one or more of the process steps included in Figure 14-1. A brief description of many of these processes is given below.

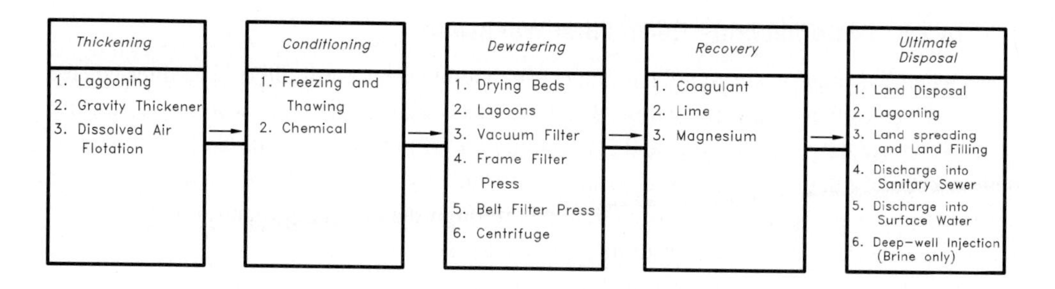

Figure 14-1 Alternative unit operations and processes for residuals processing and disposal.

14.3.1 Thickening

The general purpose for thickening is to reduce the bulk of water and to reduce the size of subsequent dewatering units. Thickening is commonly achieved in sludge lagoons or gravity thickeners. Dissolved air flotation is seldom used in water treatment.

Lagooning Lagoons are commonly used for thickening where land is available at the treatment plant site. They function quite well when designed and operated as thickeners. Alum and iron coagulated sludge without conditioning can reach 4 to 6 percent solids in a month with continuous decanting. Higher concentrations can be reached at the bottom. For design purposes, 5 to 10 percent solids concentration in two to three months of thickening is assumed.

The design involves concrete basins with overflow structure for removal of supernatant. Multiple-level decanting systems provide better operational flexibility. The thickened sludge is withdrawn from the bottom. Lagoons with 2–3 m sludge application depths are generally designed.

Many times lagoons are designed for storage, thickening, dewatering, and drying. In reality, they are used for ultimate disposal, for which they are often unsatisfactory. Further discussion of these lagoons is provided in Section 14.3.5.

Gravity Thickener Gravity thickening is used to concentrate solids by decanting excess water. The decanted water is returned to the raw water line, unless it has objectionable taste- and odor-causing compounds and microorganisms.

Gravity thickeners are usually circular settling basins with a scraper mechanism at the bottom. They may also have a hopper (without collection mechanism) for sludge accumulation and compaction. Both continuous flow and batch operation are used. In continuous-flow operation, sludge enters the thickener near the central well and the supernatant distributes radially and discharges into a peripheral wire trough.
The sludge settles and thickens at the lower depth. The scraping mechanism, if provided, moves slowly and pushes the sludge into the sludge hopper for removal.[6] The slow rotation of the scraper mechanism also prevents bridging of the sludge solids. The basin floor is sloped to a central hopper. Figure 14-2 provides details of a gravity thickener.

The thickeners on batch-fill and draw basis generally have hoppers. The sludge is allowed to settle until the desired thickening is achieved. A telescoping decant pipe removes the supernatant. The decant pipe is continuously lowered as solids settle until the maximum allowable supernatant liquid solids concentration is reached or until the sludge will not thicken further.[6] The thickened sludge is pumped out from the hopper.

Thickener design should be based on data developed from bench scale or pilot testing. Typical design parameters reported for gravity thickeners are summarized here:

- Polymer or conditioning chemicals increase the solids settling rate significantly.
- Consideration should be given to annual and seasonal variation in sludge quality and quantity.
- Typical hydraulic loading for conditioned sludge is 4 to 10 $m^3/m^2 \cdot d$ (100 to 250 gpd/ft2).[3]
- Solids loading for conditioned sludge is 20 to 80 $kg/m^2 \cdot d$ (4 to 16 $lb/ft^2 \cdot d$).[4]
- A typical depth for a gravity thickener is 4.5 to 6.5 m, and detention time might be from 8 to 24 hours.
- Solids in lime-softening sludge concentrate considerably better than solids in alum or iron coagulated sludge. Alum or iron sludges can concentrate up to from 3 to 6 percent; lime-softening sludges can concentrate up to from 9 to 20 percent. Higher solids concentrations create sludge removal problems in the thickeners.

Dissolved Air Flotation (DAF) Dissolved air flotation has been used for the thickening of wastewater treatment sludges. In water treatment, DAF has had limited application. In Scandinavian countries and in the United Kingdom, dissolved air flotation has been employed mainly for treatment of nutrient-rich reservoir water that could contain heavy algae blooms and also for low-turbidity, low-alkalinity, and colored water.[6] In a DAF unit, air is dissolved at high

Figure 14-2 Equipment details of circular gravity thickener. *(Courtesy of U.S. Filter/Link-Belt)*

pressure. The pressured flow is discharged into a flotation tank that operates at one atmosphere. Fine air bubbles rise that cause flotation of solids. The application of DAF in water treatment is limited, so this process is not covered fully in this section. Readers should refer to References 3, 4, and 6 for additional details.

14.3.2 Conditioning

Sludge conditioning is generally employed to aid in gravity thickening and mechanical dewatering. Conditioning is achieved by: (1) freezing and thawing and (2) chemical addition. In both cases, the physical and structural properties of the sludge are improved. The gelatinous structure is destroyed, water is released easily, and solids are compacted. In colder climates, the sludge is allowed to freeze in lagoons during winter months. After thawing in summer the sludge is dewatered. Excellent results have been achieved with freezing and thawing. It has been reported that alum sludge with 2 percent solids can be converted to a 20- to 30-percent-solids granular slurry that will drain readily and can be handled easily.[3,8]

Polymers are most commonly used for conditioning the water treatment sludges. Polymers improve the dewatering characteristics of the sludge. Bench-scale or pilot plant studies are needed to determine the effectiveness of the polymers and their optimum dose. The polymer dose is generally 1–10 mg per g of sludge solids.[4,6] The equipment design involves a storage facility, a feeder, and a mixer (for good dispersion of the polymer into the sludge to be conditioned).

14.3.3 Dewatering

Dewatering of water treatment sludge is necessary to remove moisture so that the sludge cake can be transported by truck to the disposal site. Sludge-dewatering systems range from simple devices to extremely complex mechanical processes. Simple devices involve natural evaporation and percolation in earthen lagoons and drying beds. Complex mechanical systems utilize conditioning and complex processes, such as vacuum filters, filter presses, belt filters, and centrifugation. The selection of any device depends on the quantity and nature of the sludge and on the disposal methods. In this section, various methods of sludge dewatering are presented.

Drying Beds Drying beds consist of perforated underdrain pipe, gravel, and sand. The sludge is applied in a depth ranging from 0.3 to 1 m. Greater sludge depths require longer drying times. Solids concentration of from 15 to 30 percent can be attained with alum sludges. The drying time may vary from three to four days. Field tests must be conducted to develop design information. Solids-loading rate can range from 100 to 300 kg of dry solids per m^2 per year (20 to 61 $lbs/ft^2{\cdot}yr$). The estimated number of bed uses may range from 1 to 20 per year. Both solids loading and the number of bed uses depend upon climate, solids concentration in the sludge, and sludge conditioning applied. Multiple beds must be provided to allow application, drainage and drying, and sludge removal cycles. The bed is usually considered dewatered when the sludge can be removed by earthmoving equipment.

Vacuum-assisted drying beds are also designed to dewater chemically conditioned sludge. The floor of the bed has porous plates, and a vacuum is applied below the plates. Cake solids of 15 percent are achieved in a one-day cycle time. These beds are attractive options, because of their low capital cost and energy requirements.

Lagoon Lagoons for thickening, dewatering, and drying are used in the United States. These lagoons have decanting systems and are designed to have sufficient filling and drying time. Three months of filling and three months of drying time are most common. Solids loading can range from 50 to 100 kg/m^2 (10 to 20 lb/ft^2) per year. For lagoon design, the net evaporation–percolation rate should be used.

Centrifuge The centrifuge uses centrifugal force to speed up the sedimentation rate of sludge solids. Sludge dewatering can be achieved by basket or solid-bowl centrifuges. Basket centrifuges operate semi-continuously. In a typical solid-bowl unit, the conditioned sludge is pumped into a horizontal or cylindrical "bowl" rotating at 1600–2000 rpm. The solids are spun to the outside of the bowl where they are scraped out by a screw conveyor. The liquid or "centrate" is returned to a sump for treatment or disposal. Centrifuges are compact, entirely enclosed, require small space, and can handle sludges that might otherwise plug the filter cloth. The disadvantages include complexity of maintenance, abrasion problems, and a centrate high in suspended solids.[15]

The sludge cake from a centrifuge contains 20–35 percent solids, and solids capture of from 85 to 98 percent is achieved. The polymer dosage for sludge conditioning prior to centrifuge may be 0.05–0.5 percent of dry solids in the feed. A centrifuge dewatering system is shown in Figure 14-3. Alum sludges require higher polymer doses. Lime-softening sludges produce cake solids, 30 to 70 percent. The bowl centrifuge can also be used to separate magnesium hydroxide selectively from calcium carbonate. Such separation is useful if lime sludge is to be recalcined.

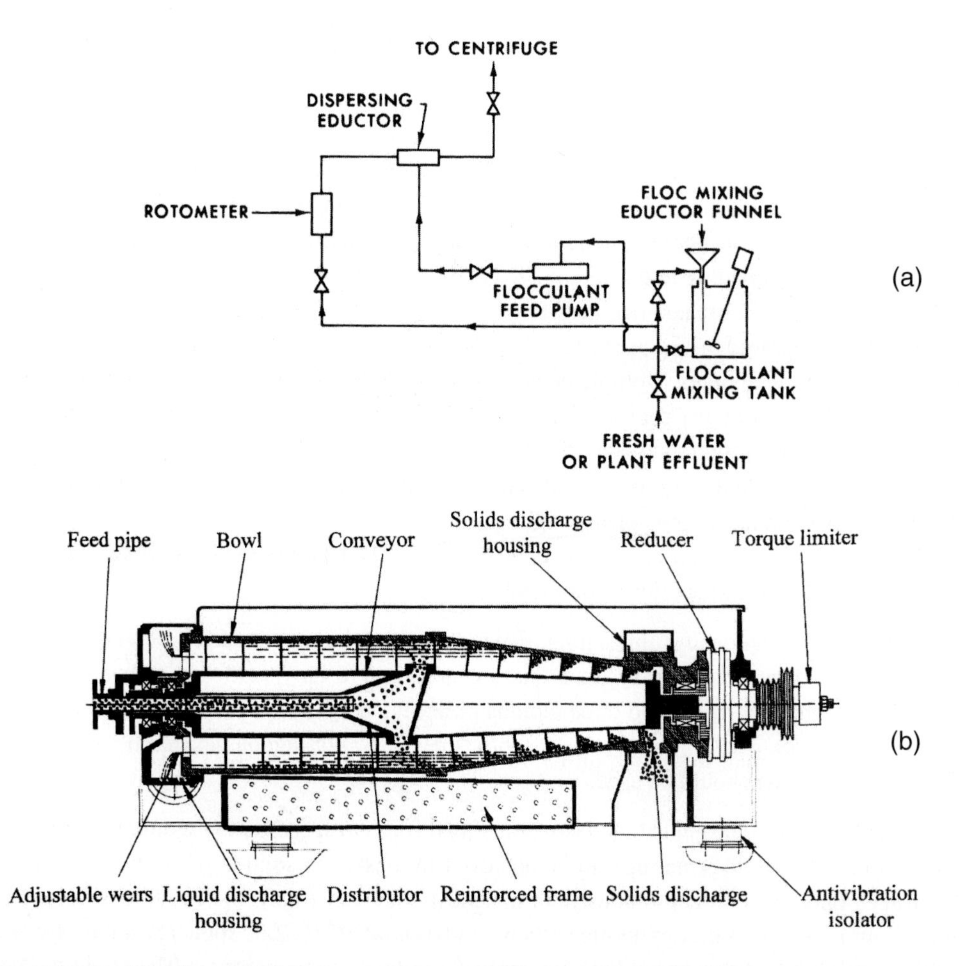

Figure 14-3 Centrifugal sludge dewatering system. (a) Typical flocculent piping diagram of centrifuge used for sludge dewatering. *(Adapted from Reference 16.)* (b) Continuous flow solid-bowl centrifuge for sludge dewatering. *(Courtesy of Andritz-Ruthner, Inc.)*

Vacuum Filter Rotary vacuum filters are widely used for dewatering coagulation and softening sludges. Vacuum filters consist of a cylindrical drum covered with cloth of natural or synthetic fabric. The drum remains partly submerged in a vat of sludge and rotates slowly. Internal vacuum that is maintained inside the drum draws the sludge to the filter medium, and water is withdrawn from the sludge. The cake-drying zone represents 40 to 60 percent of the drum surface and terminates at the cake discharge zone, where the cake is removed.[17] In a drum-type rotary vacuum filter, the sludge cake is scraped off.

Compressed air may be blown through the media to release the cake prior to scraping. A typical vacuum filter arrangement is shown in Figure 14-4.

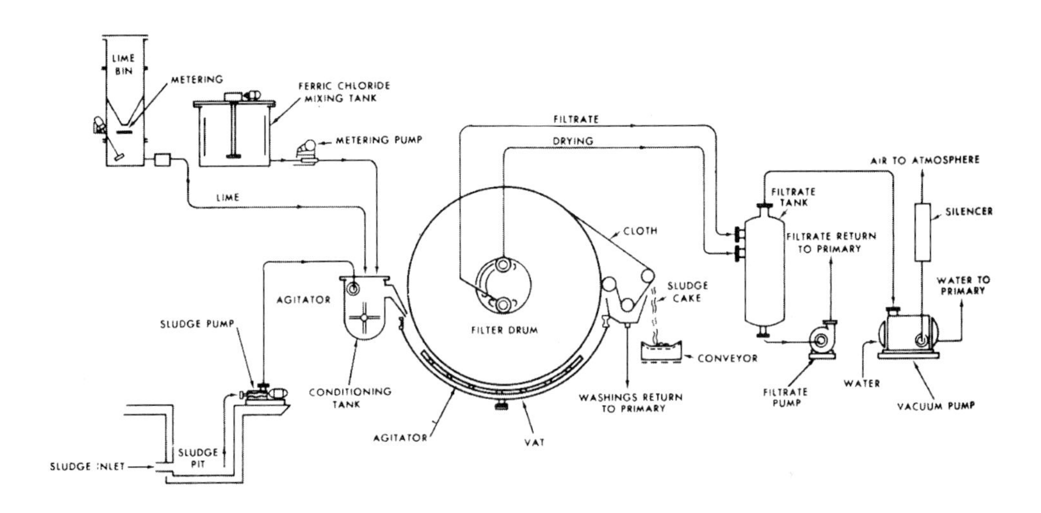

Figure 14-4 Schematic flow diagram of rotary vacuum filter system. *(Adapted from Reference 16.)*

The important design factors for rotary vacuum filters include the characteristics of the conditioned sludge, cake-formation time, viscosity, vacuum applied, specific resistance of the sludge cake, type of filter medium, and filter yield. Many equations have been developed to express the filtration rate, specific resistance of sludge, and filter yield. Basic theory of vacuum filtration may be found in some excellent publications.[6,17–19] Two test procedures used for determining the filterability of sludges are the Buchner funnel method and the filter leaf techniques. These two methods make possible the determination of the relative effects of various chemical conditioners and the calculation of the specific resistance of the sludge and filter area. Detailed procedures of the Buchner-funnel and filter-leaf tests may be found in References 6, 18–20. Generally, lime and polymers are used as sludge-conditioning chemicals. The filtration can contain less than 10 mg/L suspended solids. Typical design and operation information on vacuum filters is provided in Table 14-4.

Plate and Frame Filter Press Plate and frame presses are also called filter presses, or recessed-plate pressure filters. The design details of plate and frame presses are shown in Fig-

Table 14-4 Design and Performance Data on Vacuum Filter Sludge Dewatering

Parameter	Range
Solids	
Feed, % solids	2–6
Cake, % solids	25–35
Differential pressure (vacuum), kN/m^2 (psi)	70 (10)
Drum speed, rpm	0.2–0.5
Drying cycle, % of total period	40–60
Filter yield (dry solids)	
Alum sludge, $kg/m^2 \cdot h$ (lb/ft·h)	0.2–0.3 (0.04–0.06)
Lime softening, $kg/m^2 \cdot h$ (lb/ft·h)	2–4 (10–20)
Solids recovery, %	95–99

Source: Adapted in part from Reference 4.

ure 14-5. These consist of round or rectangular recessed plates which, when pressed together, form hollow chambers. A filter cloth is mounted on the face of each individual plate. In a fixed-volume filter press, the sludge is pumped under high pressure (350–1575 kN/m^2, or 50–225 psi) into the chamber. The water passes through the cloth, while the solids are retained and form a cake on the surface of the cloth. The sludge filling continues until the press is effectively full of cake. The entire filling operation takes up to 0–30 min. The pressure at this point is generally at the designed maximum, and it is maintained for several hours. During this time, more filtrate is removed, and the desired solids level in the cake is reached. (Filter presses can attain up to 40 percent solids.) The filter is then mechanically opened, and the dewatered cake drops from the chamber onto a conveyor belt for removal. Cake breakers are usually required to break up the rigid cake into a conveyable form.

A variation of the fixed-volume filter press (discussed above) is the variable-volume recessed-plate pressure filter. A diaphragm is placed behind the filter cloth that provides air or water pressure to squeeze the sludge injected in to the chamber.[17] Generally, a period of from 10 to 20 minutes is required to fill the press with the conditioned sludge. When the end point is reached, the sludge feed pump is automatically turned off. Water or air under high pressure is then pumped into the space between the diaphragm and the plate, thus squeezing the already formed and partially dewatered cake to the desired solids content. At the end of the cycle, the water is returned to a reservoir, the plates are automatically opened, and the sludge cake is discharged.

The filter press is applicable for dewatering difficult sludges. Metal hydroxide coagulated sludges are dewatered satisfactorily. Batch operation allows extended filtration time; typical cycle time is 8 hours. Sludge conditioning with lime or polymers increases the filter yield. Diatomaceous earth precoat is often needed.

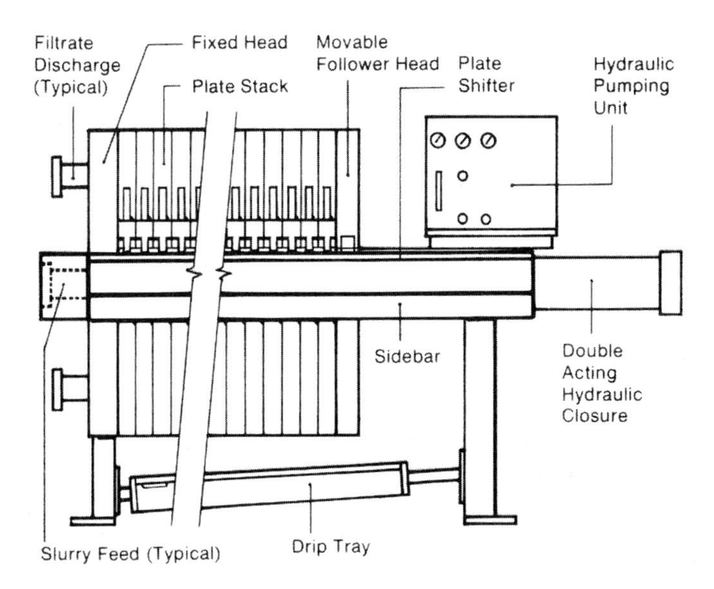

Figure 14-5 Design details of a plate and frame filter press assembly. *(Courtesy of Baker Process.)*

Belt Filter Press Belt filter presses employ single or double moving belts to dewater sludge continuously. The main advantages of a belt filter press are a drier cake, low power requirement, and continuous operation. The main disadvantages are short media life and a filtration rate sensitive to incoming sludge.

The belt filtration process involves three basic operational stages: (1) chemical conditioning, (2) gravity draining of excess water, and (3) compaction of predewatered sludge. The system components of belt filter presses are shown in Figure 14-6.

The conditioned sludge is discharged over the moving belt. Typically, 1 or 2 min. is required for drainage of the excess water. The sludge is then subject to increased pressure due to either the compression of the sludge between the carrying belt and cover belt or the application of a vacuum on the carrying belt. The sludge cake is squeezed between the two belts as it passes between various rollers. The compaction pressure can be widely varied by using a variable belt-and-roller arrangement. A continuous belt filter press typically utilizes an endless belt around a system of rollers for sludge dewatering. Belt presses applied to coagulation sludges result in cake solids from 15 to 20 percent. Polymer dosages range from 0.5 to 1 percent of the solids. Filter yield is 0.8–4.0 kg/m^2·h (0.16–0.8 lb/ft^2·h) and filter belt speed is 0.2–0.5 rpm. The solids capture efficiency is 90–99 percent.

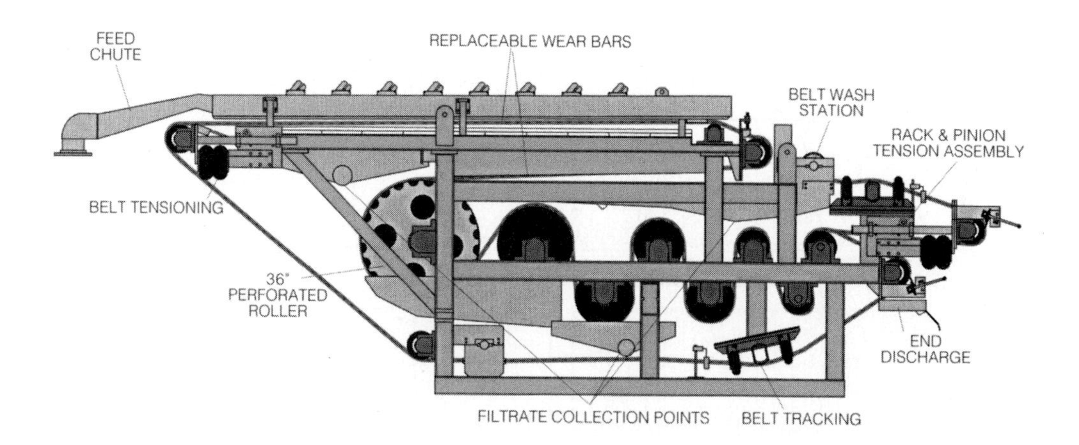

Figure 14-6 Construction details of a belt filter press used for sludge dewatering. *(Courtesy of Andritz-Ruthner, Inc.)*

14.3.4 Recovery of Chemicals

Water treatment plant sludge has been studied for recovery and reuse of many chemicals. Among these are iron, aluminum, calcium, and magnesium compounds.

Iron and aluminum recovery can be accomplished by leaching the sludge with a strong acid, normally sulfuric acid. In the 1960s, alum recovery was quite popular in Japan, when 15 plants were in operation.[4] Up to 80-percent recovery of alum was achieved at a pH of 2.5 by acidification. Acid leaching of alum is, unfortunately, accompanied by leaching of undersized heavy metals and of several organic compounds. This fact, as well as the high alum-recovery cost, has made this process less desirable. Other methods, including liquid ion exchange technique, have also been investigated.[21]

Recovery and reuse of iron coagulants in water treatment has also been investigated.[7,9] As with alum, iron recovery is possible by lowering pH by acid solution. The simplified relationship is given by Eq. (14.1).

$$2Fe(OH)_3 + 3H_2SO_4 \rightarrow Fe_2(SO_4)_3 + 6H_2O \tag{14.1}$$

Recovery of iron is 60–70 percent at pH 1.5–2. At such low pH, acid requirement is high, and other heavy metals are also released. Another possible way to reduce acid requirement is by resolubilization of iron as ferrous ion (Fe^{+2}) by adding a reducing agent to the sludge. If sulfide ion is used, the following reaction results Eq. (14.2).

$$2Fe(OH)_3 + 3Na_2S \rightarrow 2FeS + S + 6NaOH \tag{14.2}$$

This reaction is favored at low pH, so solubilization of iron for recovery is achieved by adding reducing agent as well as acid. By this method, 60 percent recovery of iron from sludge is achieved at a pH of 3. The acid requirement is significantly lowered by a reducing agent.

Also, the sludge has better settling characteristics than the sludge from extracting iron at lower pH without a reducing agent.

Magnesium precipitated as $Mg(OH)_2$ can be an effective coagulant for removal of color and turbidity from natural waters. For water high in naturally occurring magnesium, coagulation is achieved by simply adding a sufficient amount of lime to precipitate $Mg(OH)_2$. For waters low in natural magnesium, it may be necessary to add magnesium through the addition of a suitable salt. It is practical to recover both the magnesium and the lime for reuse by using suitable recovery methods. This simplifies sludge-disposal problems and improves the overall performance and economics of the process.

Separation of Mg and Ca from a sludge containing both is possible by recarbonation. The sludge from the sedimentation process is transferred to a tank for recarbonation. Carbon dioxide is dispersed in the diluted sludge to cause dissolution of magnesium hydroxide as magnesium bicarbonate ($Mg(HCO_3)_2$). The solids are separated in a thickener and $Mg(HCO_3)_2$ is decanted and stored for recycling. Thickened sludge consisting of $CaCO_3$ and inerts is vacuum-filtered to remove residual $Mg(HCO_3)_2$. The filtrate is returned to the $Mg(HCO_3)_2$ storage, and the sludge cake is transported to a reslurry tank, where conditioning and flotation agents are added to prepare the sludge for air flotation. Flotation is intended to separate the $CaCO_3$ from the remaining silt and clay. Separated $CaCO_3$ is subsequently dewatered and sent to a kiln for recalcining. The recalcining involves the heating of the dewatered sludge to about 1000°C in a kiln. Moisture and CO_2 are used in the recarbonation process. At the bottom outlet of the kiln, large pieces of recalcined lime are obtained, which are ground and taken to the storage area. A discussion of this process and other processes may be found in References 8, 22–25.

14.3.5 Ultimate Disposal

Several alternatives are available for ultimate disposal of water treatment plant residuals. Among these are disposal on land, in a sanitary sewer, and in surface waters and deep-well injection of brines.

Land Disposal Land disposal of residuals involves lagooning, land spreading, and landfilling. All these methods are governed by the Solid Waste Disposal Act (Pl 91-512) and the Resource Conservation and Recovery Act of 1976 (Pl 94-580).[26,27] These regulations require disposal of liquid or semi-liquid sludge on specific types of landfills designated for such purposes.

Lagooning Sludge lagoons are used for sludge storage, thickening, drying, and disposal. Sludge dries with time by decanting, evaporation, and percolation. These lagoons operate on an "apply and dry" rotation principle. Storage time is site-specific and depends primarily upon prevailing climatic conditions (temperature, precipitation, humidity ,and wind) and soil conditions. In dry areas, sludge dries in a relatively short time; in temperate areas, the sludge requires a longer time and also does not dewater well. In a cold climate with long periods of freezing, the drying process is aided by natural freezing and thawing. In general, six months to a year is allowed for evaporation and seepage. New sludge should not be added to the lagoon until previous applications have dried. In drying lagoons, moisture of from 70 to 50 percent can be reached. Sludge applied to a depth of 1 m will shrink to about 15 cm depth when dried to 50 percent moisture. The dried sludge can then be picked up by a front-end loader for disposal, or the cell can be caped for on-site disposal. As a general rule, where sludge can be lagooned to a depth of 3 m, a land area of 1.2–2 ha for 4000 m^3/d capacity plant will be adequate. If the sludge is applied to a depth of 1 m at a time, the berms shall be built 1–1.5 m above the maximum sludge level when filled. The berms should be 3 to 3.5 m wide to (1) permit operation of a drag line, (2) permit raising their top elevation, and (3) facilitate driving of service vehicles. Sludge dried to 50 percent moisture is easy to handle and makes excellent material for berm.

Where sufficient land is available, lagooning may be a satisfactory and economical sludge-drying and -disposal method. In view of the cost of the land, and the equipment and of the operation and maintenance cost of a lagoon, a careful evaluation must be made for process selection and design.

Land Spreading and Landfilling Residuals spreading and disposal over land have been attempted. Alum sludge solids tend to clog soil pores and prevent seed from germination; soil tilling improves this situation. Lime sludge has a beneficial effect as a soil additive for certain types of clays. Many types of clays are stabilized from shrinking or swelling and pH of acidic soils is increased.

Landfilling of dewatered sludge is achieved in regulated municipal or private landfills. Methods of landfilling include trenching (narrow and wide) and area fill (mound, layer, and dike containment). The technical considerations involved in site selection include useful life, topography, surface- and groundwater table, soil and geology, site access, and economics. A site close to the treatment plant will reduce transportation requirements. Co-disposal of sludge with municipal solid waste in sanitary landfill is also used.

Discharge into Sanitary Sewer Disposal of water treatment plant sludge into sanitary sewers is becoming more common these days. Local pretreatment guidelines generally cover the discharge requirements into the sanitary sewers.

Discharge of water treatment sludge into sanitary sewers has a low cost, because the thickening and dewatering steps may be eliminated. There may, however, be a controlled discharge requirement (to avoid slug) and continuous monitoring of flow, organic strength, and solids content of the discharge stream. Sewer-user charges are also involved. Therefore, these costs must also be considered during the preliminary evaluation of this option. The impact of solids upon the performance of a wastewater treatment plant must also be assessed. Solids loading will be increased in the primary treatment facility. Coagulants in water treatment sludges will enhance phosphorus and solids removal in the primary treatment facility of the wastewater treatment plant. Discharge of lime-softening sludge to sanitary sewers should also be carefully examined. Softening sludge may cause scaling and encrustations on weirs, channels, and piping.

Disposal into Surface Water Disposal of water treatment sludge in surface water has the advantage of relatively low capital and operational cost. In recent years, such activity is either prohibited or strictly controlled under Federal Water Pollution Control Act Amendments of 1972 and Clean Water Act of 1977.[1,2] A NPDES permit is required, as for any other industrial discharges. The outfall must be properly located and designed to minimize potential water pollution problems.

Deep-Well Injection of Brine Deep-well injection of brines from ion exchange and other demineralization processes is used in areas where geology permits and where disposal is allowed by local environmental regulations. To assess the feasibility of deep-well injection, legal aspects, site suitability, waste characteristics, and economics must be considered. Several states have, by policy, prohibited the construction of waste-injection deep wells. Other states that allow injection require a permit.

14.4 Manufacturers of Residuals-management Equipment

The equipment manufacturers of residuals-management systems are given in Appendix C. These systems include gravity thickeners, centrifuges and air flotation systems, chemical feeders and mixers for conditioning chemicals, and various types of dewatering equipment. Each equipment component must be evaluated for compatibility, operational flexibility, maintenance requirements, and design criteria. Equipment selection considerations and the responsibilities of the design engineer were covered in Section 4.10.

14.5 Information Checklist for Design of Residuals-management Facilities

Before starting the design of a residuals-management system, the design engineer must develop the necessary data and make many important decisions. The following checklist can be helpful in developing the necessary predesign data.

1. Identify the residuals-generating processes. Normally, these include sludges from coagulation–flocculation and precipitation processes and filter backwash in a conventional water treatment plant. Other process residuals may be spent brines, washwater from diatomaceous-earth filters, and spent activated carbon.
2. Conduct material mass-balance analysis to establish quality and concentrations of solids reaching the residuals streams. Such information is necessary under critical-flow and water quality conditions.
3. Evaluate various applicable methods for filter backwash recovery and sludge thickening, dewatering, and disposal. Select systems that provide the most cost-effective results and are also satisfactory to the concerned utility and regulatory agencies.
4. Develop such design parameters as solids and hydraulic loading rates, operational range, and maintenance requirements. Laboratory or pilot plant studies might be needed to develop the design parameters.

5. Obtain the design criteria from the regulatory agency concerned.
6. Select equipment manufacturers catalogs and equipment-selection guides.

14.6 Design Example

14.6.1 Design Criteria Used

The following design criteria and assumptions are used for the design of the residuals-management system.

1. Filter backwash water recovery system

   a. Provide one surge tank and one gravity settling basin for recovery of filter backwash. A constant flow is released from the surge tank into the settling basin. The overflow from the filter backwash recovery basin is returned to the rapid-mix units for reprocessing. The underflow, containing settled solids, is pumped into the thickeners for processing with the coagulation sludge from the sedimentation basins. The effluent structure will consist of V-notches and an effluent launder.

   b. The volume of filter backwash is 375 m^3 (98,700 gal) per filter unit per 24 h (Section 10.13.2, Step B.4).

   c. Provide one surge tank that will produce a constant flow into the filter backwash recovery basin.

   d. Provide a 6-h detention period and an overflow rate of 25 $m^3/m^2 \cdot d$ for the filter backwash recovery basin.

2. Gravity thickener

   Gravity thickeners shall be designed to thicken the sludge from the coagulation and flocculation facility and the settled solids from the filter backwash recovery basin.

   a. Provide two circular gravity thickeners. Minimum hydraulic and maximum solids loadings are 4 $m^3/m^2 \cdot d$ and 80 $kg/m^2 \cdot d$, respectively.

   b. The thickener overflow shall be returned into the backwash recovery sedimentation basin for reprocessing. The influent structure shall consist of an effluent well that receives sludges from sedimentation basins and sludge from the filter backwash recovery basin. The effluent structure shall consist of V-notches, an effluent launder, and a common sump.

   c. The thickened sludge shall be pumped into the sludge lagoons for dewatering and on-site disposal.

3. Sludge drying lagoons

   Provide four sludge-drying lagoons for disposal. Each lagoon shall operate on an "apply and dry" rotation principle. Storage time of lagoons shall be developed on the basis of the specific site conditions as dictated by prevailing climatic conditions. After completion of drying and filling operation, the lagoons shall be capped with local soil for disposal of residuals.

4. Unit arrangement and layout

   The residuals-management system consists of a filter backwash water surge storage tank, a filter backwash water recovery basin, gravity thickeners, sludge lagoons, and several pump stations. The general layout of residuals-management and -disposal system is illustrated in Figure 14-7. A detailed layout of filter backwash water recovery system and gravity thickeners is shown in Figure 14-8. The filter backwash water is discharged by gravity into a surge storage tank. The water from the surge storage tank is then pumped by submersible pumps at a constant rate into a flocculation box, where polymer is also added. Flocculated water then flows into a backwash water recovery basin. The reclaimed water from the recovery basin is pumped back to Junction Box D on the raw water pipeline prior to the rapid-mix basins. Provision is made to ozonate this stream, in case it becomes necessary to do so. The underflow from the recovery basin is discharged through a telescope valve into a junction box that also receives coagulation sludge from the sedimentation basins. Provision for adding polymers into this unit is also made. The combined sludge is discharged into a three-stage flocculation unit. Flocculated-sludge flow is first conveyed equally into the central feed wells of both thickeners. The thickener overflow is then discharged into a common sump and then pumped back to the surge storage tank. The underflow from the thickeners is pumped into the sludge lagoons. The sludge is allowed to dry in the lagoons. Percolation-collection piping system and decanting devices are provided to collect the liquid from the lagoons. The water collected from the lagoons can be pumped by a central pump station, either back to the surge storage tank or into the plant sanitary sewer system.

14.6.2 Design Calculations

Step A: Material Mass-Balance Calculation The sludge-processing facilities produce liquids that generally are returned to the rapid-mix units for reprocessing. The return flows increase solids in the raw water stream. In the design example, the residuals-containing streams are reprocessed in filter backwash recovery systems and sludge thickeners. The material mass-balance procedure is given in this section. These calculations are performed for maximum day and average day conditions. The filter backwash recovery system and gravity thickeners are designed for maximum day condition. The sludge lagoons are designed for average condition.

The material mass-balance calculation steps are arranged in Table 14-5. It took eight iterations to finally reach stable values of various streams. Final results after eight iterations are

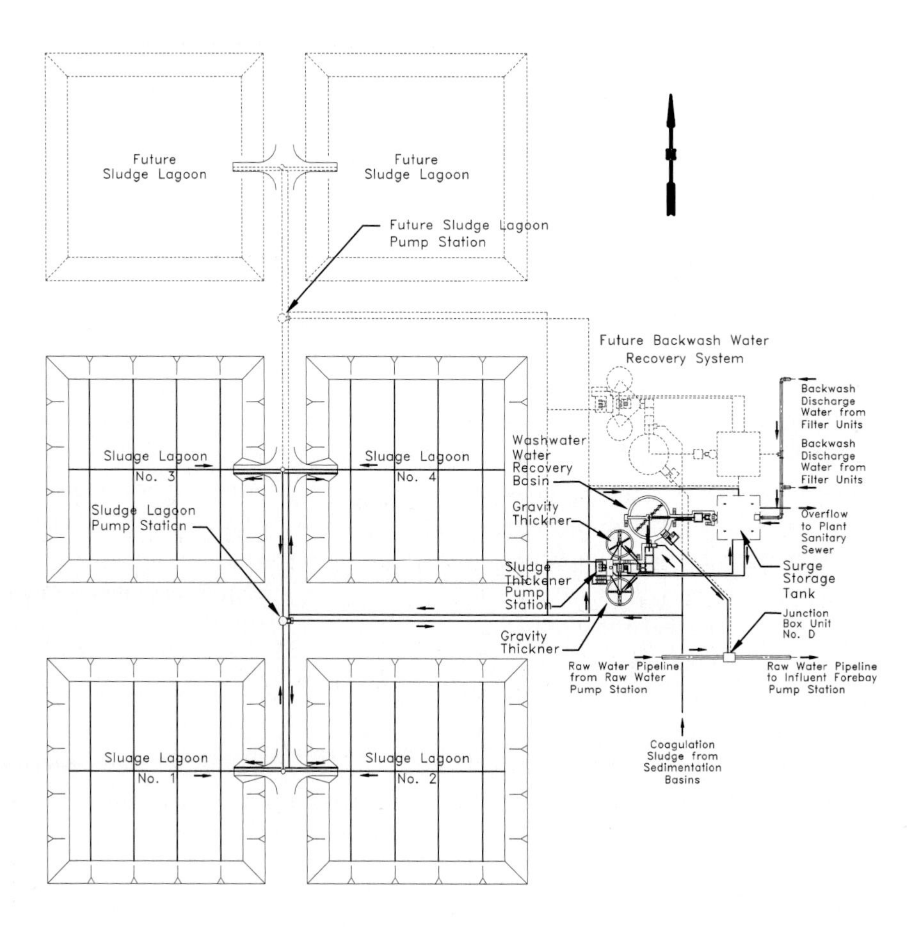

Figure 14-7 General layout of the residuals-management and -disposal system.

shown in Figure 14-9. Similarly, the material mass balance at average day design flows are provided in Table 14-6. Final results are shown in Figure 14-10.

Step B: Filter Backwash Water Recovery System

1. Characteristics of filter backwash water

   The filter backwash water recovery system receives 3000 m^3/d flow. The solids in the filter backwash water are 809.6 kg/d. Total number of filter backwash operations per day is 8 cycles. Average frequency of filter backwash cycles is therefore 3 hours.

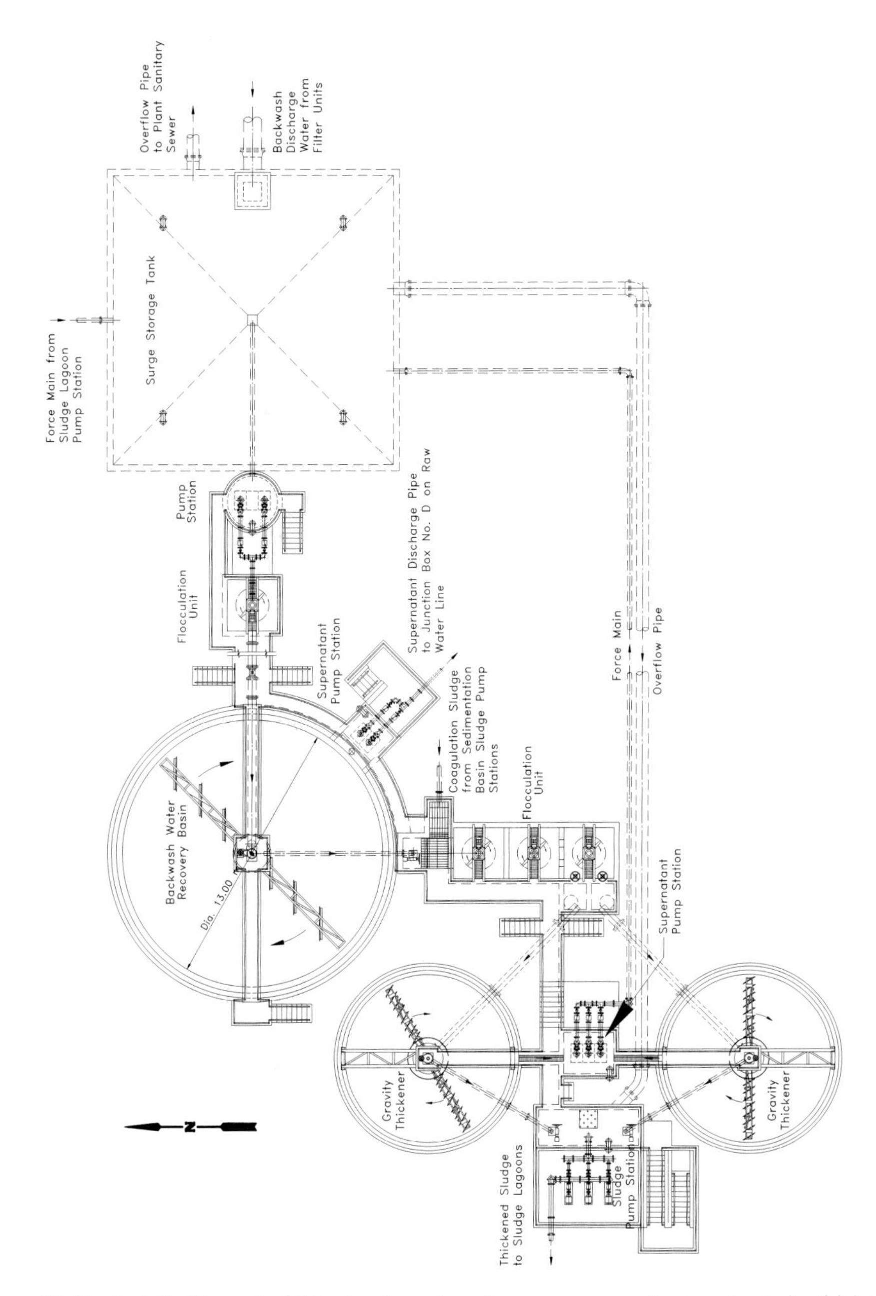

Figure 14-8 Detailed layout of filter backwash water recovery system and gravity thickeners.

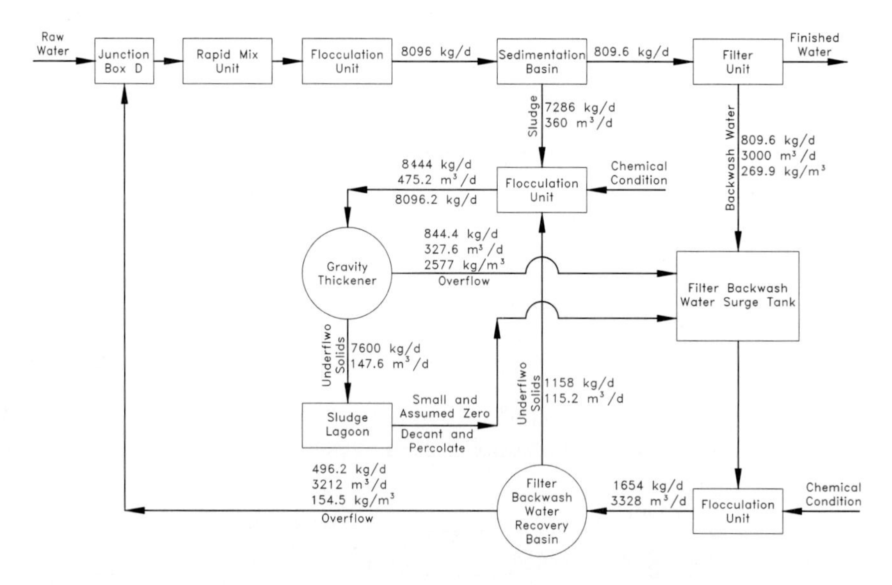

Figure 14-9 Summary of mass-balance results at design maximum day flow.

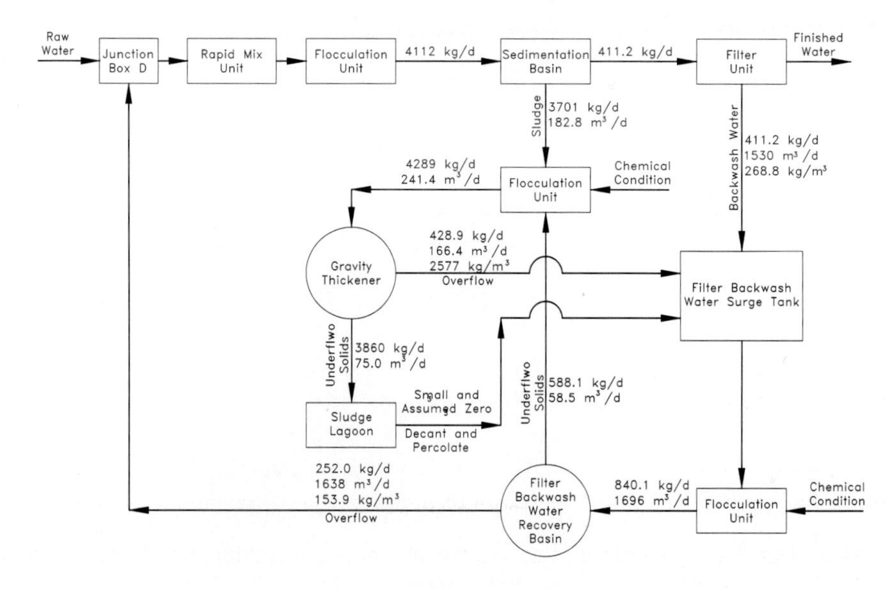

Figure 14-10 Summary of mass-balance results at design average day flow.

Table 14-5 Material Mass-Balance Analysis at Design Maximum Day Flow

Waste Stream	Number of Iterations			
	1st	2nd	7th	8th
Sedimentation basin				
Total solids produced, kg/d	7600 (Sec. 9.6.2, Step B.1)	7600 + 228 = 7828	8096	8096
Effluent solids (10% remaining), kg/d	7600 ∞ 0.1 = 760	7828 ∞ 0.1= 782.8	809.6	809.6
Sludge solids (90% efficiency), kg/d	7600 ∞ 0.9 = 6840	7828 ∞ 0.9= 7045	7286	7286
Sludge volume (2% solids, 1012 kg/m^3), m^3/d	$\frac{6840}{0.02 \times 1012} = 337.9$	$\frac{7045}{0.02 \times 1012} = 348.1$	360.0	360.0
Filter				
Solids into filter, kg/d	760	782.8	809.6	809.6
Filter backwash flow, m^3/d	3000 (Sec. 10.13.2, Step B.4)	3000	3000	3000
Solids in filter backwash, kg/d	760	782.8	809.6	809.6
Solids concentrated, mg/L	$\frac{760 \times 1000}{3000} = 253.3$	$\frac{782.8 \times 1000}{3000} = 260.9$	269.9	269.9
Filter backwash water recovery basin				
Solids into basin, kg/d	760	782.8 + 737.2 = 1520	1654	1654
Flow into basin, m^3/d	3000	3000 + 262.1 = 3262	3328	3328
Sludge solids (70% efficiency), kg/d	760 ∞ 0.7= 532	1520 ∞ 0.7 = 1064	1158	1158
Sludge volume (1% solids, 1005 kg/m^3), m^3/d	$\frac{532}{0.01 \times 1005} = 52.9$	$\frac{1064}{0.01 \times 1005} = 105.9$	115.2	115.2
Overflow, m^3/d	3000 – 52.9 = 2947	3262 – 105.9 = 3156	3212	3212
Solids in overflow, kg/d	760 ∞ 0.3 = 228	1520 ∞ 0.3 = 456	496.2	496.2
Solids concentration, mg/L	$\frac{228 \times 1000}{2947} = 77.4$	$\frac{456 \times 1000}{3156} = 144.5$	154.5	154.5
Gravity thickener				
Total solids, kg/d	6840 + 532 = 7372	7045 + 1064 = 8109	8444	8444
Flow, m^3/d	338.0 + 52.9 = 390.9	348.1 + 105.9 = 454.0	475.2	475.2
Thickener solids (90% removal), kg/d	7372 ∞ 0.9 = 6635	8109 ∞ 0.9 = 7298	7600	7600
Sludge volume (5%, 1030 kg/m^3), m^3/d	$\frac{6635}{0.05 \times 1030} = 128.8$	$\frac{7298}{0.05 \times 1030} = 141.7$	147.6	147.6
Overflow volume, m^3/d	390.9 – 128.8 = 262.1	454.0 – 141.7 = 312.3	327.6	327.6
Overflow solids, kg/d	7372 ∞ 0.1 = 737.2	8109 ∞ 0.1 = 810.9	844.4	844.4
Solids concentration, mg/L	$\frac{737.2 \times 1000}{262.1} = 2813$	$\frac{810.9 \times 1000}{312.3} = 2597$	2577	2577

Table 14-6 Material Mass Balance Analysis at Design Average Day Flow

Waste Stream	Number of Iterations			
	1st	2nd	7th	8th
Sedimentation basin				
Total solids produced, kg/d	3860 (Sec. 9.6.2, Step B.1)	3860 + 115.8 = 3976	4112	4112
Effluent solids (10% remaining), kg/d	3860 ∞ 0.1 = 386	3976 ∞ 0.1= 397.6	411.2	411.2
Sludge solids (90% efficiency), kg/d	3800 ∞ 0.9 = 3474	3976 ∞ 0.9= 3578	3701	3701
Sludge volume (2% solids, 1012 kg/m^3), m^3/d	$\frac{3474}{0.02 \times 1012} = 171.6$	$\frac{3578}{0.02 \times 1012} = 176.8$	182.8	182.8
Filter				
Solids into filter, kg/d	386	397.6	411.2	411.2
Filter backwash flow, m^3/d	1530 (Sec. 10.13.2, Step B.7)	1530	1530	1530
Solids in filter backwash, kg/d	386	397.6	411.2	411.2
Solids concentrated, mg/L	$\frac{386 \times 1000}{1530} = 252.3$	$\frac{397.6 \times 1000}{1530} = 259.9$	268.8	268.8
Filter backwash water recovery basin				
Solids into basin, kg/d	386	397.6 + 374.4 = 772.0	840.1	840.1
Flow into basin, M.D.	1530	1530 + 133.1 = 1663	1696	1696
Sludge solids (70% efficiency), kg/d	386 ∞ 0.7= 270.2	772.0 ∞ 0.7 = 540.4	588.1	588.1
Sludge volume (1% solids, 1005 kg/m^3), m^3/d	$\frac{270.2}{0.01 \times 1005} = 26.9$	$\frac{540.4}{0.01 \times 1005} = 53.8$	58.5	58.5
Overflow, m^3/d	1530 – 26.9 = 1503	1663 – 53.8 = 1609	1638	1638
Solids in overflow, kg/d	386 ∞ 0.3 = 115.8	772.0 ∞ 0.3 = 231.6	252.0	252.0
Solids concentration, mg/L	$\frac{115.8 \times 1000}{1503} = 77.4$	$\frac{231.6 \times 1000}{1609} = 143.9$	153.9	153.9
Gravity Thickener				
Total solids, kg/d	3474 + 270.2 = 3744	3578 + 540.4 = 4118	4289	4289
Flow, m^3/d	171.6 + 26.9 = 198.5	176.8 + 53.8 = 230.6	241.4	241.4
Thickener solids (90% removal), kg/d	3744 ∞ 0.9 = 3370	4118 ∞ 0.9 = 3707	3860	3860
Sludge volume (5%, 1030 kg/m^3), m^3/d	$\frac{3370}{0.05 \times 1030} = 65.4$	$\frac{3707}{0.05 \times 1030} = 72.0$	75.0	75.0
Overflow volume, m^3/d	198.5 – 65.4 = 133.1	230.6 – 72.0 = 158.6	166.4	166.4
Overflow solids, kg/d	3744 ∞ 0.1 = 374.4	4118 ∞ 0.1 = 411.8	428.9	428.9
Solids concentration, mg/L	$\frac{374.4 \times 1000}{133.1} = 2813$	$\frac{411.8 \times 1000}{158.6} = 2596$	2577	2577

2. Surge tank design

a. Compute the dimensions of the surge tank.

Design the surge tank for enough capacity to hold the volume of two filter backwashes and a six-hour overflow from the thickeners.

$$\text{Volume required} = \frac{3000 \text{ m}^3\text{/d} \times 2 \text{ backwash}}{8 \text{ backwash/d}} + \frac{327.6 \text{ m}^3\text{/d} \times 6 \text{ h}}{24 \text{ h/d}}$$

$$= 832 \text{ m}^3$$

Provide surge tank dimensions of 14 m ∞ 14 m ∞ 4 m (side water depth) plus a 1.5-m deep hopper bottom.

$$\text{Total volume provided} = 14 \text{ m} \times 14 \text{ m} \times 4 \text{ m} + \frac{(14 \text{ m})^2 \times 1.5 \text{ m}}{3}$$

$$= 882 \text{ m}$$

Provide 0.5 m freeboard. Additional capacity in the surge tank will be available in case three filters are backwashed in less than 9 hours. Two overflow pipes are provided in the surge tank to discharge the excess flow into either the sludge thickener pump station or the plant sanitary system. Ultrasonic level transducers are provided to monitor the water level in the surge tanks. The dimensions and design details of the sludge tank are given in Figure 14-11.

b. Design filter backwash water withdrawal.

Select two identical pumps for filter backwash withdrawal from the surge basin; one pump will be used as a backup. Each pump has a pumping capacity of 2.3 m^3/min (600 gpm). At this withdrawal rate, the surge tank will be emptied by one pump in approximately 6 hours. Thus, a constant flow rate of 2.3 m^3/min will be maintained from the surge tank into the backwash water recovery basin, with a second pump as a standby unit. The pump controls include a low-level pump stop. The control will also include *second pump start at high level* and *stop at normal operating level.* This will protect the surge tank against overflow.

3. Design of filter backwash recovery basin

a. Compute the dimensions.

Flow to filter backwash recovery basin = 2.3 m^3/min. The design overflow rate (per design criteria for sedimentation basins) = 25 $m^3/m^2{\cdot}d$

$$\text{Basin area} = \frac{2.3 \text{ m}^3\text{/min} \times 1440 \text{ min/d}}{25 \text{ m}^3\text{/m}^2{\cdot}\text{d}} = 132.5 \text{ m}$$

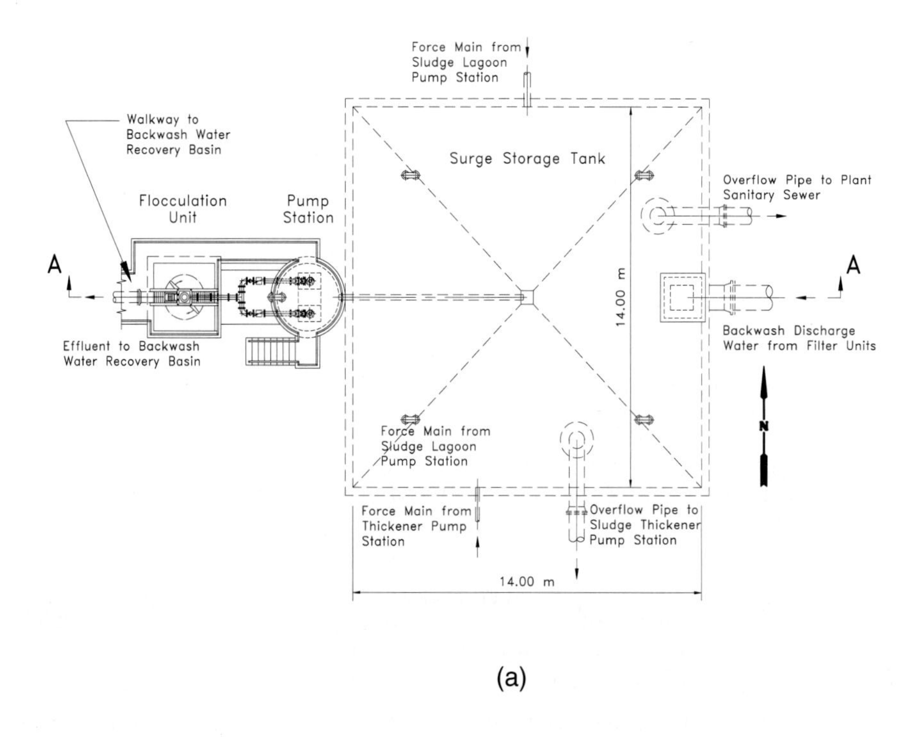

(a)

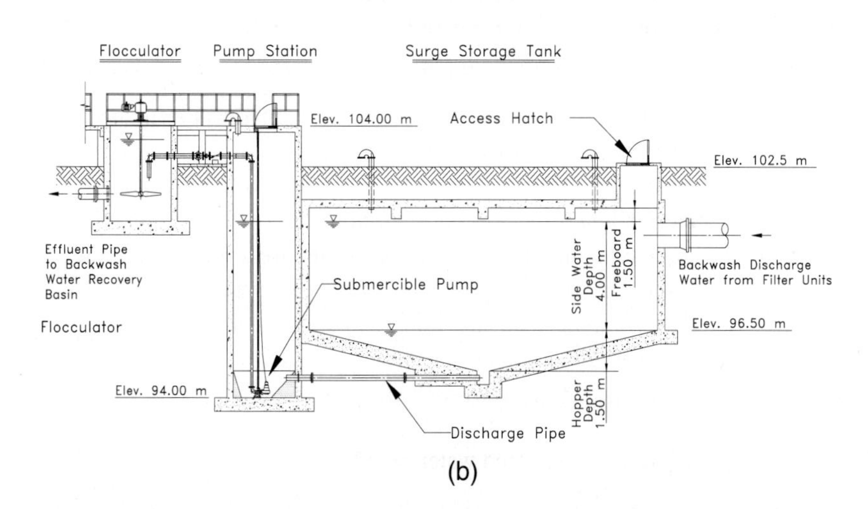

(b)

Figure 14-11 Surge tank for filter backwash. (a) Plan view. (b) Section view A-A.

Provide a 13-m (24.7-ft) diameter backwash water recovery basin. For a detention time of 6 h:

$$\text{Required volume of washwater recovery basin} = 2.3\ \text{m}^3/\text{min} \times 60\ \text{min/h} \times 6\ \text{h} = 828\ \text{m}^3$$

$$\text{Depth required} = \frac{4 \times 828\ \text{m}^3}{\pi \times (13\ \text{m})^2} = 6.24\ \text{m}$$

Provide a side-water depth of 6.3 m (20.7 ft) in the backwash water recovery basin.

b. Design the influent and effluent structures.

The influent structure of the washwater recovery basin consists of a central well. The effluent structure consists of 90°-V-notch weirs around the peripheral of an effluent launder. The effluent launder discharges into an outlet box. The entire arrangement is similar to that of a circular clarifier. The design procedure for V-notch weir, effluent launder, outlet pipe, and hydraulic profile may be found in Chapter 9. Additional details on design procedure may be found in Reference 17. The design details of the filter backwash water recovery basin are shown in Figure 14-12.

4. Filter backwash water recovery basin underflow solids

The sludge will be discharged from the filter backwash water recovery basin into a sludge collection box. The telescopic valve is used to control the solids-withdrawal rate. At an underflow concentration of 1 percent and solids capture of 70 percent, the quantity and quality of settled filter backwash solids are developed in mass-balance analysis (Table 14-5). These values areas follows:

$$\text{Quantity of backwash solids} = 1158\ \text{kg/d}$$

$$\text{Sludge-withdrawal rate} = 115.2\ \text{m}^3/\text{d}$$

5. Filter backwash water recovery basin overflow

The quantity and quality of overflow from the filter backwash water recovery basin are also developed in the mass-balance analysis (Table 14-5). These values areas follows:

$$\text{Overflow solids in recovered water} = 496.2\ \text{kg/d}$$

$$\text{Overflow of recovered water} = 3212\ \text{m}^3/\text{d}$$

$$\text{Concentration of solids in the recovered water} = 154.5\ \text{mg/L}$$

Provide two pumps, each rated at 2.3 m^3/min (600 gpm), to lift the effluent back to the raw water influent line. One pump will be in operation while the other pump is a standby unit. The pumps will be operated for equal wear.

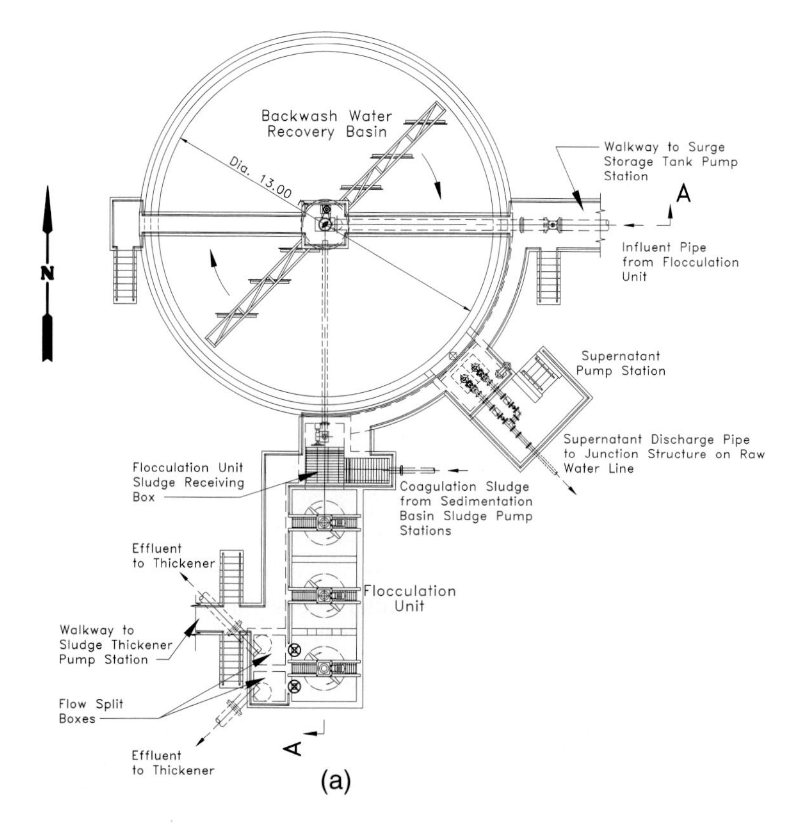

(a)

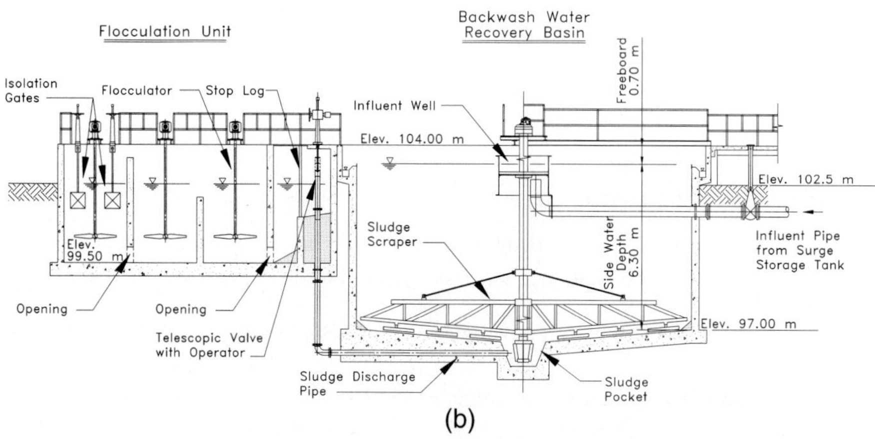

(b)

Figure 14-12 Filter backwash recovery basin and gravity thickener. (a) Plan view. (b) Section view A-A.

Step C: Gravity Thickeners

1. Characteristics of solids reaching the gravity thickeners

   The gravity thickeners solids include the underflow received from the filter backwash water recovery basin and coagulation sludge from the sedimentation basins. The quantity and quality of solids going into the thickeners are obtained in the mass-balance analysis (Figure 14-9). These values are as follows:

$$\text{Total solids} = 8444 \text{ kg/d}$$

$$\text{Total flow} = 475.2 \text{ m}^3\text{/d}$$

2. Gravity thickener design

   a. Calculate the dimensions of the thickeners.

   Based on a solids loading of 80 kg/m^2·d, the required area is as follows:

$$\text{Total area required} = \frac{8444 \text{ kg/d}}{80 \text{ kg/m}^2\cdot\text{d}} = 105.6 \text{ m}^2$$

   Provide two thickeners, then the required area for each thickener is as follows:

$$\text{Area of each thickener} = \frac{105.6 \text{ m}^2}{2} = 52.8 \text{ m}^2$$

   The required diameter of thickener is, therefore, calculated as follows:

$$\text{Required diameter of each thickener} = \frac{\sqrt{(4 \times 52.8)}}{\pi} = 8.20 \text{ m}$$

   Provide two thickeners, each of 8.2 m (27 ft) diameter.

   b. Check hydraulic loading.

$$\text{Hydraulic loading} = \frac{475.2 \text{ m}^3\text{/d}}{2 \times 52.8 \text{ m}^2} = 4.5 \text{ m}^3\text{/m}^2\cdot\text{d}$$

   The hydraulic loading of 4.5 m^3/m^2·d for the gravity thickener is above the minimum value. An increase in solids content in the receiving sludge flow could exceed the acceptable limit of solids loading of 80 kg/m^2·d. Therefore, the design should have provision for dilution of water or withdrawal of thinner sludges.

   c. Calculate the depth of the thickeners.

   The total side-water depth of a gravity thickener is composed of three separate items: the clear water zone, the settling zone, and the thickening zone. Generally, in a gravity thickener, a clear water zone of 1.0 m (3 ft), a settling zone of 1.5 m (5 ft), and a thickening zone of 3.0 m (10 ft) are considered sufficient.

$$\text{Total side-water depth} = 1.0 \text{ m} + 1.5 \text{ m} + 3.0 \text{ m} = 5.5 \text{ m}$$

A freeboard of 0.6 m (2 ft) is also provided.

d. Compute the depth of the central hopper.

The bottom of the thickener is sloped at 20 cm/m (3 in/ft).

$$\text{The depth of the central hopper} = \frac{20 \text{ cm}}{100 \text{ cm/m}} \times \frac{8.2 \text{ m}}{2} = 0.82 \text{ m}$$

Total water depth of the thickener at the central hopper is therefore 6.32 m.

e. Design of the influent and effluent structures.

The influent and effluent structures of the thickeners consist of a central well and 90°-V-notch weirs around the peripheral. These designs are similar to those discussed for the filter backwash recovery basin.

3. Thickening period

The volume of each thickener is as follows:

$$\begin{aligned}&\text{Volume of thickener}\\&= \frac{\pi}{4} \times (8.2 \text{ m})^2 \times 5.5 \text{ m} + \frac{\pi}{12} \times 0.82 \text{ m} \times (8.2 \text{ m})^2\\&= 291 \text{ m}^3 + 14 \text{ m}^3 = 305 \text{ m}^3\end{aligned}$$

$$\text{Thickening period} = \frac{2 \times 305 \text{ m}^3}{475.2 \text{ m}^3/\text{d}} = 1.3 \text{ d}$$

4. Thickened sludge withdrawal

The quantity and quality of thickened sludge are developed in mass-balance analysis (Figure 14-9). These values are as follows:

$$\text{Quantity of thickened sludge} = 7600 \text{ kg/d}$$

$$\text{Sludge-withdrawal rate} = 147.6 \text{ m}^3/\text{d}$$

The telescopic valves are used to control the sludge-blanket level. Valve position-limit switches monitor the position of the sludge valve. Motor contacts and sonic flow meters monitor the sludge pump status and sludge-withdrawal rate.

Provide three pumps, each rated at 200 L/min (53 gpm), to pump the thickened sludge to the sludge-drying lagoons. Two pumps will be in operation while the last pump is a standby unit. The pumps will be turned on and off according to the actual quantity of sludge. The pumps will also be operated alternately, for equal wear.

5. Sludge volume ratio (SAR)

The sludge volume ratio (SAR) is the volume of sludge blanket held in the thickening zone of the thickeners divided by the volume of the thickened sludge removed per day.

Volume of sludge blanket held in the of each thickener (including hopper)

$$= \frac{\pi}{4} \times (8.2\ \text{m})^2 \times 3\ \text{m} + 14\ \text{m}^3 = 158\ \text{m}^3 + 14\ \text{m}^3 = 172\ \text{m}^3$$

$$\text{SAR} = \frac{2 \times 172\ \text{m}^3}{147.6\ \text{m}^3/\text{d}} = 2.3\ \text{d}$$

6. Quality of the thickener overflow

   The quantity and quality of overflow from the thickeners are also developed in the mass-balance analysis (Figure 14-9). These values are as follows:

   Overflow solids in recovered water = 327.6 kg/d

   Overflow of recovered water = 844.4 m^3/d

   Concentration of solids in the recovered water = 2577 mg/L

   Provide three pumps, each rated at 300 L/min (80 gpm), to lift the overflow back to the surge tank of the filter backwash water recovery system. Two pumps will be in operation while the last pump is a standby unit. The pumps will be operated alternately, for equal wear.

   The design details and dimensions of the thickeners are shown in Figures 14-13.

Step D: Sludge Drying Lagoons Four sludge drying lagoons are provided for further thickening, dewatering, drying, and disposal of sludge. The sludge will dry over time by evaporation, decanting, and percolation. These lagoons will operate on an "apply and dry" rotation principle. In each rotation, the sludge is applied until the lagoon is full. After each refilling, the sludge it is allowed to concentrate and dry. Percolation, decanting, and evaporation are the mechanisms for water loss. The filling cycle is resumed only after the previous applications are completely dewatered.

The sludge lagoons are designed for maximum percolation and decanting. A 0.3-m-deep sand bed with drainage pipes is provided for water percolation. As a dewatered sludge layer builds-up in the lagoon, the percolation decreases. After build up of each 0.45-m sludge layer, it is refinished by installing stand pipes and a sand layer to increase the percolation rate.

The decant pipes are provided with multiple-level ports for removing the supernatant. The general design details of the lagoons are discussed later.

1. Site characteristics and climate

   The sludge-disposal lagoons will be constructed on the plant property. According to the Soil Conservation Service (now Soil Survey Division (SSD)) map, basket-fine sandy loam dominates the treatment plant site. The estimated permeability of the underlying soil is 5–15 cm/h. The area in general has a gentle slope of 1 percent.

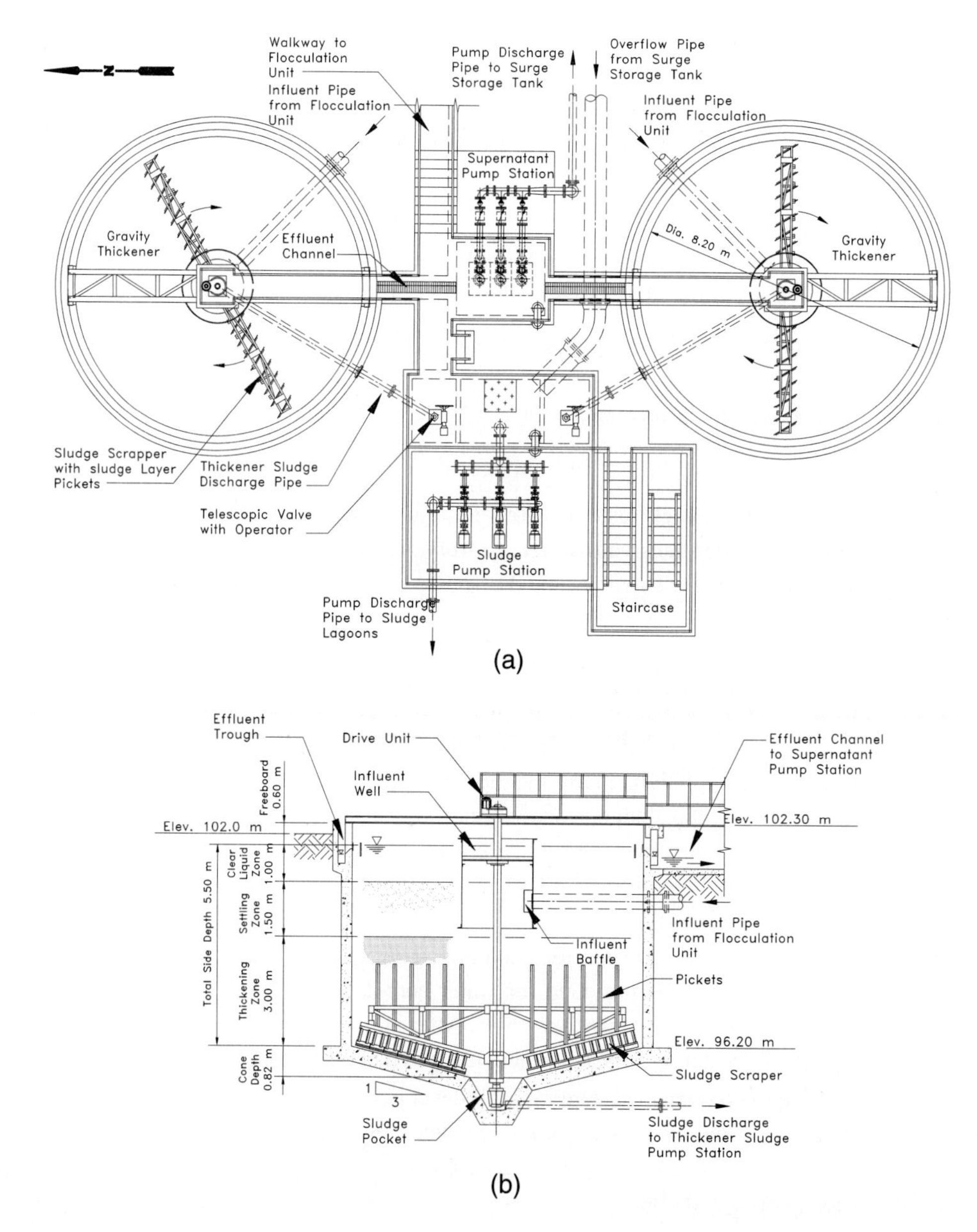

Figure 14-13 Design details of sludge gravity thickeners. (a) Plan view. (b) Conceptual section view.

The climate data are obtained from the National Oceanic and Atmosphere Administration (NOAA). The data represent the wettest in five years for monthly precipitation, temperature, and evaporation. The climatic data are displayed in Table 14-7.

Table 14-7 Climatic Data for the Wettest Year in 5 Years

Month	Mean Temperature, °C	Total Precipitation, cm	Net Lake Total Evaporation Rate, cm
January	5	3.0	6.9
February	9	11.0	5.8
March	12	5.5	8.9
April	19	12.9	8.9
May	21	8.8	14.0
June	25	14.1	16.5
July	27	7.3	21.5
August	24	12.8	21.6
September	21	2.6	20.3
October	17	12.3	13.0
November	11	1.0	10.2
December	7	3.7	5.8
	Annual mean = 16.5	Annual total = 95.0	Annual total = 153.5

2. Design details of lagoons

a. Select dimensions of the lagoons.

The selected dimensions of the lagoon are as follows:

Bottom dimensions	60 m ∞ 60 m
Side slope (horizontal : vertical)	3 : 1
Total water depth	2.55 m
Freeboard	0.60 m
Total lagoon depth	3.15 m

b. Design the base construction.

The base of the lagoon is constructed by filling an 0.3-m-deep sand layer over the bottom. Pervious tile drains are provided in the sand, to convey percolated water within the plant properly.

c. Design the intermediate-level base construction.

The lagoon is operated by filling it to a maximum water elevation, then waiting until it is completely dry before starting the next cycle. The percolation rate will decrease as solids accumulate on the base. After reaching an 0.45-m (18-in) accumulated sludge depth, the base is refinished, to enhance the percolation rate. The refinishing of the base will involve grading and sinking of sixteen 15-cm-diameter vertical pipe sections through the accumulated sludge deposit and then filling the pipes with coarse sand. A 15-cm drainage layer of sand is placed over the entire lagoon bottom surface. The construction details are shown in Figure 14-14.

d. Design the decant system.

Provide standpipes with multiple-level ports to decant the supernatant. The schematic of a standpipe is also shown in Figure 14-14.

3. Useful lagoon life

The useful life of the lagoon system is calculated from the solids-accumulation rate, running until the usable volume is used up and the facility is capped. The calculation involves application, drying time, and solids accumulation.

a. Calculate the water balance.

The mean precipitation and evaporation rate vary from month to month. Also, the filling and drying cycles of the lagoons can occur in any season of the year. Therefore, the water balance calculations are conducted on an annual-average basis.

b. Determine the precipitation.

The precipitation occurs on the total exposed area of the lagoon. The run-off from the access road over the embankments is not allowed to enter the lagoon.

$$\text{Annual average precipitation} = \frac{95.0 \text{ cm/yr}}{12 \text{ month/yr}} = 7.9 \text{ cm/month}$$

$$\text{Exposed lagoon area} = 78.9 \text{ m} \times 78.9 \text{ m} = 6225 \text{ m}^2$$

$$\text{Mean monthly precipitation} = \frac{6225 \text{ m}^2 \times 7.9 \text{ cm/month}}{100 \text{ cm/m}} = 492 \text{ m}^3\text{/month}$$

c. Determine the evaporation.

The evaporation occurs from the free-water surface. The free-water-surface area changes from a low to a high value during the filling operation and from a high to a low value during the drying cycle. Therefore, the average water-surface area is used for the evaporation calculations.

$$\text{Annual average evaporation} = \frac{153.5 \text{ cm/year}}{12 \text{ month/yr}} = 12.8 \text{ cm/month}$$

$$\text{Average water-surface area} = \left(\frac{60 \text{ m} + 75.3 \text{ m}}{2}\right)^2 = 4577 \text{ m}^2$$

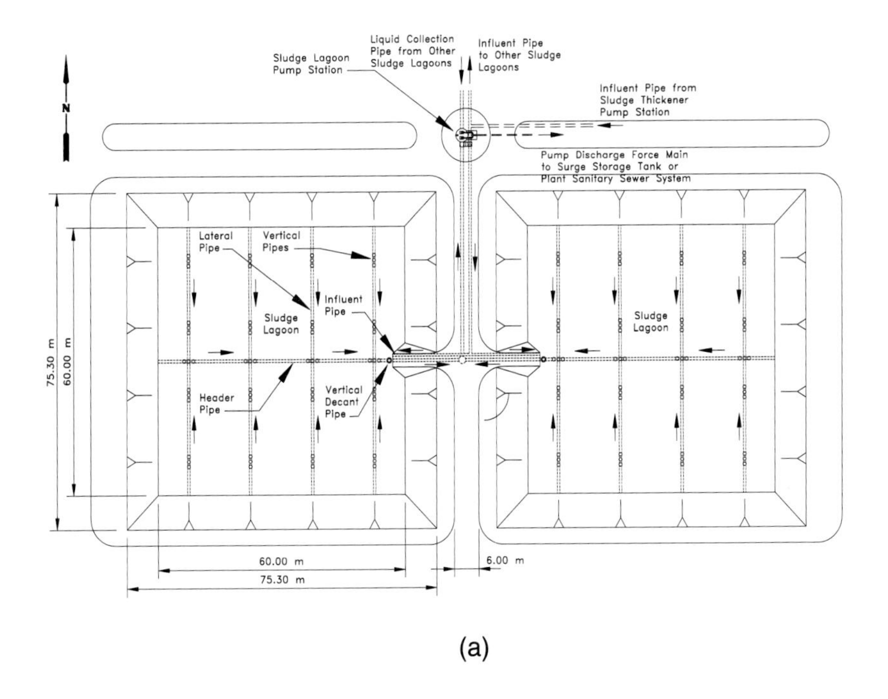

(a)

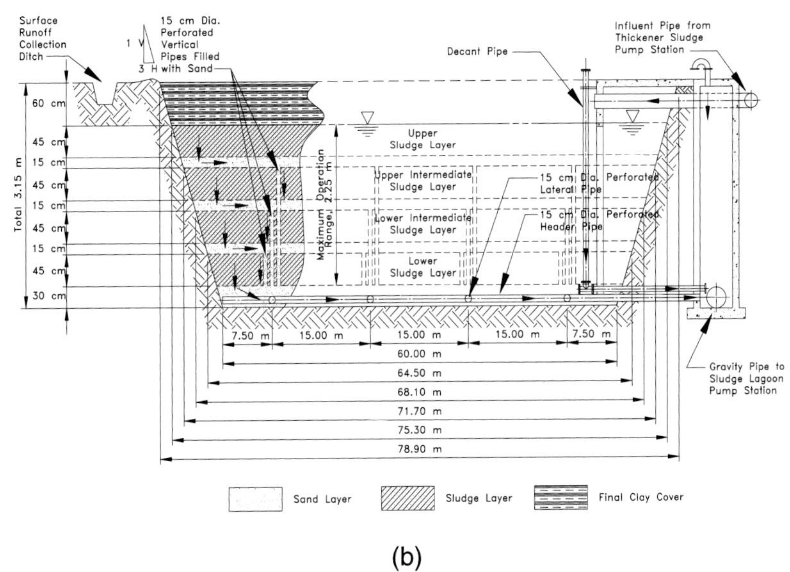

(b)

Figure 14-14 Design details of sludge-drying lagoon. (a) Plan view. (b) Section A-A, (which is not to scale).

$$\text{Average monthly evaporation} = \frac{4577\ \text{m}^2 \times 12.8\ \text{cm/month}}{100\ \text{cm/m}}$$

$$= 586\ \text{m}^3/\text{month}$$

d. Determine the percolation.

A percolation velocity of 2 m per year is assumed. This assumption can be justified as follows. At a 2-m-per-year percolation rate, a solids loading of approximately 100 kg/m^2 per year is reached at a sludge-solids concentration of 5 percent and a density of 1030 kg/m^3. This hydraulic loading is within the normal solids-loading rate used for sludge lagoons and is approximately half of the applied normal solids loading in sludge-drying beds. Furthermore, the percolation pattern through the underdrain system will involve vertical movement of water through the sludge cake, then lateral movement through a 0.15-m-thick intermediate sand bed. The water movement will further continue through the vertical pipes filled with sand, then through the 0.3-m-deep bottom sand layer before discharge into the pervious-tile drain system.

$$\text{Average monthly percolation} = \frac{200\ \text{cm/year}}{12\ \text{month/yr}} = 16.7\ \text{cm/month}$$

$$\text{Average water-surface area} = \left(\frac{60\ \text{m} + 75.3\ \text{m}}{2}\right)^2 = 4577\ \text{m}^2$$

$$\text{Average monthly evaporation} = \frac{4577\ \text{m}^2 \times 16.7\ \text{cm/month}}{100\ \text{cm/m}}$$

$$= 764\ \text{m}^3/\text{month}$$

e. Calculate the lagoon-filling cycles.

During the filling cycle, while fresh sludge is being pumped each day into the lagoon, water will be lost through evaporation and percolation. Also, precipitation will add water to the lagoon. Therefore, the filling period will be calculated by using all inputs and losses. As filling and drying cycles continue, the lagoon capacity and filling time will decrease, as a result of accumulation of sludge layers. Ultrasonic level transducers monitor and record lagoon levels all the time. The filling-cycle calculations are performed for four stages of sludge-accumulation layers: (1) lower layer, (2) lower intermediate layer, (3) upper intermediate layer, and (4) upper layer.

$$\text{Average precipitation} = 492\ \text{m}^3/\text{month}$$

$$\text{Average evaporation} = 586\ \text{m}^3/\text{month}$$

$$\text{Average percolation} = 764\ \text{m}^3/\text{month}$$

$$\text{Average sludge flow (Table 14-6)} = 75\ \text{m}^3 \times 30\ \text{d} = 2250\ \text{m}^3/\text{month}$$

$$\text{Net filling rate} = (2250 + 492 - 586 - 764)\ \text{m}^3/\text{month}$$
$$= 1392\ \text{m}^3/\text{month}$$

The total volume of the lagoon is represented by four frustums of the pyramid. The volume of each frustum is calculated from Eq. (14.3).

$$V = \frac{1}{3}D(A_1 + A_2 + \sqrt{(A_1 \times A_2)}) \tag{14.3}$$

where

V = volume of each frustum of the lagoon, m^3
D = depth of the frustum, m
A_1, A_2 = top and bottom areas of the frustum, m^2

The volumes of lagoon above each layer of deposited sludge are provided in Table 14-7.

f. Calculate the drying cycles.

The average drying period of the lagoon will decrease as sludge layers build up and lagoon capacity decreases. The rate of water loss is obtained from average precipitation, evaporation, and percolation data. The rate of water loss is 856.7 m^3/month. The average drying periods of the lagoon for each stage of lagoon operation are summarized in Table 14-7.

The volumes of a lagoon above each layer of deposited sludge are provided in Table 14-8.

g. Calculate the total life of the lagoons.

The average life of a lagoon is the time it takes to fill it with sludge solids. The sludge solids are built up in four layers. The areas and volumes of each sludge layer are provided in Table 14-9.

$$\text{Average sludge solids (Table 14-6)} = 3860\ \text{kg/d}$$

$$\text{Assume on-site sludge density} = 1450\ \text{kg/m}^3$$

$$\text{Sludge volume} = \frac{3860\ \text{kg/d}}{1450\ \text{kg/m}^3} = 2.66\ \text{m}^3/\text{d or } 80\ \text{m}^3/\text{month}$$

Average time to fill each layer is also given in Table 14-8. Average life of each lagoon = 106 month = 8.8 years. Total life of four lagoons is approximately 35 years. After the lagoons are filled with the solids, they are capped with a soil cover and the site is closed.

Table 14-8 Lagoon Filling and Drying Periods for Four Stages of Lagoon Operation

Stage	Water Surface Area , m^2	Area at Mid-Depth of Layer, m^2	Average Lagoon Depth at Mid-Layer, m	Average Lagoon Volume above Mid-Depth of Sludge Layer, m^3	Filling Period, months	Drying Period, months
Lower sludge layer	5670[a]	3988[b]	2.025[c]	9729[d]	7.0[e]	11.4[f]
Lower intermediate sludge layer	5670	4456	1.425	7197	5.2	8.4
Upper intermediate sludge layer	5670	4949	0.825	4377	3.1	5.1
Upper sludge layer	5670	5469	0.23	1253	0.9	1.5

a $75.3 \text{ m} \infty 75.3 \text{ m} = 5670 \text{ m}^2$.

b $\left(60 \text{ m} + 2 \times \left(0.3 \text{ m} + \frac{0.45 \text{ m}}{2}\right) \times 3 \text{ m/m (slope)}\right)^2 = 3988 \text{ m}^2$.

c $2.25 \text{ m} - \frac{0.45 \text{ m}}{2} = 2.205 \text{ m}$.

d $\frac{1}{3} \times 2.025 \times [3988 + 5670 + \sqrt{(3988 \times 5670)}] = 9753 \text{ m}^3$.

e $\frac{9753 \text{ m}^3}{1392 \text{ m}^3\text{/month}} = 7.0 \text{ months}$.

f $\frac{9753 \text{ m}^3}{(764 + 586 - 492) \text{ m}^3\text{/month}} = 11.4 \text{ months}$.

14.7 Operation, Maintenance, and Troubleshooting for Residuals-management Facilities

Waste streams are processed in several facilities to recover filter backwash water and concentrate solids for disposal. The filter backwash and coagulation solids are thickened in gravity thickeners and are dewatered and disposed of in the sludge lagoons. The processes involved are as follows:

Table 14-9 Top and Bottom Areas and Volumes of Each Sludge Layer

Stage	Area (B^2), m^2		Depth, m	Volume, m^3	Filling Period, months
	Bottom	Top			
Lower sludge layer	3819 (61.8^2)	4160 (64.5^2)	0.45	1795	22.4
Lower intermediate sludge layer	4277 (65.4^2)	4638 (68.1^2)	0.45	2005	25.1
Upper intermediate sludge layer	4761 (69.0^2)	5141 (71.7^2)	0.45	2227	27.8
Top sludge layer	5271 (72.6^2)	5670 (75.3^2)	0.45	2461	30.8
					107
					Total = 106

1. filter backwash water surge tank and recovery basins, to process filter backwash water and thickener overflow;
2. gravity thickener, to thicken the filter backwash solids and coagulation sludge;
3. sludge lagoons, for dewatering and on-site disposal of solids.

Proper operation and maintenance of these facilities is important for water recovery and sludge disposal. The following items are considered important for the design and operation of solids handling facilities.

14.7.1 Common Operational Problems and Troubleshooting Guide

Filter Backwash Water Surge Storage Tank The surge tank receives backwash water from the filters. Frequent *start of second pump* or *high-level alarm* in the surge tank is an indication that (a) the filter backwash cycles are too close together, or (b) the backwash withdrawal pump has a low pumping rate. A possible solution procedures includes spacing the filter backwash cycle so that not more than two filters are backwashed in less than 6 hrs. Another possible solution is to increase the capacity of the filter backwash withdrawal pumps.

Backwash Water Recovery Basin

1. The backwash water recovery basin is a circular gravity-settling basin. Poor-quality supernatant is an indication of nonsettleable colloidal particle in the filter backwash or in the thickener overflow. Polymer feed ahead of the recovery basin should be used to enhance the flocculation and sedimentation.
2. Sludge solids uniformly overflowing the weir may be due to inadequate sludge pumping. Check the sludge pumps and piping for malfunctioning or blockage. Increase sludge withdrawal to maintain at least a 2-m clear water zone in the basin.

3. Low solids in the sludge may be due to excessive sludge withdrawal, short-circuiting, or surging flow. Reduce sludge withdrawal, check and install baffles, and modify influent pumping rate.
4. Erratic operation of sludge-collection mechanism may be due to a damaged collection mechanism or to excessive sludge accumulation. Check the solids-withdrawal line and sludge-collection mechanism.
5. Sludge solids discharging over the weir in only one portion of the basin can be the result of unequal flow distribution. Level the effluent weir.
6. High algae concentration in the filter backwash return may add to taste and odor problems in the finished water. Polymer feed ahead of the backwash water recovery basin can be controlled properly to achieve low carryover of solids in the filter backwash return.

Gravity Thickeners

1. Rising sludge in the thickener is generally due to low or infrequent thickened-sludge pumping rate, low thickener overflow, or too high a depth of sludge blanket. The problem can be overcome by pumping thickened sludge more frequently or by adding dilution water for increased hydraulic loading.
2. Too thin a thickened sludge may be due to a high overflow rate, a high underflow rate, or short-circuiting through the tank. This situation is overcome by reduction in the influent sludge-pumping rate, by reduction of the dilution water, by reduction in the pumping of thickened sludge, and by maintenance of a high sludge blanket. Short-circuiting in gravity thickeners can be detected if uneven discharge of solids occurs over the effluent weir. Weir leveling or a change in baffle arrangements may be necessary.
3. Torque overload of sludge-collecting equipment can be due to accumulation of dense sludge or to a heavy foreign object jamming the scraper. The problem might be solved by agitation of the sludge blanket in front of the collector arms by rods or water jets.
4. Plugging of sludge lines and pump may be due to a too-thick sludge. The lines should be flushed and all valves should be opened fully.

Sludge-Drying Lagoons A lower percolation rate will result from build-up of the solids layer over sand. If the percolation rate is reduced significantly, surface preparation will be necessary. This is achieved by tilling the sludge layer or by drilling holes through the sludge layer into the sand.

14.7.2 Routine Operation and Maintenance

The following operation and maintenance items are necessary to keep the residuals-handling facilities in satisfactory operating condition.

Filter Backwash Water Surge Storage Tank

1. Settling of solids in the surge tank is expected under normal operating condition. Excessive build up of solids can cause reduction in pumping capacity. Check the pumps and the suction piping daily; remove all settled solids.
2. The pumps should be operated alternately, on a regular basis, for uniform wear. Both pumps will operate when the liquid level in the surge tank exceeds a preset high level. Liquid-level controller and pump operation should be checked routinely for emergency operation.
3. Check polymer-feed system and equipment in flocculation units.

Backwash Water Recovery Basin

1. Check polymer feed system and equipment in flocculation units.
2. Remove accumulations from the influent baffles, effluent weirs, scum baffles, and scum box each day.
3. Determine sludge level and adjust sludge pump as necessary.
4. Clean daily all inside exposed vertical walls and channels, with a squeegee.
5. Inspect the distribution box and clean the weirs, gates, and walls as necessary, and remove all settled solids.
6. Inspect the effluent box, and clean the weir and walls as necessary. Measure the head over the weir routinely.
7. The overflow pumps should be operated alternately, on a regular basis, for uniform wear.
8. Check electrical motors for overall operation and bearing temperature; check the overload detector twice each day.
9. Check oil level, grease reducer, and rollers on skimmer each week.
10. Check oil level for turntable bearings each week, and refill as required.
11. Grease the main bearings each week.
12. Change the oil in the gear reducer quarterly.
13. Drain the basin annually, to inspect the underwater portion of the concrete structure and mechanism. Inspect the concrete structure, and patch defective areas.

Gravity Thickener

1. The overflow pumps should be operated alternately, on a regular basis, for uniform wear.
2. The sludge pumps should be operated alternately, on a regular basis, for uniform wear.
3. Clean all vertical walls and channels by squeegee daily, and hose down walls regularly.
4. Check the sludge level daily. The sludge should be kept well below the top of the thickener. Sludge wasting should be controlled to maintain a proper sludge blanket.
5. Daily, check electrical motor for overall operation and bearing temperature; check the overload detector, and monitor unusual noises.

6. Check oil level in gear reducers weekly, fill as needed. Change oil quarterly, and lubricate worn gears weekly.
7. Drain the thickener annually, and inspect the underwater portion of the concrete structure and mechanisms. Inspect the mechanical equipment for wear and corrosion, adjust mechanisms, and set proper clearance for flights at tank walls. Patch defective concrete. Metal surfaces should be inspected for corrosion, cleaning, and painting.
8. The weirs and all surfaces should be frequently and thoroughly cleaned by a water jet.

Sludge-Drying Lagoons

1. Check the surface of the accumulated sludge layer after each drying cycle. Till the layer and level the surface for increased percolation.
2. Remove any vegetation growth.
3. Repair and redress all embankment slopes that may have been damaged by erosion and weed growth.
4. Inspect decant-removal standpipes and openings. Remove all accumulations and obstructions.
5. The percolation-return pumps should be operated alternately, on a regular basis, for uniform wear.

14.8 Specifications

The specifications for the filter backwash surge tank and recovery basin, thickeners, and lagoons designed in this chapter are briefly summarized here. The purpose of this section is to describe many components of the design that could not be covered fully in the Design Example. These specifications should be used only as a guide. Detailed specifications must be prepared for each design, in consultation with the equipment manufacturers.

14.8.1 General

Equipment Each solids-separation unit (filter backwash recovery basin and gravity thickener) shall include a complete assembly of chemical-feed, rapid-mix, flocculator, sludge-collector mechanism with drive, access bridge and walkway, influent and effluent structures, pumping facilities, and overload alarm system. The manufacturer shall furnish and deliver, ready for installation, all equipment suitable for installation in filter backwash recovery systems and gravity sludge thickeners.

Materials and Fabrication The structural steel shall conform to proper ASTM standards.[a] The minimum thickness of all submerged metal shall be not less than 6.4 mm (1/4-in) and of all above-water metal 4.8 mm (3/16 in). All iron casting shall also conform to proper

a. ASTM-American Society for Testing and Materials

ASTM standards. Design and construction shall conform to all AISC standards for structural-steel buildings.[b]

Painting All nonsubmerged ferrous materials shall be brushed clean, and submerged ferrous material shall be sandblasted to white metal. Cleaned surfaces shall be primed with approved epoxy primer and have a finished coat of approved paint or epoxy. All field welds shall be touched up with compatible paints.

14.8.2 Filter Backwash Surge Storage Tank

The filter backwash surge tank shall be 14 m ∞ 14 m, with 5.5-m side-water depth and a 1.5-m hopper, as shown in the surge storage tank design details. Overflow pipes shall be provided for emergency conditions.

Two identical constant-flow pumps, each 2.3 m^3/min (600 gpm) in capacity, will remove water from the hopper into the backwash water recovery basin. The pump control shall include low-level pump stop, high-level second pump start, and high-level alarm.

The chemical-feed system and mixer in flocculation unit, including fabricated structural frame, paddles, primary variable-speed motor reducer, a secondary worm-gear reduction unit, and motor reducer–transfer device, are similar to those covered in Chapter 8.

14.8.3 Filter Backwash Recovery Basin

The filter backwash recovery basin shall be 13 m in diameter and 6.3 m in side-water depth, with center feed and peripheral-overflow effluent launder. The general specifications of chemical feed, rapid mix, flocculator, sedimentation facility, and sludge pump have been covered in Chapters 8 and 9. The design engineers should consult these chapters to develop the design specifications of these facilities.

14.8.4 Gravity Thickener

The specifications of gravity sludge thickeners are briefly covered here to describe the general features of the equipment. Detailed specifications of similar equipment can be found in Chapters 8 and 9.

The gravity sludge thickeners shall consist of complete assembly, including chemical feed, rapid mix, flocculator, feed pumps, collector mechanism with flights, above-water center-drive mechanisms, influent central well, drive cage, access bridge, bridge support, center-pier support, and overload alarm system.

The chemical feed system and mixers in flocculation units are similar to those covered in Chapter 8.

All equipment specified herein is intended for use with combined filter backwash solids and coagulation sludge. The gravity thickener mechanisms shall be designed to handle thick-

b. AISC-American Institute of Steel Construction, Inc.

ened sludge up to a maximum of 10 percent solids concentration and to be capable of continuously plowing the thickened sludge and moving the settled sludge to the center channel for removal. Each thickener shall be of the center-feed and peripheral-overflow type, with a central driving mechanism which shall support and rotate two attached rake arms. Rake collector blades attached to the arms shall be arranged to move the settled sludge on the tank bottom to a concentric sludge channel surrounding the center column. The scrapers attached to the arms shall provide 100 percent coverage of the tank floor.

All gravity-thickening equipment shall be designed so that there will be no chains, sprockets, bearings, or operating mechanisms below the liquid surface. The drive assembly shall comprise an electric motor connected to a primary gear reducer, drive and driver sprockets with drive chain, an intermediate worm-gear reducer, pinion gear, turntable base and main spur gear, and complete automatic overload-actuating system.

14.8.5 Sludge-Drying Lagoon

The sludge lagoons are designed to dewater the sludge and to provide disposal of residuals. The lagoons are earthen basins with a bottom sand layer, drainage pipe, intermediate sand layers, and water-decant system for enhanced dewatering of sludge. Run-off-, erosion-, and sediment-control systems shall be installed.

Four sludge lagoons shall be provided for sludge dewatering and disposal. Each lagoon shall have the following design detail, summarized in Section 14.6.2 Step F.1.

Lagoon Construction Initially, the area for the lagoon shall be cleared and grubbed. The excavation of the lagoon, construction of dikes, and installation of an underdrain system shall be done simultaneously. The top surface of the embankment shall be sloped away from the lagoon to prevent surface run-off from entering the lagoon. The embankments common to two lagoons shall have corrugated metal pipe drain for proper draining of surface run-off away from the lagoon. Leftover soil shall be utilized in surface grading and in construction of levee and other embankments and shall be stockpiled for covering the lagoon surface after completion. Immediately after excavation of lagoons and construction of levees, the site shall be hydromulched to minimize erosion.

Lagoon Base and Intermediate Drainage Layers The lagoon base shall have 10-cm perforated PVC pipes embedded into 30-cm-deep sand. Two sand layers, each 15 cm deep for percolation of water, shall be provided. The lower-layer sand shall have uniformity coefficient of 1.5–2, and effective size of 0.8 mm; the perforated pipe shall be embedded completely in this layer. The upper-layer sand shall have uniformity coefficient of 1.5–2, and effective size of 0.3 mm. Both layers shall be fully stratified. Both intermediate layers shall be 15 cm deep and have uniformity coefficient and effective size of 1.5–2, and 3 mm, respectively. After filling of each layer, the surface shall be dressed. A total of 25 vertical pipes, each 15 cm in diameter, shall be embedded through the sludge layer to the lower sand layer. The vertical pipes shall be provided in a grid 15 m ∞ 15 m, filled with sand, and covered by the intermediate sand layer.

Decant System The decant system shall be a 30-cm vertical pipe having multiple ports for removal of supernatant from different layers. The vertical pipe shall have decant removal ports at 2.5 cm intervals, arranged around the vertical pipe. Each port shall have a removable stopper. The vertical pipe shall be approached by a walkway connected to the embankment.

Problems and Discussion Topics

14.1 Visit a local water treatment plant. Determine the residuals streams generated at this plant. Draw a process diagram for the residuals-management system being used at this facility.

14.2 Common processes for sludge dewatering that are generally evaluated before a final dewatering system is selected are drying beds, centrifuge, vacuum filter, plate and frame filter process, and belt filter press. Summarize the advantage and disadvantages of each process.

14.3 A water treatment plant produces 1000 kg of solids per day. The solids concentration, after thickening, is 3 percent, and specific gravity is 1.02. Calculate the area of the sludge-drying beds, if the sludge-application rate is 0.6 and the beds are used 20 times per year.

14.4 Design a gravity thickener that receives 2000 kg/d of sludge solids at 2 percent solids and specific gravity of 1.01. The thickener minimum hydraulic loading is 3.5 $m^3/m^2 \cdot d$, and maximum solid loading is 75 $kg/m^2 \cdot d$. The liquid depth in the thickener is 6 m, plus the depth of the hopper. The hopper has a side slope of 20 cm/m. Calculate the thickening period, solids concentration in the thickener overflow, and volume of thickened sludge.

The thickener efficiency is 90 percent solids capture, and solid concentration in the thickener sludge is 5 percent. Specific gravity of thickener sludge is 1.03.

14.5 A thickener receives 6000 kg/d solids from the sedimentation basin. The filter backwash system receives 250 kg/d solid from the filters. The flow diagram is such that settled solids from the filter backwash recovery basin are pumped to the thickener, and the thickener overflow is pumped to the filter backwash recovery basin. Assuming both the solid capture and the efficiency of both thickener and filter backwash recovery basin are 90 percent, conduct the mass-balance analysis and determine the solid in the thickened sludge.

14.6 Design a filter backwash recovery system that has a surge basin and a gravity sedimentation basin. There are six filters, and each is backwashed once a day. Generally, filter backwash is scheduled every four hours. It is, however, possible that two filters might need to be backwashed simultaneously. Provide three pumps: one pump for normal operation, and two pumps for emergency conditions. Calculate the dimensions of the surge tank, the pumping capacity, and the dimensions of the settling basin. Hydraulic loading on the settling basin is 20 $m^3/m^2 \cdot d$. Each filter provides 500 m^3 of backwash flow. The surge tank capacity is such that it can hold two filter backwash flows.

14.7 Assume that 2000 kg of solids reach the drying beds for dewatering each day. The total solids in the sludge are 5 percent, and specific gravity is 1.025. The sludge-drying beds are uncovered, and solids loading is 110 $kg/m^2 \cdot year$. Calculate (a) the number of beds if each paved bed is 4 m × 20 m and (b) how many beds will be filled each day if the average sludge-application rate is 25 cm.

14.8 A vacuum filter is designed for dewatering alum coagulated sludge. The sludge contains 5 percent solids, and specific gravity is 1.025. The volume of sludge is 100 m^3/d. Laboratory tests show that filter yield is 2.5 $kg/m^2 \cdot h$, and optimum chemical dosage is 9 percent lime and 3 percent polymer. The filter cake has 22 percent solids, including conditioning chemicals. If 75 percent of the added chemicals are fixed into the cake solids and solids-capture efficiency is 95 percent, calculate (a) dimensions of the vacuum filter, (b) chemical dosage (kg/d and kg/kg of solids), and (c) weight of filter cake produced per day (kg/d). Assume that filter operation is 12 h/d and that the specific gravity of the filter cake is 1.06.

14.9 Determine the volume and concentration of total solids in the filter rate from a dewatering facility using filter presses. The average volume of sludge reaching the dewatering facility is 100 m^3/d. The specific gravity of the sludge is 1.025, and total solids are 5 percent. The filters operate 8 h/d and 5 d/week. The cycle time is 60 min, and there are 5 filter units. Lime and polymer dosages are 6 and 3 percent of dry solids. Assume that 75 percent added chemicals are fixed into the sludge cake. Also, calculate the quantity and volume of sludge cake, if average moisture content in the sludge cake is 75 percent by weight and specific gravity is 1.06. Solids-capture efficiency of filter presses is 88 percent. The lime and polymer solutions used in chemical conditioning are 10 and 15 percent, respectively. The water used per cycle for filter washing and leakage through the diaphragms is 4 m^3 per cycle per filter.

14.10 Alum coagulated sludge is conditioned by using hydrated lime and polymer. Calculate the volume of conditioned sludge (m^3/d) and the capacity of the sludge-feed pump (m^3/h) used to pump conditioned sludge into the recessed-plate filter press. Use the following data:

Average flow of sludge = 50 m^3/d
Solids in sludge = 4000 kg/d
Lime dosage = 5.5 percent of dry solids
Lime slurry solution = 12 percent
Polymer = 2 percent of dry solids
Polymer solution = 5 percent

Sludge pump operates 8 cycles/d and 0.5 h/cycle.
Filter press operation is 6 d per week and 8 h/d.

14.11 1000 kg per day of sludge reaches the sludge-drying earthen cells for dewatering. The total solid in the sludge is 7 percent, and the specific gravity of the sludge is 1.025. The solids loading rate for the cells is 37 kg/m^3·year. Calculate (1) the number of cells, if each cell is 0.02 hectare in area and 1 m deep, and (2) in how many days each cell will be filled.

14.12 The material mass-balance analysis at design maximum day flow is provided in Table 14-5. Complete the material mass-balance analysis for (a) third and fourth iterations and (b) fifth and sixth iterations.

14.13 The material mass-balance analysis at design average day flow is provided in Table 14-6. Complete the material mass-balance analysis for (a) third and fourth iterations and (b) fifth and sixth iterations.

References

1. U. S. Congress. *The Federal Water Pollution Control Act Amendments of 1972* (PL 92-500), 92nd Congress, October 18, 1972.
2. U. S. Congress. *The Clean Water Act* (PL 92-217), 95th U.S. Congress, December 27, 1977.
3. ASCE and AWWA. *Water Treatment Plant Design*, 2d ed., McGraw-Hill Publishing Company, 1990.
4. James M. Montgomery, Inc. *Water Treatment Principles and Design*, John Wiley & Sons, New York, 1985.
5. USEPA. *Technology Transfer Handbook: Management of Water Treatment Plant Residuals*, EPA/625/R-95/008, USEPA, Cincinnati, OH, 1996.
6. AWWA. *Water Quality and Treatment*, 5th ed., McGraw-Hill Book Co., New York, 1999.
7. Wang, M. C., Hull, J. Q., Jao, M., Dempsey, B. A., and Cornwell, D. A. "Engineering Behavior or Water Treatment Sludge," *Jour. of Environmental Engineering, ASCE*, vol, 118, no. 6, pp. 848–864, December 1992.

8. AWWA. "American Water Works Association Research Foundation Report: Disposal of Wastes from Water Treatment Plants-Part 3," *Jour. AWWA*, vol. 61, no. 12, pp. 681–708, December 1969.

9. Pigeon, P. E., Linstedt, K. D., and Bennelt, E. R. "Recovery and Reuse of Iron Coagulants in Water Treatment," *Jour. AWWA*, vol. 70, no.7, pp. 397–403, July 1978.

10. Calkins, R. J. and Novak, J. T. "Characterization of Chemical Sludges," *Jour. AWWA*, vol. 65, no. 6, pp. 423–427, June 1973.

11. O'Connor, J. T. and Novak, J. T. "Management of Water Treatment Plant Residues," *American Water Works Association Seminar Proceedings*, AWWA Conference, June 25, 1978.

12. Weber, W. *Physicochemical Process for Water Quality Control*, Wiley-Interscience, New York, 1972.

13. Watt, R. D. and Angelbeck, D. I. "Incorporation of a Water Softening Sludge into Pozzolanic Paring Material," *Jour. AWWA*, vol. 69, no. 3, pp. 175–180, March 1977.

14. Qasim, S. R. and Lamakul, G. *Evaluation of Roll and Diatomaceous Earth Filters Used for Water Reclamation in a Commercial Carwash*, Research Report TR-3-82, Construction Research Center, The University of Texas at Arlington, Arlington, Texas, June 1982.

15. USEPA. *Process Design Manual, Sludge Treatment and Disposal,* Technology Transfer, EPA-625/1-79-011, USEPA, Cincinnati, Ohio, September 1979.

16. USEPA. *Process Design Manual for Sludge Treatment and Disposal,* Technology Transfer, EPA-625/1-74-006, USEPA, Cincinnati, Ohio, October 1974

17. Qasim, S. R. *Wastewater Treatment Plants: Planning, Design, and Operation*, 2d ed., Technomic Publishing Co., Inc., Lancaster, PA 1999.

18. Metcalf and Eddy, Inc. *Wastewater Engineering: Treatment, Disposal and Reuse*, 3d ed., McGraw-Hill Book Co., Inc., New York, 1991.

19. Reynolds, T. D. and Richards, P. A. *Unit Operations and Processes in Environmental Engineering*, PWS Publishing Co., Boston, MA, 1996.

20. Schroeder, E. D. *Water and Wastewater Treatment*, McGraw-Hill Book Co., New York, 1977.

21. Cornwell, D. A. and Lemunyon, R. M. "Feasibility Studies on Liquid Ion Exchange for Alum Recovery from Water Treatment Plant Sludges," *Jour. AWWA*, vol. 72, no. 1, pp. 64–68, January 1980.

22. Taylor, T. E., Bolding, M. E., and Stringer, C. E. "Piloting Mag-Carbonate Treatment," *Water, Engineering & Management*, pp. 50–55, April 1982.

23. Snoeyink, V. L. and Jenkins, D. *Water Chemistry*, John Wiley & Sons, New York, 1980.

24. Thompson, C. G., Singley, J. E., and Black, A. P. "Magnesium Carbonate—A Recycled Coagulant," *Jour. AWWA*, vol. 64, no.1, pp. 11–19, 1972.

25. Thompson, C. G., Singley, J. E., and Black, A. P. "Magnesium Carbonate—A Recycled Coagulant," *Jour. AWWA*, vol. 64, no. 2, pp. 93–99, 1972.

26. U. S. Congress. *The Solid Waste Disposal Act* (Pl 91-512), 91st U.S. Congress, October 26, 1970.

27. U. S. Congress. *The Resource Conservation and Recovery Act* (PL 94-580), 94th U.S. Congress, October 21, 1976.

CHAPTER 15

Plant Siting, Layout, Yard Piping, and Hydraulic Profile

15.1 Introduction

Design and construction of a water treatment plant basically involves four steps. The first step is the site selection. The second step involves development of a cost-effective process train to delineate the proper relationship of the various steps in the treatment plant. The third step involves sizing the individual treatment units and placing them in the available area to achieve proper sequence and hydraulics, to provide access for operation and maintenance of equipment, and to minimize the total space requirements. Finally, the connecting pipings are tied down with topography, process units, buildings, chemical and feed lines, roads, and pumping needs. Therefore, every effort must be made to become fully familiar with the proposed site and its neighborhood.

Yard piping includes connecting conduits, collection and division boxes, valves, and appurtenances between various units. After the treatment units, connecting pipes, and appurtenances are marked on the layout plan the head losses through pipings and treatment units are calculated. Hydraulic profile is the graphical representation of the hydraulic grade line through the plant. The elevations of treatment units and pipings are adjusted to give adequate hydraulic gradient to ensure gravity flow.

Many water treatment plants have encountered serious operational problems resulting from improper site selection, plant layout and piping, and hydraulic gradient through the plant. It is, therefore, important that the design engineers take proper care in plant siting, layout, and hydraulics. In this chapter, the basic principles of plant siting, unit location, yard piping, and preparation of hydraulic profile are presented. Furthermore, plant layout, calculation procedure, and graphical representation of the hydraulic profile are presented in the Design Example.

15.2 Plant Siting

Site selection of a water treatment plant needs careful consideration of the region's land development patterns and of engineering, environmental, and social constraints. Often, an interdisciplinary[a] team approach may be necessary to cover many aspects of site selection and planning. The following is a list of basic principles that must be considered during the site evaluation and selection process.

1. The present and future water demands, direction, and rate of growth of the service area, zoning plan and regulations, and water system master plan must be studied and developed.
2. To avoid possible misunderstandings or actual opposition to a plant siting project, it is essential that all principal groups that might be affected are consulted at an early stage. Discuss the principal jurisdictions involved with the representatives, including cities and counties, special districts, environmental groups, and local or state health departments.
3. The selected raw water sources will have a long-term influence on plant operation and performance. Therefore, site selection of a treatment plant must be integrated with the selected source of raw water. The raw water source, such as a river, lake, aqueduct, or well has a direct relationship with elevation, pumping, and transport. Also, the chemical storage and sludge disposal needs will influence the land area requirement.
4. Treatment plant sites near a river that is selected as raw water source often cause many difficult problems. These problems can include flooding, high groundwater table, or poor foundation conditions. It is often desirable to study the site selection of both intake and water treatment plant at the same time. Treatment plant sites away from the river and at higher grounds are often more advantageous than near the river.
5. Plant siting in flood plains should be avoided. If necessary, raising of the facilities and buildings above flood levels and construction of flood-protection levees should be considered. Provision must be made for collection of run-off at a central location. A stormwater management plan with outlet structure, flood gate, stormwater pumps, and outfall pipe must be developed.
6. If the service area is above the raw water source, pumping will obviously be required. It is desirable to locate the treatment plant at an elevation from which one or more pressure zones can be served directly.
7. Raw water obtained from dam or diversion intakes may offer an opportunity to utilize available head if the plant site is properly located. A gravity flow to and through the plant is desirable. Power generation from the available energy at the dam and in the pressure aqueduct has been attempted at several water treatment plants.
8. Sufficient land area must be available that can be acquired for future plant expansion and many other unforeseen needs that could develop in the future. Some of these needs might

a. The interdisciplinary team often includes professionals from sanitary and environmental engineering, urban planning, architecture, biology, government, and civic groups.

be presedimentation basins, chemical storage, roads (including truck access), on-site raw and filtered water storage, pump station, filter backwash recovery system, and residuals-management and disposal system.

9. Any restriction specific to the site should be determined. Among them are restrictions about chemical delivery, storage, and use, restrictions against building types such as chemical silos and towers, and restrictions on chemical uses and their products: lime dust, carbon dust, ozone, peroxide, ammonia, chlorine, or chlorine dioxide. Another undesired activity against which restrictions might be imposed is noisy equipment, such as drive units, compressors or blowers, ozonators, pumps, and engines.
10. Topography and accessibility of the site are important factors in site selection. Contour maps and aerial photographs should be used to evaluate the potential site. Most water treatment plants are designed to have gravity flow from one unit to the other. Therefore, topography with a moderate slope will assist in locating various treatment units in their normal sequence without excavation or fill. This will provide gravity flow, the least disruption to the normal topography, and the least in erosion control measures. The site should also have year-round, all-weather roads. Railroads will be helpful for delivery of bulk chemicals and transport residuals. The ability of the ground to support structures without expensive piling is also an important consideration in site selection. Common foundation problems are low bearing capacity, excessive settlement, differential settlement, and flotation due to a high groundwater table.
11. Land area requirement of a water treatment plant depends upon plant capacity, presedimentation basin, chemical storage, treatment processes, clearwater wells, sludge handling and disposal, and low- or high-lift pumping station. Sufficient land should also be reserved for future expansion. Typical land area requirements without presedimentation, clearwater basins, sludge-holding ponds, or drying beds are shown in Figure 15-1.[4] Availability of a suitable piece of land or necessary steps needed to acquire desired land in the area should be considered in the overall site-selection process.
12. Many factors related to a service area are also used for site selection of a water treatment plant. Some of these factors are the follows:

    a. closeness to the service;
    b. high elevation, so that most of the service area can be served by gravity;
    c. far enough distance from the first customer to provide effective time for post-chlorination requirement (a factor of little significance if the plant becomes the first customer);
    d. Closeness to existing raw water transmission pipeline;
    e. Sufficient space for clearwells and residuals-management facilities;
    f. availability of electrical power.

13. Plant sitings are directly related to overall energy requirements of raw water pumping, finished water delivery system, service-zone elevations and limits, and future land use,

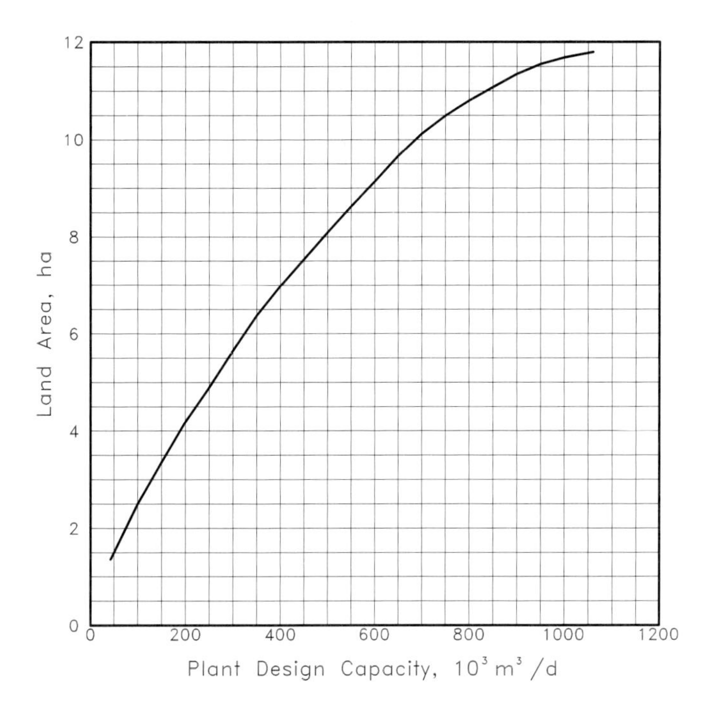

Figure 15-1 Typical land area requirement of a conventional water treatment plant (not including presedimentation, clearwater basins, sludge-holding ponds, or drying beds or lagoons). *(Source: Adapted from Reference 4.)*

including areas of peak demand. A site that eliminates unnecessary double pumping shall be preferred.

14. Most water treatment plants will require that an environmental impact statement be prepared. To prepare an environmental impact report, consult the guidelines developed by the USEPA and state agencies. Topics such as environmental setting, historic and archaeological sites, flood plains, wetlands, agricultural lands, fish and wildlife protection, endangered species protection, impact of noise, odors, chemical spills, and night lighting upon neighbors, alternatives evaluation and selection, direct and indirect impacts, and minor adverse impacts that cannot be avoided must be fully addressed.

15.3 Plant Layout Considerations

15.3.1 General Considerations in Site Development

Plant layout is the physical arrangement of designed treatment units on the selected site. Careful consideration must be given to properly locating the treatment units, connecting conduits, roads, parking facilities, administration building, and maintenance shops. The design engineer must integrate the functions of all components. Experience has shown that proper plant layout can (1) enhance the attractiveness of the plant site, (2) fit the operational needs of the processes, (3) suit the maintenance needs of the plant personnel, (4) minimize construction and operational costs, (5) offer flexibility in future process modifications and plant expansion, and (6) maintain landscaping and plant structures in harmony with the environment. A variety of factors should be considered as general design guidelines for plant layout on the selected site. Some of these factors are discussed below.

Site Topography and Geology Consideration of site topography and geology is important, because plant layout should respect existing site features, character, and topography. Site development should take advantage of the existing site topography to either emphasize or diminish the visual impact of the facility, depending on the design goals.

A site on a side-hill slope can facilitate gravity flow that will reduce pumping requirements and locate the normal sequence of units without excessive excavation and fill. Also, a side-hill location can be used to an advantage, because the chemical storage silos can have direct access both from the ground and from the upper levels.

Foundation Considerations The results of soil investigation should be utilized in locating the treatment units, buildings, and heavy equipment. Consideration should be given to load-bearing capacities, water table and flotation effects, and piling. Separation of massive concrete structures (such as basins, filters, and operations buildings) from other facilities is necessary to minimize uneven subsidence of the structure.

Length of Chemical Lines The lengths of chemical lines should be minimized to prevent clogging and operational problems. Therefore, the chemical storage and the feed equipment should be provided in close proximity to the point of application.

Access Roads Access roads should be included in the plant layout, to serve equipment in process areas. Anticipated traffic flow and parking should be considered. Employee parking at the rear and visitor parking in the front of the administration building should be provided. Trucks and service traffic should be separated from visitors and employee traffic upon site entry. Signs for visitor parking and appropriate building entry should be posted. Turning radius and proper access should be considered in the routing of chemical delivery trucks.

Noise Sources Noise should be controlled to prevent discomfort to plant personnel and neighbors. Equipment such as pumps, ejectors, generators, blowers, and compressors can produce disturbing sound levels and, therefore, should be provided with sound barriers.

Buildings Buildings are needed for personnel, process equipment, and visitors. The following considerations should be given to building design and location[6]:

1. Location of process equipment at the point of maximum usage is helpful.
2. The climate of the area should be considered in building orientation and design, to minimize heating, ventilation, lighting, and air-conditioning costs. In cold areas, buildings should not shade trucking and parking areas, to reduce snow and ice clearance problems. In mild climates, the use of an outdoor filter is acceptable. In very cold climates, filters must be covered or provided inside a building to prevent problems due to freezing.
3. The administration building should be located near the entrance and in public view. The administration building should contain offices, laboratory, instrumentation and control room, showers, lavatories, and locker rooms. In addition, the visitors' lobby should have educational displays and tour information.
4. Other building requirements include machine shops with tools and storerooms, garages, and an equipment building to house pumps, compressors, chemicals, and instrumentation. Proximity of the operation and control building to major process units is essential. Chemical system, filters, and flocculation basins require greater operator attention because of their high degree of mechanical complexity; therefore, these facilities should be located next to the operations building, for immediate operator attention and for centralization of plant control functions.

Landscaping and Site Compatibility Landscaping should reflect the character of the surrounding area. Site development should alter existing naturally stabilized site contours and drainage patterns minimally. Consideration to limiting erosion and siltation should also be given. Existing site vegetation, trees, and shrubs should be assessed and utilized. Planting should be considered for control of slope erosion, surface run-off, and enhanced attractiveness. Local soils and climatological and biological conditions should be carefully investigated by a competent landscape architect.

The developed site should be compatible with the existing land uses and the comprehensive development plan. In some instances, it may be desirable to design the buildings at the plant so as to blend the treatment plant facility with the surrounding developments.

Lighting Proper lighting at treatment facilities promotes safer operation, efficiency, and security. Consideration should be given to interior, exterior, safety, and security lighting. Illumination to highlight structural and landscape features should be provided.

Plant Utilities The utilities at the treatment plant include electrical power, natural gas, water lines, effluent lines, telephone lines, and an intercommunication system.

The design of the utility system should conform to the applicable codes and regulations of the municipality concerned and to the operating rules of the concerned utilities. All utility lines should be properly shown on the layout plans, marked on the site (if exposed), and grouped properly to facilitate repairs, modifications, and expansion.

Occupational Health and Safety A water treatment facility offers many types of occupational health hazards that must be considered as part of plant design and layout. Important factors for which proper safeguards must be provided include chemicals and chemical handling, toxic gases, fire protection, explosions, burns, electric shocks, rotating machinery parts, material and equipment handling, falls, and drowning. An excellent discussion of the subject can be found in Reference 1.

Security All accesses to the treatment plant should be controlled. Fences and other barriers should be provided to enclose the facility. Proper signs should be displayed at all accesses indicating the name and owner of the facility. The main gate should have camera-monitored access control.

Future Expansion Provisions for future plant expansion must be made. The provisions should include (1) future space requirements and (2) plant expansion and process modifications and improvements, with minimal interruption to the operation of existing plant.

15.3.2 Compact and Modular Site Development

A relatively compact site plan can minimize piping requirements. Centralization of similar process units, process equipment, personnel, and facilities may reduce total staff size, as well as optimize plant supervision and operation features.[7] The traditional linear plant layout is not ideal for chemical feed or operations, because the long length of the sedimentation basin separates the coagulation–flocculation area with filters. Heavy chemical usage and mechanical equipment areas generally are rapid mix, filters, and high-service pumps. At rapid-mix basins, chemicals such as pre-disinfectants, pH control chemicals, coagulants, and flocculants are needed. Near filters, filter aids, post-disinfectants, fluoride, and water stability control chemicals are used. The mechanical equipment associated with chemical delivery, rapid mixers, flocculator drives, and sedimentation basin sludge-collector drives is typically located in the vicinity of the rapid-mix basins. The plant control room, post-treatment chemical delivery, and high-service pumps are near the filters at the outlet end of the plant. This separation of chemical feed points and separation of mechanical equipment is undesirable from an operations viewpoint.

A compact plant layout clusters chemical feed points and mechanical equipment by locating the filters and clearwells near the head of the plant. This can be achieved by returning the settled water to the head of the plant or by providing a two-tray sedimentation basin. This concept is shown in Figure 15-2. Such an arrangement offers many advantages. Because of the compact layout achieved, many changes in the traditional design concepts applicable to linear layout must be made. The advantages and basic design considerations of a compact layout are summarized in Table 15-1.

Many important design considerations include the following:

- Modular expansion of a compact plant layout is more suitable. The designer should not only make provisions for the expected future capacity but also allow for the additional treatment units that will be required to meet more stringent future regulations.

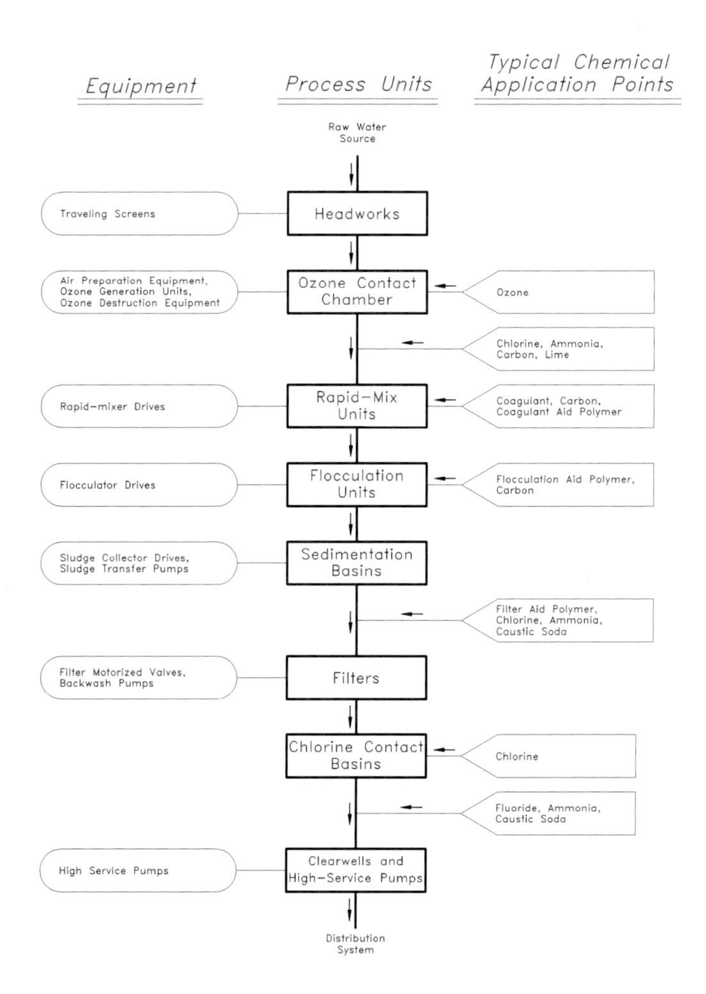

Figure 15-2 Typical service requirements of a conventional water treatment plant.

- Equipment access in a compact design may be restricted. Trucks may not be able to get as close to the process equipment as they might with a dispersed plant layout. Provisions for monorails, bridge crane, access openings, and other appropriate measures should be provided to allow for removal and replacement of equipment. Adequate clearance should be provided around all items of mechanical equipment, for maintenance and disassembly.
- In compact design and layout, the elevations of most treatment units, chemical storage and feeders, and floors of buildings become interrelated. Raising or lowering the elevation of one basin may affect the elevations of other units and building floor. Careful consideration should be given to such details at an early stage of the design.

Table 15-1 Advantages and Design Considerations of Compact Layout

Parameter	Advantages	Design Considerations
Mechanical	• The major items of mechanical equipment are consolidated into one central area near the support facilities. • Chemical feed lines are relatively short, because the chemical application points are all located close to the operations area. • Because there are fewer separate buildings, fewer HVAC systems are required.	• Removal of equipment for off-site maintenance may be more difficult because of restricted access.
Operations	• Walking distance to all equipment is short. • Mechanical equipment is close to maintenance facilities in operations building. • There is covered access by means of galleries and tunnels. • Number of plant staff may be reduced, because the equipment needing regular attention is located close to the operations building.	• Less room for maintenance activities.
Structural	• Reduced excavation volumes are possible. • Reduced concrete quantities, because of common wall construction, are possible. • Less total floor area is required.	• Because electrical and instrumentation cables are concentrated near the operations area, numerous floor openings and slab penetrations are required. • Need to make extra provisions for concrete shrinkage and expansion and for soil movement.
Electrical	• Lengths of electrical and instrumentation conduits and cables are reduced. • Cable size (determined by voltage drop) may be reduced because of shorter cable lengths.	

Table 15-1 Advantages and Design Considerations of Compact Layout (continued)

Parameter	Advantages	Design Considerations
	• Because cables can be routed indoors instead of in buried conduits between structures, cost-efficient cable trays can be used.	
	• Cables routed indoors are more accessible.	
Hydraulic	• Deleting the interconnecting piping associated with separate basins reduces hydraulic losses.	• Long channel or pipe may be required to route settled water to filters.
Architectural	• It is easier to develop an architectural theme for a single mass than for several separate structures.	• Mixed use of building (operations plus administration) complicates building's architectural design.
General	• Less land is required.	• Having one large building containing all facilities makes it more difficult to comply with fire code requirements for access–egress, because of the increased use of hallways, stairways, and so forth.
	• With fewer structures, the number of internal access roads and the amount of paving on-site is reduced.	• If a uniform top-of-wall elevation is used for all process basins (rapid mix, flocculation, sedimentation, and filters), then freeboard in the sedimentation basins may be excessive.
	• Less underground large piping is required.	• Alternatively, if the top-of-wall is varied to achieve a uniform freeboard, this will result in steps in the walkways along the tops of the basins.
		• Increased need for safety precautions arises from proximity of bulk chemical storage area to administration building.

Source: Adapted from Reference 7.

- Safety aspects need more attention with the compact design, because of the proximity of plant staff to storage, handling, and delivery of hazardous chemicals. Appropriate articles of the Uniform Fire Code (UFC) dealing with hazardous materials used in water treatment plants should be studied.[8] Minimum distances from storage tank to building, openings in buildings, air intakes, and so on, are important design considerations. Other relevant authorities about the governing codes and local regulations should be checked at an early stage of the design. Related safety issues are in compliance with the relevant building codes for fire ratings, access–egress, enclosures of stairways, and the like. Combining administration and operation into a single structure results in a building with mixed occupancy classifications. Furthermore, the filter-pipe galleries and any other lower-level portions of the operations building generally have no windows and limited means of access, because they are below ground level, are located adjacent to water-retaining structures, or both. Under such conditions, it can be difficult to meet all the requirements of the building code.[7]
- The electrical design needs require special attention in a compact layout. Electrical conduits, chemical pipes, heating-ventilating-air conditioning (HVAC), and plumbing will be concentrated into a smaller area and at two to three floors. Careful planning is needed to avoid conflicts between disciplines in both vertical and horizontal directions, room layout, floor openings, and slab penetrations.
- Compact design uses common-wall construction. Provisions for structural movement due to concrete shrinkage, expansion, soil movement, and unequal settlement must be made. The location of isolation–expansion joints should be coordinated with all disciplines, so that appropriate measures can be taken to provide for potential movement.

15.4 Yard Piping and Hydraulic Profile

Yard piping includes connecting conduits, collection and division boxes, valves and gates, and appurtenances between various treatment units. After the treatment units and connecting pipes and appurtenances are marked on the layout plan, the head losses through pipings and treatment units are calculated. Hydraulic profile is the graphical representation of the hydraulic grade line through the plant. The elevations of treatment units and pipings are adjusted to give adequate hydraulic gradient to ensure gravity flow.

Many treatment plants have encountered serious operational problems and flooding of units due to inadequate hydraulic gradient through the various units. It is, therefore, important that the design engineer take proper care in developing the hydraulic profile. In this section, the basic principles of designing yard piping and of preparation of a hydraulic profile are presented. Furthermore, the calculation procedure and graphical representation of the hydraulic profile are given in the Design Example.

15.4.1 Yard Piping

The arrangement of channels, pipelines, and appurtenances is important in transmitting flows from one treatment unit to the other. Often, flows are collected from several treatment units operating in parallel or divided into several units. Hydraulically, similar inlet piping and channels, as well as splitting boxes, are normally used for flow splitting. Hydraulic balance of flow in different treatment units without the need for throttling valves should be achieved by proper unit layout and piping arrangement. There are three basic considerations in preparing the piping layout: (1) convenience of construction and operation, (2) accessibility for maintenance, and (3) ease with which future connections can be made or lines added.

Small distances between the adjacent treatment units can provide compact plant layout and minimize the costs of connecting piping; however, it should not be forgotten that the connecting piping and space between the treatment units are the primary means of expanding a plant expeditiously. Many inexperienced designers overlook this fact. Plants with tight piping and hydraulics have resulted in serious expansion problems; often pumping was needed to route the flows through the new treatment units. In a compact layout, the need for future expansion can be easily met by modular design. Space and piping for complete modular process trains must be designated.

The connecting channels and piping might be above or below the ground level; however, all pipings must be clearly marked with reference to the treatment units on the layout plans. Provisions for future tie-ins should be made and properly indicated on the plans. Valves or gates should be provided in the connecting piping to isolate or bypass units and modules for routine servicing and maintenance. Clearly marked schematic drawings with treatment units, connecting piping, and valves will assist the designers and operator to understand the flow routing and unit isolation, and, in general, to understand the operational capability of a well-designed plant.[2–4] In many plants, underground tunnels (called pipe galleries or operating galleries) are constructed to locate piping and necessary controls. Although such galleries provide both access to piping and controls and passage between the buildings, such an expenditure can be justified only for medium and large plants. In small plants, pipes are commonly located above or below the ground. Plans showing yard piping are very helpful in reaching proper pipes and controls for operation and maintenance needs.[4]

15.4.2 Plant Hydraulics

Hydraulic profile is the graphical representation of the hydraulic grade line through the treatment plant. If the high-water level in the rapid-mix basin is known or fixed, then this level is used for hydraulic computations in the direction of the flow. If the high-water level in the clearwell or the discharge weir elevation at the head of the clearwell can be fixed, any of these levels can be used as a control point, and the head loss computations are started backward through the plant. In some plants, the hydraulic calculations are started somewhere in the middle, by using an arbitrary elevation, such as the overflow weir at the filter or the overflow weir of the sedimentation basin. At the end, the elevations of water surface are adjusted in both directions.

The total head loss through a plant is the difference between water-surface elevation in the rapid mix and the low water surface elevation in the clearwater well. If the total available head is less than the head loss through the plant, flow by gravity cannot be achieved. In this case, pumping is needed to raise the head, so that flow by gravity can occur. Intermediate pumping in a plant is considered poor planning and design. It is an expensive option and is unwarranted. The engineer must coordinate the raw water station design and treatment plant site selection to assure that the available head is not a constraint. Design experience and good judgment are required to keep raw water pumping, high-service pumping, and plant hydraulics in perfect balance.

There are many basic principles that must be considered when preparing the hydraulic profile through a plant. Some of these principles are listed here:

1. The hydraulic profiles are prepared at maximum design and at minimum initial flows.
2. The hydraulic profile is generally prepared for all main paths of flow through the plant.
3. The total head loss through a treatment plant is the sum of head losses in the treatment units and in the connecting piping and appurtenances.
4. The head losses through treatment unit include the following:

    a. head losses at the influent structure;
    b. head losses at the effluent structure;
    c. head losses through the unit;
    d. miscellaneous and free-fall surface allowance.

    The largest head loss through a treatment unit might occur at maximum design flow when the largest unit is out of service. This situation is generally prevented by scheduling maintenance during low demand seasons.

    The approximate head losses across different treatment units might be as follows:

Flume for flow measurement	0.2–1 m
Rapid mix	0.5–1 m
Flocculation basin with diffusion wall	0.01–0.5 m
Sedimentation	0.5–1.5 m
Filter	
Constant-rate	3–5 m
Declining-rate	2–5 m
Chlorination facility	0.2–2.5 m

5. The total loss through the connecting piping, channels, and appurtenances is the sum of the following:

    a. head loss due to entrance;
    b. head loss due to exit;

c. head loss due to contraction and enlargement;
d. head loss due to friction;
e. head loss due to bends, fittings, gates, valves, and meters;
f. head required over weir and other hydraulic controls;
g. free-fall surface allowance;
h. head allowance for future expansions of the treatment facility.

6. The velocity in connecting pipings and conduits is kept large enough to keep the floc in suspension. A minimum velocity of 0.6 m/s at maximum design flow is considered adequate. At minimum initial flow, a velocity of 0.2 m/s is considered necessary to transport the floc. It is desirable to provide separate lines from multiple units, so that a line can be cleaned when a unit is out of service.
7. Friction losses in pressure conduits are obtained by using the Hazen–Williams formula. (Eqs. (7.14) through (7.20))
8. The minor head losses in open channels and conduits and pressure pipes are calculated in terms of the velocity head.[9,10] Detailed discussion can be found in Section 7.2.3.
9. In channels, the depth of flow varies with the flow conditions. Therefore, the depth and grade of open channel is kept in such a way that the water surface at the design flow corresponds to the water surface across the treatment unit.

   In open channels, the flow may be either uniform or nonuniform. Uniform flow occurs in channels that have constant cross-section, flow, and velocity. Manning's equation is generally used to calculate the grade of the water surface. In design of channels, it is generally assumed that the flow is uniform. Nonuniform flow exists in channels when the cross-section changes or when flow entering the channel is not constant. Nonuniform flow generally occurs in channels that have free-fall in effluent flumes or launders or in channel junctions with surcharge. Friction formula does not apply for nonuniform flows. Backwater or drawdown analysis is necessary. Computational techniques for nonuniform flows are given in Chapter 9. Readers are referred to some excellent textbooks on hydraulics of open channel for more detailed discussion on this subject.[9,10–12]

   In water treatment plants, sufficient allowances are made for the transitions and nonuniform flows by providing invert drops. Head loss through the transitions is generally calculated by using an energy equation. Liberal amounts of free-fall are recommended, to allow for natural aeration to occur. This may enhance the stripping of volatile compounds generally associated with taste and odor problems.
10. Most of the flow-measuring devices used in water treatment plants operate with head loss. Proper head loss calculations should be made for flow-measuring devices (Venturi tube, Parshall flume, orifice plate, weir, etc.) and included in the hydraulic profile.
11. In the preparation of a hydraulic profile, the vertical scale is intentionally distorted to show the treatment facilities and the elevation of the water surface. Ground surface is also indicated, to establish the optimum elevation of the plant structures and the hydraulic controls.

15.5 Information Checklist for Plant Layout and Preparation of Hydraulic Profile

Before the layout plan of the treatment plant and the hydraulic profile through the plant can be prepared, many important design decisions should be made and the necessary data developed. A checklist of needed information is presented next.

1. Site characteristics
   a. Land area
   b. Topographical maps
   c. Geology and geotechnical data
   d. Environmental constraints
      - Plant and animal communities
      - Ecosystem
      - Endangered or locally threatened species
      - Unique or vulnerable environmental features
      - Unique archeological, historical, scientific, or cultural areas, parks, wetlands, or stream corridors
2. Permit considerations
   a. Zoning and land use
   b. National Pollutant Discharge Elimination System (NPDES) permit
      - Waste discharge
      - Stormwater discharge
   c. Clean Water Act, Section 404
   d. River and Harbor Act, Section 10
   e. Clean Air Act
   f. Building, electrical, and fire codes
3. Support system availability
   a. Transportation
      - Roads
      - Railroads

b. Power
c. Communications
d. Sanitary sewer
e. Emergency services (police, hospital, fire department, etc.)

4. Hydraulic data and plant design information

a. Maximum, average, and minimum design year and initial year flows
b. Number, dimensions, and configurations of individual treatment units and buildings
c. Hydraulic losses through individual treatment units
d. Hydraulic gradient of the raw water line from the pump station to the treatment plant site
e. Hydraulic gradient of the pipeline connecting the high-service pump station to the distribution reservoirs

15.6 Design Example

The site development plan and hydraulic profile of the designed facility are presented in this section. The basic principles of plant layout and hydraulic profile discussed earlier in this chapter are utilized in the Design Example.

15.6.1 Design Criteria Used

1. Provide a compact modular plant layout.
2. Use the hydraulic profile and flow conditions for each treatment unit presented in various chapters (Chapters 6 through 14).

15.6.2 Description of the Compact Layout Plan Developed

The proposed plant site is a 12-ha (30-acre) tract on the western end of the town. A description of the proposed plant site can be found in Section 5.4.7. The relative location of the plant site with respect to headworks, raw water line, water treatment plant, finished water transmission lines, the water distribution system of Modeltown, and so on are shown in Figure 5-4.

After careful consideration of the site development factors, an integrated compact and modular layout for the treatment units and buildings is prepared. The available land for the treatment plant is about 20.25 ha (675 m long ∞ 300 m wide). The plant layout is shown in Figure 15-3. The following discussion highlights the features of the compact layout plan.

1. General arrangement of treatment units and buildings

   In accordance with the objectives of the proposed compact layout, most treatment units and buildings are located in the middle of the plant site. The overflow weirs, equipped with sluice gates, divide the flow equally into four independently operated treatment pro-

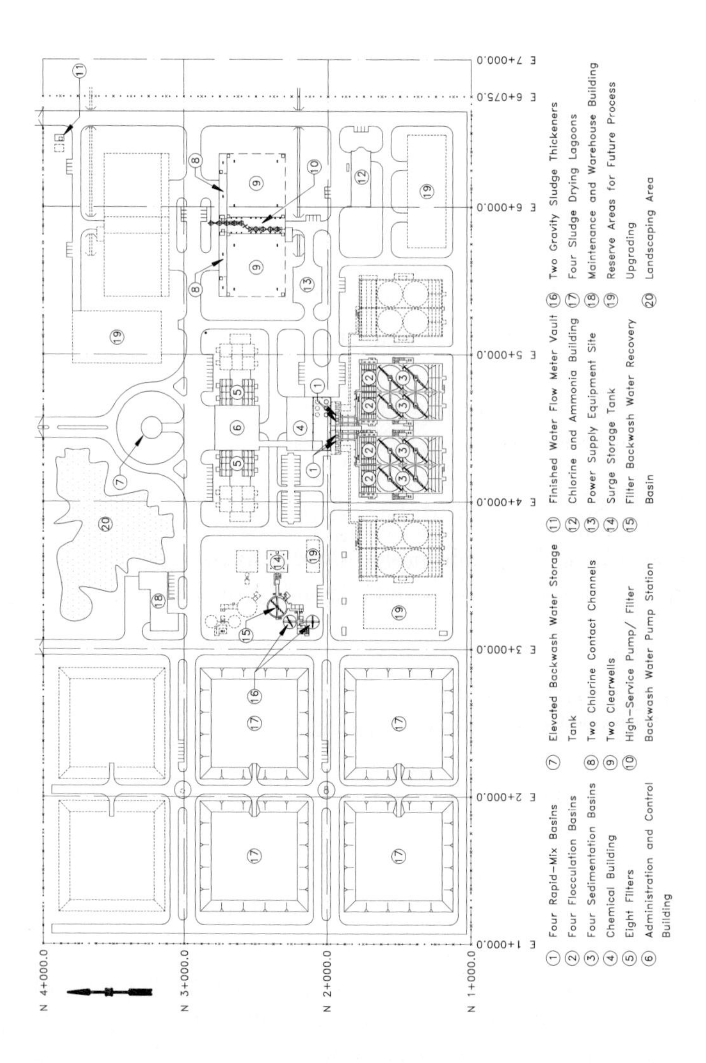

Figure 15-3 Treatment plant site plan and facility general layout.

cess trains. Each train, capable of treating one-fourth of the flow, consists of one rapid-mix basin, one flocculation basin, and one sedimentation basin. The chemical building is adjacent to the rapid-mix basins, so that the rapid mixers and flocculators will be clustered in the same area as the chemical feed equipment. The settled water is conveyed to the filter complex, which is located north of the chemical building and is integrated with the administration and control building. The filter complex includes eight filters (four on each side of the administration and control building) and two pipe galleries. An elevated filter backwash water storage tank is located on the north side of the administration and control building. The clearwater storage area is located on the east side of the filter complex and contains two chlorination channels, two underground clearwells, and the high-service/ filter backwash pump station. A finished water flow meter vault is located near the northeast corner of the plant.

The chlorine and ammonia building, which also houses the stability (Na_2CO_3) and fluoride (NaF) control chemicals and equipment, is located in the southeast part of the clear water storage area. A remote separate site is provided, because the storage and use of both chlorine and ammonia gases require special safety considerations.

The power-supply equipment site is provided southwest of the clear water storage area. This location is close to most high-power-demanding facilities, such as rapid-mix basins, chemical building, filter complex, administration and control building, and high-service/ filter-backwash pump station.

The residuals-management area is clustered together on the west side of the administration and control building. The major processing units are (1) a surge storage tank and recovery basin for waste filter backwash water recovery, (2) two sludge gravity thickeners, and (3) four lagoons for sludge dewatering and on-site ultimate disposal of residuals.

A maintenance and warehouse building is also provided on the north side of the residuals-management area. Sufficient space for the future expansion of the plant is considered in the development of site plan. (See dotted lines in Figure 15-3.) Four reserved areas are also recommended for the possibility of future process upgrading before coagulation, between sedimentation and filtration, after filtration, and for pretreatment of returned filter backwash water, respectively. These provisions can also be used as temporary storage areas for process and construction equipment and materials. A landscaping area is also provided.

2. Administration and Control Building

The ground level of the administration and control building has a visitors' lobby, reception station, operator's control center, conference room, and laboratory. The ground floor is also extended to a covered filter-control gallery on the east and west sides of the building. The filters are outside the control gallery. The upper floor of this building has offices, secretarial cubicles, library, production room, kitchen, and restrooms. A covered passage from the upper floor of the administration and control building leads to the chemical building (Figure 8-23). The lower level of the administration and control building has a filter-pipe gallery that connects the filters on the two sides of the building. The lower level has also the maintenance workshop and storage room. An opening is provided at each end of the filter-pipe gallery for convenient access and movement of valves, filtering, pumps, and other equipment. The general layout of the administration and control building is shown in Figure 15-4.

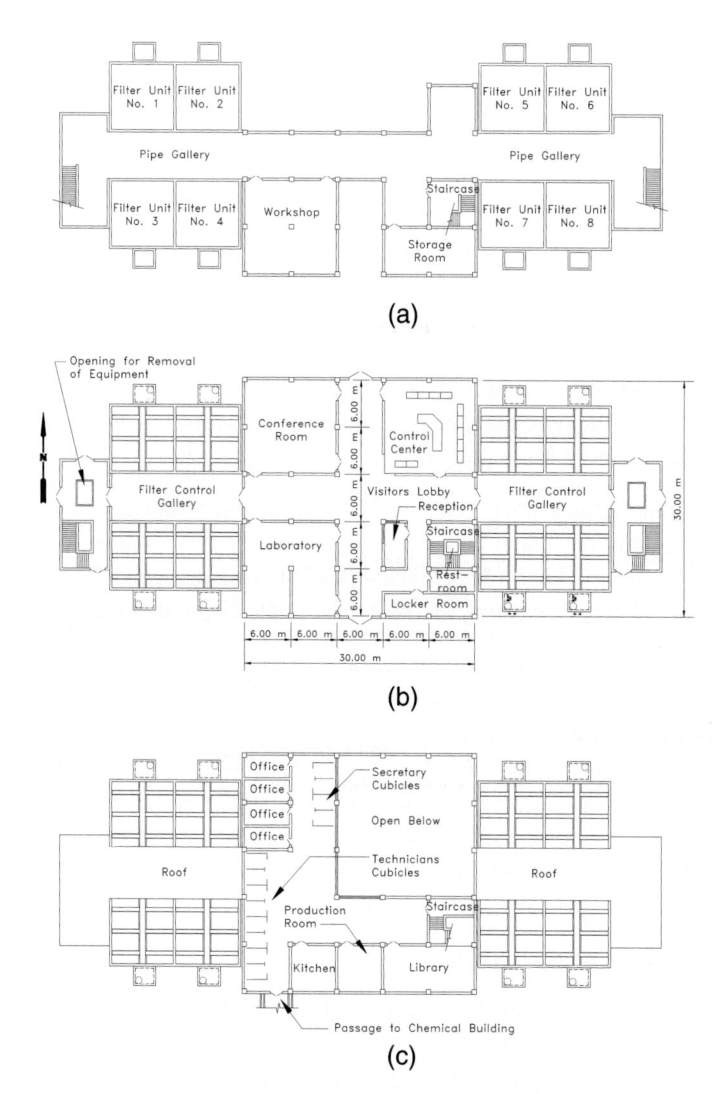

Figure 15-4 Layout plan of the Administration and Control Building. (a) Lower-level plan. (b) Ground-floor plan. (c) Upper-level plan.

3. Operation area accessibility

The filter-control and pipe galleries are directly accessible from the administration and control building. The chemical building is accessible through the covered passage. From the chemical building, the elevated walkways lead to the rapid-mix basins, Parshall flumes, flocculation basins, and sedimentation basins. Outside walkways and paved driveways lead to the clear water storage area, chlorine and ammonia buildings, electrical-equipment site, residuals-management area, and maintenance and warehouse building. All of these facilities are within a 150-m radius of the operations area.

4. Chemical handling

   The bulk chemicals for pre- and postcoagulation are housed in the chemical building. These chemicals include ferric sulfate, lime, polymers, and potassium permanganate. The ferric sulfate flows by gravity from the bulk storage tanks to the metering pumps on the ground floor. Hydrated lime storage silos are located on the roof of the building. Lime feeders and slurry tanks are located on an upper balcony, and gravity lines feed lime solution into lime-mixing chamber. The polymer systems, including drums, solution tanks, and metering pumps, are all located on the ground floor. The potassium permanganate solution tanks and feed pumps are also on the ground floor, the storage silos are located on the upper balcony. The bulk-chemical delivery is made from the covered unloading area adjacent to the building on the southeast side. Design of chemical storage and feed systems is provided in Chapter 8. Readers may refer to Section 8.8.2, Step B for design details. The chemical-feed system schematics are shown in Figure 8-22. A general layout of the chemical building is shown in Figure 8-23.

   All equipment (i.e., chlorinators, ammoniators, injectors, chemical solution tanks, and feed pumps) and other chemical storage (i.e., chlorine cylinders, back-up ammonia cylinders, stability (NA_2CO_3) and fluoride (NaF) control chemicals) are housed inside the chlorine and ammonia building. This building has docking facilities, with a monorail, for unloading and storage of chlorine and ammonia (back-up) cylinders. The 3600-kg anhydrous ammonia bulk-storage tank is located outside the building. The ammonia bulk delivery is made through an unloading station, by tank truck.

5. Clearwells and high-service pumps

   Two clearwells are located on the north side of the administration building. At this location, they will be close to the chlorine and ammonia building, from which the chlorine, ammonia, sodium carbonate, and sodium fluoride solutions can be fed directly into the chlorine contact and clearwell influent channels through short chemical delivery lines. The high-service pumps and filter backwash pumps are located between two clearwells, directly over the wet well. The high-service pipeline will exit from the plant property near the northeast corner of the plant.

6. Residuals-management area

   The residuals-management complex is on the west side of the administration building. Most processes, such as waste filter washwater surge storage tank and recovery basin, sludge thickeners, and several overflow and sludge pump stations, are located a fair distance from the administration and control building; thus, the high-maintenance area remains in relatively close proximity, to better serve operations needs. The sludge-dewatering and -disposal lagoons are located far west of the operation area.

7. Future expansion and upgrading

   A compact plant design generally poses difficulty to expansion. Modular expansion causes minimal disturbance to the earlier-constructed facilities. The future expansions are indicated by dotted lines over the layout plans. The filter-pipe gallery will serve as the connection between the expansions. Not only capacity expansions but provisions for new processes (in anticipation of stricter regulations in the future) are planned. In this design, three provisions are considered for future upgrading. The future ozone generation and application can be located in an area reserved for future expansion and upgrading. One location is adjacent to the future west flocculation/sedimentation complex. This way, the preozonation contact basins are hydraulically prior and close to the rapid-mix basins. The ozone-generation equipment will be also clustered in the same general area as the other equipment. Ozonation process is also one of the suitable alternatives if the disinfection of overflow from the filter backwash water recovery basin is required prior to its being returned back to Junction Box D on the raw water line.

   The facilities that cannot be expanded easily are constructed large enough to serve future needs. The administration and control building, chemical building, and chlorine and ammonia building are all large enough so that no expansion in the foreseeable future will be necessary.

8. Access gates, driveways, parkway, plant security, and landscaping

   Access to the water treatment plant is from the north side. The main gate is in front of the administration and control building. The chemical-delivery access gate is located near the northeast corner of the plant. Another gate, near the maintenance and warehouse building, is convenient for equipment delivery. This gate can also be used for delivery of construction materials for sludge lagoons. All gates are access-controlled, with an intercom system. In addition, a video camera is provided to monitor the activity at the gate. The entire property is fenced.

   The administration and control building is located in the front, with all treatment units on the sides and behind. Visitor parking is provided in the front, employee parking in the back. The water tower in front of the administration and control building is decorated with City's logo, to highlight the features of the plant. Paved service roads pass each treatment facility. The chemical building has roads on three sides. A main driveway from the northeast gate through the chlorine and ammonia building to the chemical building is designed for heavy vehicles, for delivery of chemicals.

   The ground surface is drained away from the main structures, which are located on a ridge. The entire site is sloped from the middle to the northwest and northeast corners of the plant. The site plan, with finished contours, is presented in Figure 15-5. As much as possible, the developed site does not alter the existing, naturally stabilized, site contour and drainage pattern. The landscaping reflects the rolling character of the surrounding area. The entire area is sodded to reduce erosion. Existing and native vegetation, trees, and shrubs are utilized throughout most of the area, except in a landscaping area located in front of the administration and control building and west side of the main gate.

15.6.3 Plant Yard Piping and Hydraulic Profile

Based on above site plan, the plant yard piping was developed. The locations of the proposed treatment units and major process pipes connecting these facilities are shown in Figures 15-6 through 15-8. The hydraulic profiles through each treatment unit have been developed individually in various chapters. Readers may refer to the following chapters for important information that is used for preparation of the hydraulic profile through the plant.

- Chapter 8 contains the hydraulic profile from rapid-mix basin to flocculation basin.
- Chapter 9 contains the hydraulic profile through the sedimentation basin.
- Chapter 10 contains the hydraulic profile from the filter unit to the filter back pressure weir common bay.
- Chapter 12 contains the hydraulic profile from the chlorine contact channel to the clearwell.

Additional information on hydraulic analysis from the raw water source to the distribution system can be found in Chapter 6 (raw water intake structure to raw water pump station), Chapter 7 (raw water pump station and raw water transmission pipeline), and Chapters 12 and 13 (high-service pump station, distribution pump stations, ground and elevated storage reservoirs, and distribution piping networks). Thus, Chapters 6 through 13 (except Chapter 11) are closely tied together and provide in-depth information on hydraulic calculations. The task of this section is simply (1) to finish hydraulic calculations for the connecting pipes that are not covered in the previous chapters, (2) to summarize the hydraulic-analysis information for all treatment units and connecting pipes, and (3) to prepare a complete hydraulic profile through the plant. The main purpose of this work is to put all hydraulic details into an overall perspective, for readers to see and realize the intricate connection between all components in this Design Example.

1. Hydraulic calculation of the head losses through the process pipes between the treatment units

   An integrated compact and modular layout for the treatment units and buildings is used to develop the site plan. As a result, the process piping between the treatment units is significantly minimized. The calculations of head loss through process piping need to be conducted for only two segments. The first one is the raw water influent pipe between Junction Box A and the influent forebay (prior to the rapid-mix basins). Several junction boxes are connected by this pipeline. The second piping carries settled water from sedimentation basin to the filter units. The hydraulic head losses through these two piping systems are calculated separately. The calculation procedures and results are summarized in Tables 15-2 and 15-3, respectively.

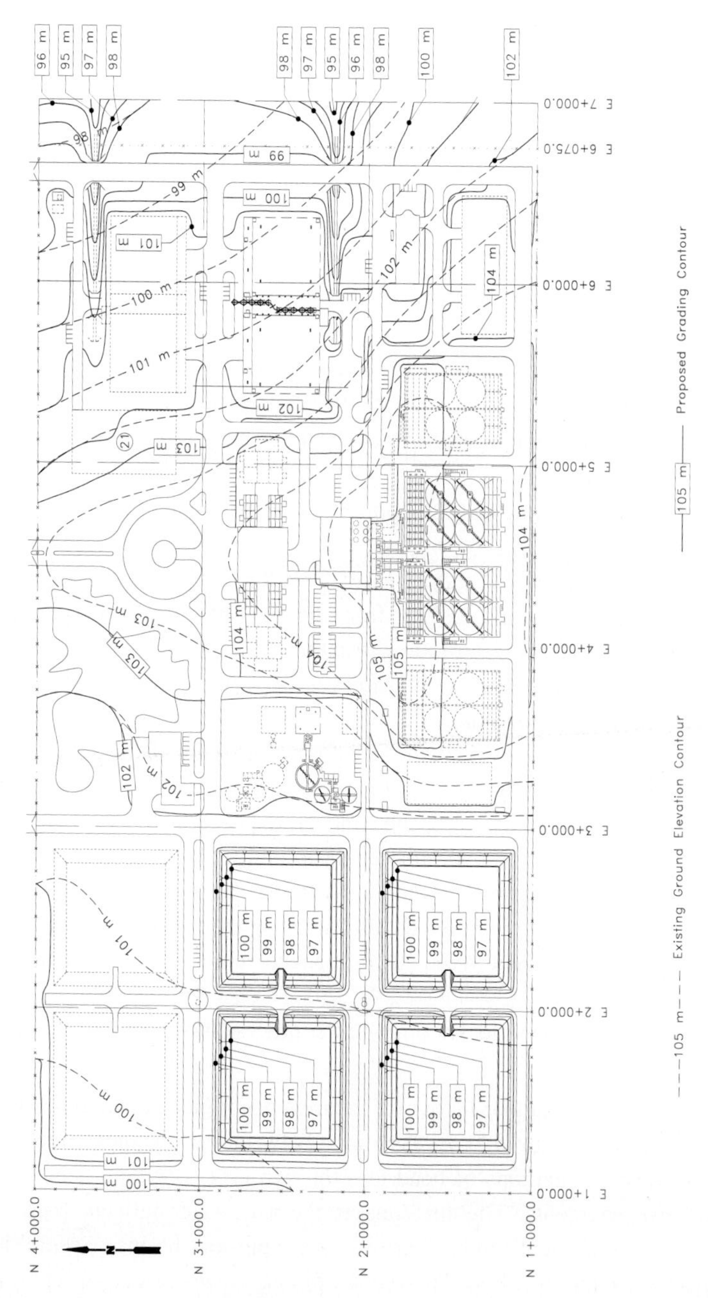

Figure 15-5 Site plan, with finished contours.

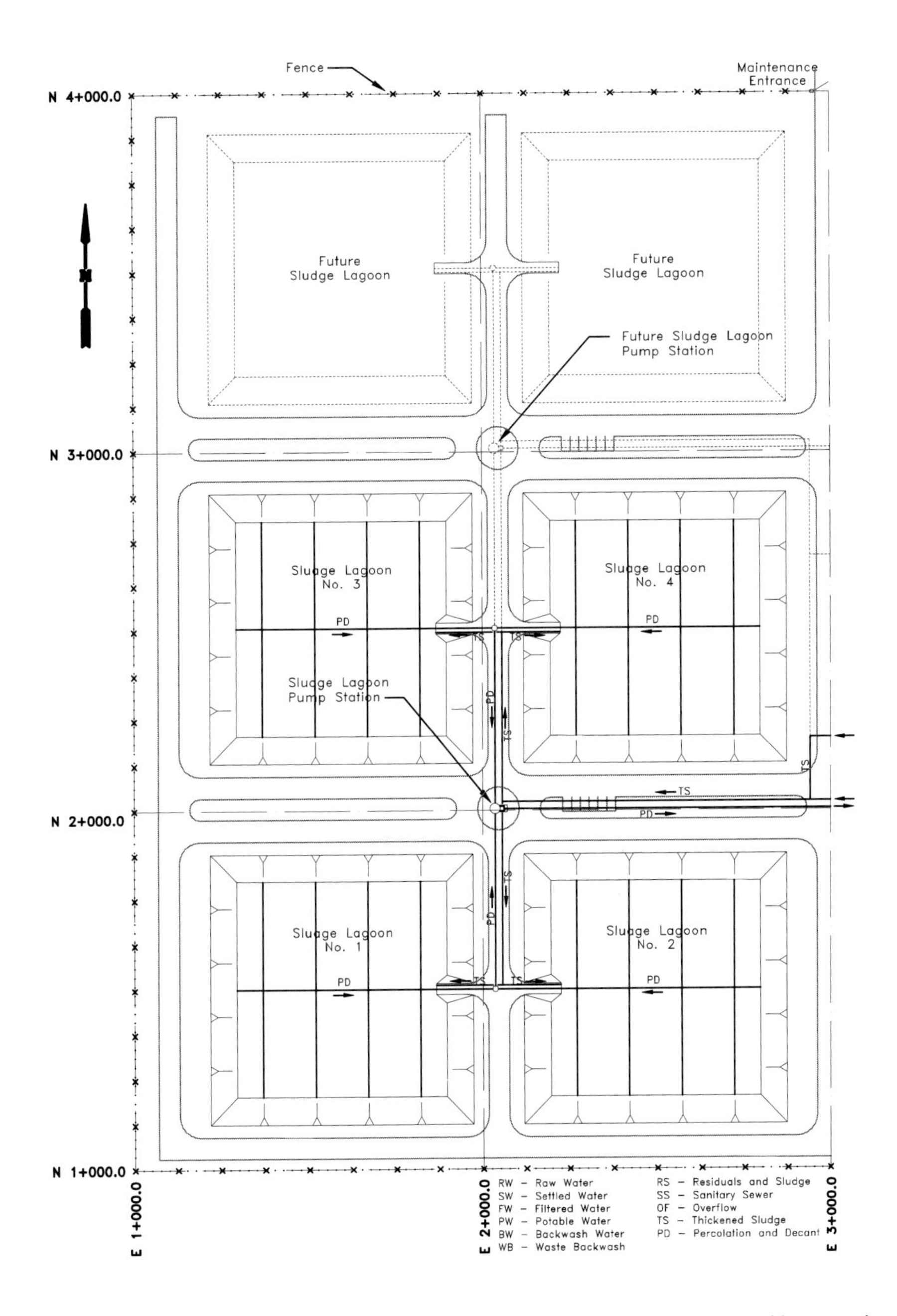

Figure 15-6 Treatment plant process piping layout–western part of the plant (the area between vertical coordinate lines E1+000.0 and E3+000.0).

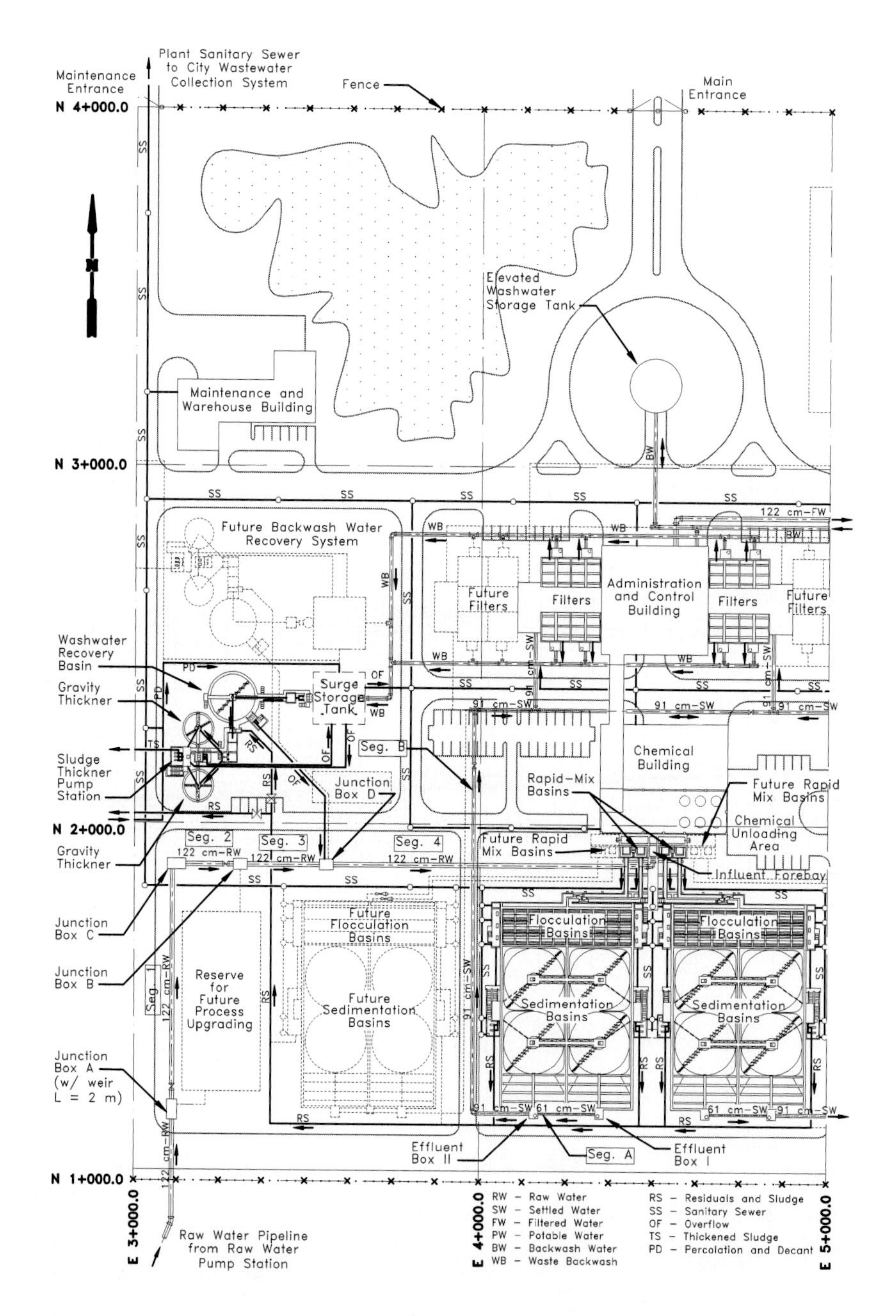

Figure 15-7 Treatment plant process piping layout–middle part of the plant (the area between vertical coordinate lines E3+000.0 and E5+000.0).

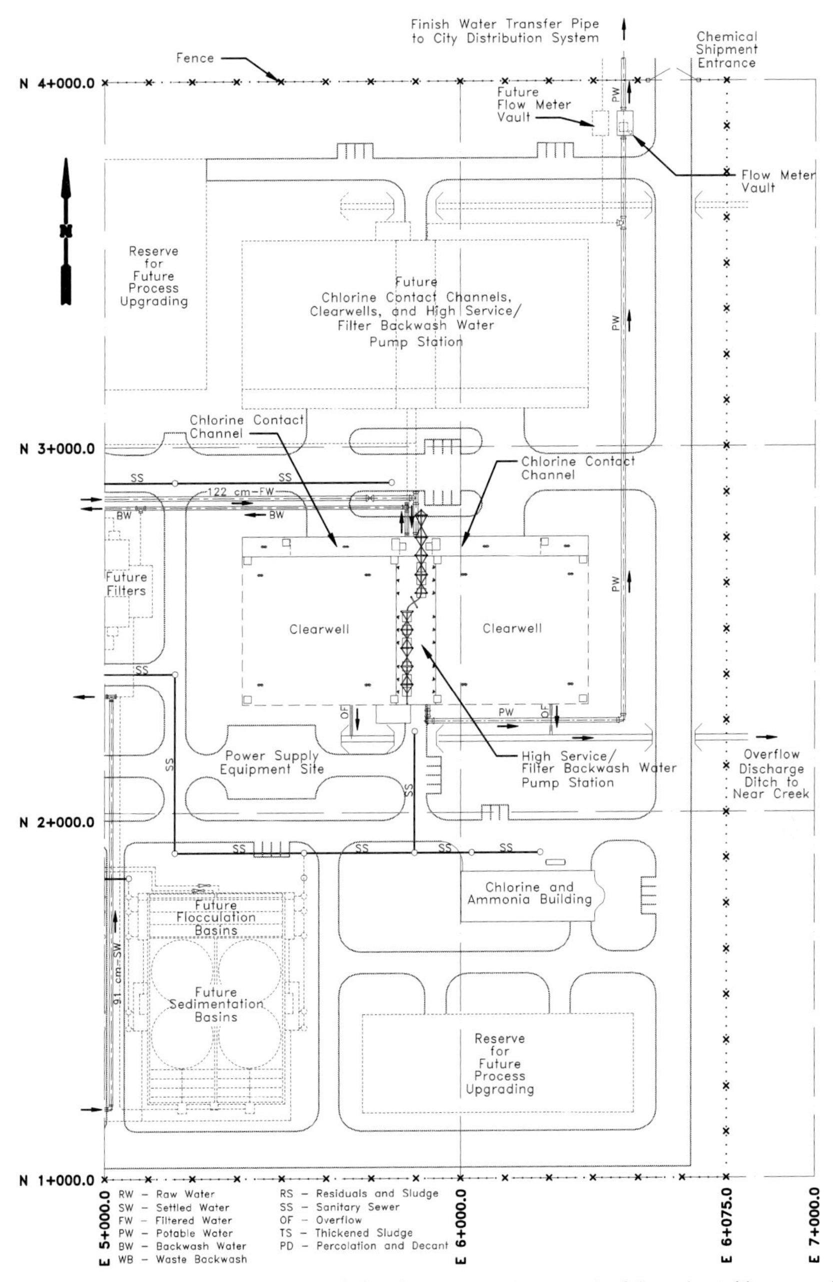

Figure 15-8 Treatment plant process piping layout–eastern part of the plant (the area between vertical coordinate lines E5+000.0 and E7+000.0).

Table 15-2 Calculations of Head Losses through the Raw Water Pipeline inside the Water Treatment Plant[a]

Appurtenance	Equation Used	Flow Rate, m^3/s	Size, m	Length, m	Velocity, m/s	Head Loss, m
Free fall over weir in Junction Box A	Eq. (8-24), $C_d = 0.6$ and $n = 0$	1.314[b]	$L = 2.00$	N/A	N/A	0.72[c]
Entrance at Junction Box A	Eq. (7.12), $K = 0.5$	1.314	1.22	N/A	1.12	0.032
Butterfly valve in Seg. 1	Eq. (7.12), $K = 1.2$	1.314	1.22	N/A	1.12	0.077
Seg. 1	Eq. (7.14), $C = 120$	1.314	1.22	70	1.12	0.067
Exit at Junction Box B	Eq. (7.12), $K = 1.0$	1.314	1.22	N/A	1.12	0.064
Entrance at Junction Box B	Eq. (7.12), $K = 0.5$	1.314	1.22	N/A	1.12	0.032
Butterfly valve in Seg. 2	Eq. (7.12), $K = 1.2$	1.314	1.22	N/A	1.12	0.077
Seg. 2	Eq. (7.14), $C = 120$	1.314	1.22	20	1.12	0.019
Exit at Junction Box C	Eq. (7.12), $K = 1.0$	1.314	1.22	N/A	1.12	0.064
Entrance at Junction Box C	Eq. (7.12), $K = 0.5$	1.314	1.22	N/A	1.12	0.032
Seg. 3	Eq. (7.14), $C = 120$	1.314	1.22	25	1.12	0.024
Exit at Junction Box D	Eq. (7.12), $K = 1.0$	1.314	1.22	N/A	1.12	0.064
Entrance at Junction Box D	Eq. (7.12), $K = 0.5$	1.314	1.22	N/A	1.12	0.032
90° elbow in Seg. 4	Eq. (7.12), $K = 0.3$	1.314	1.22	N/A	1.12	0.019
Seg. 4	Eq. (7.14), $C = 120$	1.314	1.22	100	1.12	0.096
Exit at Influent Forebay	Eq. (7.12), $K = 1.0$	1.314	1.22	N/A	1.12	0.064
					Total	1.48

a The calculations of head losses are from Junction Box A to the end of Segment 4 at Influent Forebay. (See Figure 15-7.).

b Flow in the plant influent pipe from the raw water pump station $= \dfrac{113{,}500\ m^3/d}{86{,}400\ s/d} = 1.314\ m^3/s$.

c The head over the weir $= \left(\dfrac{1.314\ m^3/s \times 3/2}{0.6 \times 2\ m \times \sqrt{(2 \times 81\ m/s^2)}}\right)^{2/3} = 0.52$ m. A free fall of 0.73 m is provided.

Table 15-3 Calculations of Head Losses through the Piping from the Sedimentation Basin to the Filter Unit[a]

Appurtenance	Equation Used	Flow Rate, m^3/s	Size, m	Length, m	Velocity, m/s	Head Loss, m
Entrance at Effluent Box I	Eq. (7.12), $K = 0.5$	0.328[b]	0.61	N/A	1.12	0.032
Seg. A (from Effluent Box I to Effluent Box II)	Eq. (7.14), $C = 120$	0.328	0.61	20	1.12	0.043
Exit at Effluent Box II	Eq. (7.12), $K = 1.0$	0.328	0.61	N/A	1.12	0.064
Entrance at Effluent Box II	Eq. (7.12), $K = 0.5$	0.657[c]	0.91	N/A	1.01	0.026
Seg. B (from Effluent Box II to Point A at the beginning of filter influent header)	Eq. (7.14), $C = 120$	0.657	0.91	180	1.01	0.199
Four 90° turns in Seg. B	Eq. (7.12), $K = 4 \infty 0.3 = 1.2$	0.657	0.91	N/A	1.01	0.062
Tee (branch-to-run) in Seg. B	Eq. (7.12), $K = 1.8$	0.657	0.91	N/A	1.01	0.094
Butterfly valves in Seg. B	Eq. (7.12), $K = 1.2$	0.657	0.91	N/A	1.01	0.062
Extra minor loss in Seg. B	Eq. (7.12), $K = 1.2$	0.657	0.91	N/A	1.01	0.062
					Total	0.64

a Location of Effluent Boxes I and II is shown in the layout of the sedimentation basins (Figure 9-14). The location of Point A at the beginning of the filter influent header is shown in the layout of the filter influent piping (Figure 10-21). The calculations of head losses are from Effluent Box I to Point A. (See Figure 15-7.).

b Flow from one sedimenation basin $= \dfrac{113{,}500\ m^3/d}{4 \times 86{,}400\ s/d} = 0.328\ m^3/s.$

c Flow from two sedimentation basin $= \dfrac{113{,}500\ m^3/d}{2 \times 86{,}400\ s/d} = 0.657\ m^3/s$

2. Summary of head loss information from individual treatment unit design

   The results of the hydraulic calculations for all the treatment units and connecting pipes are summarized in Table 15-4. The total head loss through the entire treatment plant is 10.50 m, from the Junction Box A (where the raw water from the raw water pump station is received) to the clearwell (where the finished water is stored for delivery to the City's distribution system by the high-service pumps). Out of this head loss, the largest head loss allowance is for the filters. This value is 4.63 m, which includes a 2.5-m terminal head loss allowance through dirty media. Therefore, the total head loss through the filters accounts for approximately 44 percent of the total head loss through the entire treatment plant. In addition to the liberal head loss allowance at the filters, liberal free-fall allowances are provided at the influent pipeline (Junction Box A), the influent and effluent structures of the rapid-mix basins, the influent structure of the flocculation basins, the effluent structure of the sedimentation basins, and the effluent weir of the chlorine contact channel. Such allowances become valuable tools during future expansions. The authors emphasize the importance of these considerations, which should be carefully evaluated during the initial design stages.

3. Preparation of hydraulic profile

   The hydraulic profile through the plant is prepared from a control point. In this Design Example, there are four control points. These are (1) effluent weir at the rapid mix, (2) effluent V-notch weir at the sedimentation basin, (3) filter back pressure weir at the common filter bay, and (4) weir at the chlorine contact channel, just above the clearwells. The filter back pressure weir controls the water surface elevations into the filters. Customarily, engineers have used this elevation to start hydraulic calculations. In the author's judgment, selection of a control point for start of hydraulic calculations becomes insignificant if computer programs are utilized. Calculations can be stated from any point, or from several points simultaneously. The elevations of these control points can be adjusted at any time, if necessary, because since they all use the same datum. It is therefore not required to convert elevation numbers between different datums. In the design of a new plant, the effluent weir at the clearwell has a great significance, because it is a bridge between water production (treatment units) and water distribution (high-service pumping). In an older plant that is being expanded, other reference control points could be of more significance, because the new system must be compatible with the old one. A liberal allowance for free fall at all control points must be provided, and the authors emphasize this, because extra allowances become very valuable resources when the plant is expanded in the future. Hydraulic calculations are systematically conducted to determine the head losses and prepare the hydraulic profile for various treatment units. The hydraulic calculation results presented above and in Chapters 8 through 13 are used to develop the hydraulic profile. The hydraulic profile is shown in Figures 15-9 and 15-10. Readers may refer to each chapter for the detailed procedures to obtain the important water surface and major structural elevations.

Table 15-4 Summary of Hydraulic Calculation Results for All Treatment Units

Treatment Process		Hydraulic Analysis			References	
Facility	Location	Start Point	End Point	Head Loss, m	Sec. No.	Figure
Raw water influent pipeline		Junction Box A	Influent forebay	1.48	15.6.3, Table 15-2	–
Rapid-mix basin	Influent structure	Influent forebay	Rapid-mix basin	1.17	8.8.2, Step A.5.a	8-15
	Effluent structure	Rapid-mix basin	Entry of Parshall flume	0.50	8.8.2, Step A.5.b	8-15
Flocculation basin	Parshall flume	Entry of Parshall flume	After free fall at Parshall flume	0.25	8.8.2, Step B.5.a	8-17
	Influent channel	After free fall at Parshall flume	First stage	0.23	8.8.2, Step B.5.b	8-21
	Flocculation basin	First stage	Third stage	0.00	8.8.2, Step B.5.c	8-21
	Effluent structure	Third stage	After diffusion wall	0.00	8.8.2, Step B.5.d	8-21
Sedimentation basin	Sedimentation basin	After diffusion wall	Before V-Notch weir	0.00	9.6.2, Step A.5	9-20
	Effluent structure	Before V-Notch weir	After free fall at Effluent Box I	1.11	9.6.2, Step A.5.a to d	9-20
Settled water conduit		After free fall at Effluent Box I	Beginning of filter influent header	0.64	15.6.3, Table 15-3	–
Filter	Filter system	Beginning of filter influent header	After free fall at common bay	4.63	10.13.2, Step A.5.a to h	10-23
Chlorine contact channel/clearwell		After free fall at common bay	Clearwell	0.49	12.8.2, Step A.4.a to d	12-22
Entire treatment plant		Junction Box A	Clearwell	10.50	15.6.3	15-9 & 15-10

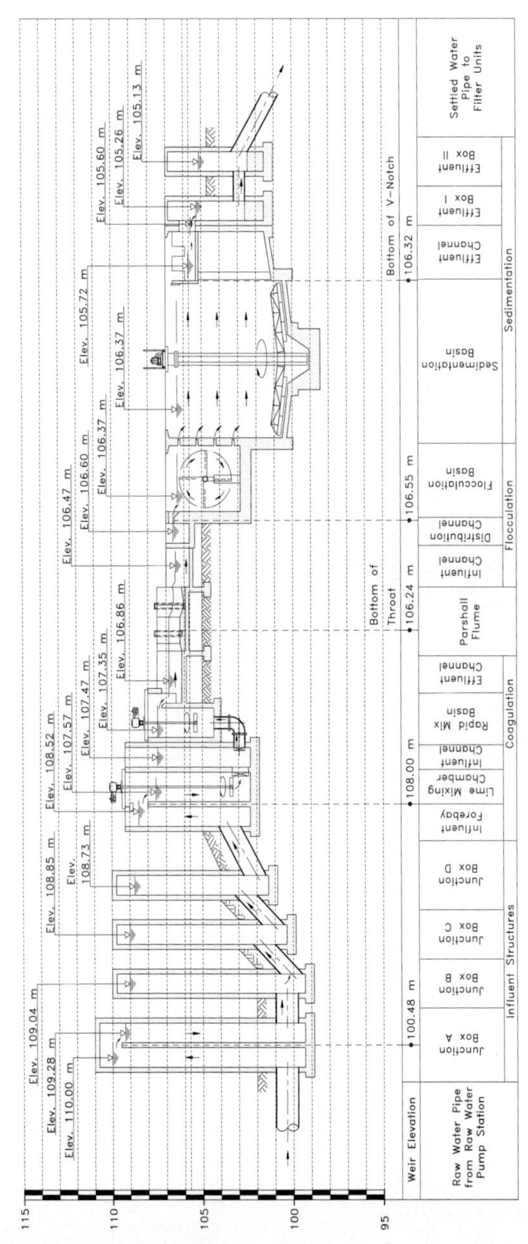

Figure 15-9 The hydraulic profile through the treatment plant. (Showing Junction Boxes A through D, the influent forebay, the lime-mixing chamber, the rapid-mix basin, the Parshall flume, the flocculation basin, and the sedimentation basin.)

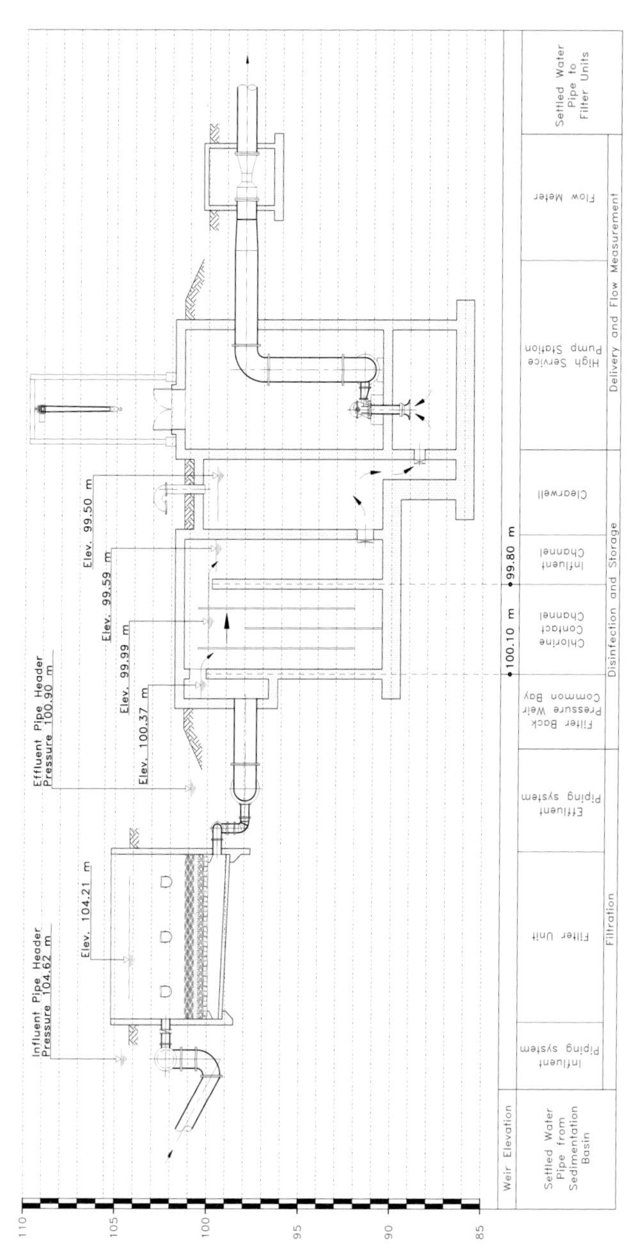

Figure 15-10 The hydraulic profile through the treatment plant. (Showing the filter unit, the filter back pressure weir common bay, the chlorine contact channel, the clearwell, the high-service pump station and the flow meter.)

The hydraulic-analysis procedures for the residuals-management system are the same as those used to calculate the head losses through the water treatment units. The hydraulic profile through the residuals-management system is therefore not presented in this Design Example. Readers could prepare a hydraulic profile for the residuals-management system by using this procedure.

Problems and Discussion Topics

15.1 As a design engineer, you have the responsibility to develop the plant layout, site plan, landscaping, and finished contours of water treatment facility. List the basic rules that you must consider in developing a compact-site plan.

15.2 Study the site plan given in figure 15-3. List the basic features that make the site plan a compact layout.

15.3 Review the plant grading plan in figure 15-5. Comment on the location of the main structure and the drainage pattern.

15.4 As a design engineer, you are required to prepare the piping and hydraulic profile of a water treatment plant. Discuss various design considerations that are necessary in developing the piping layout and hydraulic profile through the treatment plant.

15.5 The crest elevation of the effluent weir in a junction box is 61.00 m. The water flows over the weir and drops into an outlet box. The water surface elevation in the outlet box is 0.8 m below the weir crest. A 20-cm pipe connects the outlet box and the influent well of the circular sedimentation basin. The pipe is an inverted syphon and has two 45° and two 90° bends and a straight length of 68 m. Assume that K for 45° and 90° bends, for entrance and exit conditions, is 0.2, 0.3, 0.15, and 1.0, respectively; $C = 110$. Determine the water surface elevation and the total number of 90° V-notches in the sedimentation basin. Each V-notch is 15 cm deep and has a freeboard of 10 cm at the design flow of 0.015 m^3/s.

15.6 A water treatment plant is designed for a flow of 0.02 m^3/s. The water flows through a junction box that has a water surface elevation of 100.00 m. The head loss in the rapid-mix basin, the flocculation basin, the sedimentation basin, and the filter are 0.70, 0.50, 0.80, and 3.20 m, respectively. Assume that all units and the filter back pressure weir are directly connected by an inverted syphon with two 90° elbows in the connecting pipes between the units. The diameter, straight length, and C of the connecting pipes are 20 cm, 50 m, and 110, respectively. What will be the water surface elevation at the filter back pressure weir at the common bay? Also, draw the hydraulic profile through the plant. Assume that the outfall pipe into the common bay has a submerged discharge.

15.7 Obtain the hydraulic profile of the water treatment plant in your community. Tabulate the following items:
  (a) head losses in each treatment unit;
  (b) head losses in the junction boxes;
  (c) head losses in the connecting piping.

References

1. Canter, L. W. *Environmental Impact Assessment*, 2d ed., McGraw-Hill, Inc., New York, 1996.
2. Kawamura, S. *Integrated Design of Water Treatment Facilities*, John Wiley and Sons, Inc., New York, 1991.
3. ASCE and AWWA. *Water Treatment Plant Design*, 2d ed., McGraw-Hill Book Co., New York, 1990.
4. James M. Montgomery, Inc. *Water Treatment Principles and Design*, John Wiley & Sons, New York, 1985.

5. Sanks, R. L. *Water Treatment Plant Design*, Ann Arbor Science Publishers Inc., Ann Arbor, MI, 1980.
6. Qasim, S. R. *Wastewater Treatment Plants: Planning, Design, and Operation*, 2d ed., Technomic Publishing Co., Inc., Lancaster, PA, 1999.
7. Corbin, D. J., Monk, R. D. G., Hoffman, C. J., and Crumb, S. F. Jr. "Compact Treatment Plant Layout," *Jour. AWWA*, vol. 84, no. 8, pp. 36–42, August 1992.
8. International Conference of Building Officials and Western Fire Chiefs Association. *Uniform Fire Code*, Los Angeles, CA, 1991.
9. Chow, V. T. *Open Channel Hydraulics*, McGraw-Hill Book Co., New York, 1959.
10. Benefield, L. D., Judkins, J. F., and A. D. Parr, *Treatment Plant Hydraulics for Environmental Engineers*, Prentice-Hall, Inc., Englewood Cliffs, NJ, 1984.
11. Brater, E. F. and King, H. W. *Handbook of Hydraulics*, 6th ed., McGraw-Hill Co., New York, 1976.
12. French, R. H. *Open-Channel Hydraulics*, McGraw-Hill Inc., New York, 1985.

CHAPTER 16

Process Control

16.1 Introduction

In the past fifty years, advances in electronics have made possible the development of sophisticated instrumentation and computers. Consequently, many large water treatment plants utilize very sophisticated process instrumentation and controls. The plant operators are required to operate and maintain them. This trend is not surprising, because proper use of instrumentation and automatic controls can reduce labor, chemicals, and energy consumption and improve treatment process efficiency, reliability, and data acquisition and recording.

The subject of instrumentation and control is very complex, requiring an extensive background for those involved in design, analysis, and selection of instrumentation of water treatment facilities. The purpose of this chapter is to present process-control strategies, physical and chemical measuring elements and instruments, computers, and final control elements that are integral parts of the process-control systems for a water treatment facility. The type of instrumentation and control devices (selected and simplified control-loop diagrams for various treatment processes in the Design Example) are also presented in this chapter.

16.2 Benefits and Need of Instrumentation and Control Systems

The general motives for installing a sophisticated instrumentation and control system in a water treatment are numerous. The benefits may be in process improvement, equipment performance, and convenience to personnel. Many specific benefits are summarized in Table 16-1.[1,2]

Many factors are considered that influence the degree of process-control sophistication needed at a facility.[3,4] These include (a) size of the plant and future expansion needs, (b) complexity of the treatment processes and treatability aspects, (c) number of control points, (d) interdependence of process elements and controls, including process time delays, (e) number and

Table 16-1 Benefits of an Instrumentation and Control System in Water Treatment[1,2]

Purpose	Benefits
Process	• Improve process performance and process results
	• Efficient use of energy
	• Efficient use of chemicals
	• Process changes detected in a timely manner
	• Automatic execution of corrective measures
	• Greater ability to control complex processes
Equipment	• Immediate alert signal of malfunction
	• Ability to diagnose problems in remotely located equipment before malfunction occurs
	• Status known at all times
	• Automatic execution of corrective measures, and automatic response to potentially disastrous situations
	• Automatic shutdown to prevent major damage
	• Increase in running time
Personnel	• Timely and accurate process information
	• Safer operation
	• Efficient use of labor
	• Capability to quickly solve analytical problems
	• Minimize the potential for human error
	• Automated overview of plant operation
	• Decrease in manual paperwork
	• More complete records that can allow an overview of plant operation and plant behavior, and for design of future expansion
	• Increased security

capability of operating personnel available, (f) availability of qualified maintenance personnel, either owner-furnished or contracted from outside companies, and (g) useful life and process obsolescence due to rapid technological advances and unavailability of replacement parts. These factors simplify the selection of appropriate methods of control, data acquisition, and data display, and can systems may be selected.

16.3 Components of Instrumentation and Control Systems

A process-control system utilizes (1) process variables and (2) associated controls.[5] The process variables are measured by the sensors of the control systems. The measured value is transmitted, displayed, or recorded, to guide the operator in making the proper corrective process adjustments. In other automatic control systems, the process-variable signals compare the value of the variable against a preset reference and then the controller implements the selected actions for the desired results. The control variables and the associated controls are discussed next.

16.3.1 Control Variables

In a water treatment plant, many variables are measured and controlled. These variables fall into three categories: physical, chemical, and biological. Common examples of these variables are listed below.

Physical:	flow, pressure, level, temperature
Chemical:	pH, turbidity, specific conductance, dissolved oxygen
Biological:	oxygen consumption rate, TOC reduction rate, nitrate reduction rate, ammonia oxidation rate

16.3.2 Associated Controls

The automatic control system is made up of five components[6,7]:

1. the measurement section, to detect the changes in the variables;
2. signal-transmitting devices;
3. data-display readouts;
4. control systems;
5. computer and central control room.

Each of these components is discussed here.

Measurement Section or Sensing Devices The sensing devices include instruments that sense, measure, or compute the process variables. The sensing device may be on line or off line, continuous or intermittent. Common on-line process-measurement devices and their applications are summarized in Table 16-2.

Signal-Transmitting Devices The purpose of a signal-transmitting device is to transmit a process-variable signal from the sensing instrument to the readout device or a controller. The transmission may be accomplished mechanically, pneumatically, or electrically. Each type is discussed here.[1,5]

Mechanical Transmission The mechanical transmission is done by drive of a pen or indicator or by a float or cable. This method is generally limited to on-site display or location of controller.

Pneumatic Transmission The pneumatic transmission system consists of a detector and an amplifier. The detector is a flapper-nozzle unit. Regulated input air is fed to the nozzle through a reducing tube. A pressure change is transmitted as the flapper is moved away from or toward the nozzle. Small changes in back pressure at the nozzle become proportional to the movement of the flapper. The pressure change is amplified and transmitted to the receiver or controller.

The advantages of pneumatic transmission over electric transmission include absence of electrical hazard, lower effect from temperature and humidity, and lack of freezing problem. Also, they are reliable, less complicated, and easy to maintain. The limitations of pneumatic transmissions, however, are the signal lag in long tubing, the relatively short distance necessary between the detector and the controller (up to 300 m), the requirement of clean dry air, and air leakage problems. Detailed discussion on pneumatic transmission may be found in References 10 and 11.

Electronic and Radio/Microwave Transmission The electronic transmission of signals is achieved by voltage and current, pulse duration, or tone. In voltage and current transmission, the signals are transmitted by milliamp direct current (ma DC) or by voltage signals. In pulse- duration or time-pulse transmission, the length of time a voltage is transmitted is in proportion to the measured data. In tone transmission, standard telephone lines are normally used to allow signals to be transmitted. The signals are transmitted by turning the transmitter on and off, or by slightly changing the frequency or pitch of the electronic tone.[7]

Recent advances in reliable radio and microwave equipment have encouraged the use of radio/microwave transmission, where signals are transmitted over an assigned frequency per the Federal Communications Commission (FCC) regulations. The radio/microwave transmission is particularly suited where the data-gathering points are scattered over a large area and where telephone lines are not available or are too expensive. Although radio/microwave transmission is expensive at present, its use will continue to grow, particularly for large systems.

The electronics and radio/microwave control systems are rapidly gaining popularity for a number of reasons. The basic advantages are: (1) the electronic signals operate over great distances without contributing time lags, (2) the electronic signals can easily be made compatible with a digital computer, (3) the electronic units can easily handle multiple signal inputs, (4) intrinsic safety techniques have virtually eliminated electrical hazards, and (5) electronic devices can take less space, are less expensive to install, can handle almost all process instruments, and can be essentially maintenance-free.[5,10]

Data-Display Readout The transmitted information is displayed at a convenient location in a manner usable by the operating personnel. The most common types of readout device are indicators, recorders, and totalizers. These devices can be driven mechanically from the sensing devices or from the pneumatic or electronic signals. The receiving instruments might display the movement, pressure charge, or current change signals directly or be servo-operated.[7]

Control Systems Time-control systems used in water works engineering can be divided into three categories: digital, analog, and automatic control. Each of these systems is described here.[1–3]

Table 16-2 Common On-Line Process-Measurement Devices and Their Applications in Water Treatment

Measured Variables	Primary Device	Measured Signal	Common Application
Flow[a]	Venturi meter	Differential pressure	Gas, liquids
	Flow nozzle meter	Differential pressure	Gas, liquids
	Orifice meter	Differential pressure	Gas, liquids
	Electromagnetic meter	Magnetic field and voltage change	Liquids, sludges
	Turbine meter	Propeller rotation	Clean liquid
	Acoustic meter	Sound waves	Liquids, sludges
	Parshall flume	Differential elevation of water surface	Liquids
	Palmer-Bowlus flume	Differential elevation of water surface	Liquids
	Weir	Head over weir	Liquids
Pressure	Liquid-to-air diaphragm	Balance pressure across a metal diaphragm	Pressure, 0–200 kN/m^2
	Strain gauge	Dimensional change in sensor	Pressure, 0–350,000 kN/m^2
	Bellows	Displacement of mechanical linkage connected to the indicator	Pressure, 0–20,000 kN/m^2
	Bourdon tube	Uncurling motion of noncircular cross-sectional area of a curved tube	Pressure 0–35,000 kN/m^2
Liquid level	Float	Movement of a float riding on surface of liquid	Liquid head, 0–11 m
	Bubbler tube	Measurement of back pressure in a tube bubbling regulated air at slightly higher pressure than the static head outside	Liquid head, 0–56 m
	Diaphragm bulb	Pressure change in air on one side of the diaphragm caused by the liquid pressure on the other side of the diaphragm	Liquid head, 0–15 m
Sludge level	Photocell	Detection of light in a probe by a photocell across sludge blanket	Sedimentation Gravity thickener
	Ultrasonics	Detection of the ultrasonic signal transmitted between two transducers	Sedimentation Gravity thickener
Temperature	Thermocouple	Current flow in a circuit made of two different metals	Water lines, hot-water boiler
	Thermal bulb	Absolute pressure of a confined gas, proportional to the absolute temperature	Water lines

Table 16-2 Common On-Line Process-Measurement Devices and Their Applications in Water Treatment (continued)

Measured Variables	Primary Device	Measured Signal	Common Application
	Resistance temperature detector	Change in electrical resistance of temperature-sensitive element	Bearing and winding temperatures of electrical machinery, hot-water boilers
Speed	Tachometer (generator or drag-cup type)	Voltage, current	Variable-speed pump, blower, or mixer
Weight	Weight beam, hydraulic load cell, or strain gauge	Lever mechanism or spring, pressure transmitted across diaphragm, dimensional change in sensor	Chemicals
Density	Gamma radiation	Absorption of gamma rays by the liquid between radiation source and the detector	Solids concentration of thickened and dewatered sludge
	Ultrasonic sensor	Loss of ultrasonic signal by the liquid between the ultrasonic transmitter and receiver	Suspended solids concentration of thickened and dewatered sludge
pH	Selective ion electrode	Voltage produced by hydrogen ion activity	Raw water, chemical solution, rapid mix, dewatering, effluent, before chlorination, finished water
Turbidity[b]	Photocell	Measurement of light scattering or attenuation as a result of solids in suspension (in units of NTU, FTU, or JTU.)	Raw water, settled, and filtered water
Particle count	Sensors	Electrical zone sensing, light scattering (laser beam or a beam of white light), or light interruption by particles	Results supplementing turbidity data of raw water, settled, and filtered water
Color[c]	Sensor	Light absorption	Raw water, finished water
Oxidation-reduction potential	Electrode	Change in potential due to oxidation or reduction	Raw water, maintenance of proper DO in aeration basin, biofilm reactor
Total dissolved salts	Conductivity	Flow of electrical current across the solution	Raw water, filtered water
Dissolved oxygen	Membrane electrode	Electric current due to reduction of molecular oxygen	Raw water, aeration basin
Total organic carbon	Carbon analyzer	CO_2 produced from combustion of sample	Raw water, before chlorination, finished water
Chlorine residual	Sensor	Electrical output	Chlorine contact tank, clearwell

Table 16-2 Common On-Line Process-Measurement Devices and Their Applications in Water Treatment (continued)

Measured Variables	Primary Device	Measured Signal	Common Application
Gases: O_3, O_2, NH_3, H_2S, CO_2, CO, Cl_2	Sensors	Various types of sensor modules utilizing electrical impulses	Detection of hazardous condition in the room or around storage area, machine rooms
Oxygen uptake rate	Respirometers using sensor	Change in DO level with respect to time	Biofilm reactors, nitrification, taste, and odor control
Anaerobic biological condition	Sensor, combustion	Change in potential due to reduction	Biological denitrification

a Additional details may be found in Chapter 7; psi ∞ 6,895 = kN/m^2.

b See Chapter 2 for turbidity units.

c See Chapter 2 for color units.

Source: Adapted in part from References 5, 7–9.

Digital. Digital systems have two positions–on/off, open/closed, high/low, or alarm/normal. This type of signal is called status change. These signals may originate from a position, limit, float, or pressure switch.

Analog Analog information has a range of values, such as a flow rate for air or water. Level measurements or concentration measurements are also considered analog, because a range of values may result. This type of data can be reused and transmitted in its analog form, or it may be converted to digital form or combinations of analog and digital instruments, such as an analog level meter with a high-level alarm switch.

Automatic Control. There are two basic types of automatic control–discrete and continuous. The discrete control correlates equipment status (on/off) and status changes with a preset value or program of events. Discrete control could be represented by an automatic start/stop sequence, such as high pumping activities based on clearwell level. The discrete control may also be presented by a logic predefined program of events for a more complex process, such as filter backwashing. The operation might be initiated manually, by an operator using a pushbutton, or automatically, by an internal process-generated event, such as high filter head loss. Programmable controllers are commonly used to perform discrete control tasks. The PC replaces multiple components (relays and timers) in hard-wired logic networks.

Continuous control requires analog measurement for its input and manipulates a final control element as its output. Continuous control may take the form of feed-back and feed-forward control loops and of control systems or controllers that perform the control functions. Many of these control loops and control systems are illustrated in Figure 16-1.

The complexity of the control depends on such factors as treatment process, measurement required, frequency of information, environmental conditions, technical limitations, mainte-

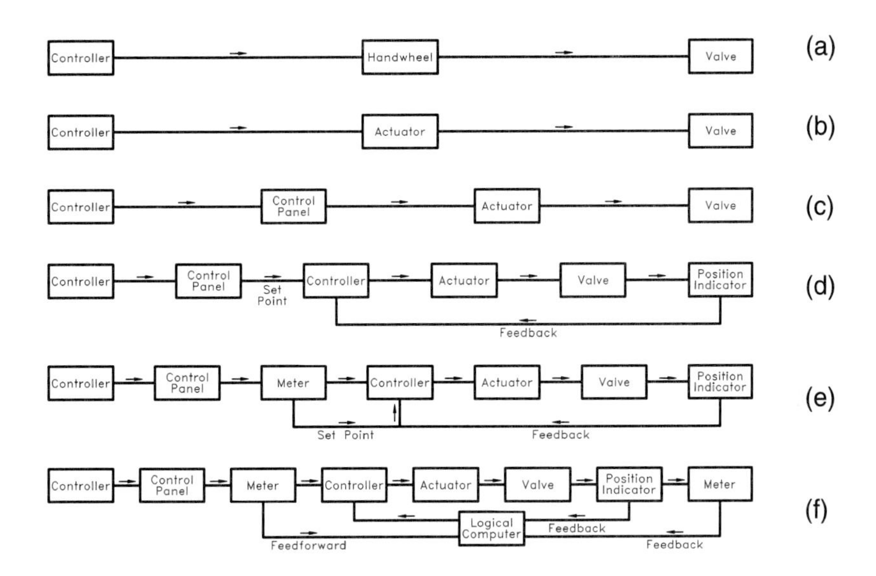

Figure 16-1 Examples of control systems. (a) Manual control. (b) Manual control made easier. (c) Remote manual control. (d) Automatic feed-back control. (e) Automatic feed-forward control. (f) Automatic feed-forward/feed-back computer control.

nance manpower, and costs. In continuous process control, there are two types of process variables: (1) the uncontrolled variable (sometimes called the wild variable), and (2) the controlled variable. For example, the chemical-feed flow in the rapid mix should be changed automatically to keep a constant ratio between plant flow and chemical flow. In this control loop, the plant flow is the uncontrolled variable, and the chemical flow is the controlled variable.

A control system is a series of related or unrelated control loops for detection of variables that manipulate the process for desired results. A typical control system and its components are illustrated in Figure 16-2. An automatic control system is made of three parts[6]:

1. a measurement section, to detect the change in the variable;
2. a reference source, to compare the value of the variable with a present reference;
3. a controller or mechanism, to manipulate the variable until the measured signal reaches the reference value.

A general description of common forms of control loops and continuous control systems follows.

Control Loop. A control loop is defined by the Instrument Society of America as a combination of one or more interconnected instruments to measure or control a process or both.[11,12] The

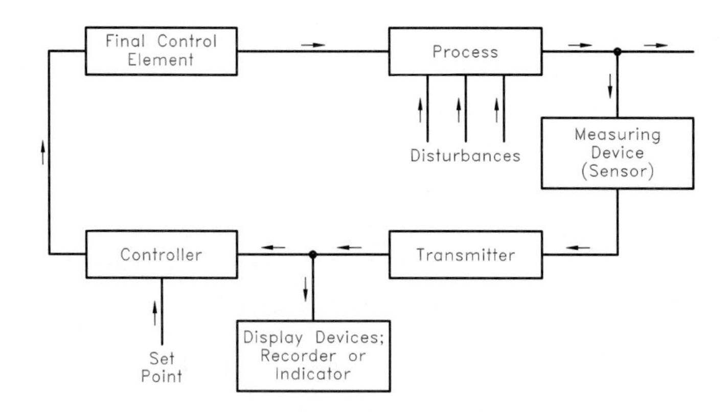

Figure 16-2 Typical Control-System Components. *(Source: Adapted from Reference 6.)*

control circuit or loop is composed of control elements and comes in two general types: feed-back, or closed loop, and feed-forward, or open loop.[1–4]

A feed-back loop in a control system is accomplished by "feeding back" the signal measuring the controlled variable to the controller, for the purpose of detecting the deviation from the setpoint. Thus, the disturbance or deviation is constantly corrected. The disadvantage is that corrections are applied only after the deviation has occurred, and feed-back elements such as meters, transmitters, and transducers are required. An example of a closed feed-back loop is the automatic chlorinator discussed in Chapter 12.

A feed-forward loop in a control system is applied without the use of feed-back information; therefore, control action is initiated before a deviation in the process occurs. This system is usually applied either to very simple control problems or to complex but predictable control loops, because there is no automatic correction. The example is chemical-feeder output controlled in proportion to the plant flow, by a ratio station. The ratio station is set manually for dosage control. The system is used for its simplicity and is best suited where accurate control and time lags are not critical.

The control system consists of a combination of individual control loops, each of which is relatively simple; however, these loops may be combined into independent, interdependent, or dependent open and closed loops, to constitute an elaborate and complex system.

Continuous-Control Systems or Controller. The device that automatically performs the control functions in regulating the control variable is a controller. It may be of different types depending upon the functional needs.[3,5]

Two-Position Control. Two-position control (also called on/off control) is the simplest form of automatic feed-back control, although it is often not referred to as feed-back control, because of its simplicity. A control band of any width can be established with on/off control. An example is pumping on/off control based on clearwell level. Other examples are the opening or closing of a valve and turning heat on or off.

Floating Control. Floating control is a variant of on/off control, with additional elements introduced: a second (negative) "on" position, and a dead band. All floating controls are three-position controls. Examples are air-conditioning systems having heating-off-cooling control action and valve systems with opening-off-closing action. Compared to two-position control, floating control allows tighter control (narrower range) and usually causes less cycling.

Proportional Control. Proportional control represents a continuous analog control, in which the output of the controller is directly proportioned to the deviation from the setpoint. The degree of proportionality is called the proportional band. The process will be unstable if the proportional band is set too narrow. Proportional control is extensively used for filter-level control in a filtration plant.

Proportional-Integral. The proportional-integral mode (also called reset) is added to compensate for the deviation from the setpoint over the entire control range of the proportional band. The rate of output change is dependent on the amount and duration of the deviation (signal error). In essence, the integral function samples the difference between the fixed setpoint of the controller and the point at which the sensor stabilizes, and adjusts the controller element in such a way as to return the parameter valve to the setpoint.

Proportional-Integral-Differential. Generally referred to as PID, the proportional-integral-differential control mode adds the differential (derivative or rate) function. The purpose of the rate function is to modify the position of the controlled element in terms of the rate at which the parameter signal deviates from the fixed setpoint. These are, perhaps, the most elaborate single means of automatic process control. In water filtration plants, PID systems are seldom used.

Sample Date Control. The control action is delayed to provide a time lag for the effect of previously applied correction to be sensed. An example is a chlorination facility where chlorine loss is controlled by the residual measured after the contact period. Another example is pH control.

Time Control. The timer control device operates on a set time schedule. An example of such a control system is the pumping of residuals from the sedimentation basin at preset time intervals.

Two- and Three-Mode Control. Two or even three modes of control may be combined with the basic controls, to achieve variable and complex control operations.

16.3.3 Computer Control Room and Data-Acquisition System

Modern medium-size water treatment plants have data acquisition systems for data logging.[6] These systems accumulate, format, record, and display large quantities of data effectively. Modern data-acquisition systems, commonly referred to as Supervisory Control and Data Acquisition (SCADA) systems, can provide accurate and impartial documentation of all process measurements and operator actions. Computers can also provide a maintenance schedule to the plant operator based on actual operating time of a particular piece of equipment, such as pump or motor.

Although SCADA systems are commonly used in many industries, they have not found wide application in small- to medium-size water treatment works. These SCADA systems accumulate the process data, display them, and also produce the necessary process corrections for optimum operation. These corrections can include control of chemical solutions and air supply, scheduling of pumps and blowers, and the like. SCADA systems that are poorly designed, or those with no previous applications, such as a newly designed system, can potentially cause numerous problems.

Central control is used to organize the plant operation in such a manner that all treatment information, important events, and alarms are displayed, indicated, and recorded at a centralized location; this location is called the *central control room*. In addition, most central facilities practice automatic or manual actuation of final control elements. Central control rooms reduce the number of personnel required to operate a large treatment facility.

Expert systems are gaining popularity in process control. This technology captures, in a computer program, the concepts, facts, and judgements used by an experienced operator in making a decision. This technology assists the operator who has limited background in complex treatment process.[5] *Neural network* techniques provide learning capabilities within the computer system.[5] This allows the computer to recognize patterns and modify its own recognition criteria to improve its performance. Readers may find additional information on this topic in References 13–16.

16.3.4 Symbols and Identification of P&I Diagram

The symbols and identification used in designing the process and instrumentation (P&I) diagrams are based on those provided by the Instrumentation Society of America (ISA), the Institute of Electrical and Electronic Engineers (IEEE), and the National Electrical Manufacturers' Association (NEMA).

16.3.5 Design of the Instrumentation and Control (I&C) System

The design of the instrumentation and control system must be conducted by I&C engineers; however, the project engineer should maintain an open line of communication with the I&C engineers and perform a thorough check of the final design. The system must be operated and maintained by the plant operators, so it is beneficial to elicit input from both the operators and the owner during the design phase.[2]

Design work usually produces the following documents in the sequence listed below. The process flow sheets (PFS) are prepared by the project engineer. A sample PFS is shown in Figure 16-3. This flow sheet then becomes the basis for the process and instrumentation diagram (P&ID).[2]

- P&ID is a functional schematic presentation of the treatment processes and the required instruments and controls. Its purpose is to illustrate the functions of the treatment plant without referring to the actual hardware.
- Process control diagram (PCD) specifies the monitoring and control loops.
- Instrumentation and input/output summaries (IIOSs) provide continuity between the P&ID and IISs. These documents specify both the quantity and the characteristics of the analog and digital instrumentation.
- Instrumentation specification sheets (ISSs) provide detailed specifications for each task and panel instrument.
- Logic diagrams for the control panels and programmable logic control (PLCs) are obtained by the instrumentation contractor.
- Panel layout drawings represent the schematic layout of P&ID.
- Loop interconnection drawings (LIDs) are primarily used in projects where the instrumentation and control systems must be interfaced with existing control equipment.
- Instrument installation details (IIDs) represent final installation details.

16.4 Manufacturers of Instrumentation and Control Systems

The names and addresses of many manufacturers and equipment suppliers of instrumentation and control systems for wastewater treatment facilities are provided in Appendix C. It is the responsibility of the process design engineer to work closely with the instrumentation and control engineer and the equipment supplier in providing data and making a final selection.[6]

16.5 Information Checklist for Design and Selection of Instrumentation and Control Systems

Specific uses of instrumentation and control, and the extent of automation provided at a water treatment facility, are decisions that must be made by the design engineer and management personnel. The extent of instrumentation could depend on the size of the plant, number of operating personnel, and available construction and operation funds. The following information must be developed and decisions made for the selection of proper instrumentation and control systems.

1. How many hours of manned operation daily will there be?
2. How available are trained instrumentation maintenance personnel?
3. Given the size and complexity of the plant, should there be no control, manual control, supervisory control, automatic control, or computer control?
4. If automatic controls are selected, decisions should be made concerning the number of control elements and loops, accuracy needed, operating range, response time of process variables, and frequency of operator input.
5. The cost of instrumentation and controls must be compared with the savings achieved by plant automation.

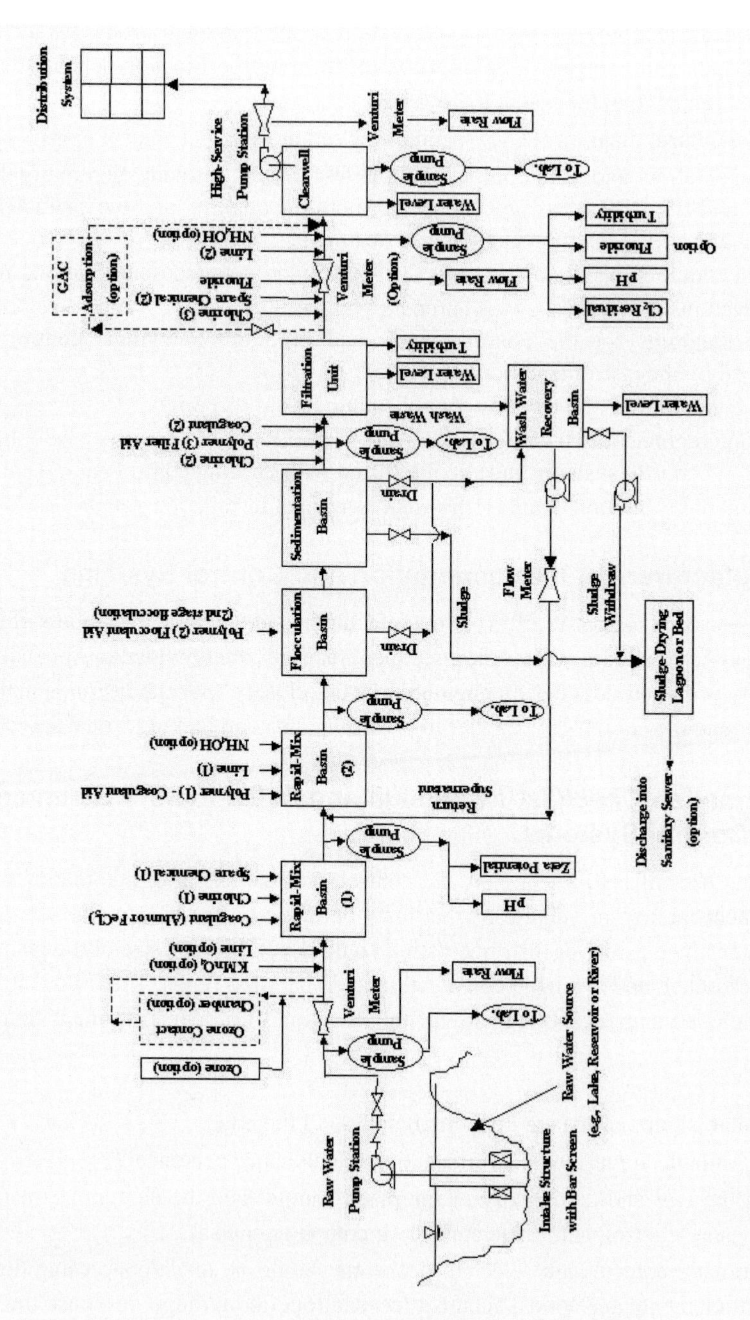

Figure 16-3 Process trains with chemical-feed and process-control points.

16.6 Design Example

16.6.1 Design Criteria Used

The basic design criteria for the instrumentation and controls selected for many treatment processes have been discussed in the respective chapters. The criteria are based on providing an economical and reliable operation of various treatment facilities. These include (1) operational range, (2) accuracy desired, (3) operational flexibility and reliability, (4) space requirements, and (5) future expansions. The selected process systems must incorporate proven industrial-control equipment and techniques of contemporary design.

16.6.2 Selected Process-Control Systems

The selected process-control system will consist of the following: (1) It shall contain automatic and continuous analyzers of sensing devices at strategic locations throughout the plant for measurement of process-control variables. Where a particular process-control variable cannot be monitored, a systematic grab-sampling device shall be instituted for piping samples directly into the laboratory. (2) The automatic control systems will be installed to offer a high degree of accuracy and permit remote control of process equipment. The instrumentation will sense flows, absolute pressure differential, and temperature. The majority of the control equipment will be solid-state electronic. (3) The central control room shall contain a combination of analog and digital-indicating controllers, annunciator alarm panels, motor control and valve operator switches, trend recorders, indicators and totalizer, converters, transducers, operator's control console, master communication equipment, telemetry transmitters and receivers, auxiliary power supply for emergency use and data-logging systems, and so on. The heating and air conditioning for this room will be designed to maintain a uniform temperature and to provide a clean atmosphere with a stable relative humidity. (4) Various control and analyzer panels shall be located throughout the plant to provide the data needed to optimize the plant efficiency.

Each of the process units and major equipment shall be monitored and, to some extent, controlled from the central control room. The general descriptions of various functions to be monitored for each treatment process are summarized in Table 16-3. Eight simplified loop diagrams are shown in Figures 16-4 through 16-7. These schematics show the operational principles of selected water supply facilities and treatment processes. Controls for many other applications, such as thickened-sludge withdrawal and backwash water recovery, can easily be developed from the examples of loops shown in Figures 16-4 through 16-7. It should, however, be mentioned that these loops are grossly simplified. Many complex loop diagrams are generally needed in the design of water treatment facilities. The instrumentation engineer should develop such diagrams in consultation with the manufacturers of various instrumentation and control devices and with the manufacturers of water treatment process equipment. Commonly used identification abbreviations are provided in Figure 16-8.

Table 16-3 Location and Functions of Major Sensing Devices Used in Process Control and in Equipment Monitoring

Treatment Unit		Sensing Device	Description	Reference
A. Raw water intake				
1. Lake elevation		Ultrasonic transducer	Display elevation	Sec. 6.8.2, Step B
2. Intake gates		Limit switch	Display green light	Sec. 6.8.2, Step B
3. Intake level		Ultrasonic transducer	Display elevation	Sec. 6.8.2, Step B
B. Fine screen				
1. Rakes	a	Time clock	Display green light	Sec. 6.8.2, Step D
	b	High-level override by upstream float	Display red light	Sec. 6.8.2, Step D
2. Motors		Torque monitor	Display red light	Sec. 6.8.2, Step D
C. Raw water pump station				
1. Gates		Limit switch	Display green light	Sec. 7.8.2, Step C
2. Wetwell		Ultrasonic transducers	Display elevation	Sec. 7.8.2, Step C
3. Pumps	a	Pressure switches	Display green light	Sec. 7.8.2, Step E
	b	Pressure indicators	Pressure-gauge reading recorded continuously	Sec. 7.8.2, Step E
	c	Motor monitors (contacts provided in motor-control center)	Lights: Green–off; Red–on; Amber–malfunction	Sec. 7.8.2, Step E Sec. 7.10.2
	e	Resistance temperature devices on pump motors	Monitored by solid-state motor protection device	Sec. 7.8.2, Step E
4. Butterfly valves		Limit switches	Lights: Green–Open; Red–Closed; Amber–Malfunction; Purple–In Transit	Sec. 7.8.2, Step E Sec. 7.10.1
D. PAC system				
1.Dust collection	a	Pressure indicator	Monitor pressure loss across filters; send alarm to control room	Sec. 11.6.2, Step B
	b	Pressure indicator	Monitor pressure loss across filters; send alarm to control room	Sec. 11.6.2, Step B
	c	Motor contact	Lights: Green–off; Red–on; Amber–malfunction	Sec. 11.6.2, Step B Sec. 11.8.2
2. Slurry tank		Ultrasonic transducer	Local and remote indication of tank level	Sec. 11.6.2, Step B Sec. 11.8.2
3. Slurry mixers		Motor contact	Lights: Green–off; Red–on; Amber–malfunction	Sec. 11.6.2, Step B Sec. 11.8.2
4. Slurry pumps		Motor contact	Lights: Green–off; Red–on; Amber–malfunction	Sec. 11.6.2, Step B Sec. 11.8.2

Table 16-3 Location and Functions of Major Sensing Devices Used in Process Control and in Equipment Monitoring (continued)

Treatment Unit		Sensing Device	Description	Reference
5. Slurry feeders		Speed controller	Remote adjustment from plant control room	Sec. 11.6.2, Step B Sec. 11.8.2
E. Motor-control center				
1. Volt meters		Voltmeters on motor-control center bus	Displays MCC bus voltage	Sec. 7.8.2, Step A
2. Current meters		Current meters in motor-control center	Current gauge reading on MCC	Sec. 7.8.2, Step E
F. Flow meter				
Venturi tube		Flow meter	Displays, records, and totalizes	Sec. 7.8.2, Step D Sec. 7.10.3
G. Raw water transmission pipeline				
Raw water	a	Turbidimeter	Display and record turbidity	Sec. 8.8.2, Step A
	b	pH meter	Display and record turbidity	Sec. 8.8.2, Step A
H. Lime-mixing chamber				
Motor		Motor contacts	Lights: Green–off; Red–on; Amber–malfunction	Sec. 8.8.2, Step A
I. Rapid-mix basin				
1. Influent	a	Turbidimeter	Display and record turbidity	Sec. 8.8.2, Step A
	b	pH meter	Display and record turbidity	Sec. 8.8.2, Step A
2. Motors		Motor contacts	Lights: Green–off; Red–on; Amber–malfunction	Sec. 8.8.2, Step A Sec. 8.10.1
3. Effluent	a	Streaming current monitor	Display and record zeta potential	Sec. 8.8.2, Step A
	b	pH meter	Display and record pH	Sec. 8.8.2, Step A
J. Parshall flumes				
Parshall flumes		Ultrasonic transducer	Digital display; flow rate converted from level data	Sec. 8.8.2, Step B Sec. 8.10.3
K. Flocculation basin				
1. Motors		Motor contact	Lights: Green–off; Red–on; Amber–malfunction	Sec. 8.8.2, Step B Sec. 8.10.2
2. Flocculators		Paddle speed	Digital display	Sec. 8.8.2, Step B Sec. 8.10.2
L. Sedimentation basin				
1. Motors		Motor contact	Lights: Green–off; Red–on; Amber–malfunction	Sec. 9.6.2, Step A Sec. 9.8

Table 16-3 Location and Functions of Major Sensing Devices Used in Process Control and in Equipment Monitoring (continued)

Treatment Unit		Sensing Device	Description	Reference
2. Effluent	a	Turbidimeter	Monitors, displays, and records settled water turbidity	Sec. 9.6.2, Step A
	b	pH meter	Monitors, displays, and records settled water pH	Sec. 9.6.2, Step A
3. Telescopic valves		Limit switch	Lights: Green–open; Red–closed; Amber–malfunction	Sec. 9.6.2, Step B
4. Sludge wetwells		Ultrasonic transducers	Display elevation	Sec. 9.6.2, Step B
5. Sludge pumps		Motor contact	Lights: Green–off; Red–on; Amber–malfunction	Sec. 9.6.2, Step B
6. Flow meter		Flow transducer	Lights: Green–off; Red–on; Amber–malfunction	Sec. 9.6.2, Step B
M. Filters				
1. Filters	a	Differential pressure gauges	Monitors, records, and totalizes	Sec. 10.13.2, Step A
	b	Ultrasonic level transducer	Displays and records filter level	Sec. 10.13.2, Step A
	c	Time clock	Records and displays run time	Sec. 10.13.2, Steps A and B
	d	Valve limit switches	Displays valve position (percent opening)	Sec. 10.13.2, Step A
	e	Valve limit switches	Lights: Green–open; Red–closed; Amber–fault; Purple–in transit	Sec. 10.13.2, Steps A and B
	f	Valve limit switch	Displays valve position (percent opening)	Sec. 10.13.2, Step B
	g	Flow meter	Displays backwash flow rate	Sec. 10.13.2, Step B
2. Effluent	a	Turbidimeter	Monitors and records	Sec. 10.13.2, Step A
	b	pH meter	Displays and records	Sec. 10.13.2, Step B
	c	Particle counter	Monitors and records	Sec. 10.13.2, Step A
3. Flow meters		Flow transducers	Lights: Green–off; Red–on; Amber–malfunction	Sec. 10.13.2, Step A
4. Backwash water		Flow transducer	Lights: Green–off; Red–on; Amber–malfunction	Sec. 10.13.2, Step A
5. Surface wash water		Flow transducer	Lights: Green–off; Red–on; Amber–malfunction	Sec. 10.13.2, Step A
6. Elevated backwash water storage tank		Ultrasonic level transducer	Displays and records water surface level	Sec. 10.13.2, Step B

Table 16-3 Location and Functions of Major Sensing Devices Used in Process Control and in Equipment Monitoring (continued)

Treatment Unit		Sensing Device	Description	Reference
N. Chlorine contact channel				
Effluent	a	Chlorine residual	Displays and records chlorine residual	Sec. 12.8.2, Step A
	b	pH meter	Displays and records	Sec. 12.8.2, Step A
O. Clearwell				
Clearwells		Ultrasonic transducer	Displays and records	Sec. 12.8.2, Step A Sec. 13.8.2, Step B
P. High-service/ backwash water pump station				
1. Pumps	a	Motor contacts	Lights: Green–off; Red–on; Amber–malfunction	Sec. 13.8.2, Step C
	b	Pressure indicator	Parameters of pump flow; used to calculate, record, and display pump station efficiency	Sec. 13.8.2, Step C
	c	Pressure indicator		
	d	Current transformer		
	e	Voltage meter (in MCC)		
	f	Resistance temperature devices on pump and motor bearings	Monitors and shuts down pump if excessive	Sec. 13.8.2, Step C
2. Valves		Valve limit switch	Lights: Green–open; Red–closed; Amber–fault; Purple–in transit	Sec. 13.8.2, Step C
3. Effluent		Chlorine residual	Displays and records	Sec. 13.10.1
Q. Flow meter vault				
Flow meter		Flow meter	Displays, records, and totalizes	Sec. 13.8.2, Step C
R. Filter backwash surge storage tank				
1. Surge storage tank		Ultrasonic level transducer	Displays and records	Sec. 14.6.2, Step B
2. Motors		Motor contacts	Lights: Green–off; Red–on; Amber–malfunction	Sec. 14.6.2, Step B
3. Pumps		Motor contacts	Lights: Green–off; Red–on; Amber–malfunction	Sec. 14.6.2, Step B
S. Backwash water recovery basin				
1. Motors	a	Motor contacts	Lights: Green–off; Red–on; Amber–malfunction	Sec. 14.6.2, Step B
	b	Motor contacts	Lights: Green–off; Red–on; Amber–malfunction	Sec. 14.6.2, Step B

Table 16-3 Location and Functions of Major Sensing Devices Used in Process Control and in Equipment Monitoring (continued)

Treatment Unit		Sensing Device	Description	Reference
2. Telescopic valve		Limit switch	Lights: Green–open; Red–closed; Amber–malfunction	Sec. 14.6.2, Step B
3. Pumps		Motor contacts	Lights: Green–off; Red–on; Amber–malfunction	Sec. 14.6.2, Step B
4. Overflow		Flow meter (sonic)	Displays, records, and totalizes	Sec. 14.6.2, Step B
T. Gravity thickeners				
1. Motors		Motor contacts	Lights: Green–off; Red–on; Amber–malfunction	Sec. 14.6.2, Step C
2. Telescopic valves		Limit switch	Lights: Green–open; Red–closed; Amber–malfunction	Sec. 14.6.2, Step C
3. Pumps		Motor contacts	Lights: Green–off; Red–on; Amber–malfunction	Sec. 14.6.2, Step C
4. Overflow		Flow meter (sonic)	Displays, records, and totalizes	Sec. 14.6.2, Step C
5. Thickened sludge		Flow meter (sonic)	Displays, records, and totalizes	Sec. 14.6.2, Step C
U. Sludge lagoons				
1. Lagoons		Ultrasonic level transducer	Displays and records	Sec. 14.6.2, Step D
2. Pumps		Motor contacts	Lights: Green–off; Red–on; Amber–malfunction	Sec. 14.6.2, Step D
3. Percolation and decant flow		Flow meter (sonic)	Displays, records, and totalizes	Sec. 14.6.2, Step D
V. Chemical storage and feed				
1. Lime	a	Ultrasonic level transducers	Displays and records	Sec. 8.8.2, Step C
	b	Galvanometric feeders	Displays and records	Sec. 8.8.2, Step C
	c	Temperature probes	Displays and records	Sec. 8.8.2, Step C
2. Coagulant	a	Ultrasonic level transducers	Displays and records	Sec. 8.8.2, Step C
	b	Meter contacts	Lights: Green–off; Red–on; Amber–malfunction	Sec. 8.8.2, Step C
	c	Diaphragm positioners	Displays, records, and adjusts	Sec. 8.8.2, Step C

Table 16-3 Location and Functions of Major Sensing Devices Used in Process Control and in Equipment Monitoring (continued)

Treatment Unit		Sensing Device	Description	Reference
3. Polymers	a	Motor controls	Lights: Green–off; Red–on; Amber–malfunction	Sec. 8.8.2, Step C
	b	Weight cells	Displays and records	Sec. 8.8.2, Step C
	c	Diaphragm positioners	Displays, records, and adjusts	Sec. 8.8.2, Step C
4. Potassium permanganate	a	Motor controls	Displays and records	Sec. 8.8.2, Step C Sec. 11.6.2, Step A
	b	Feeders	Displays and records	Sec. 8.8.2, Step C Sec. 11.6.2, Step A
	c	Ultrasonic level transducers	Displays and records	Sec. 8.8.2, Step C Sec. 11.6.2, Step A
	d	Diaphragm positioners	Displays, records, and adjusts	Sec. 8.8.2, Step C Sec. 11.6.2, Step A
5. Chlorine	a	Vacuum gauge	Displays and records; sounds alarm on loss of vacuum	Sec. 12.8.2, Step B
	b	Chlorinator	Displays and records feed rate	Sec. 12.8.2, Step B
	c	Leak detector	Displays and alarm	Sec. 12.8.2, Step B
	d	Weight cell	Displays and records	Sec. 12.8.2, Step B
6. Ammonia	a	Pressure gauge	Displays and records; sounds alarm on loss of pressure	Sec. 12.8.2, Step C
	b	Vacuum gauge	Displays and records; sounds alarm on loss of vacuum	Sec. 12.8.2, Step C
	c	Ammoniators	Displays and records feed rate	Sec. 12.8.2, Step C
	d	Leak detector	Displays and alarm	Sec. 12.8.2, Step C
7. Sodium fluoride	a	Pressure indicator	Monitors pressure loss across filters; sends alarm to control room	Sec. 12.8.2, Step D
	b	Ultrasonic level transducers	Displays and records	Sec. 12.8.2, Step D
	c	Fluoride analyzer	Displays and records	Sec. 12.8.2, Step D
	d	Diaphragm positioners	Displays, records, and adjusts	Sec. 12.8.2, Step D

Table 16-3 Location and Functions of Major Sensing Devices Used in Process Control and in Equipment Monitoring (continued)

Treatment Unit		Sensing Device	Description	Reference
8. Sodium carbonate	a	Pressure indicator	Monitors pressure loss across filters; sends alarm to control room	Sec. 13.8.2, Step A
	b	Ultrasonic level transducers	Displays and records	Sec. 13.8.2, Step A
	c	Alkalinity analyzer	Displays and records	Sec. 13.8.2, Step A
	d	Diaphragm positioners	Displays, records, and adjusts	Sec. 13.8.2, Step A

Problems and Discussion Topics

16.1 A Venturi meter is used for continuous flow-recording of industrial water. The differential pressure is transmitted and is recorded on a chart. Using Eq. (7.6), calculate the hourly flows, and totalize the flow. Use the following data.

Time:	12 Midnight	1	2	3	4	5	6	7	8	9	10	11	
h, m H_2O	0.20	0.21	0.25	0.26	0.30	0.40	0.50	0.70	1.20	1.80	2.20	2.30	
Time:	12 Noon	1	2	3	4	5	6	7	8	9	10	11	12 Midnight
h, m H_2O	2.00	1.80	1.60	2.10	2.80	3.00	2.50	2.00	1.80	0.80	0.40	0.30	0.20

16.2 An industrial plant operates from 8 am to 5 pm. The hourly water flow and TDS concentrations are given below. Totalize the TDS results and express in kg per operating day.

Time:	7	8	9	10	11	12	1	2	3	4	5	6	7
Flow, m^3/s	0	0.02	0.05	0.10	0.15	0.20	0.21	0.22	0.25	0.15	0.10	0.12	0
TOC, mg/L	0	80	120	180	300	350	300	280	250	200	150	80	0

16.3 Using the typical control system components given in Figure 16-2, list various options available at each control system component for maintaining constant temperature in an anaerobic digester. The sludge is recirculated through an external heat exchanger for heating purposes.

16.4 Match the primary flow-measuring devices with their applications. .

____Orifice meter	(a) sludge
____Flow nozzle	(b) clean liquid
____Parshall flume	(c) raw water
____Acoustic meter	
____Electromagnetic meter	

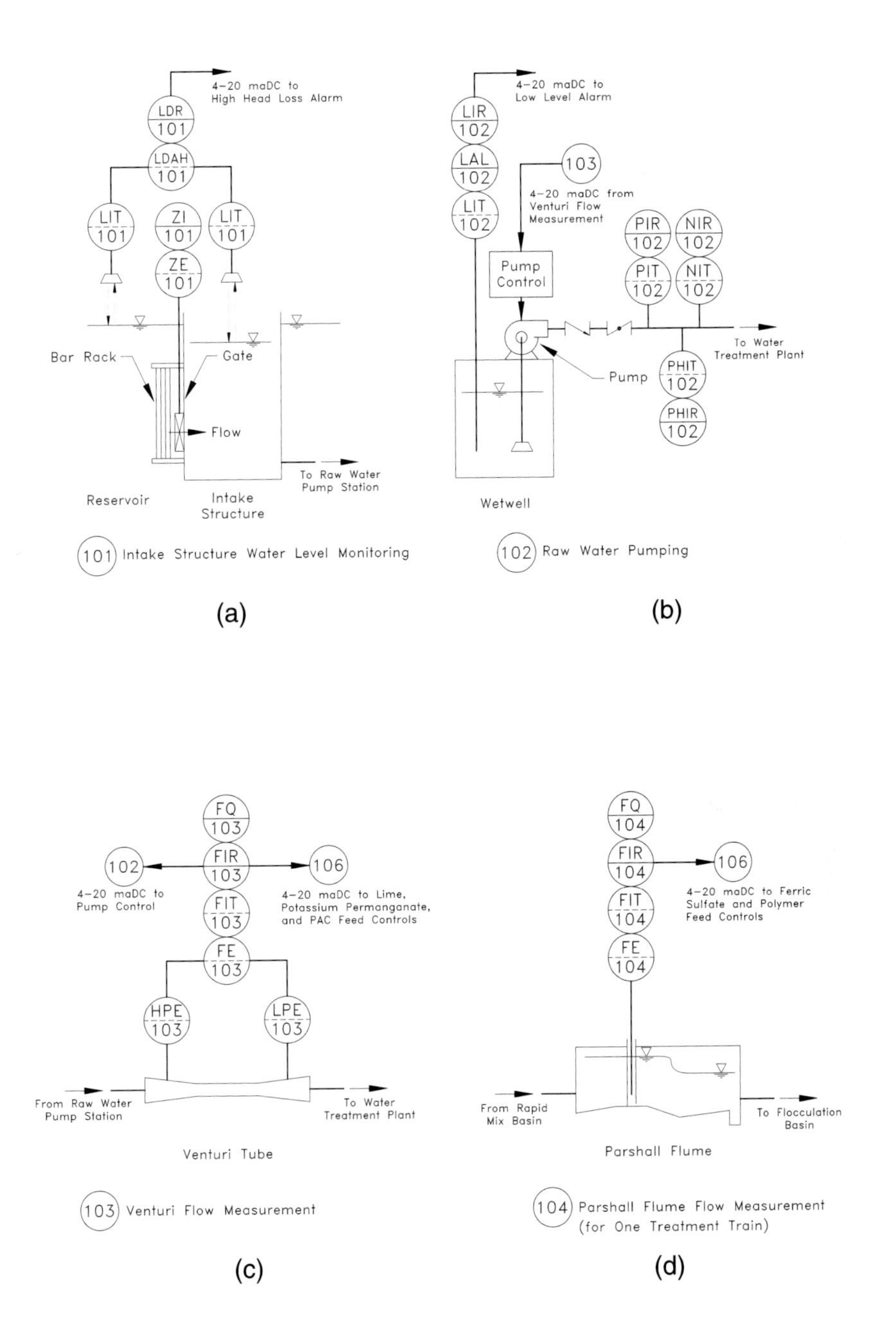

Figure 16-4 Schematics of selected control loop diagrams in the Design Example. (Showing: (a) for the intake structure water level monitoring, (b) for the raw water pumping, (c) for the Venturi flow measurement, and (d) for the Parshall flume flow measurement.)

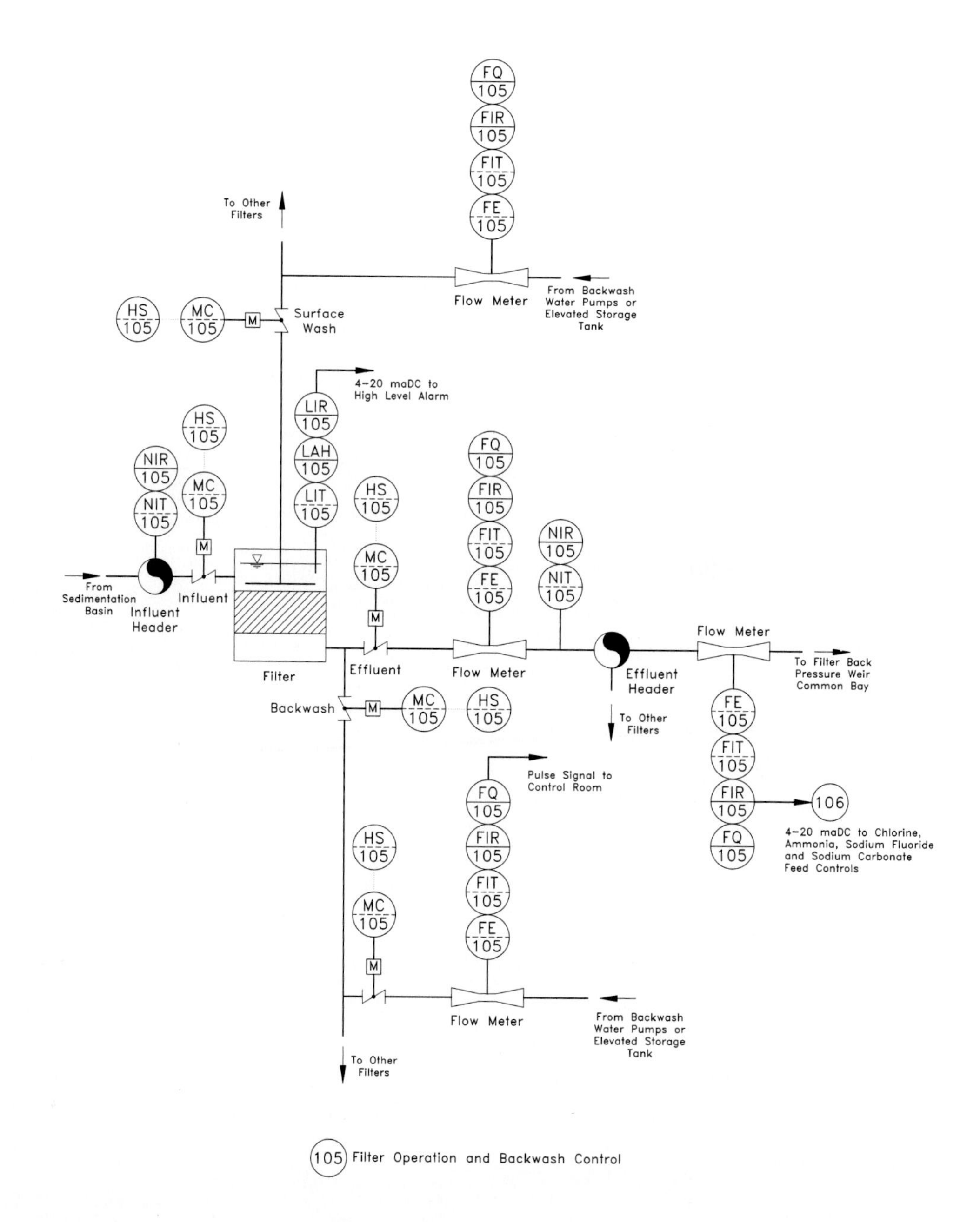

Figure 16-5 Schematics of selected control loop diagrams in the Design Example. (Showing that for the filter operation and backwash control.)

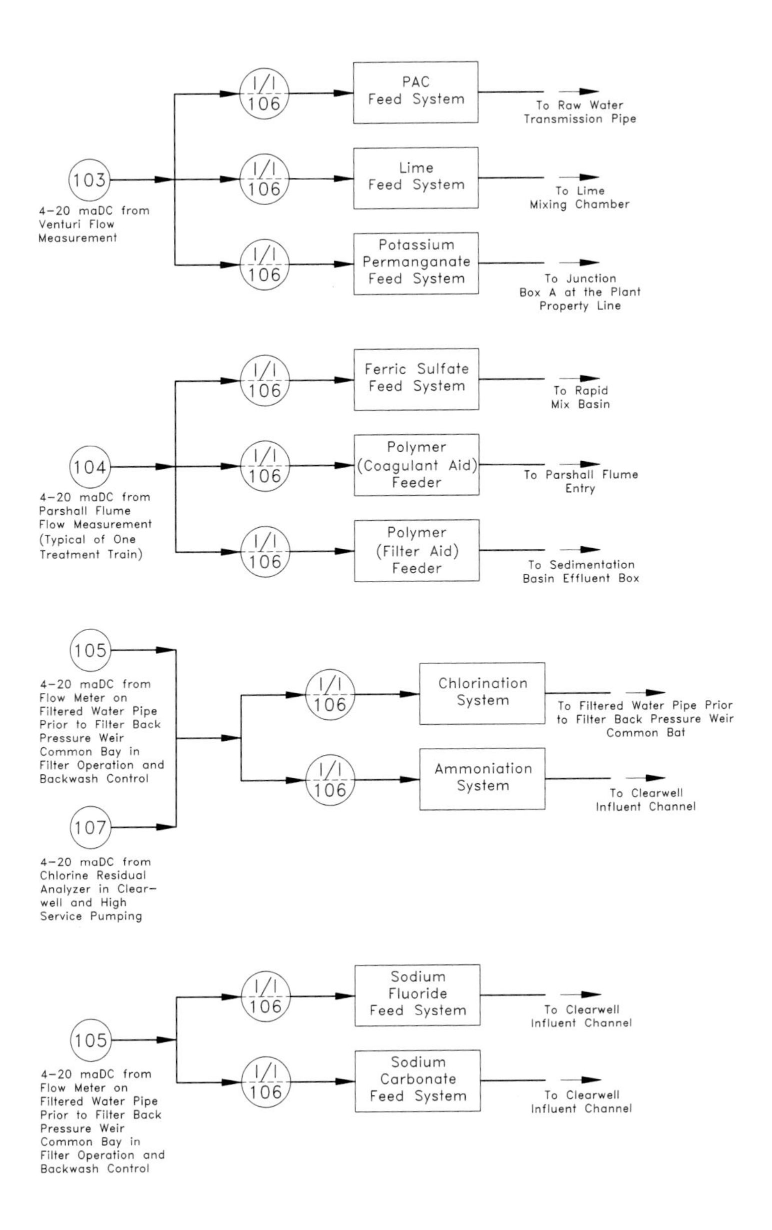

Figure 16-6 Schematics of selected control loop diagram in the Design Example. (Showing that for the chemical feed systems.)

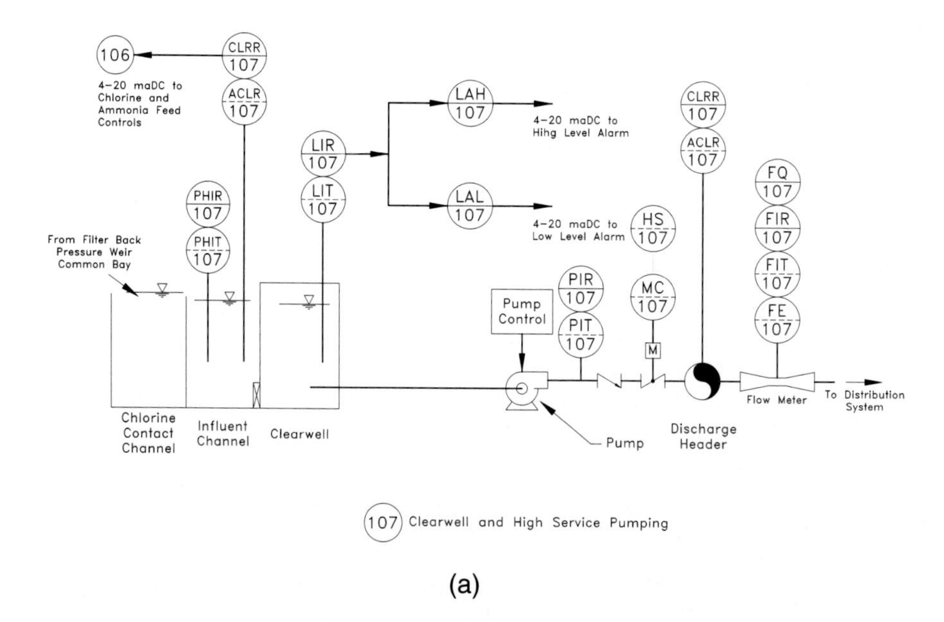

(a)

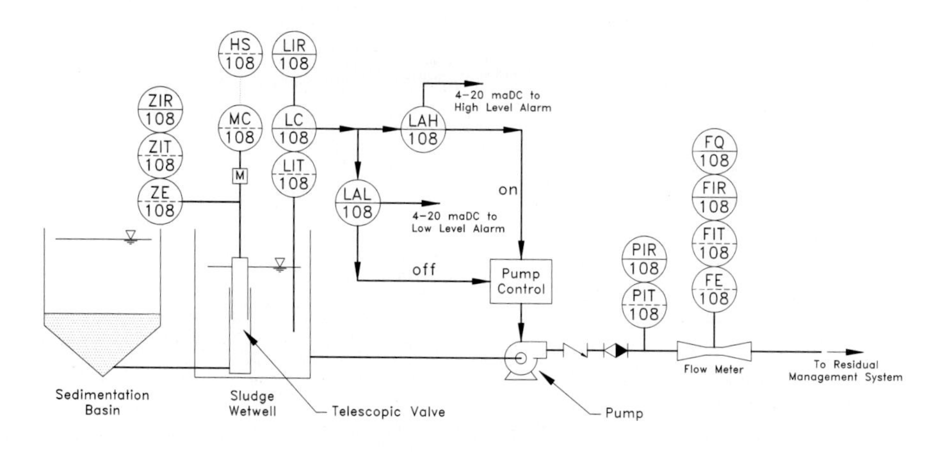

(b)

Figure 16-7 Schematics of selected control loop diagrams in the Design Example. (Showing: (a) for the clearwell and high-service pumping and (b) for the sedimentation basin sludge draw-off.)

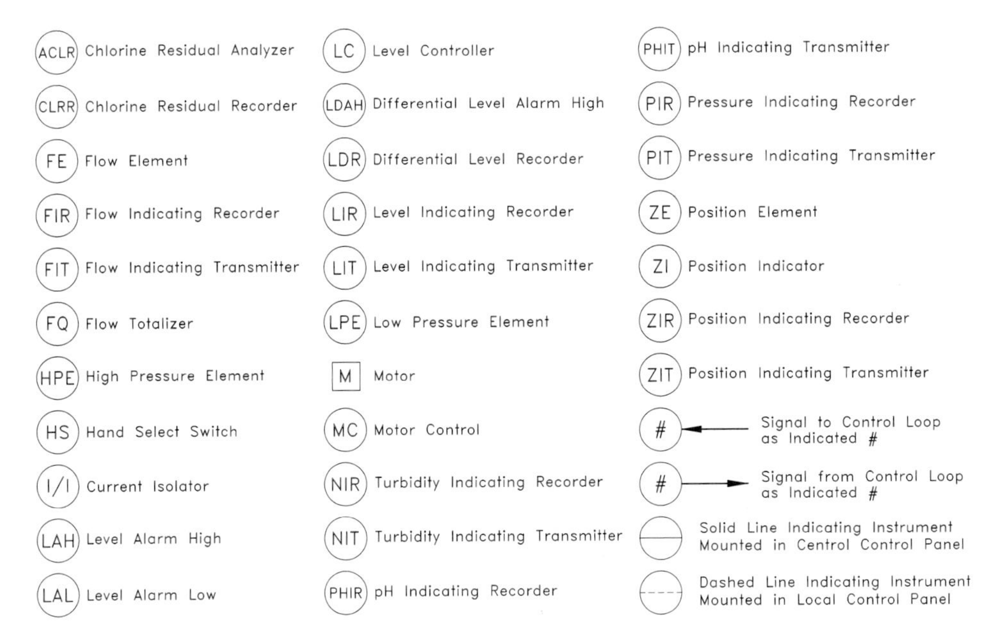

Figure 16-8 Commonly used abbreviations for P&ID.

16.5 Write measured signals for the following variables.
- **(a)** conductivity
- **(b)** pH
- **(c)** density
- **(d)** oxidation-reduction potential
- **(e)** dissolved oxygen
- **(f)** temperature
- **(g)** pressure

16.6 A process diagram of a PAC feed system is given in Figure 16-6. Draw an integrated-process mechanical and instrumentation diagram for the PAC delivery and feed system. Assume that there is a single unit of each items of equipment. Use the simplified control-loop diagrams for different units and equipment.

16.7 Discuss various types of signal-transmitting device. Give advantages, disadvantages, and applications of each type.

16.8 What is the basic difference between feed-back and feed-forward automatic-control loops? Give three examples of each in a water treatment plant.

References

1. Water Pollution Control Federation. *Process Instrumentation and Control Systems*, OM-6, Water Pollution Control Federation, Washington, DC., 1982.
2. Kawamura, S. *Integrated Design of Water Treatment Facilities*, John Wiley and Sons, Inc., New York, 1991.
3. James M. Montgomery, Inc. *Water Treatment Principles and Design*, John Wiley & Sons, New York, 1985.

4. ASCE and AWWA. *Water Treatment Plant Design*, 2d ed., McGraw-Hill Book Co., New York, 1990.
5. Qasim, S. R. *Wastewater Treatment Plants: Planning, Design, and Operation*, 2d ed., Technomic Publishing Co., Inc., Lancaster, PA, 1999.
6. Molvar, A. E. et al. *Instrumentation and Automation Experiences in Wastewater Treatment Facilities*, EPA-600/2-76-198, Municipal Environmental Research Laboratory, USEPA, Cincinnati, OH, October, 1976.
7. Committee on Instrumentation. *Instrumentation in Wastewater Treatment Plants*, MOP-21, Water Pollution Control Federation, Washington, DC., 1978.
8. Joint Committee of the WPCF and ASCE. *Wastewater Treatment Plant Design*, MOP-8, Water Pollution Control Federation, Washington, DC., 1977.
9. USEPA. *"Process Design Manual for Sludge Disposal,"* EPA-625/1-79-011, USEPA, Cincinnati, OH, September 1979.
10. Anderson, N. A. *Instrumentation for Process Measurement and Control*, Chilton Company, Radnor, PA , 1980.
11. Considine, D. M. (ed.) *Encyclopedia of Instrumentation and Control*, McGraw-Hill Publishing Co., New York, 1971.
12. Instrument Society of America. *Standard for Instrument Loop Diagram*, S5.4, 1976.
13. Brown, R. J. "An Artificial Neural Network Experiment," *Dr. Dobb's Journal*, vol. 12, no. 4, pp. 16–27, 1987.
14. Widrow, B. and Winter, R. "Neural Nets for Adaptive Filtering and Adaptive Pattern Recognition," *IEEE Computer*, vol. 21, no. 3, pp. 25–39, 1988.
15. Winston, P. H. *Artificial Intelligence*, Addison Wesley, Reading, MA, 1984.
16. Cubberly, W. H. (ed.) *Comprehensive Dictionary of Measurement and Control*, 2d ed., Instrument Society of America, Research Triangle Park, NC, 1991.

CHAPTER 17

Design Summary

In this book, the step-by-step design procedure for a medium-size water treatment facility has been developed. One Design Example has been carried through the previous twelve chapters (Chapters 5–16), to present the theory, design procedure, operation and maintenance, and equipment specifications for various components of the water treatment facility. The purpose of this chapter is to summarize the basic design information that have been developed in the Design Example. The information is summarized in Table 17-1. Reference has been made to various the sections in which the design data or details may be found.

Table 17-1 Summary of Basic Design Data and Dimensions of Water Treatment Facilities in the Design Example

Item	Design Information or Description	Reference
A. Existing water supply facility		Chapter 5
General	The existing water supply system includes seven wells, three ground storage facilities, and two elevated storage towers.	Sec. 5.4
Initial year, 2005 (including population and water demands)	Population = 67,000	Secs.5.4.3 through 5.4.5 Table 5-1 and 5-3
Design year, 2020 (including population and water demands)	Population = 76,200	Secs.5.4.3 through 5.4.5 Table 5-1 and 5-3
Existing groundwater wells	Seven wells	Figure 5-1, Table 5-4
Ground storage capacity	Three reservoirs	Figure 5-1, Table 5-5

Table 17-1 Summary of Basic Design Data and Dimensions of Water Treatment Facilities in the Design Example (continued)

Item	Design Information or Description	Reference
Elevated storage capacity	Two towers	Figure 5-1, Table 5-5
Distribution system	The existing distribution system is fed by three distribution pumping and two elevated storage reservoirs.	Figure 5-1
B. Proposed water supply facility		Chapter 5
General	The proposed capital improvement project involves construction of a new intake structure, raw water pump station, raw water pipeline, conventional water treatment plant, and high-service pumping station. The existing distribution system is expanded by adding new pipes and enlarging several existing pipelines, one ground storage and pumping, and one elevated tower.	Sec. 5.4
Surface water source	Big River Reservoir	Sec. 5.4.7.A
Water quality data	–	Secs. 5.4.8.B and D Secs. 5.4.9.A, B, and D
Project costs	–	Secs.5.4.13 through 5.4.15 Table 5-8
Environmental Impact Assessment	–	Sec. 5.4.16
C. Intake structure		Chapter 6
General	The intake structure and raw water pump station site is 0.8 km northwest of the dam. The location of intake structure is approximately 100 m from shore. The facility is a concrete box with an operation platform on the top. An access bridge is provided between the intake structure and the raw water pump station. A total of eight square gates are installed in the walls of the structure.	Sec. 6.8.1 Figures 5-1 and 6-11
Design capacity	Maximum day flow = 113,500 m^3/d	Sec. 6.8.1

Table 17-1 Summary of Basic Design Data and Dimensions of Water Treatment Facilities in the Design Example (continued)

Item	Design Information or Description	Reference
Layout	–	Sec. 6.8.2, Step A Figure 6-11
Intake structure	Include elevation design, structure dimensions, gates, and coarse screen.	Secs. 6.8.1 Sec. 6.8.2, Steps B, C, and F Figures 6-12 and 6-13
D. Raw water pump station		Chapter 7
General	The raw water pump connected to the intake structure by a suction conduit 152 cm in diameter and 135 m long. The pump station has two levels. A flow meter vault and PAC storage facility are provided adjacent to the pump station inside the property. Also see Items E and L in this table.	Sec. 7.8.1 Figures 5-1, 6-11, and 7-12 through 7-15
Design capacity	Maximum day flow = 113,500 m^3/d	Sec. 7.8.1
Layout	–	Sec. 7.8.2, Step A Figures 5-1, 6-11, 7-12 and 7-13
Pump station	Include wet well dimensions, isolation gates, fine screens, distribution channels, baffle walls, and pumps.	Secs. 7.8.1 and 7.8.3 Sec. 7.8.2, Steps C through F Figures 7-14 and 7-15
E. Venturi meter		Chapter 7
General	The Venturi-meter vault is located inside the raw water pump station property site.	Sec. 7.8.2, Step D.4 Figure 7-17
Venturi meter	–	Sec. 7.8.2, Step D.4 Figure 7-17

Table 17-1 Summary of Basic Design Data and Dimensions of Water Treatment Facilities in the Design Example (continued)

Item	Design Information or Description	Reference
F. Raw water transmission pipeline		Chapters 5 and 7
General	The raw water transmission pipeline is extended from the raw water pump station to the water treatment plant. The major portion of pipeline, 122 cm in diameter, starts from the increaser after the Venturi meter and terminated at Junction Box A, located near the south property line of the water treatment plant.	Sec. 5.4.7.D Figures 5-4 and 7-12
Design capacity	Maximum day flow = 113,500 m^3/d	Sec. 7.8.1
Pipeline	–	Sec. 5.4.7.D Figure 7-12
G. Water treatment plant location		Chapter 5
General:	The water treatment site is located on the western end of Modeltown on State Highway 101.	Sec. 5.4.7.E Figure 5-4
Plant site	–	Sec. 5.4.7.E
H. Rapid-mix and flocculation basins		Chapter 8
General:	The treatment plant receives raw water at Junction Box A near the south property boundary. The 122-cm diameter conduit then leads water into an influent forebay. Four process trains are provided for coagulation, flocculation, and sedimentation. A lime-mixing chamber is also provided prior to the rapid-mix basins for pH adjustment.	Sec. 8.8.1 Figures 8-13, 15-3 and 15-7
Design capacity	Maximum day flow = 113,500 m^3/d	Sec. 8.8.1
Layout	–	Sec. 8.8.2, Step A Figures 8-13
Lime-mixing chamber	–	Secs. 8.8.1 and 8.8.2, Step A Figures 8-14 and 8-15
Rapid-mix basins	Include mixers, influent structures, and effluent structures.	Secs. 8.8.1 and 8.8.2, Step A Figures 8-14 and 8-15

Table 17-1 Summary of Basic Design Data and Dimensions of Water Treatment Facilities in the Design Example (continued)

Item	Design Information or Description	Reference
Flocculation basins	Including Parshall flume, flocculators, influent structures, and effluent diffusion wall.	Secs. 8.8.1 and 8.8.2, Step B Figures 8-16, 8-17 and 8-19 Table 8-7
I. Chemical building		Chapter 8
General	The chemical building is located adjacent to the rapid-mix basins. The chemical building offers facilities to handle ferric sulfate, polymers (coagulant and filter aids), lime, and potassium permanganate. The lime slurry is added at the lime-mixing chamber. Ferric sulfate is fed into the rapid-mix basins. The feed points of coagulant and filter aids are the Parshall flume and the sedimentation basin effluent box, respectively. Also see Item L.	Sec. 8.8.1 Figures 8-13, 8-22, 8-23, 15-3, and 15-7
Chemical systems	–	Secs. 8.8.1 and 8.8.2, Steps C Figure 8-22 Table 8-8
Building	–	Secs. 8.8.1 and 8.8.2, Step C Figure 8-23

Table 17-1 Summary of Basic Design Data and Dimensions of Water Treatment Facilities in the Design Example (continued)

Item	Design Information or Description	Reference
J. Sedimentation basins		Chapter 9
General	Four rectangular sedimentation basins are provided. Each is an integral part of the flocculation/ sedimentation basin complex. In each basin, there are two circular sludge-collector mechanisms. The effluent-collection system consists of V-notch weirs, launder troughs, center channel, and effluent box. Settled sludge is discharged into a sludge wet well. Draw-off flows are controlled by telescopic valves. The collected sludge is transferred to a residual-management system by sludge pumps.	Sec. 9.6.1 Figures 9-14 through 9-17, 9-20, 9-22, 15-3, 15-7 and 15-8
Design capacity	Maximum day flow = 113,500 m^3/d	Sec. 9.6.1
Layout	–	Sec. 9.6.2, Step A Figure 9-14
Sedimentation basins	Include effluent collection system.	Secs. 9.6.1 and 9.6.2, Step A Figures 9-15, -16 and -20
Sludge withdraw system	Include sludge quantity, sludge collection equipment, sludge draw-off, sludge pipeline, and sludge pump station.	Secs. 9.6.1 and 9.6.2, Step B Figures 9-15, -21 and -22
K. Filters		Chapter 10
General	Eight gravity filter units are provided. There are two banks of filters, divided by the administration and control building. Each bank of filters is further divided into two groups of two filters by a pipe gallery in the middle.	Sec. 10.13.1 Figures 10-16 through 10-20, 10-23, 10-24, 15-3, 15-7, and 15-8
Capacity	Maximum day flow = 113,500 m^3/d	Sec. 10.13.1
Layout	–	Sec.10.13.2, Step A

Table 17-1 Summary of Basic Design Data and Dimensions of Water Treatment Facilities in the Design Example (continued)

Item	Design Information or Description	Reference
Filter units	Include filter media, gravel support, and underdrain system	Secs.10.13.1 and 10.13.2, Step A Figures 10-17 through 10-20, and 10-23 Tables 10-5 and 10-8
Backwash system	Include surface wash, backwash characteristics of backwash water, backwash pumps, and elevated backwash water storage tank.	Secs.10.13.1 and 10.13.2, Step B Figures 10-24, 10-25, 12-20, and 12-21
L. Taste and odor control		Chapters 7, 8 and 11
General	The potassium permanganate storage and feed system is housed in the Chemical Building. The potassium permanganate solution is fed into the raw water at Junction Box A near the south property line. The PAC storage and feed system is located close the raw water pump station. The PAC slurry is added at the raw water pump station.	Secs. 8.8.1 and 11.6.2 Figures 7-13, 8-23, 11-10, and 11-12
Potassium permanganate, $KMnO_4$	Include storage, solution tanks, and feed system.	Secs. 8.8.1 and 8.8.2, Step C Secs. 11.6.1 and 11.6.2, Step A Figures 8-22, 8-23, and 11-10 Table 8-8
Powder activated carbon (PAC)	Include storage, solution tanks, and feed system.	Secs. 11.6.1 and 11.6.2, Step B Figures 7-13 and 11-12

Table 17-1 Summary of Basic Design Data and Dimensions of Water Treatment Facilities in the Design Example (continued)

Item	Design Information or Description	Reference
M. Disinfection and clearwells		Chapters 12 and 13
General	Two chlorine contact channels are provided for disinfection. Chlorine and ammonia storage and feed systems are both housed inside the Chlorine and Ammonia Building at the south side of the clearwells. The chlorine feed point is in the filtered water pipe upstream of the filter back pressure weir common bay. The ammonia solution is fed through diffusers at each influent channel, below the weir of the chlorine contact channel.	Secs. 12.8.1 and 13.8.1 Figures 12-20 through 12-23, 15-3, and 15-8
Layout	–	Sec. 12.8.2, Steps A and B Figure 12-20
Disinfection:	Include chlorine contact channels, chlorination system and ammoniation system.	Sec. 12.8.1 Sec. 13.8.2, Steps A, B, and C Figures 12-20 through 12-22 Table 12-21
Clearwells	–	Sec. 12.8.1 Sec. 13.8.2, Step A Figures 12-20 through 12-22
N. Fluoridation		Chapter 12
General	Sodium fluoride is provided for health considerations. The sodium fluoride storage and feed system is located in the Chlorine and Ammonia Building. The sodium fluoride solution is fed through diffusers at each influent channel, below the weir of the chlorine contact channel.	Sec. 12.8.1 Figures 12-20, 12-21, 12-23, 15-3, and 15-8
Fluoridation	Include storage, solution tanks, and feed system.	Secs. 12.8.1 and 12.8.2, Step D Figures 12-20 and 12-23

Table 17-1 Summary of Basic Design Data and Dimensions of Water Treatment Facilities in the Design Example (continued)

Item	Design Information or Description	Reference
O. Stability control		Chapter 13
General	Soda ash (99.2% Na_2CO_3) is used to adjust finished stability. The sodium carbonate storage and feed system is located in the Chlorine and Ammonia Building. The sodium carbonate solution is fed through diffusers at each influent channel, below the weir of the chlorine contact channel.	Sec. 13.8.1 Figure 12-23
Stability control	Include storage, solution tanks, and feed system.	Secs. 13.8.1 and 13.8.2, Step A Figure 12-23
P. High-service pumping station		Chapters 12 and 13
General	High-service pumps are used to deliver finished water to the City's water distribution system. The high-service pump station is located between the two clearwells.	Sec. 13.8.1 Figures 12-20 and 12-21
Capacity	Maximum pumping rate = 136,200 m^3/d	Sec. 13.8.1
Layout	–	Sec. 13.8.1 Figures 12-20 and 12-21
High-service pump station:	Include wet well and pumps.	Secs. 13.8.1 and 13.8.2, Step C Figures 12-20, 12-21, 13-8, and 13-9 Table 13-5
Q. Distribution system improvement		Chapter 13
General	The improvements to the distribution system include a new distribution pump station, a new elevated storage reservoir, a number of new and enlarged pipes.	Sec. 13.8.1 Figures 13-12 and 13-13 Table 13-12
Water demand	Maximum hourly demand = 181,600 m^3/d	Secs. 13.8.1 and 13.8.2, Step G
Distribution pump stations	Total pumping capacity = 164,600 m^3/d	Secs. 13.8.1 and 13.8.2, Step D

Table 17-1 Summary of Basic Design Data and Dimensions of Water Treatment Facilities in the Design Example (continued)

Item	Design Information or Description	Reference
Ground storage capacity	Total capacity = 16,000 m^3	Secs. 13.8.1 and 13.8.2, Step E
Elevated storage capacity	Total capacity = 8,000 m^3	Secs. 13.8.1 and 13.8.2, Step F
Distribution piping system	Total number of mains = 58 pipes (30-cm diameter and up)	Secs. 13.8.1 and 13.8.2, Step D Figure 13-12 and 13-13 Table 13-12
R. Residuals-management system		Chapter 14
General	The residuals-management system consists of a filter surge storage tank, a filter backwash water recovery basin, two gravity thickeners, and four sludge-drying lagoons. Two flocculation units and several pump stations are also used to provide chemical conditioning and transfer of sludge and overflows.	Sec. 14.6.1 Figures 14-7, 14-8, 14-11 through 14-14, 15-3 and 15-6
Layout	–	Figure 14-7
Mass-balance analysis	–	Sec. 14.6.2, Step A Figures 14-9 and 14-10 Tables 14-5 and 14-6
Filter backwash water surge storage tank	Include surge storage tank, pump station, and flocculation unit.	Secs. 14.6.1 and 14.6.2, Step B Figures 14-8 and 14-11
Filter backwash water recovery basin	Include recovery basin, supernatant pump station, underflow withdraw, and flocculation unit.	Secs. 14.6.1 Sec. 14.6.2, Step B Figures 14-8 and 14-12
Gravity thickeners	Include thickeners, supernatant pump station, underflow withdraw, and sludge pump station.	Sec. 14.6.1 and 14.6.2, Step C Figures 14-8 and 14-13
Sludge-drying lagoons	Include base and intermediate-layers, decant system, and lagoon life.	Secs. 14.6.1 and 14.6.2, Step D Figure 14-14

Table 17-1 Summary of Basic Design Data and Dimensions of Water Treatment Facilities in the Design Example (continued)

Item	Design Information or Description	Reference
S. Water treatment plant siting, layout, and yard piping		Chapter 15
General	An integrated compact plant with modular layout for the treatment units and buildings is proposed.	Secs. 15.6.1 and 15.6.2 Figures 15-3 through 15-8
Layout	–	Sec. 15.6.2 Figure 15-3
Administration and Control Building	–	Sec. 15.6.2 Figure 15-4
Yard piping system	Include process piping and major plant sanitary sewers.	Sec. 15.6.3 Figures 15-6 through 15-8
T. Hydraulic profiles		Chapter 15
Hydraulic profile	An maximum head loss through the entire treatment plant is 10.5 m.	Sec. 15.6.2 Figures 15-9 through 15-10 Table 15-4
U. Process control		Chapter 16
Process control systems	Include general requirements for each major process and eight simplified loop diagrams of selected processes.	Secs. 15.6.1 and 15.6.2 Figures 16-4 through 16-7 Table 16-3

CHAPTER 18

Nonconventional Water Treatment Processes and Designs

18.1 Introduction

Many raw water supply sources contain undesired constituents that require specialized or nonconventional water treatment processes in order to produce acceptable drinking water quality. The sources of these constituents are associated both with natural processes of chemical weathering and soil leaching and with human activities, such as municipal and industrial wastewater discharges, groundwater recharge, brine disposal, leaching of land fills, and run-offs from agricultural and mining areas. The constituents of concern are both inorganic and organic. The significance of many inorganic and organic constituents commonly found in natural water supply sources have been presented in Chapter 2. Many of these constituents cannot be removed in a conventional water treatment plant and, therefore, may require special water treatment processes. The purpose of this chapter is to present the application of some special methods necessary for the removal or control of many constituents in the drinking water supply. This chapter is organized to present only those constituents that frequently occur in water sources and cannot be removed appreciably by conventional treatment processes. The material is exclusively devoted to providing an overview of the typical range and concentration, sources, chemistry, treatment technology design parameters, and equipment.

18.2 Removal of Inorganics

The presence of high concentrations of inorganic salts causes salinity. Also, certain inorganic solute can cause significant concern with respect to drinking water quality, aesthetics, and industrial uses. Their impact may be on toxicity, process operation, and product quality in industrial processes. The most common inorganic constituents of concern are total dissolved solids, nitrate, fluoride, arsenic, selenium, and other toxic constituents. Removal of these constituents involves specialized treatment processes such as demineralization, selective ion exchange, selective adsorption, or even biological treatment processes. Many of these specialized treatment methods are discussed below.

18.2.1 Demineralization or Removal of Total Dissolved Solids

The removal of total dissolved solids from water is described as *desalinization, desalting,* or *salt-water conversion*. The salinity of raw water varies with its origin, and the degree of salinity is expressed in mg/L of total dissolved solids (TDS), chloride ion Cl–, or common salt (NaCl). The salinity classification of water is based on the water source and its relative salt content. The salinity of water from different sources is summarized in Table 18-1.[1,2]

Table 18-1 Classification of Water Based on Salinity and Source

Classification	Source	TDS mg/L
Fresh or sweet	Rivers, lakes, ground	less than 500
Slightly saline	Ground, river, lake	500–1000
Mildly saline	Brackish mixture of saline and fresh water, estuary	1000–2000
Moderately saline	Inland and brackish mixture	2000–10,000
Severely saline	Inland and coastal	10,000–30,000
Sea water[a]	Offshore waters of oceans and seas	30,000–36,000

a The major elements and their typical concentrations in sea water are Cl = 19,400 mg/L, Na = 10,800 mg/L, K = 380 mg/L, Mg = 1350 mg/L, S = 885 mg/L, Ca = 422 mg/L, Br = 68 mg/L, C = 28 mg/L, Sr = 8 mg/L, and B = 4 mg/L.

Water demineralization may be achieved by different processes. Some of these processes are (1) ion exchange, (2) membrane processes, (3) distillation, and (4) freezing. These processes are described in this section. Ion exchange and membrane processes are commonly used in water treatment, so these processes are covered in greater detail in the following sections.

Ion Exchange

Process Description. Ion exchange is a unit process used to selectively remove undesired ions such as Ca^{2+}, Mg^{2+}, Fe^{2+}, Mn^{2+}, NH_4^+, NO_3^-, F^-, and other cations and anions. Ion exchange is also used to demineralize the water.[3] In both cases, the ion exchange medium consists of (1) a solid phase and (2) a mobile ion attached to an immobile functional group. The mobile ions in the resin are exchanged with ions in the water. The spent resin is regenerated and reused. The ion exchange system may be operated in either a batch or a continuous mode. In a batch process, the resin is stirred in water until the reaction is complete. The spent resin is settled, removed, and regenerated. In a continuous system, water is passed through columns packed with the desired resins.

Ion-exchange resins are *cationic* or *anionic*. Cationic resins exchange positive ions, anionic resins exchange negative ions. Sodium cation exchangers remove hardness-causing bivalent cations and replace them with sodium ions available in the bed. The bed is regenerated with sodium chloride solution. In the hydrogen cation exchange process, hydrogen ions are exchanged for various positive ions present in water (Ca^{2+}, Mg^{2+}, Na^{+}, K^{+}, etc.). The hydrogen cation exchanger is regenerated with a mineral acid. The most widely used and the cheapest regenerant is sulfuric acid.

In anion exchangers, the negative ions in water (Cl^{-}, NO_3^{-}, SO_4^{2-}, etc.) are exchanged by OH^- ions. The main types of anion exchangers are (1) the weakly-basic anion exchanger and (2) the strongly-basic anion exchanger. Weakly-basic anion exchangers remove strongly ionized acids (HCl, H_2SO_4, etc.), but they will not remove such weakly ionized acids as H_2CO_3 and H_2SiO_3. These exchangers are regenerated with a solution of soda ash (Na_2CO_3).[4]

Strongly-basic anion exchangers remove both strongly and weakly ionized acids. These resins are regenerated with a caustic soda solution. Typical ion-exchange reactions are given by the following equilibrium equations.[4]

Sodium Cation Exchanger

Softening:

$$\underset{\substack{\text{Sodium cation}\\\text{exchanger}\\\text{(insoluble)}}}{Na_2R} + \underset{\substack{\text{Hardness}\\\text{(soluble)}}}{\begin{bmatrix} Ca \\ Mg \\ Fe \end{bmatrix} \cdot \begin{bmatrix} (HCO3)_2 \\ SO_4 \\ (Cl)_2 \end{bmatrix}} \rightarrow \underset{\substack{\text{Hardness}\\\text{with resin}\\\text{(insoluble)}}}{\begin{matrix} CaR \\ MgR \\ FeR \end{matrix}} + \underset{\substack{\text{Treated}\\\text{water}\\\text{(soluble)}}}{\begin{matrix} 2\ NaHCO_3 \\ Na_2SO_4 \\ 2\ NaCl \end{matrix}} \tag{18.1}$$

Regeneration:

$$\underset{\substack{\text{Hardness}\\\text{with resin}\\\text{(insoluble)}}}{\begin{matrix} CaR \\ MgR \\ FeR \end{matrix}} + \underset{\substack{\text{Regenerant}\\\text{solution}\\\text{(soluble)}}}{2\ NaCl} \rightarrow \underset{\substack{\text{Regenerated}\\\text{bed}\\\text{(insoluble)}}}{Na_2R} + \underset{\substack{\text{Waste}\\\text{or brine}\\\text{(soluble)}}}{\begin{matrix} CaCl_2 \\ MgCl_2 \\ FeCl_2 \end{matrix}} \tag{18.2}$$

Hydrogen Cation Exchanger

Reaction with bicarbonates:

$$H_2R + \begin{matrix} Ca(HCO_3)_2 \\ Mg(HCO_3)_2 \\ Na_2Ca(HCO_3)_2 \end{matrix} \rightarrow \begin{matrix} CaR \\ MgR \\ Na_2R \end{matrix} + 2\,CO_2 + 2\,H_2O \tag{18.3}$$

Hydrogen cation exchanger (insoluble)	Cations with bicarbonate (soluble)	Cations with resin (insoluble)	Carbon dioxide (soluble)

Reaction with sulfates and chlorides:

$$H_2R + \begin{bmatrix} Ca \\ Mg \\ Na_2 \end{bmatrix} \cdot \begin{pmatrix} SO_4 \\ Cl_2 \end{pmatrix} \rightarrow \begin{matrix} CaR \\ MgR \\ Ma_2R \end{matrix} + \begin{matrix} H_2SO_4 \\ Cl_2SO_4 \end{matrix} \tag{18.4}$$

Hydrogen cation (insoluble)	Cations with sulfate and chloride (soluble)	Cations with resin (insoluble)	Sulfuric and/or hydrochloric acid (soluble)

Regeneration:

$$\begin{matrix} CaR \\ MgR \\ Na_2R \end{matrix} + H_2SO_4 \rightarrow H_2R + \begin{matrix} CaSO_4 \\ MgSO_4 \\ Na_2SO_4 \end{matrix} \tag{18.5}$$

Cations with resin (insoluble)	Regeneration solution (soluble)	Hydrogen cation exchanger (insoluble)	Waste or brine (soluble)

Weakly Basic Anion Exchanger

Reactions with acids:

$$2\ RNH_3OH + \begin{matrix} H_2SO_4 \\ 2\ HCl \\ 2\ HNO_3 \end{matrix} \rightarrow \begin{matrix} (RNH_3)_2SO_4 \\ 2\ RNH_3Cl \\ 2\ RNH_3NO_3 \end{matrix} + 2\ H_2O \tag{18.6}$$

Anion exchanger (insoluble)	Acid (soluble)	Anion exchanger (insoluble)

Regeneration:

$$\begin{matrix} (RNH_3)_2SO_4 \\ 2\ RNH_3Cl \\ 2\ RNH_3NO_3 \end{matrix} + Na_2CO_3 \rightarrow 2RNH_3OH + \begin{matrix} Na_2SO_4 \\ 2\ NaCl \\ 2\ NaCO_3 \end{matrix} + CO_2 + H_2O \tag{18.7}$$

Anion exchanger with acid (insoluble)	Regeneration solution (soluble)	Anion exchanger (insoluble)	Waste or brine (soluble)

Strongly Basic Anion Exchanger

Reactions with acids:

$$2\ RNOH + \begin{matrix} H_2SO_4 \\ 2\ HCl \\ 2\ HNO_3 \\ 2\ H_2CO_3 \\ 2\ H_2SiO_3 \end{matrix} \rightarrow \begin{matrix} (RN)_2SO_4 \\ 2\ RNCl \\ 2\ RNHCO_3 \\ 2\ RNNO_3 \\ 2RNHSiO_3 \end{matrix} + 2\ H_2O \tag{18.8}$$

Anion exchanger (insoluble)	Acids with anion (soluble)	Anion exhanger (insoluble)

Regeneration:

$$\begin{matrix} 2RNSO_4 \\ 2\ RNCl \\ 2\ RNNO_3 \\ 2\ RNHCO_3 \\ 2\ RNNSiO_3 \end{matrix} + 2\ NaOH \rightarrow 2\ RNOH + \begin{matrix} Na_2SO_4 \\ 2\ NaCl \\ 2\ NaNO_3 \\ 2\ NaHCO_3 \\ 2\ NaHSiO_3 \end{matrix} \qquad (18.9)$$

Anion exchanger with anions (insoluble)	Regeneration solution (soluble)	Anion exchanger (insoluble)	Waste or brine (soluble)

Design Considerations. Ion exchangers are often granular-packed bed columns. The water is trickled downward under pressure. When the exchange capacity of the resin is exhausted, the column is backwashed to remove any trapped solids, and the bed is then regenerated. Synthetic cation and anion resins perform better than the naturally occurring zeolites. The cation and anion resin beds for demineralization may be arranged in a series in separate exchange columns or may be mixed in a single vessel. The typical bed depths are 1–2 m (3.3–6.5 ft), and the flow rate is 0.20–0.40 $m^3/m^2{\cdot}min$ (5–10 gpm/ft^2). The total ion exchange capacity of commercially available resins is 50,000–80,000 g/m^3 (as $CaCO_3$).

Ion-exchange demineralizers produce high-quality "distilled" water. For water supply, a split treatment is generally used. A desired volume of water is demineralized, then mixed with the main stream to provide a known amount of TDS in the mixture. Organic matter and suspended solids in the feed may plug the bed and blind the resin; therefore, filtration is generally needed.[5] A schematic flow diagram of an ion-exchange bed is given in Figure 18-1(a). The ion-exchange capacity calculation is provided in Example 18-1.

Equipment and Operation. The equipment commonly needed for an ion-exchange demineralization process includes pretreatment systems, ion-exchange beds, resins, pressure pumps, regenerant solution storage and pumps, backwash, and a rinse water system. A commercial ion-exchange assembly is shown in Figure 18-1(b). The operating procedures for an ion exchanger include very equipment-specific instructions by the manufacturer. In general, the ion-exchange demineralization process is operated in a sequencing mode consisting of the following four basic steps: (1) exhaustion, (2) backwash, (3) regeneration, and (4) slow and fast rinse.

Design Parameters. Common design parameters include average design flow, total dissolved solids in the raw water, total dissolved solids required in the finished water, degree of demineralization needed, ion-exchange capacity of the resins, and flow rate through the beds.

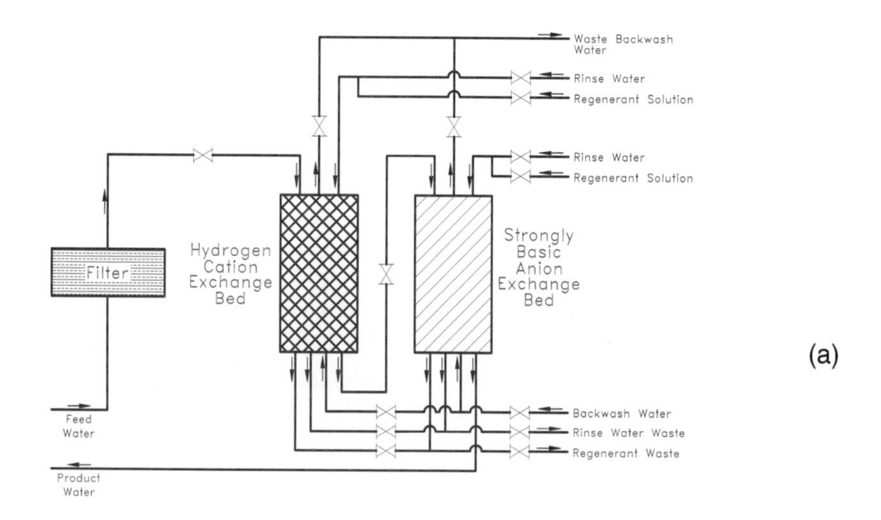

(a)

(b)

Figure 18-1 Ion-exchange demineralization system. (a) Typical flow diagram of an ion-exchange demineralization system. (b) Commercial ion-exchange equipment. *(Courtesy of Osmonics, Inc.)*

Example 18-1

Determine the volumes of hydrogen cation and strongly-basic anion exchanger beds to demineralize 100 m^3/d water that has the following chemical quality.

Cations	Anions
Ca^{2+} = 145 mg/L	HCO_{3-} = 258 mg/
Mg^{2+} = 18 mg/L	SO_4^{2-} = 220 mg/L
Na^+ = 130 mg/L	Cl^- = 214 mg/L
K^+ = 50 mg/L	NO_3^- = 50 mg/L

The ion exchange capacities of hydrogen cation and anion exchange resins are 80,000 and 50,000 g $CaCO_3/m^3$·cycle, respectively. Also, calculate the required quantities of regeneration chemicals. The regeneration cycle is once per day.

Solution

1. Compute total cations and anions in water.

   Cation–anion balance of the water sample is given below.

Cations				Anions			
Ion	Conc., mg/L	Eq. wt	mg/L as CaCO3	Ion	Conc., mg/L	Eq. wt	mg/L as $CaCO_3$
Ca^{2+}	145	20.0	362.5	HCO_{3-}	258	61.0	211.5
Mg^{2+}	18	12.2	73.8	SO_4^{2-}	220	48.0	229.2
Na^+	130	23.0	282.6	Cl^-	214	35.5	301.4
K^+	50	39.1	63.9	NO_3^-	50	62.0	40.3
			782.8				782.4

2. Compute volume of resin.

Total quantity of cations exchanged per cycle

$$= 782.8\ \text{g/m}^3 \times 100\ \text{m}^3\text{/d} \times 1\ \text{d/cycle} = 78{,}280\ \text{g as CaCO}_3\text{/cycle}$$

Required volume of hydrogen cation exchange bed

$$= \frac{78{,}280\ \text{g as CaCO}_3\text{/cycle}}{80{,}000\ \text{g as CaCO}_3\text{/m}^3\cdot\text{cycle}} = 0.98\ \text{m}^3$$

Total quantity of anions exchanged per cycle

$$= 782.4\ \text{g as CaCO}_3\text{/m}^3 \times 100\ \text{m}^3\text{/d} \times 1\ \text{d/cycle}$$

$$= 78240\ \text{g as CaCO}_3\text{/cycle}$$

Required volume of strongly-basic cation exchange bed

$$= \frac{78240 \text{ g as CaCO}_3\text{/cycle}}{50{,}000 \text{ g as CaCO}_3\text{/m}^3\cdot\text{cycle}} = 1.56 \text{ m}^3$$

In practice, because of leakage and other operational and design limitations, the required volume of resin beds is 1.2 to 1.5 times the volume theoretically required.

3. Compute regeneration chemicals per cycle.

Sulfuric acid (H_2SO_4) of 93-percent purity is used for regeneration of hydrogen cation bed.

Required quantity of sulfuric acid per cycle

$$= \frac{49.1 \text{ eq wt of H}_2\text{SO}_4 \times 78{,}280 \text{ g as CaCO}_3\text{/cycle}}{0.93 \times 50 \text{ eq wt of CaCO}_3 \times 1000 \text{ g/kg}} = 82.7 \text{ kg/cycle}$$

A dilute solution of 2-percent strength is generally used. Actual quantity of acid used is two to three times the amount theoretically required.

Sodium hydroxide (NaOH) of 95-percent purity is used for regeneration of a strongly basic anion exchange bed.

Required quantity of sodium hydroxide per cycle

$$= \frac{40 \text{ eq wt of NaOH} \times 78{,}240 \text{ g as CaCO}_3\text{/cycle}}{0.95 \times 50 \text{ eq wt of CaCO}_3 \times 1000 \text{ g/kg}} = 65.9 \text{ kg/cycle}$$

Actual quantity of NaOH solution used is two to three times the amount theoretically required.

Membrane Processes Membrane processes are those in which a membrane is used to permeate high-quality water while rejecting the passage of dissolved and suspended solids. These processes can be classified into two groups based on the process-driving forces: (1) processes driven by pressure–reverse osmosis, nanofiltration, ultrafiltration, and microfiltration, and (2) processes driven by electric current–electrodialysis, and electrodialysis reversal. In the water industry, the membrane processes have been used for demineralization and for removal of both dissolved and suspended particles. Tremendous improvements have been made in recent years, and the utilization of membrane technology has dramatically increased in potable water treatment. It is expected that membrane processes will be used more and more in the future as more stringent drinking water quality standards will likely become enforced.

Pressure-Driven Membrane Processes

Process Description. These membrane processes are originally developed on the principle of reverse osmosis. Osmosis is the natural passage of water through a semi-permeable membrane from a weaker solution to a stronger solution, to equalize the chemical potentials in the membrane-separated solution. Osmotic pressure is the driving force for osmosis to occur. In reverse osmosis, an external pressure greater than the osmotic pressure is applied to the solution, causing water to flow against the natural direction through the membrane, thus producing high-quality demineralized water. The principles of osmosis and reverse osmosis are shown in Figure 18-2.[6] Reverse osmosis (RO) processes have the highest material-rejection capacity (0.0001–0.001 μm range). Reverse osmosis processes are mostly used for demineralization in industrial water supply or for desalinization of seawater in drinking water supply.

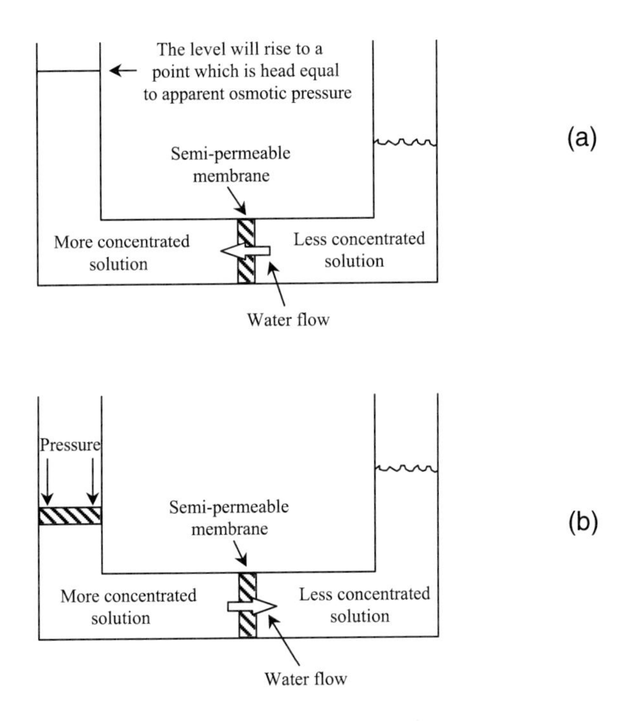

Figure 18-2 Principle of osmosis and reverse osmosis. *(Reprinted from Reference 6, with permission of Technomic Publishing Co., Inc., copyright 1999.)* (a) Osmosis. (b) Reverse osmosis.

Nanofiltration (NF), ultrafiltration (UF), and microfiltration (MF) are other membrane processes, which are commonly used not for demineralization but for removal of particles, color, NOMs, and other inorganic and organic contaminates. The particle-removal ranges of these processes are quite different from that of reverse osmosis. Nanofiltration processes reject molecules in the range from 0.001 to 0.01 μm. Such a process can be used for softening and for removal of TDS and TOC. Ultrafiltration is the membrane process that is designated to remove colloidal-sized particles, in the range from 0.005 to 0.1 μm. It is not effective for demineralization purpose; however, most turbidity-causing particles, viruses, and most organic substances (such as large molecular NOMs and proteins) are within the removal range of ultrafiltration. Ultrafiltration is actually a physical screening process using a relatively coarse membrane. Microfiltration utilizes membranes that typically have a pore size of 0.45 μm. The particle-removal range is between 0.05 μm and 1 μm. In general, this process is not applicable in potable water treatment. It is mostly used in industrial water supply and in pretreatment prior to a reverse osmosis process. Depending on the type of process, the membrane may reject suspended solids, molecules, or ions, while permitting acceptable rates of water passage. The applications of different membrane processes for separation of common materials are shown in Figure 18-3.[7]

Equipment and Operation. Reverse osmosis modules suitable for water treatment involve arrangement of membranes and their supporting structures so that feed water under high pressure (up to 10,000 kN/m^2 (1500 lb/in^2)) can pass through the membrane surface, while product water is collected from the opposite face without brine contamination. The applied pressure for an ultrafiltration system is normally below 1000 kN/m^2 (150 psig).[8] It is much lower than that required for a reverse osmosis process. A schematic of a reverse osmosis system is shown in Figure 18-4 (a).

Reverse osmosis modules have been developed in four different types of designs: plate and frame, large tube, spiral wound, and hollow fine fiber.

Many types of membranes for reverse osmosis have been developed, but cellulose acetate and polyamide (nylon) are currently the most widely used membrane materials in reverse osmosis system. Typical membranes are approximately 100 μm thick having a surface skin of about 0.2 μm thick that serves as the rejecting surface. The remaining layer is porous and spongy, and serves as backing material. The hollow fine fibers have outer and inner diameters of 50 and 25 μm. Basic design and operational information of reverse osmosis systems are summarized in Table 18-2.[9] A commercially available membrane, a spiral wound or scroll type, is illustrated in Figures 18-4 (b).

The major equipment for a nanofiltration process is similar to that for reverse osmosis system, and includes membrane module with support system, high-pressure pump and piping, and a brine-handling and disposal system. The equipment for an ultrafiltration system is different from that for reverse osmosis or nanofiltration process. For instance, the pressure requirement for the feed pump is much lower. There is no brine-handling and disposal system required because inorganic ions and molecules can not be effectively removed by nultrafiltration process. Therefore, the nultrafiltration is only periodically backwashed by using either water or air similar to that of gravity filtration system. Occasional maintenance is required to clean the membranes. This is done by soaking the membrane in a weak acidic or basic solution. The membrane processes are highly automated systems. They are usually easy to operate with the help of control software. The operating procedures is provided by the manufacturer of the equipment.

Design Parameters. The design of a membrane system to handle demineralization requires consideration of the following variables: plant capacity, salinity of raw water, pretreatment needs, product-recovery rate, rejection rate, applied pressure, feed-water temperature, and

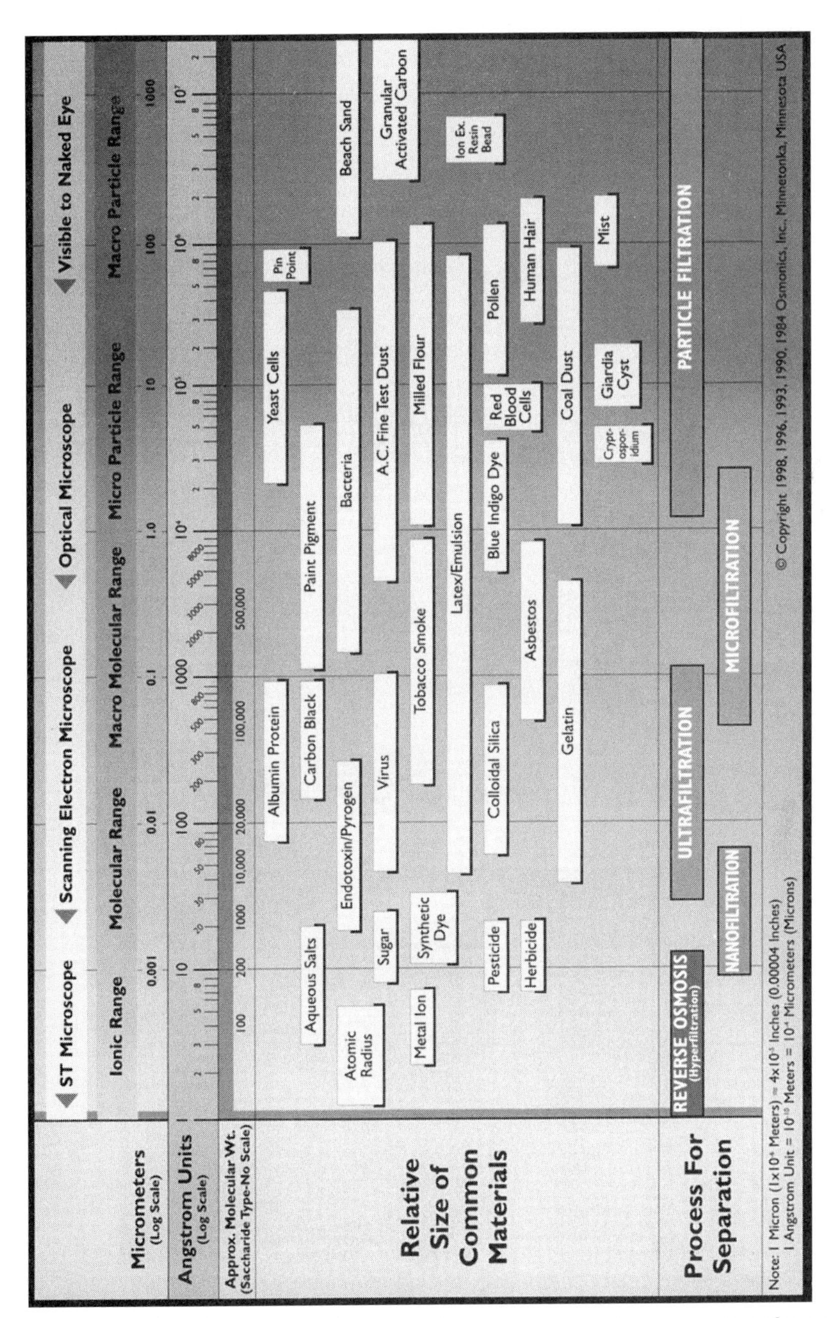

Figure 18-3 Application of pressure-driven membrane processes for separation of common materials. *(Courtesy of Osmonics, Inc.)*

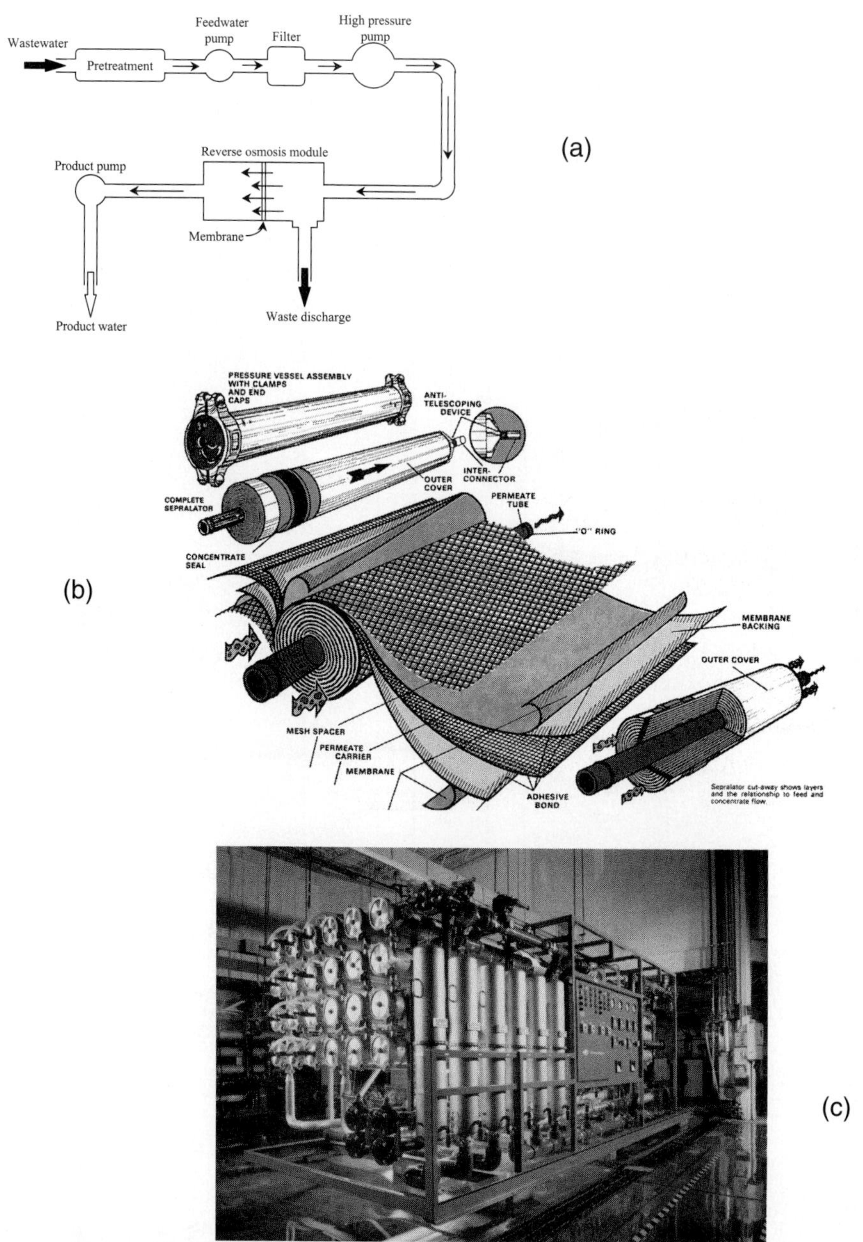

Figure 18-4 Reverse osmosis schematic and commercially available membrane material and equipment. (a) Schematic of reverse osmosis system. (b) Spiral wound or scroll configuration membrane material. *(Courtesy of Osmosis, Inc.)* (c) A commercial membrane equipment. *(Courtesy of Osmosis, Inc.)*

Table 18-2 Summary of Design and Operation Parameters of a Reverse Osmosis System

Parameters	Descriptions	Range	Typical
Pressure	Higher pressure gives greater flux, but pressure capability of membranes is limited.	2000–7000 kN/m^2 (300–1000 psig)	4140 kN/m^2 (600 psig)
Temperature	Higher temperatures give greater flux. The life of membrane is reduced at temperatures greater than 38°C.	16–38°C	21°C
Packing density	It is the area of the membrane that can be placed per unit volume of the pressure vessel.	164–1640 m^2/m^3 (50–500 ft^2/ft^3)	820 m^2/m^3 (250 ft^2/ft^3)
Flux	Flux is recovery of product water per unit area of the membrane. Flux tends to decrease with length of run. Flux may decrease by 50 percent after 2 years of operation.	0.1–5 $m^3/m^2{\cdot}d$ (2.5–120 gpd/ft^2)	0.5–1.5 $m^3/m^2{\cdot}d$ (12–35 gpd/ft^2)
Recovery factor	Recovery factor is the ratio of the volume of product water to the total volume treated. At a higher recovery factor, there may be more salt in the product water.	75–95 percent	80 percent
Salt rejection	Salt rejection depends on the type of membrane and the salt-concentration gradient.	85–99 percent	95 percent
Pre-treatment and feed water	The feed water must be free of scale forming constituents (calcium, magnesium, iron, manganese, silicon, etc.). Concentrations of such ions should be reduced by pretreatment. TDS should be less than 10,000 mg/L. Organics, microorganisms, and oil and grease must be removed to prevent coating and fouling of membranes. Cellulose-acetate membranes are subject to hydrolysis at high and low pH.		
	Turbidity and particle size must be controlled:		
	pH	4.5–5.5	4.7
	turbidity	–	< 1 NTU
	particle size	–	< 25 μm
Feed water stream velocity	High velocities and turbulent flow are necessary to minimize concentration polarization at the membrane surface.	0.01–0.8 m/s (0.04–2.5 ft/s)	–
Life	Life of membrane depends on quality of feed water.	–	2 years
Cleaning	Membranes must be cleaned periodically by depressurization, high-velocity water, air-water mixture, and chemical solutions.	periodic	–
Power	Power is needed for pumping at high operational pressures.	2–5 kWh/m^3 (9–17 kWh/1000 gal)	–

method of brine disposal. The ultrafiltration and macrofiltration processes are not suitable for demineralization so the design variables for these systems are plant capacity, turbidity of raw water, flux rate of treatment module, transmembrane pressure, removal of turbidity and particles, backwash quantity and quality, and average cleaning frequency.

The design procedure for a membrane plant to handle demineralization includes (1) determining raw water quality and selecting pretreatment processes, (2) selecting reverse osmosis system after consultation with the equipment manufacturers, (3) selecting operating parameters (flux rate, rejection factor, applied pressure, system life, performance level, etc.), (4) calculating system size, (5) determining power requirement, (6) selecting brine-disposal system, and (7) estimating system economics, including amortization of capital cost, labor, supplies and chemicals, and brine disposal. A commercial membrane assembly are shown in Figures 18-4 (c).

Example 18-2

A municipal water supply source has a total dissolved solids (TDS) concentration of 1000 mg/L. Develop the design, and size the various components of a reverse osmosis system, to produce finished water having a TDS concentration of less than 300 mg/L. The plant capacity is 19,000 m^3/d. Use the following data:

Plant design capacity, Q	19,000 m^3/d (5 MGD)
Recovery factor, R	75 percent
Salt-rejection factor, S	95 percent
Design pressure, P	4140 kN/m^2 (600 psig)
Feed water temperature, T	27°C
TDS in raw water, C_0	1000 mg/L
TDS in finished water, C	300 mg/L
Flux rate, f	0.82 $m^3/m^2 \cdot d$ (20 gpd/ft^2)

Solution

1. Compute the system size for when using split treatment (Figure 18-5).

$$C_e = \frac{C_0(1-S)}{R} = \frac{1000 \text{ mg/L} \times (1-0.95)}{0.75} = 67 \text{ mg/L}$$

Using the flow and concentration for the split treatment in Figure 18-5 (a), calculate the capacity "q_e" of the RO unit.

$$(19{,}000 \text{ m}^3/\text{d} - q_e \text{ m}^3/\text{d}) \times 1000 \text{ mg/L} + q_e \text{ m}^3/\text{d} \times 67 \text{ mg/L}$$
$$= 19000 \text{ m}^3/\text{d} \times 300 \text{ mg/L}$$

Solve for q_e = 14,255 m^3/d.

$$\text{Feed water flow to RO system, } q_f = \frac{q_e}{R} = \frac{14{,}255}{0.75} = 19007 \text{ m}^3/\text{d}$$

$$\text{Brine flow } q_b = (q_f - q_e) = 19{,}007 \text{ m}^3/\text{d} - 14{,}255 \text{ m}^3/\text{d} = 4752 \text{ m}^3/\text{d}$$

$$\text{TDS in brine } C_b = \frac{0.95 \times 1000 \text{ mg/L}}{1 - 0.75} = 3800 \text{ mg/L}$$

$$\text{Total raw water flow } Q_0 = q_f \text{ m}^3/\text{d} + (19{,}000 - q_e) \text{ m}^3/\text{d}$$
$$= 19{,}007 \text{ m}^3/\text{d} + (19{,}000 - 14{,}255) \text{ m}^3/\text{d} = 23{,}752 \text{ m}^3/\text{d}$$

The final results of the material mass-balance analysis is shown in Figure 18-5(b).

For a flux rate of 0.82 $m^3/m^2{\cdot}d$ (20 gpd/ft^2), calculate the required area of the RO membrane.

$$A = \frac{14255 \text{ m}^3/\text{d}}{0.82 \text{ m}^3/\text{m}^2{\cdot}\text{d}} = 17384 \text{ m}^2$$

Assuming a packing density of 820 m^2/m^3, calculate the required total volume.

$$V = \frac{17384 \text{ m}^2}{820 \text{ m}^2/\text{m}^3} = 22 \text{ m}^3$$

Assuming unit volume of each module is 0.03 m^3/module, calculate the total number of modules required.

$$n = \frac{22 \text{ m}^3}{0.03 \text{ m}^3/\text{module}} = 733 \text{ modules}$$

If 10 modules are used per pressure vessel, calculate the total number of pressure vessels needed.

$$N = \frac{733 \text{ module}}{10 \text{ module/pressure vessel}} = 73.3 \text{ pressure vessels}$$

Provide 75 pressure vessels.

2. Compute power consumption.

$$\text{Water power} = \text{design pressure} \times \text{flow rate}$$
$$= \frac{4140 \text{ kN/m}^2 \times 19{,}007 \text{ m}^3/\text{d}}{(60 \text{ s/min} \times 60 \text{ min/h} \times 24 \text{ h/d})} = 911 \text{ kN·m/s or kW}$$

Assuming 95 percent pump efficiency,

$$\text{Brake power} = \frac{911 \text{ kW}}{0.95} = 959 \text{ kW}$$

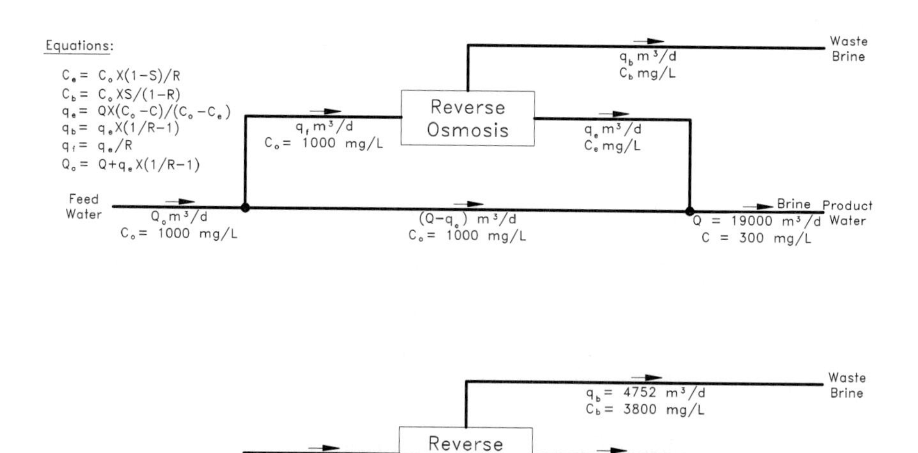

Figure 18-5 Mass-balance analysis around the RO unit in Example 18-2. (a) Mass balance with concentrations and flows of RO system. (b) Final result of mass-balance analysis.

Assuming 88 percent motor efficiency,

$$\text{Motor power} = \frac{959 \text{ kW}}{0.88} = 1090 \text{ kW}$$

If provision is desired to operate the system at higher design pressure, the power calculations must be corrected.

Electric-Driven Membrane Processes

Process Description. Electrodialysis (ED) is also a membrane process, and it is effective for demineralization. In this process, an electrical potential is applied to push the ions through ion-selective membranes.[9,10] This process is most practical and is widely used for treatment of brackish waters. The energy requirement is directly proportional to the concentration of salts in the water being treated. An electrical voltage is applied in cells (also called stacks). The cells contain alternatively arranged cation- and anion-permeable membranes. The cations and anions in the feed liquid migrate to the opposite poles. They pass the cation and anion-selective membranes, thus producing clean water and brine streams. To avoid membrane fouling, the water must be pretreated as is done for reverse osmosis. The schematic flow diagram of an electrodialysis system is given in Figure 18-6. The electrodialysis reversal (EDR) process is similar to electrodialysis, except that the polarity of the direct current is periodically reversed. The purpose of reversing the polarity is to automatically flush scale-forming materials from the membrane surface and significantly reduce the requirements for pretreatment of feedwater. Unlike pressure-driven membrane processes, the electrodialysis processes are not effective for removal of dissolved organics and particulates.

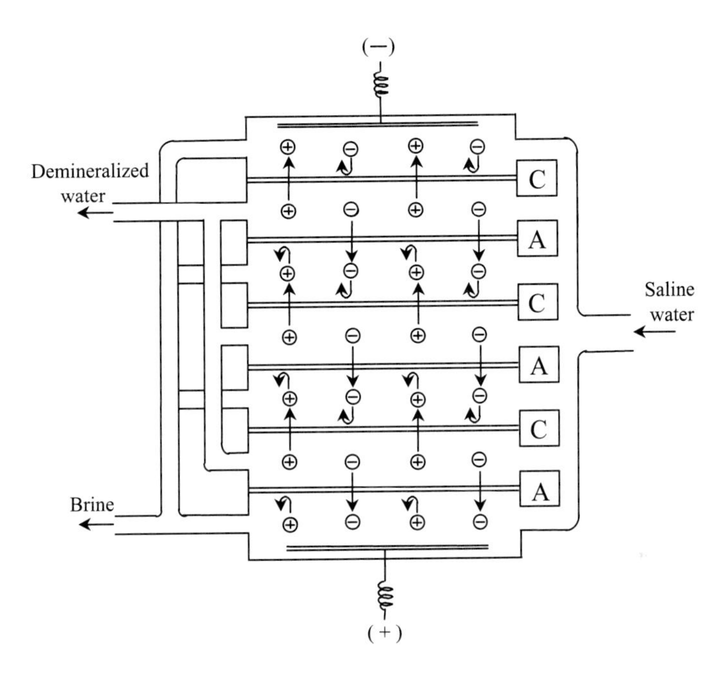

Figure 18-6 Mechanism of the electrodialytic process. *(Reprinted from Reference 6, with permission of Technomic Publishing Co., Inc., copyright 1999.)*

Equipment and Operation. The major equipment for an electrodialysis system includes a membrane stack with support system, a pretreatment system, and a brine-handling and -disposal system.

Design Parameters . The design parameters for electrodialysis systems are average design flow, salinity, product-recovery rate, and influent quality to the membrane, and pretreatment requirements.

Distillation Processes

Process Description . Distillation is the oldest demineralization process. It consists of evaporating part or all of the water from a saline solution and the subsequent condensation of the mineral-free vapor. The energy requirements are very high for the system. One modification of the simple distillation process is multiple-effect evaporation, where water is evaporated in different stages under a small vacuum created by the condensed steam from the previous effect.[11] The latent heat of condensation is used to preheat the incoming water. This method has been under extensive commercial development for many years and is used for production of fresh water from the seawater.[11] This process is shown in Figure 18-7(a). Another modification is distillation with vapor compression (Figure 18-7(b)). This system utilizes the latent heat of the compressed steam to preheat and evaporate the incoming water. Both modifications conserve energy, but the energy requirements are still quite large.

Solar distillation offers the advantage of using energy otherwise wasted. The basic principles of solar distillation had been known for many years before the first significant installation unit was made in Chile in 1872 for recovery of fresh water from saline water. The simplest solar-distillation unit consists of a shallow blackened pan with sloping glass, exposed to the sun (Figure 18-7(c)).[12] The evaporated water condenses on the sloping glass sheets (because they are cooler) and runs down to the collecting channels for recovery of condensate. With present techniques, about 3 kg of distillate can be produced per day per square meter of evaporating pan. This gives an approximate operating efficiency of 35 percent at an average solar radiation of 47.5 calories/cm^2·d. Such yields have been achieved in practice with experimental and full-scale large units operating over long periods.[13,14]

Equipment and Operation. The equipment commonly needed for distillation includes a raw liquid feed system, a still or boiler, liquid-heating and -evaporating systems, vapor-separation and -condensation systems, and condensate- and brine-recovery systems.

Design Parameters. Common design parameters for the distillation process are the design flow, the raw water temperature, the number of stages, the temperature drop allowed in each stage, the efficiency of heat exchangers, and the brine-handling and -disposal systems.

Freeze Process

Process Description. If saline solution is cooled to the freezing point and ice crystals are formed, the dissolved salts remain in the solution. Thus, the brine contains a higher concentration of salts than the initial solution, and the ice contains pure water. This process has been used in colder climates, where ice is formed in saline water during the winter. The ice is separated from the brine and melted in summer to produce fresh water. It should be noted that the freezing point of a solution and the quality of the ice produced depends on the salinity of the solution.[11] Freezing temperature is lower for higher-salinity water, and ice crystals produced become coated with the brine at excessively high salt content.

Equipment and Operation. The equipment commonly needed for the freeze process includes a cooling system to freeze the ice slab (not applicable in cold climate), ice-separation and -melting systems, and systems for the collection and removal of brine and fresh water.

Design Parameter. Common design parameters for the freeze process are design flow, salt concentration, cooling rate, land area, depth of impoundment needed (if natural cooling is used), and brine disposal.

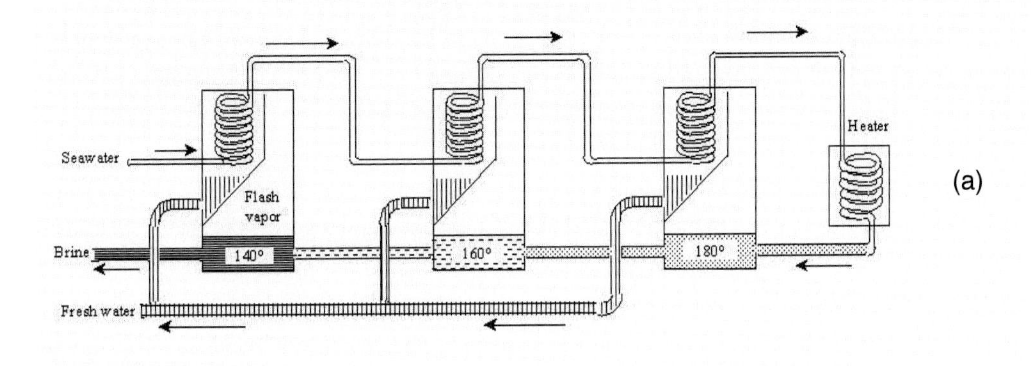

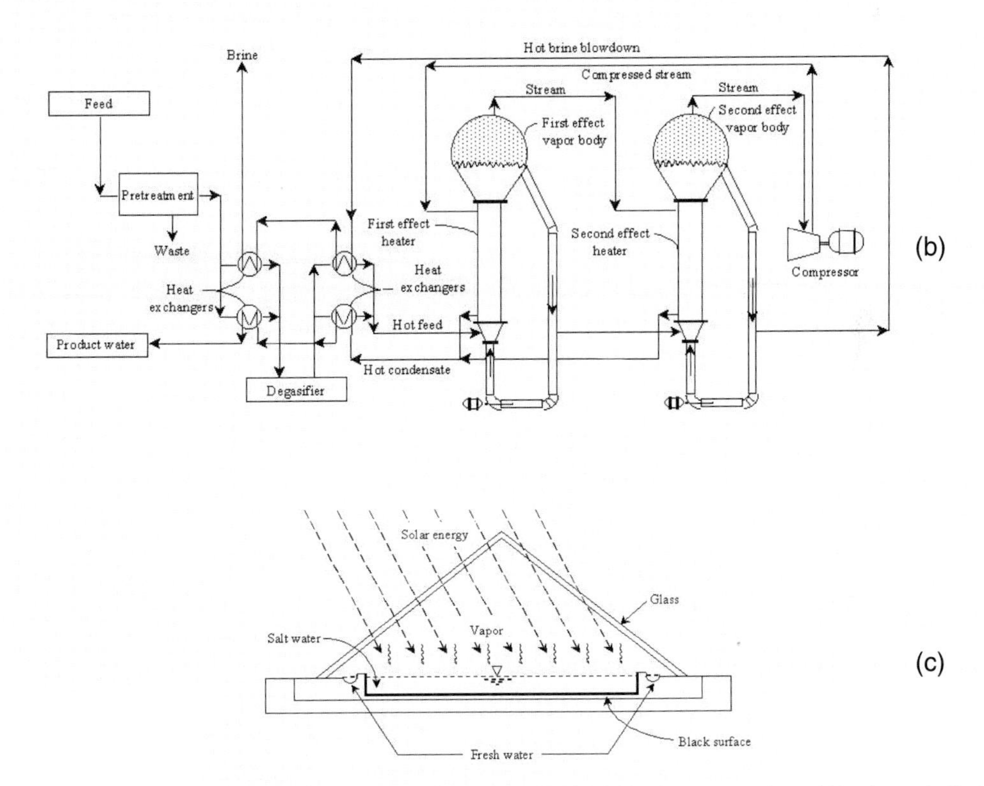

Figure 18-7 Water distillation. *(Reprinted from Reference 6, with permission of Technomic Publishing Co., Inc., copyright 1999.)* (a) Flash distillation. (b) Forced-circulation vapor-compression distillation. (c) Solar distillation.

18.2.2 Nitrate Removal

The presence of nitrate in high concentrations in drinking water constitutes an indirect health hazard. Nitrate is considered acutely toxic, because it can be reduced in the stomach or by saliva to nitrite and so induce methemoglobinemia in infants. The National Primary Drinking Water Standard for nitrate is 10 mg/L as nitrogen.

Process Description Several nitrate-removal techniques have been studied. These are biological denitrification, membrane, and ion exchange.[5,15] Each process is briefly described next.

Biological Denitrification. Biological denitrification of nitrate is an effective means of nitrate removal from water. In this process, nitrate is reduced to nitrogen gas by bacteria under anoxic conditions. The bacteria require a suitable carbon source, such as methanol or ethanol (Eq. (18.10). This process is effective, but it has not gained widespread acceptability for potable water, because both the bacteria and the organic feed source are water contaminants that will also require removal; however, this process is gaining acceptance in Europe.

$$\underset{\substack{\text{Methanol} \\ \text{(carbon source)}}}{6\,HNO_3 + 5CH_3OH} \rightarrow 5\,CO_2 + 3\,N_2 + 13\,H_2O \qquad (18.10)$$

The biological denitrification process commonly used in water treatment employs fixed media bed and fluidized media bed anaerobic filters. The clarifier is typically needed to separate the organisms grown on the media for nitrate reduction. The effluent must be aerated to remove nitrogen gas and volatile organics. Methanol or ethanol must be of "food grade". Special care is needed to remove microorganisms and residual organics from the carbon source and the microbiological end-products.

Membrane Process. Reverse osmosis has received consideration as an alternative for nitrate removal. Nitrate rejection has been 65 to 90 percent, depending on the influent water quality and other constituents in the feed water. Because other cations and anions are also rejected, reverse osmosis systems are used primarily for treating high-TDS or saline water where demineralization is necessary.

Ion Exchange Process. Ion exchange is another effective and economical process for removal of nitrate from drinking water. Strong-base anionic resins are used that remove nitrate ion in competition with other major anions.[15] If water has a TDS less than 1000 mg/L, nitrate will be the ion second most preferred by strong-base anionic resins in competition with sulfate, chloride, and bicarbonate (sulfate being the most preferred). The development of a selective anionic exchange resin (tributylamine, strong base) that prefers nitrate over sulfate allowed the construction of a 1-mgd (3800 m^3/d) treatment facility in McFarland, California. The plant operates in semi-automatic mode, blending 70 percent treated water with 30 percent untreated.[16]

The design of a nitrate-removal column will involve selection of the resin based on its nitrate preference, exchange capacity, and regeneration requirements. Because nitrate removal is sensitive to the feed rate, the loading rate is generally kept below 300 L/min·m^3 (2.5 gpm/ft^3). The depth of the resin bed should be greater than 0.6 m. A pilot plant study could be necessary to develop design and operation criteria for the nitrate-removal facility.[5] It is important that an accurate and reliable continuous nitrate-monitoring instrumentation be used to determine the nitrate concentration in the treated water. This is necessary, because, in some resins, once nitrate breakthrough occurs, the concentration of nitrate in the effluent could far exceed that in the influent for some period of time.

Equipment and Operation. The major equipment for nitrate removal from drinking water greatly depends upon the method used. A biological denitrification system requires a suspended or attached growth biological reactor, clarifier, methanol or ethanol feed system, and apparatus to remove residual organics (activated carbon). The major equipment for reverse osmosis and ion exchange systems is the same as given under demineralization.

Design Parameters. The design parameters for a nitrate-removal system include design flow rate, concentration of nitrate, concentration of competing anions, and TDS.

18.2.3 Defluoridation

Fluoride has long been recognized as a preventive agent against tooth decay. The optimum concentration of fluoride in drinking water for the prevention of dental caries is 1.0 mg/L; however, concentrations above 1.5 mg/L increase the occurrence and severity of tooth mottling. High concentrations of fluoride are toxic and can cause fluorosis, mottled tooth and enamel, kidney and thyroid injury, and death. The National Primary and Secondary Drinking Water Standards for fluoride are 4.0 and 2.0 mg/L, respectively.

Process Description. Both lime softening and alum coagulation are capable of reducing fluoride levels to some degree; however, the only acceptable method of defluoridation is adsorption onto activated alumina (Al_2O_3) or bauxite.[a] Adsorption onto activated alumina can reduce fluoride levels to below drinking water standards.[5]

Activated alumina is used in much the same way as ion-exchange resins. The water is filtered through a bed of activated alumina. The exchange capacity of alumina for fluoride depends on three important factors: (1) the initial fluoride concentration, (2) the pH of the feed water, and (3) the particle size of the alumina.[17] The capacity of activated alumina for fluoride removal increases with increasing fluoride concentration in the influent. The optimum pH of the influent for defluoridation is around 5.5. In a packed-bed column, a smaller particle size of alumina provides the most rapid uptake of fluoride; however, the smaller particles have a greater tendency to be washed out during backwash and are subject to dissolution by NaOH solution during regener-

a. Alum coagulation at high alum dosages will reduce fluoride level in a conventional coagulation process. Alum coagulation, may, however not reduce fluoride levels to acceptable drinking water standards.

ation. Activated alumina of 28–48 mesh is most commonly used. The fluoride capacity of activated alumina is 6–8 kg/m^3 (0.4–0.5 lb/ft^3). Other contaminants, particularly arsenic, may reduce the effective removal rate of fluoride. The minimum recommended empty-bed contact time for fluoride removal is 5 minutes. For example, a 0.028-m^3 (1-ft^3) bed of alumina should have a maximum flow rate of 0.095 L/s (1.5 gpm), to provide at least 5 minutes of empty-bed contact time between the water and the alumina.[5,17,18]

The regeneration of the alumina bed involves backwashing, regeneration, and neutralization. Backwashing is done to remove trapped suspended material, break up the packed bed, and redistribute the medium. It is done at 50-percent bed expansion for 10–15 minutes. Regeneration of an alumina bed is achieved by NaOH solution. Typically, NaOH solution of from 0.5 to 2 percent at a flow rate of 0.56–6.7 L/s·m^3 (0.25–3 gpm/ft^3) for 1–1.5 hours are used. Activated alumina dissolves to some degree during regeneration. Neutralization of activated alumina after regeneration is necessary to return the bed into acidic condition. To achieve this, the bed is rinsed with water to remove excess caustic. The bed is then rinsed with either H_2SO_4 or HCl solution that has a pH of 2.0–2.5. The acid solution is percolated through the bed at the treatment flow rate under monitoring of the effluent pH. When the pH of the effluent drops to the desired treatment level, the bed resumes operation.[5] The flow from backwash, regeneration, neutralization, rinsing, and acidification operations should be collected as brine. The brine should be handled in a way similar to the brine obtained from the demineralization processes.

The standard design and operating procedures for full-scale fluoride-removal facilities are not fully developed. Pilot testing may be necessary to develop design and operating criteria.

Equipment and Operation. The major equipment for defluoridation of a municipal water supply by activated alumina include an activated-alumina bed, acid and base feed, a pH adjustment and control system, raw water filtration and backwash system, and alumina-bed regeneration and neutralization system.

Design Parameters. Important design parameters for designing the alumina bed for defluoridation include design flow, fluoride concentration in raw water, pH of the feed water, filtration rate, and fluoride uptake rate by the alumina. The design procedure is similar to the ion-exchange system.

18.2.4 Arsenic Removal

Arsenic accumulates in the human body over a relatively short period of time. The major characteristics of acute arsenic poisoning are profound gastrointestinal damage and cardiac abnormalities. Arsenic in drinking water is at the forefront for risk of cancer of the skin, liver, lungs, kidneys, and bladder. The information on health risks with arsenic is expected to drive the current standard of 50 μg/L down to 5 μg/L, or less.

Arsenic can occur in water in four oxidation states (+5, +3, 0, −3). Most common forms in natural water are pentavalent (arsenate, As(V)) or trivalent (arsenite, As(III)), the later being the most toxic.[5] The source of arsenic may natural events or human activities. The concentration of arsenic in natural fresh water sources varies from 0.3 to 230 μg/L. In general, high concentration of arsenic is restricted mostly to groundwaters, in the form of As(III).[19–21]

Process Description. Arsenic can be removed by enhanced or conventional coagulation with iron or aluminum salts. Other removal methods include activated alumina, ion exchange, and membrane.[22]

Enhanced Coagulation. Arsenic in pentavalent from (As(V)) is readily removed by coagulation at a high dosage of iron or aluminum. The optimum dosage depends on the raw water quality and should be determined by jar tests. Research has suggested that the optimum pH for arsenic removal is either below 5.5 or between 8.5 and 9. Lime softening is also very effective in removing As(V) at pH above 10.5. Conventional coagulation methods are not very effective in removing As(III). Lime softening above pH 11 can remove As(III) up to 80 percent. To achieve higher removal efficiency, an oxidation process to convert As(III) to As(V) prior to coagulation is required. Chlorine, ozone, and potassium permanganate have been successfully used for oxidation.[22–26] The coagulation step is commonly needed to remove turbidity from the surface waters. Most arsenic problems are generally encountered in groundwater (which is free of turbidity), so coagulation is not a viable method for arsenic removal.

Activated Alumina. Arsenic is better retained than fluoride in an activated-alumina column. Because arsenic is strongly adsorbed/exchanged by activated alumina, a higher concentration of NaOH solution is needed to elute arsenic from the alumina bed. Some investigators suggested that a 4-percent solution of NaOH is required for arsenic, as compared to the 1-percent solution typically needed for fluoride. The activated-alumina system for arsenic removal is designed and operated similar to that for fluoride removal.[5,27,28]

Ion Exchange. Anion exchange resins have been used to remove arsenic ion from water supply. Some strong-base anion exchange resins remove As(V) present as the divalent anion $HASO_4^{2-}$ over other monovalent anions in the water supply. Design considerations for an anion exchange resin bed are the same as were discussed under nitrate removal (Section 18.2.2). Arsenic removal by anion exchange resins is currently in development.[5,22,28,29]

Membrane. Reverse osmosis and nanofiltration processes have been studied for arsenic removal from drinking water supply. In high-pressure reverse osmosis system, arsenic was successfully removed with a relative low volume of reject water flow. Similar removal efficiencies for both As(V) and As(III) were also observed with membrane processes.[5,22,28,30]

Equipment and Operation. The major equipment for arsenic removal by enhanced coagulation, activated alumina, ion exchange, and membrane are similar to those that have been presented in coagulation–flocculation (Chapter 8), defluoridation (Section 18.2.3), nitrate removal (Section 18.2.2), and demineralization (Section 18.2.1), respectively. Readers should refer to these chapters and sections for details.

Design Parameters. The major design parameters for arsenic removal by activated alumina, ion exchange or membrane processes are the design flow, concentration of arsenic, pH, other constituents in the raw water, removal capacity of the resin or alumina, flow rate, backwash, regeneration, neutralization, and rinse cycles. The enhanced coagulation process is designed through procedures similar to those for conventional coagulation process.

18.2.5 Heavy Metals Removal

Heavy metals are micro-inorganic pollutants that can be present in higher concentrations in some raw water sources.

Coagulation by aluminum sulfate alone very successfully removes silver, lead, and copper and reduces selenium, vanadium, and mercury by about 50 percent; however, this treatment has little effect on nickel, cobalt, manganese, and chromium.[31] Conventional coagulation using alum brings about an efficient reduction of many heavy metals, and the use of a sand filter followed by coagulation reduces the levels of silver, mercury, and copper to virtually zero; sand filtration, however, does not appreciably change the contents of chromium, cadmium, vanadium, and cobalt. Zinc and nickel are reduced by coagulation under pressure of chlorine.[31]

High-pH coagulation, in conjunction with softening, is very effective in removing most of the heavy metals. An adequate reduction in the content of undesirable or toxic ions can also be obtained by filtration through granular activated carbon. Silver and mercury are completely removed, and content of zinc, nickel, and other heavy metals is reduced significantly. Activated alumina is also very effective in removing many heavy metals.[31]

18.3 Removal of Organics

Organic compounds in the water supply affect water quality. Many organic chemicals cause disagreeable tastes and odors.

Other anthropogenic pollutants, although generally present in low concentrations, are of great concern. For example, vinyl chloride, benzene, PCBs, some cleaning solvents, many pesticides and herbicides, and many other organic compounds are known to be carcinogenic. Other substances, such as chloroforms, THMs, and other substitution and organohalogen oxidation products are cancer-suspect agents. In this section, the sources of organic compounds and applicable processes for removal of these compounds are presented.

18.3.1 Sources

Organic compounds reach water supply sources through natural and man-made sources. These sources are discussed below.

Natural Sources The source of natural organic matter (NOM) is dissolution of naturally-occurring organic materials that originate from plant and animal residues. These include aquatic plants and animals and drainage through decaying vegetation such as grass, leaves, and trees.[32] Other sources are algae, bacteria, and actinomycetes.

Mostly, NOMs are humic substances containing decomposed proteins and carbohydrates and phenolic polymers derived from lignins and tannins. The organic carbon concentration in ground and surface waters due to natural sources may range from 0.1 to 2 mg/L and 1 to 20 mg/L, respectively. Highly colored water draining from most swamps can contain TOC concentrations as high as 200 mg/L.[5] Discussion on organic compounds in natural waters, measurement techniques, and effects on water quality can be found in Chapters 2, 11, and 12.

Man-Made Sources Synthetic organic compounds reach the water supply sources through municipal and industrial discharges, agriculture, spills, and intentional dumping. Other sources of man-made organics are those that are produced as by-products during disinfection of water. Most of these are anthropogenic organic pollutants, and their concentration in aquatic environment is quite low (parts-per-billion level). These compounds include synthetic detergents, pesticides, cleaning solvents, PCBs, trihalomethanes, and others.[33,34] Detailed discussion on these organic compounds can be found in Chapters 2, 11, and 12.

18.3.2 Treatment Technologies

Applicable Processes Organic matter is removed from the water supply by many conventional and nonconventional water treatment processes. Commonly used processes are aeration or air stripping, oxidation, enhanced coagulation, adsorption, biologically active carbon (BAC), ion exchange, and membrane. Each method is best suited for specific organic compounds or a group of compounds.

Organic compounds resulting from natural sources are generally high-molecular-weight humic and fulvic acids. These and other naturally occurring compounds are effectively removed by coagulation and oxidation. Low-pH coagulation at a high coagulant dose (enhanced coagulation) and coagulation in conjunction with softening has shown enhanced removal of organic matter resulting from natural sources. Air stripping effectively removes volatile organics of low solubility and low molecular weight. Examples of such compounds are chloroform, benzene, bromoform, carbon tetrachloride, and chlorobenzene. Carbon adsorption is effectively used for removal of synthetic organics, such as pesticides and aromatic compounds.[35,36] Carbon adsorption is also used to remove taste- and odor-causing organic compounds from natural sources. The readers are referred to Chapter 11 for detailed discussion of this subject. Ion-exchange technology for removal of organics from water is gaining popularity.[37,38] In reality, most organic substances react differently with resins, in accordance to their nature. Some pass through the bed unaffected; others are reversibly fixed and are released by regeneration, while others are irreversibly fixed and tend to blind the resins. In general, the strongly-basic anion exchange resins are blinded and cation exchange resins are unaffected, while weakly-basic anion exchange resins retain organics in a reversible manner.[31] Ion exchange effectively removes polar and ionized organic compounds such as phenols and amines. Certain macroporous exchangers have the ability to remove color and TOC from water supplies. Other carbonaceous resins are effective in removing nonpolar organics like the trihalomethanes, phenols, pesticides, and certain other chlorinated organic compounds.[5] The biologically-active carbon beds (BAC) are also extensively researched for removal of NOM and other micropollutants. These are activated carbon or anthracite beds that are allowed to develop microbiological film. Organic compounds coming in contact with the film are decomposed by the microorganisms. Ozonation enhances the biodegradability of organic compounds, and they are more readily removed by BAC.

Removal of organics by membrane techniques is currently a growing area of research and application. Applied to surface water treatment, ultrafiltration removes very many pollutants, including organics with molecular mass in excess of about 70 g/mole. Many pesticides, including lindane, parathion, and others, are effectively removed.[31]

Removal of Micropollutants The treatment methods for most organic compounds in natural waters have been presented. In this section, the special treatment methods for elimination of many organic micropollutants are discussed. These organic micropollutants are trihalomethanes (or chlorination by-products), phenols, hydrocarbons, detergents, and pesticides.

Trihalomethanes

Phenols and Phenol Compounds. Phenols and phenol compounds are not removed by conventional treatments such as coagulation and filtration. Special treatments, such as oxidation with chlorine dioxide, ozone, activated carbon, and ozone in combination with activated carbon are needed to eliminate these compounds.

High dosage of chlorine dioxide will oxidize chlorophenol and other phenolic compounds. Removal of excess chlorine may, however, be necessary. Ozone at proper dosage and pH destroys phenols and phenolic compounds. Lower ozone concentration is needed at higher pH, which doubles when the pH is dropped from 12 to 7.[31] At normal pH range of the natural water (7 to 8.5), high ozone dosage is needed for destruction of phenolic compounds. The ozone dose also varies with the concentration and types of phenolic compounds.

Phenols and phenolic compounds are effectively removed by activated carbon. Granular activated carbon is very effective in removing all forms of phenolic compounds. Removal by powdered activated carbon depends on the type of phenolic compounds and their concentrations and the dose of carbon. A combined treatment by ozone and activated carbon is exceedingly effective in removing large concentrations of phenolic compounds. BAC beds are also very effective in removing phenolic compounds.

Hydrocarbons. Hydrocarbons reach raw water sources from accidental spills and leaking of underground storage tanks. A large portion is removed by conventional treatment processes (coagulation–flocculation, sedimentation, and filtration). Traces are removed by carbon adsorption and oxidation by ozone.

Pesticides. Pesticides and herbicides are generally unaffected by conventional water treatment processes. Slow sand filtration has limited effect with some pesticides. Ozone oxidizes many pesticides and herbicides. Chlorine has no effect on chlorine-based pesticides but partially oxidizes some herbicides. Potassium permanganate partially destroys some pesticides and herbicides; however, the effectiveness is low and cannot be accepted for removal of pesticides generally found in surface waters.[31] To date, carbon adsorption is the method most often used for the removal of pesticides. Activated carbon in powdered as well as granular forms has been used for removal of pesticides from water supplies. Both the pH and temperature have little effect on removal efficiencies of carbon.[31,39] BAC beds following ozonation are quite effective in removing pesticides and herbicides.

Detergents. Detergents, surfactants, or surface-active agents are large organic molecules that are slightly soluble in water and cause foaming at the air-water interface. Coagulation–flocculation clarification generally does not remove detergents from the water. Filtration removes little or no detergent. Prechlorination does not break down detergent molecules. The commonly used removal methods are foam separation, ozone oxidation, and carbon adsorption.

Foam separation of detergent is used where the concentration is high. The process involves blowing air into water, slightly below the surface. The foam produced is separated. The process efficiency depends on air-flow rate, size of air bubbles, and nature and concentration of detergent.

Ozone breaks down the detergent molecule. Ozone dose in the range of from 1.5 to 3 mg/L may be necessary to eliminate 50 percent of detergent from the water.[31] There is always a residual detergent concentration that cannot be destroyed by ozonation.

Carbon adsorption is a very effective process for detergent removal. A 50-percent reduction in detergent content can be expected with a dose of from 12.5 to 25 mg/L.[31,35] Of course, the removal will depend on the concentration and nature of detergent.

Granular activated carbon is very effective in removing detergents from water. Ozone in combination with activated carbon or BAC is valuable when detergent content reaches occasional peaks or when a low residual content is required.[31] Ozone and activated carbon treatment is generally justified if other micropollutants are also of concern. For detergent removal alone, granular activated is considered sufficient.

Equipment and Design Parameters The above discussion indicates that conventional treatment processes remove organics that originate from natural sources. Microconcentrations of natural and synthetic organics require ozone and/or carbon adsorption. Equipment and design parameters for ozone treatment and carbon adsorption have been discussed in Chapters 11 and 12. Readers are referred to these chapters for details.

When ozone and granular activated carbon beds are used, the combination must be carefully evaluated in terms of equipment and of technical and economic factors. There may be an advantage to providing ozonation prior to gravity filters but after carbon beds. This arrangement will remove oxidation products resulting from ozonation and prechlorination. If chlorine residual is to be maintained in the distribution system, chlorination and ammoniation steps must follow carbon beds.

Problems and Discussion Topics

18.1 Discuss the advantages and disadvantages of various demineralization processes (ion exchange, distillation, and membrane processes) if used for recovery of potable water from a brackish water source.

18.2 Well water having a total hardness of 300 mg/L as $CaCO_3$ is softened by split treatment. A zeolite bed takes 3/4 of the total flow and produces water at a hardness of 3 mg/L as $CaCO_3$. One-fourth of flow is directly mixed with the softened water. The zeolite bed has 28.3 m^3 resin capacity. The water softening capacity of the resin is 62 kg/m^3 total hardness as $CaCO_3$. Exhausted resin is regenerated by 98 percent pure NaCl solution. The salt consumption is three times the theoretical amount and used as 4 percent salt solution. Calculate (a) regeneration period if 950 m^3 per day of water is passed through the softener, (b) the hardness of the finished water, and (c) salt consumption per regeneration cycle.

18.3 A hydrogen cation exchange bed has a capacity of 288 kg/m^3. Regeneration cycle is 5 days and flow is 1900 m^3/d. Calculate the volume of bed and theoretical quantity of acid needed for regeneration. The concentration of major cations are Ca^{2+} = 28 mg/L, Mg^{2+} = 21 mg/L, Na^+ = 56 mg/L, and K^+ = 20 mg/L.

18.4 Design an anion exchange system. The resin bed has a capacity of 320 kg/m^3 as $CaCO_3$. Regeneration cycle is 5 days, and flow is 1900 m^3/d. Calculate the theoretical amount of NaOH needed for regeneration. The concentrations of major anions in water are SO_4^{2-} = 35 mg/L, Cl^- = 68 mg/L, HCO_3^- = 37 mg/L, CO_3^{2-}= 19 mg/L, NO_3^-= 6 mg/L, CO_2 = 12 mg/L, and SiO_2 = 2 mg/L.

18.5 Describe the reverse osmosis process. What pretreatment is required before applying a reverse-osmosis module? Explain the construction and operation of the following membrane module designs: plate and frame, large tube, spiral wound, and hollow fine fiber.

18.6 Design the components of a reverse osmosis system to produce demineralized water from brackish water that has TDS of 2000 mg/L. The finished water production capacity of the plant is 2000 m^3/d. Use the following data: recovery factor R = 75 percent, salt-rejection factor S = 95 percent, and flux rate = 0.65 $m^3/m^2\cdot d$ at the design pressure of 4000 kN/m^2.

18.7 Water is withdrawn from a lake that has high TDS concentration. The water is demineralized by a reverse osmosis system. Describe the pretreatment processes and draw a process diagram.

18.8 Discuss the advantages and disadvantages of distillation process over membrane process.

18.9 Calculate the methanol consumption (kg/d) for denitrification of 0.01 m^3/s well water that contains 20 mg/L nitrate nitrogen and 1.0 mg/L dissolved oxygen.

18.10 The minimum recommended empty-bed contact time for fluoride removal in an activated alumina is 5 min. What is the volume of the alumina bed if the maximum flow rate is 0.15 L/s?

18.11 Sodium hydroxide solution is needed to regenerate activated alumina bed. A 1.5-percent solution at a flow rate of 0.6 $L/s\cdot m^3$ is used for 1 hour. Calculate the volume of the bed if total quantity of NaOH solution consumed is 5 L.

18.12 Discuss various methods of arsenic removal from well water. How does arsenic-removal mechanism differ from fluoride removal in an activated alumina column?

18.13 Discuss the health effects of high concentrations of selenium in a municipal water supply. Review selenium-removal methods presented in Reference 5, and discuss them.

18.14 Review Reference 26 and discuss sources and removal methods of seasonal pesticides in surface waters.

18.15 A water treatment plant receives water from a long raw water line that is periodically chlorinated to reduce the growth in the transmission line. As a result, the concentration of trihalomethanes in the raw water is high. Discuss the methods of removing trihalomethanes from the water supply.

References

1. Fair, G. M., Geyer, J. C., and Okun, D. A. *Water and Wastewater Engineering*, Vol. 2, John Wiley & Sons, Inc., New York, 1968.

2. Bishop, P. L. *Marine Pollution and its Control*, McGraw-Hill Book Co., New York, 1983.
3. Dorfner, K. *Ion Exchangers Properties and Applications*, 3d ed., Ann Arbor Science, Ann Arbor, MI, 1977.
4. Tchobanoglous, G. and Schroeder, E. D. *Water Quality: Characteristics, Modeling, and Modification*, Addison-Wesley Publishing Co., Reading, Massachusetts, 1985.
5. James M. Montgomery, Inc. *Water Treatment Principles and Design*, John Wiley & Sons, New York, 1985.
6. Qasim, S. R. *Wastewater Treatment Plants: Planning, Design, and Operation*, 2d ed., Technomic Publishing Co., Inc., Lancaster, PA, 1999.
7. Osmonics, Inc. *The Filtration Spectrum*, Osmonics, Inc., Minnetonka, MN, 1998.
8. Cheryan, M. *Ultrafiltration Handbook*, Technomic Publishing Co., Inc., Lancaster, PA, 1986.
9. Cruver, J. E. *Membrane Processes in Physicochemical Process for Water Quality Control*, (W.G. Weber, Jr., ed.), Wiley Interscience, New York, 1972.
10. Farrell, J. B. and Smith, R. N. "Process Applications of Electrodialysis," *Ind. Eng. Chem.*, vol. 54, no. 6, p. 29, June 1962.
11. AWWA. *Water Quality and Treatment*, 5th ed., McGraw-Hill Book Co., New York, 1999.
12. Viessman, W. and Hammer, M. J. *Water Supply and Pollution Control*, 6th ed., Addison-Wesley, Menlo Park, CA, 1998.
13. United Nations. *Solar Distillation as a Means of Meeting Small-Scale Water Demand*, Publication No. 70.II.B.1, United Nations, Department of Economic and Social Affairs, New York, 1970.
14. Qasim, S. R. "Treatment of Domestic Sewage by Using Solar Distillation and Plant Culture," *Journal Environ. Sci. Health*, vol. A13, no. 88, pp. 615–627, 1978.
15. Guter, G. A. *Removal of Nitrate from Contaminated Water Supplies for Public Use*, Interim Report, EPA-6000/2-81-029, USEPA, Cincinnati, OH, 1981.
16. Evans, S. "Nitrate Removal by Ion Exchange," *Journal Water Pollution Control Federation*, vol. 45, no. 4, pp. 632–636, April 1973.
17. Choi, W. W. and Chen, K. Y. "The Removal of Fluoride from Waters by Adsorption," *Jour. AWWA*, vol. 71, no. 10, pp. 562–570, October 1979.
18. Robel, F. and Williams, F. S. *Pilot Plant Study of Fluoride and Arsenic Removal from Potable Water*, EPA-600/2-80-100, USEPA, Cincinnati, OH, August 1980.
19. Fergusson, J. F. *The Heavy Element–Chemistry, Environmental Impact and Health Effects*, Pergamon Press, Oxford, NY, 1990.
20. Fergusson, J. F. and Anderson, M. A. "Chemical forms of Arsenic in Water Supplies and Their Removal," *Chemistry of Water Supply, Treatment, and Distribution* (A. J. Rubin, ed.), Ann Arbor Science Publisher, Inc., Ann Arbor, Michigan, 1974.
21. Fergusson, J. F. and Davis, J. A. "Review of the Arsenic Cycle in Natural Waters," *Water Research*, vol. 6, no. 11, pp. 1259–1274, November 1972.
22. Jekel, M. R. "Removal of Arsenic in Drinking Water Treatment," *Arsenic in the Environment, Part I: Cycling and Characterization* (J. O. Nriagu, ed.), John Wiley & Sons, New York, 1994.
23. Zhu, G., et al. "Coagulation Diagrams for Arsenic Removal by Enhanced Coagulation," *Proceedings of 1997 Fall Meeting of Texas Section–American Society of Civil Engineers*, pp. 32–42, October 1–4, 1997.
24. Hering, J. G., et al. "Arsenic Removal by Ferric Chloride," *Jour. AWWA*, vol. 88, no. 4, pp. 155–167, April 1996.
25. Cheng, R. C., et al. "Enhanced Coagulation for Arsenic Removal," *Jour. AWWA*, vol. 86, no. 9, pp. 79–90, September 1994.
26. Edwards, M. "Chemistry of Arsenic Removal during Coagulation and Fe-Mn Oxidation," *Jour. AWWA*, vol. 86, no. 9, pp. 64-78, September 1994.
27. Ghosh, M. M. and Yuan, J. R. "Adsorption of Inorganic Arsenic and Organoarsenicals on Hydrous Oxides," *Environ. Prog.*, vol. 6, no. 3, pp. 150–157, March 1987.
28. Fox, K. R. and Sorg, T. J. "Controlling Arsenic, Fluoride, and Uranium by Point-of-Use Treatment," *Jour. AWWA*, vol. 79, no. 10, pp. 81–84, October 1987.

29. Clifford, D. A. "Ion Exchange and Inorganic Adsorption," *Water Quality and Treatment* (F. W. Pontius, ed.), 4th ed., McGraw-Hill, Inc., New York, 1990.
30. Hering, J. G. and Elimelech, M. *Arsenic Removal by Enhanced Coagulation and Membrane Processes*, Final Report, Department Of Civil and Environmental Engineering, UCLA, Los Angeles, CA, 1995.
31. Degremont. *Water Treatment Handbook*, 5th ed., Degremont, Rueil-Malmaison, France, 1979.
32. Gjessing, E.T. *Physical and Chemical Characteristics of Aquatic Humus*, Ann Arbor Science, Ann Arbor, Michigan, 1976.
33. Donaldson, W. T. "Trace Organics in Water," *Environmental Science and Technology*, vol. 11, no. 4, pp. 348–351, April 1977.
34. Rook, J. J. "Haloforms in Drinking Water," *Jour. AWWA*, vol. 68, no. 3, pp. 168–172, March 1976.
35. Randtke, S. J. "Organic Contaminant Removal by Coagulation and Related Process Combinations," *Jour. AWWA*, vol. 80, no. 5, pp. 40–56, May 1988.
36. Hand, O. W., et al. "Designing Fixed-Bed Adsorbers to Remove Mixtures of Organics," *Jour. AWWA*, vol. 81, no. 1, pp. 67–77, January 1989.
37. Anderson, C. and Maier, W. G. *The Removal of Organic Material from Water Supplies by Ion Exchange*, Report for Office of Water Research and Technology, Washington, D.C., 1977.
38. Chen. A. S. C., et al. "Activated Alumina for Removing Dissolved Organic Compounds," *Jour. AWWA*, vol. 81, no. 1, pp. 53–60, January 1989.
39. Miltner, R. C., et al. "Treatment of Seasonal Pesticides in Surface Waters," *Jour. AWWA*, vol. 81, no. 1, pp. 43–52, January 1989.

CHAPTER 19

Avoiding Design Errors

19.1 Introduction

Design errors do happen, and often they pass through undetected by the project team and the review process. In fact, in the midst of complex details in plans and specifications, little things can be easily overlooked, and any designer can become the victim. As the saying goes, mistakes once committed are often difficult to detect in spite of a regress and careful review of plans and specifications.

The purpose of this chapter is to present many examples of design errors and design deficiencies that have been discovered during the construction, operation, and maintenance phases of various projects. Also, in this chapter, steps have been presented that can be implemented by the design and review teams to reduce or eliminate many types of design errors.

19.2 Examples of Design Errors and Deficiencies

Some blunders are senseless, and their occurrence is perhaps due to an oversight or omission while details of the drawings are being worked out. Others may occur because a designer simply forgot to include some items in the design calculations, made a poor judgment or assumption, or perhaps overlooked the implications of the design criteria. Many other poor design provisions may not be the errors in the real sense but may be least desirable from the point of view of plant construction or expansion or may not provide the desired flexibility and convenience in plant operation and routine maintenance. Following is a list of many types of design errors or blunders that have happened in many designs.

1. While preparing the unit layout, a design engineer lowered the foundation elevation of one treatment unit to avoid construction above the ground; however, he forgot to make proper

changes in the hydraulic profile and relative elevations of the other units. As a result, the free water surface dropped below that of the other units that followed it.

Many similar incidents of design blunders can be cited where a change in the design assumption criteria or critical elevation was made, but the resulting effects were not carried throughout the entire design.

2. Many times the designed units flooded because the designer did not check the hydraulics at design flow under emergency conditions. In one instance, a filter flooded during backwashing, because the effluent piping for backwash was not adequate. In another instance, the filter media got washed off, because the backwash gutters were not high enough.
3. In many designs, the common wall between two units failed when one unit was drained. On many other occasions, the bottom slab heaved upward because of hydrostatic pressure, or the side wall collapsed due to the earth pressure when the units were drained. In all cases, the designer forgot to check the structural safety under the most critical loading conditions.
4. Many designs have been prepared with incorrect design criteria or design assumptions. As a result, extensive revisions had to be made later to meet the specific criteria supplied by the concerned regulatory agency. Examples of such criteria are detention time, solids loading, standby units, maximum capacity, and chemical dosages.

   In one instance, the designer misinterpreted the design criteria or inadvertently oversized the variable-speed pumps. Naturally, the unit flooded when the pumps operated at that flow.
5. Wright reported that, once, a flume was constructed to carry a raw water supply across a coulee. The piers were tall and unstable, but the structure would have been quite stable once the flume was in place and tied to the piers. During the construction phase, however, several piers fell over.[1]
6. In one design of a clarifier, Wright reported that the detailer neglected to show the steel bars from cantilevered effluent launders hooked and embedded into the basin wall. No one detected this, and, when the facility was first put into service, the launder dropped off.[1]
7. In another settling basin incident, Wright cited that the designer correctly provided the corbels to support the weight of the effluent launder full of water. This condition could occur when the side gates on the effluent lines were closed, and the basin drained; however, the designer inadvertently did not anchor the launder to the corbel against uplift. When the basin was first filled, the launders floated off the support from buoyant force.[1]
8. In one case, Wright reported that a package-pumping station was installed in a flood plain. The station was provided with a submarine-type hatch and all electrical connections and leads were sealed against flooding. The pumps were controlled through a pneumatic system employing a bubbler tube for level sensing in the wet well. The bubbler tube was not extended above the flood stage. As a result, the station was flooded through the bubbler tube.[1]

9. Flooding of treatment facilities is a common occurrence in spite of flood-protection levees. Reported reasons for flooding have included the following: check valves or gate valves were not provided in the stormwater, so the water backed into the plant area; stormwater or effluent pumps were not provided; the storm drainage within the plant site was inadequately designed due to incorrect assumptions of rainfall intensity or run-off coefficient, causing flash flooding; or flood protection levees were not high enough at some sections.
10. In one instance, Wright reported that the discharge piping of a pump included a harnessed dresser coupling arrangement. The harnessed connection was to take the hydraulic thrust, and therefore the bolts were designed and specified to be high-tensile bolts; instead, mild steel harness bolts were installed. When the pumps were tested, the harness bolts failed, resulting in pump base-bending, and the anchor bolts failed with consequent movement of the entire pumping unit away from the header.
11. In one instance, the suction piping was not properly anchored and secured. Due to vibrations, the opening around the wall cracked open. As a result, the dry pit flooded due to high water level in the wet well.
12. In one design, the pump station was built with a check valve in the suction line.
13. A chemical-feed pump was connected to the chemical solution tank. The tank was not vented to the atmosphere. When the chemical feed pump withdrew the liquid, the tank collapsed from atmospheric pressure.
14. The routine detailing of design plans may be mislabeled, simply due to an oversight. An example may be that 40 joists on 30-cm centers may incorrectly be labeled "30 joists on 40-cm centers."
15. In a chlorinator room, the blower switch and gas masks were located inside the room, which could not be reached in case of an accident.
16. Sanks reported that, in one plan, several fabricated vertical filters were spaced so closely that the operator had to squeeze sideways to get between them. In another plant, he reported, the control panel required removal of the face plate to replace the desiccant.[2]
17. Many designs have been prepared with little consideration of plant construction and future expansion. Other designs have overlooked the basic operation and maintenance needs of the facility and safety considerations. Common examples of such design deficiencies reported by Wright, Sanks, and others include the following:[1–4]

    a. Not enough space was left for future expansion, or another unit blocked the expansion.
    b. A tee of a blind flange was not provided in a pipe at an appropriate location for future connection.
    c. Provisions were not made for future expansions of a building.
    d. Enough work space was not left around a piece of machinery for servicing or placing a small hoist for lifting heavy pieces.

e. Bolted pipe fittings were cast into walls without enough clearance to install the bolt or put a wrench.
f. Valves were provided in lines without sufficient clearance for the handwheels or for the operators to reach them.
g. Ceiling hooks were not cast into concrete at appropriate places for a chain hoist.
h. Dimensions of the equipment did not fit the unit size.
i. Floor drains were not provided, or the floor was not properly sloped.
j. A door in the building was too small to bring in the equipment (or take it out, if the equipment was installed before the structure was completed).
k. Enough space for chemical storage was not provided, or chemical-feeding equipment was not properly designed.
l. Isolation of noisy or heat-producing equipment was not considered.
m. Freezing of pipes, valves, and equipment in cold climates was disregarded, with disastrous consequences.
n. Safety aspects are critical, because a bulk-chemical storage area offers potential chemical spills. Consideration was not given to items such as a chemical spill or gas leak. Gas lighter than air (ammonia) travels upward through open stairways or other floor penetrations, and gas heavier than air (chlorine) would disperse horizontally and affect the rooms and tunnels on lower levels.[5]
o. Adequate safety measures to protect not only the plant operators but also the support personnel working in the administration–operations building was not given. Among these are (1) leak detector, (2) deluge systems, (3) fume scrubbers, (4) minimum distances from storage tanks to buildings, (5) opening in buildings, and (6) air intakes.[5] Most Uniform Fire Codes (UFC) contain provisions relating to such considerations.[6]
p. Electrical conduits, chemical pipes, heating-ventilating-air conditioning (HVAC), and plumbing are generally concentrated on different floors of the buildings. Often, careful planning to avoid conflicts between disciplines in both vertical and horizontal directions was not given.
q. The room layout in the operations building was not properly planned. As a result, the laboratory and restrooms, with numerous drains, were located directly over electrical and computer rooms.
r. Greater concentration of electrical conduits and chemical piping and plumbing resulted in excessive slab penetrations and openings. Proper allowances for structural safety of the slab was not given.
s. The location of isolation–expansion joints were not coordinated with all disciplines for potential movement and location of embedded piping across isolation joints.

There is a long list of such small design deficiencies. The designer should be conscious of them, and proper considerations should be given during early stages of the design.

19.3 Procedure to Avoid or Reduce Common Design Errors and Deficiencies

Once an error is made in plans and specifications, there is a good chance that it will escape detection. To keep such incidents from occurring, many design engineering firms have developed a comprehensive design-review process. Although such review processes may vary from job to job, based on availability of personnel within the organization, there are some basic items that must be considered under all conditions. Some of these items are listed here.

1. Several reviews of the design and checks of the plans and specifications are generally necessary at various stages of the design to eliminate errors and inconsistencies.
2. The preliminary design data of the water treatment facility is developed as part of the planning process. (See Chapter 5.) This includes determination of design capacity, process selection, number of process trains, and preliminary unit sizes. Justification is necessary if major changes are expected. In the authors' opinion, the predesign report must be used as a guide during the entire design calculation phase and during preparation of plans and specifications; however, a thorough review of the predesign report should be made by the designers, checkers, project engineer, and project manager. At this review, the work plan, responsibilities, time schedules, budgets, etc., should be discussed. Any changes in the design data developed in the predesign report (including process train) must be approved by everyone concerned. Afterward, the design data should not be changed.
3. The second review should be made after the design of the individual treatment units are complete. At this review, the basic design criteria of the concerned regulatory agency, material mass-balance analyses, flows, and loadings under normal and emergency conditions (when largest unit is out of service), unit details, and the mechanical equipment for each unit should be checked. Consideration should be given to equipment compatibility, minimum and maximum chemical feed rates, operating pressures, ratings of the pumps, chemical feeders, etc. Any changes in meeting the necessary space, flexibility requirements, and equipment compatibility and performance should be made at this time.
4. The third review should be made after the layout plan, yard pipings, and hydraulic profile are complete. At this stage, the review team should include the design engineer and checker, project manager or principals, and perhaps one or two experienced designers who have not been involved in the preliminary reviews. Any changes made during this review should be listed and documented in a separate checklist. These changes should be carried through in all drawings. Small consulting engineering firms may not have enough in-house expertise. In such circumstances, a competent consultant may be used to check the design.

This review should be made with a limited number of design items at a time. These items include (1) plant schematics and unit layout, (2) future expansion, (3) hydraulic profile and elevations of the treatment units, (4) space around equipment, (5) flexibility, (6) freezing, and (7) plant operation and maintenance and safety. Each of these items and review considerations is discussed here.

1. The plant schematics and unit layout should be thoroughly checked. Special consideration should be given to the following items:

   Forward and return flow lines

   Material mass balance

   Basin volumes and hydraulic loadings

   Pumps, piping, valves, gates, bypasses, collection and splitter boxes, etc.

2. Space for future expansion of the plants should be identified on the layout. This includes future basins, buildings, appurtenances, major pipings, etc. Most designers prefer to mark these on the layout plan in dotted lines, so that any unit or building obstructing the future expansion can be easily detected.

   Pipe connections for future expansion should also be checked. Many designers prefer to provide tees and crosses at appropriate places; the blocked end is used for future connections.

3. Hydraulic calculations and profile should be thoroughly checked for all major paths through the plant, including the filter backwash and residuals train. Unit elevations, ground elevations, building floor elevations, and water surface elevations at critical flow conditions should be marked on the hydraulic profile sheet. Omissions and errors in the hydraulic profiles and units and building elevations can have disastrous consequences.

4. Adequate space around the equipment should be provided to go around and place small hoist and other tools for equipment maintenance. Headroom, doors and accesses, ceiling hooks for chain hoists, etc., should be carefully checked.

5. Plant flexibility is important to operate the plant properly under high- and low-demand conditions. Pipe channels, valves, and structures should be sized for design flow when the largest unit is out of operation. The chemical feeders should be sized for extreme conditions. Space for chemical storage and future feeders should be properly planned.

6. Freeze protection of equipment and pipes should be checked if freezing is expected. Channels, pipes, sludge lines, and drains that flow intermittently should have enough earth cover or be designed to drain completely. Moving parts, such as sprockets, chain, etc., should be kept below freeze depth in the basin.

7. Operation and maintenance and safety considerations should not be overlooked by the reviewing team. Some designers prefer writing an O&M manual during the design phase, to include the general operational flexibility of the plant. Following is a list of basic operations and maintenance and safety items that should be checked by the review team:

Arrangements to bypass a unit should be made for routine maintenance.

Floor drains and proper slope should be provided for draining and flushing the unit. Chlorine-solution lines should be provided to wash the basin walls, weirs, and channels for routine operational needs.

Sampling point for each process should be clearly identified in the O&M manual. In large plants, samples from different units are piped directly to the laboratory, so the representative fresh samples can be obtained by the laboratory staff directly.

The review team should evaluate the equipment and controls for complexity and maintenance needs. A simple system should always be preferred over a complex system.

Hand wheels and valves should have operating clearance.

The operating panels should be accessible, and strip charts, recording pens, and desiccant containers should be easily replaceable.

Hazardous chemicals (chlorine, ozone, ammonia, etc.) should be isolated. Noisy and heat-producing equipment should also be isolated. Arrangement should be made for dry-feeder equipment for easy filling. All storage tanks must have secondary containment. Proper clearance between bulk storage chemicals and administration and control building and air intakes and scrubbers must be made. UFC must be followed throughout the checking process.

All stairs and walkways should have proper handrails.

Laboratories and workshops should have adequate space, equipment and tools, lavatory, and showers.

As mentioned earlier, each review process may vary from project to project, and with the availability of resources of the design firm. Each firm should develop the review procedure that may best fit its resources and the needs of the project. The review process given above should be used only as a guide, as it is neither complete nor definitive.

Problems and Discussion Topics

19.1 Information checklists have been presented in Chapters 6 through 15 for the design of various treatment processes. These checklists were highly specialized for individual processes. Using this information, develop a generalized design checklist for an entire water treatment plant that a designer could use as a guide during the entire design phase of a project.

19.2 Develop a comprehensive checklist for design review that could be used by a review team. Arrange the checklist items under the following headings:
- **(a)** predesign information that must be carried out in the plans and specifications;
- **(b)** process train and instrumentation;
- **(c)** plant layout and piping;
- **(d)** buildings and equipment;
- **(e)** plant hydraulics;
- **(f)** roads, drainage, and landscaping;

(g) operation and maintenance flexibility;
(h) expansion flexibility.

19.3 You have been hired as an environmental engineer by a small consulting engineering company to complete the final plans and specifications of a medium-sized water treatment plant. There is one additional civil/environmental engineer in your organization. The predesign for the project was prepared by another engineer who is no longer with this organization. Discuss how you would organize the project components, including revisions to the predesign report, design task and review process, review team, and project timetable. The project must be completed within 18 months. The project involves doubling the existing water treatment facility and high-service pumping station.

19.4 Complete problem 19-3 again, but instead assuming that a large consulting engineering company has hired you as a junior design engineer to assist a project manager who prepared the predesign report.

References

1. Wright, J. R. "Design Blunders," *Water Treatment Design for Practicing Engineers* (R. L. Sanks, ed.), Ann Arbor Science Publisher, Ann Arbor, MI, 1978.
2. Sanks, R. L. "How to Avoid Design Blunders," *Water Treatment Plant Design for Practicing Engineers* (R. L. Sanks, ed.), Ann Arbor Science Publisher, Ann Arbor, MI, 1978.
3. Regan, T. M. "O&M Manuals and Operator Training," *Water Treatment Plant Design for Practicing Engineers* (R. L. Sanks, ed.), Ann Arbor Science Publisher, Ann Arbor, MI, 1978.
4. Qasim, S. R. *Wastewater Treatment Plants: Planning, Design, and Operation*, 2d ed., Technomic Publishing Co., Inc., Lancaster, PA, 1999.
5. Corbin, D. J. et al. "Compact Treatment Plant Layout," *Jour. AWWA*, vol. 84, no. 8, pp. 36–42, August 1992.
6. International Fire Code Institute. *Uniform Fire Code*, International Conference of Building Officials and Western Fire Chiefs Association, CA, 1997.

APPENDIX A

Physical and Chemical Properties of Water

This appendix summarizes the principal physical and chemical properties of water commonly used in water treatment plant design.

Table A-1 Basic Properties of Water

Property	Value
Molecular formula	H_2O
Molecular weight, g/mol	18
Ionization constant (K_w) at 25°C	10^{-14}
Specific weight, γ at 25°C, kN/m^3 (62.43 lb$_f$/ft^3)	9.81
1 U.S. gallon weight (average), lb	8.34
Density at 4°C, g/cm^3 (1.94 slug/ft^3)	1
Specific heat, cal/g·°C (1 Btu/lb·°F)	1
Dynamic viscosity (μ), N·s/m^2 (2.089×10^{-5} lb$_f$·s/ft^2)	1×10^{-3}
Kinematic viscosity (ν), m^2/s (1.078×10^{-5} ft^2/s)	1.002×10^{-6}
One atm (at 760 mm Hg or 14.7 psi), m (ft) water column	10.33 (33.9)
Boiling point at 1 atm, °C (°F)	100 (212)
Melting point at 1 atm, °C (°F)	0 (32)
Heat of fusion at 0°C, cal/g (Btu/lb)	80 cal/g (144)
Heat of vaporization at 100°C, cal/g (Btu/lb)	540 (973)
Modulus of elasticity at 15°C, kN/m^2 (lb$_f$/in^2)	2.15 (312×10^3)

Source: Adapted in part from Reference 4.

Table A-2 Density, Viscosities, Surface Tension, and Vapor Pressure of Water

Temperature, °C	Temperature, °F	Density (ρ), g/cm^3	Dynamic Viscosity (μ), $\times 10^3$ N·s/m^2	Kinematic Viscosity (ν), $\times 10^6$ m^2/s	Surface Tension against Air (σ), $\times 10^3$ N/m	Vapor Pressure (p), mm Hg
0	32	0.9998	1.787	1.787	75.60	4.579
3.98	39.2	1.0000	1.568	1.568	75.00	6.092
5	41	0.9999	1.519	1.519	74.90	6.543
10	50	0.9997	1.307	1.307	74.22	9.209
15	59	0.9991	1.139	1.140	73.49	12.788
20	68	0.9982	1.002	1.004	72.75	17.535
25	77	0.9970	0.890	0.893	71.97	23.756
30	86	0.9957	0.798	0.801	71.18	31.824
35	95	0.9941	0.719	0.724	70.37	42.175
40	104	0.9922	0.653	0.658	69.56	55.324
45	113	0.9903	0.596	0.602	68.74	71.880
50	122	0.9881	0.547	0.553	67.91	92.510

Source: Adapted in part from Reference 4.

Table A-3 Solubility of Oxygen and Other Gases in Water

Temperature		Dissolved Oxygen Saturation[a] (C_s), mg/L					Decrease in Oxygen Concentration, mg oxygen per 100 mg Chlorine	Solubility of Other Gases, mg/L		
		Chlorine Concentration, mg/L								
°C	°F	0	5000	10,000	15,000	20,000		Nitrogen	Carbon Dioxide	Air
0	32	14.62	13.79	12.97	12.14	11.32	0.017	22.81	1.00	37.50
5	41	12.80	12.09	11.39	10.70	10.01	0.014	20.16	0.83	32.94
10	50	11.33	10.73	10.13	9.55	8.98	0.012	17.93	0.70	29.27
15	59	10.15	9.65	9.14	8.63	8.14	0.010	16.15	0.59	26.25
20	68	9.17	8.73	8.30	7.86	7.42	0.009	14.72	0.51	23.74
25	77	8.38	7.96	7.56	7.15	6.74	0.008	13.55	0.43	21.58
30	86	7.63	7.25	6.86	6.49	6.13	0.008	12.53	0.38	19.60

a. The solubility of oxygen (C_s) exposed to air containing 20.9 percent oxygen by volume at 1 atmosphere (760 mm Hg). Under any other barometric pressure, the solubility can be calculated from the equation

$$C'_s = \frac{C_s(P - p)}{(760 - p)}$$

where

p = vapor pressure at the temperature of water, mm Hg

P = atmosphere pressure, mm Hg

Source: Adapted in part from Reference 4.

Table A-4 Barometric Pressure with Altitude

Elevation above Sea Level,		Absolute Pressure in Head of Water (H_{abs}),	
m	ft	m	ft
0	0	10.3	33.9
305	1000	10.0	32.8
457	1500	9.8	32.1
610	2000	9.6	31.5
1,219	4000	8.9	29.2
1,829	6000	8.3	27.2
2,438	8000	7.7	25.2
3,048	10,000	7.1	23.4
4,572	15,000	5.9	19.2

Source: Adapted in part from Reference 4.

References

1. APHA, AWWA, and WEF. *Standard Methods for Examination of Water and Wastewater*, American Public Health Association, 18th ed., Washington, D.C., 1992.
2. Fay, J. A. *Introduction to Fluid Mechanics*, MIT Press, Cambridge, MA, 1994.
3. Lange, N. A. (Editor) *Lange's Handbook of Chemistry*, 13th ed., McGraw-Hill Book Co., New York, 1985.
4. Qasim, S. R. *Wastewater Treatment Plants: Planning, Design, and Operation*, 2d ed., Technomic Publishing Co., Inc., Lancaster, PA, 1999.
5. Zipparro, V. J. (Editor) *Davis's Handbook of Applied Hydraulics in Engineering*, 4th ed., McGraw-Hill Book Co., New York, 1993.

APPENDIX B

Constants and Coefficients Used for Hydraulic Head Loss Calculation

This appendix deals with the constants commonly used to calculate minor head losses in open channels and pressure conduits. Many of these constants have been used in the Design Examples to calculate the losses caused by contraction and expansion and exit and entrance in the appurtenances. The values of most of these constants can be obtained in many hydraulics texts and handbooks. The readers are referred to the references cited at the end of this appendix.

B.1 Minor Head Losses in Open Channels

B.1.1 Sudden Contraction or Inlet Losses

Sharp-cornered	Round-cornered	Bell-mouthed	Equation
0.5	0.25	0.05	$\left(\frac{v_2^2}{2g} - \frac{v_1^2}{2g}\right)$

where

v_1 and v_2 = the velocities upstream and downstream of the contraction or inlet

B.1.2 Sudden Enlargement or Outlet Losses

Sharp-cornered	Bell-mouthed	Equation
0.2–1.0	0.1	$\frac{v_1^2}{2g} - \frac{v_2^2}{2g}$

where

v_1 and v_2 = the velocities upstream and downstream of the enlargement or outlet

B.1.3 Junction and Diversion Boxes

a. In a small junction or diversion box where there is no change in channel size and direction,

Flow-through box	Terminal box	Equation
0.05	1.0	$\frac{v^2}{2g}$

where

v = the velocity upstream or downstream of the box

b. In a small junction or diversion box where there is no change in channel size, but a change in channel direction,

45° turn	90° turn	Equation
0.25–0.3	1.0–1.2	$\frac{v^2}{2g}$

where

v = the velocity upstream or downstream of the box

c. In a large junction or diversion box in which the velocity is small,

Entrance	Exit	Equation
0.5	1.0	$\frac{v_2^2}{2g}$

where

v_1 and v_2 = the velocities upstream and downstream of the box

B.1.4 Syphon

Head loss at syphon	2.78	$\frac{v^2}{2g}$

where

v = the velocity through the syphon

B.1.5 Sluice Gate

Head loss at gate	0.2–0.8	$\frac{v^2}{2g}$

where

v = the velocity through the gate opening

B.2 Minor Head Losses in Pressure Conduits

B.2.1 Gate Valve

Full open	One-fourth closed	One-half closed	Three-forths closed	Typical value	Equation
0.19	1.15	5.6	24.0	1.0	$\frac{v^2}{2g}$

where

v = the velocity in the pipe upstream or downstream of the valve

B.2.2 Butterfly Valve

Full open	20°	Angle closed 40°	60°	Typical value	Equation
0.3	1.4	10	94	1.2	$\frac{v^2}{2g}$

where

v = the velocity in the pipe upstream or downstream of the valve

B.2.3 Check Valve

Head loss at valve	1.5–2.5	$\frac{v^2}{2g}$

where

v = the velocity in the pipe upstream or downstream of the valve

B.2.4 Plug Valve

Head loss at valve	1.0	$\frac{v^2}{2g}$

where

v = the velocity in the pipe upstream or downstream of the valve

B.2.5 Entrance

Pipe projecting into tank	End of pipe flushed with tank	Slightly rounded	Bell-mouthed	Equation
0.83	0.50	0.23	0.04	$\frac{v^2}{2g}$

where

v = the velocity in the pipe after the entrance

B.2.6 Exit

Head loss from conduit to still water	1.0	$\frac{v^2}{2g}$

where

v = the velocity in the pipe before the exit

B.2.7 Elbow (45–61 cm diameter)

22.5°	45°	90°	Equation
0.1–0.2	0.2–0.3	0.25–0.6	$\frac{v^2}{2g}$

where

v = the velocity in the pipe upstream or downstream of the elbow

B.2.8 Tee

Run-to run	**Branch-to-run**	**Run-to-branch**	**Equation**
0.25–0.6	0.6–1.8	0.6–1.8	$\frac{v^2}{2g}$

where

v = the velocity in the pipe (run or branch) upstream of the tee

B.2.9 Reducer

Head loss through reducer (with angle of divergence 10°–20°	0.15–0.2	$\frac{v^2}{2g}$

where

v = the velocity in the pipe downstream of the reducer

B.2.10 Increaser

Head loss through increaser (with angle of divergence 10°–20°)	0.05–0.3	$\frac{v^2}{2g}$

where

v = the velocity in the pipe upstream of the reducer

References

1. Benefield, L. D., Judkins, J. F. and Parr, A. D. *Treatment Plant Hydraulics for Environmental Engineers*, Prentice-Hall, Englewood Cliffs, NJ, 1984.
2. Brater, E. F. and King, H. W. *Handbook of Hydraulics*, McGraw-Hill Book Co., New York, 1976.
3. FMC Corporation. *Hydraulics and Useful Information*, Chicago Pumps, Chicago, IL, 1973.
4. Heald, C. C. (Editor) *Cameron Hydraulic Data*, 18th ed., Ingersoll-Dresser Pumps, Liberty Corner, NJ, 1995.
5. Qasim, S. R. *Wastewater Treatment Plants: Planning, Design, and Operation*, 2d ed., Technomic Publishing Co., Inc., Lancaster, PA, 1999.
6. Zipparro, V. J. (Editor) *Davis's Handbook of Applied Hydraulics in Engineering*, 4th ed., McGraw-Hill Book Co., New York, 1993.

APPENDIX C

Manufacturers and Suppliers of Water Treatment Process Equipment

A list of the equipment manufacturers and suppliers of water treatment processes is provided in this appendix. The alphabetical listings are based on the keywords of major treatment processes, facilities, contaminants, chemicals, and class esof equipment. Information for each listing includes the complete address, phone number, and fax number. Toll-free phone number and web site addresses are also provided when available. This list is prepared from the *1999 AWWA Sourcebook*, *1997-98 Buyers' Guide of Water Environmental Federation*, and *1997-98 Buyers' Guide of Water and Waste Digest*. Other references include catalogs of equipment manufacturers, and buyers' information from such sources as *Water, Industrial Wastes*, *Water Engineering and Management*, *Industrial Water Engineering*, *Plant Engineering*, *Specifying Engineering*, *Instrument Society of America*, *Thomas Register*, and many others.

C.1 LISTING BY TYPE OF PROCESS OR EQUIPMENT

Aeration Systems
Aqua-Aerobic Systems, Inc.
Envirex/U.S. Filter Corp.
EnviroQuip, Inc.
Infilco Degremont, Inc.
Lakeside Equipment Co.
Lightnin Aerators
Philadelphia Mixers
Sanitaire/ITT Industries

Blowers and Compressors
Dresser Industries Roots Operations
Hoffman Air & Filtration Systems
Lamson Centrifugal Blowers
Semblex, Inc.
Spencer Turbine Co.
Turblex, Inc.

Chemical Feed Equipment
Acrison, Inc.
Alldos, Inc.
Aqua Smart, Inc.
Bailey-Fischer & Porter Co.
C & E Services, Inc.
Calgon Corp.
Capital Controls Company,Inc.
Chemco Equipment, Inc.
Fluid Dynamics, Inc.
FMC Corporation
LMT Milton Roy
Merrick Industries, Inc.
Penn Process Technologies
Semblex, Inc.
Smith & Loveless, Inc.
Stockhausen, Incorporated
Sweetwater Technologies
Wallace & Tiernan, Inc./U.S. Filter

Chemicals
American InternationalChemical
BOC Gases
Carus Chemical Company
Chemical Lime company
The Dow Chemical Company
Fe_3, Inc.
General Alum & Chemical
Polydyne, Inc.
Tetra Technologies, Inc.

Chlorinators and Chloramination
Alldos, Inc.
Bailey-Fischer & Porter Co.
Capital Controls Company, Inc.
Chlorinators, Inc.
Penn Process Technologies, Inc.
PPG Industries
Wallace & Tiernan, Inc./U.S. Filter

Clarifiers and Sedimentation Basins
EIMCO Process Equipment
FMC Corp.
Envirex/ U.S. Filter
EnviroQuip, Inc.
F. B. Leopold, Inc.
Infilco Degremont, Inc.
Kruger Inc.
Parkson Corporation
Smith & Loveless, Inc.
Walker Process Equipment
Westech IWS

Coagulants and Flocculants
Fe_3, Inc.
General Alum & Chemical
General Chemical Corp.
Midland Resources, Inc.
Nalco Chemical Co.
Stockhausen, Incorporated
Summit Research Labs.
Tetra Technologies, Inc.
Walker Process Equipment
Westwood Chemical

Controls
Bailey-Fischer & Porter Co.
Bristol Babcock
Capital Controls Company, Inc.
F. B. Leopold, Inc.
Fluid Solutions, Inc.
Fuji Electric Corp. of America
Kruger Inc.
Micro-Comm, Inc.
RACO Manufacturing & Engineering Co.
Systems Integrated
Tesco Controls Inc.
U.S. Filter

Corrosion Control and Softening Equipment
C & E Services, Inc.
Calgon Corp.
Carus Chemical Company
Chemical Lime Company
Infilco Degremont, Inc.

Dewatering Equipment Services
Alfa-Laval Separation, Inc. of Sharples Division
Andritz Ruthnor, Inc.
Ashbrook Corp.
EIMCO Process Equipment
Envirex/U.S. Filter
EnviroQuip, Inc.
Humboldt Decanter, Inc.
Naro Separation, Inc.
Roediger Pittsburgh, Inc.

Dissolved Air Flotation Equipment
EIMCO Process Equipment
Envirex/U.S. Filter
F. B. Leopold, Inc.
Infilco Degremont
Polychem Corp.
Walker Process Equipment
Westech IWS

Disinfection Equipment
Alldos, Inc.
Bailey-Fischer & Porter Co.
Capital Controls Company, Inc.
Chlorinators, Inc.
Ozonia North America
Penn Process Technologies, Inc.

Trojan Technologies, Inc.
Ultra Tech International, Inc.
Wallace & Tiernan, Inc./U.S. Filter

Distribution Systems (Pipes and Fitting)

A-1 Pipe Inc.
AMERON INTERNATIONAL Concrete & Steel Pipe Group
Gifford-Hill-American, Inc.
Mueller Co.
Northwest Pipe company
Plexco
Price Brothers Company
U.S. Pipe & Foundry Co.

Filter Media, Equipment and Controls

American Materials Company
Calgon Carbon Corporation
EIMCO Process Equipment
Enviroquip, Inc.
F. B. Leopold, Inc.
Infilco Degremont
Lang Filter Media Co.
Pall Corporation
Parkson Corporated
Tetra Technologies, Inc.
U.S. Filter
U.S. Silica
Valcan Industries, Inc.

Flow and Level Measurement

ABB Water Meters, Inc.
AMR Data Corporation
Badger Meter, Inc.
Bailey-Fischer & Porter Co.
Dynasonics, Inc.
Krohne, Inc.
Milltronics, Inc.
Mueller Co., A Tyco International Ltd. Company
Panametrics, Inc.
Polysonics, Inc.
Sensus Technologies, Inc.
Sparling Instruments Co.

Instruments (Analyzing, Recording and Control)

Bran & Luebbe, Inc.
Badger Meter, Inc.
Bailey-Fischer & Porter Co.
Bristol Babcock, Inc.
Capital Controls Company, Inc.
Hatch Company
Honeywell, Inc.
Thermo environmental Instruments, Inc.
Penn Process Technologies, Inc.
Orion Research Incorporated
Wallace & Tiernan, Inc./U.S. Filter

Intakes and Screens

Antritz Ruthner, Inc.
Envirex/U.S. Filter
FMC Corporation
Infilco Degremont, Inc.
Johns & Attwood, Inc.
Parkson Corporated
Valcan Industries, Inc.

Ion Exchange

C & E services, Inc.
Ionics, Incorporated
Osmonics, Inc.
Smith & Loveless, Inc.
Tetra Technologies, Inc.
U.S. Filter

Laboratory Supply

HF Scientific
HACH Company
Phipps & Bird
Watson-Marlow, Inc.

Ozonation

Fuji Electric Corp. of America
Infilco Degremont, Inc.
Ionics Incorporated
Osmonics, Inc.
Ozonia North America
Ozotech, Inc.

Membrane and Reverse Osmosis

Aqua Care Systems, Inc.
Aquasource North America
The Dow Chemical Company
Ionics Incorporated
Osmonics, Inc.
Pall Corporation
Smith & Loveless, Inc.
U.S. Filter
Westech IWS
Zenon Environmental, Inc.

Polymers

General Alum & Chemical
Polydyne Inc.
Stockhausen, Incorporated

Pumps

A-C Pump/ITT Industries
Alldos, Inc.
Aurora Pump
C & E Services, Inc.
Capital Controls Company, Inc.
Davis EMU
EBARA International, Inc.
Fairbanks Morse Pump
Gorman-Rupp Co.
Goulds Pumps, Inc.
Ingersoll-Dresser Pump Co.
Johnston Pump Company
Milton Roy Company
PACO Pumps, Inc.
Patterson Pump
Pulsa Feeder, Inc.
Smith & Loveless, Inc.
Wallace & Tiernan, Inc./U.S. Filter
Watson-Marlow, Inc.

Rapid Mixing and Flocculation Equipment

Chemineer, Inc.
EIMCO Process Equipment
FMC Corporation
Kruger, Inc.
Lightnin Mixers Equipment
Philadelphia Mixers
Walker Process Equipment
Westech Engineering, Inc.

Residuals Handling and Control

Ashbrook Corp.

Bird Machine Co.
Envirex/U.S. Filter
FMC Corporation
Gravity Flow Systems
Serpentix Conveyor Corp.
Walker Process Equipment
Westech IWS

Scada Systems

Alpine Technology, Inc.
Bristol Babcock Inc.
Control Microsystems
HSQ Technology
Micro-Comm, Inc.
Motorola, Inc.
Northern Digital, Inc.
QEI, Inc.
Systems Integrated
Tesco Controls, Inc.

Sludge Concentration and Thickening Equipment

Alfa-Laval Separation, Inc. of Sharples Division
Andritz Ruthner, Inc.
Ashbrook Corp.
Bird Machine Co.
EIMCO Process Equipment
Envirex/u.s. fILTER
EnviroQuip, Inc.
Lakeside Equipment Corp.
Parkson Corporated
Roediger Pittsburg, Inc.
Walker Process Equipment
Westech IWS

Water Storage Tanks (Ground and Elevated)

Caldwell Tanks, Inc.
Chicago Bridge & Iron Co.
DYK Incorporated
Landmark Structures, Inc.
Natgun Corporation
Pittsburg Tank & Tower Company
TEMCOR

Taste and Odor Control

Alldos, Inc.
Calgon Carbon Corporation
Carus Chemical Company
FE_3, Inc. (FINI)
FMC Corporation
Ionics Incorporated
Northwestern Carbon
Norit Americas, Inc.
Osmonics, Inc.
Ozonia North America
Ozotech, Inc.
Pall Corporation

Water Treatment Package Plants

EIMCO Process Equipment
Infilco Degremont, Inc.
EnviroQuip, Inc.
Kinetico Incorporated
Osmonics, Inc.
Ozotech, Inc.
Pall Corporation
Smith & Loveless, Inc.
Tonka Equipment Company
U.S. Filter
Westech IWS

Valves and Gates

APCO Willamette Valve and Primer Corporation
DeZurik
Flomatic Valves
GA Industries Inc.
Mueller Co., A Tyco International Ltd. Company
OCV Control Valves
Red Valve Company Inc.
Rodney Hunt Co.
Ross Valve Manufacturing Co., Inc.
Val-Matic Valve & Manufacturing Corp.

C.2 Addresses, Phone Numbers, and Other Information

A-1 PIPE INC.
P.O. Box 398
Bridgeport, NJ 08014-0398
Ph: (609) 467-1144
Fax: (609) 467-3860

ABB WATER METERS, INC.
P.O. Box 1852
Ocala, FL 34478
Ph: (352) 732-4670
Fax: (352) 732-1950
www.abb.com

A-C PUMP/ITT INDUSTRIES
1150 Tennessee Avenue
Cincinnati, OH 45229
Ph: (513) 482-2500
Fax: (513) 482-2569
www.ittacpump.com

ACRISON, INC.
20 Empire Blvd.
Moonachie, NJ 07074
Ph: (201) 440-8300
Fax: (201) 440-4939
www.acrison.com

ALFA LAVAL SEPARATION, INC., SHARPLES DIVISION
955 Mearns Road
Warminster, PA 18974-0556
Ph: (215) 443-4000 or (215) 443-4012
Fax: (215) 443-4139 or (215) 443-4011

ALLDOS, INC.
2220 Northwest Parkway
Suite 180
Marietta, GA 30067
Ph: (770) 956-7996
Fax: (770) 956-7836

AQUA SMART, INC.
800 Wendell Court
Atlanta, GA 30336
Ph: (800) 278-2762 or (404) 696-4406
Fax: (404) 696-3712

ALPINE TECHNOLOGY, INC.
2 Cielo Center, Suite 300
1250 Capitol of Texas Highway
Austin, TX 78746
Ph: (512) 329-2809
Fax: (512) 328-4792

AMERICAN INTERNATIONAL CHEMICAL
17 Strathmore Road
Natick, MA 01760
Ph: (800) 238-0001 or (508) 655-5805
Fax: (508) 655-0927
www.aicma.com

AMERICAN MATERIALS CORPORATION
717 Short St.
P.O. Box 388
Eau Claire, WI 54701
Ph: (800) 238-9139 or (715) 835-2251
Fax: (715) 835-0662
www.redflint.com

AMERON INTERNATIONAL CONCRETE & STEEL PIPE GROUP
10681 Foothill Blvd., #450
Rancho Cucamonga, CA 91730
Ph: (909) 944-4100
Fax: (909) 944-4112
www.ameron-intl.com

AMR DATA CORP.
62 2nd Ave.
Burlington, MA 01803
Ph: (781) 359-2713
Fax: (781) 359-2723
www.amrdata.com

ANDRITZ RUTHNER, INC.
1010 Commercial Blvd.,
Arlington, TX 76017
Ph: (817) 465-5611
Fax: (817) 468-3961

APCO WILLAMETTE VALVE AND PRIMER CORPORATION
1100 Via Callejon
San Clemente, CA 92673-6230
Ph: (949) 361-9900
Fax: (949) 361-3414
www.apcovalves.com

AQUA AEROBIC SYSTEMS, INC.
6306 N. Alpine Road
P. O. Box 93-0785
Rockford, IL 61130
Ph: (815) 654-2501
Fax: (815) 654-2501
www.aqua-aerobic.com

AQUA CARE SYSTEMS, INC.
9542 Hardpan Road
Angola, NY 14141
Ph: (716) 549-2500
Fax: (716) 549-3950

AQUASOURCE NORTH AMERICA
P.O. Box 70295
Richmond, VA 23255-0295
Ph: (804) 672-8160
Fax: (804) 672-8135

ASHBROOK CORP.
11600 E. Hardy
Houston, TX 77093
Ph: (713) 449-0322
Fax: (713) 449-1324

AURORA PUMP
800 Airport Road
North Aurora, IL 60542
Ph: (630) 859-7000
Fax: (630) 859-7060
www.gspump.com

BADGER METER, INC.
4545 W. Brown Deer Road
Milwaukee, WI 53223
Ph: (414) 355-0400
Fax: (414) 371-5980
www.badgermeter.com

BAILEY-FISCHER & PORTER CO.
125 E. Country Line Road
Warminster, PA 18974
Ph: (215) 674-6339
Fax: (215) 674-6740
www.ebpa.com/bfp/bfp home.htm

BIRD MACHINE CO.
100 Neponset Street
P.O. Box 9103
South Walpole, MA 02071-9103
Ph: (508) 668-0400
Fax: (508) 668-6855

BOC GASES
575 Mountain Ave.
Murray Hill, NJ 07974
Ph: (800) 742-4726 or (201) 984-2622
Fax: (201) 984-3633
www.boc.com

BRAN & LUEBBE, INC.
1025 Busch Parkway
Buffalo Grove, IL 60089
Ph: (847) 520-0700
Fax: (847) 520-0855
www.bran-luebbe.cle

BRISTOL BABCOCK, INC.
1100 Buckingham Street
Watertown, CT 06795
Ph: (860) 945-2200
Fax: (860) 945-2278
www.bristolbabcock.com

CALDWELL TANKS, INC.
4000 Tower Road
Louisville, KY 40219
Ph: (502) 964-3361
Fax: (502) 966-8732
www.caldwelltanks.com

CALGON CARBON CORPORATION
400 Calgon Carbon Drive
Pittsburgh, PA 15205
Ph: (412) 787-6700
Fax: (412) 787-4511
www.calgoncarbon.com

CALGON CORP.
P.O. Box 1346
Pittsburgh, PA 15230
Ph: (412) 494-8765
Fax: (412) 494-8224
www.calgon.com
CAPITAL CONTROLS COMPANY, INC.
3000 Advance Lane
Colmar, PA 18915
Ph: (215) 997-4000
Fax: (215) 997-4062
www.capitalcontrols.com

CARUS CHEMICAL CO.
315 Fifth Street
Peru, IL 61354
Ph: (815) 223-1500
Fax: (815) 224-6697
www.caruschem.com

C & E SERVICES, INC.
1950 Old Gallows Road
Suite 550
Vienna, VA 22182
Ph: (703) 506-1960
Fax: (703) 506-1957

CHEMCO EQUIPMENT, INC.
1500 Industrial Drive
Monongahela, PA 15063
Ph: (800) 501-5113 or (724) 258-7333
Fax: (724) 258-7350
www.chemcoequipment.com

CHEMINEER, INC.
P.O. Box 1123
Dayton, OH 45401
Ph: (513) 454-3200
Fax: (513) 454-3379
www.chemineer.com

CHICAGO BRIDGE & IRON CO.
601 W. 143rd Street
Plainfield, IL 60544
Ph: (815) 439-3100
Fax: (815) 439-3130
www.chicago-bridge.com

CHLORINATORS, INC.
P.O. Box 1822
Palm City, FL 34990-6822
Ph: (561) 288-4854
Fax: (561) 287-3238

CONTROL MICROSYSTEMS
28 Steacie Drive
Kanata, ON K2K 2A9
Canada
Ph: (613) 591-1943
Fax: (613) 591-1022
www.controlmicrosystems.com

DAVIS EMU
1706 Metcalf Avenue
Thomasville, GA 31792
Ph: (912) 226-5733
Fax: (912) 226-4793

DEZURIK
A UNIT OF GENERAL SIGNAL
250 Riverside Avenue N
Sartell, MN 56377-1743
Ph: (320) 259-2000
Fax: (320) 259-2227

THE DOW CHEMICAL CO.
2020 Dow Center
Midland, MI 48675
Ph: (800) 447-4369
Fax: (517) 832-1465
www.dow.com

DRESSER INDUSTRIES, ROOTS OPERATIONS
5310 Taneytown Pike
Taneytown, MD 21287
Ph: (301) 756-2602
Fax: (301) 756-2615

DYK INCORPORATED
P.O. Box 696
1214 Pioneer Way
El Cajon, CA 92022-0696
Ph: (800) 227-8181 or (619) 440-8181
Fax: (619) 440-8653
www.dyk.com

DYNASONICS, INC.
3535 S.W. Corporate Parkway
Palm City, FL 34990
Ph: (800) 535-3569
Fax: (407) 287-0011

EBARA INTERNATIONAL CORP.
1651 Cedar Line Drive
Rock Hill SC 29730
Ph: (803) 327-5005
Fax: (803) 327-5097

EIMCO PROCESS EQUIPMENT
P.O. Box 300
Salt Lake City, UT 84110
Ph: (801) 526-2000
Fax: (801) 526-2425

ENVIREX/U.S. FILTER
P.O. Box 1604
Waukesha, WI 53187-1604
Ph: (414) 547-0141
Fax: (414) 547-4120

ENVIROQUIP, INC.
P.O. Box 9069
Austin, TX 78728-8519
Ph: (512) 834-6010
Fax: (512) 834-6039

FAIRBANKS MOORSE PUMP
3601 Fairbanks Avenue
Kansas City, KS 66106
Ph: (913) 371-5000
Fax: (913) 371-2272
www.pentairpump.com

F. B. LEOPOLD CO., INC.
227 S. Division Street
Zelienople, PA 16063
Ph: (412) 452-6300
Fax: (412) 452-1377

FE_3, INC.
P.O. Box 808
Celina, TX 75009
Ph: (800) 441-2659 or (972) 382-2381
Fax: (972) 382-3211
www.fe3.com

FLOMATIC VALVES
145 Murray Street
Glens Falls, NY 12801
Ph: (518) 761-9797
Fax: (518) 761-9798
www.flomatic.com

FLUID DYNAMICS, INC.
6595 Odell Place, Suite E
Boulder, CO 80301
Ph: (303) 530-7300

FLUID SOLUTIONS
2650 Napa Valley Corporate Drive
Napa, CA 94558
Ph: (707) 255-0123
Fax: (707) 255-8465

FMC CORP.
1735 Market Street
Philadelphia, PA 19103
Ph: (800) 530-5927
Fax: (215) 299-5998
www.h2o2.com

FUJI ELECTRIC CORP. OF AMERICA
Park 80 West, Plaza II
Saddle Brook, NJ 07663
Ph: (201) 712-0555
Fax: (201) 368-8258
www.fujielectric.com

GA INDUSTRIES, INC.
9025 Marshall Road
Cranberry Township, PA 16066
Ph: (724) 776-1020
Fax: (724) 776-1254

GENERAL ALUM & CHEMICAL
P.O. Box 819
Holland, OH 43528
Ph: (419) 865-8000
Fax: (419) 865-6413

GENERAL CHEMICAL CORP.
90 E. Halsey Road
Parsippany, NJ 07054
Ph: (973) 515-0900
Fax: (973) 515-4461
www.genchem.com

GIFFORD-HILL-AMERICAN, INC.
P.O. Box 569470
Dallas, TX 75356-9470
Ph: (972) 262-3600
Fax: (972) 264-6236
www.ghainc.com

GORMAN RUPP CO.
305 Bowman Street
Mansfield, OH 44902
Ph: (419) 755-1011
Fax: (419) 755-1404
www.gormanrupp.com

GOULDS PUMPS, INC.
East Bayard Street
Seneca Falls, NY 13148
Ph: (315) 568-7647
Fax: (315) 568-7644

GRAVITY FLOW SYSTEMS, INC.
34 North Church Street
P.O. Box 525
Carbondale, PA 18407-0525
Ph: (800) 237-7500 or (716) 549-2500
Fax: (716) 282-3081

HACH CO.
P.O. Box 389
Loveland, CO 80539-0389
Ph: (800) 227-4224
Fax: (907) 669-2932
www.hach.com

HF SCIENTIFIC
3170 Metro Parkway
Fort Myers, FL 33916
Ph: (941) 337-2116
Fax: (941) 332-7643
www.hfscientific.com

HOFFMAN AIR & FILTRATION SYSTEMS
6181 Thompson Road
P.O. Box 548
East Syracuse, NY 13057-0548
Ph: (800) 258-8008 or (315) 432-8600
Fax: (315) 432-8682
www.hoffmanair.com

HONEYWELL, INC.
1100 Virginia Drive
Fort Washington, PA 19034
Ph: (215) 641-3050
Fax: (215) 641-3599
www.honeywell.com

HSQ TECHNOLOGY
1435 Huntington Avenue
San Francisco, CA 94080-5999
Ph: (650) 952-4310
Fax: (650) 952-7212
www.hsq.com

HUMBOLDT DECANTER, INC.
3883 Steve Reynolds Blvd.
Norcross, GA 30093-3051
Ph: (404) 564-7300

INFILCO DEGREMONT, INC.
P.O. Box 71390
Richmond, VA 23255-1390
Ph: (804) 756-7600
Fax: (804) 756-7643
www.wateronline.com/companie s/infilco
www.infilcodegremont.com

INGERSOLL-DRESSER PUMP CO.
MD Highway 97
P.O. Box 91
Taneytown, MD 21787
Ph: (410) 756-3278
Fax: (410) 756-2275

IONICS, INCORPORATED
65 Grove Street
Watertown, MA 02472
Ph: (617) 926-2500
Fax: (617) 926-4304
www.ionics.com

JONES & ATTWOOD, INC.
1931 Industrial Drive
Libertyville, IL 60048
Ph: (847) 367-5480
Fax: (847) 367-8983

JOHNSTON PUMP COMPANY
800 Koomey Road
Brookshire, TX 77423
Ph: (281) 934-6009
Fax: (281) 934-6090
www.johnstonpump.com

KINETICO INCORPORATED
10845 Kinsman Road
Newbury, OH 44065-9701
Ph: (440) 564-9111
Fax: (440) 564-7696
www.kinetico.com

KROHNE, INC.
7 Dearborn Road
Peabody, MA 01960
Ph: (978) 535-6060
Fax: (978) 535-1720
www.krohneamerica.com

KRUGER, INC.
401 Harrison Oaks Blvd., Suite 100
Cary, NC 27513-2412
Ph: (919) 677-8310
Fax: (919) 677-0082
www.krugerworld.com

LAKESIDE EQUIPMENT CO.
1022 E. Devon Avenue
P.O. Box 8448
Bartlett, IL 60103
Ph: (708) 837-5640
Fax: (708) 837-5647

LAMSON CENTRIFUGAL BLOWERS
Gardner Denver Machinery, Inc.
1800 Gardner Expressway
Quincy, IL 62301
Ph: (800) 682-9868 or (217) 228-5400
Fax: (217) 228-8247
www.gardnerdenver.com

LANDMARK STRUCTURES, INC.
1665 Harmon Road
Fort Worth, TX 76177
Ph: (817) 439-8888
Fax: (817) 439-9001
www.ldmkusa.com

LIGHTNIN AERATORS
135 Mt. Read Blvd.
Rochester, NY 14611
Ph: (716) 436-5550
Fax: (716) 527-1720
www.lightnin-mixers.com

LIGHTNIN MIXERS EQUIPMENT
135 Mt. Read Blvd.
Rochester, NY 14611
Ph: (716) 436-5550
Fax: (716) 527-1720

MILTON ROY COMPANY
201 Ivyland Road
Ivyland, PA 18974
Ph: (215) 441-0800
Fax: (215) 441-8620
www.miltonroy.com

MERRICK INDUSTRIES, INC.
10 Arthur Drive
Lynn Haven, FL 32444
Ph: (800) 345-8440
Fax: (904) 265-9768

MICRO-COMM, INC.
6980 W. 153rd St.
Overland Park, KS 66223
Ph: (913) 681-6349
Fax: (913) 681-2753

MIDLAND RESOURCES, INC.
3211 Clinton Parkway Ct.,
Suite 1
Lawrence, KS 66047
Ph: (785) 842-7424
Fax: (785) 842-3150
www.cjnetworks/~midland.com

MILLTRONICS, INC.
709 Stadium Drive
Arlington, TX 76011
Ph: (817) 277-3543
Fax: (817) 277-3894
www.milltronics.com

MOTOROLA, INC.
1301 E. Algonquin Road
Schaumburg, IL 60196
Ph: (800) 247-2346
Fax: (847) 538-7163
www.mol.com/moscad

MUELLER CO., A TYCO INTERNATIONAL LTD. COMANY
500 W. Eldorado Street
Decatur, IL 62525
Ph: (217) 423-4471
Fax: (217) 425-7537
www.uellerflo.com

NALCO CHEMICAL CO.
One Nalco Center
Naperville, IL 60563
Ph: (708)305-1000

NARO SEPARATION, INC.
5202 Brittmore
Houston, TX 77041
Ph: (713) 849-2181

NATGUN CORPORATION
11 Teal Road
Wakefield, MA 01880
Ph: (781) 246-1133
Fax: (781) 245-3279
www.natgun.com

NORIT AMERICAS, INC.
1050 Crown Pointe Parkway
Suite 1500
Atlanta, GA 30338
Ph: (770) 512-4610
Fax: (770) 512-4622
www.norit.com

NORTHERN DIGITAL, INC.
4520 California Ave., Suite 200
Bakersfield, CA 93309
Ph: (805) 322-6044
Fax: (805) 322-1209
www.northerndigital.com

NORTHWESTERN CARBON
11711 Reading Road
P.O. Box 130
Red Bluff, CA 96080-0130
Ph: (530) 527-2664
Fax: (530) 527-0544
www.nwcarbon.com

NORTHWEST PIPE COMPANY
12005 N. Burgard
P.O. Box 83149
Portland, OR 97283-0149
Ph: (800) 824-9824 or (503) 285-1400
Fax: (503) 285-2913
www.nwpipe.com

OCV CONTROL VALVES
7400 E. 42nd Place
Tulsa, OK 74145
Ph: (800) 331-4113
Fax: (918) 622-8916
www.controlvalves.com

ORION RESEARCH INCORPORATED
500 Cummings Center
Beverly, MA 01915
Ph: (800) 225-1480
Fax: (978) 922-4426

OSMONICS, INC.
5951 Clearwater Drive
Minnetonka, MN 55343
Ph: (612) 933-2277 or (800) 848-1750
Fax: (612) 933-9141
www.osmonics.com

OZONIA NORTH AMERICA
491 Edward H. Ross Drive
Elmwood Park, NJ 07407
Ph: (201) 794-3100
Fax: (201) 794-3358

OZOTECH, INC.
2401 Oberlin Road
Yreka, CA 96097
Ph: (530) 842-4189
Fax: (530) 842-3238
www.snowcrest.net/ozotech

PACO PUMPS, INC.
800 Koomey Road
Brookshire, TX 77423
Ph: (281) 934-6012
Fax: (281) 934-6090
www.paco-pumps.com

PALL CORPORATION
25 Harbor Park Drive
Port Washington, NY 11050
Ph: (516) 484-3600
Fax: (516) 484-3628
www.pall.com

PANAMETRICS, INC.
221 Harbor Park Drive
Waltham, MA 02154-3418
Ph: (800) 833-9438 or (781) 899-2719
Fax: (781) 891-8582
www. panametrics.com

PARKSON CORP.
2727 N.W. 62nd Street
P. O. Box 408399
Fort Lauderdale, FL 33340-8399
Ph: (954) 974-6610

PATTERSON PUMP
P.O. Box 790
Toccoa, GA 30577
Ph: (706) 886-2101
Fax: (706) 886-0023
www.pattersonpumps.com

PENN PROCESS TECHNOLOGIES
6100 Easton Road
Plumsteadville, PA 18949
Ph: (215) 766-7766
Fax: (215) 766-8290

PHILADELPHIA MIXERS
1221 E. Main
Palmyra, PA 17078
Ph: (717) 838-1341
Fax: (717) 832-8802

PHIPPS & BIRD
1519 Summit Avenue
Richmond, VA 23230-4511
Ph: (804) 254-2737
Fax: (804) 254-2955
www.phippsbird.com

PITTSBURGH TANK & TOWER COMPANY
515 Pennel Street
P.O. Box 913
Henderson, KY 42420
Ph: (502) 826-9000
Fax: (502) 827-4417
www.watertank.com

PLEXCO
1050 Illinois Route # 83
Suite #200
Bensenville, IL 60106-1048
Ph: (630) 350-3700
Fax: (630) 350-2894
www.plexco.com

POLYCHEM CORP.
P.O. Box 527
Phoenixville, PA 19460-0527
Ph: (610) 935-0225
Fax: (610) 935-7151
www.polychemcorp.com

POLYDYNE, INC.
7150 Granite Circle
Toledo, OH 43617
Ph: (419) 843-8066
Fax: (419) 843-6041

POLYSONICS, INC.
10335 Landsbury, Suite 300
Houston, TX 77099-7531
Ph: (713) 530-0885

PPG INDUSTRIES
One PPG Place
Pittsburgh, PA 15272
Ph: (800) 245-2974
Fax: (412) 434-3695

PRICE BROTHERS CO.
367 W. 2nd Street
P.O. Box 825
Dayton, OH 45401-0825
Ph: (937) 226-8700
Fax: (937) 226-8711

PULSA FEEDER, INC.
2883 Brighton-Henrietta Town Line Road
Rochester, NY 14623
Ph: (716) 292-8000
Fax: (716) 424-5619

QEI, INC.
60 Fadem Road
Springfield, NJ 07081
Ph: (973) 379-7400
Fax: (973) 379-2138
www.qeiinc.com

RACO MANUFACTURING & ENGINEERING CO.
1400 62nd Street
Emeryville, CA 94608
Ph: (510) 658-6713
Fax: (281) 658-3153
www.racoman.com

RED VALVE COMPANY INC.
700 N. Bell Avenue
Pittsburgh, PA 15106
Ph: (412) 279-0044
Fax: (412) 279-7878
www.redvalve.com

RODNEY HUNT CO.
46 Mill Street
Orange, MA 01364
Ph: (978) 544-2511
Fax: (978) 544-7204
www.rodneyhunt.com

ROEDIGER PITTSBURGH, INC.
3812 Route 8
Allison Park, PA 15101
Ph: (412) 487-6010
Fax: (412) 487-6005

ROSS VALVE MANUFACTURING CO., INC.
6 Oakwood Ave.
Troy, NY 12180
Ph: (518) 274-0961
Fax: (518) 274-0210
www.rossvalve.com

SANITAIRE/ITT INDUSTRIES
9333 N. 49th Street
Brown Deer, WI 53223
Ph: (414) 365-2200
Fax: (414) 365-2210
www.sanitaire.com

SEMBLEX, INC.
1635 W. Walnut Street
Springfield, MO 65806
Ph: (417) 866-1035
Fax: (417) 866-0235

SENSUS TECHNOLOGIES, INC.
450 N. Gallatin Ave.
P.O. Box 487
Uniontwon, PA 15401-0487
Ph: (800) 638-3748 or (724) 439-7894
Fax: (800) 888-2403
www.sensus.com

SERPENTIX CONVEYOR CORP.
9085 Marshall Court
Westminster, CO 80030
Ph: (303) 430-8427
Fax: (303) 430-7337
www.serpentix.com

SMITH & LOVELESS, INC.
14040 Santa Fe Trail Drive
Lenexa, KS 66215
Ph: (913) 888-5201
Fax: (913) 888-2173
www.smithandloveless.com

SPARLING INSTRUMENTS CO.
4097 N. Temple City Blvd.
El Monte, CA 91731
Ph: (818) 444-0571
Fax: (818) 452-0723

SPENCER TURBINE CO.
600 Day Hill Road
Windsor, CT 06095
Ph: (860) 688-8361
Fax: (860) 688-0098

STOCKHAUSEN, INC.
3408 Doyle Street
Greensboro, NC 27406
Ph: (910) 333-3564
Fax: (910) 333-3518

SUMMIT RESEARCH LABS.
45 River Road, Suite 300
Flemington, NJ 08822
Ph: (908) 782-9500
Fax: (908) 469-2932

SWEETWATER TECHNOLOGIES
P.O. Box 1473
Temecula, CA 92593
Ph: (909) 676-2443
Fax: (909) 676-2913

SYSTEMS INTEGRATED
8080 Dagget Street
San Diego, CA 92111
Ph: (619) 277-0700
Fax: (619) 277-1490

TEMCOR
24724 S. Wilmington Avenue
Carson, CA 90745
Ph: (310) 549-4311
Fax: (310) 549-4588
www.temcor.com

TESCO CONTROLS, INC.
P.O. Box 239012
Sacramento, CA 95823-9012
Ph: (916) 395-8800
Fax: (916) 429-2817

TETRA TECHNOLOGIES, INC.
6302 Benjamin Road, Suite 408
Tampa, FL 33634
Ph: (813) 886-9331
Fax: (813) 886-0651

THERMO ENVIRONMENTAL INSTRUMENTS, INC.
8 W. Forge Parkway
Carson, CA 90745
Ph: (310) 549-4311
Fax: (310) 549-4588
www.thermoei.com

TONKA EQUIPMENT COMPANY
13305 Watertower Circle
Plymouth, MN 55441
Ph: (612) 559-2837
Fax: (612) 559-1979

TROJAN TECHNOLOGIES, INC.
3020 Gore Road
London, ON N5V 4T7
Canada
Ph: (519) 457-3400
Fax: (519) 457-3030
www.trojanuv.com

TURBLEX, INC.
1635 W. Walnut Street
Springfield, MO 65806
Ph: (417) 864-5599
Fax: (417) 866-0235

ULTRA TECH INTERNATIONAL, INC.
9454-9 Phillips Highway
Jacksonville, FL 32256
Ph: (904) 292-1611

U.S. FILTER CORP.
40-004 Cook St.
Palm Desert, CA 92211
Ph: (760) 340-0098
Fax: (760) 341-9368
www.usfilter.com

U.S. PIPE & FOUNDRY CO.
P.O. Box 10406
Birmingham, AL 35202-0406
Ph: (205) 254-7442
Fax: (205) 254-7165
www.uspipe.com

U.S. SILICA
P.O. Box 187
Berkeley Springs, WV 25411
Ph: (304) 258-2500
Fax: (304) 258-8295
www.u-s-silica.com

VAL-MATIC VALVE & MANUFACTURING CORP.
905 Riverside Drive
Elmhurst, IL 60126
Ph: (630) 941-7600
Fax: (630) 941-8042
www.walker-process.com

VULCAN INDUSTRIES, INC.
212 S. Kirlin Street
Missouri Valley, IA 51555
Ph: (712) 642-2755
Fax: (712) 642-4256
www.vulcanindustries.com

WALKER PROCESS EQUIPMENT
DIVISION OF McNISH CORP.
840 N. Russell Avenue
Aurora, IL 60506
Ph: (630) 892-7921
Fax: (630) 892-7951
www.walker-process.com

WALLACE & TIERNAN, INC./U.S. FILTER
25 Main Street
Belleville, NJ 07109-3057
Ph: (201) 759-8000
Fax: (201) 759-0348

WATSON-MARLOW, INC.
220 Ballardvale Street
Wilmington, MA 01887
Ph: (800) 282-8823
Fax: (907) 658-0041
www.watson-marlow.com/watson-marlow/

WESTECH IWS
3625 South West Temple
Salt Lake City, UT 84115
Ph: (801) 265-1000
Fax: (801) 265-1080
www.westech-inc.com

WESTWOOD CHEMICAL
46 Tower Drive
Middletown, NY 10941
Ph: (914) 692-6721
Fax: (914) 695-1906

ZENON ENVIRONMENTAL, INC.
845 Harrington Court
Burlington, ON L7N 3P3 Canada
Ph: (905) 639-6320, Ext. 202
Fax: (905) 639-1812
www.zenonenv.com

APPENDIX D

Design Parameters, Abbreviations, Symbols, Constants, Conversion Factors, and Other Useful Information for Water Works Engineering

Table D-1 Design Parameters, Units of Expression, and Typical Design Values for Water Treatment Processes

Design Parameter	Typical Design Values (SI units)	SI Units	Factor k Divide by k to Obtain ← Multiply by k to Obtain →	U.S. Customary Units
Intake and raw water pump station				
Velocity through coarse screen	0.05–0.08	m/s	3.28	ft/s
Velocity through fine screen	0.4–0.8	m/s	3.28	ft/s
Rapid mix				
Detention time	0.2–5	min	–	min
Velocity gradient	700–1000	s^{-1}	–	s^{-1}
Gt	3×10^4–6×10^4	–	–	–
Flocculation				
Detention time	20–60	min	–	min
Velocity gradient	15–60	s^{-1}	–	s^{-1}
Gt	1×10^4–15×10^4	–	–	–
Peripheral velocity of paddle	0.3–0.6	m/s	3.28	ft/s
Shaft rotational speed	1.5–5	rpm	–	rpm
Sedimentation				
Coagulation				
Detention time	2–8	h	–	h
Surface loading	20–40	$m^3/m^2{\cdot}d$	24.55	gpd/ft^2
Weir loading	200–300	$m^3/m{\cdot}d$	80.52	gpd/ft
Softening				
Detention time	1–6	h	–	h
Surface loading	40–60	$m^3/m^2{\cdot}d$	24.55	gpd/ft^2
Weir loading	250–350	$m^3/m{\cdot}d$	80.52	gpd/ft
Filtration				
Hydraulic application rate				
Slow sand filter	10	$m^3/m^2{\cdot}d$	0.017	gpm/ft^2
Rapid sand filter	120–240	$m^3/m^2{\cdot}d$	0.017	gpm/ft^2
High-rate filter	240–600	$m^3/m^2{\cdot}d$	0.017	gpm/ft^2
Media Design				

Table D-1 Design Parameters, Units of Expression, and Typical Design Values for Water Treatment Processes (cont'd)

Design Parameter	Typical Design Values (SI units)	SI Units	Factor k Divide by k to Obtain ← Multiply by k to Obtain →	U.S. Customary Units
Single medium				
Depth	50–150	cm	0.394	inch
Uniformity coefficient	1.2–1.7	–	–	–
Effective size	0.5–1.5	mm	0.394	inch
Dual media				
Sand layer				
Depth	20–40	cm	0.394	inch
Uniformity coefficient	1.2–1.7	–	–	–
Effective size	0.45–0.6	mm	0.0394	inch
Anthracite layer				
Depth	30–60	cm	0.394	inch
Uniformity coefficient	1.3–1.8	–	–	–
Effective size	0.7–2	mm	0.0394	inch
Backwash system				
Backwash low rate	37–50	$m^3/m^2 \cdot h$	0.409	gpm/ft^2
Rotating arm surface wash	1.2–2.4	$m^3/m^2 \cdot h$	0.409	gpm/ft^2
Fixed arm surface wash	5–10	$m^3/m^2 \cdot h$	0.409	gpm/ft^2
Air scour system	37–73	$m^3/m^2 \cdot h$	0.409	gpm/ft^2
Taste and odor control				
Potassium permanagate dose	0.5–2.5	mg/L	–	mg/L
PAC	0.5–5	mg/L	–	mg/L
Disinfection				
Chlorine dose	1–5	mg/L	–	mg/L
Chlorine residual	0.5– 1	mg/L	–	mg/L
Ozone dose	1– 5	mg/L	–	mg/L
Fluoridation				
Fluoride dose	0.7–1.2	mg/L	–	mg/L
Distribution system				

Table D-1 Design Parameters, Units of Expression, and Typical Design Values for Water Treatment Processes (cont'd)

Design Parameter	Typical Design Values (SI units)	SI Units	Factor k Divide by k to Obtain ← Multiply by k to Obtain →	U.S. Customary Units
Velocity in mains	1–2	m/s	0.3048	ft/s
Pressure	138–1000	kpa	0.145	psi
Residuals-management				
Quantity (coagulation)	8–210	$kg/1000m^3$	8.346	$lb/10^6$ gal
Density (dry wt.)	1200–1520	kg/m^2	0.0624	lb/ft^3
Quantity (softening)	43–2230	$kg/1000m^3$	8.346	$lb/10^6$ gal
Gravity thickener				
Hydraulic loading	4–10	$m^3/m^2{\cdot}d$	24.54	gpd/ft^2
Solids loading	20–80	$kg/m^2{\cdot}d$	0.205	$lb/ft^2{\cdot}d$
Dewatering				
Drying beds	100–300	kg dry solids/$m^2{\cdot}yr$	0.205	$lb/ft^2{\cdot}yr$
Drying lagoon	50–100	$kg/m^2{\cdot}yr$	0.205	$lb/ft^2{\cdot}yr$
Belt filter press	0.8–4	kg/m^2h	0.205	$lb/ft^2{\cdot}h$

Table D-2 Abbreviations and Symbols

Abbreviations	Symbols	Abbreviations	Symbols
Acceration due to gravity (LT^{-2})	g	Gallon (L^3)	gal
Area (L^2)	A	Gallons per minute (L^3T^{-1})	gal/min or gpm
Atmosphere	atm	Gallons per second (L^3T^{-1})	gal/s or gps
British thermal uni	Btu	Horsepower	hp
Calorie	cal	Horsepower-hour	hp-h
Centimeter	cm	Hour	h
Centipoise	cp	Inch	in
Cubic centimeter	cm^3	Joule	J
Degree Celsius	°C	Kilocalorie	kcal
Degree Fahrenheit	°F	Kilonewton per cubic meter	kN/m^3
Degree Kelvin	°K	Liter (L^3)	L
Degree Rankine	°R	Million gallons per day (L^3T^{-1})	mgd
Density (ML^{-3})	p	Pound force	lb_f
Discharge (L^3T^{-1})	Q	Pound mass	lb_m
Energy	E	Standard temperature and pressure	STP @ 273°K and 760 mmHg
Feet per minute (LT^{-1})	fpm or ft/min	Watt	W
Feet per second (LT^{-1})	fps or ft/s		

M = mass; L = length; T = time

Source: Reprinted from Reference 1, with permission of Technomic Publishing Co., Inc., copyright 1999.

Table D-3 Values of Useful Constants

Acceleration caused by gravity, $g = 9.807\ m/s^2$ ($32.174\ ft/s^2$)
Standard atmosphere = 101.325 kN/m^2 (14.696 lb_f/in^2)
= 101.325 kPa (1.013 bars)
= 10.333 m (33.899 ft) of water
Molecular weight of air = 29.1
Density of air (15°C and 1 atm) = 1.23 kg/m^3 (0.0766 lb_m/ft^3)
Dynamic viscosity of air, μ (15°C and 1 atm) = 17.8 x 10^{-6} $N{\cdot}s/m^2$ (0.372 x 10^{-6} $lb_f{\cdot}s/ft^2$)
Kinematic viscosity of air, ν (15°C and 1 atm) = 14.4 x 10^{-6} m^2/s (155 x 10^{-6} ft^2/s)
Specific heat of air = 1005 J/kg·°K (0.24 $Btu/lb_m{\cdot}°R$)

Source: Reprinted from Reference 1, with permission of Technomic Publishing Co., Inc., copyright 1999.

Table D-4 Conversion Factors, Mixed Units

Length (L)					
mile	yard	ft	in.	m	cm
1	1760	5280	6.336×10^4	1.609×10^3	1.609×10^5
5.68×10^{-4}	1	3	36	0.9144	91.44
1.894×10^{-4}	0.333	1	12	0.3048	30.48
1.578×10^{-5}	0.028	0.083	1	0.0254	2.54
6.214×10^{-4}	1.094	3.281	39.37	1	100

Area (A)					
mile^2	acre	yard^2	ft^2	in^2	m^2
1	640[a]	3.098×10^6	2.788×10^7	4.014×10^9	2.59×10^6
1.563×10^{-3}	1	4840	43,560	6.27×10^6	4047
3.228×10^{-7}	2.066×10^{-4}	1	9	1296	0.836
3.587×10^{-8}	2.3×10^{-5}	0.111	1	144	0.093
2.491×10^{-10}	1.59×10^{-7}	7.716×10^{-4}	6.944×10^{-3}	1	6.452×10^{-4}
3.861×10^{-7}	2.5×10^{-4}	1.196	10.764	1550	1

Volume (V)						
acre·ft	U.S. Gal	ft^3	in.^{3}	L	m^3	cm^3
1	325,851	43,560	75.3×10^{-6}	1.23×10^6	1230	1.23×10^9
3.07×10^{-6}	1	0.134	231.6	3.785	3.875×10^{-3}	3875
2.3×10^{-5}	7.481	1	1728	28.317	0.028	28,317
1.33×10^{-8}	4.329×10^{-3}	5.787×10^{-4}	1	0.016	1.639×10^{-5}	16.39
8.1×10^{-7}	0.264	0.035	61.02	1	1×10^{-3}	1000
8.13×10^{-4}	264.2	35.31	6.10×10^4	1000	1	10^6

Time (T)					
yr	mon	d	h	min	s
1	12	365	8760	525,600	3.1536×10^7

Table D-4 Conversion Factors, Mixed Units (cont'd)

Velocity (L/T)				
ft/s	ft/min	m/s	m/min	cm/s
1	60	0.3048	18.29	30.48
0.017	1	5.08×10^{-3}	0.3048	0.5080
3.281	196.8	1	60	100
0.055	3.28	0.017	1	1.70
0.032	1.969	0.01	0.588	1

Discharge (L^3/T)					
mgd	gpm	ft^3/s	ft^3/min	L/s	m^3/d
1	694.4	1.547	92.82	43.75	3.78×10^3
1.44×10^{-3}	1	2.228×10^{-3}	0.134	0.063	5.45
0.646	448.9	1	60	28.32	2447
0.011	7.481	0.017	1	0.472	40.78
0.023	15.85	0.035	2.119	1	86.41
2.64×10^{-4}	0.183	4.09×10^{-4}	0.025	0.012	1

Mass (M)					
ton	lb_m	grain	ounce (oz)	kg	g
1	2000	1.4×10^7	32,000	907.2	907,185
0.0005	1	7000	16	0.454	454
7.14×10^{-8}	1.429×10^{-4}	1	2.29×10^{-3}	6.48×10^{-5}	0.065
3.125×10^{-5}	0.0625	437.6	1	0.028	28.35
1.10×10^{-3}	2.205	1.54×10^4	35.27	1	1000
1.10×10^{-6}	2.20×10^{-3}	15.43	0.035	10^{-3}	1

Table D-4 Conversion Factors, Mixed Units (cont'd)

Temperature (T)			
°F	°C	°K	°R
°F	$\frac{5}{9}$(°F – 32)	$\frac{5}{9}$°F + 255.38	°F + 459.69
$\frac{9}{5}$°C + 32	°C	°C + 273.16	$\frac{9}{5}$°C + 491.69
$\frac{9}{5}$°K – 459.69	°K – 273.16	°K	$\frac{9}{5}$°K
°R – 459.69	$\frac{5}{9}$°R – 273.16	$\frac{5}{9}$°R	°R

Density (M/L^3)				
lb/ft^3	lb/gal (U.S.)	kg/m^3	kg/L	g/cm^3
1	0.1337	16.019	0.01602	0.01602
7.48	1	119.8	0.1198	0.1198
0.0624	8.345×10^{-3}	1	0.001	0.001
62.43	8.345	1000	1	1

Pressure (F/L^2)						
lb/in.^{2}	ft water	in Hg	atm	mm Hg	kg/cm^2	N/m^2
1	2.307	2.036	0.068	51.71	0.0703	6895
0.4335	1	0.8825	0.0295	22.41	0.0305	2989
0.4912	1.133	1	0.033	25.40	0.035	3386
14.70	33.93	29.92	1	760	1.033	1.013×10^5
0.019	0.045	0.039	1.30×10^{-3}	1	1.36×10^{-3}	133.3
14.23	32.78	28.96	0.968	744.7	1	98,070
1.45×10^{-4}	3.35×10^{-4}	2.96×10^{-4}	9.87×10^{-6}	7.50×10^{-3}	1.02×10^{-5}	1

Table D-4 Conversion Factors, Mixed Units (cont'd)

Force (F)		
lb_f	N	dyne
1	4.448	4.448×10^5
0.225	1	10^5
2.25×10^{-6}	10^{-5}	1

Energy (E)					
kW·h	hp·h	Btu	J	kJ	calories
1	1.341	3412	3.6×10^6	3600	8.6×10^5
0.7457	1	2545	2.684×10^6	2685	6.4×10^5
2.930×10^{-6}	3.929×10^{-4}	1	1055	1.055	252
2.778×10^{-7}	3.72×10^{-7}	9.48×10^{-4}	1	0.001	0.239
2.778×10^{-4}	3.72×10^{-4}	0.948	1000	1	239
1.16×10^{-6}	1.56×10^{-6}	3.97×10^{-3}	4.186	$4.18 \text{ x } 10^{-3}$	1

Power (P)					
kW	Btu/min	hp	ft·lb/s	kg·m/s	cal/min
1	56.89	1.341	737.6	102	14,330
0.018	1	0.024	12.97	1.793	252
0.746	42.44	1	550	76.09	10,690
1.35×10^{-3}	0.077	1.82×10^{-3}	1	0.138	19.43
9.76×10^{-3}	0.558	0.013	7.233	1	137.6
6.98×10^{-5}	3.97×10^{-3}	9.355×10^{-5}	0.0514	7.12×10^{-3}	1

Table D-4 Conversion Factors, Mixed Units (cont'd)

Viscosity

Dynamic Absolute Viscosity (μ)

cp	$lb_f \cdot s/ft^2$	$lb_m/ft \cdot s$	g/cm·s	$N \cdot s/m^2$	kg/m·s	dp
1	2.09×10^{-5}	6.72×10^{-4}	0.01	1×10^{-3}	1×10^{-3}	1×10^{-3}
4.78×10^4	1	32.15	478.5	47.85	47.85	47.85
1488	0.031	1	14.88	1.488	1.488	1.488
100	2.09×10^{-3}	0.0672	1	0.10	0.10	0.10
1000	0.021	0.672	10	1	1	1

Kinematic Viscosity (ν)

centistoke	ft^2/s	cm^2/s	m^2/s	myriastoke
1	1.076×10^{-5}	0.01	1.0×10^{-6}	1.0×10^{-6}
9.29×10^4	1	929.4	0.093	0.093
100	1.076×10^{-3}	1	1.0×10^{-4}	1.0×10^{-4}
10^6	10.76	10^4	1	1

Table D-5 Commonly Used Commercial Pipe Sizes (4–96 inches Diameter)

Pipe Size Available in the U.S. Market,		
inch	**mm**	**Equivalent Pipe Size in Metric Unit, mm**
4	101.6	100
6	152.4	150
8	203.2	200
10	254.0	250
12	304.8	300
14	355.6	350
16	406.4	400
18	457.2	450
20	508.0	500
24	609.6	600
27	685.8	700
30	762.0	750
36	914.4	900
42	1066	1050
48	1219	1200
54	1371	1400
60	1524	1500
72	1829	1800
78	1981	2000
84	2134	2100
90	2286	2250
96	2438	2400

Table D-6 Chemical Symbols, Atomic Numbers, Atomic Weights, and Valences of Common Elements

Element	Symbol	Atomic Number	Atomic Weight	Valence
Aluminum	Al	13	26.9815	3
Antimony	Sb	51	121.75	3,5
Argon	Ar	18	39.948	—
Arsenic	As	33	74.9216	3,5
Barium	Ba	56	137.34	2
Beryllium	Be	4	9.0122	—
Bismuth	Bi	83	208.980	3,5
Boron	B	5	10.811	3
Bromine	Br	35	79.909	1,3,5,7
Cadmium	Cd	48	112.40	2
Calcium	Ca	20	40.08	2
Carbon	C	6	12.01115	4
Chlorine	Cl	17	35.453	1,3,5,7
Chromium	Cr	24	51.996	2,3,6
Cobalt	Co	27	58.9332	2,3
Copper	Cu	29	63.54	1,2
Fluorine	F	9	18.9984	1
Gold	Au	79	196.967	1,3
Helium	He	2	4.0026	—
Hydrogen	H	1	1.00797	1
Iodine	I	53	126.9044	1,3,5,7
Iron	Fe	26	55.847	2,3
Lead	Pb	82	207.19	2,4
Lithium	Li	3	6.939	1
Magnesium	Mg	12	24.312	2
Manganese	Mn	25	54.9380	2,3,4,6,7
Mercury	Hg	80	200.59	1,2
Molybdenum	Mo	42	94.94	3,4,6
Neon	Ne	10	20.183	—
Nickel	Ni	28	58.71	2,3
Nitrogen	N	7	14.0067	3,5

Table D-6 Chemical Symbols, Atomic Numbers, Atomic Weights, and Valences of Common Elements (cont'd)

Element	Symbol	Atomic Number	Atomic Weight	Valence
Oxygen	O	8	15.9994	2
Phosphorus	P	15	30.9738	3,5
Platinum	Pt	78	195.09	2,4
Potassium	K	19	39.102	1
Radium	Ra	88	226.05	—
Silicon	Si	14	28.086	4
Silver	Ag	47	107.870	1
Sodium	Na	11	22.9898	1
Strontium	Sr	38	87.62	2
Sulfur	S	16	32.064	2,4,6
Tin	Sn	50	118.69	2,4
Titanium	Ti	22	47.90	—
Tungsten	W	74	183.85	3,4
Uranium	U	92	238.03	4,6
Zinc	Zn	30	65.37	2

References

1. Qasim, S. R. *Wastewater Treatment Plants: Planning, Design, and Operation*, 2d ed., Technomic Publishing Co., Inc., Lancaster, PA, 1999.

Index

C

D

E

F

G

H

I

J

K

L

M

N

O

P

R

S

T

About the Authors

SYED R. QASIM, Professor of Civil and Environmental Engineering at The University of Texas at Arlington, has over 30 years of design, research and teaching experience in water and wastewater treatment processes. He has written three books, and published over 100 technical papers.

EDWARD M. MOTLEY is Vice President and Director of Engineering with Chiang, Patel & Yerby Inc., a consulting engineering firm in Dallas, TX. He has over 20 years of consulting experience in planning, design, and construction of advanced water, wastewater treatment, and conveyance projects.

GUANG ZHU has ten years of consulting experience with Beijing Municipal Engineering Design and Research Institute, Beijing, China, and is author of many technical papers in environmental engineering. He joined Chiang, Patel & Yerby Inc. as an environmental engineering process design engineer after completing his Ph.D. in 1996.